AN INTRODUCTION TO COMPUTATIONAL STOCHASTIC PDES

GABRIEL J. LORD
Heriot-Watt University, Edinburgh

CATHERINE E. POWELL
University of Manchester

TONY SHARDLOW
University of Bath

CAMBRIDGE
UNIVERSITY PRESS

CAMBRIDGE
UNIVERSITY PRESS

32 Avenue of the Americas, New York, NY 10013-2473, USA

Cambridge University Press is part of the University of Cambridge.

It furthers the University's mission by disseminating knowledge in the pursuit of education, learning and research at the highest international levels of excellence.

www.cambridge.org
Information on this title: www.cambridge.org/9780521728522

First published 2014

Printed in the United States of America

A catalog record for this publication is available from the British Library.

Library of Congress Cataloging in Publication Data
Lord, Gabriel J., author.
An introduction to computational stochastic PDEs / Gabriel J. Lord, Heriot-Watt University, Edinburgh, Catherine E. Powell, University of Manchester, Tony Shardlow, University of Bath.
pages cm – (Cambridge texts in applied mathematics; 50)
Includes bibliographical references and index.
ISBN 978-0-521-89990-1 (hardback) – ISBN 978-0-521-72852-2 (paperback)
1. Stochastic partial differential equations. I. Powell, Catherine E., author.
II. Shardlow, Tony, author. III. Title.
QA274.25.L67 2014
519.2′2–dc23 2014005535

ISBN 978-0-521-89990-1 Hardback
ISBN 978-0-521-72852-2 Paperback

Additional resources for this publication at www.cambridge.org/9780521728522

Contents

Preface

Techniques for solving many of the differential equations traditionally used by applied mathematicians to model phenomena such as fluid flow, neural dynamics, electromagnetic scattering, tumour growth, telecommunications, phase transitions, etc. are now mature. Parameters within those models (e.g., material properties, boundary conditions, forcing terms, domain geometries) are often assumed to be known exactly, even when it is clear that is not the case. In the past, mathematicians were unable to incorporate noise and/or uncertainty into models because they were constrained both by the lack of computational resources and the lack of research into stochastic analysis. These are no longer good excuses. The rapid increase in computing power witnessed in recent decades allows the extra level of complexity induced by uncertainty to be incorporated into numerical simulations. Moreover, there are a growing number of researchers working on stochastic partial differential equations (PDEs) and their results are continually improving our theoretical understanding of the behaviour of stochastic systems. The transition from working with purely deterministic systems to working with stochastic systems is understandably daunting for recent graduates who have majored in applied mathematics. It is perhaps even more so for established researchers who have not received any training in probability theory and stochastic processes. We hope this book bridges this gap and will provide training for a new generation of researchers — that is, you.

This text provides a friendly introduction and practical route into the numerical solution and analysis of stochastic PDEs. It is suitable for mathematically grounded graduates who wish to learn about stochastic PDEs and numerical solution methods. The book will also serve established researchers who wish to incorporate uncertainty into their mathematical models and seek an introduction to the latest numerical techniques. We assume knowledge of undergraduate-level mathematics, including some basic analysis and linear algebra, but provide background material on probability theory and numerical methods for solving differential equations. Our treatment of model problems includes analysis, appropriate numerical methods and a discussion of practical implementation. MATLAB is a convenient computer environment for numerical scientific computing and is used throughout the book to solve examples that illustrate key concepts. We provide code to implement the algorithms on model problems, and sample code is available from the authors' or the publisher's website*. Each chapter concludes with exercises, to help the reader study and become more

* http://www.cambridge.org/9780521728522

familiar with the concepts involved, and a section of notes, which contains pointers and references to the latest research directions and results.

The book is divided into three parts, as follows.

Part One: Deterministic Differential Equations We start with a deterministic or non-random outlook and introduce preliminary background material on functional analysis, numerical analysis, and differential equations. Chapter 1 reviews linear analysis and introduces Banach and Hilbert spaces, as well as the Fourier transform and other key tools from Fourier analysis. Chapter 2 treats elliptic PDEs, starting with a two-point boundary-value problem (BVP), and develops Galerkin approximation and the finite element method. Chapter 3 develops numerical methods for initial-value problems for ordinary differential equations (ODEs) and a class of semilinear PDEs that includes reaction–diffusion equations. We develop finite difference methods and spectral and finite element Galerkin methods. Chapters 2 and 3 include not only error analysis for selected numerical methods but also MATLAB implementations for test problems that illustrate numerically the theoretical orders of convergence.

Part Two: Stochastic Processes and Random Fields Here we turn to probability theory and develop the theory of stochastic processes (one parameter families of random variables) and random fields (multi-parameter families of random variables). Stochastic processes and random fields are used to model the uncertain inputs to the differential equations studied in Part Three and are also the appropriate way to interpret the corresponding solutions. Chapter 4 starts with elementary probability theory, including random variables, limit theorems, and sampling methods. The Monte Carlo method is introduced and applied to a differential equation with random initial data. Chapters 5–7 then develop theory and computational methods for stochastic processes and random fields. Specific stochastic processes discussed include Brownian motion, white noise, the Brownian bridge, and fractional Brownian motion. In Chapters 6 and 7, we pay particular attention to the important special classes of stationary processes and isotropic random fields. Simulation methods are developed, including a quadrature scheme, the turning bands method, and the highly efficient FFT-based circulant embedding method. The theory of these numerical methods is developed alongside practical implementations in MATLAB.

Part Three: Stochastic Differential Equations There are many ways to incorporate stochastic effects into differential equations. In the last part of the book, we consider three classes of stochastic model problems, each of which can be viewed as an extension to a deterministic model introduced in Part One. These are:

Chapter 8	ODE (3.6)	+	white noise forcing
Chapter 9	Elliptic BVP (2.1)	+	correlated random data
Chapter 10	Semilinear PDE (3.39)	+	space–time noise forcing

Note the progression from models for time t and sample variable ω in Chapter 8, to models for space $\boldsymbol{x}$ and ω in Chapter 9, and finally to models for $t, \boldsymbol{x}, \omega$ in Chapter 10. In each case, we adapt the techniques from Chapters 2 and 3 to show that the problems are well posed and to develop numerical approximation schemes. MATLAB implementations are also discussed. Brownian motion is key to developing the time-dependent problems with white noise forcing considered in Chapters 8 and 10 using the Itô calculus. It is these types of differential equations that are traditionally known as stochastic differential equations (SODEs and SPDEs). In Chapter 9, however, we consider elliptic BVPs with both a forcing term and coefficients that are represented by random fields *not* of white noise type. Many authors prefer to reserve the term 'stochastic PDE' only for PDEs forced by white noise. We interpret it more broadly, however, and the title of this book is intended to incorporate PDEs with data and/or forcing terms described by both white noise (which is uncorrelated) and correlated random fields. The analytical tools required to solve these two types of problems are, of course, very different and we give an overview of the key results.

Chapter 8 introduces the class of stochastic ordinary differential equations (SODEs) consisting of ODEs with white noise forcing, discusses existence and uniqueness of solutions in the sense of Itô calculus, and develops the Euler–Maruyama and Milstein approximation schemes. Strong approximation (of samples of the solution) and weak approximation (of averages) are discussed, as well as the multilevel Monte Carlo method. Chapter 9 treats elliptic BVPs with correlated random data on two-dimensional spatial domains. These typically arise in the modelling of fluid flow in porous media. Solutions are also correlated random fields and, here, do not depend on time. To begin, we consider log-normal coefficients. After sampling the input data, we study weak solutions to the resulting deterministic problems and apply the Galerkin finite element method. The Monte Carlo method is then used to estimate the mean and variance. By approximating the data using Karhunen–Loève expansions, the stochastic PDE problem may also be converted to a deterministic one on a (possibly) high-dimensional parameter space. After setting up an appropriate weak formulation, the stochastic Galerkin finite element method (SGFEM), which couples finite elements in physical space with global polynomial approximation on a parameter space, is developed in detail. Chapter 10 develops stochastic parabolic PDEs, such as reaction–diffusion equations forced by a space–time Wiener process, and we discuss (strong) numerical approximation in space and in time. Model problems arising in the fields of neuroscience and fluid dynamics are included.

The number of questions that can be asked of stochastic PDEs is large. Broadly speaking, they fall into two categories: *forward problems* (sampling the solution, determining exit times, computing moments, etc.) and *inverse problems* (e.g., fitting a model to a set of observations). In this book, we focus on forward problems for specific model problems. We pay particular attention to elliptic PDEs with coefficients given by correlated random fields and reaction–diffusion equations with white noise forcing. We also focus on methods to compute individual samples of solutions and to compute moments (means, variances) of functionals of the solutions. Many other stochastic PDE models are neglected (hyperbolic problems, random domains, forcing by Levy processes, to name a few) as are many important questions (exit time problems, long-time simulation, filtering). However, this book covers a wide range of topics necessary for studying these problems and will leave the reader well prepared to tackle the latest research on the numerical solution of a wide range of stochastic PDEs.

1
Linear Analysis

This chapter introduces theoretical tools for studying stochastic differential equations in later chapters. §1.1 and §1.2 review Banach and Hilbert spaces, the mathematical structures given to sets of random variables and the natural home for solutions of differential equations. §1.3 reviews the theory of linear operators, especially the spectral theory of compact and symmetric operators, and §1.4 reviews Fourier analysis.

1.1 Banach spaces C^r and L^p

Banach and Hilbert spaces are fundamental to the analysis of differential equations and random processes. This section treats Banach spaces, reviewing first the notions of norm, convergence, and completeness before giving Definition 1.7 of a Banach space. We assume readers are familiar with real and complex vector spaces.

Definition 1.1 (norm) A *norm* $\|\cdot\|$ is a function from a real (respectively, complex) vector space X to $\mathbb{R}^+$ such that

(i) $\|u\| = 0$ if and only if $u = 0$,
(ii) $\|\lambda u\| = |\lambda|\,\|u\|$ for all $u \in X$ and $\lambda \in \mathbb{R}$ (resp., $\mathbb{C}$), and
(iii) $\|u + v\| \le \|u\| + \|v\|$ for all $u, v \in X$ (triangle inequality).

A *normed vector space* $(X, \|\cdot\|)$ is a vector space X with a norm $\|\cdot\|$. If only conditions (ii) and (iii) hold, $\|\cdot\|$ is called a *semi-norm* and denoted $|\cdot|_X$.

Example 1.2 $(\mathbb{R}^d, \|\cdot\|_2)$ is a normed vector space with

$$\|\boldsymbol{u}\|_2 := \left(|u_1|^2 + \cdots + |u_d|^2\right)^{1/2}$$

for the column vector $\boldsymbol{u} = [u_1, \ldots, u_d]^{\mathsf{T}} \in \mathbb{R}^d$, where $|\cdot|$ denotes absolute value. More generally, $\|\boldsymbol{u}\|_\infty := \max\{|u_1|, \ldots, |u_d|\}$ and $\|\boldsymbol{u}\|_p := (|u_1|^p + \cdots + |u_d|^p)^{1/p}$ for $p \ge 1$ is a norm and $(\mathbb{R}^d, \|\cdot\|_p)$ is a normed vector space. When $d = 1$, these norms are all equal to the absolute value.

Definition 1.3 (domain) A domain D is a non-trivial, connected, open subset of $\mathbb{R}^d$ and a domain is bounded if $D \subset \{\boldsymbol{x} \in \mathbb{R}^d : \|\boldsymbol{x}\|_2 \le R\}$ for some $R > 0$. The boundary of a domain is denoted ∂D and we always assume the boundary is piecewise smooth (e.g., the boundary of a polygon or a sphere).

Example 1.4 (continuous functions) For a subset $D \subset \mathbb{R}^d$, let $\mathrm{C}(D)$ denote the set of real-valued continuous functions on D. If D is a domain, functions in $\mathrm{C}(D)$ may be unbounded. However, functions in $\mathrm{C}(\bar{D})$, where $\bar{D}$ is the closure of D, are bounded and $\big(\mathrm{C}(\bar{D}), \|\cdot\|_\infty\big)$ is a normed vector space with the supremum norm,

$$\|u\|_\infty := \sup_{x \in \bar{D}} |u(\boldsymbol{x})|, \qquad u \in \mathrm{C}(\bar{D}).$$

A norm $\|\cdot\|$ on a vector space X measures the size of elements in X and provides a notion of convergence: for $u, u_n \in X$, we write $u = \lim_{n\to\infty} u_n$ or $u_n \to u$ as $n \to \infty$ in $(X, \|\cdot\|)$ if $\|u_n - u\| \to 0$ as $n \to \infty$. For example, the notion of convergence on $\mathrm{C}(\bar{D})$ is known as uniform convergence.

Definition 1.5 (uniform and pointwise convergence) We say $u_n \in \mathrm{C}(\bar{D})$ converges *uniformly* to a limit u if $\|u_n - u\|_\infty \to 0$ as $n \to \infty$. Explicitly, for every $\epsilon > 0$, there exists $N \in \mathbb{N}$ such that, for all $\boldsymbol{x} \in \bar{D}$ and all $n \geq N$, $|u_n(\boldsymbol{x}) - u(\boldsymbol{x})| < \epsilon$. In uniform convergence, N depends only on ϵ. This should be contrasted with the notion of pointwise convergence, which applies to all functions $u_n : D \to \mathbb{R}$. We say $u_n \to u$ *pointwise* if, for every $\boldsymbol{x} \in D$ and every $\epsilon > 0$, there exists $N \in \mathbb{N}$ such that for all $n \geq N$, $|u_n(\boldsymbol{x}) - u(\boldsymbol{x})| < \epsilon$. In pointwise convergence, N may depend both on ϵ and $\boldsymbol{x}$.

There are many techniques, both computational and analytical, for finding approximate solutions $u_n \in X$ to mathematical problems posed on a vector space X. When u_n is a Cauchy sequence and X is complete, u_n converges to some $u \in X$, the so-called limit point, and this is often key in showing a mathematical model is well posed and proving the existence of a solution.

Definition 1.6 (Cauchy sequence, complete) Consider a normed vector space $(X, \|\cdot\|)$. A sequence $u_n \in X$ for $n \in \mathbb{N}$ is called a *Cauchy sequence* if, for all $\epsilon > 0$, there exists an $N \in \mathbb{N}$ such that

$$\|u_n - u_m\| < \epsilon \qquad \text{for all } n, m \geq N.$$

A normed vector space $(X, \|\cdot\|)$ is said to be *complete* if every Cauchy sequence u_n in X converges to a limit point $u \in X$. In other words, there exists a $u \in X$ such that $\|u_n - u\| \to 0$ as $n \to \infty$.

Definition 1.7 (Banach space) A *Banach space* is a complete normed vector space.

Example 1.8 $(\mathbb{R}, |\cdot|)$ and $(\mathbb{R}^d, \|\cdot\|_p)$ for $1 \leq p \leq \infty$ are Banach spaces.

Example 1.9 $\big(\mathrm{C}(\bar{D}), \|\cdot\|_\infty\big)$ is a Banach space if D is bounded. See Exercise 1.1. If D is unbounded, the set of bounded continuous functions $\mathrm{C}_\mathrm{b}(D)$ on D gives a Banach space.

The contraction mapping theorem is used in Chapters 3, 8, and 10 to prove the existence and uniqueness of solutions to initial-value problems.

Theorem 1.10 (contraction mapping) *Let Y be a non-empty closed subset of the Banach space $(X, \|\cdot\|)$. Consider a mapping $\mathfrak{J} : Y \to Y$ such that, for some $\mu \in (0,1)$,*

$$\|\mathfrak{J}u - \mathfrak{J}v\| \leq \mu \|u - v\|, \qquad \text{for all } u, v \in Y. \tag{1.1}$$

There exists a unique fixed point of $\mathfrak{J}$ in Y; that is, there is a unique $u \in Y$ such that $\mathfrak{J}u = u$.

Proof Fix $u_0 \in Y$ and consider $u_n = \mathcal{J}^n u_0$ (the nth iterate of u_0 under application of $\mathcal{J}$). The sequence u_n is easily shown to be Cauchy in Y using (1.1) and therefore converges to a limit $u \in Y$ because Y is complete (as a closed subset of X). Now $u_n \to u$ and hence $u_{n+1} = \mathcal{J}u_n \to \mathcal{J}u$ as $n \to \infty$. We conclude that u_n converges to a fixed point of $\mathcal{J}$.

If $u, v \in Y$ are both fixed points of $\mathcal{J}$, then $\mathcal{J}u - \mathcal{J}v = u - v$. But (1.1) holds and hence $u = v$ and the fixed point is unique. □

Spaces of continuously differentiable functions

The smoothness, also called regularity, of a function is described by its derivatives and we now define spaces of functions with a given number of continuous derivatives. For a domain $D \subset \mathbb{R}^d$ and Banach space $(Y, \|\cdot\|_Y)$, consider a function $u\colon D \to Y$. We denote the partial derivative operator with respect to x_j by $\mathcal{D}_j := \frac{\partial}{\partial x_j}$. Given a multi-index $\boldsymbol{\alpha} = (\alpha_1, \ldots, \alpha_d)$, we define $|\boldsymbol{\alpha}| := \alpha_1 + \cdots + \alpha_d$ and $\mathcal{D}^{\boldsymbol{\alpha}} := \mathcal{D}_1^{\alpha_1} \cdots \mathcal{D}_d^{\alpha_d}$, so that

$$\mathcal{D}^{\boldsymbol{\alpha}} u = \frac{\partial^{|\boldsymbol{\alpha}|} u}{\partial x_1^{\alpha_1} \cdots \partial x_d^{\alpha_d}}.$$

Definition 1.11 (continuous functions)

(i) $\mathrm{C}(D, Y)$ is the set of continuous functions $u\colon D \to Y$. If D is bounded, we equip $\mathrm{C}(\bar{D}, Y)$ with the norm

$$\|u\|_\infty := \sup_{\boldsymbol{x} \in \bar{D}} \|u(\boldsymbol{x})\|_Y. \tag{1.2}$$

(ii) $\mathrm{C}^r(D, Y)$ with $r \in \mathbb{N}$ is the set of functions $u\colon D \to Y$ such that $\mathcal{D}^{\boldsymbol{\alpha}} u \in \mathrm{C}(D, Y)$ for $|\boldsymbol{\alpha}| \le r$; that is, functions whose derivatives up to and including order r are continuous. We equip $\mathrm{C}^r(\bar{D}, Y)$ with the norm

$$\|u\|_{\mathrm{C}^r(\bar{D}, Y)} := \sum_{0 \le |\boldsymbol{\alpha}| \le r} \|\mathcal{D}^{\boldsymbol{\alpha}} u\|_\infty.$$

We abbreviate the notation so that $\mathrm{C}(D, \mathbb{R})$ is denoted by $\mathrm{C}(D)$ and $\mathrm{C}^r(D, \mathbb{R})$ by $\mathrm{C}^r(D)$.

Proposition 1.12 *If the domain D is bounded, $\mathrm{C}(\bar{D}, Y)$ and $\mathrm{C}^r(\bar{D}, Y)$ are Banach spaces.*

Proof The case of $\mathrm{C}(\bar{D})$ is considered in Exercise 1.1. □

The following sets of continuous functions, which are not provided with a norm, are also useful.

Definition 1.13 (infinitely differentiable functions)

(i) $\mathrm{C}^\infty(D, Y)$ is the set $\bigcap_{r \in \mathbb{N}} \mathrm{C}^r(D, Y)$ of infinitely differentiable functions from D to Y.
(ii) $\mathrm{C}_\mathrm{c}^\infty(D, Y)$ is the set of $u \in \mathrm{C}^\infty(D, Y)$ such that $\operatorname{supp} u$ is a compact subset of D, where the support $\operatorname{supp} u$ denotes the closure of $\{\boldsymbol{x} \in D\colon u(\boldsymbol{x}) \neq 0\}$. (The definition of compact is recalled in Definition 1.66).

The spaces $\mathrm{C}^r(D,Y)$ specify the regularity of a function via the number, r, of continuous derivatives. More refined concepts of regularity include Hölder and Lipschitz regularity.

Definition 1.14 (Hölder and Lipschitz) Let $(X, \|\cdot\|_X)$ and $(Y, \|\cdot\|_Y)$ be Banach spaces. A function $u \colon X \to Y$ is *Hölder continuous* with constant $\gamma \in (0,1]$ if there is a constant $L > 0$ so that

$$\|u(x_1) - u(x_2)\|_Y \le L\|x_1 - x_2\|_X^{\gamma}, \qquad \forall x_1, x_2 \in X.$$

If the above holds with $\gamma = 1$, then u is *Lipschitz continuous* or *globally* Lipschitz continuous to stress that L is uniform for $x_1, x_2 \in X$. The space $\mathrm{C}^{r,\gamma}(D)$ is the set of functions in $\mathrm{C}^r(D)$ whose rth derivatives are Hölder continuous with exponent γ.

Lebesgue integrals and measurability

The Lebesgue integral is an important generalisation of the Riemann integral. For a function $u \colon [a,b] \to \mathbb{R}$, the Riemann integral $\int_a^b u(x)\,dx$ is given by a limit of sums $\sum_{j=0}^{N-1} u(\xi_j)(x_{j+1} - x_j)$ for points $\xi_j \in [x_j, x_{j+1}]$, with respect to refinement of the partition $a = x_0 < \cdots < x_N = b$. In other words, u is approximated by a piecewise constant function, whose integral is easy to evaluate, and a limiting process defines the integral of u. The Lebesgue integral is also defined by a limit, but instead of piecewise constant approximations on a partition of $[a,b]$, approximations constant on *measurable sets* are used.

Let 1_F be the indicator function of a set F so

$$1_F(x) := \begin{cases} 1, & x \in F, \\ 0, & x \notin F. \end{cases}$$

Suppose that $\{F_j\}$ are measurable sets in $[a,b]$ (see Definition 1.15) and $\mu(F_j)$ denotes the measure of F_j (e.g., if $F_j = [a,b]$ then $\mu([a,b]) = |b-a|$). The Lebesgue integral of u with respect to the measure μ is defined via

$$\int_a^b u(x)\,d\mu(x) = \lim \sum_j u_j\, \mu(F_j),$$

where the limit is taken as the functions $\sum_j u_j 1_{F_j}(x)$ converge to $u(x)$. The idea is illustrated in Figure 1.1, where the function $u(x)$ is approximated by $\sum_{i=1}^3 u_j 1_{F_j}(x)$ for

$$\begin{aligned} F_1 &= u^{-1}([-1,-0.5]), & F_2 &= u^{-1}((-0.5,0.5]), & F_3 &= u^{-1}((0.5,1]), \\ u_1 &= -0.8, & u_2 &= 0, & u_3 &= 0.8. \end{aligned}$$

Here, $u^{-1}([a,b]) := \{x \in \mathbb{R} \colon u(x) \in [a,b]\}$. To precisely define the Lebesgue integral, we must first form a collection of subsets $\mathcal{F}$ that we can measure.

Definition 1.15 (σ-algebra) A set $\mathcal{F}$ of subsets of a set X is a *σ-algebra* if

(i) the empty set $\{\} \in \mathcal{F}$,
(ii) the complement $F^{\mathrm{c}} := \{x \in X \colon x \notin F\} \in \mathcal{F}$ for all $F \in \mathcal{F}$, and
(iii) the union $\cup_{j \in \mathbb{N}} F_j \in \mathcal{F}$ for $F_j \in \mathcal{F}$.

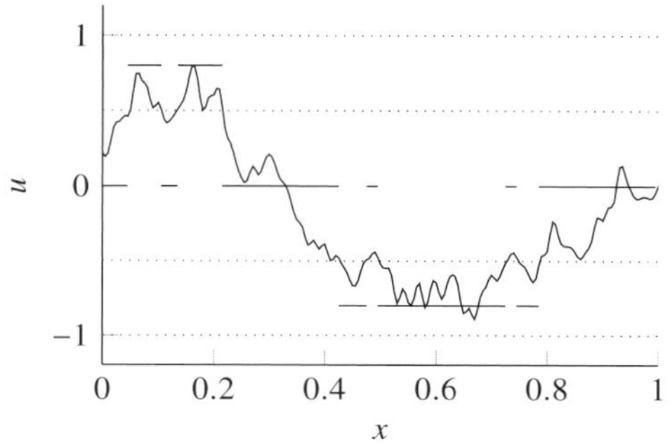

Figure 1.1 We approximate a function $u(x)$ by the simple function $\sum_{j=1}^{3} u_j 1_{F_j}(x)$. The sets F_j arise by partitioning the y-axis and using the pullback sets, which are measurable when u is measurable. In contrast to the Riemann integral, F_j are not simple intervals.

Thus, a σ-algebra is a collection of subsets of X that contains the empty set and is closed under forming complements and countable unions. Any $F \in \mathcal{F}$ is known as a *measurable set* and the pair $(X, \mathcal{F})$ is known as a *measurable space*.

It is natural to ask why the power set of X (i.e., the set of all subsets of X) is not chosen for $\mathcal{F}$. It turns out, for example using Vitali sets or in the Banach–Tarski paradox, that non-intuitive effects arise through measuring every set.

For topological spaces like $\mathbb{R}^d$ or Banach spaces, $\mathcal{F}$ is chosen to be the smallest σ-algebra that contains all open sets; this is known as the *Borel σ-algebra*.

Definition 1.16 (Borel σ-algebra) For a topological space Y, $\mathcal{B}(Y)$ denotes the Borel σ-algebra and equals the smallest σ-algebra containing all open subsets of Y.

Definition 1.17 (measure) A measure μ on a measurable space $(X, \mathcal{F})$ is a mapping from $\mathcal{F}$ to $\mathbb{R}^+ \cup \{\infty\}$ such that

(i) the empty set has measure zero, $\mu(\{\}) = 0$, and
(ii) $\mu(\cup_{j \in \mathbb{N}} F_j) = \sum_{j \in \mathbb{N}} \mu(F_j)$ if $F_j \in \mathcal{F}$ are disjoint (i.e., $F_k \cap F_j = \{\}$ for $k \neq j$).

Together $(X, \mathcal{F}, \mu)$ form a *measure space*. We say the measure space is *σ-finite* if $X = \cup_{j=1}^{\infty} F_j$ for some $F_j \in \mathcal{F}$ with $\mu(F_j) < \infty$. A set $F \in \mathcal{F}$ such that $\mu(F) = 0$ is known as a *null set* and the measure space is said to be *complete* if all subsets of null sets belong to $\mathcal{F}$.

Any measure space $(X, \mathcal{F}, \mu)$ can be extended to a complete measure space by adding in all subsets of null sets. In this book, we always implicitly assume this extension to a complete measure space has been made.

Example 1.18 (Lebesgue measure) The usual notion of volume of subsets of $\mathbb{R}^d$ gives rise to Lebesgue measure, which we denote Leb, on the Borel sets $\mathcal{B}(\mathbb{R}^d)$. The measure space $(\mathbb{R}^d, \mathcal{B}(\mathbb{R}^d), \text{Leb})$ is σ-finite and (once subsets of null sets are included) complete.

Integrals are defined for measurable functions. Let $(X, \mathcal{F}, \mu)$ be a measure space and $(Y, \|\cdot\|_Y)$ be a Banach space.

Definition 1.19 (measurable) (i) A function $u\colon X \to \mathbb{R}$ is *$\mathcal{F}$-measurable* if $\{x \in X\colon u(x) \le a\} \in \mathcal{F}$ for every $a \in \mathbb{R}$. A function $u\colon X \to Y$ is *$\mathcal{F}$-measurable* if the pullback set $u^{-1}(G) \in \mathcal{F}$ for any $G \in \mathcal{B}(Y)$. We simply write *u is measurable* if the underlying σ-algebra is clear and write *u is Borel measurable* if $\mathcal{F}$ is a Borel σ-algebra.

(ii) A measurable function $u = 0$ *almost surely* (a.s.) or *almost everywhere* (a.e.) if $\mu(\{x \in X : u(x) \ne 0\}) = 0$. We sometimes write μ-a.s. to stress the measure.

The following lemma requires the underlying measure space $(X, \mathcal{F}, \mathbb{P})$ to be *complete* (as we assume throughout the book). See Exercise 1.2.

Lemma 1.20 *If $u_n : X \to Y$ is measurable and $u_n(x) \to u(x)$ for almost all $x \in X$, then $u\colon X \to Y$ is measurable.*

Finally, we define the integral of a measurable function $u\colon X \to Y$ with respect to $(X, \mathcal{F}, \mu)$.

Definition 1.21 (integral)

(i) A function $s\colon X \to Y$ is *simple* if there exist $s_j \in Y$ and $F_j \in \mathcal{F}$ for $j = 1, \ldots, N$ with $\mu(F_j) < \infty$ such that

$$s(x) = \sum_{j=1}^{N} s_j 1_{F_j}(x), \qquad x \in X.$$

The integral of a simple function s with respect to a measure space $(X, \mathcal{F}, \mu)$ is

$$\int_X s(x)\, d\mu(x) := \sum_{j=1}^{N} s_j\, \mu(F_j). \tag{1.3}$$

(ii) We say a measurable function u is *integrable with respect to μ* if there exist simple functions u_n such that $u_n(x) \to u(x)$ as $n \to \infty$ for almost all $x \in X$ and u_n is a Cauchy sequence in the sense that, for all $\epsilon > 0$,

$$\int_X \|u_n(x) - u_m(x)\|_Y\, d\mu(x) < \epsilon,$$

for any n, m sufficiently large. Notice that $s(x) = \|u_n(x) - u_m(x)\|_Y$ is a simple function from X to $\mathbb{R}$ and the integral is defined by (1.3).

(iii) If u is integrable, define

$$\int_X u(x)\, d\mu(x) := \lim_{n\to\infty} \int_X u_n(x)\, d\mu(x).$$

Note that $\int_X \|u(x)\|_Y\, d\mu(x) < \infty$ (see Exercise 1.3). If $F \in \mathcal{F}$, define

$$\int_F u(x)\, d\mu(x) := \int_X u(x) 1_F(x)\, d\mu(x).$$

When $Y = \mathbb{R}^d$, the above definition gives the *Lebesgue integral* and, for a Borel set $D \subset \mathbb{R}^d$, the integral with respect to $(D, \mathcal{B}(D), \mathrm{Leb})$ is denoted

$$\int_D u(\boldsymbol{x})\, d\mathrm{Leb}(\boldsymbol{x}) = \int_D u(\boldsymbol{x})\, d\boldsymbol{x}.$$

This corresponds to the usual notion of volume beneath a surface. However, the Lebesgue integral is much more general than the Riemann integral, as the partitions essential for the definition of the Riemann integral do not make sense for the probability spaces of Chapter 4. Definition 1.21 includes the case where Y is a Banach space, the so-called *Bochner integral*.

The following elementary properties hold in general for integrable functions.

(i) For integrable functions $u, v\colon X \to Y$ and $a, b \in \mathbb{R}$,

$$\int_X \big(au(x) + bv(x)\big)\, d\mu(x) = a \int_X u(x)\, d\mu(x) + b \int_X v(x)\, d\mu(x).$$

(ii) For $u\colon X \to Y$,

$$\left\| \int_X u(x)\, d\mu(x) \right\|_Y \le \int_X \|u(x)\|_Y \, d\mu(x). \tag{1.4}$$

We state without proof the dominated convergence theorem on limits of integrals and Fubini's theorem for integration on product spaces.

Theorem 1.22 (dominated convergence) *Consider a sequence of measurable functions $u_n\colon X \to Y$ such that $u_n(x) \to u(x)$ in Y as $n \to \infty$ for almost all $x \in X$. If there is a real-valued integrable function $\bar{U}$ such that $\|u_n(x)\|_Y \le |\bar{U}(x)|$ for $n \in \mathbb{N}$ and almost all $x \in X$, then*

$$\lim_{n\to\infty} \int_X u_n(x)\, d\mu(x) = \int_X \lim_{n\to\infty} u_n(x)\, d\mu(x) = \int_X u(x)\, d\mu(x).$$

Now suppose that $(X_k, \mathcal{F}_k, \mu_k)$ for $k = 1,2$ are a pair of σ-finite measure spaces.

Definition 1.23 (product measure) Denote by $\mathcal{F}_1 \times \mathcal{F}_2$ the smallest σ-algebra containing all sets $F_1 \times F_2$ for $F_k \in \mathcal{F}_k$. The *product measure* $\mu_1 \times \mu_2$ on $\mathcal{F}_1 \times \mathcal{F}_2$ is defined by $(\mu_1 \times \mu_2)(F_1 \times F_2) := \mu_1(F_1)\, \mu_2(F_2)$.

Theorem 1.24 (Fubini) *Suppose that $(X_k, \mathcal{F}_k, \mu_k)$ for $k = 1,2$ are σ-finite measure spaces and consider a measurable function $u\colon X_1 \times X_2 \to Y$. If*

$$\int_{X_2} \left(\int_{X_1} \|u(x_1, x_2)\|_Y \, d\mu_1(x_1) \right) d\mu_2(x_2) < \infty, \tag{1.5}$$

then u is integrable with respect to the product measure $\mu_1 \times \mu_2$ and

$$\begin{aligned}\int_{X_1 \times X_2} u(x_1, x_2)\, d(\mu_1 \times \mu_2)(x_1, x_2) &= \int_{X_2} \left(\int_{X_1} u(x_1, x_2)\, d\mu_1(x_1) \right) d\mu_2(x_2) \\ &= \int_{X_1} \left(\int_{X_2} u(x_1, x_2)\, d\mu_2(x_2) \right) d\mu_1(x_1).\end{aligned}$$

To understand the necessity of (1.5), see Exercise 1.7.

Example 1.25 For Borel sets $X \subset \mathbb{R}^n$ and $Y \subset \mathbb{R}^m$, both measure spaces $(X, \mathcal{B}(X), \mathrm{Leb})$ and $(Y, \mathcal{B}(Y), \mathrm{Leb})$ are σ-finite. Consider $u\colon X \times Y \to \mathbb{R}$ such that $\int_X \int_Y |u(\boldsymbol{x}, \boldsymbol{y})|\, d\boldsymbol{y}\, d\boldsymbol{x} < \infty$. Then Fubini's theorem applies and

$$\int_{X \times Y} u(\boldsymbol{x}, \boldsymbol{y})\, d(\mathrm{Leb} \times \mathrm{Leb})(\boldsymbol{x}, \boldsymbol{y}) = \int_X \int_Y u(\boldsymbol{x}, \boldsymbol{y})\, d\boldsymbol{y}\, d\boldsymbol{x} = \int_Y \int_X u(\boldsymbol{x}, \boldsymbol{y})\, d\boldsymbol{x}\, d\boldsymbol{y}.$$

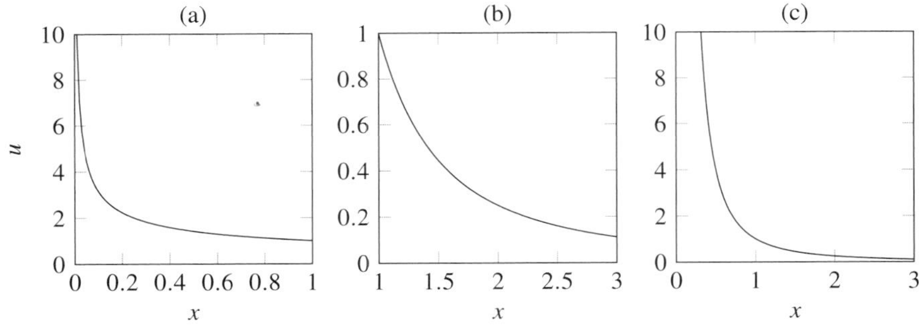

Figure 1.2 (a) If $u(x) = x^{-1/2}$ then $u \in L^p(0,1)$ for $1 \le p < 2$; (b) if $u(x) = x^{-2}$ then $u \in L^p(1,\infty)$ for $p > 2$; (c) if $u(x) = x^{-2}$ then $u \notin L^p(0,\infty)$ for any $p \ge 1$.

L^p spaces

With Lebesgue integrals now defined, we introduce spaces of functions that have finite integrals.

Definition 1.26 (L^p spaces) Let $(Y, \|\cdot\|_Y)$ be a Banach space and $1 \le p < \infty$.

(i) For a domain D, $L^p(D)$ is the set of Borel measurable functions $u\colon D \to \mathbb{R}$ with $\|u\|_{L^p(D)} < \infty$, where

$$\|u\|_{L^p(D)} := \left(\int_D |u(\boldsymbol{x})|^p \, d\boldsymbol{x}\right)^{1/p}. \tag{1.6}$$

$L^1(D)$ comprises the real-valued functions on D that are integrable with respect to Lebesgue measure.

(ii) Let $(X, \mathcal{F}, \mu)$ be a measure space. The space $L^p(X,Y)$ is the set of $\mathcal{F}$-measurable functions $u\colon X \to Y$ such that $\|u\|_{L^p(X,Y)} < \infty$, where

$$\|u\|_{L^p(X,Y)} := \left(\int_X \|u(x)\|_Y^p \, d\mu(x)\right)^{1/p}.$$

$L^1(X,Y)$ comprises the integrable (with respect to μ) functions $X \to Y$. We write $L^p(X)$ for $L^p(X,\mathbb{R})$.

(iii) $L^\infty(X,Y)$ is the set of $\mathcal{F}$-measurable functions $u\colon X \to Y$ such that

$$\|u\|_{L^\infty(X,Y)} := \operatorname*{ess\,sup}_{x \in X} \|u(x)\|_Y < \infty$$

and $\operatorname{ess\,sup}_{x\in X} \|u(x)\|_Y$ denotes the essential supremum or the smallest number that bounds $\|u\|_Y$ almost everywhere. We write $L^\infty(X)$ for $L^\infty(X,\mathbb{R})$.

We examine further the case $p = 2$ in §1.2. The following example illustrates $L^p(D)$ for different choices of p and D.

Example 1.27 Let $D = (0,1)$ and $u(x) = x^\alpha$ with $\alpha < 0$. Then $u \in L^p(D)$ if and only if

$$\int_0^1 x^{\alpha p} \, dx = \left[\frac{x^{1+\alpha p}}{1+\alpha p}\right]_0^1 < \infty \iff 1 + \alpha p > 0.$$

Hence, $u \in L^p(D)$ for $1 \le p < -1/\alpha$. See Figure 1.2(a) for the case $\alpha = -1/2$.

If $D = (1,\infty)$, then $u \in L^p(D)$ if and only if $1 + \alpha p < 0$ or $p > -1/\alpha$. See Figure 1.2(b) for the case $\alpha = -2$. If $D = (0,\infty)$ then $u \notin L^p(D)$ for any $p \ge 1$, as u grows too quickly near the origin or decays too slowly at infinity; see Figure 1.2(c).

More generally, it can be shown $L^q(D) \subset L^p(D)$ for $1 \le p \le q$ for any bounded domain D. See Exercise 1.9.

The basic inequalities for working on Lebesgue spaces are the following:

Hölder's inequality: Suppose that $1 \le p,q \le \infty$ with $1/p + 1/q = 1$ (p,q are said to be *conjugate exponents* and this includes $\{p,q\} = \{1,\infty\}$). For $u \in L^p(X)$ and $v \in L^q(X)$,

$$\|uv\|_{L^1(X)} \le \|u\|_{L^p(X)} \|v\|_{L^q(X)}. \tag{1.7}$$

Minkowski's inequality: For $u,v \in L^p(X,Y)$,

$$\|u+v\|_{L^p(X,Y)} \le \|u\|_{L^p(X,Y)} + \|v\|_{L^p(X,Y)}.$$

Minkowski's inequality provides the triangle inequality needed to establish that $\|\cdot\|_{L^p(X,Y)}$ is a norm. Note however that, as functions in $L^p(X,Y)$ in Definition 1.26 are unique only up to sets of measure zero, $\|\cdot\|_{L^p(X,Y)}$ fails the first axiom in Definition 1.1. The trick is to work on the set of equivalence classes of functions that are equal almost everywhere with respect to the given measure on X. Then, when we refer to functions in the $L^p(X,Y)$ spaces, we mean functions that are representative of those equivalence classes. Using a rigorous definition, it can be shown that $L^p(X,Y)$ is a complete normed vector space.

Lemma 1.28 *$L^p(X,Y)$ is a Banach space for $1 \le p \le \infty$.*

1.2 Hilbert spaces L^2 and H^r

Hilbert spaces are Banach spaces with the additional structure of an inner product.

Definition 1.29 (inner product) An *inner product* on a real (resp., complex) vector space X is a function $\langle\cdot,\cdot\rangle \colon X \times X \to \mathbb{R}$ (resp., $\mathbb{C}$) that is

(i) positive definite: $\langle u,u\rangle \ge 0$ and $\langle u,u\rangle = 0$ if and only if $u = 0$,
(ii) conjugate symmetric: $\langle u,v\rangle = \overline{\langle v,u\rangle}$, where $\bar{\ }$ denotes the complex conjugate of u, and
(iii) linear in the first argument: $\langle \lambda u + \mu v, w\rangle = \lambda\langle u,w\rangle + \mu\langle v,w\rangle$ for real (resp., complex) λ,μ and $u,v,w \in X$.

Definition 1.30 (Hilbert space) Let H be a real (resp., complex) vector space with inner product $\langle\cdot,\cdot\rangle$. Then, H is a real (resp., complex) *Hilbert space* if it is complete with respect to the induced norm $\|u\| := \langle u,u\rangle^{1/2}$. In particular, any Hilbert space is a Banach space.

Example 1.31 $\mathbb{R}^d$ and $\mathbb{C}^d$ are Hilbert spaces with the inner product

$$\langle \boldsymbol{x},\boldsymbol{y}\rangle := \boldsymbol{x}^{\mathsf{T}}\bar{\boldsymbol{y}} = x_1\bar{y}_1 + x_2\bar{y}_2 + \cdots + x_d\bar{y}_d.$$

In this book, the norm $\|\cdot\|_2$ on $\mathbb{R}^d$ denotes $\|\boldsymbol{x}\|_2 := \langle \boldsymbol{x},\boldsymbol{x}\rangle^{1/2} = \left(x_1^2 + \cdots + x_d^2\right)^{1/2}$.

The following inequality is frequently used when working with Hilbert spaces.

Lemma 1.32 (Cauchy–Schwarz inequality) *Let H be a Hilbert space. Then*

$$\left|\left\langle u,v\right\rangle\right| \le \left\|u\right\| \left\|v\right\|, \qquad \forall u,v \in H. \tag{1.8}$$

Proof See Exercise 1.10. □

Hilbert space $L^2(D)$

Recall the space $L^p(D)$ introduced in Definition 1.26. When $p = 2$, there is an inner product and $L^2(D)$ is a Hilbert space. Functions in this space are called *square integrable*.

Proposition 1.33 *For any domain D, $L^2(D)$ is a real Hilbert space with inner product*

$$\left\langle u,v\right\rangle_{L^2(D)} := \int_D u(\boldsymbol{x})v(\boldsymbol{x})\,d\boldsymbol{x}.$$

Proof $L^2(D)$ is a Banach space by Lemma 1.28. It is elementary to verify that $\langle\cdot,\cdot\rangle_{L^2(D)}$ is an inner product and that $\langle u,u\rangle_{L^2(D)}^{1/2}$ agrees with (1.6) with $p = 2$. □

More generally, we may work with the L^2 space of functions taking values in any Hilbert space H.

Proposition 1.34 *Let $(X,\mathcal{F},\mu)$ be a measure space and H be a Hilbert space with inner product $\langle\cdot,\cdot\rangle$. Then, $L^2(X,H)$ is a Hilbert space with inner product*

$$\left\langle u,v\right\rangle_{L^2(X,H)} := \int_X \left\langle u(x),v(x)\right\rangle d\mu(x).$$

Example 1.35 Later on, we meet the complex Hilbert space $L^2(\mathbb{R}^d,\mathbb{C})$ with inner product $\langle u,v\rangle_{L^2(\mathbb{R}^d,\mathbb{C})} := \int_{\mathbb{R}^d} u(\boldsymbol{x})\overline{v(\boldsymbol{x})}\,d\boldsymbol{x}$. For stochastic PDEs and a second measurable space $(\Omega,\mathcal{F},\mathbb{P})$, we use $L^2(\Omega,L^2(D))$, the square integrable functions $u(\boldsymbol{x},\omega)$ on $D\times\Omega$ such that $u(\cdot,\omega) \in L^2(D)$ for almost every $\omega \in D$, with inner product

$$\left\langle u,v\right\rangle_{L^2(\Omega,L^2(D))} := \int_\Omega \int_D u(\boldsymbol{x},\omega)v(\boldsymbol{x},\omega)\,d\boldsymbol{x}\,d\mathbb{P}(\omega).$$

Orthogonal projections and orthonormal bases

Consider a Hilbert space H with inner product $\langle\cdot,\cdot\rangle$ and let G be a subspace of H (i.e., $G \subset H$ and is a vector space).

Definition 1.36 (orthogonal complement) The *orthogonal complement* of a closed subspace G of H is

$$G^\perp := \left\{u \in H : \left\langle u,v\right\rangle = 0 \text{ for all } v \in G\right\}.$$

Note that $G \cap G^\perp = \{0\}$. For a non-zero vector $\boldsymbol{u} \in \mathbb{R}^d$, $G = \operatorname{span}\{\boldsymbol{u}\}$ is a subspace of $\mathbb{R}^d$ and $G^\perp$ is the hyperplane through the origin orthogonal to $\boldsymbol{u}$.

Theorem 1.37 *If G is a closed subspace of a Hilbert space H, every $u \in H$ may be uniquely written $u = p^* + q$ for $p^* \in G$ and $q \in G^\perp$. Further, $\|u - p^*\| = \inf_{p\in G}\|u - p\|$.*

Proof We assume that H is a real Hilbert space. Consider $p_n \in G$ such that $\|u - p_n\| \to \inf_{p\in G}\|u - p\|$ as $n \to \infty$. As G is closed, it may be shown that p_n is a Cauchy sequence that has a limit point p^* in G (see Exercise 1.16). Clearly $\|u - p^*\| = \inf_{p\in G}\|u - p\|$. For $v \in G$, let $\phi(\epsilon) := \|u - p^* + \epsilon v\|^2$. Then $\phi(\epsilon)$ has a minimum at $\epsilon = 0$ and $\phi'(0) = 2\langle u - p^*, v\rangle = 0$ for all $v \in G$, which means that $u - p^* \in G^\perp$. To show uniqueness, suppose $u = p_i + q_i$ for $i = 1, 2$ with $p_i \in G$, $q_i \in G^\perp$. Because $G \cap G^\perp = \{0\}$, we see that $p_1 - p_2 = q_1 - q_2 = 0$ and the decomposition is unique. □

Definition 1.38 (orthogonal projection) The orthogonal projection $P\colon H \to G$ is defined by $Pu = p^*$, where p^* is given in Theorem 1.37. Note that $P^2 = P$.

We can represent the orthogonal projection using an orthonormal basis of H.

Definition 1.39 (orthonormal basis) Let H be a Hilbert space. A set $\{\phi_j : j \in \mathbb{N}\} \subset H$ is *orthonormal* if

$$\langle \phi_j, \phi_k \rangle = \begin{cases} 1, & j = k, \\ 0, & j \neq k. \end{cases}$$

This is also written as $\langle \phi_j, \phi_k \rangle = \delta_{jk}$, where δ_{jk} is known as the *Kronecker delta* function. If the span of $\{\phi_j\}$ is additionally dense in H then we have a *complete* orthonormal set and $\{\phi_j\}$ is said to be an *orthonormal basis* for H. In that case, $u = \sum_{j=1}^\infty \langle u, \phi_j\rangle \phi_j$ for any $u \in H$ and

$$\|u\|^2 = \sum_{j=1}^\infty |\langle u, \phi_j \rangle|^2 = \sum_{j=1}^\infty |\langle \phi_j, u \rangle|^2. \tag{1.9}$$

A Hilbert space is *separable* if it contains a countable dense subset and every separable Hilbert space has an orthonormal basis (as can be proved by applying the Gram–Schmidt orthogonalisation procedure to the countable dense subset).

Example 1.40 For any domain D, $L^2(D)$ is separable and has an orthonormal basis. For example,

$$\left\{ \sqrt{2}\sin(j\pi x) : j \in \mathbb{N} \right\} \qquad \text{is an orthonormal basis for } L^2(0,1),$$
$$\left\{ \frac{1}{\sqrt{2\pi}} \mathrm{e}^{ikx} : k \in \mathbb{Z} \right\} \qquad \text{is an orthonormal basis for } L^2((-\pi,\pi),\mathbb{C}).$$

We frequently work with the orthogonal projection from H to the span of the first N basis functions.

Lemma 1.41 *Consider a separable Hilbert space H with an orthonormal basis $\{\phi_j : j \in \mathbb{N}\}$. Let $G_N :=$ span $\{\phi_1, \dots \phi_N\}$ and denote the orthogonal projection from H to G_N by P_N. Then,*

$$P_N u = \sum_{j=1}^N \langle u, \phi_j \rangle \phi_j. \tag{1.10}$$

In particular, $\|P_N u\| \le \|u\|$ and $\|u - P_N u\| \to 0$ as $N \to \infty$.

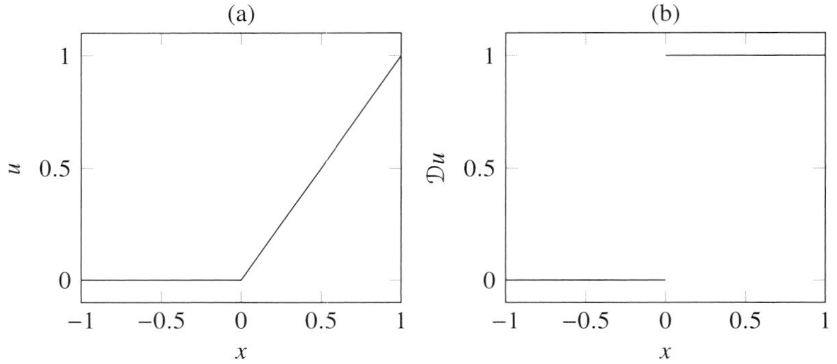

Figure 1.3 (a) Plot of $u(x)$ defined by (1.11) and (b) the derivative of $u(x)$ defined for $x \neq 0$.

Proof Let $u \in H$. For any $p \in G_N$,

$$u - p = \sum_{j=1}^{N} \langle u - p, \phi_j \rangle \phi_j + \sum_{j=N+1}^{\infty} \langle u, \phi_j \rangle \phi_j$$

and $\|u - p\|^2 = \sum_{j=1}^{N} |\langle u - p, \phi_j \rangle|^2 + \sum_{j=N+1}^{\infty} |\langle u, \phi_j \rangle|^2$. Hence, $\|u - p\|$ is minimized by choosing $p = \sum_{j=1}^{N} \langle u, \phi_j \rangle \phi_j$ and the orthogonal projection P_N is given by (1.10). We have $\|P_N u\| \le \|u\|$ because

$$\|u\|^2 = \sum_{j=1}^{\infty} |\langle \phi_j, u \rangle|^2 \qquad \text{and} \qquad \|P_N u\|^2 = \sum_{j=1}^{N} |\langle \phi_j, u \rangle|^2.$$

Finally, $\|u - P_N u\|^2 = \sum_{j=N+1}^{\infty} |\langle \phi_j, u \rangle|^2 \to 0$ as $N \to \infty$. □

Sobolev spaces

Just as $C^r(D, Y)$ describes the regularity of continuous functions, Sobolev spaces describe the regularity of Lebesgue-integrable functions. We assign a meaning to the derivative $\mathcal{D}u$, even when u may not have a derivative in the traditional sense. For example,

$$u(x) = \begin{cases} 0, & x < 0, \\ x, & x \ge 0 \end{cases} \tag{1.11}$$

is piecewise differentiable but, at $x = 0$, the derivative is not defined. See Figure 1.3. We define the derivative

$$\mathcal{D}u(x) = \begin{cases} 0, & x < 0, \\ 0, & x = 0, \\ 1, & x > 0, \end{cases} \tag{1.12}$$

where we have made a choice for $\mathcal{D}u(0)$. To make a formal definition of the derivative and show that this choice for $\mathcal{D}u$ makes sense, we integrate $\mathcal{D}u$ by parts against a test function in $C_c^\infty(D)$ (see Definition 1.13). The space $C_c^\infty(D)$ is used because the functions are differentiable and the space is dense in $L^2(D)$ (any function in $L^2(D)$ is approximated arbitrarily well by one in $C_c^\infty(D)$). Let $D = (a,b)$ and $\phi \in C_c^\infty(D)$. Integration by parts gives

$$\int_a^b \mathcal{D}u(x)\phi(x)\,dx = \big[u(x)\phi(x)\big]_a^b - \int_a^b u(x)\mathcal{D}\phi(x)\,dx.$$

Because ϕ has support in D, ϕ is identically zero outside D and ϕ and all its derivatives are zero on the boundary of D. In particular, $\phi(a) = \phi(b) = 0$ and

$$\int_a^b \mathcal{D}u(x)\phi(x)\,dx = -\int_a^b u(x)\mathcal{D}\phi(x)\,dx, \qquad \forall \phi \in C_c^\infty(\mathbb{R}). \tag{1.13}$$

It is easy to show that (1.12) satisfies this condition and, because the integral is independent of the choice for $\mathcal{D}u(0)$, it does not matter what value is chosen for $\mathcal{D}u(0)$ in (1.12) (see Exercise 1.11). Repeatedly integrating by parts, for $\alpha \in \mathbb{N}$,

$$\int_a^b \mathcal{D}^\alpha u(x)\phi(x)\,dx = (-1)^\alpha \int_a^b u(x)\mathcal{D}^\alpha \phi(x)\,dx.$$

This calculation for functions u of $x \in \mathbb{R}$ also extends to functions of $\boldsymbol{x} \in \mathbb{R}^d$ by using the multi-index notation.

Definition 1.42 (weak derivative) Let Y be a Banach space. We say a measurable function $\mathcal{D}^\alpha u \colon D \to \mathbb{R}$ is the αth *weak derivative* of a measurable function $u \colon D \to Y$ if

$$\int_D \mathcal{D}^\alpha u(\boldsymbol{x})\phi(\boldsymbol{x})\,d\boldsymbol{x} = (-1)^{|\alpha|}\int_D u(\boldsymbol{x})\mathcal{D}^\alpha \phi(\boldsymbol{x})\,d\boldsymbol{x}, \qquad \forall \phi \in C_c^\infty(D).$$

Example 1.43 (Dirac delta function) $\mathcal{D}u$ defined by (1.12) is constant for $x \neq 0$ and hence $\mathcal{D}^2 u(x)$, the derivative of $\mathcal{D}u$, is zero for all $x \neq 0$. However, the zero function in $L^2(\mathbb{R})$ is not the derivative of $\mathcal{D}u$. In fact, $\mathcal{D}^2 u(x) = \delta(x)$, the Dirac delta function, defined by

$$\delta(x) = 0 \quad \text{for } x \neq 0, \qquad \int_{\mathbb{R}} \delta(x)\phi(x)\,dx = \phi(0),$$

for any continuous function $\phi(x)$. In particular, $\int_{-\epsilon}^{\epsilon} \delta(x)\,dx = 1$ for any $\epsilon > 0$ and the delta function has unit mass at $x = 0$, representing the derivative of $\mathcal{D}u$ at $x = 0$. The delta function is frequently used, even though it is not a function in the classical sense and, strictly speaking, requires the theory of distributions.

Definition 1.44 (Sobolev spaces) Let D be a domain and Y be a Banach space. For $p \geq 1$, the Sobolev space $W^{r,p}(D,Y)$ is the set of functions whose weak derivatives up to order $r \in \mathbb{N}$ are in $L^p(D,Y)$. That is,

$$W^{r,p}(D,Y) := \big\{u \colon \mathcal{D}^\alpha u \in L^p(D,Y) \text{ if } |\alpha| \leq r\big\}.$$

If $p = 2$ and H is a Hilbert space, $H^r(D,H)$ is used to denote $W^{r,2}(D,H)$.

Proposition 1.45 *$W^{r,p}(D,Y)$ is a Banach space with norm*

$$\|u\|_{W^{r,p}(D,Y)} := \Bigg(\sum_{0\leq|\alpha|\leq r} \|\mathcal{D}^\alpha u\|_{L^p(D,Y)}^p\Bigg)^{1/p}$$

and $H^r(D,H)$ is a Hilbert space with inner product

$$\langle u,v\rangle_{H^r(D,H)} := \sum_{0\le|\alpha|\le r} \langle \mathcal{D}^\alpha u, \mathcal{D}^\alpha v\rangle_{L^2(D,H)}.$$

We abbreviate $H^r(D,\mathbb{R})$ to $H^r(D)$ and the norm on $H^r(D)$ is

$$\|u\|_{H^r(D)} := \left(\sum_{0\le|\alpha|\le r} \|\mathcal{D}^\alpha u\|^2_{L^2(D)}\right)^{1/2}. \tag{1.14}$$

It is also convenient to define the Sobolev semi-norm (recall Definition 1.1):

$$|u|_{H^r(D)} := \left(\sum_{|\alpha|=r} \|\mathcal{D}^\alpha u\|^2_{L^2(D)}\right)^{1/2}. \tag{1.15}$$

By using fractional powers (see Definition 1.87), it is possible to define fractional derivatives and define Sobolev spaces for any $r \in \mathbb{R}$.

Sobolev spaces that incorporate boundary conditions are needed to analyse solutions to boundary-value problems. For one-dimensional domains $D = (a,b)$, the most frequently used in our studies are, for homogeneous Dirichlet conditions,

$$H_0^1(a,b) := \{u \in H^1(a,b) : u(a) = u(b) = 0\},$$

and, for periodic conditions,

$$H^2_{\mathrm{per}}(a,b) := \{u \in H^2(a,b) : u(a) = u(b) \text{ and } u_x(a) = u_x(b)\}.$$

Precise definitions are given below.

Definition 1.46 (completion) X is the completion of Y with respect to $\|\cdot\|$ if X equals the union of the space Y and the limit points of sequences in Y with respect to $\|\cdot\|$.

Definition 1.47 (Sobolev spaces H_0^1, H^2_{per})

(i) For a domain D, $H_0^1(D)$ is the completion of $C_c^\infty(D)$ with respect to the $H^1(D)$ norm and is a Hilbert space with the $H^1(D)$ inner product.

(ii) $H^2_{\mathrm{per}}(a,b)$ is the completion with respect to the $H^2(a,b)$ norm of the set of $u \in C^\infty([a,b])$ such that the pth derivative $u^{(p)}(a) = u^{(p)}(b)$ for $p = 0,1,\ldots$. It is a Hilbert space with the $H^2(D)$ inner product.

(iii) Let $D := (a_1,b_1)\times(a_2,b_2) \subset \mathbb{R}^2$ be a domain. $H^2_{\mathrm{per}}(D)$ is the completion with respect to the $H^2(D)$ norm of the set of $u \in C^\infty(\bar{D})$ such that

$$\frac{\partial^p}{\partial x_1^p}u(a_1,y) = \frac{\partial^p}{\partial x_1^p}u(b_1,y), \qquad \frac{\partial^p}{\partial x_2^p}u(x,a_2) = \frac{\partial^p}{\partial x_2^p}u(x,b_2)$$

for $x \in (a_1,b_1)$, $y \in (a_2,b_2)$, and $p = 0,1,\ldots$. It is a Hilbert space with the $H^2(D)$ inner product.

Below, we introduce a norm on $H_0^1(D)$ that is equivalent to $\|\cdot\|_{H^1(D)}$. First, we state Poincaré's inequality. Recall the definition of $|u|_{H^1(D)}$ from (1.15).

Theorem 1.48 (Poincaré's inequality) *For a bounded domain D, there exists a constant $K_{\mathrm{p}} > 0$ such that $\|u\|_{L^2(D)} \le K_{\mathrm{p}}|u|_{H^1(D)}$ for any $u \in H_0^1(D)$.*

Proof See Exercise 1.14 for the case $D \subset \mathbb{R}$. □

Definition 1.49 (equivalence of norms) Consider two norms $\|\cdot\|_A$ and $\|\cdot\|_B$ on a Banach space X. We say the norms are *equivalent* if there exist constants $K_1, K_2 > 0$ such that

$$K_1\|u\|_A \le \|u\|_B \le K_2\|u\|_A, \qquad \forall u \in X.$$

For equivalent norms, a sequence u_n converges to u with respect to $\|\cdot\|_A$ or equivalently $\|u_n - u\|_A \to 0$ as $n \to \infty$ if and only if $\|u_n - u\|_B \to 0$ as $n \to \infty$.

Lemma 1.50 *Suppose that D is a bounded domain. Then $\|u\|_{H_0^1(D)} := |u|_{H^1(D)}$ is a norm. Further, $\|u\|_{H^1(D)}$ and $\|u\|_{H_0^1(D)}$ are equivalent norms on $H_0^1(D)$. In particular,*

$$\|u\|_{H_0^1(D)} \le \|u\|_{H^1(D)} \le \left(1 + K_p^2\right)^{1/2}\|u\|_{H_0^1(D)}, \qquad \forall u \in H_0^1(D).$$

Proof This is a simple consequence of the definitions of $\|\cdot\|_{H^1(D)}$ and $\|\cdot\|_{H_0^1(D)}$ and the Poincaré inequality (Theorem 1.48). □

The relationship between the space $H^r(D)$ of functions with Lebesgue-integrable weak derivatives and the space $C^r(\bar{D})$ of functions with classical derivatives is described by the Sobolev embedding theorems.

Theorem 1.51 (Sobolev embedding) *Let D be a bounded domain in $\mathbb{R}^d$ and $u \in H^r(D)$ for $r \in \mathbb{N}$. If $r > d/2$, then $u\colon D \to \mathbb{R}$ can be extended to a continuous function $u\colon \bar{D} \to \mathbb{R}$. Further, $u \in C(\bar{D})$ and there exists $K > 0$ such that $\|u\|_{C(\bar{D})} \le K\|u\|_{H^r(D)}$.*

If $r > s + d/2$ for $s \in \mathbb{N}$, then $u \in C^s(\bar{D})$ and there exists $K > 0$ such that $\|u\|_{C^s(\bar{D})} \le K\|u\|_{H^r(D)}$.

The theorem is dimension dependent. Any function in $H^1(D)$ for $D = (a, b)$ is continuous and bounded, but in higher dimensions this is not true. For example in dimension $d = 2$, the function $u(\boldsymbol{x}) = \log|\log r|$, for $r = \|\boldsymbol{x}\|_2$, belongs to $H^1(D)$ for $D = \{\boldsymbol{x} \in \mathbb{R}^2 : \|\boldsymbol{x}\|_2 < 1/2\}$. However, $u(\boldsymbol{x})$ is unbounded at $\boldsymbol{x} = \boldsymbol{0}$ and does not belong to $C(\bar{D})$. See Exercise 1.12.

1.3 Linear operators and spectral theory

Definition 1.52 (linear operator) Let X and Y be two real (resp., complex) vector spaces. A function $L\colon X \to Y$ is said to be a *linear operator* if, for any $u, v \in X$ and any scalar $a \in \mathbb{R}$ (resp., $\mathbb{C}$),

$$L(u + v) = L(u) + L(v), \qquad L(av) = aL(v).$$

We usually drop the brackets and write Lu for $L(u)$.

Examples include matrix multiplication of vectors in $\mathbb{R}^d$, the projection operator P_N of Lemma 1.41, and differentiation of functions in $C^r(D)$ or $H^r(D)$. We first discuss operators L that are bounded and compact, before looking at the Hilbert–Schmidt spectral theory.

Definition 1.53 (bounded linear operator) Let $(X, \|\cdot\|_X), (Y, \|\cdot\|_Y)$ be Banach spaces. A linear operator $L\colon X \to Y$ is *bounded* if for some $K > 0$

$$\|Lu\|_Y \le K\|u\|_X, \qquad \forall u \in X.$$

We write $\mathcal{L}(X,Y)$ for the set of bounded linear operators $L\colon X \to Y$. In the case $X = Y$, we write $\mathcal{L}(X) = \mathcal{L}(X,X)$.

Bounded linear operators are important because they are exactly the continuous linear operators. (see Exercise 1.15).

Lemma 1.54 *The set $\mathcal{L}(X,Y)$ is a Banach space with the operator norm*

$$\|L\|_{\mathcal{L}(X,Y)} := \sup_{u\neq 0} \frac{\|Lu\|_Y}{\|u\|_X}.$$

If $L \in \mathcal{L}(X,Y)$, then $\|Lu\|_Y \le \|L\|_{\mathcal{L}(X,Y)}\|u\|_X$ for any $u \in X$.

Proof See Exercise 1.15. □

Example 1.55 (matrices) With $X = \mathbb{R}^d = Y$, multiplication by a $d\times d$ matrix is a bounded linear operation. We characterise the $\mathcal{L}(\mathbb{R}^d)$ norm of symmetric matrices in Example 1.71.

Again, the case $Y = \mathbb{R}$ deserves special attention.

Definition 1.56 (bounded linear functional) A *bounded linear functional* on X is a member of $\mathcal{L}(X,\mathbb{R})$. The set of bounded linear functionals on a Banach space is known as the *dual space* and is itself a Banach space with the norm $\|\cdot\|_{\mathcal{L}(X,\mathbb{R})}$.

Bounded linear functionals are characterised by the Riesz representation theorem and the Lax–Milgram lemma.

Theorem 1.57 (Riesz representation) *Let H be a Hilbert space with inner product $\langle\cdot,\cdot\rangle$ and let ℓ be a bounded linear functional on H. There exists a unique $u_\ell \in H$ such that*

$$\langle u_\ell, x\rangle = \ell(x), \qquad \forall x \in H.$$

Definition 1.58 (bilinear form) Let X, Y and Z be three real vector spaces. A function $a\colon X \times Y \to Z$ is a *bilinear form* if, for any $y \in Y$, $x \mapsto a(x,y)$ is a linear operator from X to Z and, for any $x \in X$, $y \mapsto a(x,y)$ is a linear operator from Y to Z.

If X, Y, Z have norms $\|\cdot\|_X$, $\|\cdot\|_Y$, and $\|\cdot\|_Z$, we say a is *bounded* if, for some $\alpha > 0$,

$$\|a(x,y)\|_Z \le \alpha\|x\|_X\|y\|_Y, \qquad \forall x \in X, y \in Y.$$

If $a\colon X \times X \to \mathbb{R}$, we say a is *coercive* if, for some $\beta > 0$,

$$a(x,x) \ge \beta\|x\|_X^2, \qquad \forall x \in X.$$

Lemma 1.59 (Lax–Milgram) *Let H be a real Hilbert space with norm $\|\cdot\|$ and let ℓ be a bounded linear functional on H. Let $a\colon H \times H \to \mathbb{R}$ be a bilinear form that is bounded and coercive. There exists a unique $u_\ell \in H$ such that $a(u_\ell, x) = \ell(x)$ for all $x \in H$.*

The Lax–Milgram lemma implies the Riesz representation theorem, by using the inner product $\langle\cdot,\cdot\rangle$ on H for a. Lax–Milgram, however, is more general because a need not be symmetric or an inner product. Both theorems provide existence and uniqueness of a u_ℓ that represents the bounded linear functional ℓ and are frequently used for the proof of existence and uniqueness of solutions to PDEs (see Chapter 2).

Hilbert–Schmidt operators

An important class of bounded linear operators are the Hilbert–Schmidt operators.

Definition 1.60 Let H, U be separable Hilbert spaces with norms $\|\cdot\|$, $\|\cdot\|_U$ respectively. For an orthonormal basis $\{\phi_j : j \in \mathbb{N}\}$ of U, define the Hilbert–Schmidt norm

$$\|L\|_{\mathrm{HS}(U,H)} := \left(\sum_{j=1}^{\infty} \|L\phi_j\|_U^2\right)^{1/2}.$$

The set $\mathrm{HS}(U,H) := \{L \in \mathcal{L}(U,H) : \|L\|_{\mathrm{HS}(U,H)} < \infty\}$ is a Banach space with the Hilbert–Schmidt norm. An $L \in \mathrm{HS}(U,H)$ is known as a *Hilbert–Schmidt* operator. We write $\|L\|_{\mathrm{HS}} = \|L\|_{\mathrm{HS}(H,H)}$ if $U = H$.

Example 1.61 (matrix) The linear operators from $\mathbb{R}^d$ to $\mathbb{R}^d$ are the $\mathbb{R}^{d\times d}$ matrices, $A = (a_{ij})$. Denote by $\boldsymbol{e}_j$ the standard basis of $\mathbb{R}^d$. The Hilbert–Schmidt norm is $\|A\|_{\mathrm{HS}} = \left(\sum_{i,j=1}^d a_{ij}^2\right)^{1/2}$ and coincides with the Frobenius norm $\|A\|_{\mathrm{F}} := \left(\sum_{i,j=1}^d a_{ij}^2\right)^{1/2}$.

Example 1.62 Define $L \in \mathcal{L}(L^2(0,1))$ by

$$(Lu)(x) = \int_0^x u(y)\,dy, \qquad u \in L^2(0,1), \quad 0 < x < 1. \tag{1.16}$$

Consider the orthonormal basis $\phi_j(x) = \sqrt{2}\sin(j\pi x)$. We have

$$L\phi_j(x) = \sqrt{2}(1 - \cos(j\pi x))\frac{1}{j\pi}.$$

Therefore, $\|L\phi_j\|_{L^2(0,1)} \le 2\sqrt{2}/j\pi$ and L is a Hilbert–Schmidt operator because

$$\|L\|_{\mathrm{HS}}^2 = \sum_{j=1}^{\infty} \|L\phi_j\|_{L^2(0,1)}^2 \le \sum_{j=1}^{\infty} \frac{8}{j^2\pi^2} < \infty.$$

Lemma 1.63 *Let H be separable Hilbert space. If $L \in \mathrm{HS}(H,H)$, then $\|L\|_{\mathcal{L}(H)} \le \|L\|_{\mathrm{HS}}$. In particular, Hilbert–Schmidt operators are bounded.*

Proof Let $\{\phi_j : j \in \mathbb{N}\}$ be an orthonormal basis for H and write $u = \sum_{j=1}^\infty u_j\phi_j$ for $u \in H$. Then, $\|u\| = \left(\sum_{j=1}^\infty u_j^2\right)^{1/2}$ and, using the Cauchy–Schwarz inequality,

$$\|L\|_{\mathcal{L}(H)} = \sup_{\|u\|=1} \|Lu\| = \sup_{\|u\|=1} \left\|\sum_{j=1}^{\infty} Lu_j\phi_j\right\| \le \sup_{\sum_{j=1}^\infty |u_j|^2=1} \sum_{j=1}^{\infty} |u_j|\,\|L\phi_j\|$$
$$\le \left(\sum_{j=1}^{\infty} \|L\phi_j\|^2\right)^{1/2} = \|L\|_{\mathrm{HS}}.$$
□

We now focus on integral operators and their relationship with the Hilbert–Schmidt operators on $L^2(D)$.

Definition 1.64 (integral operator with kernel G) For a domain D and $G \in L^2(D \times D)$, the integral operator L on $L^2(D)$ with *kernel* G is defined by

$$(Lu)(\boldsymbol{x}) := \int_D G(\boldsymbol{x},\boldsymbol{y})u(\boldsymbol{y})\,d\boldsymbol{y}, \qquad \boldsymbol{x} \in D, \quad u \in L^2(D). \tag{1.17}$$

For example, the operator L in (1.16) is an integral operator on $L^2(0,1)$ with kernel $G(x,y) = 1_{(0,x)}(y)$.

Theorem 1.65 (integral operators, Hilbert–Schmidt) *Any integral operator with kernel $G \in L^2(D \times D)$ is a Hilbert-Schmidt operator on $L^2(D)$.*

Conversely, any Hilbert–Schmidt operator L on $L^2(D)$ can be written as (1.17) *with* $\|L\|_{\mathrm{HS}} = \|G\|_{L^2(D\times D)}$.

Proof First, we show that L defined by (1.17) is Hilbert–Schmidt. $L^2(D)$ is a separable Hilbert space and has an orthonormal basis $\{\phi_j\}$. Further, $\{\phi_j(\boldsymbol{x})\phi_k(\boldsymbol{y})\colon j,k = 1,\ldots,\infty\}$ is an orthonormal basis for $L^2(D \times D)$ and

$$G(\boldsymbol{x},\boldsymbol{y}) = \sum_{j,k=1}^{\infty} g_{jk}\phi_j(\boldsymbol{x})\phi_k(\boldsymbol{y}) \tag{1.18}$$

with coefficients $g_{jk} \in \mathbb{R}$ defined by

$$g_{jk} := \left\langle G, \phi_j\phi_k \right\rangle_{L^2(D\times D)} = \int_D\int_D G(\boldsymbol{x},\boldsymbol{y})\phi_j(\boldsymbol{x})\phi_k(\boldsymbol{y})\,d\boldsymbol{x}\,d\boldsymbol{y} = \left\langle \phi_j, L\phi_k \right\rangle_{L^2(D)}.$$

Using (1.9), we obtain

$$\|G\|_{L^2(D\times D)} = \left(\sum_{j,k=1}^{\infty}|g_{jk}|^2\right)^{1/2}, \qquad \|L\phi_k\| = \left(\sum_{j=1}^{\infty}\left|\left\langle\phi_j, L\phi_k\right\rangle\right|^2\right)^{1/2} \tag{1.19}$$

and so

$$\begin{aligned}\|L\|_{\mathrm{HS}} &= \left(\sum_{k=1}^{\infty}\|L\phi_k\|^2\right)^{1/2} = \left(\sum_{j,k=1}^{\infty}\left|\left\langle\phi_j, L\phi_k\right\rangle\right|^2\right)^{1/2} = \left(\sum_{j,k=1}^{\infty}|g_{jk}|^2\right)^{1/2} \\ &= \|G\|_{L^2(D\times D)}.\end{aligned} \tag{1.20}$$

Hence, the integral operator L with kernel G is a Hilbert–Schmidt operator with norm $\|L\|_{\mathrm{HS}} = \|G\|_{L^2(D\times D)} < \infty$.

Hilbert–Schmidt operators on $L^2(D)$ can be written as integral operators, as we now show. Write $L\phi_k = \sum_{j=1}^{\infty} g_{jk}\phi_j$ for $g_{jk} = \langle L\phi_k, \phi_j\rangle_{L^2(D)}$. For $u = \sum_{k=1}^{\infty} u_k\phi_k$,

$$Lu = L\sum_{k=1}^{\infty}u_k\phi_k = \sum_{k=1}^{\infty}u_kL\phi_k = \sum_{k=1}^{\infty}u_k\sum_{j=1}^{\infty}\left\langle L\phi_k,\phi_j\right\rangle_{L^2(D)}\phi_j = \sum_{j,k=1}^{\infty}g_{jk}\phi_j u_k.$$

Now, substitute $u_k = \langle u, \phi_k\rangle_{L^2(D)}$, to find

$$(Lu)(\boldsymbol{x}) = \int_D \sum_{j,k=1}^{\infty} g_{jk}\phi_j(\boldsymbol{x})\phi_k(\boldsymbol{y})u(\boldsymbol{y})\,d\boldsymbol{y},$$

so that L is an integral operator with kernel G defined by (1.18). As L is a Hilbert–Schmidt operator and (1.20) holds, $\sum_{j,k=1}^{\infty}|g_{jk}|^2 < \infty$ and G is well defined in $L^2(D \times D)$. □

We mention one more property of the integral operators.

Definition 1.66 (compact set) A set B in a Banach space X is *compact* if every sequence u_n in B has a subsequence $\tilde{u}_n$ converging to a limit $u \in B$.

The easiest sets to work with in finite dimensions are the closed and bounded sets, because they are compact (the Heine–Borel theorem). Closed and bounded sets are not compact in infinite dimensions; for example, the sequence defined by an orthonormal basis of a Hilbert space has no convergent subsequences (see Exercise 1.18). However, compact sets are still easy to find and important, because the integral operators map bounded sets to compact ones.

Definition 1.67 (compact linear operator) A linear operator $L\colon X \to Y$, where X and Y are Banach spaces, is *compact* if the image of any bounded set B in X has compact closure in Y. In other words, the closure of $L(B) = \{Lu\colon u \in B\}$ is compact in Y.

Theorem 1.68 *For $G \in L^2(D \times D)$, the integral operator L on $L^2(D)$ defined by* (1.17) *is compact.*

Spectral theory of compact symmetric operators

Any real symmetric matrix A can be transformed into a diagonal matrix $\Lambda = Q^{\mathsf{T}} A Q$, by an orthogonal matrix Q of eigenvectors. We now look at the generalisation of this result to infinite-dimensional Hilbert spaces.

Definition 1.69 (symmetric operator) $L \in \mathcal{L}(H)$ is *symmetric* on a Hilbert space H if

$$\langle Lu, v \rangle = \langle u, Lv \rangle \qquad \text{for any } u, v \in H.$$

For the Hilbert space $\mathbb{R}^d$ with inner product $\langle \boldsymbol{u}, \boldsymbol{v} \rangle = \boldsymbol{u}^{\mathsf{T}} \boldsymbol{v}$, the symmetric linear operators in $\mathcal{L}(\mathbb{R}^d, \mathbb{R}^d)$ are the $d \times d$ symmetric matrices.

Definition 1.70 (eigenvalue) If $L \in \mathcal{L}(H)$, λ is an *eigenvalue* of L if there exists a non-zero $\phi \in H$ such that $L\phi = \lambda\phi$. Here, ϕ is called an *eigenvector* or *eigenfunction.*

Example 1.71 ($\mathcal{L}(\mathbb{R}^d)$ norm) Let A be a $d \times d$ matrix and recall that

$$\|A\|_{\mathcal{L}(\mathbb{R}^d)} := \sup_{\boldsymbol{x} \neq \boldsymbol{0}} \frac{\|A\boldsymbol{x}\|_2}{\|\boldsymbol{x}\|_2},$$

where we treat $\mathbb{R}^d$ as the Hilbert space with norm $\|\cdot\|_2$. By differentiating

$$\frac{\|A\boldsymbol{x}\|_2^2}{\|\boldsymbol{x}\|_2^2} = \frac{\boldsymbol{x}^{\mathsf{T}} A^{\mathsf{T}} A \boldsymbol{x}}{\boldsymbol{x}^{\mathsf{T}} \boldsymbol{x}}$$

with respect to $\boldsymbol{x}$, we find that the maximum occurs at an eigenvector of $A^{\mathsf{T}} A$. Further, $\|A\boldsymbol{u}\|_2^2 / \|\boldsymbol{u}\|_2^2 = \lambda$, for an eigenvector $\boldsymbol{u}$ of $A^{\mathsf{T}} A$ with eigenvalue λ. Hence $\|A\|_{\mathcal{L}(\mathbb{R}^d)} = \left(\rho(A^{\mathsf{T}} A)\right)^{1/2}$, where $\rho(\cdot)$ denotes the spectral radius (the magnitude of the largest eigenvalue). When A is symmetric, we obtain $\|A\|_{\mathcal{L}(\mathbb{R}^d)} = \rho(A)$.

Lemma 1.72 *For a domain D, suppose that $G \in L^2(D \times D)$ is symmetric (i.e., $G(\boldsymbol{x}, \boldsymbol{y}) = G(\boldsymbol{y}, \boldsymbol{x})$ for all $\boldsymbol{x}, \boldsymbol{y} \in D$). The integral operator L in* (1.17) *with kernel G is symmetric.*

Proof Let L denote the integral operator defined in (1.17). If $u, v \in L^2(D)$, then $|u|, |v| \in L^2(D)$. Further, $|G| \in L^2(D \times D)$ and hence

$$\int_D \int_D \left| u(\boldsymbol{x}) G(\boldsymbol{x}, \boldsymbol{y}) v(\boldsymbol{y}) \right| d\boldsymbol{x}\, d\boldsymbol{y} < \infty.$$

Therefore, Fubini's theorem applies and

$$\begin{aligned} \langle u, Lv \rangle_{L^2(D)} &= \int_D u(\boldsymbol{x}) \left(\int_D G(\boldsymbol{x}, \boldsymbol{y}) v(\boldsymbol{y})\, d\boldsymbol{y} \right) d\boldsymbol{x} \\ &= \int_D \left(\int_D G(\boldsymbol{y}, \boldsymbol{x}) u(\boldsymbol{x})\, d\boldsymbol{x} \right) v(\boldsymbol{y})\, d\boldsymbol{y} = \langle Lu, v \rangle_{L^2(D)}. \end{aligned}$$ □

The major theorem concerning eigenvalues of compact symmetric operators, such as integral operators on $H = L^2(D)$, is the Hilbert–Schmidt spectral theorem.

Theorem 1.73 (Hilbert–Schmidt spectral) *Let H be a countably infinite-dimensional Hilbert space and let $L \in \mathcal{L}(H)$ be symmetric and compact. Denote the eigenvalues of L by λ_j, ordered so that $|\lambda_j| \ge |\lambda_{j+1}|$, and denote the corresponding eigenfunctions by $\phi_j \in H$. Then,*

(i) *all eigenvalues λ_j are real and $\lambda_j \to 0$ as $j \to \infty$,*
(ii) *ϕ_j can be chosen to form an orthonormal basis for the range of L, and*
(iii) *for any $u \in H$,*

$$Lu = \sum_{j=1}^{\infty} \lambda_j \langle u, \phi_j \rangle \phi_j. \tag{1.21}$$

Example 1.74 (symmetric matrix) If A is a $d \times d$ symmetric matrix, there exists an orthogonal matrix Q of eigenvectors and a diagonal matrix Λ of eigenvalues such that $AQ = Q\Lambda$. Denote the jth entry of Λ by λ_j and the jth column of Q by $\boldsymbol{\phi}_j$. We may assume that $|\lambda_{j+1}| \le |\lambda_j|$. As $\{\boldsymbol{\phi}_j\}$ form an orthonormal basis for $\mathbb{R}^d$, $\boldsymbol{u} = \sum_{j=1}^{d} \langle \boldsymbol{u}, \boldsymbol{\phi}_j \rangle \boldsymbol{\phi}_j$ for any $\boldsymbol{u} \in \mathbb{R}^d$ and the finite-dimensional analogue of (1.21) is

$$A\boldsymbol{u} = \sum_{j=1}^{d} \lambda_j \langle \boldsymbol{u}, \boldsymbol{\phi}_j \rangle \boldsymbol{\phi}_j.$$

Non-negative definite operators

Definition 1.75 (non-negative definite, positive definite) A function $G \colon D \times D \to \mathbb{R}$ is *non-negative definite* if, for any $\boldsymbol{x}_j \in D$ and $a_j \in \mathbb{R}$ for $j = 1, \dots, N$, we have

$$\sum_{j,k=1}^{N} a_j a_k G(\boldsymbol{x}_j, \boldsymbol{x}_k) \ge 0.$$

A linear operator $L \in \mathcal{L}(H)$ on a Hilbert space H is *non-negative definite* (sometimes called positive semi-definite) if $\langle u, Lu \rangle \ge 0$ for any $u \in H$ and *positive definite* if $\langle u, Lu \rangle > 0$ for any $u \in H \setminus \{0\}$.

Example 1.76 (matrices) Let A be a $d \times d$ real symmetric matrix. If $\boldsymbol{u}^{\mathsf{T}} A \boldsymbol{u} > 0$ (respectively, ≥ 0) for all $\boldsymbol{u} \in \mathbb{R}^d \setminus \{\boldsymbol{0}\}$ or, equivalently, if all the eigenvalues of A are positive (resp., non-negative), then A is *positive* (resp., *non-negative*) definite.

Non-negative definite operators arise frequently in the following chapters. We now study non-negative definite operators L on $H = L^2(D)$.

Lemma 1.77 *For a domain D, if $G \in \mathrm{C}(D \times D)$ is a non-negative definite function, then the integral operator L on $L^2(D)$ with kernel G is non-negative.*

Proof For $u, v \in L^2(D)$, we must show

$$\langle u, Lv \rangle = \int_D \int_D u(\boldsymbol{x}) G(\boldsymbol{x}, \boldsymbol{y}) v(\boldsymbol{y}) \, d\boldsymbol{x} \, d\boldsymbol{y} \geq 0.$$

We may choose simple functions G_N of the form

$$G_N(\boldsymbol{x}, \boldsymbol{y}) = \sum_{j,k=1}^{N} G(\boldsymbol{x}_j, \boldsymbol{y}_k) 1_{F_j}(\boldsymbol{x}) 1_{F_k}(\boldsymbol{y}), \qquad \text{for } F_j \in \mathcal{B}(D \times D)$$

and $\boldsymbol{x}_j, \boldsymbol{y}_k \in D$, such that $G_N \to G$ pointwise as $N \to \infty$ (see Exercise 1.4). Then,

$$\langle u, Lv \rangle = \lim_{N \to \infty} \sum_{j,k=1}^{N} \int_{F_j} u(\boldsymbol{x}) \, d\boldsymbol{x} \; G(\boldsymbol{x}_j, \boldsymbol{y}_k) \int_{F_k} v(\boldsymbol{y}) \, d\boldsymbol{y},$$

which is non-negative by the definition of non-negative definite function. □

Lemma 1.78 (Dini) *For a bounded domain D, let $f_n \in \mathrm{C}(\bar{D})$ be such that $f_n(\boldsymbol{x}) \leq f_{n+1}(\boldsymbol{x})$ for $n \in \mathbb{N}$ and $f_n(\boldsymbol{x}) \to f(\boldsymbol{x})$ as $n \to \infty$ for all $\boldsymbol{x} \in \bar{D}$. Then $\|f - f_n\|_\infty \to 0$ as $n \to \infty$.*

Proof Suppose for a contradiction that $\|f - f_n\|_\infty \geq \epsilon$ as $n \to \infty$ for some $\epsilon > 0$. Let $D_n := \{\boldsymbol{x} \in \bar{D} : f(\boldsymbol{x}) - f_n(\boldsymbol{x}) < \epsilon\}$. Then, there exists $\boldsymbol{x}_n \in \bar{D}$ with $\boldsymbol{x}_n \notin D_n$ for all n sufficiently large. Because $\bar{D}$ is compact, $\boldsymbol{x}_n$ has a limit point $\boldsymbol{x} \in \bar{D}$. As f_n is increasing, the set D_n is increasing, $D_n \subset D_{n+1}$, and $\boldsymbol{x} \notin D_n$ for any n. Hence, $f(\boldsymbol{x}) - f_n(\boldsymbol{x}) \geq \epsilon$ for all n. This is a contradiction as $f_n(\boldsymbol{x}) \to f(\boldsymbol{x})$. □

Definition 1.79 (trace class) For a separable Hilbert space H, a non-negative definite operator $L \in \mathcal{L}(H)$ is of *trace class* if $\operatorname{Tr} L < \infty$, where the trace is defined by $\operatorname{Tr} L := \sum_{j=1}^{\infty} \langle L\phi_j, \phi_j \rangle$, for an orthonormal basis $\{\phi_j : j \in \mathbb{N}\}$. See Exercise 1.19.

Theorem 1.80 (Mercer) *For a bounded domain D, let $G \in \mathrm{C}(\bar{D} \times \bar{D})$ be a symmetric and non-negative definite function and let L be the corresponding integral operator* (1.17). *There exist eigenfunctions ϕ_j of L with eigenvalues $\lambda_j > 0$ such that $\phi_j \in \mathrm{C}(\bar{D})$ and*

$$G(\boldsymbol{x}, \boldsymbol{y}) = \sum_{j=1}^{\infty} \lambda_j \phi_j(\boldsymbol{x}) \phi_j(\boldsymbol{y}), \qquad \boldsymbol{x}, \boldsymbol{y} \in \bar{D},$$

where the sum converges in $\mathrm{C}(\bar{D} \times \bar{D})$. Furthermore,

$$\sup_{\boldsymbol{x}, \boldsymbol{y} \in \bar{D}} \left| G(\boldsymbol{x}, \boldsymbol{y}) - \sum_{j=1}^{N} \lambda_j \phi_j(\boldsymbol{x}) \phi_j(\boldsymbol{y}) \right| \leq \sup_{\boldsymbol{x} \in \bar{D}} \sum_{j=N+1}^{\infty} \lambda_j |\phi_j(\boldsymbol{x})|^2. \tag{1.22}$$

The operator L is of trace class and

$$\operatorname{Tr} L = \int_D G(\boldsymbol{x},\boldsymbol{x})\,d\boldsymbol{x}.$$

Proof By Theorem 1.68, L is a compact linear operator on $L^2(D)$. Then, G and hence L is symmetric and Theorem 1.73 applies with $H = L^2(D)$. There exists an orthonormal basis of eigenfunctions ϕ_j for the range of L. By Lemma 1.77, L is non-negative definite and the eigenvalues $\lambda_j > 0$ (eigenfunctions with zero eigenvalues do not contribute to the range of L). Now $\langle \phi_j, L\phi_k\rangle_{L^2(D)} = \lambda_j$ for $j = k$ and $\langle \phi_j, L\phi_k\rangle_{L^2(D)} = 0$ for $j \neq k$, so that (1.18) gives $G(\boldsymbol{x},\boldsymbol{y}) = \sum_{j=1}^{\infty} \lambda_j \phi_j(\boldsymbol{x})\phi_j(\boldsymbol{y})$. Here $\phi_j(\boldsymbol{x})$ is a continuous function of $\boldsymbol{x}$ for $\lambda_j \neq 0$, because $\phi_j(\boldsymbol{x}) = (1/\lambda_j)\int_D G(\boldsymbol{x},\boldsymbol{y})\phi_j(\boldsymbol{y})\,d\boldsymbol{y}$ and the kernel G is continuous.

Define $G_N(\boldsymbol{x},\boldsymbol{y})$ via $G_N(\boldsymbol{x},\boldsymbol{y}) = \sum_{j=1}^{N} \lambda_j\phi_j(\boldsymbol{x})\phi_j(\boldsymbol{y})$. The function $G - G_N$ is non-negative definite, as the corresponding integral operator has non-negative eigenvalues λ_j for $j = N+1,\dots,\infty$. Under the non-negative definite condition (Definition 1.75 with $a_j = 1$ and $\boldsymbol{x}_j = \boldsymbol{x}$), $G(\boldsymbol{x},\boldsymbol{x}) - G_N(\boldsymbol{x},\boldsymbol{x}) \geq 0$ for all $\boldsymbol{x} \in \bar{D}$. Let $g_N(\boldsymbol{x}) := G_N(\boldsymbol{x},\boldsymbol{x})$; then g_N is increasing and bounded and has a limit, say g. The convergence is uniform by Lemma 1.78 and $\|g - g_N\|_\infty \to 0$ as $N \to \infty$.

By the Cauchy–Schwarz inequality (Lemma 1.32),

$$\sum_{j=N+1}^{\infty} \lambda_j\phi_j(\boldsymbol{x})\phi_j(\boldsymbol{y}) \leq \left(\sum_{j=N+1}^{\infty} \lambda_j|\phi_j(\boldsymbol{x})|^2 \sum_{j=N+1}^{\infty} \lambda_j|\phi_j(\boldsymbol{y})|^2\right)^{1/2},$$

so that $\|G - G_N\|_\infty \leq \|g - g_N\|_\infty$, which is (1.22). Then, G is a uniform limit of G_N in $\mathrm{C}(\bar{D}\times\bar{D})$. Finally, $G(\boldsymbol{x},\boldsymbol{x}) = \sum_{j=1}^{\infty}\lambda_j\phi_j(\boldsymbol{x})^2$. Integrating over D and using the fact that $\int_D \phi_j(\boldsymbol{x})^2\,d\boldsymbol{x} = 1$, we see

$$\int_D G(\boldsymbol{x},\boldsymbol{x})\,d\boldsymbol{x} = \sum_{j=1}^{\infty}\lambda_j = \operatorname{Tr} L.$$

As $G \in \mathrm{C}(\bar{D}\times\bar{D})$ and D is bounded, G and the integral of G over D are bounded. Hence, $\operatorname{Tr} L < \infty$ and L has trace class. □

Unbounded operators

The Hilbert–Schmidt and projection operators are examples of bounded operators. The differential operators that arise in the study of PDEs are unbounded and usually defined on a strict subset of an underlying Hilbert space H. The following example with the Laplace operator illustrates the situation well.

Example 1.81 (Laplacian with Dirichlet conditions) Consider the Dirichlet problem for the Poisson equation: for a given $f \in L^2(0,1)$, find u such that

$$-u_{xx} = f, \qquad u(0) = u(1) = 0. \tag{1.23}$$

We discuss this problem in detail in Chapter 2. Here, rather than consider (1.23) with classical derivatives, we use weak derivatives and seek $u \in H^2(0,1)$. We also assume that $u \in H_0^1(0,1)$, the Sobolev space given in (1.47). The Sobolev embedding theorem gives

$H_0^1(0,1) \subset C([0,1])$ and hence the Dirichlet conditions $u(0) = u(1) = 0$ are well defined. We recast (1.23) in terms of an operator A defined by

$$Au := -u_{xx}, \qquad u \in H^2(0,1) \cap H_0^1(0,1). \tag{1.24}$$

Note that A maps from a strict subset of $L^2(0,1)$ to $L^2(0,1)$ and that A is unbounded because, for $\phi_j(x) = \sqrt{2}\sin(j\pi x)$, we have

$$\|A\phi_j\|_{L^2(0,1)} = \pi^2 j^2 \|\phi_j\|_{L^2(0,1)} = \pi^2 j^2 \to \infty \qquad \text{as } j \to \infty.$$

Now (1.23) becomes

$$\text{find } u \in H^2(0,1) \cap H_0^1(0,1) \text{ such that } Au = f. \tag{1.25}$$

Definition 1.82 (domain of linear operator) For a linear operator $A\colon \mathcal{D}(A) \subset H \to H$, the set $\mathcal{D}(A)$ is known as the *domain of* A.

For the operator A in Example 1.81, the domain is $\mathcal{D}(A) = H^2(0,1) \cap H_0^1(0,1)$.

The inverse of an unbounded operator often has much nicer properties. For the operator A in (1.24), A^{-1} is symmetric and compact and the Hilbert–Schmidt spectral theorem applies. We investigate this in Example 1.83. We adopt the convention that L is a bounded operator with eigenvalues ν_j and A is an unbounded operator with eigenvalues λ_j.

Example 1.83 Continuing with Example 1.81 for fixed $y \in (0,1)$, let $G(x,y)$ be the solution of

$$-G_{xx} = \delta(x-y), \qquad G(0,y) = G(1,y) = 0,$$

where δ is the Dirac delta function, introduced in Example 1.43. Here, G is known as the *Green's function* for the operator A. In this case (see Exercise 1.20),

$$G(x,y) = \begin{cases} x(1-y), & x < y, \\ y(1-x), & x \ge y. \end{cases} \tag{1.26}$$

Define

$$(Lf)(x) := \int_0^1 G(x,y)f(y)\,dy, \qquad \forall f \in L^2(0,1).$$

Then

$$A\int_0^1 G(x,y)f(y)\,dy = \int_0^1 AG(x,y)f(y)\,dy = \int_0^1 \delta(x-y)f(y)\,dy = f(x)$$

and $A(Lf) = f$, so that L is the inverse of A. The second derivatives of Lf are well defined in $L^2(0,1)$ and it is easy to check that Lf obeys the Dirichlet boundary conditions. Hence, $Lf \in H^2(0,1) \cap H_0^1(0,1) = \mathcal{D}(A)$ and $u = Lf$ is the solution to (1.25).

Observe that G is a symmetric function and is square integrable. Then, the operator L is symmetric (see Lemma 1.72) and compact (see Theorem 1.68) on $L^2(0,1)$. By Theorem 1.73, there exists a decreasing set of eigenvalues ν_j corresponding to an orthonormal basis of eigenfunctions $\phi_j \in L^2(0,1)$ of L. Now $L\phi_j = \nu_j\phi_j$. Hence, $\phi_j \in \mathcal{D}(A)$ and, applying the operator A, we have $\phi_j = \nu_j A\phi_j$. Then,

$$-A\phi_j + \frac{1}{\nu_j}\phi_j = \frac{d\phi_j^2}{dx^2} + \frac{1}{\nu_j}\phi_j = 0, \qquad \phi_j(0) = 0 = \phi_j(1).$$

The solution is $\nu_j = 1/(\pi^2 j^2)$ for $j \in \mathbb{N}$ and the corresponding normalised eigenfunctions are $\phi_j(x) = \sqrt{2}\sin(j\pi x)$.

From (1.21),

$$Lf = \sum_{j=1}^{\infty} \frac{1}{\pi^2 j^2} \langle f, \phi_j \rangle_{L^2(0,1)} \phi_j, \qquad \forall f \in L^2(0,1).$$

Because $\{\phi_j\}$ is an orthonormal basis for $L^2(0,1)$ and A is symmetric,

$$Au = \sum_{j=1}^{\infty} \langle Au, \phi_j \rangle_{L^2(0,1)} \phi_j = \sum_{j=1}^{\infty} \langle u, A\phi_j \rangle_{L^2(0,1)} \phi_j$$

and, since $A\phi_j = \pi^2 j^2 \phi_j$, $\lambda_j = \pi^2 j^2$ are the eigenvalues of A and

$$Au = \sum_{j=1}^{\infty} \pi^2 j^2 \langle u, \phi_j \rangle_{L^2(0,1)} \phi_j, \qquad \forall u \in \mathcal{D}(A).$$

Hence, we have derived a representation for A in terms of the eigenfunctions ϕ_j.

Example 1.84 (periodic boundary conditions) The Laplace operator with periodic boundary conditions maps constant functions to zero and does not have a well-defined inverse or Green's function. Instead, we consider the following boundary-value problem: find u such that

$$u - u_{xx} = f, \qquad u(-\pi) = u(\pi), \qquad u_x(-\pi) = u_x(\pi), \tag{1.27}$$

for a given $f \in L^2(-\pi,\pi)$. Here, we choose the space $H^2_{\text{per}}(-\pi,\pi)$ for the domain. Again, $H^2_{\text{per}}(-\pi,\pi) \subset \mathrm{C}^1([-\pi,\pi])$ by the Sobolev embedding theorem and the boundary conditions are well defined. Define the operator A by

$$Au := u - u_{xx} \qquad \text{for } u \in \mathcal{D}(A) = H^2_{\text{per}}(-\pi,\pi),$$

so that (1.27) becomes: find $u \in \mathcal{D}(A)$ such that $Au = f$. The Green's function for this problem is the solution of $G - G_{xx} = \delta(x-y)$, subject to the periodic boundary conditions, and is a symmetric and square integrable function (see Exercise 1.20).

Once again, the Hilbert–Schmidt spectral theory applies to $L = A^{-1}$ and the eigenvalues ν of L are found by solving

$$-A\phi + \frac{1}{\nu}\phi = -\phi + \frac{d\phi^2}{dx^2} + \frac{1}{\nu}\phi = 0,$$

subject to $\phi(-\pi) = \phi(\pi)$ and $\phi_x(-\pi) = \phi_x(\pi)$. Using complex notation, the eigenvalues of L are $\nu_k = 1/(1+k^2)$ for $k \in \mathbb{Z}$ and the corresponding eigenfunctions are $\psi_k(x) = \frac{1}{\sqrt{2\pi}} e^{ikx}$. Therefore, A has eigenvalues $\lambda_k = 1/\nu_k = 1 + k^2$ and, for $u \in \mathcal{D}(A)$,

$$Au = \sum_{k \in \mathbb{Z}} (1 + k^2) u_k \psi_k, \qquad u_k = \langle u, \psi_k \rangle_{L^2((-\pi,\pi),\mathbb{C})}.$$

When u is a real-valued function, $u_{-k} = \bar{u}_k$ and Au is also real valued.

In Theorem 1.73, we index the eigenvalues λ_j of L by $j \in \mathbb{N}$ and the corresponding eigenfunctions $\phi_j \in L^2(-\pi,\pi)$ and are real-valued functions. Here, we still have the representation

$$Au = \sum_{j=1}^{\infty} \lambda_j u_j \phi_j, \qquad u_j = \langle u, \phi_j \rangle_{L^2(-\pi,\pi)}, \quad u \in \mathcal{D}(A)$$

if, we set for $k \in \mathbb{N}$,

$$\begin{aligned} \phi_1(x) &= \frac{1}{\sqrt{2\pi}}, & \phi_{2k}(x) &= \frac{1}{\sqrt{\pi}} \sin(kx), & \phi_{2k+1}(x) &= \frac{1}{\sqrt{\pi}} \cos(kx), \\ \lambda_1 &= 1, & \lambda_{2k} &= 1 + k^2, & \lambda_{2k+1} &= 1 + k^2. \end{aligned} \tag{1.28}$$

Powers of a linear operator

Fractional powers, A^α for $\alpha \in \mathbb{R}$, of a linear operator A are useful for understanding Sobolev spaces and analysing PDEs. Here, we take a simple approach and assume the following.

Assumption 1.85 A is a linear operator from $\mathcal{D}(A) \subset H$ to a Hilbert space H with an orthonormal basis of eigenfunctions $\{\phi_j : j \in \mathbb{N}\}$ and eigenvalues $\lambda_j > 0$ ordered so that $\lambda_{j+1} \ge \lambda_j$.

In Examples 1.83 and 1.84, we have already met operators satisfying Assumption 1.85 with $H = L^2(D)$. All operators of this type are self-adjoint.

Lemma 1.86 (self-adjoint) *Let A obey Assumption 1.85. Then, $\langle Au, v \rangle = \langle u, Av \rangle$ for all $u, v \in \mathcal{D}(A)$. We say A is* self-adjoint.

Proof If the assumption holds then $Au = \sum_{j=1}^{\infty} \lambda_j \langle u, \phi_j \rangle \phi_j$ for $u \in \mathcal{D}(A)$. Hence,

$$\langle Au, v \rangle = \sum_{j=1}^{\infty} \lambda_j \langle u, \phi_j \rangle \langle v, \phi_j \rangle = \langle u, Av \rangle.$$ □

As the eigenvalues are strictly positive, λ_j^α is well defined in $\mathbb{R}^+$ for any $\alpha \in \mathbb{R}$ and we have the following definition.

Definition 1.87 (fractional powers of a linear operator) Under Assumption 1.85, define the fractional power $A^\alpha u$ for $\alpha \in \mathbb{R}$ by

$$A^\alpha u := \sum_{j=1}^{\infty} \lambda_j^\alpha u_j \phi_j, \qquad \text{for } u = \sum_{j=1}^{\infty} u_j \phi_j, \quad u_j \in \mathbb{R},$$

and let the domain $\mathcal{D}(A^\alpha)$ be the set of $u = \sum_{j=1}^{\infty} u_j \phi_j$ such that $A^\alpha u \in H$.

Theorem 1.88 *Under Assumption 1.85, the domain $\mathcal{D}(A^\alpha)$ for $\alpha \in \mathbb{R}$ is a Hilbert space with inner product $\langle u, v \rangle_\alpha := \langle A^\alpha u, A^\alpha v \rangle$ and corresponding induced norm $\|u\|_\alpha := \|A^\alpha u\|$.*

We list some elementary properties of $\mathcal{D}(A^\alpha)$.

Lemma 1.89 *Let Assumption 1.85 hold.*

(i) $\|u\|_0 = \|u\|$ *for* $u \in \mathcal{D}(A^0) = H$.
(ii) $\langle u, v\rangle_{1/2} = \langle A^{1/2}u, A^{1/2}v\rangle = \langle Au, v\rangle$ *for* $u \in \mathcal{D}(A)$ *and* $v \in \mathcal{D}(A^{1/2})$.
$\langle u, v\rangle = \langle A^{1/2}u, A^{-1/2}v\rangle$ *for* $u \in \mathcal{D}(A^{1/2})$ *and* $v \in H$.
(iii) *For* $\alpha > 0$, $\|u\|_\alpha \geq \lambda_1^\alpha \|u\|$ *for* $u \in \mathcal{D}(A^\alpha) \subset H$.
(iv) *For* $\alpha < 0$, $H \subset \mathcal{D}(A^\alpha)$ *and* $A^\alpha \in \mathcal{L}(H)$ *with* $\|A^\alpha\|_{\mathcal{L}(H)} = \lambda_1^\alpha$.

Proof The result is an elementary consequence of Definition 1.87 and the fact that λ_j is a non-decreasing sequence. For example, for $\alpha < 0$, A^α has non-increasing eigenvalues λ_j^α bounded from above by λ_1^α and

$$\|A^\alpha u\| = \left(\sum_{j=1}^{\infty} \lambda_j^{2\alpha} |u_j|^2\right)^{1/2} \leq \lambda_1^\alpha \|u\|$$

with equality if $u = \phi_1$. Then, $\|A^\alpha\|_{\mathcal{L}(H)} = \sup_{u\neq 0}\|A^\alpha u\|/\|u\| = \lambda_1^\alpha$. □

Example 1.90 (fractional powers for Laplacian with Dirichlet conditions) Continuing with Example 1.83, the Laplace operator $Au = -u_{xx}$ with domain $\mathcal{D}(A) = H^2(0,1) \cap H_0^1(0,1)$ has eigenfunctions $\phi_j(x) = \sqrt{2}\sin(j\pi x)$ and eigenvalues $\lambda_j = j^2\pi^2$ for $j \in \mathbb{N}$. For $u = \sum_j u_j \phi_j \in \mathcal{D}(A^\alpha)$, the fractional power A^α and fractional power norm $\|\cdot\|_\alpha$ are then given by

$$A^\alpha u = \sum_{j=1}^{\infty} (j\pi)^{2\alpha} u_j \phi_j, \qquad \|u\|_\alpha = \left(\sum_{j=1}^{\infty} (j\pi)^{4\alpha} u_j^2\right)^{1/2}.$$

Example 1.91 (fractional power with periodic conditions) Consider the operator $Au = u - u_{xx}$ with domain $\mathcal{D}(A) = H^2_{\text{per}}(-\pi,\pi)$ from Example 1.84. We write the definition of A^α in complex notation using the eigenfunctions $\psi_k(x) = \frac{1}{\sqrt{2\pi}}\mathrm{e}^{\mathrm{i}kx}$ and eigenvalues $\lambda_k = 1 + k^2$ indexed by $k \in \mathbb{Z}$. Writing $u = \sum_{k\in\mathbb{Z}} u_k \psi_k$, we have

$$A^\alpha u = \sum_{k\in\mathbb{Z}} (1 + k^2)^\alpha u_k \psi_k, \qquad \|u\|_\alpha = \left(\sum_{k\in\mathbb{Z}} (1 + k^2)^{2\alpha} |u_k|^2\right)^{1/2}.$$

Once again, when $u_{-k} = \bar{u}_k$, both u and $A^\alpha u$ are real valued.

Example 1.92 (Dirac delta function) Continuing from Example 1.91, we show that the Dirac delta function belongs to $\mathcal{D}(A^\alpha)$ for $\alpha < -1/4$ and that $\mathcal{D}(A^\alpha)$ is strictly larger than $L^2(-\pi,\pi)$. Although the delta function is not a function, formally we can write

$$\delta = \sum_{k\in\mathbb{Z}} \langle \delta, \psi_k\rangle_{L^2((-\pi,\pi),\mathbb{C})} \psi_k = \sum_{k\in\mathbb{Z}} \overline{\psi_k(0)} \psi_k \tag{1.29}$$

because, for continuous functions ϕ,

$$\langle \delta, \phi\rangle_{L^2((-\pi,\pi),\mathbb{C})} = \int_{-\pi}^{\pi} \delta(x)\,\overline{\phi(x)}\,dx = \overline{\phi(0)}.$$

Substituting for ψ_k, $\delta(x) = \frac{1}{2\pi}\sum_{k\in\mathbb{Z}} \mathrm{e}^{\mathrm{i}kx}$ and

$$A^\alpha \delta(x) = \frac{1}{2\pi} \sum_{k\in\mathbb{Z}} (1 + k^2)^\alpha \mathrm{e}^{\mathrm{i}kx}.$$

The sum is real valued as the imaginary parts of $e^{\pm ikx}$ cancel. Although δ and the series (1.29) are not well defined in $L^2(-\pi,\pi)$, the fractional power $A^\alpha\delta$ is well defined in $L^2(-\pi,\pi)$ if $\alpha < -1/4$, as

$$\|A^\alpha \delta\|_{L^2(-\pi,\pi)} = \left(\frac{1}{2\pi}\sum_{k\in\mathbb{Z}}(1+k^2)^{2\alpha}\right)^{1/2}.$$

We conclude that the delta function $\delta \in \mathcal{D}(A^\alpha)$ for $\alpha < -1/4$.

Fractional powers have a number of important applications. In §3.3, they are used to analyse the exponential e^{-At} of a linear operator. We now give two applications: the first shows a relationship between the Sobolev spaces $H^r(-\pi,\pi)$ and fractional powers of the Laplacian. The second describes the approximation of u by its projection onto the first N eigenfunctions.

Proposition 1.93 *For $D = (-\pi,\pi)$, let $Au = u - u_{xx}$ with domain $\mathcal{D}(A) = H^2_{\mathrm{per}}(-\pi,\pi)$ and let $\|\cdot\|_\alpha$ denote the corresponding fractional power norm. For $r \in \mathbb{N}$, the $H^r(-\pi,\pi)$ norm (see Definition 1.44) is equivalent to the fractional power norm $\|u\|_{r/2}$.*

Proof For $u \in L^2(-\pi,\pi)$, write

$$u(x) = \frac{1}{\sqrt{2\pi}}\sum_{k\in\mathbb{Z}} u_k e^{ikx}$$

for some $u_k \in \mathbb{C}$ with $u_{-k} = \bar{u}_k$. Differentiation of u with respect to x results in the coefficient u_k being multiplied by ik. In terms of the Sobolev norms, this means

$$\|u\|^2_{H^r(-\pi,\pi)} = \sum_{0\le|\alpha|\le r}\|\mathcal{D}^\alpha u\|^2_{L^2(-\pi,\pi)} = \sum_{k\in\mathbb{Z}}(1+k^2+\cdots+k^{2r})|u_k|^2. \tag{1.30}$$

The condition $u \in H^r(-\pi,\pi)$ now concerns the decay of the coefficients $|u_k|$ as $|k| \to \infty$. Note that $1 + k^2 \le 2k^2$ for $k \ne 0$ and

$$\frac{1}{2^r}(1+k^2)^r \le (1+k^2+\cdots+k^{2r}) \le (1+k^2)^r.$$

As $\|u\|^2_{r/2} = \sum_{k\in\mathbb{Z}}(1+k^2)^r|u_k|^2$, we obtain $\frac{1}{2^r}\|u\|^2_{r/2} \le \|u\|^2_{H^r(-\pi,\pi)} \le \|u\|^2_{r/2}$, which gives the required norm equivalence. □

We have treated, for simplicity, the case $r \in \mathbb{N}$, but the proposition can be extended when $H^r(-\pi,\pi)$ is defined for negative and non-integer values r. In that case, following Example 1.92, we can say the delta function $\delta \in H^r(-\pi,\pi)$ for $r < -1/2$.

Finally, we study the orthogonal projection P_N from H onto the first N eigenfunctions of A (see Lemma 1.41). $P_N u$ is often used to approximate u and, when $u \in \mathcal{D}(A^\alpha)$ for some $\alpha > 0$, we have the following estimate on the approximation error $\|P_N u - u\|$. We see smoother functions (where α is large) are easier to approximate.

Lemma 1.94 *Let Assumption 1.85 hold. For $\alpha > 0$, $\|u - P_N u\| \le \lambda_{N+1}^{-\alpha}\|u\|_\alpha$ for $u \in \mathcal{D}(A^\alpha)$ and $\|A^{-\alpha}(I - P_N)\|_{\mathcal{L}(H)} = \lambda_{N+1}^{-\alpha}$.*

Proof With $u_j = \langle u, \phi_j \rangle$, we have

$$u - P_N u = \sum_{j=N+1}^{\infty} u_j \phi_j = \sum_{j=N+1}^{\infty} \lambda_j^{-\alpha} (\lambda_j^{\alpha} u_j) \phi_j.$$

As $\lambda_j^{-\alpha}$ is non-increasing, $\|u - P_N u\| \le \lambda_{N+1}^{-\alpha} \|u\|_\alpha$. Further,

$$\|A^{-\alpha}(I - P_N)u\| = \left(\sum_{j=N+1}^{\infty} \lambda_j^{-2\alpha} |u_j|^2 \right)^{1/2} \le \lambda_{N+1}^{-\alpha} \|(I - P_N)u\| \le \lambda_{N+1}^{-\alpha} \|u\|,$$

with equality if $u = \phi_{N+1}$. The result now follows from the definition of the operator norm (Lemma 1.54). □

1.4 Fourier analysis

In Fourier analysis, a function is decomposed into a linear combination of elementary waves of different frequencies. These waves are typically sines or cosines (often written using the complex exponential) and coincide with the eigenfunctions of the Laplacian (see Examples 1.83 and 1.84). A given function u in a Hilbert space (the so-called physical space) can be written as a linear combination of waves, and the set of coefficients in this expansion belong to a second Hilbert space, known as the frequency domain or Fourier space. The mapping between the physical space and the frequency domain is known as a Fourier transform, and it is an invertible mapping that preserves the inner products and hence the norms on the Hilbert spaces up to a scalar constant. In other words, the Fourier transform is an isometry. Many problems admit elegant solutions in the frequency domain, as we see when studying differential equations and stochastic processes in later chapters. We briefly review the following four tools from Fourier analysis:

The discrete Fourier transform (DFT) which maps vectors in $\mathbb{C}^d$ to vectors in $\mathbb{C}^d$.

Fourier series which maps functions $u \in L^2((a,b),\mathbb{C})$ to a set of coefficients $\{u_k\} \in \ell^2 := L^2(\mathbb{Z},\mathbb{C})$.

The Fourier transform which maps functions in $L^2(\mathbb{R}^d,\mathbb{C})$ to functions in $L^2(\mathbb{R}^d,\mathbb{C})$.

The Hankel transform which maps isotropic functions in $L^2(\mathbb{R}^d,\mathbb{C})$ to isotropic functions in $L^2(\mathbb{R}^d,\mathbb{C})$.

Discrete Fourier Transform

The discrete Fourier transform (DFT) is a key computational tool for applying Fourier transforms. For example, it is used in Chapter 3 for the numerical solution of PDEs and in Chapters 6 and 7 to generate approximations of Gaussian random fields. The DFT gives an expression for a vector $\boldsymbol{u} \in \mathbb{C}^d$ in terms of the columns of the following matrix.

Definition 1.95 (Fourier matrix) The Fourier matrix W is the $d \times d$ matrix with complex entries $w_{\ell m} = \omega^{(\ell-1)(m-1)}/\sqrt{d}$, where $\omega := \mathrm{e}^{-2\pi \mathrm{i}/d}$ is the dth root of unity.

The Fourier matrix is unitary (i.e., $W^*W = I$ where * denotes conjugate transpose), as is verified by the following calculation. Let the mth column of W be $\boldsymbol{w}_m$. Then,

$$\boldsymbol{w}_i^* \boldsymbol{w}_m = \frac{1}{d} \sum_{\ell=1}^{d} \omega^{-(\ell-1)(i-1)} \omega^{(\ell-1)(m-1)} = \frac{1}{d} \sum_{\ell=1}^{d} \omega^{(\ell-1)(m-i)} = \delta_{im},$$

where δ_{im} is the Kronecker delta function. In other words, the vectors $\{\boldsymbol{w}_m : m = 1, \ldots, d\}$ form an orthonormal basis for $\mathbb{C}^d$, which is known as the Fourier basis. It is the expression of vectors in $\mathbb{C}^d$ in the Fourier basis that is of interest.

Definition 1.96 (discrete Fourier transform) The DFT of $\boldsymbol{u} \in \mathbb{C}^d$ is the vector $\hat{\boldsymbol{u}} = [\hat{u}_1, \ldots, \hat{u}_d]^{\mathsf{T}} \in \mathbb{C}^d$ given by

$$\hat{u}_\ell := \sum_{m=1}^{d} u_m \omega^{(\ell-1)(m-1)}, \qquad \ell = 1, \ldots, d, \quad \omega := \mathrm{e}^{-2\pi \mathrm{i}/d}. \tag{1.31}$$

In many applications, $\boldsymbol{u} \in \mathbb{R}^d$ (is real valued), in which case $\hat{u}_\ell = \bar{\hat{u}}_{d-\ell}$. The *inverse discrete Fourier transform (IDFT)* of $\hat{\boldsymbol{u}} \in \mathbb{C}^d$ is the vector $\boldsymbol{u} \in \mathbb{C}^d$ defined by

$$u_\ell := \frac{1}{d} \sum_{m=1}^{d} \hat{u}_m \omega^{-(m-1)(\ell-1)}, \qquad \ell = 1, \ldots, d. \tag{1.32}$$

The coefficients $\hat{u}_\ell$ of the DFT are *not* the coefficients of $\boldsymbol{u}$ in the Fourier basis $\boldsymbol{w}_m$ and in fact $\hat{\boldsymbol{u}} = \sqrt{d}\, W \boldsymbol{u}$. The choice of scaling means that

$$\|\hat{\boldsymbol{u}}\|_2 = \sqrt{d} \|\boldsymbol{u}\|_2, \qquad \hat{\boldsymbol{u}}^{\mathsf{T}} \hat{\boldsymbol{v}} = d\, \boldsymbol{u}^{\mathsf{T}} \boldsymbol{v}, \tag{1.33}$$

which are known as the Plancherel and Parseval identities respectively. This scaling is used in computing environments such as Matlab and (1.31)–(1.32) agree with the Matlab commands `fft` and `ifft` respectively. Routines such as `fft` implement the fast Fourier transform (FFT) algorithm and are very efficient, especially when d is a power of 2, running in $\mathcal{O}(d \log d)$ operations. Compare this to a standard matrix vector multiplication, which requires $\mathcal{O}(d^2)$ operations.

Fourier series $L^2((a,b),\mathbb{C}) \longleftrightarrow L^2(\mathbb{Z},\mathbb{C})$

As in Example 1.84, the functions $\frac{1}{\sqrt{2\pi}} \mathrm{e}^{\mathrm{i}kx}$ form an orthonormal basis for $L^2((-\pi,\pi),\mathbb{C})$ and, similarly, $\psi_k(x) := \frac{1}{\sqrt{b-a}} \mathrm{e}^{2\pi \mathrm{i}kx/(b-a)}$ form an orthonormal basis for $L^2((a,b),\mathbb{C})$. Hence, any $u \in L^2((a,b),\mathbb{C})$ can be written as

$$u(x) = \frac{1}{\sqrt{b-a}} \sum_{k=-\infty}^{\infty} u_k \mathrm{e}^{2\pi \mathrm{i}kx/(b-a)}, \qquad u_k = \langle u, \psi_k \rangle_{L^2((a,b),\mathbb{C})}.$$

In the real-valued case, $u \in L^2(a,b)$ and $u_{-k} = \bar{u}_k$. This expansion is called a Fourier series and provides a mapping from the space $L^2((a,b),\mathbb{C})$ to the set of Fourier coefficients $\{u_k\} \in L^2(\mathbb{Z},\mathbb{C})$. Here, $L^2(\mathbb{Z},\mathbb{C})$ is the Hilbert space of sequences $\{u_k : k \in \mathbb{Z}\}$ that are square integrable, which we denote ℓ^2, and has inner product $\langle \{u_k\}, \{v_k\} \rangle_{\ell^2} := \sum_k u_k \bar{v}_k$.

It is convenient to rescale the above equations when defining the Fourier series.

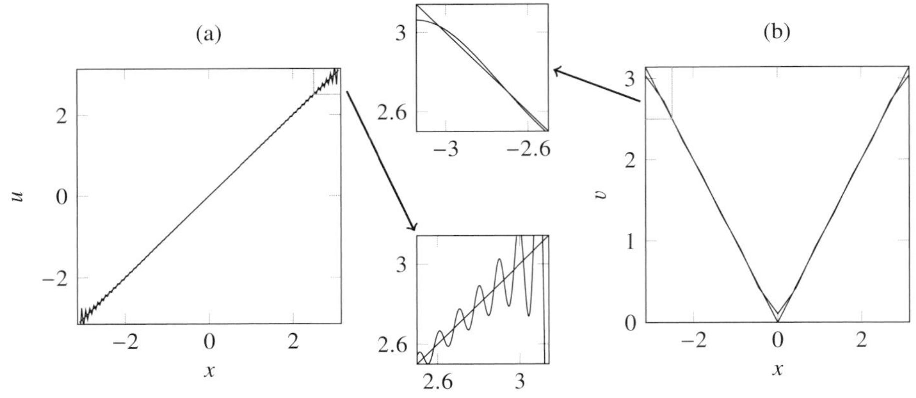

Figure 1.4 Plots of (a) $u(x) = x$ and the series (1.37) truncated to 64 terms and (b) $v(x) = |x|$ and the series (1.38) truncated to 8 terms, with an enlargement of the behaviour near to the boundary. The large oscillations seen in (a) are called the Gibbs phenomenon and occur because $u(-\pi) \neq u(\pi)$.

Definition 1.97 (Fourier series) Let $u \in L^2((a,b),\mathbb{C})$. The Fourier series is

$$u(x) = \sum_{k=-\infty}^{\infty} u_k e^{2\pi i k x/(b-a)}, \tag{1.34}$$

where the Fourier coefficients $\{u_k\} \in \ell^2$ are defined by

$$u_k := \frac{1}{b-a} \int_a^b u(x) e^{-2\pi i k x/(b-a)}\, dx. \tag{1.35}$$

With this scaling, the Parseval identity says that

$$\langle u, v \rangle_{L^2((a,b),\mathbb{C})} = (b-a)\langle \{u_k\}, \{v_k\} \rangle_{\ell^2}. \tag{1.36}$$

Example 1.98 (explicit calculations of coefficients) The function $u(x) = x$ on $(-\pi,\pi)$ has Fourier coefficients $u_0 = 0$ and

$$u_k = \frac{1}{2\pi} \int_{-\pi}^{\pi} x e^{-ikx}\, dx = -\frac{i}{2\pi} \int_{-\pi}^{\pi} x \sin(kx)\, dx = \frac{i}{k}(-1)^k, \quad \text{for } k \neq 0.$$

The function $v(x) = |x|$ on $(-\pi,\pi)$ has Fourier coefficients $v_0 = \pi/2$ and

$$v_k = \frac{1}{\pi} \int_0^{\pi} x e^{-ikx}\, dx = \frac{1}{\pi} \int_0^{\pi} x \cos(kx)\, dx = \frac{1}{\pi k^2}(\cos(k\pi) - 1), \qquad \text{for } k \neq 0.$$

The coefficients v_k are real because $v(x)$ is an even function. We also have

$$u(x) = \sum_{k=-\infty}^{\infty} u_k e^{ikx} = \sum_{j=1}^{\infty} \frac{2}{j}(-1)^{j+1} \sin(jx), \tag{1.37}$$

$$v(x) = \sum_{k=-\infty}^{\infty} v_k e^{ikx} = \frac{\pi}{2} + \sum_{j=1}^{\infty} \frac{2(\cos(j\pi) - 1)}{\pi j^2} \cos(jx). \tag{1.38}$$

Odd functions can always be written as Fourier sine series and even functions can be written as Fourier cosine series. See Figure 1.4 and Exercise 1.28. Notice that $|u_k| = \mathcal{O}(1/k)$ and

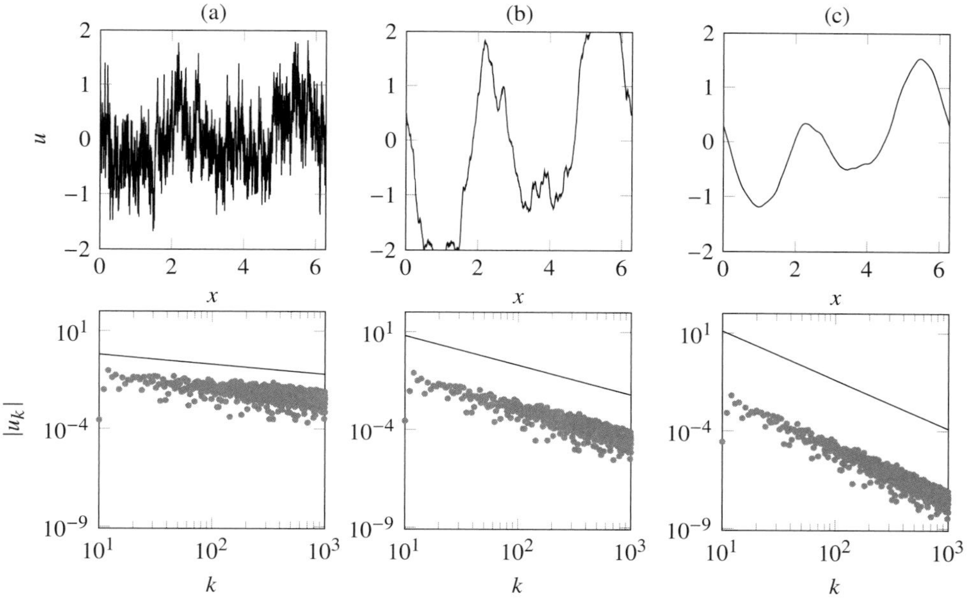

Figure 1.5 For three sets of Fourier coefficients u_k, the lower plots show $|u_k|$ against k and a line with slope (a) $-1/2$, (b) $-3/2$, (c) $-5/2$ showing the rate of decay of u_k. The upper plots show the corresponding function $u(x)$ on $[0, 2\pi]$ given by (1.34) and we know that (a) $u \in L^2(0, 2\pi)$, (b) $u \in H^1(0, 2\pi)$, and (c) $\in H^2(0, 2\pi)$. The function $u(x)$ is smoother when u_k decays more rapidly.

$|v_k| = \mathcal{O}(1/k^2)$. The rapid rate of decay of $|v_k|$ compared to $|u_k|$ results from fact that the periodic extension of v is continuous while that of u is discontinuous. The rate of decay of Fourier coefficients is of keen interest in later chapters.

Example 1.99 (smoothness) To illustrate how the rate of decay of u_k determines the smoothness of the function $u(x)$, Figure 1.5 gives three sets of Fourier coefficients u_k with rates of decay (a) $-1/2$, (b) $-3/2$, and (c) $-5/2$. The corresponding functions $u(x)$, defined by (1.34) on $[0, 2\pi]$, are shown in the upper plots of Figure 1.5 and belong to (a) $L^2(0, 2\pi)$, (b) $H^1(0, 2\pi)$, and (c) $H^2(0, 2\pi)$ respectively. Visually, we see a change in the smoothness of $u(x)$ as the decay rate of u_k is varied and in fact $u \in H^r(0, 2\pi)$ if $u_k = \mathcal{O}(|k|^{-(2r+1+\epsilon)/2})$ for some $\epsilon > 0$ (see Exercise 1.25).

In general, to use the Fourier series of a function $u(x)$, we find the coefficients u_k via a numerical approximation and truncate the Fourier series (1.34) to a finite number of terms. We now describe how to approximate u_k. For an even integer J, consider the set of points $x_j = a + jh$ for $h = (b - a)/J$ and $j = 0, \dots, J$. We approximate u_k by U_k, the composite trapezium rule approximation (see (A.1)) to the integral (1.35). That is,

$$U_k := \frac{h}{b-a}\left(\frac{u(a)e^{-2\pi i k a/(b-a)}}{2} + \sum_{j=1}^{J-1} u(x_j)e^{-2\pi i k x_j/(b-a)} + \frac{u(b)e^{-2\pi i k b/(b-a)}}{2}\right). \tag{1.39}$$

Given U_k, we define $I_J u(x)$ by

$$I_J u(x) := \sum_{k=-J/2+1}^{J/2} U_k \mathrm{e}^{2\pi \mathrm{i} k x/(b-a)} \tag{1.40}$$

and use it to approximate $u(x)$. To evaluate U_k efficiently, we write

$$U_k = \frac{h}{b-a} \mathrm{e}^{-2\pi \mathrm{i} k a/(b-a)} \left(\frac{u(a)+u(b)}{2} + \sum_{j=1}^{J-1} u(x_j) \mathrm{e}^{-2\pi \mathrm{i} j k/J} \right). \tag{1.41}$$

A DFT (1.31) of the vector $[(u(a)+u(b))/2, u(x_1), \dots, u(x_{J-1})]^{\mathsf{T}} \in \mathbb{R}^J$ gives the bracketed term and it is efficiently evaluated for $k = 0, \dots, J-1$ by applying the FFT. Using the symmetry $u_{-k} = \bar{u}_k$, we compute the required values of U_k with Algorithm 1.1.

If $u \in H^2_{\mathrm{per}}(a,b)$, $I_J u$ is a good approximation to u in the following sense.

Lemma 1.100 *With $I_J u$ defined by* (1.40)*, there exists $K > 0$ such that*

$$\|u - I_J u\|_{H^1(a,b)} \le \frac{K}{J^{1/2}} \|u\|_{H^2(a,b)}, \qquad \forall u \in H^2_{\mathrm{per}}(a,b).$$

Proof See Exercise 1.23. □

Example 1.101 (estimating a Sobolev norm) As we know all the coefficients of $I_J u$ in Fourier space, $\|I_J u\|_{L^2(a,b)}$ and $\|I_J u\|_{H^1(a,b)}$ are easy to evaluate from U_k and provide estimates of $\|u\|_{H^1(a,b)}$ and $\|u\|_{H^2(a,b)}$. In particular, from (1.14) and (1.36),

$$\|u\|_{L^2(a,b)} \approx \|I_J u\|_{L^2(a,b)} = \left((b-a) \sum_{k=-J/2+1}^{J/2} |U_k|^2 \right)^{1/2},$$

$$\|u\|_{H^1(a,b)} \approx \|I_J u\|_{H^1(a,b)} = \left(\|u\|^2_{L^2(a,b)} + (b-a) \sum_{k=-J/2+1}^{J/2} \frac{(2\pi k)^2}{(b-a)^2} |U_k|^2 \right)^{1/2}.$$

See Algorithm 1.2. If $u \in H^2_{\mathrm{per}}(a,b)$, these approximations converge with $\mathcal{O}(J^{-2})$ (see Exercise 1.24). In Figure 1.6, we show the error in approximating the norms of the functions $u_1(x) = x^2(x-1)^2$ and $u_2(x) = \cos(\pi x)$ on $[0,1]$. For $u_1(x)$, the correct answer is computed to machine accuracy with $J \approx 10^3$. However, because $u_2(0) \neq u_2(1)$, the approximation to $\|u_2\|_{H^1(0,1)}$ diverges.

Algorithm 1.1 Code to find U_k as defined by (1.39) for $k = -J/2+1, \dots, J/2$. The inputs are a vector `u` of values $u(x)$ on the grid $x_j = a + jh$ for $j = 0, \dots, J$ and $h = (b-a)/J$; the grid spacing `h`; and `a`, `b` to define the interval (a,b). The outputs are a vector `Uk` of U_k and a vector `nu` of the corresponding frequencies $2\pi k/(b-a)$.

```
function [Uk, nu] = get_coeffs(u, a, b)
J=length(u)-1; h=(b-a)/J; u1=[(u(1)+u(J+1))/2; u(2:J)];
Uk=(h/(b-a))*exp(-2*pi*sqrt(-1)*[0:J-1]'*a/(b-a)).*fft(u1);
assert(mod(J,2)==0); % J must be even
nu=2*pi/(b-a)*[0:J/2,-J/2+1:-1]';
```

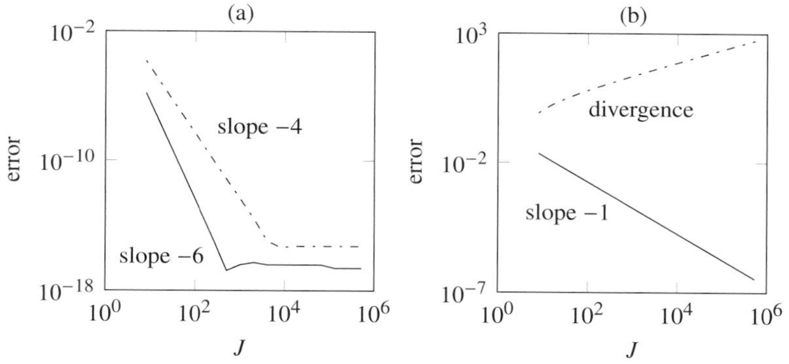

Figure 1.6 The error in approximating $\|u\|$ with $\|I_J u\|$ for the $L^2(0,1)$ (solid) and $H^1(0,1)$ (dash-dot) norm of (a) $u_1(x) = x^2(x-1)^2$ and (b) $u_2(x) = \cos(x\pi/2)$ against the number of points J. We see divergence of the approximation to the $H^1(0,1)$ norm in (b), because $u_2(x)$ is not in $H^2_{\text{per}}(0,1)$.

Algorithm 1.2 Code to approximate the $L^2(a,b)$ and $H^1(a,b)$. The inputs are a function handle `fhandle` to evaluate u, the domain `a`, `b`, and discretisation parameter `J`. The outputs are the $L^2(a,b)$ and $H^1(a,b)$ norms of $I_J u$.

```
function [l2_norm, h1_norm] = get_norm(fhandle, a, b, J)
grid=[a:(b-a)/J:b]';
% evaluate the function on the grid
u=fhandle(grid);
[Uk,nu]=get_coeffs(u, a, b);
l2_norm =sqrt(b-a)* norm(Uk);
dUk=nu.*Uk;
h1_norm = norm([l2_norm, sqrt(b-a)*norm(dUk)]);
```

Fourier transform $L^2(\mathbb{R}^d, \mathbb{C}) \longleftrightarrow L^2(\mathbb{R}^d, \mathbb{C})$

The Fourier series on an interval (a,b) arises from the orthonormal basis provided by the eigenfunctions of the Laplacian with periodic boundary conditions. On the whole of $\mathbb{R}^d$, $e^{i\lambda^T x}$ is an eigenfunction of the Laplacian for every $\lambda \in \mathbb{C}^d$ and, to include every wave number λ in the Fourier representation of $u \in L^2(\mathbb{R}^d, \mathbb{C})$, integrals are used.

Definition 1.102 (Fourier transform) For $u \in L^2(\mathbb{R}^d, \mathbb{C})$, the Fourier transform

$$\hat{u}(\lambda) := \frac{1}{(2\pi)^{d/2}} \int_{\mathbb{R}^d} u(\boldsymbol{x}) e^{-i\lambda^T \boldsymbol{x}} \, d\boldsymbol{x}, \qquad \lambda \in \mathbb{R}^d.$$

The inverse mapping $\hat{u} \mapsto u$ is denoted $\check{u}$ and defined via

$$\check{u}(\boldsymbol{x}) := \frac{1}{(2\pi)^{d/2}} \int_{\mathbb{R}^d} u(\lambda) e^{i\lambda^T \boldsymbol{x}} \, d\lambda. \tag{1.42}$$

The Parseval identity says that, for $u, v \in L^2(\mathbb{R}^d, \mathbb{C})$,

$$\langle u, v \rangle_{L^2(\mathbb{R}^d,\mathbb{C})} = \langle \hat{u}, \hat{v} \rangle_{L^2(\mathbb{R}^d,\mathbb{C})}, \qquad \|u\|_{L^2(\mathbb{R}^d,\mathbb{C})} = \|\hat{u}\|_{L^2(\mathbb{R}^d,\mathbb{C})}. \tag{1.43}$$

Further properties of the Fourier transform are considered in Exercise 1.21.

Example 1.103 (Fourier transform of an exponential function) Let $d = 1$ and consider $u(x) = e^{-a|x|}$ for some $a > 0$. The Fourier transform is

$$\hat{u}(\lambda) = \sqrt{\frac{2}{\pi}} \frac{a}{a^2 + \lambda^2}, \tag{1.44}$$

as is easily verified by integrating or using a computer algebra system.

Example 1.104 (Fourier transform of Gaussian functions) For $d = 1$ and $a > 0$, the Gaussian function $u(x) = e^{-ax^2}$ has Fourier transform

$$\hat{u}(\lambda) = \frac{1}{\sqrt{2a}} e^{-\lambda^2/4a}. \tag{1.45}$$

For a $d \times d$ symmetric positive-definite matrix C, the function $u(\boldsymbol{x}) = e^{-\boldsymbol{x}^{\mathsf{T}} C \boldsymbol{x}}$ has Fourier transform

$$\hat{u}(\boldsymbol{\lambda}) = \frac{1}{2^{d/2}\sqrt{\det C}} e^{-\boldsymbol{\lambda}^{\mathsf{T}} C^{-1} \boldsymbol{\lambda}/4}. \tag{1.46}$$

Example 1.105 (Fourier transform of derivatives) For $u \in L^2(\mathbb{R}, \mathbb{C})$, the Fourier transform of the derivative $\frac{du}{dx} = \mathcal{D}u$ is found by integrating by parts,

$$\widehat{\mathcal{D}u}(\lambda) = \frac{1}{(2\pi)^{1/2}} \int_{\mathbb{R}} e^{-i\lambda x} \mathcal{D}u(x)\, dx = -\frac{1}{(2\pi)^{1/2}} \int_{\mathbb{R}} \frac{d}{dx} e^{-i\lambda x} u(x)\, dx = (i\lambda)\hat{u}(\lambda),$$

where the boundary term disappears since $|u(x)| \to 0$ as $|x| \to \infty$. We see that the derivative of the Fourier transform equals $i\lambda$ times the Fourier transform $\hat{u}(\lambda)$.

Fourier integrals for isotropic functions in $L^2(\mathbb{R}^d, \mathbb{C})$

Definition 1.106 (isotropic) A function $u\colon \mathbb{R}^d \to \mathbb{C}$ is *isotropic* if $u(\boldsymbol{x})$ depends only on $r := \|\boldsymbol{x}\|_2$.

The Fourier transform of an isotropic function has a special form, known as the Hankel transform, which is used in Chapter 7 to develop the turning bands algorithm. To derive this special form, we make use of the Bessel function of the first kind $J_p(r)$ of order p. For $p = -1/2, 0, 1/2$, we have

$$J_{-1/2}(r) = \sqrt{\frac{2}{\pi r}} \cos(r), \quad J_0(r) = \frac{1}{2\pi} \int_{-\pi}^{\pi} e^{ir\cos\theta}\, d\theta, \quad J_{1/2}(r) = \sqrt{\frac{2}{\pi r}} \sin(r), \tag{1.47}$$

and these functions feature in the Fourier transform of isotropic functions in dimensions $d = 1, 2, 3$ respectively. The case of general d is considered in Appendix A.3.

Theorem 1.107 (Hankel transform) *If $u \in L^2(\mathbb{R}^d, \mathbb{C})$ is isotropic, the Fourier transform $\hat{u}(\boldsymbol{\lambda})$ only depends on $s := \|\boldsymbol{\lambda}\|_2$ and, in the cases $d = 1, 2, 3$, we have*

$$\hat{u}(s) = \frac{1}{(2\pi)^{d/2}} \int_{\mathbb{R}^d} e^{-i\boldsymbol{\lambda}^{\mathsf{T}}\boldsymbol{x}} u(\boldsymbol{x})\, d\boldsymbol{x} = \begin{cases} \sqrt{\frac{2}{\pi}} \int_0^\infty \cos(rs) u(r)\, dr, & d = 1, \\ \int_0^\infty J_0(rs) u(r) r\, dr, & d = 2, \\ \sqrt{2\pi} \int_0^\infty \frac{\sin(rs)}{rs} u(r) r^2\, dr, & d = 3. \end{cases}$$

Proof To demonstrate the general principle that an isotropic function $u(\boldsymbol{x})$ has an isotropic Fourier transform $\hat{u}(\boldsymbol{\lambda})$, let Q be a rotation on $\mathbb{R}^d$. For the change of variable $\boldsymbol{x} = Q\boldsymbol{z}$, $u(\boldsymbol{x}) = u(\boldsymbol{z})$ and

$$(2\pi)^{d/2}\hat{u}(Q\boldsymbol{\lambda}) = \int_{\mathbb{R}^d} e^{-i(Q\boldsymbol{\lambda})^{\mathsf{T}}\boldsymbol{x}} u(\boldsymbol{x})\, d\boldsymbol{x} = \int_{\mathbb{R}^d} e^{-i\boldsymbol{\lambda}^{\mathsf{T}} Q^{\mathsf{T}} Q\boldsymbol{z}} u(\boldsymbol{z}) \left|\det Q\right| d\boldsymbol{z}.$$

As $Q^{\mathsf{T}}Q = I$, $\hat{u}(Q\boldsymbol{\lambda}) = \hat{u}(\boldsymbol{\lambda})$. Because $\hat{u}(\boldsymbol{\lambda})$ is unchanged by rotation, it is a function of $s = \|\boldsymbol{\lambda}\|_2$ and the Fourier transform is isotropic.

We treat the case $d = 3$ and leave $d = 1, 2$ for Exercise 1.26. Consider spherical coordinates $(x_1, x_2, x_3) = r(\cos\psi, \sin\psi\cos\theta, \sin\psi\sin\theta)$ with $-\pi \le \theta_1 \le \pi$ and $0 \le \psi \le \pi$. Pick $\boldsymbol{\lambda} = [s, 0, 0]^{\mathsf{T}}$ so that

$$\int_{\mathbb{R}^3} e^{-i\boldsymbol{\lambda}^{\mathsf{T}}\boldsymbol{x}} u(\boldsymbol{x})\, d\boldsymbol{x} = \int_0^\infty \int_{-\pi}^{\pi} \int_0^{\pi} e^{-isr\cos\psi} (\sin\psi)\, d\psi\, d\theta\, u(r) r^2\, dr.$$

Let $t = -\cos\psi$ so that $dt = \sin\psi\, d\psi$ and

$$\int_{-\pi}^{\pi} \int_0^{\pi} e^{-isr\cos\psi} \sin\psi\, d\psi\, d\theta = \int_{-\pi}^{\pi} \int_{-1}^{1} e^{irst}\, dt\, d\theta = \frac{2\pi}{rs} 2\sin(rs).$$

Hence,

$$\int_{\mathbb{R}^3} e^{-i\boldsymbol{\lambda}^{\mathsf{T}}\boldsymbol{x}} u(\boldsymbol{x})\, d\boldsymbol{x} = 4\pi \int_0^\infty \frac{\sin(rs)}{rs} u(r) r^2\, dr = (2\pi)^{3/2} \int_0^\infty \frac{J_{1/2}(rs)}{\sqrt{rs}} u(r) r^2\, dr. \qquad \square$$

1.5 Notes

The Lebesgue integral is usually developed from limits of non-negative functions, for example, Rudin (1987). Our approach follows Zeidler (1995) for Lebesgue integrals and (Dudley, 1999; Yosida, 1995) for Bochner integrals. Neither the Borel σ-algebra nor the product measure (even of two complete measures) is complete in general; our convention is to always use the extension to the complete measure space. Lemma 1.20 is given for example by Dudley (2002, Theorem 4.2.2) for pointwise convergence and this extends to almost sure convergence if the measure space is complete (Exercise 1.2). The Lebesgue integral is important in probability theory, as we see in Chapter 4, and is developed in many probability text books including (Fristedt and Gray, 1997; Williams, 1991).

Hilbert spaces and the spaces $H^r(D)$ are developed in (Renardy and Rogers, 2004; Robinson, 2001; Temam, 1988; Zeidler, 1995). An excellent reference for the Sobolev embedding theorem is Adams and Fournier (2003). Our treatment of non-negative integral operators is more detailed as they are important for the Karhunen–Loève theorem in Chapter 5. (Lax, 2002; Riesz and Sz.-Nagy, 1990) treat Mercer's theorem. To study fractional powers fully, the operator A should be developed as a *sectorial operator*, which says that the spectrum of L lies in a sector of the complex plane; see Pazy (1983). We have given applications to the Laplacian with periodic and Dirichlet conditions and this is suitable for the reaction–diffusion equations studied in later chapters.

The Fourier transform is covered in detail by Sneddon (1995) as a method. The FFT is discussed in Brigham (1988) and mathematical theory is discussed in Koerner (1989). FFT and spectral methods for MATLAB are discussed in Trefethen (2000).

Exercises

1.1 For a bounded domain D, prove that $C(\bar{D})$ is complete with respect to the supremum norm $\|u\|_\infty = \sup_{\boldsymbol{x}\in\bar{D}} |u(\boldsymbol{x})|$. By constructing a suitable Cauchy sequence, show that $C(\bar{D})$ is not complete with respect to the $L^2(D)$ norm.

1.2 Let $(X, \mathcal{F}, \mu)$ be a complete measure space, Y be a Banach space, and $f: X \to Y$ a measurable function. Show that $f = g$ a.s. implies that g is measurable. Give an example to show that the completeness assumption on the measure space is necessary.

1.3 Using Definition 1.21, prove that if $u: X \to Y$ is integrable then

$$\int_X \|u(x)\|_Y \, d\mu(x) < \infty.$$

1.4 a. If $u: \mathbb{R} \to \mathbb{R}$ is measurable, define simple functions $u_n: \mathbb{R} \to \mathbb{R}$ such that $u_n(x) \to u(x)$ for all $x \in \mathbb{R}$.

b. If $G \in C(\mathbb{R} \times \mathbb{R})$, define simple functions $G_N(x, y)$ of the form

$$\sum_{j,k=1}^{N} G(x_j, x_k) 1_{F_j}(x) 1_{F_k}(y),$$

for $x_j \in \mathbb{R}$ and $F_j \in \mathcal{B}(\mathbb{R})$, such that $G_N \to G$ pointwise as $N \to \infty$.

1.5 Give an example of a Lebesgue integrable function $u: \mathbb{R} \to \mathbb{R}$ that is not Riemann integrable.

1.6 By introducing a measure on $\mathbb{Z}$, prove that

$$\lim_{n\to\infty} \sum_{j\in\mathbb{Z}} u_{n,j} = \sum_{j\in\mathbb{Z}} u_j, \tag{1.48}$$

for sequences $u_{n,j}$ converging to u_j as $n \to \infty$ and such that $|u_{n,j}| \le \bar{U}_j$ for a sequence $\bar{U}_j$ with $\sum_{j\in\mathbb{Z}} \bar{U}_j < \infty$.

1.7 Evaluate

$$\int_0^b \int_0^a f(x, y)\, dx\, dy, \qquad \int_0^a \int_0^b f(x, y)\, dy\, dx,$$

taking care to observe the order of integration, for

$$f(x, y) = \begin{cases} \dfrac{xy(y^2 - x^2)}{(x^2 + y^2)^3}, & 0 < x < a, \quad 0 < y < b, \\ 0, & x = 0, \quad y = 0. \end{cases}$$

Why does Fubini's theorem *not* apply?

1.8 Using Fubini's theorem, prove that

$$\frac{1}{\sqrt{2\pi}} \int_{-\infty}^{\infty} e^{-x^2/2}\, dx = 1. \tag{1.49}$$

Hint: first square both sides.

1.9 For $f \in \mathrm{C}^1([0,1])$ and $p \geq 1$, prove that

$$\left(\int_0^1 |f(x)|\,dx\right)^p \leq \int_0^1 |f(x)|^p\,dx.$$

This is a version of Jensen's inequality (see Lemma 4.56 and §A.4). For $a < b$, show that $L^q(a,b) \subset L^p(a,b)$ for $1 \leq p \leq q$.

1.10 On a real Hilbert space H, prove the Cauchy–Schwarz inequality $|\langle u, v\rangle| \leq \|u\|\|v\|$ for $u, v \in H$. Hint: find the minimum of $\|u + \alpha v\|^2$ with respect to $\alpha \in \mathbb{R}$.

1.11 Prove that (1.12) is the weak derivative of (1.11) (by showing (1.13) holds) and that $\mathcal{D}u$ does not have a weak derivative in the set of measurable functions from $\mathbb{R} \to \mathbb{R}$.

1.12 a. Let $u(r) = \log|\log r|$ for $r > 0$. Sketch a graph of $u(r)$ and $u'(r)$. Show that $u \in L^2(0,1/2)$ but $u \notin \mathrm{C}([0,1/2])$ and $u' \notin L^2(0,1/2)$.

b. Let $u(\boldsymbol{x}) = \log|\log r|$ for $r = \|\boldsymbol{x}\|_2$ and $D = \{\boldsymbol{x} \in \mathbb{R}^2 \colon \|\boldsymbol{x}\|_2 < 1/2\}$. Show that u has a well-defined weak derivative $\mathcal{D}_i u$ in $L^2(D)$ for $i = 1,2$. Hence show that $u \in H^1(D)$ and $u \notin \mathrm{C}(\bar{D})$.

1.13 For $\ell > 0$ and $u \in \mathrm{C}^{n+1}([-\ell,\ell])$, show that there exists a polynomial p of degree at most n such that

$$\|u - p\|_{L^2(-\ell,\ell)} \leq \frac{1}{(n+1)!}\|D^{n+1}u\|_{L^2(-\ell,\ell)}\ell^{n+1}.$$

Hint: use Theorem A.1 and the Cauchy–Schwarz inequality.

1.14 Prove the Poincaré inequality for the domain $D = (0,\ell)$ by the following methods:

a. With $K_{\mathrm{p}} = \ell$, by writing $u(x) = \int_0^x u'(s)\,ds$.

b. With $K_{\mathrm{p}} = \ell/\pi$, by writing u in the basis $\{\sin(k\pi x/\ell) \colon k \in \mathbb{N}\}$.

Show that $K_{\mathrm{p}} = \ell/\pi$ is the optimal constant.

1.15 For Banach spaces X, Y, prove that $L \in \mathcal{L}(X,Y)$ if and only if L is continuous from X to Y. Prove that $\|Lu\|_Y \leq \|L\|_{\mathcal{L}(X,Y)}\|u\|_X$ for $u \in X$.

1.16 Let H be a real Hilbert space. Show that

$$2\|u\|^2 + 2\|v\|^2 = \|u + v\|^2 + \|u - v\|^2, \qquad u, v \in H.$$

For a closed subspace G of H, let $p_n \in G$ be such that $\|u - p_n\| \to \inf_{p \in G}\|u - p\|$. Prove that p_n is a Cauchy sequence.

1.17 Prove the Riesz representation theorem (Theorem 1.57). Hint: consider $G = \{u \in H \colon \ell(u) = 0\}$. Show that $G^{\perp}$ is one dimensional and apply Theorem 1.37.

1.18 Let H be a separable Hilbert space and $B = \{u \in H \colon \|u\| \leq 1\}$. If H is infinite dimensional, prove the following:

a. B is not compact.

b. If $L \colon H \to H$ is compact, $L(B)$ is bounded and $L \in \mathcal{L}(H)$.

c. Let μ be a measure on H such that $\mu(F) = \mu(x + F)$ for $F \in \mathcal{B}(H)$, $x \in H$, and $x + F := \{x + y \colon y \in F\}$ (we say the measure is *translation invariant*). If $\mu(B)$ is finite, then $\mu(B) = 0$.

1.19 Let $L \in \mathcal{L}(H)$ be non-negative. Prove that $\operatorname{Tr} L = \sum_{j=1}^{\infty}\langle L\phi_j, \phi_j\rangle$ is independent of the orthonormal basis $\{\phi_j \colon j = 1, \ldots, \infty\}$.

1.20 Calculate the Green's functions for the operators in Examples 1.81 and 1.84.

1.21 Prove that the Fourier transform of a convolution $u \star v$, defined by

$$(u \star v)(\boldsymbol{x}) := \int_{\mathbb{R}^d} u(\boldsymbol{x} - \boldsymbol{y}) v(\boldsymbol{y}) \, d\boldsymbol{y},$$

for $u, v \in L^1(\mathbb{R}^d)$, is given by

$$\widehat{u \star v}(\boldsymbol{\lambda}) = (2\pi)^{d/2} \hat{u}(\boldsymbol{\lambda}) \hat{v}(\boldsymbol{\lambda}). \tag{1.50}$$

1.22 With u_k and U_k defined by (1.35) and (1.39) in the case $(a,b) = (0,\ell)$, show that

$$U_k = \sum_{n \in \mathbb{Z}} u_{k+nJ}, \qquad \forall u \in H^2_{\text{per}}(0,\ell).$$

1.23 Let $Au := u - u_{xx}$ for $u \in \mathcal{D}(A) := H^2_{\text{per}}(0,2\pi)$ and denote the associated fractional power norm by $\|\cdot\|_r$. Let J be even and I_J be defined by (1.40). Let P_J denote the orthogonal projection from $L^2(0,2\pi)$ onto $\operatorname{span}\{e^{2\pi i j x} : j = -J/2+1, \ldots, J/2\}$.

a. Using Exercise 1.22, prove that for $r > 1/4$ there exists $K > 0$ such that

$$\left\|I_J u - P_J u\right\|_{L^2(0,2\pi)} \le \frac{K}{J^{2r}} \left\|u\right\|_r, \qquad \forall u \in \mathcal{D}(A^r).$$

b. Prove Lemma 1.100 by showing that, for some $K > 0$,

$$\left\|u - I_J u\right\|_{1/2} \le K \frac{1}{J} \left\|u\right\|_1.$$

Hint: use Lemma 1.94 and Proposition 1.93.

c. Let u_n be given by (1.35) and suppose there exists $K, p > 0$ such that $|u_n| \le K|n|^{-(1+p)}$ for all $n \in \mathbb{Z}$. Show that

$$\left|\left\|u\right\|_{r/2} - \left\|I_J u\right\|_{r/2}\right| = \mathcal{O}\big(J^{-(1+p)}\big), \qquad 0 < 2r \le p.$$

Now prove that $\|u_1\|_{H^1(0,2\pi)} - \|I_J u_1\|_{H^1(0,2\pi)} = \mathcal{O}(J^{-3} \log J)$ for the function $u_1(x)$ shown in Figure 1.6.

1.24 a. Prove that the trapezium rule approximation $T_J(u^2)$ to $\int_0^1 u(x)^2 \, dx$ can be written as $T_J(u^2) = \|I_J u\|^2_{L^2(0,1)}$ for any $u \in C(0,1)$ such that $u(0) = u(1)$.

b. By using Theorem A.6, explain the higher rates of convergence shown in Figure 1.6(a) for the $L^2(0,1)$ norm.

1.25 Consider $u \in L^2(0,2\pi)$. Suppose that the Fourier coefficient $u_k = \mathcal{O}\big(|k|^{-(2r+1+\epsilon)/2}\big)$, for some $r, \epsilon > 0$. Show that $u \in H^r(0,2\pi)$.

1.26 Derive the formula for the Hankel transform given in Theorem 1.107 for $d = 1, 2$.

1.27 The discrete cosine transform (DCT-1) $\boldsymbol{y} \in \mathbb{R}^N$ of a vector $\boldsymbol{x} \in \mathbb{R}^N$ is defined by

$$y_k := \frac{1}{2}\big(x_1 + (-1)^{k-1} x_N\big) + \sum_{j=2}^{N-1} x_j \cos\left(\frac{\pi(k-1)(j-1)}{N-1}\right),$$

for $k = 1, \ldots, N$. Show how to evaluate $\boldsymbol{y}$ by using the FFT on a vector of length $2N-2$, which you need to specify in terms of x_j.

The discrete sine transform (DST-1) $\boldsymbol{y} \in \mathbb{R}^N$ of $\boldsymbol{x} \in \mathbb{R}^N$ is defined by

$$y_k := \sum_{j=1}^{N} x_j \sin\left(\frac{\pi j k}{N+1}\right), \qquad k = 1, \ldots, N.$$

Show how to evaluate $\boldsymbol{y}$ by using the FFT on a vector of length $2N + 2$. Implement your algorithms in MATLAB.

1.28 Define the truncated series for (1.37) by

$$u^N(x) = \sum_{j=1}^{N} \frac{2}{j}(-1)^{j+1} \sin(jx).$$

Using DST-1, evaluate $u^N(x)$ at $x = k\pi/(N+1)$ for $k = -N, \ldots, N$ and plot $u^N(x)$ against x for different values of N. Define the truncated series for (1.38) by

$$v^N(x) = \frac{\pi}{2} + \sum_{j=1}^{N-1} \frac{2(\cos j\pi - 1)}{\pi j^2} \cos(jx).$$

Making use of DCT-1, evaluate $u^N(x)$ at $x = (k-1)\pi/(N-1)$ for $k = -N+2, \ldots, N$ and plot $v^N(x)$ against x for different values of N.

2

Galerkin Approximation and Finite Elements

Let $D \subset \mathbb{R}^2$ be a two-dimensional domain with boundary ∂D. We consider the partial differential equation

$$-\nabla \cdot \big(a(\boldsymbol{x})\nabla u(\boldsymbol{x})\big) = f(\boldsymbol{x}), \qquad \boldsymbol{x} \in D, \tag{2.1}$$

where $a(\boldsymbol{x}) > 0$ for all $\boldsymbol{x} \in D$, together with the Dirichlet boundary condition

$$u(\boldsymbol{x}) = g(\boldsymbol{x}), \qquad \boldsymbol{x} \in \partial D. \tag{2.2}$$

Given a source term f, a diffusion coefficient a and boundary data g, our aim is to find a function u that satisfies the differential equation in D and boundary conditions on ∂D. Together, (2.1) and (2.2) represent an important class of elliptic boundary-value problems (BVPs). The differential equation (2.1) provides a model for steady-state diffusion and is ubiquitous in models of, for example, heat transfer, fluid flows and electrostatic potentials.

In this chapter, we review classical numerical methods for solving elliptic BVPs based on so-called Galerkin approximation. For now, we assume the inputs a and f are *deterministic* functions of $\boldsymbol{x} \in D$ but in Chapter 9 we extend the Galerkin framework to accommodate elliptic BVPs with *random* a and f. The first step is to interpret the derivatives in (2.1) in the weak sense (recall Definition 1.42) and introduce a variational formulation that has a weak solution in a suitable Sobolev space, V. The Galerkin approximation is then sought in an appropriate finite-dimensional subspace $\widetilde{V} \subset V$. In §2.1, as an introduction, we study a two-point BVP for a linear second-order ODE on $D \subset \mathbb{R}$. We use this simple one-dimensional problem to illustrate the variational approach and consider Galerkin approximation of two types: the spectral Galerkin method and the Galerkin finite element method. In §2.2, we outline the variational formulation of the two-dimensional problem (2.1)–(2.2). Finally, in §2.3, we apply the finite element method.

2.1 Two-point boundary-value problems

Fix a domain $D = (a,b)$ and let $p,q,f\colon D \to \mathbb{R}$ be given functions. We want to find $u\colon D \to \mathbb{R}$ that satisfies

$$-\frac{d}{dx}\left(p(x)\frac{du(x)}{dx}\right) + q(x)u(x) = f(x), \qquad a < x < b, \tag{2.3}$$

and the homogeneous Dirichlet boundary conditions

$$u(a) = u(b) = 0. \tag{2.4}$$

Equation (2.3) provides a model for steady-state reaction–diffusion; p is the diffusion coefficient and qu is the reaction term. Note that (2.1) has no reaction term and so (2.3) is more general. Before developing approximation schemes for the BVP (2.3)–(2.4), we define precisely what we mean by a *solution*. Below, we develop the concepts of classical, strong and weak solutions. To begin, we make the following simplifying assumption on p and q.

Assumption 2.1 (regularity of coefficients) The following conditions hold:

(diffusion) $p \in \mathrm{C}^1([a,b])$ and $p(x) \geq p_0 > 0$ for all $x \in (a,b)$, for some p_0.
(reaction) $q \in \mathrm{C}([a,b])$ and $q(x) \geq 0$ for all $x \in (a,b)$.

If we interpret derivatives in (2.3) in the classical sense then each term in the equation needs to be continuous. In that case, Assumption 2.1 must hold and a solution is a smooth function with continuous first and second derivatives.

Definition 2.2 (classical solution) Let $f \in \mathrm{C}(a,b)$. A function $u \in \mathrm{C}^2(a,b) \cap \mathrm{C}([a,b])$ that satisfies (2.3) for all $x \in (a,b)$ and the boundary conditions (2.4) is called a *classical solution*.

In Definition 2.2, we have $u \in \mathrm{C}([a,b])$, where the interval is closed, and $u \in \mathrm{C}^2(a,b)$, where the interval is open. This means that u (but not necessarily its derivatives) has a well-defined limit at the boundary. Hence the boundary conditions $u(a) = u(b) = 0$ are well defined.

Example 2.3 Suppose $D = (0,1)$. If p, q, and f are constants then the function

$$u(x) = \frac{f}{q}\left[1 - \left(\frac{\exp(\sqrt{s}x) + \exp(\sqrt{s}(1-x))}{1 + \exp(\sqrt{s})}\right)\right], \qquad s := \frac{q}{p},$$

satisfies (2.3)–(2.4) for every $x \in D$. Hence, u is a classical solution.

Often, we want to solve problems with $f \in L^2(a,b)$. The notion of a classical solution is then too restrictive, so we interpret the derivatives in (2.3) in the weak sense (see Definition 1.42) and seek solutions in a Sobolev space.

Definition 2.4 (strong solution) Let $f \in L^2(a,b)$. A function $u \in H^2(a,b) \cap H_0^1(a,b)$ that satisfies (2.3) for almost every $x \in (a,b)$ (with derivatives interpreted in the weak sense) is called a *strong solution*.

Example 2.5 Suppose $D = (0,1)$, $p = 1$, $q = 0$ and let

$$f(x) = \begin{cases} -1, & 0 \leq x < \dfrac{1}{2}, \\ 1, & \dfrac{1}{2} \leq x \leq 1. \end{cases} \tag{2.5}$$

We can construct a function $u \in \mathrm{C}^1([0,1])$ that satisfies (2.3) in the classical sense on the separate intervals $(0,1/2)$ and $(1/2,1)$, and the boundary conditions (2.4), as follows:

$$u(x) = \begin{cases} \dfrac{x^2}{2} - \dfrac{x}{4}, & 0 \leq x < \dfrac{1}{2}, \\ -\dfrac{x^2}{2} + \dfrac{3x}{4} - \dfrac{1}{4}, & \dfrac{1}{2} \leq x \leq 1. \end{cases} \tag{2.6}$$

See Figure 2.1. Note that $u \notin \mathrm{C}^2(0,1)$ as the second derivative jumps from $+1$ to -1 at $x = 1/2$. However, $u \in H^2(0,1) \cap H_0^1(0,1)$ and is a strong solution by Definition 2.4.

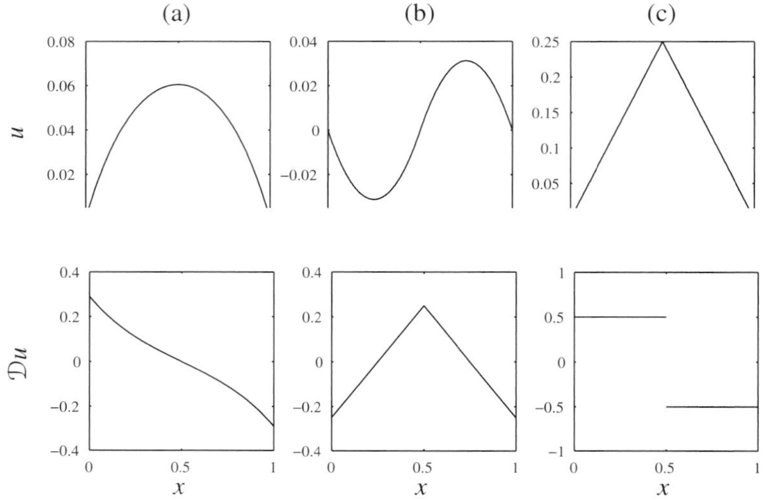

Figure 2.1 The top line shows (a) $u \in C^2(0,1) \cap C([0,1])$ from Example 2.3 with $p = 1$, $q = 10$ and $f = 1$, (b) $u \in H^2(0,1) \cap H_0^1(0,1)$ from Example 2.5 and (c) $u \in H_0^1(0,1)$ from Example 2.6. The bottom line shows the corresponding (weak) first derivatives $\mathcal{D}u$.

We can go one step further and find solutions to the BVP (2.3)–(2.4) when $f \notin L^2(a,b)$. Recall the definition of the delta function from Example 1.43.

Example 2.6 Let $D = (0,1)$, $p = 1$, $q = 0$ and $f(x) = \delta(x - \frac{1}{2})$. The function

$$u(x) = \begin{cases} \dfrac{x}{2}, & 0 \le x < \dfrac{1}{2}, \\ \dfrac{1}{2} - \dfrac{x}{2}, & \dfrac{1}{2} \le x \le 1 \end{cases}$$

satisfies (2.3) in the classical sense on the separate intervals $(0,1/2)$ and $(1/2,1)$ and also (2.4). If we admit weak derivatives, u solves (2.3) a.e. on the interval $(0,1)$ but the weak derivative $\mathcal{D}^1 u$ does not have a well-defined weak derivative in $L^2(0,1)$. In other words, $u \notin H^2(0,1)$. In fact, we have only $u \in H_0^1(0,1)$ (see Figure 2.1) and so u is neither a classical nor a strong solution.

Variational formulation of two-point BVPs

We need a mathematical framework that accommodates problems like the one defined in Example 2.6. To this end, we now introduce the variational formulation of the BVP (2.3)–(2.4). Let $\phi \in C_c^\infty([a,b])$ (see Definition 1.13). Multiplying (2.3) by ϕ and integrating over $[a,b]$ gives

$$\int_a^b -\phi(x)\big(p(x)u'(x)\big)' + \phi(x)q(x)u(x)\,dx = \int_a^b f(x)\phi(x)\,dx. \tag{2.7}$$

Integrating by parts gives

$$-\left[\phi(x)p(x)u'(x)\right]_a^b + \int_a^b \left(\phi'(x)p(x)u'(x) + \phi(x)q(x)u(x)\right) dx = \int_a^b f(x)\phi(x)\, dx$$

and the first term is zero as $\phi(a) = \phi(b) = 0$. Hence, any solution u to (2.3)–(2.4) satisfies the variational problem

$$a(u,\phi) = \ell(\phi), \qquad \forall \phi \in \mathrm{C}_c^\infty(a,b) \tag{2.8}$$

where

$$a(u,\phi) := \int_a^b \left(p(x)u'(x)\phi'(x) + q(x)u(x)\phi(x)\right) dx, \tag{2.9}$$

$$\ell(\phi) := \left\langle f,\phi\right\rangle_{L^2(a,b)}. \tag{2.10}$$

The integrals in (2.8) are actually well defined for $u \in H_0^1(a,b)$ and any $\phi \in H_0^1(a,b)$. By borrowing smoothness from the *test function* ϕ, u only needs to have weak first derivatives. This leads to the notion of a weak solution.

Definition 2.7 (weak solution) A *weak solution* to the BVP (2.3)–(2.4) is a function $u \in H_0^1(a,b)$ that satisfies

$$a(u,v) = \ell(v), \qquad \forall v \in H_0^1(a,b), \tag{2.11}$$

where a and ℓ are defined in (2.9) and (2.10). Equation (2.11) is known as the *weak form* of (2.3)–(2.4) or the *variational problem*.

We do not need Assumption 2.1 to have a well-defined variational problem. Indeed, we can replace it with the following weaker conditions.

Assumption 2.8 (regularity of coefficients) The following conditions hold:

(diffusion) $p \in L^\infty(a,b)$ and $p(x) \geq p_0 > 0$ a.e. in (a,b) for some p_0.
(reaction) $q \in L^\infty(a,b)$ and $q(x) \geq 0$ a.e. in (a,b).

This means, for example, that p and q can now be piecewise constant functions.

From Theorem 1.51 for a bounded domain in one dimension, we know that $H_0^1(a,b) \subset \mathrm{C}([a,b])$. The weak solution is therefore *continuous* but, unlike the classical solution, need not be differentiable. In the weak formulation (2.11), the space in which we seek u is called the *solution* or *trial space* and the space from which we select the functions v is the *test space*. The trial and test spaces for (2.3)–(2.4) coincide because the boundary conditions are homogeneous. However, if we replace (2.4) with

$$u(a) = \alpha, \qquad u(b) = \beta, \tag{2.12}$$

with non-zero α, β, then we must replace the solution space with

$$\left\{u \in H^1(a,b) \colon u(a) = \alpha \text{ and } u(b) = \beta\right\},$$

while the test space is still $H_0^1(a,b)$. The variational formulation of elliptic BVPs with *non-homogeneous* boundary conditions is discussed in §2.2.

When solving differential equations by the variational method, the solution space is often denoted V. For (2.3)–(2.4), we have

$$V := \{v \in H^1(a,b)\colon v(a) = 0 = v(b)\} = H_0^1(a,b), \tag{2.13}$$

and on V we define the so-called *energy norm*

$$\|u\|_{\mathrm{E}} := a(u,u)^{1/2} = \left(\int_a^b p(x)u'(x)^2 + q(x)u(x)^2\,dx\right)^{1/2}.$$

Observe that $a(\cdot,\cdot)$ is a symmetric bilinear form (see Definition 1.58). In fact, $a(\cdot,\cdot)\colon V\times V \to \mathbb{R}$ is an inner product (see Exercise 2.1) on V and $\|\cdot\|_{\mathrm{E}}$ is the induced norm. Recall that $\|u\|_{H_0^1(a,b)} := |u|_{H^1(a,b)}$ is also a norm on V. The next result shows that it is equivalent to the energy norm.

Lemma 2.9 *Let Assumption 2.8 hold. Then $\|\cdot\|_{\mathrm{E}}$ is a well-defined norm on V and is equivalent (see Definition 1.49) to the norm $\|\cdot\|_{H_0^1(a,b)}$.*

Proof Using the positivity conditions on p and q, it is easy to show that $\|\cdot\|_{\mathrm{E}}$ is a norm and

$$p_0^{1/2}\|u\|_{H_0^1(a,b)} \le \|u\|_{\mathrm{E}} \le \left(\|p\|_{L^\infty(a,b)} + \|q\|_{L^\infty(a,b)}K_{\mathrm{p}}^2\right)^{1/2}\|u\|_{H_0^1(a,b)} \tag{2.14}$$

by applying Poincaré's inequality (see Exercise 2.2). □

We can use the energy norm together with the Lax–Milgram lemma (Lemma 1.59) to prove existence and uniqueness of a weak solution, as follows.

Theorem 2.10 *Let Assumption 2.8 hold and suppose $f \in L^2(a,b)$. Then (2.11) has a unique solution $u \in V = H_0^1(a,b)$.*

Proof V is a Hilbert space, $a(\cdot,\cdot)\colon V \times V \to \mathbb{R}$ is a bilinear form, and $\|\cdot\|_{H_0^1(a,b)}$ is a norm on V. The Cauchy–Schwarz and Poincaré inequalities yield

$$|\ell(v)| \le \|f\|_{L^2(a,b)}\|v\|_{L^2(a,b)} \le K_{\mathrm{p}}\|f\|_{L^2(a,b)}\|v\|_{H_0^1(a,b)}, \qquad \forall v \in V.$$

Since $f \in L^2(a,b)$, ℓ is bounded from V to $\mathbb{R}$. In addition,

$$a(v,v) = \|v\|_{\mathrm{E}}^2 \ge p_0\|v\|_{H_0^1(a,b)}^2, \qquad \forall v \in V,$$

and, using Lemma 2.9,

$$|a(w,v)| \le \|w\|_{\mathrm{E}}\|v\|_{\mathrm{E}} \le \left(\|p\|_{L^\infty(a,b)} + K_{\mathrm{p}}^2\|q\|_{L^\infty(a,b)}\right)\|w\|_{H_0^1(a,b)}\|v\|_{H_0^1(a,b)}$$

for every $w, v \in V$. $a(\cdot,\cdot)$ is therefore bounded and coercive on V with respect to the norm $\|\cdot\|_{H_0^1(a,b)}$ and the Lax–Milgram lemma proves the result. □

In the proof of Theorem 2.10, we deduce that ℓ is a bounded linear operator on V by applying the Cauchy–Schwarz inequality and assuming $f \in L^2(a,b)$. To apply the Lax–Milgram lemma, however, we only need $\ell \in \mathcal{L}(V,\mathbb{R})$, the dual space of V. For $V = H_0^1(a,b)$, $f \in H^{-1}(a,b)$ suffices.

Example 2.11 Let $f(x) = \delta(x - \frac{1}{2})$ and $D = (0, 1)$, as in Example 2.6. For any $v \in V$, we have $|\ell(v)| \le \|v\|_{H_0^1(0,1)}$ since Jensen's inequality gives

$$|\ell(v)| = |v(1/2)| = \left|\int_0^{1/2} v'(x)\,dx\right| \le \int_0^1 |v'(x)|\,dx \le \left(\int_0^1 |v'(x)|^2\,dx\right)^{1/2}.$$

Thus, ℓ is a bounded linear operator on V and the resulting BVP (2.3)–(2.4) has a unique weak solution. This happens because $f(x) = \delta(x - \frac{1}{2}) \in H^{-1}(0, 1)$ (see Example 1.92 and the discussion after Proposition 1.93).

For the BVP (2.3)–(2.4), there is a much simpler way to establish uniqueness of solutions; see Exercise 2.3. Since $a(\cdot, \cdot)$ is symmetric, the Riesz representation theorem (Theorem 1.57) can also be applied. However, the strategy used in Theorem 2.10 is more general and will be useful in Chapter 9. We now clarify the relationship between a strong solution and a weak solution.

Theorem 2.12 *Let Assumption 2.1 hold and let $f \in L^2(a, b)$. Any strong solution to* (2.3)–(2.4) *is a weak solution. If the weak solution u belongs to $H^2(a, b)$, then u is a strong solution.*

Proof If u is a strong solution then $u \in H_0^1(a, b)$ by Definition 2.4. The integration by parts calculation (2.7)–(2.8) is valid when we replace ϕ with $v \in H_0^1(a, b)$ and so $a(u, v) = \ell(v)$ for any $v \in H_0^1(a, b)$. Hence, u is a weak solution. On the other hand, if u is the weak solution then $u \in H_0^1(a, b)$ but may not possess square-integrable second derivatives. If we assume $u \in H^2(a, b)$, we can reverse the integration by parts calculation that led to (2.11) to obtain

$$\int_a^b \left(-(p(x)u'(x))' + q(x)u(x) - f(x)\right)v(x)\,dx = 0, \qquad \forall v \in V.$$

This implies $-(p\,u')' + q\,u - f = 0$ a.e in (a, b) and so u is a strong solution. □

We sometimes assume that the weak solution belongs to $H^2(a, b)$ to simplify analysis and so we state this formally for later reference.

Assumption 2.13 (H^2-regularity) Let u satisfy (2.11). There exists a constant $K_2 > 0$ such that, for every $f \in L^2(a, b)$, the solution $u \in H^2(a, b) \cap H_0^1(a, b)$ and

$$|u|_{H^2(a,b)} \le K_2 \|f\|_{L^2(a,b)}.$$

Although Assumption 2.13 does not hold for arbitrary choices of p and q, it does hold, in particular, when Assumption 2.1 holds (e.g., as in Example 2.5).

Galerkin approximation

The variational problem (2.11) is the starting point for a large class of numerical methods. The idea is to introduce a *finite-dimensional* subspace $\widetilde{V} \subset V$ and solve (2.11) on $\widetilde{V}$. In general, we require *two* subspaces, one for the solution space and one for the test space. For the homogeneous boundary conditions (2.4), however, the spaces coincide and we have the following definition.

Definition 2.14 (Galerkin approximation) The *Galerkin approximation* for (2.3)–(2.4) is the function $\tilde{u} \in \widetilde{V} \subset V = H_0^1(a,b)$ satisfying

$$a(\tilde{u}, v) = \ell(v), \qquad \forall v \in \widetilde{V}. \tag{2.15}$$

When the trial and test spaces do not coincide in (2.11), the subspaces must be constructed in a compatible way. See Exercise 2.7 and §2.2 for more details.

Theorem 2.15 *Let Assumption 2.8 hold and suppose $f \in L^2(a,b)$. Then (2.15) has a unique solution $\tilde{u} \in \widetilde{V}$.*

Proof Since $\widetilde{V} \subset H_0^1(a,b)$, the proof follows exactly as for Theorem 2.10. □

Let $\widetilde{V} = \operatorname{span}\{\phi_1, \phi_2, \ldots, \phi_J\}$, where the functions ϕ_j are linearly independent. We can then express the Galerkin solution as

$$\tilde{u} = \sum_{j=1}^{J} u_j \phi_j \tag{2.16}$$

for coefficients $u_j \in \mathbb{R}$ to be determined. Substituting (2.16) into (2.15) gives

$$a\left(\sum_{j=1}^{J} u_j \phi_j, v\right) = \sum_{j=1}^{J} u_j a(\phi_j, v) = \ell(v), \qquad \forall v \in \widetilde{V},$$

and setting $v = \phi_i \in \widetilde{V}$ gives

$$\sum_{j=1}^{J} u_j a(\phi_j, \phi_i) = \ell(\phi_i), \qquad i = 1, \ldots, J. \tag{2.17}$$

If we define the matrix $A \in \mathbb{R}^{J \times J}$ and the vector $\boldsymbol{b} \in \mathbb{R}^J$ via

$$a_{ij} = a(\phi_i, \phi_j), \quad b_i = \ell(\phi_i), \qquad i, j = 1, \ldots, J, \tag{2.18}$$

then the coefficients in (2.16) are obtained by solving the linear system

$$A\boldsymbol{u} = \boldsymbol{b} \tag{2.19}$$

where $\boldsymbol{u} = [u_1, \ldots, u_J]^{\mathsf{T}}$. A is clearly symmetric, as $a(\cdot,\cdot)$ is symmetric. The next result tells us that the linear system (2.19) is uniquely solvable.

Theorem 2.16 *Under Assumption 2.8, the matrix A with entries a_{ij} in (2.18) is positive definite.*

Proof Let $\boldsymbol{v} \in \mathbb{R}^J \setminus \{\boldsymbol{0}\}$. Then $v = \sum_{i=1}^{J} v_i \phi_i \in \widetilde{V}$ and

$$\boldsymbol{v}^{\mathsf{T}} A \boldsymbol{v} = \sum_{i,j=1}^{J} v_j a(\phi_i, \phi_j) v_i = \sum_{j=1}^{J} v_j a\left(\sum_{i=1}^{J} v_i \phi_i, \phi_j\right) = a\left(\sum_{i=1}^{J} v_i \phi_i, \sum_{j=1}^{J} v_j \phi_j\right).$$

Hence, $\boldsymbol{v}^{\mathsf{T}} A \boldsymbol{v} = a(v,v) = \|v\|_E^2 > 0$. We obtain a strict inequality because

$$\boldsymbol{v}^{\mathsf{T}} A \boldsymbol{v} = 0 \iff \|v\|_E = 0 \iff v = 0 \iff \boldsymbol{v} = \boldsymbol{0}$$

since the basis functions ϕ_j are linearly independent. □

The Galerkin approximation has a very desirable property. It is the *best approximation* in $\widetilde{V}$ to the weak solution, with respect to the energy norm.

Theorem 2.17 (best approximation) *If $u \in V$ and $\tilde{u} \in \widetilde{V}$ satisfy* (2.11) *and* (2.15) *respectively and $\widetilde{V} \subset V$ then*

$$\|u - \tilde{u}\|_{\mathrm{E}} \le \|u - v\|_{\mathrm{E}}, \qquad \forall v \in \widetilde{V}. \tag{2.20}$$

Proof Since $\widetilde{V} \subset V$, (2.11) implies $a(u,v) = \ell(v)$ for all $v \in \widetilde{V}$. Subtracting (2.15) then gives

$$a(u - \tilde{u}, v) = 0, \qquad \forall v \in \widetilde{V}. \tag{2.21}$$

Property (2.21) is known as *Galerkin orthogonality* and is key. Since $\tilde{u} \in \widetilde{V}$,

$$a(u - \tilde{u}, u - \tilde{u}) = a(u - \tilde{u}, u) - a(u - \tilde{u}, \tilde{u}) = a(u - \tilde{u}, u) = a(u - \tilde{u}, u - v)$$

for any $v \in \widetilde{V}$. Applying the Cauchy–Schwarz inequality now gives

$$\|u - \tilde{u}\|_{\mathrm{E}}^2 = a(u - \tilde{u}, u - \tilde{u}) = a(u - \tilde{u}, u - v) \le \|u - \tilde{u}\|_{\mathrm{E}} \|u - v\|_{\mathrm{E}},$$

which proves the result. □

Theorem 2.17 is known as Céa's lemma. Using Lemma 2.9, we can also show that $\tilde{u}$ is the best approximation with respect to the norm $\|\cdot\|_{H_0^1(a,b)}$, up to a constant K that depends on the data.

Theorem 2.18 (quasi-optimal approximation) *Let Assumption 2.8 hold. If $u \in V$ and $\tilde{u} \in \widetilde{V}$ satisfy* (2.11) *and* (2.15) *respectively and $\widetilde{V} \subset V$ then*

$$\|u - \tilde{u}\|_{H_0^1(a,b)} \le K \|u - v\|_{H_0^1(a,b)}, \qquad \forall v \in \widetilde{V}, \tag{2.22}$$

where $K := p_0^{-1/2} \left(\|p\|_{L^\infty(a,b)} + \|q\|_{L^\infty(a,b)} K_{\mathrm{p}}^2 \right)^{1/2}$.

Since $\tilde{u} \in \widetilde{V}$, (2.20) says that the Galerkin approximation satisfies

$$\|u - \tilde{u}\|_{\mathrm{E}} = \inf_{v \in \widetilde{V}} \|u - v\|_{\mathrm{E}}.$$

Definition 1.38 then tells us that $\tilde{u} := P_G u$, where $P_G \colon V \to \widetilde{V}$ is the orthogonal projection from $V = H_0^1(a,b)$ to the chosen finite-dimensional subspace $\widetilde{V}$.

Definition 2.19 (Galerkin projection) The *Galerkin projection* $P_G \colon V \to \widetilde{V}$ is the orthogonal projection satisfying $\|u - P_G u\|_{\mathrm{E}} = \|u - \tilde{u}\|_{\mathrm{E}}$ or, equivalently,

$$a(P_G u, v) = a(u, v), \qquad \forall v \in \widetilde{V},\ u \in V. \tag{2.23}$$

We now consider two specific ways to construct $\widetilde{V} \subset V = H_0^1(a,b)$.

Spectral Galerkin method

First, we rewrite (2.3)–(2.4) as $Au = f$, using the differential operator

$$Au := -\frac{d}{dx}\left(p(x)\frac{du}{dx}\right) + q(x)u. \tag{2.24}$$

If we interpret derivatives in the weak sense then the domain of A is $\mathcal{D}(A) = H^2(a,b) \cap H_0^1(a,b)$ (see Definition 2.4). The *spectral Galerkin method* uses the eigenfunctions of A to construct the subspace $\widetilde{V}$. Let G denote the Green's function associated with A, so that $AG = \delta(x-y)$ for $a < x < b$ and $G(a,y) = G(b,y) = 0$. We now show that when G is square integrable and symmetric (see Exercise 2.4), there is a complete set of eigenfunctions.

Lemma 2.20 *Suppose that $G \in L^2((a,b)\times(a,b))$ and G is symmetric (i.e., $G(x,y) = G(y,x)$ for $x,y \in (a,b)$). The eigenfunctions ϕ_j of A provide a complete orthonormal basis of $L^2(a,b)$ and the eigenvalues satisfy $\lambda_j > 0$.*

Proof The solution to $Au = f$ can be written as

$$u(x) = \int_a^b G(x,y) f(y)\,dy =: (Lf)(x), \tag{2.25}$$

where L is a symmetric, compact integral operator on $L^2(a,b)$. The result follows by applying the Hilbert–Schmidt spectral theorem (see Theorem 1.73) to L and noting that A and $L = A^{-1}$ have the same eigenfunctions. These functions form a complete orthonormal basis for the range of L. As $\langle u, Au\rangle_{L^2(a,b)} = a(u,u) > 0$ for $u \neq 0$, the kernel of A is trivial and the result follows. □

Assume $|\lambda_{j+1}| \ge |\lambda_j|$ and choose $\widetilde{V} = \operatorname{span}\{\phi_1,\dots,\phi_J\}$, where ϕ_j is the jth normalised eigenfunction of A associated with eigenvalue λ_j. Then,

$$a(\phi_i,\phi_j) = \langle \phi_i, A\phi_j\rangle_{L^2(a,b)} = \lambda_j \langle \phi_i, \phi_j\rangle_{L^2(a,b)} = \lambda_j \delta_{ij},$$

where δ_{ij} is the Kronecker delta function, and the matrix A in (2.19) is *diagonal*. The Galerkin approximation is trivial to compute at the level of the linear system and, in (2.16),

$$u_j = \frac{\ell(\phi_j)}{a(\phi_j,\phi_j)} = \lambda_j^{-1}\langle f, \phi_j\rangle_{L^2(a,b)}.$$

This is the main advantage of the spectral Galerkin method.

We write u_J for $\tilde{u}$ to stress the dependence on J. The spectral Galerkin approximation is

$$u_J := \sum_{j=1}^{J} \lambda_j^{-1}\langle f, \phi_j\rangle_{L^2(a,b)} \phi_j. \tag{2.26}$$

Rearranging, we see that

$$u_J = \sum_{j=1}^{J} \lambda_j^{-1} a(u,\phi_j)\phi_j = \sum_{j=1}^{J} \lambda_j^{-1}\langle u, A\phi_j\rangle_{L^2(a,b)}\phi_j = \sum_{j=1}^{J}\langle u, \phi_j\rangle_{L^2(a,b)}\phi_j.$$

Since the basis functions are orthonormal, $u_J = P_J u$ where P_J is the orthogonal projection (see Lemma 1.41) from $L^2(a,b)$ onto $\widetilde{V} := \operatorname{span}\{\phi_1,\dots,\phi_J\}$. We now study the error in approximating u by u_J using fractional powers of linear operators. Recall the meaning of A^α from Definition 1.87 and fractional power norms $\|\cdot\|_\alpha$ from Theorem 1.88.

Lemma 2.21 (convergence of spectral Galerkin) *Let u be the solution to* (2.11) *and let u_J be the solution to* (2.15) *constructed via* (2.26). *For $\beta + 1 > \alpha$,*

$$\|u - u_J\|_\alpha \le \lambda_{J+1}^{-(\beta+1-\alpha)} \|f\|_\beta, \tag{2.27}$$

where λ_{J+1} is the $(J+1)$st eigenvalue of A and, in particular,

$$\|u - u_J\|_E \le \lambda_{J+1}^{-1/2} \|f\|_{L^2(a,b)}. \tag{2.28}$$

Proof First note that

$$\|u - u_J\|_\alpha = \|A^\alpha (u - u_J)\|_{L^2(a,b)} = \|A^{\alpha-1} A(u - u_J)\|_{L^2(a,b)}.$$

As $u - u_J = (I - P_J)u$ and $Au = f$, we have

$$\|u - u_J\|_\alpha = \|A^{\alpha-1}(I - P_J) f\|_{L^2(a,b)} \le \|A^{-(\beta+1-\alpha)}(I - P_J)\|_{\mathcal{L}(L^2(a,b))} \|A^\beta f\|_{L^2(a,b)}$$

for any $\beta \in \mathbb{R}$. By Lemma 1.94, with $H = L^2(a,b)$ and $\beta + 1 > \alpha$, we have

$$\|A^{-(\beta+1-\alpha)}(I - P_J)\|_{\mathcal{L}(L^2(a,b))} \le \lambda_{J+1}^{-(\beta+1-\alpha)}.$$

This proves (2.27). By Lemma 1.89(ii), we have $\|u\|_E = \|u\|_{1/2}$ so that choosing $\alpha = 1/2$ and $\beta = 0$ gives (2.28). □

The estimate (2.27) shows that the convergence rate of the energy norm error increases with the regularity of $f \in \mathcal{D}(A^\beta)$.

Example 2.22 (Laplace's equation) For $Au = -u_{xx}$ on $D = (0,1)$ with $u(0) = 0 = u(1)$, we have $\lambda_j = j^2\pi^2$ (see Example 1.81). Hence, (2.28) gives

$$\|u - u_J\|_E \le \frac{1}{(J+1)\pi} \|f\|_{L^2(0,1)}.$$

If f is analytic (see Definition A.2), then it belongs to every fractional power space and

$$\|u - u_J\|_E \le \frac{1}{((J+1)\pi)^{2\beta+1}} \|f\|_\beta, \qquad \forall \beta > -1/2,$$

so that the error in the energy norm converges faster than any power of $1/J$.

The eigenfunctions ϕ_j are known explicitly for the Laplace operator on $D = (a,b)$ and computations can be performed efficiently using the fast Fourier transform. For most practical problems of interest, however, the eigenfunctions are not known exactly and must be computed numerically.

In Chapter 3, we study spectral Galerkin approximation in space for a time-dependent PDE, examining convergence theoretically and giving MATLAB code (see Example 3.39). We also use the spectral Galerkin approximation for a PDE with time-dependent stochastic forcing in §10.7.

The Galerkin finite element method

The finite element method (FEM) is an alternative approach to constructing $\widetilde{V}$ in (2.15). The idea is to partition $\bar{D} = [a, b]$ into subintervals or *elements* and choose $\widetilde{V}$ to be a set of polynomials that are defined piecewise on the elements and globally continuous (since we need $\widetilde{V} \subset H_0^1(a, b) \subset C([a, b])$).

To begin, we partition $\bar{D}$ into n_e elements $e_k = [z_{k-1}, z_k]$ of uniform width

$$h := \frac{b-a}{n_e}, \tag{2.29}$$

with $n_e + 1$ vertices $z_k = a + kh$, $k = 0, \dots, n_e$. This partition is known as the finite element *mesh*. After fixing the mesh, we fix $r \in \mathbb{N}$ and choose

$$\widetilde{V} = \left\{ v \in C([a, b]) \colon v|_{e_k} \in \mathbb{P}_r(e_k) \text{ for } k = 1, \dots, n_e \text{ and } v(a) = 0 = v(b) \right\} \tag{2.30}$$

where $\mathbb{P}_r(e_k)$ is the set of polynomials of degree r or less on element e_k.

Definition 2.23 (global basis functions) In the Galerkin finite element method, we have $\widetilde{V} = \operatorname{span}\{\phi_1, \dots, \phi_J\}$ where the *global basis functions* ϕ_j satisfy

$$\phi_j\big|_{e_k} \in \mathbb{P}_r(e_k) \quad \text{and} \quad \phi_j(x_i) = \delta_{ij}, \tag{2.31}$$

(where δ_{ij} is the Kronecker delta function) and $\{x_1, \dots, x_J\} \subset [a, b]$ are a set of *nodes*, positioned so that $\widetilde{V} \subset C([a, b])$.

The Galerkin FEM approximation is usually denoted u_h to stress the dependence on the element width h in (2.29). We also write V^h instead of $\widetilde{V}$ and P_h instead of P_G for the Galerkin projection, so that $u_h = P_h u$. Using (2.31),

$$u_h = \sum_{j=1}^{J} u_j \phi_j \qquad \text{where } u_j = u_h(x_j). \tag{2.32}$$

The coefficients u_j are the values of u_h at the nodes and so $\{\phi_1, \dots, \phi_J\}$ is often called a *nodal* basis. The number of basis functions, J, depends on the chosen polynomial degree r in (2.30) and on the number of elements in the mesh, or, equivalently on the length h in (2.29). As the polynomial degree is increased and the mesh width is decreased, the number of basis functions increases.

Example 2.24 (piecewise linear) If $r = 1$, the nodes x_j are necessarily the (interior) vertices z_j of the mesh. We have $V^h = \operatorname{span}\{\phi_1, \dots, \phi_J\}$ where $J = n_e - 1$. Each ϕ_j is piecewise linear and satisfies

$$\phi_j(x_i) = \delta_{ij}.$$

See Figure 2.2. These functions are called *hat* functions and each one has support on two elements (i.e., $\operatorname{supp}(\phi_j) = e_j \cup e_{j+1}$). Any $v \in V^h$ is continuous on $[a, b]$ and belongs to $H^1(a, b)$. By excluding ϕ_0 and ϕ_{J+1} (associated with the vertices $z_0 = a$ and $z_{J+1} = b$), any $v \in V^h$ also satisfies $v(a) = 0$ and $v(b) = 0$.

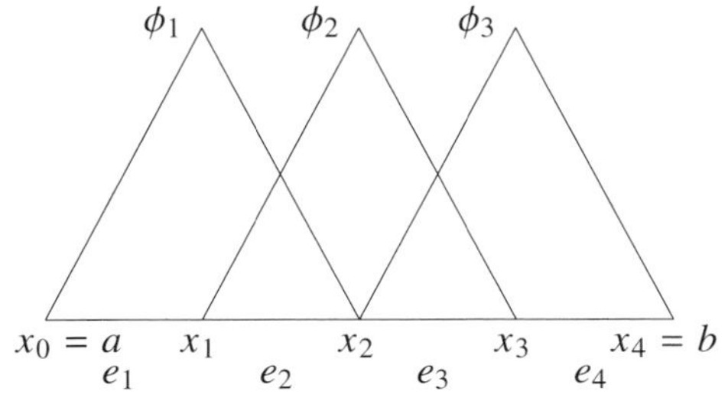

Figure 2.2 Piecewise linear ($r = 1$) global basis functions as defined in (2.31). The mesh on $[a, b]$ consists of $n_e = 4$ finite elements and has $J = 3$ interior vertices.

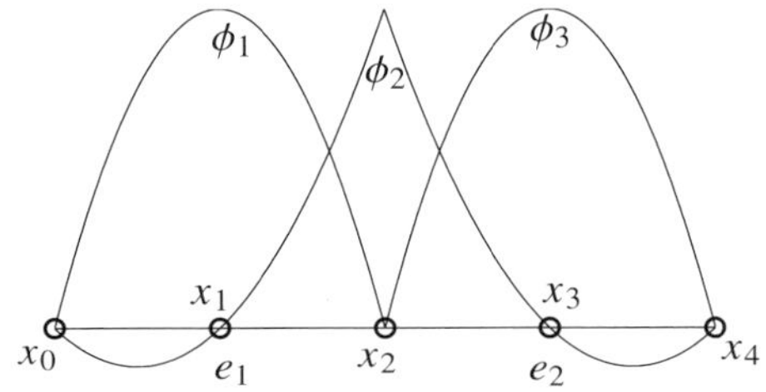

Figure 2.3 Piecewise quadratic ($r = 2$) global basis functions as defined in (2.31). The mesh on $[a, b]$ consists of $n_e = 2$ finite elements and has $J = 3$ interior nodes.

Example 2.25 (piecewise quadratic) To define quadratic basis functions ϕ_j, we place three nodes in each element, two at the vertices to ensure functions in V^h are globally continuous on $[a, b]$ and the third at the midpoint (see Figure 2.3). With $r = 2$, we now have $J = 2n_e - 1$ interior nodes. Once again, excluding basis functions corresponding to boundary nodes ensures $V^h \subset H_0^1(a, b)$.

Having selected global basis functions, we can assemble A and $\boldsymbol{b}$ in (2.19). For (2.3), the Galerkin matrix has the form $A = K + M$, where for $i, j = 1, \dots, J$

$$k_{ij} := \int_a^b p(x) \frac{d\phi_i(x)}{dx} \frac{d\phi_j(x)}{dx} \, dx, \qquad m_{ij} := \int_a^b q(x)\phi_i(x)\phi_j(x) \, dx.$$

K is called the *diffusion matrix* and M is the *mass matrix* (see Exercise 2.5). In (2.19),

$$a_{ij} := a(\phi_i, \phi_j) = \sum_{k=1}^{n_e} \int_{e_k} p(x) \frac{d\phi_i(x)}{dx} \frac{d\phi_j(x)}{dx} + q(x)\phi_i(x)\phi_j(x) \, dx, \tag{2.33}$$

$$b_i := \ell(\phi_i) = \sum_{k=1}^{n_e} \int_{e_k} f(x)\phi_i(x) \, dx. \tag{2.34}$$

Breaking the integrals up over the elements is key for computational efficiency. For fixed ϕ_i and ϕ_j, only a small number of element integrals in (2.33)–(2.34) are non-zero. Indeed, $a(\phi_i, \phi_j) = 0$ unless ϕ_i and ϕ_j are both non-zero on at least one element. If the polynomial degree r is small, ϕ_i overlaps with few other ϕ_js. The matrix A is thus sparse, a major attraction of Galerkin FEMs.

Example 2.26 (piecewise linear) For $r = 1$, the global basis functions are the hat functions in Figure 2.2. The functions ϕ_i and ϕ_j are both non-zero only when $x_i = z_i$ and $x_j = z_j$ are vertices of the same element. Hence, $a(\phi_i, \phi_j) \neq 0$ only when $j = i - 1, i$ or $i + 1$ and so A is tridiagonal.

Writing the Galerkin solution in terms of the global basis functions is useful for analysis but, in computations, we always work with *local* basis functions. Observe that if we restrict $v \in V^h$ to an element e_k we have

$$v|_{e_k} = \sum_{i=1}^{r+1} v_i^k \phi_i^k, \qquad k = 1, \ldots, n_e, \tag{2.35}$$

where $\operatorname{span}\{\phi_1^k, \ldots, \phi_{r+1}^k\} = \mathbb{P}_r(e_k)$. Thus, at element level, $v \in V^h$ can be expanded using $r + 1$ *local* basis functions defined by

$$\phi_i^k \in \mathbb{P}_r(e_k) \quad \text{and} \quad \phi_i^k(x_j) = \begin{cases} 1, & x_j = x_i^k, \\ 0, & x_j \neq x_i^k, \end{cases} \tag{2.36}$$

where $\{x_1^k, \ldots, x_{r+1}^k\} \subset [z_{k-1}, z_k]$ is an appropriate set of element nodes.

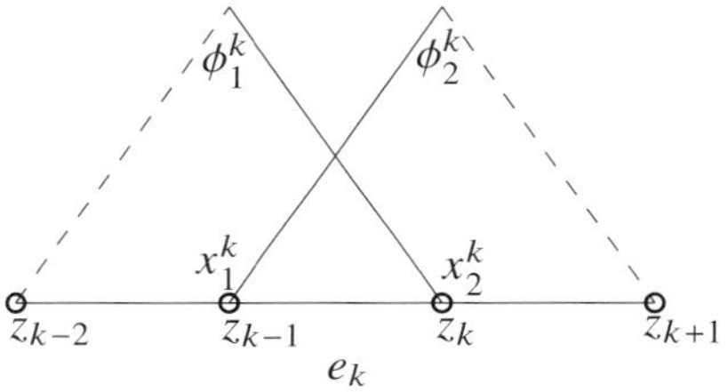

Figure 2.4 Local basis functions for piecewise linear approximation ($r = 1$). Each element has two local nodes (the vertices) ordered from left to right.

Example 2.27 (piecewise linear) When $r = 1$, only the global basis functions ϕ_{k-1} and ϕ_k (see Figure 2.4) are non-zero on a fixed element e_k. For $v \in V^h$, $v|_{e_k} = v_1^k \phi_1^k + v_2^k \phi_2^k$, where the local basis functions are $\phi_1^k = \phi_{k-1}|_{e_k}$ and $\phi_2^k = \phi_k|_{e_k}$, or equivalently, using (2.36) with $x_1^k = z_{k-1}$ and $x_2^k = z_k$,

$$\phi_1^k(x) = \frac{z_k - x}{h}, \qquad \phi_2^k(x) = \frac{x - z_{k-1}}{h}.$$

Example 2.28 (piecewise quadratic) When $r = 2$, three global basis functions are non-zero on a fixed element e_k and for each $v \in V^h$, $v|_{e_k} = v_1^k \phi_1^k + v_2^k \phi_2^k + v_3^k \phi_3^k$. See Figure 2.5 and Exercise 2.6.

Working with *local* basis functions means we can construct A and $\boldsymbol{b}$ in (2.19) from easy-to-compute small *element arrays*. On a fixed element e_k, only $r + 1$ global basis functions are non-zero. The local supports of these functions are, by definition, the local basis functions. We define the *element matrix* $A^k \in \mathbb{R}^{(r+1)\times(r+1)}$ by

$$a_{st}^k := \int_{e_k} p(x) \frac{d\phi_s^k(x)}{dx} \frac{d\phi_t^k(x)}{dx} + q(x)\phi_s^k(x)\phi_t^k(x)\, dx, \tag{2.37}$$

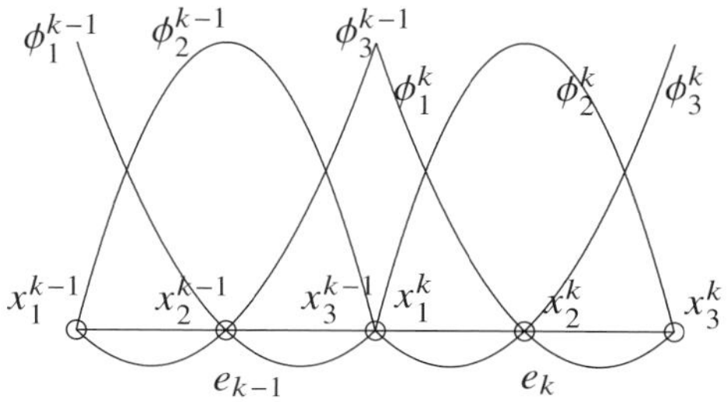

Figure 2.5 Local basis functions for piecewise quadratic approximation ($r = 2$) on adjoining elements. Each element has three local nodes — the vertices and the midpoint.

for $s,t = 1,\ldots,r+1$ and the *element vector* $\boldsymbol{b}^k \in \mathbb{R}^{r+1}$ by

$$b_s^k := \int_{e_k} f(x)\phi_s^k(x)\,dx. \tag{2.38}$$

Note that we can also write the element matrix as $A^k = K^k + M^k$, where

$$k_{st}^k := \int_{e_k} p(x)\frac{d\phi_s^k(x)}{dx}\frac{d\phi_t^k(x)}{dx}\,dx, \qquad m_{st}^k := \int_{e_k} q(x)\phi_s^k(x)\phi_t^k(x)\,dx. \tag{2.39}$$

When we solve time-dependent problems in Chapter 3, it will be convenient to assemble the diffusion and mass matrices separately.

Now, from (2.33), we observe that each entry of the global matrix A is a sum of entries in a few element matrices A^k, and similarly for $\boldsymbol{b}$.

Example 2.29 (piecewise linear) Consider two elements e_i and e_{i+1} meeting at vertex z_i (see Figure 2.6). If $r = 1$, we obtain 2×2 element matrices

$$A^i = \begin{pmatrix} a_{11}^i & a_{12}^i \\ a_{21}^i & a_{22}^i \end{pmatrix}, \qquad A^{i+1} = \begin{pmatrix} a_{11}^{i+1} & a_{12}^{i+1} \\ a_{21}^{i+1} & a_{22}^{i+1} \end{pmatrix}.$$

The global basis function ϕ_i corresponding to z_i is non-zero only on $e_i \cup e_{i+1}$. Using the numbering system in Figure 2.6, the local basis functions on e_i are $\phi_1^i = \phi_{i-1}|_{e_i}$ and $\phi_2^i = \phi_i|_{e_i}$ and the local basis functions on e_{i+1} are $\phi_1^{i+1} = \phi_i|_{e_{i+1}}$ and $\phi_2^{i+1} = \phi_{i+1}|_{e_{i+1}}$. Comparing (2.33)–(2.34) and (2.37), we see that the entries in A associated with ϕ_i are

$$a_{i,i-1} = a_{21}^i, \qquad a_{i,i} = a_{22}^i + a_{11}^{i+1}, \qquad a_{i,i+1} = a_{12}^{i+1},$$

and similarly for the other ϕ_js.

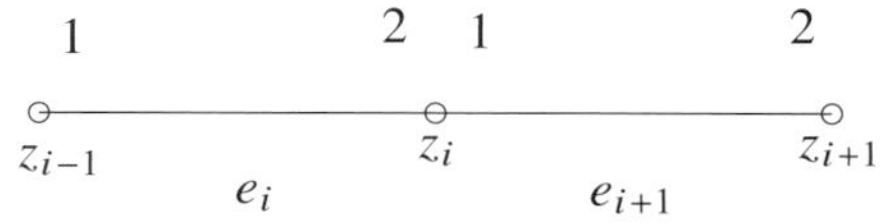

Figure 2.6 Two elements e_i and e_{i+1} meeting at vertex z_i and the local vertex numbering convention $(1,2)$ for piecewise linear approximation.

Algorithm 2.1 Piecewise linear FEM for the BVP (2.3)–(2.4) on $D = (0, 1)$ with piecewise constant p, q, f. The inputs are `ne`, the number of elements and `p,q,f`, vectors of length n_e containing the data values in each element. The outputs are `uh`, a vector containing the coefficients of the finite element approximation; `A`, the Galerkin matrix; `b`, the right-hand side vector; `K` the diffusion matrix; and `M` the mass matrix.

```
function [uh,A,b,K,M]=oned_linear_FEM(ne,p,q,f)
% set-up 1d FE mesh
h=(1/ne); xx=0:h:1; nvtx=length(xx);
J=ne-1; elt2vert=[1:J+1;2:(J+2)]';
% initialise global matrices
K = sparse(nvtx,nvtx); M = sparse(nvtx,nvtx); b=zeros(nvtx,1);
% compute element matrices
[Kks,Mks,bks]=get_elt_arrays(h,p,q,f,ne);
% Assemble element arrays into global arrays
for row_no=1:2
    nrow=elt2vert(:,row_no);
    for col_no=1:2
        ncol=elt2vert(:,col_no);
        K=K+sparse(nrow,ncol,Kks(:,row_no,col_no),nvtx,nvtx);
        M=M+sparse(nrow,ncol,Mks(:,row_no,col_no),nvtx,nvtx);
    end
    b = b+sparse(nrow,1,bks(:,row_no),nvtx,1);
end
% impose homogeneous boundary condition
K([1,end],:)=[]; K(:,[1,end])=[]; M([1,end],:)=[]; M(:,[1,end])=[];
A=K+M; b(1)=[]; b(end)=[];
% solve linear system for interior degrees of freedom;
u_int=A\b; uh=[0;u_int;0]; plot(xx,uh,'-k');
```

Algorithm 2.1 demonstrates how to amalgamate the element arrays to form A and $\boldsymbol{b}$ for the case $r = 1$ and the numbering system shown in Figure 2.6. We also assume that p, q and f are piecewise constant functions so that the integrals in (2.37)–(2.38) are simple. Three-dimensional arrays are a nice feature of MATLAB that can store *all* the element arrays as a single data structure. Once assembled, (2.19) is solved (see Exercise 2.9) to provide the values of the Galerkin finite element solution u_h at the J interior nodes. Some care is needed with boundary conditions. Recall,

$$u_h = \sum_{j=1}^{J} u_j \phi_j = 0\,\phi_0 + \sum_{j=1}^{J} u_j \phi_j + 0\,\phi_{J+1}. \tag{2.40}$$

If we assemble *every* element array then we include $e_1 = [a, z_1]$ and $e_{J+1} = [z_J, b]$ and so $A \in \mathbb{R}^{(J+2)\times(J+2)}$ and $\boldsymbol{b} \in \mathbb{R}^{J+2}$. We effectively include ϕ_0 and ϕ_{J+1} in the basis for V^h. Since they don't play a role in (2.40), we may 'throw away' entries with contributions from ϕ_0 and ϕ_{J+1}. See also Exercise 2.7.

For piecewise linear approximation in one dimension and constant p, q and f, we can construct the Galerkin matrix A and vector $\boldsymbol{b}$ directly (see Exercise 2.8). However, Algorithm 2.1 is easily modified to handle $r > 1$, general data and two-dimensional domains (see §2.3). The key task is to provide a subroutine `get_elt_arrays` to form the appropriate element arrays.

Example 2.30 Let $r = 1$. If p, q and f are piecewise constant then, for each element e_k of length h, straightforward integration in (2.37)–(2.38) gives

$$A^k = \underbrace{p_k \begin{pmatrix} 1/h & -1/h \\ -1/h & 1/h \end{pmatrix}}_{K^k} + \underbrace{q_k \begin{pmatrix} h/3 & h/6 \\ h/6 & h/3 \end{pmatrix}}_{M^k}, \qquad \boldsymbol{b}^k = f_k \begin{pmatrix} h/2 \\ h/2 \end{pmatrix},$$

where p_k, q_k and f_k denote the values of p, q and f in element e_k. Algorithm 2.2 computes these arrays for use in Algorithm 2.1.

Algorithm 2.2 Subroutine to compute element arrays (with $r = 1$) for the BVP (2.3)–(2.4) with piecewise constant p, q, f. The inputs `p`,`q` and `f` are as in Algorithm 2.1, `h` is the mesh width and `ne` is the number of elements. The outputs are `Kks` and `Mks`, the element diffusion and mass matrices and `bks`, the element vectors.

```
function [Kks,Mks,bks]=get_elt_arrays(h,p,q,f,ne)
Kks = zeros(ne,2,2);   Mks=zeros(ne,2,2);
Kks(:,1,1)=(p./h);     Kks(:,1,2)=-(p./h);
Kks(:,2,1)=-(p./h);    Kks(:,2,2)=(p./h);
Mks(:,1,1)=(q.*h./3);  Mks(:,1,2)=(q.*h./6);
Mks(:,2,1)=(q.*h./6);  Mks(:,2,2)=(q.*h./3);
bks=zeros(ne,2); bks(:,1)= f.*(h./2); bks(:,2) = f.*(h./2);
```

Note that if p, q and f are simply constants, then all the element arrays are identical. For general meshes, coefficients and polynomial degrees, the element arrays are different for each element and quadrature rules are usually needed to compute the integrals in (2.37) and (2.38). See Exercise 2.6 and §2.3.

Example 2.31 Consider (2.3)–(2.4) on $D = (0, 1)$ with constant data $f = 1, p = 1$ and $q = 10$. The piecewise linear finite element approximation associated with $h = 1/8$ can be computed using Algorithm 2.1 with the commands:

```
>> ne=8; [uh,A,b]=oned_linear_FEM(ne,ones(ne,1),10.*ones(ne,1),ones(ne,1));
```

Approximations generated with varying mesh size h are shown in Figure 2.7.

The accuracy of the Galerkin FEM approximation is clearly affected by the mesh width (see Figure 2.7), but also by the degree of polynomials used to construct V^h. Indeed, there are two strategies for improving the approximation: h-refinement (decreasing the mesh size) and r-refinement (increasing the polynomial degree). We consider only h-refinement and assume that $r = 1$ is fixed.

The best approximation property (2.20) is the starting point for error analysis. We know that u_h yields a smaller energy norm error than any other approximation in V^h. The next result shows that, when $r = 1$, the error is $\mathcal{O}(h)$ and hence converges to zero as $h \to 0$. The proof makes use of $I_h u$, the *piecewise linear interpolant* of u, which is defined on the element $e_k = [z_{k-1}, z_k]$ by

$$I_h u(x) := \left(\frac{x - z_{k-1}}{z_k - z_{k-1}}\right) u(z_k) + \left(\frac{x - z_k}{z_{k-1} - z_k}\right) u(z_{k-1}), \qquad z_{k-1} \le x \le z_k, \tag{2.41}$$

and, crucially, also belongs to V^h.

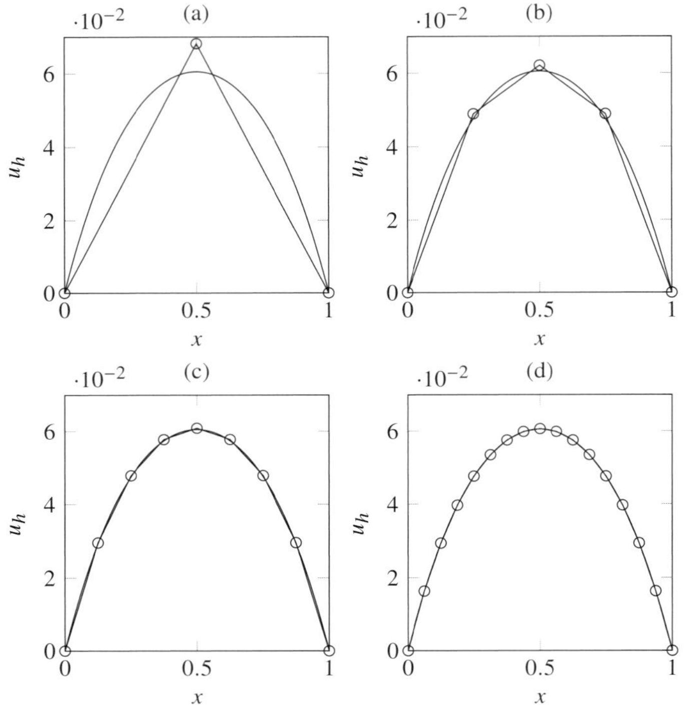

Figure 2.7 Piecewise linear FEM approximations (line with circles) for Example 2.31 with mesh widths (a) $h = 1/2$, (b) $h = 1/4$, (c) $h = 1/8$ and (d) $h = 1/16$ generated with the command `[uh,A,b]=oned_linear_FEM(ne,ones(ne,1),10*ones(ne,1),ones(ne,1));` where `ne` $= h^{-1}$. The exact solution is also shown (solid line).

Theorem 2.32 *Assume $f \in L^2(a,b)$ and let Assumption 2.13 hold. Let $u \in H^2(a,b) \cap H_0^1(a,b)$ be the solution to (2.11) and $u_h \in V^h$ be the solution to (2.15), constructed using piecewise linear finite elements. Then,*

$$\|u - u_h\|_{\mathrm{E}} \le K\, h\, \|f\|_{L^2(a,b)}$$

where $K > 0$ is a constant that depends on the coefficients p and q.

Proof Using (2.20), we know $\|u - u_h\|_{\mathrm{E}} \le \|u - v\|_{\mathrm{E}}$ for any $v \in V^h$. Let $\{z_k\}_{k=0}^{J+1}$ denote the vertices of the mesh and consider $v = I_h u$, the piecewise linear interpolant (2.41) of u. Note that $e = u - I_h u \in H_0^1(a,b)$ and satisfies

$$\begin{aligned} \|e\|_{\mathrm{E}}^2 &= \sum_{k=1}^{n_e} \int_{z_{k-1}}^{z_k} \left(p(x)\big(e(x)''\big)^2 + q(x) e(x)^2\right) dx \\ &\le \sum_{k=1}^{n_e} \|p\|_{L^\infty(a,b)} \int_{z_{k-1}}^{z_k} \big(e(x)'\big)^2\, dx + \|q\|_{L^\infty(a,b)} \int_{z_{k-1}}^{z_k} e(x)^2\, dx. \end{aligned}$$

Our task is then to bound the following norms of the element errors:

$$\|e'\|_{L^2(e_k)}^2 = \int_{z_{k-1}}^{z_k} \big(e(x)'\big)^2\, dx, \qquad \|e\|_{L^2(e_k)}^2 = \int_{z_{k-1}}^{z_k} e(x)^2\, dx. \tag{2.42}$$

Since $e(z_{k-1}) = 0 = e(z_k)$, it can be shown using Fourier techniques that

$$\|e'\|^2_{L^2(e_k)} \le \frac{h^2}{\pi^2}\|e''\|^2_{L^2(e_k)}, \qquad \|e\|^2_{L^2(e_k)} \le \frac{h^4}{\pi^4}\|e''\|^2_{L^2(e_k)}, \tag{2.43}$$

(see Exercise 2.10). Observe now that $e'' = u''$ since $I_h u$ is linear. Combining the above bounds and summing over the elements gives

$$\begin{aligned}\|u-u_h\|^2_{\mathrm{E}} \le \|e\|^2_{\mathrm{E}} &\le \sum_{k=1}^{n_e}\left(\frac{\|p\|_{L^\infty(a,b)}h^2}{\pi^2} + \frac{\|q\|_{L^\infty(a,b)}h^4}{\pi^4}\right)\|u''\|^2_{L^2(e_k)} \\ &\le h^2\left(\frac{\|p\|_{L^\infty(a,b)}}{\pi^2} + \frac{\|q\|_{L^\infty(a,b)}}{\pi^4}\right)|u|^2_{H^2(a,b)}.\end{aligned}$$

Since Assumption 2.13 holds,

$$K := K_2\left(\frac{\|p\|_{L^\infty(a,b)}}{\pi^2} + \frac{\|q\|_{L^\infty(a,b)}}{\pi^4}\right)^{1/2} \tag{2.44}$$

and we have $\|u-u_h\|_{\mathrm{E}} \le Kh\|f\|_{L^2(a,b)}$ as required. □

It should be noted that in the case of piecewise linear approximation, for some BVPs of the form (2.3)–(2.4), we actually obtain $u_h = I_h u$. See Exercises 2.12 and 2.13. An alternative method of proof for Theorem 2.32 that leads to a different constant K is also considered in Exercise 2.11.

Algorithm 2.3 Code to compute the error in the piecewise linear FEM approximation for Example 2.33. The inputs are `ne`, the number of elements and `uh`, the approximation generated by Algorithm 2.1. The output `error` is the energy norm error $\|u-u_h\|_{\mathrm{E}}$.

```
function error=test_FEM_error(ne,uh)
h=(1/ne); xx=0:h:1; Ek2=zeros(ne,1);
u1s=uh(1:end-1); u2s=uh(2:end);
% quadrature weights and points
weights=h.*[1/6;2/3;1/6];
x_quad=[xx(1:end-1)',[xx(1:end-1)+h/2]',xx(2:end)'];
for i=1:3
    Ek2=Ek2+weights(i).*Ek2_eval(x_quad(:,i),u1s./h,u2s./h);
end
error=sqrt(sum(Ek2));

function Ek2=Ek2_eval(x,u1,u2);
Ek2=(0.5-x+u1-u2).^2;
return
```

Example 2.33 Consider (2.3)–(2.4) with $D = (0,1)$, $f = 1$, $p = 1$ and $q = 0$. The exact solution is $u = \frac{1}{2}x(1-x)$ and the error is

$$\|u-u_h\|^2_{\mathrm{E}} = \sum_{k=1}^{n_e}\int_{z_{k-1}}^{z_k}\left(u'(x) - u_h'(x)\right)^2\,dx = \sum_{k=1}^{n_e}\int_{z_{k-1}}^{z_k}\left(\frac{1}{2} - x - u_h'(x)\right)^2\,dx.$$

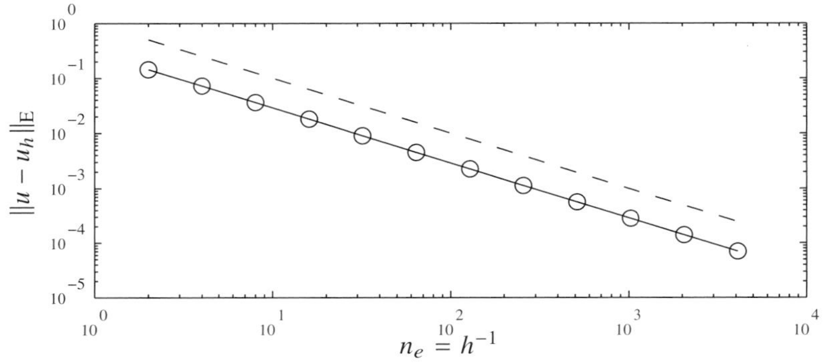

Figure 2.8 Plot of the error $\|u - u_h\|_{\mathrm{E}}$ for the piecewise linear finite element approximation of the BVP (2.3)–(2.4) on $D = (0, 1)$ with $f = 1, p = 1$ and $q = 0$. The horizontal axis displays the number of elements or the reciprocal of the mesh size. The solid line with circles is the computed error and the dashed line has a reference slope of -1.

Expanding $u_h\big|_{e_k} = u_1^k \phi_1^k + u_2^k \phi_2^k$ as in Example 2.27 gives

$$\|u - u_h\|_{\mathrm{E}}^2 = \sum_{k=1}^{n_e} \int_{z_{k-1}}^{z_k} \left(\frac{1}{2} - x + \frac{u_1^k}{h} - \frac{u_2^k}{h} \right)^2 dx =: \sum_{k=1}^{n_e} E_k^2.$$

The element errors E_k^2 can be computed exactly using Simpson's rule:

$$\int_{z_{k-1}}^{z_k} f(x)\, dx \approx \frac{h}{6} \big(f(z_{k-1}) + 4 f(z_{k-1/2}) + f(z_k) \big).$$

This is implemented in Algorithm 2.3. The energy norm errors computed for increasing numbers of elements $n_e = h^{-1}$ are plotted in Figure 2.8. The theoretical $\mathcal{O}(h)$ convergence rate predicted by Theorem 2.32 is clearly observed.

For example, with $n_e = 16$ elements, the error can be computed using Algorithms 2.1 and 2.3 with the commands:

```
>> ne=16; [uh,A,b]=oned_linear_FEM(ne,ones(ne,1),zeros(ne,1),ones(ne,1));
>> error=test_FEM_error(ne,uh)
error =
    0.0180
```

See also Exercise 2.12.

2.2 Variational formulation of elliptic PDEs

Having studied Galerkin approximation and the finite element method for the two-point BVP (2.3)–(2.4), we are ready to develop analogous theory for two-dimensional elliptic PDEs. In this section, we begin with an introduction to the analysis of weak solutions. Let $D \subset \mathbb{R}^2$ be a bounded domain with piecewise smooth boundary ∂D (recall Definition 1.3). Given suitable functions $g \colon \partial D \to \mathbb{R}$ and $a, f \colon D \to \mathbb{R}$ with $a(\boldsymbol{x}) > 0$ for all $\boldsymbol{x}$ in D, we want to find $u(\boldsymbol{x})$ that satisfies

$$-\nabla \cdot \big(a(\boldsymbol{x})\nabla u(\boldsymbol{x})\big) = -\sum_{j=1}^{2} \frac{\partial}{\partial x_j}\left(a(\boldsymbol{x})\frac{\partial u(\boldsymbol{x})}{\partial x_j}\right) = f(\boldsymbol{x}), \qquad \boldsymbol{x} \in D, \tag{2.45}$$

and the Dirichlet boundary condition

$$u(\boldsymbol{x}) = g(\boldsymbol{x}), \qquad \boldsymbol{x} \in \partial D. \tag{2.46}$$

If $D \subset \mathbb{R}$ and $g = 0$ then (2.45)–(2.46) is a special case of the BVP (2.3)–(2.4) with $p = a$ and $q = 0$. For brevity, we now omit the reaction term qu but we do allow for non-homogeneous boundary data. This leads to subtle changes in the variational formulation and corresponding analysis.

The notion of a *classical solution* (see Definition 2.2) generalises easily for (2.45)–(2.46). In particular, classical solutions exist only if $a \in \mathrm{C}^1(\bar{D})$. In many practical applications, however, a is discontinuous and we are content with weak solutions. We make the following assumption.

Assumption 2.34 (regularity of coefficients) The diffusion coefficient $a(x)$ satisfies

$$0 < a_{\min} \le a(\boldsymbol{x}) \le a_{\max} < \infty, \qquad \text{for almost all } \boldsymbol{x} \in D,$$

for some real constants $a_{\min}$ and $a_{\max}$. In particular, $a \in L^\infty(D)$.

Our first task is to set up a variational formulation of (2.45)–(2.46). Following our approach in §2.1, we multiply both sides of (2.45) by a test function $\phi \in \mathrm{C}_c^\infty(D)$ and integrate over D, to obtain

$$-\int_D \nabla \cdot \big(a(\boldsymbol{x})\nabla u(\boldsymbol{x})\big)\phi(\boldsymbol{x})\, d\boldsymbol{x} = \int_D f(\boldsymbol{x})\phi(\boldsymbol{x})\, d\boldsymbol{x}. \tag{2.47}$$

The product rule for differentiation gives

$$\nabla \cdot (\phi a \nabla u) = \nabla \cdot (a\nabla u)\phi + a\nabla u \cdot \nabla \phi$$

and so we can split the integral on the left-hand side of (2.47) to obtain

$$\int_D a(\boldsymbol{x})\nabla u(\boldsymbol{x}) \cdot \nabla\phi(\boldsymbol{x})\, d\boldsymbol{x} - \int_D \nabla \cdot \big(\phi(\boldsymbol{x})a(\boldsymbol{x})\nabla u(\boldsymbol{x})\big)\, d\boldsymbol{x} = \int_D f(\boldsymbol{x})\phi(\boldsymbol{x})\, d\boldsymbol{x}.$$

Applying the divergence theorem (in two space dimensions) gives

$$\int_D \nabla \cdot \big(\phi(\boldsymbol{x})a(\boldsymbol{x})\nabla u(\boldsymbol{x})\big)\, d\boldsymbol{x} = \int_{\partial D} \phi(\boldsymbol{x})\big(a(\boldsymbol{x})\nabla u(\boldsymbol{x})\big) \cdot \boldsymbol{n}\, ds,$$

where $\boldsymbol{n}$ denotes the unit outward pointing normal vector on ∂D and the integral on the right-hand side is a line integral. Since $\phi \in \mathrm{C}_c^\infty(D)$, we have $\phi(\boldsymbol{x}) = 0$ for $\boldsymbol{x} \in \partial D$ and the line integral is zero. Hence,

$$\int_D a(\boldsymbol{x})\, \nabla u(\boldsymbol{x}) \cdot \nabla\phi(\boldsymbol{x})\, d\boldsymbol{x} = \int_D f(\boldsymbol{x})\phi(\boldsymbol{x})\, d\boldsymbol{x}. \tag{2.48}$$

Any solution u to (2.45) therefore satisfies the variational problem

$$a(u, \phi) = \ell(\phi), \qquad \forall \phi \in \mathrm{C}_c^\infty(D),$$

where, now,

$$a(u,\phi) := \int_D a(\boldsymbol{x})\, \nabla u(\boldsymbol{x}) \cdot \nabla \phi(\boldsymbol{x})\, d\boldsymbol{x}, \tag{2.49}$$

$$\ell(\phi) := \langle f, \phi \rangle_{L^2(D)}. \tag{2.50}$$

If Assumption 2.34 holds, the symmetric bilinear form $a(\cdot,\cdot)$ is bounded from $V \times V$ to $\mathbb{R}$, where $V = H_0^1(D)$ (see Definition 1.58). However, if $g \neq 0$ in (2.46), we cannot look for u in V. The correct solution space is often written as

$$W = \{w \in H^1(D) \colon w|_{\partial D} = g\}, \tag{2.51}$$

but is defined precisely as

$$W := H_g^1(D) := \{w \in H^1(D) \colon \gamma w = g\}. \tag{2.52}$$

In (2.52), $\gamma \colon H^1(D) \to L^2(\partial D)$ is a so-called *trace operator* that maps functions on D onto functions on the boundary ∂D. In Chapter 1, we defined the space $L^2(D)$ associated with a domain $D \subset \mathbb{R}^d$. The definition of $L^2(\partial D)$ follows naturally from Definition 1.26(ii).

Definition 2.35 ($L^2(\partial D)$ spaces) For a domain $D \subset \mathbb{R}^d$ with boundary ∂D, consider the measure space $(\partial D, \mathcal{B}(\partial D), V)$, where $V(B)$ is the volume (i.e., arc length for $d = 2$, area for $d = 3$) of a Borel set $B \in \mathcal{B}(\partial D)$. $L^2(\partial D)$ is the Hilbert space $L^2(\partial D, \mathbb{R})$ equipped with the norm

$$\|g\|_{L^2(\partial D)} := \left(\int_{\partial D} g(\boldsymbol{x})^2\, dV(\boldsymbol{x}) \right)^{1/2}.$$

Example 2.36 ($d = 2$) If $D \subset \mathbb{R}^2$, we parameterise ∂D by the arc length s of $\boldsymbol{x} \in \partial D$ from a fixed reference point $\boldsymbol{x}_0 \in \partial D$. We can then write $g \in L^2(\partial D)$ as a function of s so that $g \colon [0, L] \to \mathbb{R}$, where L is the length of ∂D, and

$$\|g\|_{L^2(\partial D)} = \left(\int_0^L g(s)^2 ds \right)^{1/2}.$$

The need for precision in (2.52) must be stressed. For $w \in \mathrm{C}(\bar{D})$, the restriction $w|_{\partial D}$ is well defined. When $D \subset \mathbb{R}^2$, however, we have no guarantee that $w \in H^1(D)$ is continuous on $\bar{D}$. Since ∂D is a set of measure zero, we can assign any value to w on ∂D without changing its meaning in the L^2 sense. By observing that $w \in H^1(D)$ can be approximated by an appropriate sequence of functions $\{w_n\}$ that *are* continuous on $\bar{D}$, and that the restrictions of these functions form a Cauchy sequence, classical theory shows that

$$\gamma w := \lim_{n \to \infty} w_n|_{\partial D} \tag{2.53}$$

is well defined. For our work, it suffices to know that the operator γ exists and belongs to $\mathcal{L}(H^1(D), L^2(\partial D))$ (see Definition 1.53).

Lemma 2.37 (trace operator on $H^1(D)$) *Consider a bounded domain $D \subset \mathbb{R}^2$ with sufficiently smooth boundary ∂D. There exists a bounded linear operator $\gamma \colon H^1(D) \to L^2(\partial D)$ such that*

$$\gamma w = w|_{\partial D}, \qquad \forall w \in \mathrm{C}^1(\bar{D}).$$

Lemma 2.37 says that when w is smooth enough, γw coincides with our usual interpretation of $w|_{\partial D}$. Our earlier assumption that a domain D has a piecewise smooth boundary is important. Such domains are *sufficiently smooth* in the sense of Lemma 2.37 and this facilitates the analysis. When we look for an approximation in a finite-dimensional subspace, we can always choose $W^h \subset W$ so that $w|_{\partial D}$ is well defined for any $w \in W^h$. We must also give some thought to the choice of g in (2.46). The image of γ is not the whole of $L^2(\partial D)$ but a certain subspace. If the weak solution is to satisfy the boundary condition (in the sense of trace) then g must belong to this subspace.

Definition 2.38 ($H^{1/2}(\partial D)$ space) Let $D \subset \mathbb{R}^2$ be a bounded domain. The *trace space* $H^{1/2}(\partial D)$ is defined as

$$H^{1/2}(\partial D) := \gamma(H^1(D)) = \{\gamma w \colon w \in H^1(D)\},$$

where γ is the trace operator on $H^1(D)$. $H^{1/2}(\partial D)$ is a Hilbert space and is equipped with the norm

$$\|g\|_{H^{1/2}(\partial D)} := \inf\{\|w\|_{H^1(D)} \colon \gamma w = g \text{ and } w \in H^1(D)\}.$$

If $g \in H^{1/2}(\partial D)$, we can find an $H^1(D)$ function whose trace is g and this implies that the solution space W in (2.52) is non-empty.

Lemma 2.39 *There exists $K_\gamma > 0$ such that, for all $g \in H^{1/2}(\partial D)$, we can find $u_g \in H^1(D)$ with $\gamma u_g = g$ and*

$$\|u_g\|_{H^1(D)} \le K_\gamma \|g\|_{H^{1/2}(\partial D)}. \tag{2.54}$$

We can now state the weak form of (2.45)–(2.46).

Definition 2.40 (weak solution) A *weak solution* to the BVP (2.45)–(2.46) is a function $u \in W$ that satisfies

$$a(u,v) = \ell(v), \qquad \forall v \in V, \tag{2.55}$$

where a and ℓ are defined as in (2.49) and (2.50).

Most of the theory from §2.1 can be applied to analyse (2.55). Some modifications are needed however to accommodate non-zero boundary data. Since $W = H^1_g(D)$ and $V = H^1_0(D)$, we have $W \ne V$ whenever $g \ne 0$ and we cannot apply the Lax–Milgram lemma directly as in Theorem 2.10. Moreover, the analogous *energy norm*

$$|u|_{\mathrm{E}} := a(u,u)^{1/2} = \left(\int_D a\nabla u \cdot \nabla u \, dx\right)^{1/2} \tag{2.56}$$

is only a norm on V. In much of the analysis below, it acts as a semi-norm.

Lemma 2.41 *Let Assumption 2.34 hold. The bilinear form $a(\cdot,\cdot)$ in (2.49) is bounded from $H^1(D) \times H^1_0(D)$ to $\mathbb{R}$. Further, the semi-norm $|\cdot|_{\mathrm{E}}$ in (2.56) is equivalent to the semi-norm $|\cdot|_{H^1(D)}$ on $H^1(D)$.*

Proof For any $w \in H^1(D)$ and $v \in H^1_0(D)$,

$$\begin{aligned}|a(w,v)| &\le a_{\max}\|\nabla w\|_{L^2(D)}\|\nabla v\|_{L^2(D)} = a_{\max}|w|_{H^1(D)}|v|_{H^1(D)} \\ &\le a_{\max}\|w\|_{H^1(D)}|v|_{H^1(D)} = a_{\max}\|w\|_{H^1(D)}\|v\|_{H^1_0(D)}.\end{aligned}$$

Hence, $a(\cdot,\cdot)$ is bounded from $H^1(D) \times H_0^1(D)$ to $\mathbb{R}$. Now, for any $w \in H^1(D)$, the assumption on the diffusion coefficient gives

$$a_{\min}\|\nabla w\|_{L^2(D)}^2 \le a(w,w) \le a_{\max}\|\nabla w\|_{L^2(D)}^2$$

and so we have the equivalence

$$\sqrt{a_{\min}}|w|_{H^1(D)} \le |w|_E \le \sqrt{a_{\max}}|w|_{H^1(D)}. \tag{2.57}$$

□

Let $g \in H^{1/2}(\partial D)$ and $u_g \in H^1(D)$ be such that $\gamma u_g = g$ and consider the variational problem: find $u_0 \in V$ such that

$$a(u_0, v) = \hat{\ell}(v), \qquad \forall v \in V, \tag{2.58}$$

where, on the right-hand side,

$$\hat{\ell}(v) := \ell(v) - a(u_g, v). \tag{2.59}$$

The variational problems (2.58) and (2.55) are equivalent. Indeed, if $u_0 \in V$ solves (2.58) then $u = u_0 + u_g \in W$ and solves (2.55). On the other hand, if $u \in W$ solves (2.55) then $u_0 = 0 \in V$ and solves (2.58) with $u_g = u$. Establishing the well-posedness of the weak problem (2.55) is therefore equivalent to establishing the existence and uniqueness of $u_0 \in V$ satisfying (2.58).

Theorem 2.42 *Let Assumption 2.34 hold, $f \in L^2(D)$ and $g \in H^{1/2}(\partial D)$. Then* (2.55) *has a unique solution $u \in W = H_g^1(D)$.*

Proof Use the Lax–Milgram lemma (see Exercise 2.14) to show that there exists a unique $u_0 \in V = H_0^1(D)$ satisfying (2.58). The result follows. □

The next result gives an upper bound for the $H^1(D)$ semi-norm of the weak solution in terms of the data a, f and g.

Theorem 2.43 *Let $u \in W$ satisfy* (2.55). *If the conditions of Theorem 2.42 hold then*

$$|u|_{H^1(D)} \le K\left(\|f\|_{L^2(D)} + \|g\|_{H^{1/2}(\partial D)}\right)$$

where $K := \max\left\{K_p a_{\min}^{-1}, K_\gamma\left(1 + a_{\max}a_{\min}^{-1}\right)\right\}$.

Proof We can write $u = u_0 + u_g$ for some $u_0 \in V$ satisfying (2.58) and $u_g \in H^1(D)$ satisfying $\gamma u_g = g$. Since $a(u_0,u_0) = \hat{\ell}(u_0)$, we have

$$|u_0|_{H^1(D)}^2 \le a_{\min}^{-1} a(u_0,u_0) = a_{\min}^{-1}\hat{\ell}(u_0).$$

Applying the Cauchy–Schwarz inequality and Assumption 2.34 in (2.59) gives

$$|\hat{\ell}(u_0)| \le \|f\|_{L^2(D)}\|u_0\|_{L^2(D)} + a_{\max}|u_g|_{H^1(D)}|u_0|_{H_0^1(D)}.$$

Next, applying Poincaré's inequality and (2.54) gives

$$|\hat{\ell}(u_0)| \le \left(K_p\|f\|_{L^2(D)} + a_{\max}K_\gamma\|g\|_{H^{1/2}(\partial D)}\right)|u_0|_{H^1(D)}. \tag{2.60}$$

Hence,

$$|u_0|_{H^1(D)} \le a_{\min}^{-1}\big(K_p\|f\|_{L^2(D)} + a_{\max}\|g\|_{H^{1/2}(\partial D)}\big).$$

Now note that $|u|_{H^1(D)} \le |u_0|_{H^1(D)} + |u_g|_{H^1(D)}$ and apply (2.54) once more. □

In practice, we often approximate the data. Sometimes this is done at the outset because the true data is too complex and sometimes at the level of implementation. Approximating a and f leads to a modified weak problem with a solution $\tilde{u} \in W$ that differs from the solution of the weak form of (2.45)–(2.46). We now study the resulting weak solution error $u - \tilde{u}$.

Variational formulation with approximate data

Replacing a and f in (2.45) by approximations $\tilde{a}$ and $\tilde{f}$, respectively, leads to the perturbed weak problem: find $\tilde{u} \in W$ such that

$$\tilde{a}(\tilde{u}, v) = \tilde{\ell}(v), \qquad \forall v \in V, \tag{2.61}$$

where $\tilde{a}\colon W \times V \to \mathbb{R}$ and $\tilde{\ell}\colon V \to \mathbb{R}$ are now defined by

$$\tilde{a}(w, v) := \int_D \tilde{a}(\boldsymbol{x})\nabla w(\boldsymbol{x}) \cdot \nabla v(\boldsymbol{x})\, d\boldsymbol{x}, \qquad \tilde{\ell}(v) := \int_D \tilde{f}(\boldsymbol{x}) v(\boldsymbol{x})\, d\boldsymbol{x}. \tag{2.62}$$

If $\tilde{f} \in L^2(D)$ and $\tilde{a}$ satisfies Assumption 2.44 then it is straightforward to prove the analogue of Theorem 2.42 using the Lax–Milgram lemma.

Assumption 2.44 (regularity of coefficients) The diffusion coefficient $\tilde{a}(x)$ satisfies

$$0 < \tilde{a}_{\min} \le \tilde{a}(\boldsymbol{x}) \le \tilde{a}_{\max}, \qquad \text{for almost all } \boldsymbol{x} \in D,$$

for some real constants $\tilde{a}_{\min}$ and $\tilde{a}_{\max}$. In particular, $\tilde{a} \in L^\infty(D)$.

Theorem 2.45 *Let Assumption 2.44 hold, $\tilde{f} \in L^2(D)$ and $g \in H^{1/2}(\partial D)$. Then* (2.61) *has a unique solution $\tilde{u} \in W = H^1_g(D)$.*

The next result gives an upper bound for the error $|u - \tilde{u}|_{H^1(D)}$ in terms of the data errors $\|f - \tilde{f}\|_{L^2(D)}$ and $\|a - \tilde{a}\|_{L^\infty(D)}$.

Theorem 2.46 *Let the conditions of Theorems 2.42 and 2.45 hold so that $u \in W$ satisfies* (2.55) *and $\tilde{u} \in W$ satisfies* (2.61). *Then,*

$$|u - \tilde{u}|_{H^1(D)} \le K_p \tilde{a}_{\min}^{-1}\|f - \tilde{f}\|_{L^2(D)} + \tilde{a}_{\min}^{-1}\|a - \tilde{a}\|_{L^\infty(D)}|u|_{H^1(D)}. \tag{2.63}$$

Proof Since $u, \tilde{u} \in W$, we have $\tilde{u} - u \in V$ and

$$\begin{aligned}
|\tilde{u} - u|^2_{H^1(D)} &\le \tilde{a}_{\min}^{-1}\big(\tilde{a}(\tilde{u}, \tilde{u} - u) - \tilde{a}(u, \tilde{u} - u)\big)\\
&= \tilde{a}_{\min}^{-1}\big(\tilde{\ell}(\tilde{u} - u) - \tilde{a}(u, \tilde{u} - u) + a(u, \tilde{u} - u) - \ell(\tilde{u} - u)\big)\\
&= \tilde{a}_{\min}^{-1}\Big(\big(\tilde{\ell}(\tilde{u} - u) - \ell(\tilde{u} - u)\big) + \big(a(u, \tilde{u} - u) - \tilde{a}(u, \tilde{u} - u)\big)\Big).
\end{aligned}$$

Applying the Cauchy–Schwarz and Poincaré inequalities gives

$$\tilde{\ell}(\tilde{u} - u) - \ell(\tilde{u} - u) \le \|f - \tilde{f}\|_{L^2(D)}\|\tilde{u} - u\|_{L^2(D)} \le K_p\|f - \tilde{f}\|_{L^2(D)}|\tilde{u} - u|_{H^1(D)}.$$

Note also that

$$a\,(u,\tilde{u}-u) - \tilde{a}\,(u,\tilde{u}-u) \le \|a-\tilde{a}\|_{L^\infty(D)} |u|_{H^1(D)} |\tilde{u}-u|_{H^1(D)}.$$

Combining all these bounds gives the result. □

We did not make an assumption about the type of approximations $\tilde{a}$ and $\tilde{f}$. However, the bound in (2.63) is only useful if we have some guarantee that $\|f-\tilde{f}\|_{L^2(D)}$ and $\|a-\tilde{a}\|_{L^\infty(D)}$ are small.

Galerkin approximation

We return now to the original BVP (2.45)–(2.46) with data a and f. Once again, the weak problem (2.55) provides the starting point for Galerkin approximation. If $g \ne 0$, the solution space and test space are different. We have to choose *two* finite-dimensional subspaces, $W^h \subset W = H^1_g(D)$ and $V^h \subset V = H^1_0(D)$. There are several possibilities. Clearly, $v - w \in V$ for any v and w in W^h. In *Galerkin* approximation, we construct the subspaces so that

$$v - w \in V^h, \qquad \forall v, w \in W^h. \tag{2.64}$$

Definition 2.47 (Galerkin approximation) Let $W^h \subset W$ and $V^h \subset V$ and suppose that (2.64) holds. The *Galerkin approximation* for (2.45)–(2.46) is the function $u_h \in W^h$ satisfying

$$a(u_h, v) = \ell(v), \qquad \forall v \in V^h. \tag{2.65}$$

To ensure (2.64), we assume that W^h is constructed so that each $w \in W^h$ has the form $w = w_0 + w_g$, where $w_0 \in V^h$ and $w_g \in H^1(D)$ is a *fixed* function satisfying the boundary conditions (see §2.3 for some practical advice).

Theorem 2.48 *If the conditions of Theorem 2.42 hold then* (2.65) *has a unique solution* $u_h \in W^h$.

Proof Since $g \in H^{1/2}(\partial D)$, there exists $w_g \in H^1(D)$ such that $\gamma w_g = g$. Following the proof of Theorem 2.42, we can apply the Lax–Milgram lemma on $V^h \times V^h$ to show that there exists a unique $u_h^0 \in V^h$ satisfying

$$a(u_h^0, v) = \hat{\ell}(v), \qquad \forall v \in V^h, \tag{2.66}$$

where $\hat{\ell}(v) = \ell(v) - a(w_g, v)$. By construction of W^h, $u_h = u_h^0 + w_g$ is then the unique function in W^h satisfying (2.65). □

The next two results are the analogues of Theorems 2.17 and 2.18.

Theorem 2.49 (best approximation) *Suppose* $V^h \subset V$ *and* $W^h \subset W$ *and let* $u \in W$ *and* $u_h \in W^h$ *satisfy* (2.55) *and* (2.65), *respectively. If* (2.64) *holds then*

$$|u - u_h|_E = \inf_{w \in W^h} |u - w|_E. \tag{2.67}$$

Proof Since $V^h \subset V$, we obtain $a(u - u_h, v) = 0$ for all $v \in V^h$ (Galerkin orthogonality). If (2.64) holds then for, any $w \in W^h$, we have $w - u_h \in V^h$ and

$$\begin{aligned}|u - u_h|_E^2 &= a(u - u_h, u - u_h) = a(u - u_h, u - w) + a(u - u_h, w - u_h)\\ &= a(u - u_h, u - w).\end{aligned}$$

Now apply the Cauchy–Schwarz inequality and note that $u_h \in W^h$. □

It is crucial that (2.64) holds in the above proof. Otherwise, $u_h \in W^h$ is not guaranteed to be the best approximation.

Theorem 2.50 (quasi-optimal approximation) *Let Assumption 2.34 hold and suppose the conditions of Theorem 2.49 hold. Then*

$$|u - u_h|_{H^1(D)} \le \sqrt{\frac{a_{\max}}{a_{\min}}} |u - w|_{H^1(D)}, \qquad \forall w \in W^h. \tag{2.68}$$

Proof This follows easily from the equivalence (2.57) and (2.67). □

Galerkin approximation with approximate data

Finally, we consider the accuracy of the Galerkin approximation $\tilde{u}_h$ when the original data a and f are approximated.

Definition 2.51 (Galerkin approximation) Let $W^h \subset W$ and $V^h \subset V$ and suppose that (2.64) holds. The *Galerkin approximation* for (2.45)–(2.46) with approximate data $\tilde{a}$ and $\tilde{f}$ is the function $\tilde{u}_h \in W^h$ satisfying

$$\tilde{a}(\tilde{u}_h, v) = \tilde{\ell}(v), \qquad \forall v \in V^h, \tag{2.69}$$

where $\tilde{a}(\cdot,\cdot)$ and $\tilde{\ell}(\cdot)$ are defined in (2.62).

If $\tilde{a}$ and $\tilde{f}$ satisfy the same assumptions as a and f, respectively, we can extend Theorem 2.48 to prove the well-posedness of (2.69).

Theorem 2.52 *Let Assumption 2.44 hold, $\tilde{f} \in L^2(D)$ and $g \in H^{1/2}(\partial D)$. Then* (2.69) *has a unique solution $\tilde{u}_h \in W^h$.*

Now, by the triangle inequality,

$$|u - \tilde{u}_h|_{H^1(D)} \le |u - \tilde{u}|_{H^1(D)} + |\tilde{u} - \tilde{u}_h|_{H^1(D)}. \tag{2.70}$$

This splitting is natural if we first approximate the data and then solve a finite-dimensional version of the perturbed problem. The first term is the error in the weak solution caused by approximating the data. We obtained a bound for this in Theorem 2.46. The second term is the Galerkin error associated with approximating the perturbed solution $\tilde{u} \in W$ by $\tilde{u}_h \in W^h$. Since $\tilde{u}$ and $\tilde{u}_h$ are solutions to weak problems with the same bilinear form $\tilde{a}(\cdot,\cdot)$ and linear functional $\tilde{\ell}(\cdot)$, Theorem 2.50 can be extended to show that

$$|\tilde{u} - \tilde{u}_h|_{H^1(D)} \le \sqrt{\frac{\tilde{a}_{\max}}{\tilde{a}_{\min}}} \inf_{w \in W^h} |\tilde{u} - w|_{H^1(D)}. \tag{2.71}$$

A precise bound can be found after choosing specific subspaces V^h and W^h.

Alternatively, we have

$$|u - \tilde{u}_h|_{H^1(D)} \leq |u - u_h|_{H^1(D)} + |u_h - \tilde{u}_h|_{H^1(D)}. \tag{2.72}$$

This splitting is more natural if we make approximations only at the finite-dimensional level. For deterministic BVPs, this is usually the case. The second term is the error in the *Galerkin* solution caused by modifying $a(\cdot,\cdot)$ and $\ell(\cdot)$ (e.g., by approximating a and f by piecewise constant functions and then integrating exactly as we did in our finite element experiments in §2.1). If W^h is chosen so that $u_h - \tilde{u}_h \in V_h$, it is straightforward to modify the proof of Theorem 2.46 to show that

$$|u_h - \tilde{u}_h|_{H^1(D)} \leq K_{\mathrm{p}} \tilde{a}_{\min}^{-1} \|f - \tilde{f}\|_{L^2(D)} + \tilde{a}_{\min}^{-1} \|a - \tilde{a}\|_{L^\infty(D)} |u_h|_{H^1(D)}. \tag{2.73}$$

Note that u_h now appears on the right-hand side. The first term in (2.72) is the Galerkin error associated with approximating $u \in W$ by $u_h \in W^h$.

In Chapter 9, we consider BVPs with random data. One approach is to approximate the data by spatially smooth random functions and then apply the so-called stochastic Galerkin method. The error splitting (2.70) is useful for the analysis. Another approach is to apply the standard finite element method to the deterministic elliptic BVPs that arise when approximations of spatially rough realisations of the random data are generated using numerical methods. In that case, the bound (2.72) will be useful. To analyse the Galerkin error $u - u_h$ (or, similarly, $\tilde{u} - \tilde{u}_h$), we need to choose a common set of basis functions to construct both V^h and W^h so that (2.64) holds. We now focus on finite element approximation.

2.3 The Galerkin finite element method for elliptic PDEs

Following the ideas in §2.1, we partition $\bar{D}$ into n_e *elements* and choose W^h and V^h to be sets of piecewise polynomials. This is less trivial in two dimensions. There are many possible ways to partition $\bar{D} \subset \mathbb{R}^2$ (some good and some bad) and hence, many possible choices for V^h and W^h.

Triangular finite elements

Rectangular domains D can be divided up into quadrilaterals, and any two-dimensional domain with a polygonal boundary can be partitioned with triangles. Since they offer more flexibility in terms of the domain, we focus exclusively on triangular elements. In numerical experiments, we fix $D = (0,1) \times (0,1)$ for convenience but this is not a restriction. Two meshes of triangular elements on the unit square are shown in Figure 2.9.

Definition 2.53 (admissible mesh, characteristic mesh size) Consider a set of n_e non-overlapping triangles $\mathcal{T}_h := \{\triangle_1, \ldots, \triangle_{n_e}\}$ such that $\bar{D} = \cup_{k=1}^{n_e} \bar{\triangle}_k$. Distinct triangles must meet only at a single vertex or else share an entire edge. Let h_k be the length of the longest edge of $\triangle_k$ and let $h := \max_k h_k$. We say $\mathcal{T}_h$ is an *admissible mesh* with *characteristic mesh size* h.

Note that h is a *representative* mesh size and an admissible mesh may contain elements of quite different sizes (as in the left plot in Figure 2.9). To study the Galerkin finite element error as $h \to 0$, we require a *sequence* of meshes and, to ensure that the mesh quality does not deteriorate as refinements are made, we work with shape-regular meshes.

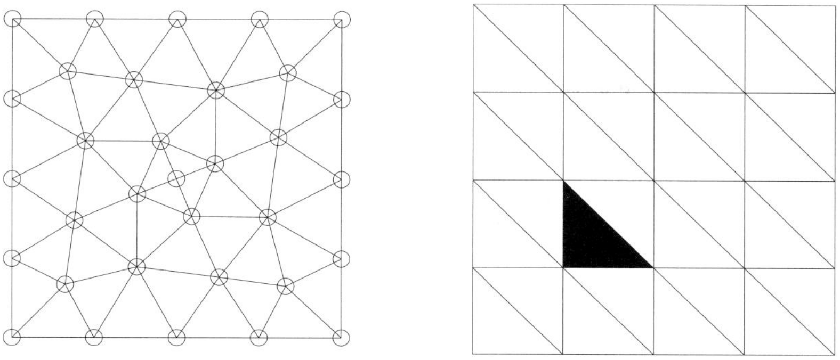

Figure 2.9 Admissible triangular meshes $\mathcal{T}_h$ on $D = (0,1) \times (0,1)$. The mesh on the left has $n_e = 48$ elements, $J = 17$ interior vertices, and $J_b = 16$ boundary vertices. The mesh on the right has $n_e = 32$ elements, $J = 9$ interior vertices, and $J_b = 16$ boundary vertices.

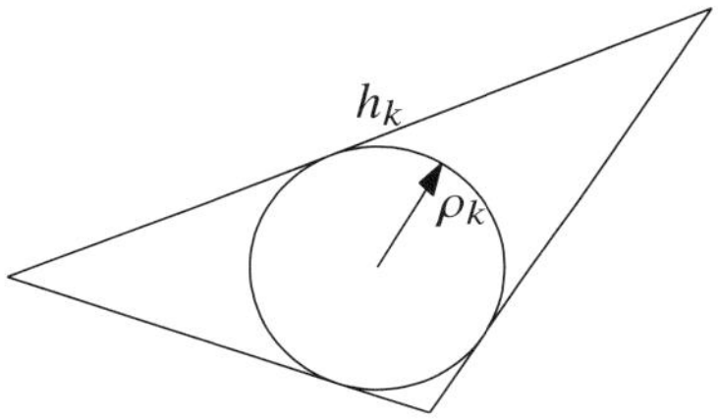

Figure 2.10 A triangular element $\triangle_k$ of size h_k, with inscribed circle of radius ρ_k. For shape-regular meshes, ρ_k/h_k never becomes too small and is bounded away from zero as the mesh is refined.

Definition 2.54 (shape-regular mesh) A sequence of meshes $\{\mathcal{T}_h\}$ is *shape-regular* if there exists a constant $\kappa > 0$ independent of h such that

$$\frac{\rho_k}{h_k} \geq \kappa, \qquad \forall \triangle_k \in \mathcal{T}_h, \tag{2.74}$$

where ρ_k is the radius of the largest inscribed circle and h_k is the size of $\triangle_k$ (see Figure 2.10).

If (2.74) holds for a sequence of increasingly fine meshes $\{\mathcal{T}_h\}$, then no triangles with small interior angles appear as $h \to 0$. A simple refinement strategy that leads to shape-regular mesh sequences is illustrated in Figure 2.11.

Algorithm 2.4 generates uniform meshes of right-angled triangles on the unit square (as shown on the right in Figure 2.9). The mesh is created by partitioning the domain into `ns` × `ns` squares and dividing each square into two triangles. Increasing `ns` refines the mesh. For such simple meshes, Definition 2.53 says that the characteristic mesh size h (which appears later in the analysis) is $h = \sqrt{2}\,\texttt{ns}^{-1}$. In our numerical experiments, we use the parameter $h = \texttt{ns}^{-1}$ to represent the mesh size; see Figure 2.12.

In addition to creating a mesh, Algorithm 2.4 imposes a global numbering system on the vertices and elements. Starting at the bottom left corner of D, the vertices are labelled lexicographically (from left to right and bottom to top). The elements are divided into two groups (see Figure 2.12). Triangles of type A are numbered first, from left to right and bottom to top, followed by triangles of type B. The output `elt2vert` lists the labels of the vertices of each element.

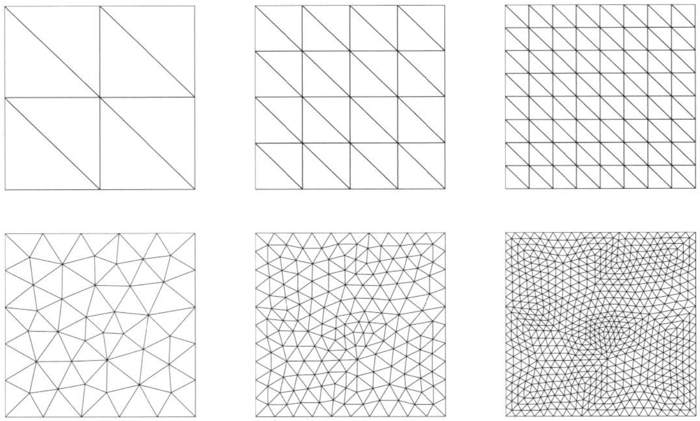

Figure 2.11 Two sequences of shape-regular meshes. Starting from the mesh on the left, each triangle is divided into four equally sized smaller triangles by joining the midpoints of the edges. At each refinement step, n_e increases fourfold and h is halved.

Figure 2.12 The meshes generated by Algorithm 2.4 contain triangles of type A and type B. The vertices are labelled in an anti-clockwise direction, starting at the right angle.

Example 2.55 We can generate the uniform mesh on the right in Figure 2.9 using Algorithm 2.4 with the command:

```
>> [xv,yv,elt2vert,nvtx,ne,h]=uniform_mesh_info(4);
```

The global labels of the vertices of the sixth triangle (the shaded triangle in Figure 2.9) can be found using the command:

```
>> elt2vert(6,:)
ans =
     7     8    12
```

These are the labels of the first, second and third vertices, respectively, in Figure 2.12. The coordinates of the vertices can be found with the command:

```
>> [xv(elt2vert(6,:)), yv(elt2vert(6,:))]
ans =
    0.2500    0.2500
    0.5000    0.2500
    0.2500    0.5000
```

Algorithm 2.4 Code to generate uniform meshes of right-angled triangles on $D = (0,1) \times (0,1)$. The input `ns` is the number of squares in the partition. The outputs are `xv,yv`, vectors of coordinates of the vertices; `elt2vert`, an array whose ith row contains the global labels of the ith element; `nvtx`, the number of vertices; `ne`, the number of elements and `h`, the mesh size.

```
function [xv,yv,elt2vert,nvtx,ne,h]=uniform_mesh_info(ns)
h=1/ns; x=0:h:1; y=x; [xv,yv]=meshgrid(x,y);
% co-ordinates of vertices
xv=xv'; xv=xv(:); yv=yv'; yv=yv(:);
n2=ns*ns;nvtx=(ns+1)*(ns+1); ne=2*n2;
% global vertex labels of individual elements
elt2vert=zeros(ne,3); vv=reshape(1:nvtx, ns+1,ns+1);
v1=vv(1:ns,1:ns); v2=vv(2:end,1:ns); v3=vv(1:ns,2:end);
elt2vert(1:n2,:)=[v1(:),v2(:),v3(:)];
v1=vv(2:end,2:end); elt2vert(n2+1:end,:)=[v1(:),v3(:),v2(:)];
triplot(elt2vert,xv,yv,'k'); axis square;
```

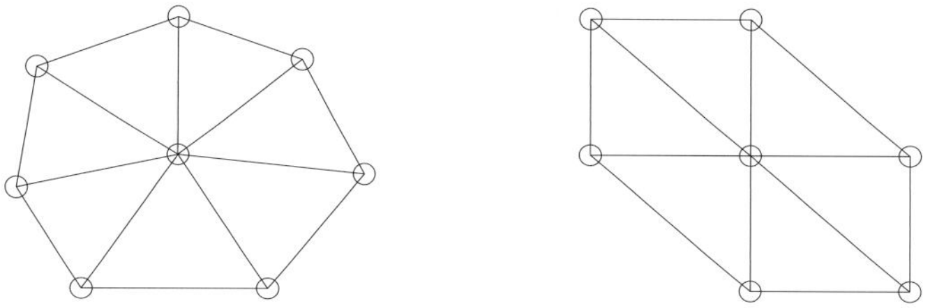

Figure 2.13 Patches of triangles. The piecewise linear basis function $\phi_j(\boldsymbol{x})$ associated with the central vertex $\boldsymbol{x}_j$ takes the value one at $\boldsymbol{x}_j$ and is zero at all other vertices.

Global basis functions

Given a triangular finite element mesh $\mathcal{T}_h$, we now choose $V^h \subset H_0^1(D)$ to be a set of continuous piecewise polynomials of fixed total degree. That is,

$$V^h := \left\{ v \in C(\bar{D}) \text{ with } v = 0 \text{ on } \partial D \text{ and } v|_{\triangle_k} \in \mathbb{P}_r(\triangle_k) \text{ for all } \triangle_k \in \mathcal{T}_h \right\}, \tag{2.75}$$

where $\mathbb{P}_r(\triangle_k)$ denotes polynomials in $\boldsymbol{x} = (x,y)$ of total degree r or less on the triangle $\triangle_k$. We are not obliged to have $V^h \subset C(\bar{D})$ but this ensures $v|_{\partial D}$ is well defined for any $v \in V^h$. Once again, we use nodal bases; $V^h = \text{span}\{\phi_1(\boldsymbol{x}),\dots,\phi_J(\boldsymbol{x})\}$ where ϕ_j satisfies

$$\phi_j(\boldsymbol{x}_i) = \delta_{ij}, \tag{2.76}$$

and $\{\boldsymbol{x}_1,\dots,\boldsymbol{x}_J\}$ is a set of nodes positioned at appropriate locations in D to ensure $V^h \subset C(\bar{D})$. The polynomial degree r determines the functions ϕ_j as well as the number and the positions of the nodes.

Example 2.56 (piecewise linear elements) Let $r = 1$. The global basis functions are defined by

$$\phi_j|_{\triangle_k} \in \mathbb{P}_1(\triangle_k) \quad \text{and} \quad \phi_j(\boldsymbol{x}_i) = \delta_{ij}, \tag{2.77}$$

where $\boldsymbol{x}_1,\dots,\boldsymbol{x}_J$ are the J *interior vertices* of the mesh (see Figure 2.9). The support $\text{supp}(\phi_j)$ is a *patch* of elements (see Figure 2.13) meeting at vertex $\boldsymbol{x}_j$.

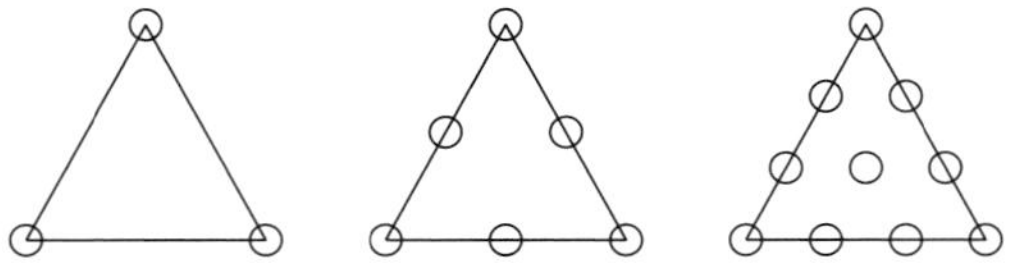

Figure 2.14 $\mathbb{P}_1$ (left), $\mathbb{P}_2$ (middle) and $\mathbb{P}_3$ (right) elements with 3, 6 and 10 nodes, respectively. When $r = 2$, the nodes are the vertices and the midpoints of the edges. When $r = 3$, there are four nodes on each edge and one node at the centre of the element.

To understand why we choose the nodes to be the vertices, consider two elements $\triangle_1$ and $\triangle_2$ that share an edge $\mathcal{E}$. As ϕ_j is *linear* on both elements and $\phi_j|_{\triangle_1}$ and $\phi_j|_{\triangle_2}$ agree at *two* points on $\mathcal{E}$ (the vertices), $\phi_j(\boldsymbol{x})$ is continuous across $\mathcal{E}$. This is true for all edges and so $v \in V^h$ belongs to $C(\bar{D})$. Since no nodes are placed on the boundary, any $v \in V^h = \text{span}\{\phi_1, \ldots, \phi_J\}$ is also zero on ∂D.

Higher-order elements are analogously defined. On each triangle, the basis functions are composed of the monomials $x^a y^b$ with $0 \le a + b \le r$. For example, polynomials of degree $r \le 2$ are linear combinations of $1, x, y, x^2, xy$ and y^2. For a given r, the number of terms is

$$n_r := \binom{r+2}{r} = \frac{(r+2)(r+1)}{2}.$$

When $r = 1$, $n_r = 3$ and $\phi_j(x, y)|_{\triangle_k} = ax + by + c$. We can determine the coefficients a, b and c because, by construction, we know the values $\phi_j(\boldsymbol{x}_i)$ at *three* nodes $\boldsymbol{x}_i$ in $\triangle_k$ (i.e., the vertices). In general, we need n_r nodes in each element. To obtain a globally continuous approximation, we also need to configure the nodes so that $r + 1$ of them lie on each edge; see Figure 2.14.

To construct W^h, we need to include the boundary nodes, $\boldsymbol{x}_{J+1}, \ldots, \boldsymbol{x}_{J+J_b}$. In the case of piecewise linear elements, these are the *vertices* that lie on ∂D (see Figure 2.9) and we number these nodes *after* the interior ones. We construct W^h by enriching the basis for V^h with the polynomials ϕ_j associated with the J_b boundary nodes. Specifically, each $w \in W^h$ has the form

$$w(\boldsymbol{x}) = \sum_{i=1}^{J} w_i \phi_i(\boldsymbol{x}) + \sum_{i=J+1}^{J+J_b} w_i \phi_i(\boldsymbol{x}) =: w_0(\boldsymbol{x}) + w_g(\boldsymbol{x}). \tag{2.78}$$

By construction, $w_0 \in V^h$ is zero on the boundary and we *fix* the coefficients

$$\boldsymbol{w}_{\mathrm{B}} := [w_{J+1}, \ldots, w_{J+J_b}]^\mathsf{T}.$$

In this way, (2.64) is satisfied. Although we need $W^h \subset H^1_g(D)$ for the analysis in §2.2, we often relax this in practice. With the construction (2.78), we have

$$w|_{\partial D} = w_g|_{\partial D} = \sum_{i=J+1}^{J+J_b} w_i \phi_i|_{\partial D}.$$

Unless g is a linear combination of the chosen polynomials, the boundary condition will only be satisfied *approximately* and $W^h \not\subset H^1_g(D)$. One possibility is to assume that w

interpolates g at the boundary nodes and, for this, we fix

$$w_i := g(\boldsymbol{x}_i), \qquad \text{for } i = J+1, \dots, J+J_b. \tag{2.79}$$

The Galerkin finite element approximation can now be written as

$$u_h(\boldsymbol{x}) = \sum_{i=1}^{J} u_i \phi_i(\boldsymbol{x}) + \sum_{i=J+1}^{J+J_b} w_i \phi_i(\boldsymbol{x}) = u_0(\boldsymbol{x}) + w_g(\boldsymbol{x}). \tag{2.80}$$

Substituting (2.80) into (2.65) (or into (2.61)) and setting $v = \phi_j \in V^h$ gives

$$\sum_{i=1}^{J} u_i a(\phi_i, \phi_j) = \ell(\phi_j) - \sum_{i=J+1}^{J+J_b} w_i a(\phi_i, \phi_j), \qquad j = 1, \dots, J. \tag{2.81}$$

To express these equations in matrix-vector notation, we define the Galerkin matrix $A \in \mathbb{R}^{(J+J_b)\times(J+J_b)}$ and the vector $\boldsymbol{b} \in \mathbb{R}^{J+J_b}$ via

$$a_{ij} := a(\phi_i, \phi_j) = \int_D a(\boldsymbol{x}) \nabla\phi_i(\boldsymbol{x}) \cdot \nabla\phi_j(\boldsymbol{x})\, d\boldsymbol{x}, \qquad i,j = 1, \dots, J+J_b, \tag{2.82}$$

$$b_i := \ell(\phi_i) = \int_D f(\boldsymbol{x}) \phi_i(\boldsymbol{x})\, d\boldsymbol{x}, \qquad i = 1, \dots, J+J_b. \tag{2.83}$$

Note that A coincides with the diffusion matrix here. If a reaction term qu is added in (2.45), we also need the mass matrix M defined via

$$m_{ij} := \int_D q(\boldsymbol{x}) \phi_i(\boldsymbol{x}) \phi_j(\boldsymbol{x})\, d\boldsymbol{x}, \qquad i,j = 1, \dots, J+J_b. \tag{2.84}$$

From (2.82), we see that $a_{ij} \neq 0$ only when $\operatorname{supp}(\phi_i)$ and $\operatorname{supp}(\phi_j)$ intersect. When $r = 1$, this happens only if $\boldsymbol{x}_i$ and $\boldsymbol{x}_j$ are vertices of the same triangle (see Figure 2.13) and so A is sparse. Partitioning A and $\boldsymbol{b}$ as

$$A = \begin{pmatrix} A_{\mathrm{II}} & A_{\mathrm{IB}} \\ A_{\mathrm{BI}} & A_{\mathrm{BB}} \end{pmatrix}, \qquad \boldsymbol{b} = \begin{pmatrix} \boldsymbol{b}_{\mathrm{I}} \\ \boldsymbol{b}_{\mathrm{B}} \end{pmatrix},$$

where $A_{\mathrm{II}} \in \mathbb{R}^{J\times J}$, $A_{\mathrm{IB}} \in \mathbb{R}^{J\times J_b}$, $\boldsymbol{b}_{\mathrm{I}} \in \mathbb{R}^{J}$ (and similarly for the other blocks), the Galerkin equations (2.81) can finally be written as

$$A_{\mathrm{II}} \boldsymbol{u}_{\mathrm{I}} = \boldsymbol{b}_{\mathrm{I}} - A_{\mathrm{IB}} \boldsymbol{w}_{\mathrm{B}}, \tag{2.85}$$

where $\boldsymbol{w}_{\mathrm{B}}$ is the discrete boundary data, as defined for example in (2.79).

Local basis functions

The integrals in (2.82) and (2.83) can be broken up over the elements $\triangle_k$ in $\mathcal{T}_h$. Following the arguments in §2.1, we then assemble A and $\boldsymbol{b}$ using the element arrays $A^k \in \mathbb{R}^{n_r \times n_r}$ and vectors $\boldsymbol{b}^k \in \mathbb{R}^{n_r}$ defined by

$$a^k_{pq} := \int_{\triangle_k} a(\boldsymbol{x})\, \nabla\phi^k_p(\boldsymbol{x}) \cdot \nabla\phi^k_q(\boldsymbol{x})\, d\boldsymbol{x}, \qquad p,q = 1, \dots n_r, \tag{2.86}$$

$$b^k_p := \int_{\triangle_k} f(\boldsymbol{x}) \phi^k_p(\boldsymbol{x})\, d\boldsymbol{x}, \qquad p = 1, \dots n_r, \tag{2.87}$$

where $\{\phi^k_1, \dots, \phi^k_{n_r}\}$ are the *local basis functions* for element $\triangle_k$.

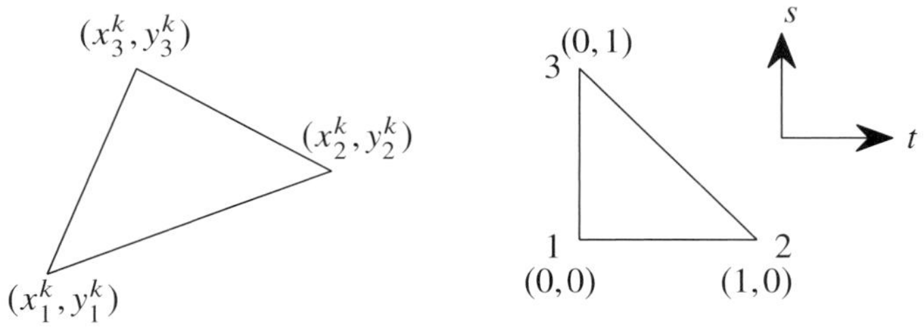

Figure 2.15 Arbitrary element $\triangle_k$ (left) and reference element $\triangle_*$ (right). Under the mapping (2.92)–(2.93), $(0,0) \to (x_1^k, y_1^k)$, $(1,0) \to (x_2^k, y_2^k)$ and $(0,1) \to (x_3^k, y_3^k)$.

Example 2.57 (piecewise linear elements) Let $r = 1$ so that $n_r = 3$. For an arbitrary triangle $\triangle_k$ with vertices $(x_1^k, y_1^k), (x_2^k, y_2^k), (x_3^k, y_3^k)$, $v \in V^h$ satisfies

$$v|_{\triangle_k} = v_1^k \phi_1^k + v_2^k \phi_2^k + v_3^k \phi_3^k$$

where the local basis functions $\phi_1^k, \phi_2^k, \phi_3^k$ are defined by

$$\phi_i^k \in \mathbb{P}_1(\triangle_k), \qquad \phi_i^k(\boldsymbol{x}) = \begin{cases} 1, & \boldsymbol{x} = (x_i^k, y_i^k), \\ 0, & \boldsymbol{x} = (x_j^k, y_j^k) \text{ and } i \neq j. \end{cases} \tag{2.88}$$

To work out these functions explicitly, we write $\phi_i^k(\boldsymbol{x}) = a_i x + b_i y + c_i$, for $i = 1,2,3$, and use (2.88) to obtain

$$\begin{pmatrix} x_1^k & y_1^k & 1 \\ x_2^k & y_2^k & 1 \\ x_3^k & y_3^k & 1 \end{pmatrix} \begin{pmatrix} a_i \\ b_i \\ c_i \end{pmatrix} = \boldsymbol{e}_i, \tag{2.89}$$

where $\boldsymbol{e}_i$ is the ith column of the 3×3 identity matrix. The coefficients a_i, b_i, c_i depend on the vertices and also on the area of the element, denoted $|\triangle_k|$. See Exercise 2.15 for details and the quadratic case ($r = 2$).

Given the local basis functions, we *could* compute A^k and $\boldsymbol{b}^k$ by using (2.86)–(2.87) and integrating over each $\triangle_k$ separately. However, this is inefficient. For shape-regular meshes, all the elements of $\mathcal{T}_h$ can be constructed from transformations of a representative element $\triangle_*$. With a suitable change of variables, all the element arrays in (2.86)–(2.87) can then be computed by performing integration on a single element $\triangle_*$.

Definition 2.58 (reference element) For triangular meshes, the *reference element* $\triangle_*$ is the right-angled triangle in the s–t plane with vertices $(0,0)$, $(1,0)$ and $(0,1)$, labelled one to three, respectively (see Figure 2.15).

Definition 2.59 (reference element basis functions) The reference element basis functions $\psi_1, \dots, \psi_{n_r}$ of degree r are defined by

$$\psi_i \in \mathbb{P}_r(\triangle_*), \qquad \psi_i(s,t) = \begin{cases} 1, & (s,t) = (s_i, t_i), \\ 0, & (s,t) = (s_j, t_j) \text{ and } i \neq j, \end{cases} \tag{2.90}$$

where $(s_i, t_i), i = 1, \dots, n_r$, are the reference element nodes (see Figure 2.14).

Example 2.60 (piecewise linear) Let $r = 1$. The local piecewise linear basis functions associated with the three vertices of $\triangle_*$ are

$$\psi_1(\boldsymbol{x}) := 1 - s - t, \qquad \psi_2(\boldsymbol{x}) := s, \qquad \psi_3(\boldsymbol{x}) := t. \tag{2.91}$$

See Exercise 2.16.

The linear functions in (2.91) can be used to define a simple map from the reference element to an arbitrary triangle $\triangle_k \in \mathcal{T}_h$ as follows:

$$x(s,t) = x_1^k(1 - s - t) + x_2^k s + x_3^k t, \tag{2.92}$$

$$y(s,t) = y_1^k(1 - s - t) + y_2^k s + y_3^k t, \tag{2.93}$$

where (x_i^k, y_i^k), $i = 1,2,3$, are the vertices of $\triangle_k$. This is an affine mapping that transforms edges of $\triangle_*$ into edges of $\triangle_k$. See Exercise 2.17 and Figure 2.15. We use the linear polynomials to define the mapping regardless of the polynomial degree r actually used to define V^h and W^h. Applying the transformation (2.92)–(2.93) in (2.86) gives

$$a_{pq}^k = \int_{\triangle_*} a(x(s,t),y(s,t))\, \nabla\phi_p^k(x(s,t),y(s,t)) \cdot \nabla\phi_q^k(x(s,t),y(s,t)) \det(J_k)\, ds\, dt,$$

for $p,q = 1,\ldots,n_r$, where $\nabla := \left(\frac{\partial}{\partial x}, \frac{\partial}{\partial y}\right)^{\mathsf{T}}$ and J_k is the Jacobian matrix

$$J_k := \begin{pmatrix} \dfrac{\partial x}{\partial s} & \dfrac{\partial y}{\partial s} \\ \dfrac{\partial x}{\partial t} & \dfrac{\partial y}{\partial t} \end{pmatrix}. \tag{2.94}$$

Now, since $\phi_p^k(x(s,t),y(s,t)) = \psi_p(s,t)$, for all elements $\triangle_k$, we have

$$a_{pq}^k = \int_{\triangle_*} a(x(s,t),y(s,t))\, \nabla\psi_p(s,t) \cdot \nabla\psi_q(s,t) \det(J_k)\, ds\, dt. \tag{2.95}$$

Similarly, from (2.87) we obtain

$$b_q^k = \int_{\triangle_*} f(x(s,t),y(s,t))\, \psi_q(s,t) \det(J_k)\, ds\, dt, \qquad q = 1,\ldots,n_r. \tag{2.96}$$

The Jacobian matrices J_k in (2.94) are important. They allow us to map derivatives between coordinate systems and perform the integration in (2.95) and (2.96) efficiently. For each of the n_r reference element basis functions ψ_p,

$$\nabla\psi_p = \begin{pmatrix} \dfrac{\partial\psi_p}{\partial x} \\ \dfrac{\partial\psi_p}{\partial y} \end{pmatrix} = \begin{pmatrix} \dfrac{\partial s}{\partial x} & \dfrac{\partial t}{\partial x} \\ \dfrac{\partial s}{\partial y} & \dfrac{\partial t}{\partial y} \end{pmatrix} \begin{pmatrix} \dfrac{\partial\psi_p}{\partial s} \\ \dfrac{\partial\psi_p}{\partial t} \end{pmatrix} = J_k^{-1} \begin{pmatrix} \dfrac{\partial\psi_p}{\partial s} \\ \dfrac{\partial\psi_p}{\partial t} \end{pmatrix}. \tag{2.97}$$

For triangular elements, we have the mapping (2.92)–(2.93) and so

$$J_k = \begin{pmatrix} x_2 - x_1 & y_2 - y_1 \\ x_3 - x_1 & y_3 - y_1 \end{pmatrix}, \qquad J_k^{-1} = \frac{1}{\det(J_k)} \begin{pmatrix} y_3 - y_1 & y_1 - y_2 \\ x_1 - x_3 & x_2 - x_1 \end{pmatrix}, \tag{2.98}$$

where, recall, $(x_1,y_1),(x_2,y_2)$ and (x_3,y_3) are the vertices of $\triangle_k$. Note that the entries of these matrices do not depend on s and t. In addition,

$$\det(J_k) = (x_2 - x_1)(y_3 - y_1) - (y_2 - y_1)(x_3 - x_1) = 2|\triangle_k|.$$

Algorithm 2.5 computes the matrices J_k and J_k^{-1} for any triangular mesh $\mathcal{T}_h$.

Algorithm 2.5 Code to compute Jacobian matrices for any triangular mesh. The inputs are defined in the same way as the outputs of Algorithm 2.4. The outputs are `Jks`, `invJks`, three-dimensional arrays containing the 2×2 matrices J_k and J_k^{-1} and `detJks`, a vector containing the determinants $\det(J_k)$.

```
function [Jks,invJks,detJks]=get_jac_info(xv,yv,ne,elt2vert);
Jks=zeros(ne,2,2); invJks=zeros(ne,2,2);
% all vertices of all elements
x1=xv(elt2vert(:,1)); x2=xv(elt2vert(:,2)); x3=xv(elt2vert(:,3));
y1=yv(elt2vert(:,1)); y2=yv(elt2vert(:,2)); y3=yv(elt2vert(:,3));
% Jk matrices,determinants and inverses
Jks(:,1,1)=x2-x1;Jks(:,1,2)=y2-y1; Jks(:,2,1)=x3-x1;Jks(:,2,2)=y3-y1;
detJks=(Jks(:,1,1).*Jks(:,2,2))-(Jks(:,1,2).*Jks(:,2,1));
invJks(:,1,1)=(1./detJks).*(y3-y1);invJks(:,1,2)=(1./detJks).*(y1-y2);
invJks(:,2,1)=(1./detJks).*(x1-x3);invJks(:,2,2)=(1./detJks).*(x2-x1);
```

Example 2.61 (uniform meshes) For the meshes generated by Algorithm 2.4, we have $\det(J_k) = 2|\triangle_k| = h^2 = \texttt{ns}^{-2}$ and, since the triangles have edges aligned with the x and y axes, each J_k is diagonal. For triangles of type A,

$$J_k = \begin{pmatrix} h & 0 \\ 0 & h \end{pmatrix}, \qquad J_k^{-1} = \begin{pmatrix} 1/h & 0 \\ 0 & 1/h \end{pmatrix},$$

(see Figure 2.12) and, for triangles of type B,

$$J_k = \begin{pmatrix} -h & 0 \\ 0 & -h \end{pmatrix}, \qquad J_k^{-1} = \begin{pmatrix} -1/h & 0 \\ 0 & -1/h \end{pmatrix}.$$

The Jacobian matrix for the shaded triangle (of type A) in Figure 2.9 can be computed using Algorithms 2.4 and 2.5 with the following commands:

```
>> [xv,yv,elt2vert,nvtx,ne,h]=uniform_mesh_info(4);
>> [Jks, invJks detJks]=get_jac_info(xv,yv,ne,elt2vert);
>> squeeze(Jks(6,:,:))
ans =
    0.2500         0
         0    0.2500
```

Given the Jacobian matrices, we can compute the element arrays. For triangular elements, we have shown that

$$a_{pq}^k = 2|\triangle_k| \int_{\triangle_*} a(x(s,t), y(s,t)) \nabla\psi_p(s,t) \cdot \nabla\psi_q(s,t)\, ds\, dt, \tag{2.99}$$

$$b_q^k = 2|\triangle_k| \int_{\triangle_*} f(x(s,t), y(s,t)) \psi_q(s,t)\, ds\, dt, \tag{2.100}$$

for $p, q = 1, \dots, n_r$, where $\nabla\psi_p$ and $\nabla\psi_q$ are as in (2.97) with J_k^{-1} defined as in (2.98). The integrals in (2.99)–(2.100) are all over the reference element $\triangle_*$ and can be performed efficiently using quadrature. Hence, for a given set of Q weights w_i and nodes (s_i, t_i) on $\triangle_*$, we compute

$$b_q^k = 2|\triangle_k| \sum_{i=1}^{Q} w_i f(x(s_i,t_i), y(s_i,t_i))\, \psi_q(s_i,t_i), \tag{2.101}$$

and similarly for a_{pq}^k. If a and f are simple functions (e.g., piecewise constants or polynomials) then the integrands in (2.99)–(2.100) are polynomials and we can find quadrature rules that are exact for the chosen r. Otherwise, we are solving a problem of the form (2.69) where $\tilde{a}(\cdot,\cdot)$ and $\tilde{\ell}(\cdot)$ arise from inexact integration. Algorithm 2.6 returns the element arrays for $r = 1$ and piecewise constant data. In that case, quadrature is not needed; see Exercise 2.18.

Algorithm 2.6 Code to compute the element arrays for piecewise linear finite elements and piecewise constant data. The first six inputs are as for Algorithms 2.4 and 2.5 and `a` and `f` are vectors of length n_e containing the values of a and f in each element. The outputs `Aks`, `bks` are the element arrays.

```
function [Aks,bks]=get_elt_arrays2D(xv,yv,invJks,detJks,ne,...
                              elt2vert,a,f)
bks=zeros(ne,3); Aks=zeros(ne,3,3);
dpsi_ds=[-1,1,0]; dpsi_dt=[-1,0,1]; % for r=1
for i=1:3
    for j=1:3
        grad=[dpsi_ds(i) dpsi_ds(j); dpsi_dt(i) dpsi_dt(j)];
        v1=squeeze([invJks(:,1,1:2)])*grad;
        v2=squeeze([invJks(:,2,1:2)])*grad;
        int=prod(v1,2)+prod(v2,2);
        Aks(:,i,j)=Aks(:,i,j)+a.*detJks.*int./2;
    end
    bks(:,i)=bks(:,i)+f.*detJks./6;
end
```

Example 2.62 (piecewise linear elements) Let $r = 1$ and suppose $a = 1$ and $f = 1$. We can compute the arrays $A^k \in \mathbb{R}^{3\times 3}$ and $\boldsymbol{b}^k \in \mathbb{R}^3$ for the shaded triangle in Figure 2.9 using Algorithms 2.4–2.6 with the following commands:

```
>> [xv,yv,elt2vert,nvtx,ne,h]=uniform_mesh_info(4);
>> [Jks,invJks,detJks]=get_jac_info(xv,yv,ne,elt2vert);
>> [Aks,bks]=get_elt_arrays2D(xv,yv,invJks,detJks,ne,...
                              elt2vert,ones(ne,1),ones(ne,1));
>> squeeze(Aks(6,:,:))
ans =
    1.0000   -0.5000   -0.5000
   -0.5000    0.5000         0
   -0.5000         0    0.5000
>> bks(6,:)
ans =
    0.0104    0.0104    0.0104
```

Algorithm 2.7 assembles and then solves the Galerkin system (2.85) for piecewise linear elements and piecewise constant data, using MATLAB's backslash command. An iterative solver is also investigated in Exercise 2.19.

Algorithm 2.7 Piecewise linear FEM for the BVP (2.45)–(2.46) on $D = (0,1) \times (0,1)$ with piecewise constant a and f and $g = 0$. The inputs are defined as in Algorithms 2.4 and 2.6. The outputs are the Galerkin matrix `A_int` associated with the interior nodes, the right-hand side vector `rhs` and a vector `u_int` containing the approximation at the interior nodes.

```
function [u_int,A_int,rhs]=twod_linear_FEM(ns,xv,yv,elt2vert,...
                                       nvtx,ne,h,a,f)
[Jks,invJks,detJks]=get_jac_info(xv,yv,ne,elt2vert);
[Aks,bks]=get_elt_arrays2D(xv,yv,invJks,detJks,ne,elt2vert,a,f);
A = sparse(nvtx,nvtx); b = zeros(nvtx,1);
for row_no=1:3
    nrow=elt2vert(:,row_no);
    for col_no=1:3
        ncol=elt2vert(:,col_no);
        A=A+sparse(nrow,ncol,Aks(:,row_no,col_no),nvtx,nvtx);
    end
    b = b+sparse(nrow,1,bks(:,row_no),nvtx,1);
end
% get discrete Dirichlet boundary data
b_nodes=find((xv==0)|(xv==1)|(yv==0)|(yv==1));
int_nodes=1:nvtx; int_nodes(b_nodes)=[]; b_int=b(int_nodes);
wB=feval('g_eval',xv(b_nodes),yv(b_nodes));
% solve linear system for interior nodes;
A_int=A(int_nodes,int_nodes); rhs=(b_int-A(int_nodes,b_nodes)*wB);
u_int=A_int\rhs;
uh=zeros(nvtx,1); uh(int_nodes)=u_int; uh(b_nodes)=wB;
m=ns+1;mesh(reshape(xv,m,m),reshape(yv,m,m),reshape(uh,m,m));
axis square; title('finite element solution');
end
function g=g_eval(x,y)
g=zeros(size(x));
end
```

Example 2.63 Consider (2.45)–(2.46) on $D = (0,1)\times(0,1)$ with constant data $a = 1, f = 1$ and $g = 0$. The piecewise linear approximation associated with the uniform mesh on the right in Figure 2.9 can be generated using Algorithm 2.7 with the commands:

```
>> ns=4;[xv,yv,elt2vert,nvtx,ne,h]=uniform_mesh_info(ns);
>> [u_int,A_int,rhs]=twod_linear_FEM(ns,xv,yv,elt2vert,nvtx,...
                              ne,h,ones(ne,1),ones(ne,1));
```

Approximations on four successively refined meshes are shown in Figure 2.16. Similarly, approximations generated with the piecewise constant coefficient

$$a(\boldsymbol{x}) = \begin{cases} 10^{-2}, & \boldsymbol{x} \in [0.25, 0.75] \times [0.25, 0.75], \\ 1, & \text{otherwise}, \end{cases} \tag{2.102}$$

are shown in Figure 2.17. The correct calling sequence in this case is

```
>> [u_int,A_int,rhs]=twod_linear_FEM(ns,xv,yv,elt2vert,nvtx,...
                              ne,h,a,ones(ne,1));
```

where the kth entry of the vector `a` is one or 10^{-2}. See Exercise 2.20 for details and an example with non-homogeneous boundary conditions.

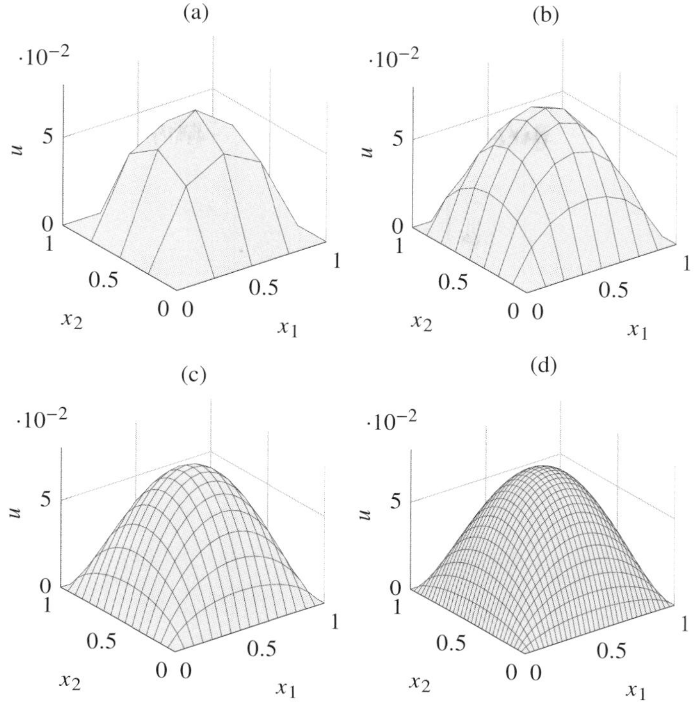

Figure 2.16 Piecewise linear FEM approximation for Example 2.63 with $a = 1$, $f = 1$ and $g = 0$ on uniform meshes generated by Algorithm 2.4 with (a) `ns` $= 4$ ($h = 1/4$), (b) `ns` $= 8$ ($h = 1/8$), (c) `ns` $= 16$ ($h = 1/16$) and (d) `ns` $= 32$ ($h = 1/32$).

Algorithm 2.7 does not need uniform meshes. Any mesh generator can be used, as long as the element and vertex information is specified correctly. The subroutine `g_eval` can be edited to accommodate different boundary conditions (see Exercise 2.20), and other polynomial degrees and data approximations can be handled by providing the appropriate element arrays.

Finite element error analysis

Mapping to and from $\triangle_*$ not only makes computations efficient but also facilitates error analysis. We return now to the finite-dimensional Galerkin problem (2.65) and focus on piecewise linear elements. That is, we choose

$$V^h = \left\{ v \in \mathrm{C}(\bar{D}) \text{ with } v = 0 \text{ on } \partial D \text{ and } v|_{\triangle_k} \in \mathbb{P}_1(\triangle_k) \text{ for all } \triangle_k \in \mathcal{T}_h \right\},$$

and W^h to be the set of functions of the form (2.78) that interpolate the boundary data. We also assume now that $g\colon \partial D \to \mathbb{R}$ is a suitable linear polynomial so that $W^h \subset H^1_g(D)$. In that case, Theorem 2.49 gives

$$|u - u_h|_E \le |u - w|_E, \qquad \forall w \in W^h, \tag{2.103}$$

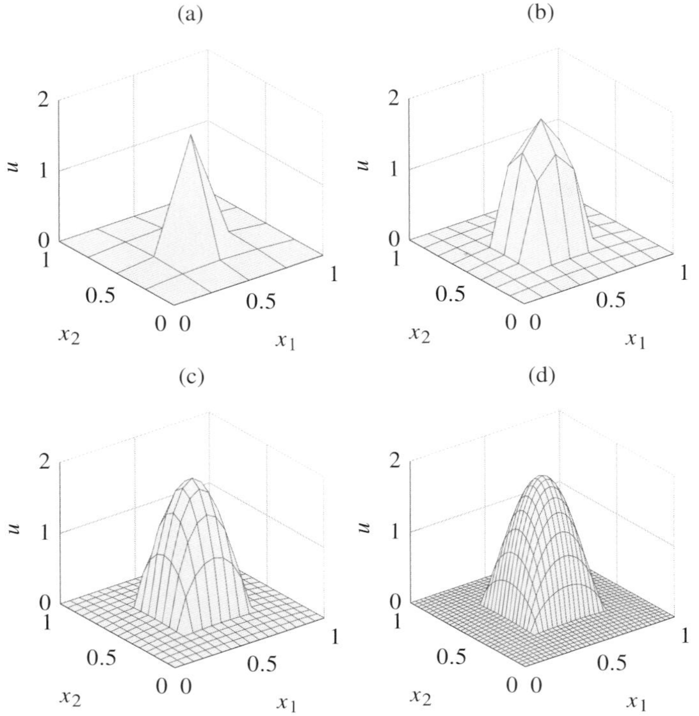

Figure 2.17 Piecewise linear FEM approximation for Example 2.63 with a given by (2.102), $f = 1$ and $g = 0$ on uniform meshes generated by Algorithm 2.4 with (a) `ns` = 4 ($h = 1/4$), (b) `ns` = 8 ($h = 1/8$), (c) `ns` = 16 ($h = 1/16$) and (d) `ns` = 32 ($h = 1/32$).

and using (2.57) gives

$$|u - u_h|_E^2 \le a_{\max} |u - w|_{H^1(D)}^2 = a_{\max} \sum_{k=1}^{n_e} |u - w|_{H^1(\triangle_k)}^2, \qquad \forall w \in W^h. \tag{2.104}$$

This is the starting point for error analysis.

Consider the interpolant $I_h u$ of the weak solution u, satisfying

$$I_h u(\boldsymbol{x}_i) = u(\boldsymbol{x}_i), \qquad \text{for } i = 1, \ldots, J + J_b, \tag{2.105}$$

where $\boldsymbol{x}_i$ denotes a vertex of the mesh, and

$$I_h u\big|_{\triangle_k} \in \mathbb{P}_1(\triangle_k), \qquad \text{for all } \triangle_k \in \mathcal{T}_h. \tag{2.106}$$

In (2.104), we need to choose a specific function w that belongs to our finite element space. The assumption on g ensures $w = I_h u$ satisfies the correct boundary condition. However, u is not guaranteed to be continuous and so neither is $I_h u$. To work with the interpolant, we assume the weak solution has some extra smoothness. Theorem 1.51 tells us that if $u \in H^2(D)$ then $u \in C(\bar{D})$ and, in that case, $I_h u \in W^h$. We make the following assumption (recall also Assumption 2.13).

Assumption 2.64 (H^2-regularity) There exists a constant $K_2 > 0$ such that, for every $f \in L^2(D)$, the solution u to (2.55) belongs to $H^2(D)$ and satisfies

$$|u|_{H^2(D)} \le K_2 \|f\|_{L^2(D)}.$$

Now, let $e := u - w = u - I_h u$ where $I_h u$ is defined by (2.105)–(2.106). Recall (2.92)–(2.93) and denote by $\hat{e}$ the mapped interpolation error from $\triangle_k$ onto the reference element so that $e(x(s,t), y(s,t)) = \hat{e}(s,t)$ for $(x,y) \in \triangle_k$. We need to derive bounds for the element errors $|e|_{H^1(\triangle_k)}$. To consider all elements simultaneously, we first relate $|e|_{H^1(\triangle_k)}$ to $|\hat{e}|_{H^1(\triangle_*)}$ (see Lemma 2.65). Working on the reference element, we relate $|\hat{e}|_{H^1(\triangle_*)}$ to $|\hat{u}|_{H^2(\triangle_*)}$ (where $\hat{u}$ is the mapped u) and then, finally, we map this quantity back to $|u|_{H^2(\triangle_k)}$ (see Lemma 2.66).

Lemma 2.65 *For all triangles $\triangle_k \in \mathcal{T}_h$,*

$$|e|^2_{H^1(\triangle_k)} \le 2\left(\frac{h_k^2}{|\triangle_k|}\right)|\hat{e}|^2_{H^1(\triangle_*)}. \tag{2.107}$$

Proof Recall that $\det(J_k) = 2|\triangle_k|$ for all elements and so

$$|e|^2_{H^1(\triangle_k)} = \int_{\triangle_k} \|\nabla e(x,y)\|_2^2 \,dx\,dy = 2|\triangle_k| \int_{\triangle_*} \|\nabla \hat{e}(s,t)\|_2^2 \,ds\,dt.$$

Using (2.94) to map the derivatives gives the result. See Exercise 2.21. □

Each time we map to and from the reference element using (2.92)–(2.93), it is important to know whether powers of h_k enter our bounds. If the mesh is shape-regular (recall Definition 2.54 and Figure 2.10), we have

$$\frac{h_k^2}{|\triangle_k|} \le \frac{\rho_k^2}{\kappa^2|\triangle_k|} \le \frac{\rho_k^2}{\kappa^2(\pi\rho_k^2)} \le \frac{1}{\pi\kappa^2}, \qquad \forall \triangle_k \in \mathcal{T}_h.$$

Hence, the constant in (2.107) is bounded independently of h_k and does not blow up as $h = \max_k h_k \to 0$ and the mesh is refined. In the next result, we pick up a power of h_k when we move from $\triangle_*$ back to the elements of $\mathcal{T}_h$.

Lemma 2.66 *For all triangles $\triangle_k \in \mathcal{T}_h$,*

$$|\hat{e}|^2_{H^1(\triangle_*)} \le K h_k^2 \left(\frac{h_k^2}{|\triangle_k|}\right)|u|^2_{H^2(\triangle_k)}, \tag{2.108}$$

where K is a constant independent of h_k.

Proof First, working on the reference element, it can be shown that there exists a constant c_1, independent of h_k, such that

$$|\hat{u} - I_h\hat{u}|^2_{H^1(\triangle_*)} \le c_1 |\hat{u} - I_h\hat{u}|^2_{H^2(\triangle_*)}. \tag{2.109}$$

We omit the details. As $I_h\hat{u}$ is linear, its second derivatives are zero and so

$$|\hat{e}|^2_{H^1(\triangle_*)} = |\hat{u} - I_h\hat{u}|^2_{H^1(\triangle_*)} \le c_1 |\hat{u}|^2_{H^2(\triangle_*)}. \tag{2.110}$$

Next, since $\det(J_k)^{-1} = 1/2|\triangle_k|^{-1}$, we have

$$|\hat{u}|^2_{H^2(\triangle_*)} = \frac{1}{2|\triangle_k|}\int_{\triangle_k} \left(\frac{\partial^2 u(x,y)}{\partial s^2}\right)^2 + \left(\frac{\partial^2 u(x,y)}{\partial s \partial t}\right)^2 + \left(\frac{\partial^2 u(x,y)}{\partial t^2}\right)^2 dx\,dy.$$

A messy (but straightforward) calculation to map each of the derivatives back to second derivatives with respect to x and y then gives

$$|\hat{u}|^2_{H^2(\triangle_*)} \le c_2 h_k^2 \left(\frac{h_k^2}{|\triangle_k|} \right) |u|^2_{H^2(\triangle_k)}, \tag{2.111}$$

for another constant c_2 independent of h_k. See Exercise 2.22 for details. The result now follows with $K = c_1 c_2$. □

Combining (2.107) and (2.108) with (2.104), the next result states that the energy norm of the finite element error is $\mathcal{O}(h)$.

Theorem 2.67 *Let u be the solution to* (2.55) *and let u_h be the piecewise linear finite element approximation satisfying* (2.65). *If Assumption 2.64 holds and the finite element mesh $\mathcal{T}_h$ is shape regular, then*

$$|u - u_h|_E \le K \sqrt{a_{\max}}\, h \, \|f\|_{L^2(D)}, \tag{2.112}$$

where $K > 0$ is a constant independent of h.

Proof See Exercise 2.23. □

Corollary 2.68 *If the assumptions of Theorem 2.67 hold, then*

$$|u - u_h|_{H^1(D)} \le K\, h \sqrt{\frac{a_{\max}}{a_{\min}}} \|f\|_{L^2(D)}. \tag{2.113}$$

Finally, note that when we implemented the finite element method, we assumed that a and f were piecewise constant functions. If the data are not piecewise constant then returning to (2.72) and using (2.68), we have

$$\begin{aligned} |u - \tilde{u}_h|_{H^1(D)} \le K h \sqrt{\frac{a_{\max}}{a_{\min}}} \|f\|_{L^2(D)} + K_p \tilde{a}_{\min}^{-1} \|f - \tilde{f}\|_{L^2(D)} \\ + \tilde{a}_{\min}^{-1} \|a - \tilde{a}\|_{L^\infty(D)} |u_h|_{H^1(D)}. \end{aligned}$$

For approximations that are constant on the elements, we have

$$\tilde{a}_{\min} := \min_{\triangle_k \in \mathcal{T}_h} a_k, \qquad \tilde{a}_{\max} := \max_{\triangle_k \in \mathcal{T}_h} a_k,$$

where a_k is the value chosen to represent a in $\triangle_k$. To balance the terms in the error bound when piecewise linear finite elements are used, the approximations $\tilde{a}$ and $\tilde{f}$ should ideally be chosen so that the errors $\|f - \tilde{f}\|_{L^2(D)}$ and $\|a - \tilde{a}\|_{L^\infty(D)}$ are $\mathcal{O}(h)$. We return to this in Chapter 9 when we consider some specific approximations for random diffusion coefficients.

2.4 Notes

In this chapter, we described Galerkin approximation, for which the finite-dimensional trial and test spaces are the same (up to boundary conditions). Choosing different polynomial spaces for V^h and W^h leads to the so-called Petrov–Galerkin method. The subject of finite element approximation is vast and we have barely scratched the surface. Indeed, we restricted our discussion to one- and two-dimensional domains, piecewise linear elements, uniform meshes, and conforming approximations.

There are many excellent textbooks that deal with the theory of Galerkin approximation and finite element methods (e.g., Braess, 1997; Brenner and Scott, 2008; Elman, Silvester, and Wathen, 2005b; Hackbusch, 1992; Strang and Fix, 1973). The two-point boundary-value problem (2.3)–(2.4) is discussed in (Strang and Fix, 1973; Süli and Mayers, 2003). Hackbusch (1992) deals specifically with the variational treatment of elliptic PDEs and contains rigorous proofs and many of the technical details that we have omitted. Brenner and Scott (2008) is accessible for graduate students and discusses some of the variational crimes typically committed when finite element methods are implemented (e.g., as a result of non-trivial boundary conditions or domains with curved boundaries). Elman et al. (2005b) focuses on practical implementation and covers a range of PDE problems. Iterative solvers and preconditioning schemes are also discussed.

Each boundary-value problem encountered provides different challenges for variational analysis due to the geometry of the domain, the smoothness of the data, the nature of the boundary conditions, the order of the derivatives in the PDE(s), and the function spaces involved in the weak form. Indeed, few real-life PDE problems are as nice as the elliptic model problem. At the level of implementation, it is easy to forget about many of the theoretical issues. Finite element methods can be applied for a wide variety of problems that are intractable for analysis, and numerical results obtained with FEM software packages must always be interpreted with care. The variational analysis of *systems* of PDEs, where there is more than one solution variable, is more challenging. However, the variational analysis of saddle-point problems (that arise, e.g., in the variational formulation of the Stokes and Navier–Stokes equations) and the associated theory of mixed finite elements is well developed; see Brezzi and Fortin (1991).

The spectral Galerkin approximation is particularly useful for simple domains and smooth data or in any situation where the eigenfunctions ϕ_j are explicitly known. In that case, computations can be performed efficiently by the fast Fourier transform. For further information on spectral methods, see (Canuto et al., 1988; Gottlieb and Orszag, 1977). The spectral method that we discussed is closely related to pseudo-spectral methods (Fornberg, 1996).

Some remarks are warranted about the technical details that we skipped in §2.2. Theorem 2.46 is similar to Strang's first lemma (see Ciarlet, 1978, Theorem 4.1.1), which provides an abstract error estimate for $\|u - u_h\|_V$ where $u \in V$ satisfies a variational problem

$$a(u,v) = \ell(v), \qquad \forall v \in V,$$

and $u_h \in V^h \subset V$ satisfies a variational problem

$$\tilde{a}(u_h,v) = \tilde{\ell}(v), \qquad \forall v \in V^h.$$

Although the $\tilde{a}(\cdot,\cdot)$ and $\tilde{\ell}(\cdot)$ we considered arose from approximating the data, Strang's first and second lemmas (see Ciarlet, 1978, Theorem 4.2.2) can be used to analyse errors arising from any approximation to the bilinear form $a(\cdot,\cdot)$ and linear functional $\ell(\cdot)$.

The space $H^{1/2}(\partial D)$ can also be defined as a fractional power Sobolev space. For details of this interpretation, see Renardy and Rogers (2004). More details about *trace operators* and a rigorous analysis of the operator $\gamma\colon H^1(D) \to L^2(\partial D)$ can be found, for example, in Renardy and Rogers (2004) and Hackbusch (1992, Section 6.2.5). Lemma 2.37 is taken from Temam (1988) and Lemma 2.39 is a consequence of the open mapping theorem. We assumed at the outset that domains $D \subset \mathbb{R}^2$ satisfy Definition 1.3. This is adequate for most of the results in this book. Domains of this sort are indeed sufficiently smooth in the context of Lemma 2.37 and lead to a well-defined and bounded trace operator. A more precise statement of Lemma 2.37 involves the assumption that D is a domain whose boundary is of class C^1.

For the error analysis in §2.3, we focused on the energy norm (or, equivalently, the $H^1(D)$ semi-norm). Error estimates in the $L^2(D)$ norm are discussed in Hackbusch (1992, Chapter 8). The proof of the result

$$|\hat{u} - I_h\hat{u}|^2_{H^1(\triangle_*)} \le c_1 |\hat{u}|^2_{H^2(\triangle_*)},$$

in (2.109)–(2.110), can be found in Braess (1997, Chapter 6) and is a special case of a more general result known as the Bramble–Hilbert lemma. In the proof of Theorem 2.67, we assumed the weak solution is H^2-regular (i.e., we used Assumption 2.13). To get more smoothness out of the solution naturally requires more smoothness from the input data. For elliptic problems, if $a \in \mathrm{C}^{0,1}(\bar{D})$ (i.e., is Lipschitz continuous) and D is a convex polygon then $u \in H^2(D)$ (see Hackbusch, 1992, Chapter 9). This covers domains such as squares and rectangles. Non-convex domains, like L-shaped domains, lead to weak solutions with singularities that do not belong to $H^2(D)$. See Elman et al. (2005b, Chapter 1) for a discussion of a test problem on an L-shaped domain for which the piecewise linear finite element error is shown to be $\mathcal{O}(h^{2/3})$. In general, if degree r polynomials are used, it is possible to obtain $\mathcal{O}(h^r)$ estimates for the energy norm error, provided the weak solution belongs to $H^{r+1}(D)$. Of course, this requires additional smoothness from the input data. We also assumed $W^h \subset W := H^1_g$. Whenever our finite element space does not satisfy this requirement, we require non-conforming error analysis techniques (see Brenner and Scott, 2008).

We attempted, wherever possible, to quantify the dependence of the constants K in our bounds on $a_{\min}, a_{\max}$ (or $\|p\|_{L^\infty(D)}$ and p_0 for the two-point BVP in §2.1). Note, however, that our use of these constants in moving between semi-norms via the equivalence (2.57) is quite crude. In some applications, a may vary by several orders of magnitude across the domain and bounds that depend on the ratio $a_{\max}/a_{\min}$ can be pessimistic with respect to the coefficients.

The MATLAB code developed in this chapter has been designed in a modular way so that, by providing alternative subroutines, readers can experiment with other polynomial degrees, meshes and data approximations. The code has also been designed to interface easily with the random field generators discussed in Chapter 7. It is used in Chapter 9 to perform Monte Carlo finite element computations for elliptic BVPs with *random* data. We mention two other freely available MATLAB finite element codes for the model elliptic PDE in two dimensions. The first, `ifiss` (Elman, Ramage, and Silvester, 2007), accompanies Elman et al. (2005b) and applies standard finite elements. We have followed a similar programming style. The second, `pifiss` (Powell and Silvester, 2007), applies a more sophisticated *mixed* finite element method and approximates, in addition to the scalar variable u, the flux $\boldsymbol{q} := -\nabla u$.

Exercises

2.1 Let $V = H_0^1(a,b)$. Show that if Assumption 2.8 holds, then $a(\cdot,\cdot)\colon V \times V \to \mathbb{R}$ in (2.9) is an inner product on V.

2.2 Let p, q satisfy the conditions of Assumption 2.8. Prove the norm equivalence (2.14). In addition, show that the energy norm $\|\cdot\|_E$ is equivalent to the norm $\|\cdot\|_{H^1(a,b)}$.

2.3 Let Assumption 2.8 hold. Assume, for contradiction, that there are two weak solutions u_1, u_2 satisfying (2.11) and show that $u_1 = u_2$. Hence, deduce that (2.11) has at most one solution.

2.4 Let Assumption 2.1 hold. Consider the operator A given by (2.24) and let $G(x,y)$ denote the Green's function satisfying (2.25).

a. Show that

$$G'(x,y_+) - G'(x,y_-) = \frac{-1}{p(y)}, \tag{2.114}$$

where $G'(x,y_+)$ denotes the limit of $G'(x,y+\epsilon)$ as ϵ approaches 0 from above and $G'(x,y_-)$ is the limit as ϵ approaches 0 from below.

b. Suppose that $u_1, u_2 \in C^2([a,b])$ are non-zero functions such that $Au_i = 0$ for $i = 1,2$ with $u_1(a) = u_2(b) = 0$. Show that

$$G(x,y) = \begin{cases} Cu_1(x)u_2(y), & x \le y, \\ Cu_2(x)u_1(y), & x > y, \end{cases}$$

for a constant C that you should determine.

c. Show that $G(x,y)$ is symmetric and belongs to $L^2((a,b)\times(a,b))$.

2.5 Fix the coefficients $p = 1$ and $q = 1$. Using Algorithms 2.1 and 2.2, investigate the condition number of the Galerkin matrix A. Determine the dependence of the condition numbers of the matrices A, K and M on the finite element mesh width h.

2.6 Using (2.36), show that the local quadratic basis functions associated with the nodes $x_1^k = 0$, $x_2^k = h/2$, $x_3^k = h$ on the element $e_k = [0,h]$ are

$$\phi_1^k(x) = \frac{2x^2}{h^2} - \frac{3x}{h} + 1, \qquad \phi_2^k(x) = \frac{4x}{h} - \frac{4x^2}{h^2}, \qquad \phi_3^k(x) = \frac{2x^2}{h^2} - \frac{x}{h}.$$

Assume that $q = 0$ and p and f are constants. Compute the element vector $\boldsymbol{b}^k \in \mathbb{R}^3$ and show that the 3×3 element matrix A^k for this specific element is

$$A^k = \frac{1}{3h}\begin{pmatrix} 7 & -8 & 1 \\ -8 & 16 & -8 \\ 1 & -8 & 7 \end{pmatrix}.$$

2.7 Consider the BVP (2.3) on $D = (0,1)$ with non-homogeneous boundary conditions $u(0) = \alpha$ and $u(1) = \beta$. By writing

$$u_h(x) = \alpha\phi_0(x) + \sum_{i=1}^{J} u_i\phi_i(x) + \beta\phi_{J+1}(x)$$

where $\phi_0(x)$ and $\phi_{J+1}(x)$ are the hat functions associated with $x = 0$ and $x = 1$ respectively, derive the linear system $A\boldsymbol{u} = \hat{\boldsymbol{b}}$ to be solved for the interior values $u_i = u_h(z_i)$, $i = 1, \dots, J$. Relate this linear system to the one obtained for homogeneous conditions. Edit Algorithm 2.1 and compute the finite element approximation to (2.3) with the boundary conditions $u(0) = 1$ and $u(1) = 0$, for $p = 1$, $q = 10$ and $f = 1$.

2.8 For the BVP (2.3)–(2.4) on $D = (0,1)$ with $p = 1, q = 0$, and constant f, show that the Galerkin finite element matrix associated with polynomials of degree $r = 1$ is the tridiagonal matrix

$$A = \begin{pmatrix} 2/h & -1/h & & & \\ -1/h & 2/h & -1/h & & \\ & \ddots & \ddots & \ddots & \\ & & \ddots & \ddots & -1/h \\ & & & -1/h & 2/h \end{pmatrix}.$$

Hence, determine that the finite element approximation is equivalent to a centred finite difference approximation.

2.9 Algorithm 2.1 uses the Matlab backslash command to solve the piecewise linear Galerkin finite element system $A\boldsymbol{u} = \boldsymbol{b}$. By successively refining the mesh, investigate the efficiency of this solver for the linear systems associated with the BVP in Example 2.31, with respect to the dimension J.

2.10 By writing $e(x)\big|_{e_k} = \sum_{n=1}^{\infty} a_n \sin\left(\frac{n\pi(x-z_{k-1})}{h}\right)$ in Theorem 2.32 as a Fourier sine series, show that

$$\int_{z_{k-1}}^{z_k} e(x)^2\,dx = \frac{h}{2}\sum_{n=1}^{\infty} |a_n|^2$$

$$\int_{z_{k-1}}^{z_k} e'(x)^2\,dx = \frac{h}{2}\sum_{n=1}^{\infty} \left(\frac{n\pi}{h}\right)^2 |a_n|^2$$

$$\int_{z_{k-1}}^{z_k} e''(x)^2\,dx = \frac{h}{2}\sum_{n=1}^{\infty} \left(\frac{n\pi}{h}\right)^4 |a_n|^2$$

and hence prove (2.43).

2.11 Let $D = (0,1)$ and define $e = u - I_h u$ as in Theorem 2.32.

a. Use Exercise 1.14(a) and apply a change of variable, to show that

$$\int_0^h e(y)^2\,dy \le \int_0^h h^2 e'(y)^2\,dy.$$

b. Similarly, show that

$$\int_0^h e'(y)^2\, dy \le h^2 \int_0^h e''(y)^2\, dy.$$

c. Hence, prove Theorem 2.32 without using Exercise 2.10. Compare the constant K in your error bound with the one in (2.44).

2.12 Modify Algorithm 2.1 and compute the piecewise linear finite element approximation to the solution of the BVP (2.3)–(2.4) with data chosen as in Example 2.5. Compare the approximation with the exact solution at the vertices of the mesh. What do you observe?

2.13 a. Let $z_k \in [0,1]$ and find the Green's function $G_k = G(x, z_k)$ satisfying

$$-\frac{d^2 G_k}{dx^2} = \delta(x - z_k), \qquad 0 \le x \le 1,$$

and boundary conditions

$$G(0, z_k) = 0, \qquad G(1, z_k) = 0.$$

b. Let z_k be any vertex of the mesh associated with a piecewise linear finite element discretization of the BVP in Example 2.5. Use your answer to part (a) to explain your observations in Exercise 2.12.

2.14 Let Assumption 2.34 hold. Use the Lax–Milgram Lemma to prove the existence and uniqueness of $u_0 \in V$ satisfying (2.58) and hence prove Theorem 2.42.

2.15 Solve (2.89) and hence find the local piecewise linear finite element basis functions ϕ_1^k, ϕ_2^k and ϕ_3^k for an element $\triangle_k$ in two dimensions. Similarly, find the six piecewise quadratic basis functions $\phi_i^k, i = 1, \dots, 6$, associated with $r = 2$.

2.16 Show that the local piecewise linear basis functions ψ_1, ψ_2 and ψ_3 associated with the reference element $\triangle_*$ in Figure 2.15 are given by (2.91). Similarly, find the six quadratic reference element basis functions associated with $r = 2$.

2.17 Show that the mapping (2.92)–(2.93) is affine and show that, for the triangle $\triangle_k$ with vertices (h,h), $(2h,h)$ and $(h,2h)$, the mapping corresponds to a certain shift and scaling of the reference element.

2.18 Let $r = 1$ and assume the diffusion coefficient a and function f are piecewise constant on the elements of the mesh.

a. Derive the element matrix $A^k \in \mathbb{R}^{3\times 3}$ and the element vector $\boldsymbol{b}^k \in \mathbb{R}^3$ for a general triangle $\triangle_k$ with vertices $(x_1^k, y_1^k), (x_2^k, y_2^k), (x_3^k, y_3^k)$.

b. Now fix $D = (0,1) \times (0,1)$ and assume the mesh $\mathcal{T}_h$ is generated using Algorithm 2.4. Show that

$$A^k = a_k A^*, \qquad \boldsymbol{b}^k = h^2 f_k \boldsymbol{b}^*, \qquad \forall \triangle_k \in \mathcal{T}_h,$$

where a_k and f_k are the values of a and f in $\triangle_k$ and

$$A^* = \begin{pmatrix} 1 & -\frac{1}{2} & -\frac{1}{2} \\ -\frac{1}{2} & \frac{1}{2} & 0 \\ -\frac{1}{2} & 0 & \frac{1}{2} \end{pmatrix}, \qquad \boldsymbol{b}^* = \begin{pmatrix} \frac{1}{6} \\ \frac{1}{6} \\ \frac{1}{6} \end{pmatrix}.$$

2.19 Use Algorithms 2.4 and 2.7 to generate the Galerkin finite element systems associated with the BVP in Example 2.63 (for both choices of a) on a sequence of successively refined meshes.

a. Investigate the efficiency of the MATLAB backslash solver with respect to the dimension J.
b. Solve the systems using MATLAB's conjugate gradient solver `pcg`. Record the number of iterations required to solve the systems to a fixed tolerance as J is increased. What do you observe?

2.20 Let $D = (0,1) \times (0,1)$ and write $\partial D = \partial D_{\mathrm{D}} \cup \partial D_{\mathrm{N}}$ where $\partial D_{\mathrm{D}} = \{0,1\} \times [0,1]$ and $\partial D_{\mathrm{N}} = \partial D \setminus \partial D_{\mathrm{D}}$. Consider the BVP

$$\begin{aligned} -\nabla \cdot \big(a(\boldsymbol{x})\nabla u(\boldsymbol{x})\big) &= f(\boldsymbol{x}), && \boldsymbol{x} \in D, \\ u(\boldsymbol{x}) &= g(\boldsymbol{x}), && \boldsymbol{x} \in \partial D_{\mathrm{D}}, \\ a(\boldsymbol{x})\nabla u(\boldsymbol{x}) \cdot \boldsymbol{n}(\boldsymbol{x}) &= 0, && \boldsymbol{x} \in \partial D_{\mathrm{N}}. \end{aligned}$$

Here $\boldsymbol{n}(\boldsymbol{x})$ is the unit normal vector at $\boldsymbol{x} \in \partial D$. Modify Algorithm 2.7 for these boundary conditions. Let $f = 0$ and, for $\boldsymbol{x} = [x, y]^{\mathsf{T}}$, let

$$g(\boldsymbol{x}) = \begin{cases} 1, & x = 0, \\ 0, & x = 1. \end{cases}$$

Compute the piecewise linear finite element approximation on uniform meshes generated by Algorithm 2.4 with diffusion coefficient (a) $a = 1$ and (b) a given by (2.102).

2.21 Prove Lemma 2.65.

2.22 Prove (2.111) in Lemma 2.66.

2.23 Use Lemmas 2.65 and 2.66 to prove Theorem 2.67.

3

Time-dependent Differential Equations

This chapter addresses the numerical approximation of initial-value problems (IVPs) for a class of ordinary differential equations (ODEs) and a class of semilinear partial differential equations (PDEs). We study systems of ODEs of the form

$$\frac{d\boldsymbol{u}}{dt} = \boldsymbol{f}(\boldsymbol{u}), \qquad t > 0, \tag{3.1}$$

where $\boldsymbol{f} : \mathbb{R}^d \to \mathbb{R}^d$ is, in general, a nonlinear function and we specify initial data $\boldsymbol{u}_0 \in \mathbb{R}^d$, also called an initial condition. In §2.1, we studied two-point boundary-value problems for ODEs, with $u(x)$ a function of a spatial variable x. Now we focus on $\boldsymbol{u}(t)$ as a function of time t and always specify $\boldsymbol{u}(0) = \boldsymbol{u}_0$. Such IVPs are used to model the dynamics of a wide variety of phenomena in physics, biology, chemistry and economics. For example, we can think of (3.1) as modelling reactions between d chemicals. In that setting, each component of $\boldsymbol{u}$ is the time-varying concentration of one chemical and the nonlinear function $\boldsymbol{f}$ provides the reaction rates. Later, in Chapter 8, we include effects of random forcing on ODEs and study IVPs for stochastic ODEs (SODEs).

For spatio-temporal modelling, PDEs are often used and we focus on reaction–diffusion equations for a scalar quantity of interest $u = u(t, \boldsymbol{x})$, for $t > 0$ and $\boldsymbol{x}$ in a domain D. That is, we consider

$$u_t = \Delta u + f(u), \tag{3.2}$$

where $\Delta = \nabla^2$ is the Laplacian, $f : \mathbb{R} \to \mathbb{R}$ is a possibly nonlinear function, and u_t denotes the *partial* derivative with respect to t. We specify an initial condition $u(0, \boldsymbol{x}) = u_0(\boldsymbol{x})$ for a given function $u_0 : D \to \mathbb{R}$. Note that (3.2) is a parabolic PDE.

As we see in §5.1, the Laplacian Δ models diffusion. In the context of chemical reactions, we then have diffusion of one chemical and the nonlinearity f again models reaction rates. A concrete example is the Allen–Cahn equation

$$u_t = \varepsilon\, \Delta u + u - u^3, \qquad u(0, \boldsymbol{x}) = u_0(\boldsymbol{x}), \tag{3.3}$$

which models phase transitions in a binary alloy (where $\varepsilon > 0$ is a modelling parameter that controls the amount of diffusion). We usually work on a bounded domain D and prescribe conditions on the boundary ∂D, such as homogeneous Dirichlet boundary conditions $u(t, \boldsymbol{x}) = 0$ for $\boldsymbol{x} \in \partial D$ and $t > 0$. Chapter 10 shows how to add random fluctuations in space and time to the PDE (3.2).

In §3.1, we prove existence and uniqueness of the solution to the IVP for ODEs and develop numerical time-stepping methods, such as the well-known Euler methods. We provide MATLAB implementations and review convergence and stability theory.

The study of PDEs begins in §3.2 with the linear PDE $u_t = \Delta u$, the so-called *heat equation*. Our approach is to treat the heat equation and other semilinear PDEs as an ODE taking values in a Hilbert space H of functions of the spatial variable $\boldsymbol{x}$. That is, we write the IVP for the heat equation as

$$\frac{du}{dt} = -Au, \qquad u(0) = u_0 \in H = L^2(D) \tag{3.4}$$

where $Au = -\Delta u$ for u in the domain $\mathcal{D}(A)$ of A (see Definition 1.82). We introduce the semigroup of linear operators e^{-tA} and write the solution of (3.4) as $u(t) = \mathrm{e}^{-tA} u_0$. Written in this form, the solution $\mathrm{e}^{-tA} u_0$ looks very much like the solution of a linear ODE on $\mathbb{R}$. We extend this approach in §3.3 and write semilinear PDEs like (3.2) as ODEs on a function space, viz.

$$\frac{du}{dt} = -Au + f(u), \qquad u(0) = u_0, \tag{3.5}$$

for a class of unbounded operators A on $\mathcal{D}(A) \subset H$ that satisfy Assumption 1.85. We define the *mild solution* for (3.5) and prove its existence and uniqueness.

In §3.4, we introduce the method of lines and finite differences for the approximation of reaction–diffusion equations. The method of lines is a general technique for approximating PDEs by ODEs and we use it again in §3.5 and §3.6 with Galerkin approximation in space. Both spectral and finite element Galerkin methods (see Chapter 2) are studied. These methods are combined with the semi-implicit Euler approximation in time, which is a practical method because of its good stability properties. We analyse the convergence of the spectral Galerkin method for a linear problem. This result requires an assumption on the smoothness of the initial data. The analysis is complemented by a numerical study of convergence. To conclude, we prove a non-smooth error estimate for a linear problem and obtain convergence with a much weaker regularity assumption on the initial data. An extended example is given for the finite element approximation in space. The non-smooth analysis is also applied in Chapter 10, where we consider PDEs with space–time random fluctuations.

3.1 Initial-value problems for ODEs

We start by reviewing the theory of existence, uniqueness, and numerical approximation of solutions to the following initial-value problem (IVP): find $\boldsymbol{u} : [0,T] \to \mathbb{R}^d$ such that

$$\frac{d\boldsymbol{u}}{dt} = \boldsymbol{f}(\boldsymbol{u}), \qquad \boldsymbol{u}(0) = \boldsymbol{u}_0 \in \mathbb{R}^d, \tag{3.6}$$

where $\boldsymbol{f} : \mathbb{R}^d \to \mathbb{R}^d$. This is a system of d first-order ODEs; many ODEs of higher (differential) order can be converted to (3.6) as in Exercise 3.1. First, we reformulate (3.6) as an integral equation.

Lemma 3.1 *Assume* $\boldsymbol{f} \in \mathrm{C}(\mathbb{R}^d, \mathbb{R}^d)$. *Then,* $\boldsymbol{u}$ *is a solution of* (3.6) *on* $[0,T]$ *if and only if*

$$\boldsymbol{u}(t) = \boldsymbol{u}_0 + \int_0^t \boldsymbol{f}(\boldsymbol{u}(s))\, ds, \qquad 0 \le t \le T. \tag{3.7}$$

Proof By integrating (3.6), we obtain

$$u(t) - u(0) = \int_0^t \frac{du(s)}{ds}\,ds = \int_0^t f(u(s))\,ds.$$

Hence, $u(t)$ satisfies (3.7). Conversely, we derive (3.6) from (3.7) by differentiation. □

By writing the solution of (3.6) as the fixed point of a mapping on a Banach space, we show that the contraction mapping theorem (Theorem 1.10) provides existence and uniqueness of the solution. We work with the Banach space $X := C([0,T],\mathbb{R}^d)$ with norm

$$\|u\|_X := \sup_{0\le t\le T} \|u(t)\|_2 \tag{3.8}$$

and the mapping $\mathcal{J}$ defined, for $u \in X$, by

$$(\mathcal{J}u)(t) := u_0 + \int_0^t f(u(s))\,ds. \tag{3.9}$$

If $\mathcal{J}u = u$ then, by Lemma 3.1, u is a solution of (3.6) on the interval $[0,T]$. We assume f is globally Lipschitz continuous (see Definition 1.14).

Theorem 3.2 *Suppose that* $f\colon \mathbb{R}^d \to \mathbb{R}^d$ *satisfies, for some* $L > 0$,

$$\begin{aligned} \|f(u)\|_2 &\le L(1 + \|u\|_2), \\ \|f(u_1) - f(u_2)\|_2 &\le L\|u_1 - u_2\|_2, \qquad \forall u, u_1, u_2 \in \mathbb{R}^d. \end{aligned} \tag{3.10}$$

Then, there exists a unique solution u *of* (3.6) *for* $t \ge 0$.

Proof First notice that $\mathcal{J}$ defined by (3.9) maps into the space $X := C([0,T],\mathbb{R}^d)$ because $(\mathcal{J}u)(t)$ is a continuous function on a compact interval $[0,T]$ and hence is bounded. Next, we show that $\mathcal{J}$ is a contraction if $LT < 1$. By the Lipschitz condition (3.10),

$$\begin{aligned} \|(\mathcal{J}u)(t) - (\mathcal{J}v)(t)\|_2 &\le \int_0^t L\|u(s) - v(s)\|_2\,ds \\ &\le Lt \sup_{0\le s\le t} \|u(s) - v(s)\|_2 \le LT\,\|u - v\|_X. \end{aligned}$$

The contraction mapping theorem therefore applies for T sufficiently small and provides a unique fixed point $u \in X$ of $\mathcal{J}$. By Lemma 3.1, u is a solution of the IVP (3.6) on $[0,T]$. Repeating the argument on the time intervals $[T,2T]$, $[2T,3T]$, …, we find a unique solution for all $t > 0$. □

Numerical time-stepping methods

For most choices of nonlinearity f, explicit solutions $u(t)$ are unavailable and numerical techniques are necessary. We examine the numerical approximation of $u(t_n)$ by u_n, where $t_n = n\Delta t$ for $n = 0,1,2,\ldots$ for a time step $\Delta t > 0$.

The simplest numerical approximation to the IVP (3.6) is the explicit Euler method. The integral equation (3.7) gives

$$u(t_{n+1}) - u(t_n) = \int_0^{t_{n+1}} f(u(s))\,ds - \int_0^{t_n} f(u(s))\,ds = \int_{t_n}^{t_{n+1}} f(u(s))\,ds. \tag{3.11}$$

If we approximate $\boldsymbol{f}(\boldsymbol{u}(s))$ by the constant $\boldsymbol{f}(\boldsymbol{u}(t_n))$ for $s \in [t_n, t_{n+1})$, then

$$\boldsymbol{u}(t_{n+1}) \approx \boldsymbol{u}(t_n) + \Delta t\, \boldsymbol{f}(\boldsymbol{u}(t_n)).$$

This yields the *explicit Euler method* and it is implemented in Algorithm 3.1.

Algorithm 3.1 Code to apply the explicit Euler method. The inputs are the initial vector `u0`, final time `T`, number of steps `N` (so that $\Delta t = T/N$), state space dimension `d`, and a function handle `fhandle` to evaluate $\boldsymbol{f}(\boldsymbol{u})$. The outputs are a vector `t` of times $t_0, \ldots, t_N$ and a matrix `u` with columns $\boldsymbol{u}_0, \ldots, \boldsymbol{u}_N$.

```
function [t,u]=exp_euler(u0,T,N,d,fhandle)
Dt=T/N;              % set time step
u=zeros(d,N+1);      % preallocate solution u
t=[0:Dt:T]';         % make time vector
u(:,1)=u0; u_n=u0; % set inital data
for n=1:N, % time loop
  u_new=u_n+Dt*fhandle(u_n); % explicit Euler step
  u(:,n+1)=u_new; u_n=u_new;
end
```

Definition 3.3 (explicit Euler method) For a time step $\Delta t > 0$ and initial condition $\boldsymbol{u}_0 \in \mathbb{R}^d$, the explicit Euler approximation $\boldsymbol{u}_n$ to the solution $\boldsymbol{u}(t_n)$ of (3.6) for $t_n = n\Delta t$ is defined by

$$\boldsymbol{u}_{n+1} = \boldsymbol{u}_n + \Delta t\, \boldsymbol{f}(\boldsymbol{u}_n). \tag{3.12}$$

We have chosen a uniform time step (i.e., Δt is independent of n). The word 'explicit' indicates that $\boldsymbol{u}_{n+1}$ is expressed in terms of known quantities. Using the initial condition $\boldsymbol{u}(0) = \boldsymbol{u}_0$, we have an approximation $\boldsymbol{u}_1$ to $\boldsymbol{u}(\Delta t) = \boldsymbol{u}(t_1)$ that we can compute directly. From $\boldsymbol{u}_1$, we then compute an approximation $\boldsymbol{u}_2$ to $\boldsymbol{u}(2\Delta t)$ and so on. Contrast this method with the implicit Euler method, where $\boldsymbol{u}_{n+1}$ is found by solving a nonlinear system of equations.

Definition 3.4 (implicit Euler method) For a time step $\Delta t > 0$ and initial condition $\boldsymbol{u}_0 \in \mathbb{R}^d$, the implicit Euler approximation $\boldsymbol{u}_n$ to the solution $\boldsymbol{u}(t_n)$ of (3.6) for $t_n = n\Delta t$ is defined by

$$\boldsymbol{u}_{n+1} = \boldsymbol{u}_n + \Delta t\, \boldsymbol{f}(\boldsymbol{u}_{n+1}). \tag{3.13}$$

Notice that (3.13) can be derived from (3.11) by approximating $\boldsymbol{f}(\boldsymbol{u}(s))$ by $\boldsymbol{f}(\boldsymbol{u}(t_{n+1}))$ for $s \in (t_n, t_{n+1}]$. The implicit Euler method is implemented in Algorithm 3.2. The MATLAB optimization toolbox function `fsolve` is used to solve the nonlinear system at each step.

Example 3.5 (population model) To illustrate the implicit and explicit Euler methods, we consider the following system of two ODEs:

$$\frac{d}{dt}\begin{pmatrix} u_1 \\ u_2 \end{pmatrix} = \begin{pmatrix} u_1(1-u_2) \\ u_2(u_1-1) \end{pmatrix}, \qquad \begin{pmatrix} u_1(0) \\ u_2(0) \end{pmatrix} = \boldsymbol{u}_0. \tag{3.14}$$

This is a simplified model of an interacting population of u_2 predators and u_1 prey. The

Algorithm 3.2 Code to apply the implicit Euler method. The inputs and outputs are defined as in Algorithm 3.1.

```
function [t,u]=imp_euler(u0,T,N,d,fhandle)
Dt=T/N;              % set time step
u=zeros(d,N+1);      % preallocate solution u
t=[0:Dt:N*Dt]';      % set time
options=optimset('Display','off');
u(:,1)=u0; u_n=u0; % set initial condition
for n=1:N,           % time loop
  u_new=fsolve(@(u) impeuler_step(u,u_n,Dt,fhandle),u_n,options);
  u(:,n+1)=u_new; u_n=u_new;
end
function step=impeuler_step(u,u_n,Dt,fhandle)
step=u-u_n-Dt*fhandle(u);
return
```

explicit Euler method gives

$$\boldsymbol{u}_{n+1} = \begin{pmatrix} u_{1,n+1} \\ u_{2,n+1} \end{pmatrix} = \begin{pmatrix} u_{1,n} \\ u_{2,n} \end{pmatrix} + \Delta t \begin{pmatrix} u_{1,n}(1 - u_{2,n}) \\ u_{2,n}(u_{1,n} - 1) \end{pmatrix}, \qquad \begin{pmatrix} u_{1,0} \\ u_{2,0} \end{pmatrix} = \boldsymbol{u}_0,$$

where $\boldsymbol{u}_{n+1} \approx \boldsymbol{u}(t_{n+1})$. Given initial data $\boldsymbol{u}_0 = [0.5, 0.1]^{\mathsf{T}}$, we generate an approximation to $\boldsymbol{u}(t)$ for $t \in [0, 10]$ with $\Delta t = 0.01$ (so $N = 1000$) using the following Matlab commands to call Algorithm 3.1:

```
>> u0=[0.5;0.1]; T=10; N=1000; d=2;
>> [t,u]=exp_euler(u0,T,N,d,@(u) [u(1)*(1-u(2));u(2)*(u(1)-1)]);
```

Figure 3.1(a) shows a plot of the resulting approximations to the two components u_1, u_2 of the solution. Similarly, the implicit Euler method gives

$$\boldsymbol{u}_{n+1} = \begin{pmatrix} u_{1,n+1} \\ u_{2,n+1} \end{pmatrix} = \begin{pmatrix} u_{1,n} \\ u_{2,n} \end{pmatrix} + \Delta t \begin{pmatrix} u_{1,n+1}(1 - u_{2,n+1}) \\ u_{2,n+1}(u_{1,n+1} - 1) \end{pmatrix}, \qquad \begin{pmatrix} u_{1,0} \\ u_{2,0} \end{pmatrix} = \boldsymbol{u}_0,$$

where we must solve a nonlinear system at each step to determine $\boldsymbol{u}_{n+1}$. We generate a second approximation by using the following commands to call Algorithm 3.2:

```
>> u0=[0.5;0.1]; T=10; N=1000; d=2;
>> [t,u]=imp_euler(u0,T,N,d,@(u) [u(1)*(1-u(2));u(2)*(u(1)-1)]);
```

Figure 3.1(b) shows a plot of the resulting approximation. Comparing both figures, we observe that, although the Euler approximations are similar, they are not identical. For the same time step, the implicit method takes longer to execute than the explicit method, owing to the nonlinear solve.

The Euler methods are examples of numerical methods called numerical integrators. Both methods are known to converge on finite time intervals with first order, so the error satisfies $\|\boldsymbol{u}(t_n) - \boldsymbol{u}_n\|_2 \le K \Delta t$ for $0 \le t_n \le T$ for some constant K independent of Δt, but dependent on T. They are examples of low-order methods.

Definition 3.6 (convergence/order) An approximation $\boldsymbol{u}_n$ *converges* to $\boldsymbol{u}(t_n)$ on $[0, T]$ as $\Delta t \to 0$ if

$$\max_{0 \le t_n \le T} \|\boldsymbol{u}(t_n) - \boldsymbol{u}_n\|_2 \to 0,$$

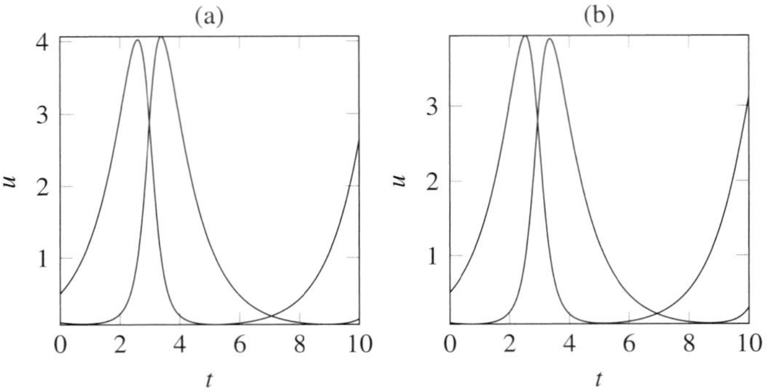

Figure 3.1 Plot of (a) the explicit Euler and (b) the implicit Euler approximations to u_1 and u_2 in (3.14) computed with $\Delta t = 0.01$. In each case, the approximations satisfy the initial conditions $u_1(0) = 0.5$ and $u_2(0) = 0.1$.

and has pth *order* if, for some constant $K > 0$,

$$\left\|\boldsymbol{u}(t_n) - \boldsymbol{u}_n\right\|_2 \le K\Delta t^p, \qquad \text{for } 0 \le t_n \le T.$$

Higher-order methods can be derived, for example, by taking higher-order polynomial approximations for $\boldsymbol{f}$ in (3.11). We restrict attention to low-order methods here, as it is the generalisations of such methods that are important for stochastic differential equations (as considered in Chapter 8). We prove that the explicit Euler method converges with first order.

Theorem 3.7 (convergence of explicit Euler) *Assume that $\boldsymbol{f} : \mathbb{R}^d \to \mathbb{R}^d$ satisfies (3.10) and that the solution $\boldsymbol{u}$ of (3.6) belongs to $\mathrm{C}^2([0,T],\mathbb{R}^d)$. Let $\boldsymbol{u}_n$ be the explicit Euler approximation (3.12) to $\boldsymbol{u}(t_n)$. Then, $\boldsymbol{u}_n$ converges to $\boldsymbol{u}(t_n)$ with first order on $[0,T]$. In particular,*

$$\max_{0 \le t_n \le T} \left\|\boldsymbol{u}(t_n) - \boldsymbol{u}_n\right\|_2 \le \frac{K}{L}(\mathrm{e}^{TL} - 1)\Delta t,$$

where $K := \|\boldsymbol{u}\|_{\mathrm{C}^2([0,T],\mathbb{R}^d)}/2$.

Proof Taylor's theorem (Theorem A.1) gives

$$\boldsymbol{u}(t) = \boldsymbol{u}(s) + (t - s)\boldsymbol{u}'(s) + \boldsymbol{R}_1,$$

where the remainder term is

$$\boldsymbol{R}_1 := (t - s)^2 \int_0^1 (1 - r)\boldsymbol{u}''(s + r(t - s))\, dr.$$

Clearly, $\|\boldsymbol{R}_1\|_2 \le \|\boldsymbol{u}\|_{\mathrm{C}^2([0,T],\mathbb{R}^d)}\Delta t^2/2$. If $\boldsymbol{u}(t)$ satisfies (3.6), then

$$\boldsymbol{u}(t) = \boldsymbol{u}(s) + (t - s)\,\boldsymbol{f}(\boldsymbol{u}(s)) + \boldsymbol{R}_1. \tag{3.15}$$

Setting $t = t_{n+1}$ and $s = t_n$ and subtracting (3.12) gives

$$\boldsymbol{u}(t_{n+1}) - \boldsymbol{u}_{n+1} = \boldsymbol{u}(t_n) - \boldsymbol{u}_n + \Delta t\left[\boldsymbol{f}(\boldsymbol{u}(t_n)) - \boldsymbol{f}(\boldsymbol{u}_n)\right] + \boldsymbol{R}_1. \tag{3.16}$$

Denote the error at time t_n by $\boldsymbol{e}_n = \boldsymbol{u}(t_n) - \boldsymbol{u}_n$. With the Lipschitz constant denoted L and $K := \|\boldsymbol{u}\|_{C^2([0,T],\mathbb{R}^d)}/2$, we have

$$\|\boldsymbol{e}_{n+1}\|_2 \le \|\boldsymbol{e}_n\|_2 + \Delta t\, L \|\boldsymbol{e}_n\|_2 + \Delta t^2\, K = (1 + \Delta t\, L)\|\boldsymbol{e}_n\|_2 + \Delta t^2 K. \tag{3.17}$$

If we apply (3.17) recursively, we obtain

$$\|\boldsymbol{e}_{n+1}\|_2 \le (1 + \Delta t\, L)^{n+1}\|\boldsymbol{e}_0\|_2 + \sum_{k=0}^{n}(1 + \Delta t\, L)^k \Delta t^2\, K.$$

Using $\sum_{k=0}^{n} \lambda^k = (\lambda^{n+1} - 1)/(\lambda - 1)$ for $\lambda \ne 1$ gives

$$\|\boldsymbol{e}_{n+1}\|_2 \le (1 + \Delta t L)^{n+1}\|\boldsymbol{e}_0\|_2 + \frac{\Delta t^2 K}{\Delta t L}\left((1 + \Delta t L)^{n+1} - 1\right).$$

Note that $1 + x \le \mathrm{e}^x$ for $x \ge 0$ and so

$$\|\boldsymbol{e}_{n+1}\|_2 \le (\mathrm{e}^{\Delta t L})^{n+1}\|\boldsymbol{e}_0\|_2 + \frac{K}{L}\Delta t(\mathrm{e}^{\Delta t L(n+1)} - 1).$$

Reducing the indices by one and substituting $t_n = n\Delta t$ gives

$$\|\boldsymbol{e}_n\|_2 \le \mathrm{e}^{t_n L}\|\boldsymbol{e}_0\|_2 + \frac{K}{L}\Delta t\left(\mathrm{e}^{t_n L} - 1\right).$$

Finally, $\boldsymbol{e}_0 = \boldsymbol{u}_0 - \boldsymbol{u}(0) = \boldsymbol{0}$ and hence

$$\max_{0 \le t_n \le T}\|\boldsymbol{e}_n\|_2 \le \frac{K}{L}\Delta t\left(\mathrm{e}^{TL} - 1\right). \qquad \square$$

We now examine the convergence of time-stepping methods numerically.

Testing convergence numerically

To investigate convergence of a time-stepping method from simulations, we compute a sequence of approximations $\boldsymbol{u}_{N(\kappa)}$ to $\boldsymbol{u}(T)$ with time steps $\Delta t_\kappa := T/N(\kappa)$ for some $N = N(\kappa)$ depending on $\kappa \in \mathbb{N}$. We then estimate the error $\|\boldsymbol{u}(T) - \boldsymbol{u}_{N(\kappa)}\|_2$ not knowing the exact solution $\boldsymbol{u}(T)$, for each κ considered. We discuss three possibilities.

Method A: Approximate $\boldsymbol{u}(T)$ by a reference solution $\boldsymbol{u}_{N_{\mathrm{ref}}}$ obtained with time step $\Delta t_{\mathrm{ref}} = T/N_{\mathrm{ref}}$ using the same numerical method used to generate the approximations $\boldsymbol{u}_{N(\kappa)}$. To compute $\boldsymbol{u}_{N(\kappa)}$, take $N(\kappa) := N_{\mathrm{ref}}/\kappa$ steps so that $\Delta t_\kappa = \kappa\Delta t_{\mathrm{ref}}$ and choose κ so that Δt_{ref} is much smaller than each Δt_κ. Then examine $\|\boldsymbol{u}_{N_{\mathrm{ref}}} - \boldsymbol{u}_{N(\kappa)}\|_2$ for each κ.

Method B: Consider pairs of approximations generated with the same method that correspond to κ_i and κ_{i+1}. That is, consider $\boldsymbol{u}_{N(\kappa_i)}$ (computed with $\Delta t_{\kappa_i} = T/N(\kappa_i)$) and $\boldsymbol{u}_{N(\kappa_{i+1})}$ (computed with $\Delta t_{\kappa_{i+1}} = T/N(\kappa_{i+1})$), where the number of steps $N(\kappa_i) > N(\kappa_{i+1})$. Then examine the error $\|\boldsymbol{u}_{N(\kappa_i)} - \boldsymbol{u}_{N(\kappa_{i+1})}\|_2$ for $\kappa_1, \kappa_2, \dots, \kappa_K$, taking $\boldsymbol{u}_{N(\kappa_i)}$ as an approximation to $\boldsymbol{u}(T)$.

Method C: Use a more accurate numerical method to approximate $\boldsymbol{u}(T)$ by $\hat{\boldsymbol{u}}_{N_{\mathrm{ref}}}$ using a time step Δt_{ref} that is much smaller than Δt_κ, for each κ considered. As in method A, choose N_{ref} and κ such that $T = N_{\mathrm{ref}}\Delta t_{\mathrm{ref}}$ and $\Delta t_\kappa = \kappa\Delta t_{\mathrm{ref}}$. Then examine $\|\hat{\boldsymbol{u}}_{N_{\mathrm{ref}}} - \boldsymbol{u}_{N(\kappa)}\|_2$.

Example 3.8 In Figure 3.2, we plot the results of a numerical investigation into solving the population model (3.14) using the explicit Euler method. Algorithm 3.3 uses methods A and B above to estimate the error. Method C is used in Exercise 3.8. The following MATLAB commands call Algorithm 3.3 to estimate the error at $T = 1$ with $\Delta t_{\text{ref}} = 10^{-6}$ (so $N_{\text{ref}} = 10^6$). We take $\kappa = 10, 10^2, 10^3, 10^4, 10^5$ corresponding to $\Delta t_\kappa = 10^{-5}, 10^{-4}, 10^{-3}, 10^{-2}, 10^{-1}$.

```
>> u0=[0.5;0.1]; T=1; Nref=1e6; d=2;
>> kappa=[10,1e2,1e3,1e4,1e5];
>> [dt,errA,errB]=exp_euler_conv(u0,T,Nref,d,...
                              @(u) [u(1)*(1-u(2));u(2)*(u(1)-1)],kappa);
```

The plot in Figure 3.2 is on a log-log scale and the slope indicates the order of convergence. We observe the theoretical rate of convergence of order 1.

Algorithm 3.3 Code to examine convergence of the explicit Euler method. The inputs (similar to Algorithm 3.1) are the initial vector `u0`, final time `T`, number of steps `Nref` (so that $\Delta t_{\text{ref}} = T/N_{\text{ref}}$), state space dimension `d`, a function handle `fhandle` to evaluate $\boldsymbol{f}(\boldsymbol{u})$, and a vector of values for `kappa`. The outputs are a vector `dt` of time steps $\Delta t_\kappa = \kappa \Delta t_{\text{ref}}$ and vectors `errA`, `errB` of errors estimated using methods A and B, respectively.

```
function [dt,errA,errB]=exp_euler_conv(u0,T,Nref,d,fhandle,kappa)
[t,uref]=exp_euler(u0,T,Nref,d,fhandle); % compute reference soln
uTref=uref(:,end); dtref=T/Nref; uTold=uTref;
for k=1:length(kappa)
  N=Nref/kappa(k); dt(k)=T/N;
  [t,u]=exp_euler(u0,T,N,d,fhandle); % compute approximate soln
  uT=u(:,end);
  errA(k)=norm(uTref-uT); % error by method A
  errB(k)=norm(uTold-uT); uTold=uT; % error by method B
end
```

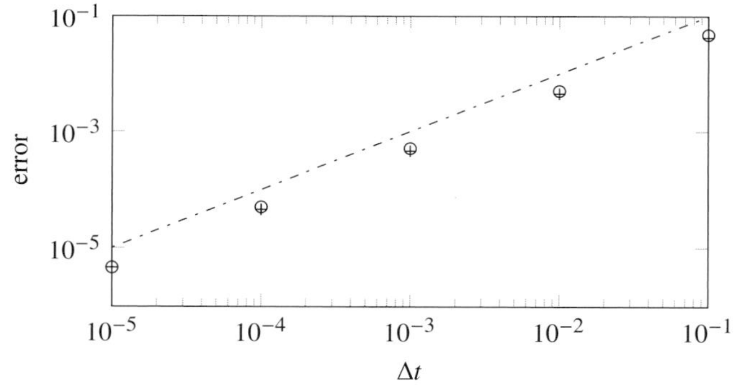

Figure 3.2 Approximation errors for the explicit Euler scheme, estimated by methods A (○) and B (+). For method A, the reference solution was computed with $\Delta t_{\text{ref}} = 10^{-6}$. A reference line (dash-dot) of slope 1 shows the theoretical order of convergence.

Stability

Even though a single step of the implicit Euler method is more expensive than an explicit Euler step, the implicit method is often preferred. The implicit method is more *stable* and the approximations behave better when large time steps Δt are used. Consequently, the implicit Euler method is often able to achieve a good approximation with many fewer steps and in less overall time than the explicit method. To make the notion of stability precise, we introduce the scalar linear test IVP.

Example 3.9 (linear test IVP) For $\lambda \in \mathbb{C}$, consider the IVP

$$\frac{du}{dt} = \lambda u, \qquad u(0) = u_0.$$

A key feature for $\operatorname{Re} \lambda < 0$ is that solutions decay to zero; that is, $u(t) \to 0$ as $t \to \infty$. We now ask, when do the Euler methods produce approximations with the same behaviour?

Applying the explicit scheme to the test equation gives

$$u_n = (1 + \lambda\Delta t)u_{n-1} = \left(1 + \lambda\Delta t\right)^n u_0$$

and applying the implicit scheme gives

$$u_n = u_{n-1} + \lambda\Delta t\, u_n \qquad \text{or} \qquad u_n = \left(1 - \lambda\Delta t\right)^{-n} u_0.$$

The conditions under which the Euler methods provide approximate solutions with the decay property are very different. Assume that $u_0 \neq 0$.

Explicit Euler method: $u_n \to 0$ as $n \to \infty$ if and only if $|1 + \lambda\Delta t|^n \to 0$. This is true if and only if $\lambda\Delta t$ lies *inside* a circle of radius one with centre -1 in the complex plane.

Implicit Euler method: $u_n \to 0$ as $n \to \infty$ if and only if $|1 - \lambda\Delta t|^{-n} \to 0$. This is true if and only if $\lambda\Delta t$ lies *outside* the circle of radius one with centre 1 in the complex plane.

These regions in the complex plane are called *regions of absolute stability* and are plotted in Figure 3.3. Consider the case when λ is real and negative. The explicit Euler method converges to zero if $\Delta t < -2/\lambda$, which places a restriction on the time step. However, there is no such restriction for the implicit Euler method. For *stiff* differential equations, where there are fast and slow time scales, the explicit Euler method must use a time step determined by the fastest time scale in the system, while the dynamics of interest may take place over much longer time scales. This means a large number of time steps are required and the calculation becomes expensive compared to the implicit Euler method, which often provides accurate answers with a much larger time step.

To end this section, we briefly mention three further numerical integrators. First, consider a semilinear system of ODEs (see also §3.3) given by

$$\frac{d\boldsymbol{u}}{dt} = -M\boldsymbol{u} + \boldsymbol{f}(\boldsymbol{u}), \tag{3.18}$$

where M is a $d \times d$ positive-definite matrix and we are given initial data $\boldsymbol{u}(0) = \boldsymbol{u}_0$. If we treat the linear term $M\boldsymbol{u}$ in (3.18) implicitly and $\boldsymbol{f}(\boldsymbol{u})$ explicitly, then we obtain the *semi-implicit Euler method*,

$$\boldsymbol{u}_{n+1} = \boldsymbol{u}_n - \Delta t\, M\boldsymbol{u}_{n+1} + \Delta t\, \boldsymbol{f}(\boldsymbol{u}_n), \qquad \boldsymbol{u}(0) = \boldsymbol{u}_0,$$

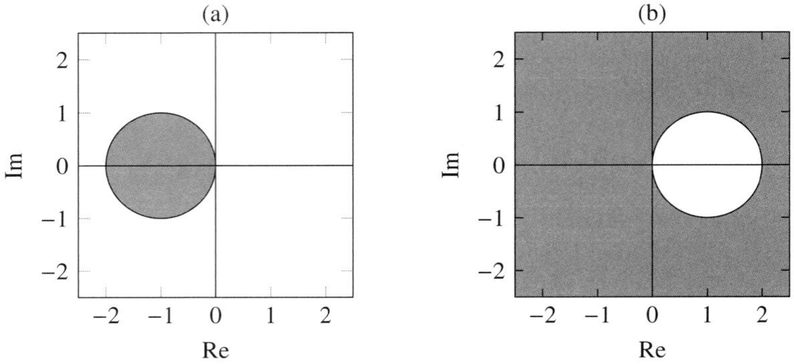

Figure 3.3 The shaded regions indicate the regions of absolute stability for (a) the explicit Euler method and (b) the implicit Euler method. Notice that the region of absolute stability for (b) includes the whole of the left half-plane.

or, equivalently,

$$(I + \Delta t\, M)\boldsymbol{u}_{n+1} = \boldsymbol{u}_n + \Delta t\, \boldsymbol{f}(\boldsymbol{u}_n), \qquad \boldsymbol{u}(0) = \boldsymbol{u}_0. \tag{3.19}$$

At each step, we need to solve a *linear* system of equations. Compared to the explicit Euler method, the linear solve is an additional cost, but small compared to solving the nonlinear equations associated with the implicit Euler method. For many models, it is the $M\boldsymbol{u}$ term that contains the fast and slow time scales. Treating this term implicitly means that the method is more stable and we can take larger time steps and achieve a better approximation than when using an explicit method. It can be shown that the semi-implicit method (3.19) is of first order (the same order as the implicit and explicit Euler methods).

The *trapezoidal method* for (3.6) is given by

$$\boldsymbol{u}_{n+1} = \boldsymbol{u}_n + \frac{1}{2}\Delta t \left[\boldsymbol{f}(\boldsymbol{u}_n) + \boldsymbol{f}(\boldsymbol{u}_{n+1})\right]. \tag{3.20}$$

It is implicit and second-order accurate, so that the error is proportional to Δt^2. For many problems, it provides greater accuracy than the explicit Euler method for a given time step. The method is considered further in Exercises 3.5 and 3.6. To avoid the cost of the nonlinear solve, the trapezoidal method can be modified by using an intermediate step $\tilde{\boldsymbol{u}}_{n+1}$ as follows:

$$\begin{aligned} \tilde{\boldsymbol{u}}_{n+1} &= \boldsymbol{u}_n + \Delta t\, \boldsymbol{f}(\boldsymbol{u}_n) \\ \boldsymbol{u}_{n+1} &= \boldsymbol{u}_n + \frac{1}{2}\Delta t \left[\boldsymbol{f}(\boldsymbol{u}_n) + \boldsymbol{f}(\tilde{\boldsymbol{u}}_{n+1})\right]. \end{aligned} \tag{3.21}$$

This is known as *Heun's method* and is also of second order. See Exercise 3.7. We meet a stochastic version of Heun's method in Chapter 8 when we discuss numerical methods for Stratonovich SODEs.

3.2 Semigroups of linear operators

Theorem 3.2 establishes existence and uniqueness of the solution $\boldsymbol{u}(t)$ to (3.6). This leads naturally to the idea of a *solution operator* or *semigroup* $S(t)$ defined for initial data $\boldsymbol{u}_0 \in \mathbb{R}^d$ by

$$S(t)\boldsymbol{u}_0 = \boldsymbol{u}(t), \qquad t \geq 0. \tag{3.22}$$

Starting at $t = 0$, we have $\boldsymbol{u}(0) = \boldsymbol{u}_0$ and so $S(0)\boldsymbol{u}_0 = \boldsymbol{u}_0$. Theorem 3.2 tells us that we obtain $\boldsymbol{u}(s+t)$ by first flowing forward by time s and then flowing forward by time t (using $\boldsymbol{u}(s)$ as initial data). Hence, the solution operator for (3.6) satisfies

$$S(0) = I \qquad \text{and} \qquad S(t)S(s) = S(t+s). \tag{3.23}$$

The space $\mathbb{R}^d$ where $\boldsymbol{u}(t)$ exists is called the *phase space* and $S(t)\colon \mathbb{R}^d \to \mathbb{R}^d$ is a one-parameter family of operators on the phase space $\mathbb{R}^d$. The pair $(\mathbb{R}^d, S(t))$ is termed a *dynamical system*. We treat semigroups on an abstract phase space X.

Definition 3.10 (semigroup) Let X be a vector space. $S(t)\colon X \to X$ is a *semigroup* if

(i) $S(0) = I$ (the identity operator on X) and
(ii) $S(t+s) = S(t)S(s)$ for $s, t \geq 0$.

We say $S(t)$ is a semigroup of *linear operators* if $S(t)\colon X \to X$ is linear for each $t \geq 0$.

When X is a Banach space, an important class of semigroups of linear operators is characterised by the infinitesimal generator.

Definition 3.11 (infinitesimal generator) Let $S(t)$ be a semigroup of linear operators on a Banach space X. The *infinitesimal generator* $-A$ is defined by

$$(-A)u := \lim_{h \downarrow 0} \frac{S(h)u - u}{h}, \tag{3.24}$$

where $\lim_{h\downarrow 0}$ denotes the limit from above with respect to the norm on X. The domain of A, denoted $\mathcal{D}(A)$, is the set of $u \in X$ where the limit exists.

Intuitively, the limit in (3.24) is the time derivative of $S(t)u$ at $t = 0$.

Example 3.12 (linear ODE) Consider the linear ODE

$$\frac{du}{dt} = -\lambda u$$

with initial condition $u(0) = u_0 \in \mathbb{R}$ and $\lambda > 0$. Then, the phase space is $\mathbb{R}$ and $S(t)u_0 := u(t) = \mathrm{e}^{-\lambda t}u_0$. It is clear that $S(t) = \mathrm{e}^{-\lambda t}$ satisfies both conditions in Definition 3.10 and is a semigroup of linear operators. The infinitesimal generator of $S(t)$ is $-\lambda$ and $-\lambda$ is said to *generate* the semigroup $S(t) = \mathrm{e}^{-\lambda t}$.

Our aim is to apply semigroup theory to study PDEs where the solutions vary in space and time. In that case, the phase space X contains functions of the spatial coordinate $\boldsymbol{x} \in D$ and solutions $u(t, \boldsymbol{x})$ of the PDE are interpreted as functions $u(t)$ of t taking values in the function space X, thereby suppressing the dependence on $\boldsymbol{x}$. Using this trick, PDEs can be written as ODEs (with only one derivative made explicit) on X. Existence and uniqueness of the solution leads naturally to a semigroup $S(t)$ and we identify the generator $-A$ for several linear PDEs.

Example 3.13 (heat equation) Consider the heat equation

$$u_t = u_{xx},$$

with $x \in \mathbb{R}$ and the initial condition $u(0,x) = u_0(x)$. To find the solution, we use the Fourier transform (see Definition 1.102) and recall that $\widehat{u_x}(\lambda) = -\mathrm{i}\lambda\hat{u}$ (as in Example 1.105). Thus, the heat equation becomes

$$\frac{d}{dt}\hat{u} = -\lambda^2\hat{u} \tag{3.25}$$

so that $\hat{u}(\lambda,t) = \mathrm{e}^{-\lambda^2 t}\hat{u}_0(\lambda)$. By (1.45), $\hat{v}(\lambda,t) = \mathrm{e}^{-\lambda^2 t}$ is the Fourier transform of $v(x) = \mathrm{e}^{-x^2/4t}/\sqrt{2t}$. The identity (1.50) turns the inverse Fourier transform of the product $\hat{u}(\lambda,t) = \mathrm{e}^{-\lambda^2 t}\hat{u}_0(\lambda)$ into the following convolution:

$$u(t,x) = \frac{1}{\sqrt{2\pi}}(v \star u_0)(t,x) = \frac{1}{\sqrt{4\pi t}}\int_{\mathbb{R}} \mathrm{e}^{-(x-y)^2/4t}u_0(y)\,dy. \tag{3.26}$$

The mapping from $u_0(x)$ to $u(t,x)$ given by (3.26) for any $t > 0$ is a linear operator and forms a semigroup of linear operators. If $u_0 \in L^2(\mathbb{R})$, then $u(t) \in L^2(\mathbb{R})$ by the Parseval identity (1.43). Here, we write $u(t,x)$ as a function $u\colon [0,\infty) \to L^2(\mathbb{R})$ and so interpret $u(t)$ as a function in $L^2(\mathbb{R})$. The solution given in (3.26) then defines the so-called heat semigroup $S(t)u_0 = u(t)$ on $L^2(\mathbb{R})$. Using Definition 3.11, the infinitesimal generator satisfies

$$(-Au_0)(x) = \lim_{h\downarrow 0}\frac{(S(h)u_0)(x) - u_0(x)}{h} = \lim_{h\downarrow 0}\frac{u(h,x) - u(0,x)}{h} = u_t(0,x).$$

As $u_t = u_{xx}$, we see $-Au_0 = (u_0)_{xx}$, which is well defined in $L^2(\mathbb{R})$ for $u_0 \in \mathcal{D}(A) = H^2(\mathbb{R})$. We can now rewrite the heat equation $u_t = u_{xx}$ as the following ODE on $L^2(\mathbb{R})$:

$$\frac{du}{dt} = -Au, \qquad u(0) = u_0 \in L^2(\mathbb{R}).$$

Example 3.14 (advection equation) Consider the advection equation

$$u_t = u_x, \qquad u(0,x) = u_0(x),$$

with $x \in \mathbb{R}$, which has solution $u(t,x) = u_0(x+t)$. Notice that $u(t,x)$ depends linearly on the initial condition and if $u_0 \in \mathrm{C}(\mathbb{R})$ then $u(t,x)$ is continuous as a function of x. We write $u(t,x)$ as $u(t)$ and interpret $u(t)$ as a function in $\mathrm{C}(\mathbb{R})$. We can use the solution $u(t)$ to define a semigroup on $\mathrm{C}(\mathbb{R})$; that is, we take $(S(t)u_0)(x) := u(t)$. It is straightforward to see that $S(t)$ satisfies the conditions of Definition 3.10. Here, the infinitesimal generator is given by

$$(-Au_0)(x) = \lim_{h\downarrow 0}\frac{(S(h)u_0)(x) - u_0(x)}{h} = \lim_{h\downarrow 0}\frac{u_0(x+h) - u_0(x)}{h} = (u_0)_x(x). \tag{3.27}$$

This provides a well-defined limit for each $x \in \mathbb{R}$ for $u_0 \in \mathrm{C}^1(\mathbb{R})$. Note, however, that Definition 3.11 requires convergence in the sense of a Banach space norm and not a pointwise norm. With the operator $-A$ in hand, we can rewrite the advection equation $u_t = u_x$ as the ODE $\frac{du}{dt} = -Au$.

Semigroups of linear operators often have poor regularity, such as the semigroup on $C(\mathbb{R})$ defined by the advection equation in Example 3.14, where any discontinuity (in the first derivatives, say) of the initial condition $u_0(x)$ propagates forward and remains a discontinuity of $u(t,x) = u_0(t+x)$ for all $t \ge 0$. We focus on semigroups with a smoothing property. That is, for $t > 0$, $S(t)u_0$ is continuous and differentiable as a function of x even when u_0 is not. The simplest form of regularity is to ask that $S(t)u$ be continuous and this leads to the C_0 semigroups.

Definition 3.15 (C_0 semigroup) Let X be a Banach space. $S(t)$ is a C_0 *semigroup* (or a strongly continuous semigroup) on X if it satisfies Definition 3.10 and, in addition, $S(t)u$ is a continuous function from $[0,\infty)$ to X for any $u \in X$.

We present three examples: the first looks at the heat equation on $\mathbb{R}$ and the following two look at the heat equation on the bounded domains $D = (0,\pi)$ and $D = (0,2\pi)^2$ with Dirichlet and periodic boundary conditions, respectively.

Example 3.16 (heat semigroup) Continuing from Example 3.13,

$$\left\|u(t) - u_0\right\|_{L^2(\mathbb{R})} = \left\|\hat{u}(t) - \hat{u}_0\right\|_{L^2(\mathbb{R})} = \left(\int_{\mathbb{R}} \left(e^{-\lambda^2 t} - 1\right)^2 \hat{u}_0(\lambda)^2\, d\lambda\right)^{1/2}$$

by (3.25) and the Parseval identity (1.43). This converges to zero as $t \downarrow 0$ by Theorem 1.22 for any $u_0 \in L^2(\mathbb{R})$. The heat semigroup obeys $\|S(t)u_0 - u_0\|_{L^2(\mathbb{R})} \to 0$ as $t \downarrow 0$ and $S(t)$ is a C_0 semigroup on $L^2(\mathbb{R})$.

In addition, the heat semigroup is smoothing so that $u(t,x)$ is differentiable for $t > 0$ even if u_0 is not. For example, differentiating (3.26) gives

$$u_x(t,x) = \frac{\partial u}{\partial x}(t,x) = \frac{1}{\sqrt{4\pi t}} \int_{\mathbb{R}} \frac{-2(x-y)}{4t} e^{-(x-y)^2/4t} u_0(y)\, dy.$$

This is well defined for $t > 0$ as the integrand behaves like e^{-y^2} at infinity. This argument can be repeated to show that $u(t,x)$ is differentiable to any degree in t and x. More generally, it can be shown that $u(t,x)$ is analytic.

Example 3.17 (heat equation with Dirichlet conditions) Let $D = (0,\pi)$ and consider the heat equation

$$u_t = u_{xx}, \qquad x \in D, \quad t > 0,$$

with boundary conditions $u(t,0) = u(t,\pi) = 0$ and initial data $u(0,x) = u_0(x)$. We can write this as an ODE on the Hilbert space $H = L^2(0,\pi)$ using the operator A defined in Example 1.81. That is, $Au = -u_{xx}$ with $\mathcal{D}(A) = H^2(0,\pi) \cap H_0^1(0,\pi)$ and we have the IVP

$$\frac{du}{dt} = -Au, \qquad u(0) = u_0 \in L^2(0,\pi). \tag{3.28}$$

In this example, A has orthonormal eigenfunctions $\phi_j(x) = \sqrt{2/\pi}\sin(jx)$ and it is straightforward to verify that the eigenvalues are $\lambda_j = j^2$ (see Example 1.81) for $j \in \mathbb{N}$.

It is also instructive to write $\hat{u}_j(t) = \langle u(t,\cdot), \phi_j\rangle_{L^2(0,\pi)}$ and

$$u(t,x) = \sum_{j=1}^{\infty} \hat{u}_j(t)\phi_j(x). \tag{3.29}$$

Substituting the expansion (3.29) into (3.28), we see that the Fourier coefficients $\hat{u}_j$ satisfy the decoupled linear ODEs

$$\frac{d\hat{u}_j}{dt} = -\lambda_j \hat{u}_j, \qquad \hat{u}_j(0) = \left\langle u_0, \phi_j \right\rangle_{L^2(0,\pi)}$$

with solution $\hat{u}_j(t) = \mathrm{e}^{-\lambda_j t}\langle u_0, \phi_j\rangle_{L^2(0,\pi)}$. Then,

$$u(t,x) = \sum_{j=1}^{\infty} \hat{u}_j(0)\,\sqrt{2/\pi}\,\sin(jx)\mathrm{e}^{-j^2 t} \tag{3.30}$$

and, in particular, $u(t,x)$ is a smooth function of t and x. For $u_0 \in L^2(D)$, we see that the solution $u(t,\cdot) \in H^r(D)$ for any $t, r > 0$.

Let's examine what happens when we change to periodic boundary conditions.

Example 3.18 (heat equation with periodic boundary conditions) Consider the heat equation on $D = (0,2\pi)^2$,

$$u_t = \Delta u, \qquad u(0) = u_0 \in L^2(D),$$

with periodic boundary conditions. We rewrite this equation as an ODE on the Hilbert space $H = L^2(D)$ by taking $A = -\Delta$ with $\mathcal{D}(A) = H^2_{\mathrm{per}}(D)$ (see Definition 1.47) to get

$$\frac{du}{dt} = -Au, \qquad u(0) = u_0.$$

As in Example 1.84, we use complex notation to write the eigenfunctions of the operator A as $\phi_{j_1,j_2}(\boldsymbol{x}) = \mathrm{e}^{\mathrm{i}j_1x_1}\mathrm{e}^{\mathrm{i}j_2x_2}/2\pi$ and the eigenvalues as $\lambda_{j_1,j_2} = j_1^2 + j_2^2$ for $j_1, j_2 \in \mathbb{Z}$. Then,

$$u(t,\boldsymbol{x}) = \sum_{j_1,j_2\in\mathbb{Z}} \hat{u}_{j_1,j_2}(t)\phi_{j_1,j_2}(\boldsymbol{x}), \tag{3.31}$$

with $\hat{u}_{j_1,j_2}(t) = \langle u(t,\cdot), \phi_{j_1,j_2}\rangle_{L^2(D)}$. The Fourier coefficients $\hat{u}_{j_1,j_2}$ satisfy a system of decoupled linear ODEs

$$\frac{d}{dt}\hat{u}_{j_1,j_2} = -\lambda_{j_1,j_2}\hat{u}_{j_1,j_2}, \qquad \hat{u}_{j_1,j_2}(0) = \left\langle u_0, \phi_{j_1,j_2}\right\rangle_{L^2(D)}$$

with solution $\hat{u}_{j_1,j_2}(t) = \mathrm{e}^{-\lambda_{j_1,j_2}t}\langle u_0, \phi_{j_1,j_2}\rangle_{L^2(D)}$. Again $u(t,\boldsymbol{x})$ is a smooth function of t and $\boldsymbol{x}$. The Laplacian with periodic boundary conditions has a zero eigenvalue, which can be awkward in the analysis. An alternative is to analyse $v(t) := \mathrm{e}^{-t}u(t)$, which satisfies $v_t = \Delta v - v = -(I - \Delta)v$ and the operator $I - \Delta$ is positive definite.

Semigroup theory

In Examples 3.16–3.18, the PDEs all have smooth solutions, even when the initial data is rough. In Examples 3.17 and 3.18 where the domain is bounded, A yields a basis of eigenfunctions for $L^2(D)$. Furthermore in Example 3.17, all the eigenvalues are positive and we focus on this case. Specifically, we work in situations where A satisfies Assumption 1.85 and the solution $u(t)$ can be written as a series. We restate the assumption for convenience.

Assumption 3.19 Suppose that H is a Hilbert space with inner product $\langle\cdot,\cdot\rangle$ and that the linear operator $-A\colon \mathcal{D}(A) \subset H \to H$ has a complete orthonormal set of eigenfunctions $\{\phi_j : j \in \mathbb{N}\}$ and eigenvalues $\lambda_j > 0$, ordered so that $\lambda_{j+1} \ge \lambda_j$.

With $\lambda_j > 0$, A has an inverse and this is convenient for the theory that follows. With extra work, much of the theory can be extended to non-negative definite operators A, such as the A in Example 3.18.

For this general class of operators satisfying Assumption 3.19, we define and give properties of the exponential.

Definition 3.20 (exponential) Suppose Assumption 3.19 holds. Then, for $t \ge 0$, the exponential of $-tA$ is defined by

$$e^{-tA}u := \sum_{j=1}^{\infty} e^{-\lambda_j t}\langle u,\phi_j\rangle\phi_j.$$

The right-hand side already appeared in (3.30) and we may write the solution (3.30) as $u(t) = e^{-tA}u_0$. Unlike the one-dimensional exponential e^{-t}, we define e^{-tA} for $t \ge 0$ only. This is because, for $t < 0$, $e^{-\lambda_j t}$ blows up as $\lambda_j \to \infty$. Thus, we deal with semigroups, which are defined only for positive time. Lemma 3.21 shows that e^{-tA} is indeed a semigroup and is generated by $-A$.

Lemma 3.21 *Suppose Assumption 3.19 holds. Then, $S(t) := e^{-tA}$ is a semigroup of linear operators on H and its infinitesimal generator is $-A$.*

Proof Clearly, $S(t)$ is a linear operator. To show $S(t)$ is a semigroup, notice that $S(0)u = u$ so that $S(0) = I$. Further,

$$\begin{aligned}S(t+s)u &= \sum_{j=1}^{\infty} e^{-\lambda_j(t+s)}\langle u,\phi_j\rangle\phi_j \\ &= \sum_{j=1}^{\infty} e^{-\lambda_j s}e^{-\lambda_j t}\langle u,\phi_j\rangle\phi_j = S(s)S(t)u,\end{aligned}$$

because $\langle S(t)u,\phi_j\rangle = e^{-\lambda_j t}\langle u,\phi_j\rangle$. Hence, $S(t)$ satisfies Definition 3.10. Its infinitesimal generator is the operator $-A$, since

$$\begin{aligned}\frac{S(h)u - u}{h} &= \sum_{j=1}^{\infty} \frac{(e^{-\lambda_j h} - 1)}{h}\langle u,\phi_j\rangle\phi_j \\ &\to \sum_{j=1}^{\infty} -\lambda_j\langle u,\phi_j\rangle\phi_j = -Au, \qquad \text{as } h \downarrow 0,\end{aligned}$$

where the limit is well defined for any $u \in \mathcal{D}(A)$. □

Under Assumption 3.19, the fractional powers A^α and the associated fractional power norms $\|\cdot\|_\alpha = \|A^\alpha\cdot\|$ (where $\|\cdot\|$ is the norm on H) are well defined for $\alpha \in \mathbb{R}$ (see Theorem 1.88). By using fractional power norms, we make precise statements about the regularity of solutions and initial data in this chapter and also in Chapter 10. To do this, we need the following lemma.

Lemma 3.22 *Suppose that Assumption 3.19 holds. Then*

(i) *For* $u \in \mathcal{D}(A^\alpha)$ *and* $\alpha \in \mathbb{R}$, *we have* $A^\alpha e^{-tA} u = e^{-tA} A^\alpha u$ *for* $t \geq 0$.

(ii) *For each* $\alpha \geq 0$, *there exists a constant* K *such that*

$$\|A^\alpha e^{-tA}\|_{\mathcal{L}(H)} \leq K t^{-\alpha}, \qquad t > 0.$$

(iii) *For* $\alpha \in [0,1]$, *there is a constant* K *such that*

$$\|A^{-\alpha}(I - e^{-tA})\|_{\mathcal{L}(H)} \leq K t^\alpha, \qquad t \geq 0.$$

Proof (i) From Definition 3.20,

$$A^\alpha e^{-tA} u = A^\alpha \sum_{j=1}^\infty e^{-\lambda_j t} \langle u, \phi_j \rangle \phi_j = \sum_{j=1}^\infty e^{-\lambda_j t} \langle u, \phi_j \rangle A^\alpha \phi_j,$$

and the result follows as ϕ_j is an eigenfunction of A.

(ii) For $x \geq 0$, $x^\alpha e^{-x}$ is bounded by $K_\alpha := (\alpha/e)^\alpha$ for $\alpha > 0$. Then,

$$\lambda_j^\alpha e^{-\lambda_j t} = \frac{1}{t^\alpha} (\lambda_j t)^\alpha e^{-\lambda_j t} \leq \frac{K_\alpha}{t^\alpha} \tag{3.32}$$

and

$$\|A^\alpha e^{-tA} u\|^2 = \sum_{j=1}^\infty \lambda_j^{2\alpha} e^{-2\lambda_j t} \langle u, \phi_j \rangle^2 \leq \frac{K_\alpha^2}{t^{2\alpha}} \sum_{j=1}^\infty \langle u, \phi_j \rangle^2.$$

Hence, $\|A^\alpha e^{-tA} u\| \leq K_\alpha t^{-\alpha} \|u\|$, which implies the stated bound with $K = K_\alpha$ on the operator norm of $A^\alpha e^{-tA}$. The case $\alpha = 0$ holds as $\|e^{-tA}\|_{\mathcal{L}(H)} \leq 1$.

(iii) Using (3.32) for $\alpha > 0$,

$$\lambda^{-\alpha}(I - e^{-\lambda t}) = \int_0^t \lambda^{1-\alpha} e^{-\lambda s}\, ds \leq \int_0^t \frac{K_{1-\alpha}}{s^{1-\alpha}}\, ds = \frac{K_{1-\alpha}}{\alpha} t^\alpha.$$

Arguing as in (ii), we see $\|A^{-\alpha}(I - e^{-At})u\| \leq K t^\alpha \|u\|$ for $K := K_{1-\alpha}/\alpha$. □

3.3 Semilinear evolution equations

Semilinear evolution equations can be analysed by exploiting the linear theory and the variation of constants formula. We first review variation of constants for semilinear ODEs.

Example 3.23 (variation of constants formula for ODEs) Consider the IVP for the semilinear system of ODEs

$$\frac{d\boldsymbol{u}}{dt} = -M\boldsymbol{u} + \boldsymbol{f}(\boldsymbol{u}), \qquad \boldsymbol{u}(0) = \boldsymbol{u}_0 \in \mathbb{R}^d, \tag{3.33}$$

where M is a $d \times d$ positive-definite matrix (see also (3.18)). The right-hand side is said to be *semilinear* as it is a sum of a linear term $-M\boldsymbol{u}$ and a nonlinear term $\boldsymbol{f}(\boldsymbol{u})$. To find an expression for the solution, multiply (3.33) by the integrating factor e^{tM}, to find

$$\frac{d}{dt}\left[e^{tM} \boldsymbol{u}(t)\right] = e^{tM} \frac{d\boldsymbol{u}}{dt} + M e^{tM} \boldsymbol{u} = e^{tM} \boldsymbol{f}(\boldsymbol{u}).$$

A simple integration yields

$$e^{tM}\boldsymbol{u}(t) = \boldsymbol{u}(0) + \int_0^t e^{sM}\boldsymbol{f}(\boldsymbol{u}(s))\,ds.$$

We have derived the *variation of constants* formula

$$\boldsymbol{u}(t) = e^{-tM}\boldsymbol{u}_0 + \int_0^t e^{-(t-s)M}\boldsymbol{f}(\boldsymbol{u}(s))\,ds. \tag{3.34}$$

We see that the semigroup $S(t) = e^{-tM}$ generated by the linear operator $-M$ is a key element of the variation of constants formula (3.34).

Many PDEs, such as reaction–diffusion equations, have a semilinear structure and a solution that can be written using a variation of constants formula.

Example 3.24 (PDE with Dirichlet boundary conditions) For $f\colon \mathbb{R} \to \mathbb{R}$, consider the reaction–diffusion equation

$$u_t = \Delta u + f(u), \tag{3.35}$$

with homogeneous Dirichlet boundary conditions on a bounded domain D and an initial condition $u(0,\boldsymbol{x}) = u_0(\boldsymbol{x})$ for $\boldsymbol{x} \in D$. The solution $u(t,\boldsymbol{x})$ is a function of $t > 0$ and $\boldsymbol{x} \in D$ and, for the semigroup theory, we interpret the solution as a function $u\colon [0,\infty) \to L^2(D)$ and write (3.35) as a semilinear ODE on the space $L^2(D)$. That is,

$$\frac{du}{dt} = -Au + f(u), \tag{3.36}$$

where $A = -\Delta$ with $\mathcal{D}(A) = H^2(D) \cap H_0^1(D)$. Here, $-A$ is the infinitesimal generator of the semigroup $S(t) = e^{-tA}$ on $H = L^2(D)$ (see Example 3.17).

Example 3.25 (PDE with periodic boundary conditions) Consider the reaction–diffusion equation (3.35) on $D = (0,a)$ or $D = (0,a_1) \times (0,a_2)$ with periodic boundary conditions instead of Dirichlet conditions. Using Example 3.18, we may rewrite (3.35) as an ODE on $H = L^2(D)$ by taking $A = -\Delta$ with domain $\mathcal{D}(A) = H^2_{\text{per}}(D)$. This gives the semilinear ODE

$$\frac{du}{dt} = -Au + f(u), \tag{3.37}$$

on $L^2(D)$. Even though A does not satisfy Assumption 3.19, $S(t) = e^{-tA}$ is a well-defined semigroup on $L^2(D)$.

Example 3.26 (Langmuir reaction) A specific model of reaction kinetics and diffusion of a chemical is given by

$$u_t = \varepsilon\,\Delta u - \frac{\alpha u}{1+u^2}, \qquad u(0) = u_0 \in L^2(D), \tag{3.38}$$

where $\varepsilon > 0$ and $\alpha \in \mathbb{R}$ are modelling parameters. In the notation of (3.36) with Dirichlet boundary conditions, $A = -\varepsilon\,\Delta$ with $\mathcal{D}(A) = H^2(D) \cap H_0^1(D)$ and $f(u) = -\alpha u/(1+u^2)$.

The above reaction–diffusion equations are examples of semilinear evolution equations on a Hilbert space. Boundary conditions are not stated explicitly, rather incorporated into the operator A.

Definition 3.27 (semilinear evolution equation) A semilinear evolution equation on a Hilbert space H is a differential equation of the form

$$\frac{du}{dt} = -Au + f(u), \qquad u(0) = u_0 \in H \tag{3.39}$$

where A satisfies Assumption 3.19 and $f\colon H \to H$.

The semigroup $S(t) = \mathrm{e}^{-tA}$ generated by $-A$ gives rise to the following variation of constants formula for (3.39):

$$u(t) = \mathrm{e}^{-tA}u_0 + \int_0^t \mathrm{e}^{-(t-s)A} f(u(s))\,ds \tag{3.40}$$

(see also (3.34) for ODEs). The variation of constants formula is central to the idea of *mild solutions*.

Definition 3.28 (mild solution) A *mild solution* of (3.39) on $[0,T]$ is a function $u \in \mathrm{C}([0,T],H)$ such that (3.40) holds for $0 \le t \le T$.

The following theorem establishes existence and uniqueness of the mild solution to (3.39) and mirrors Theorem 3.2 for ODEs. Indeed, the method of proof is similar and is based on the contraction mapping theorem (Theorem 1.10) and Gronwall's inequality (see Exercise 3.4).

Theorem 3.29 *Suppose that Assumption 3.19 holds, the initial data $u_0 \in H$, and $f\colon H \to H$ satisfies, for a constant $L > 0$,*

$$\begin{aligned} \|f(u)\| &\le L\big(1 + \|u\|\big), \\ \|f(u_1) - f(u_2)\| &\le L\|u_1 - u_2\|, \qquad \forall u, u_1, u_2 \in H. \end{aligned} \tag{3.41}$$

Then, there exists a unique mild solution $u(t)$ to (3.39) for $t \ge 0$. Further, for $T > 0$, there exists $K_T > 0$ such that, for any $u_0 \in H$,

$$\|u(t)\| \le K_T\big(1 + \|u_0\|\big), \qquad 0 \le t \le T. \tag{3.42}$$

Proof Consider the Banach space $X := \mathrm{C}([0,T],H)$ with norm $\|u\|_X := \sup_{0\le t\le T}\|u(t)\|$ and define, for $u \in X$,

$$(\mathcal{J}u)(t) := \mathrm{e}^{-tA}u_0 + \int_0^t \mathrm{e}^{-(t-s)A} f(u(s))\,ds.$$

Fixed points of $\mathcal{J}$ are mild solutions (see Definition 3.28). We show that $\mathcal{J}$ maps into X and is a contraction. By the triangle inequality, (1.4), and Lemma 1.54,

$$\begin{aligned} \|(\mathcal{J}u)(t)\| &\le \|\mathrm{e}^{-tA}u_0\| + \left\|\int_0^t \mathrm{e}^{-(t-s)A} f(u(s))\,ds\right\| \\ &\le \|\mathrm{e}^{-tA}\|_{\mathcal{L}(H)}\|u_0\| + \int_0^t \|\mathrm{e}^{-(t-s)A}\|_{\mathcal{L}(H)}\|f(u(s))\|\,ds. \end{aligned}$$

As $\|\mathrm{e}^{-tA}\|_{\mathcal{L}(H)} \le 1$ and $\|f(u(s))\| \le L(1 + \|u(s)\|)$, we see

$$\|(\mathcal{J}u)(t)\| \le \|u_0\| + \int_0^t L\big(1 + \|u(s)\|\big)\,ds. \tag{3.43}$$

For $u \in X$, we have that $\|\mathcal{J}u\|_X := \sup_{t\in[0,T]}\|(\mathcal{J}u)(t)\|$ is finite. Similarly, we can show that $\mathcal{J}u\colon [0,T] \to H$ is continuous and hence $\mathcal{J}u \in X$.

To show $\mathcal{J}$ is a contraction, note that for $u_1, u_2 \in X$

$$\begin{aligned}\|(\mathcal{J}u_1)(t) - (\mathcal{J}u_2)(t)\| &= \left\|\int_0^t \mathrm{e}^{-(t-s)A}\big(f(u_1(s)) - f(u_2(s))\big)\, ds\right\| \\ &\le \int_0^t L\|u_1(s) - u_2(s)\|\, ds \\ &\le LT\|u_1 - u_2\|_X, \qquad 0 \le t \le T.\end{aligned}$$

We have used (1.4), (3.41), and $\|\mathrm{e}^{-tA}\|_{\mathcal{L}(H)} \le 1$. Hence, $\|\mathcal{J}u_1 - \mathcal{J}u_2\|_X \le LT\|u_1 - u_2\|_X$ and $\mathcal{J}$ is a contraction if $LT < 1$.

For small T, the contraction mapping theorem (Theorem 1.10) applies to $\mathcal{J}$ on X and $\mathcal{J}$ has a unique fixed point in X and, in particular, there exists a unique mild solution on $[0,T]$. As f satisfies the Lipschitz condition (3.41) uniformly over H, the argument can be repeated over the time intervals $[T,2T]$, $[2T,3T]$, etc., and a unique mild solution exists for all time $t > 0$. Finally, by (3.43),

$$\|u(t)\| \le \|u_0\| + \int_0^t L(1 + \|u(s)\|)\, ds.$$

Gronwall's inequality (see Exercise 3.4) completes the proof of (3.42). □

The most common class of nonlinearities $f\colon H \to H$ are the so-called *Nemytskii operators* defined for $H = L^2(D)$. Here Lipschitz continuous means globally Lipschitz continuous, as in Definition 1.14.

Lemma 3.30 (Nemytskii) *Suppose that $\tilde{f}\colon \mathbb{R} \to \mathbb{R}$ is Lipschitz continuous. If the domain D is bounded, $f(u)(\boldsymbol{x}) := \tilde{f}(u(\boldsymbol{x}))$ is a well-defined Lipschitz function $f\colon L^2(D) \to L^2(D)$. It is common to use the same notation f for both $\tilde{f}$ and f.*

Proof We must show f is well defined by showing $f(u) \in L^2(D)$ for $u \in L^2(D)$. Since $\tilde{f}$ is Lipschitz, we can find L so that

$$\|f(u)\|^2_{L^2(D)} = \int_D f(u)(\boldsymbol{x})^2\, d\boldsymbol{x} = \int_D \tilde{f}(u(\boldsymbol{x}))^2\, d\boldsymbol{x} \le L^2 \int_D \big(1 + u(\boldsymbol{x})\big)^2\, d\boldsymbol{x} < \infty.$$

Hence $f\colon L^2(D) \to L^2(D)$ and satisfies the linear growth condition (3.41). A similar argument shows that f is Lipschitz continuous. □

The Langmuir nonlinearity f in (3.38) is Lipschitz (see Exercise 3.15) and gives a well-defined Nemytskii operator. However, the nonlinearity in the Allen–Cahn equation (3.3) does not and different techniques are required.

We finish this section by describing the regularity of solutions to (3.39).

Proposition 3.31 (regularity in time) *Let the assumptions of Theorem 3.29 hold and the initial data $u_0 \in \mathcal{D}(A^\gamma)$ for some $\gamma \in [0,1]$. For $T > 0$ and $\epsilon > 0$, there exists a constant $K_{RT} > 0$ such that the solution $u(t)$ of (3.39) satisfies*

$$\|u(t_2) - u(t_1)\| \le K_{RT}(t_2 - t_1)^\theta\big(1 + \|u_0\|_\gamma\big),$$

for $0 \le t_1 \le t_2 \le T$ and $\theta := \min\{\gamma, 1 - \epsilon\}$.

Proof Write $u(t_2) - u(t_1) = \mathtt{I} + \mathtt{II}$, for

$$\mathtt{I} := \left(e^{-t_2 A} - e^{-t_1 A}\right)u_0, \qquad \mathtt{II} := \int_0^{t_2} e^{-(t_2-s)A} f(u(s))\,ds - \int_0^{t_1} e^{-(t_1-s)A} f(u(s))\,ds.$$

We estimate each term. For the first term, $\mathtt{I} = e^{-t_1 A} A^{-\gamma}\left(e^{-(t_2-t_1)A} - I\right)A^{\gamma}u_0$ from Lemma 3.22(i) and $e^{-A(t+s)} = e^{-At}e^{-As}$. Using properties of the operator norm in Lemma 1.54,

$$\|\mathtt{I}\| \le \left\|A^{-\gamma}\left(I - e^{-(t_2-t_1)A}\right)\right\|_{\mathcal{L}(H)}\left\|A^{\gamma}u_0\right\|$$

and, by Lemma 3.22(iii), there exists $C_{\mathtt{I}} > 0$ such that $\|\mathtt{I}\| \le C_{\mathtt{I}}(t_2 - t_1)^{\gamma}\|u_0\|_{\gamma}$.

For the second term, write $\mathtt{II} = \mathtt{II}_1 + \mathtt{II}_2$ for

$$\mathtt{II}_1 := \int_0^{t_1} \left(e^{-(t_2-s)A} - e^{-(t_1-s)A}\right) f(u(s))\,ds, \qquad \mathtt{II}_2 := \int_{t_1}^{t_2} e^{-(t_2-s)A} f(u(s))\,ds.$$

We estimate the norms of each term $\mathtt{II}_1$ and $\mathtt{II}_2$. For $\mathtt{II}_1$,

$$\begin{aligned}\|\mathtt{II}_1\| &= \left\|\int_0^{t_1} \left(e^{-(t_2-s)A} - e^{-(t_1-s)A}\right) f(u(s))\,ds\right\| \\ &\le \int_0^{t_1} \left\|e^{-(t_2-s)A} - e^{-(t_1-s)A}\right\|_{\mathcal{L}(H)} ds \sup_{0\le s\le t_1}\|f(u(s))\|.\end{aligned}$$

Note that $e^{-(t_2-s)A} - e^{-(t_1-s)A} = A^{-(1-\epsilon)}\left(e^{-(t_2-t_1)A} - I\right)A^{1-\epsilon}e^{-(t_1-s)A}$. Then,

$$\begin{aligned}&\int_0^{t_1} \left\|e^{-(t_2-s)A} - e^{-(t_1-s)A}\right\|_{\mathcal{L}(H)} ds \\ &\quad\le \int_0^{t_1} \left\|A^{-(1-\epsilon)}\left(I - e^{-(t_2-t_1)A}\right)\right\|_{\mathcal{L}(H)}\left\|A^{1-\epsilon}e^{-(t_1-s)A}\right\|_{\mathcal{L}(H)} ds \\ &\quad\le K_1(t_2-t_1)^{1-\epsilon}\int_0^{t_1} K_2(t_1-s)^{-(1-\epsilon)}\,ds \le K_1K_2\,\frac{t_1^{\epsilon}}{\epsilon}\,(t_2-t_1)^{1-\epsilon},\end{aligned}$$

where K_1, K_2 are constants given by Lemma 3.22(ii)–(iii). Thus, for $C_0 := K_1\,K_2\,T^{\epsilon}/\epsilon$,

$$\|\mathtt{II}_1\| \le C_0(t_2-t_1)^{1-\epsilon}\sup_{0\le s\le t_1}\|f(u(s))\|.$$

For $\mathtt{II}_2$,

$$\begin{aligned}\|\mathtt{II}_2\| &= \left\|\int_{t_1}^{t_2} e^{-(t_2-s)A} f(u(s))\,ds\right\| \\ &\le \int_{t_1}^{t_2} \left\|e^{-(t_2-s)A} f(u(s))\right\| ds \le (t_2-t_1)\sup_{t_1\le s\le t_2}\|f(u(s))\|.\end{aligned}$$

As f satisfies the Lipschitz condition (3.41) and $0 \le t_1 \le t_2 \le T$, we have

$$\begin{aligned}\|\mathtt{II}\| \le \|\mathtt{II}_1\| + \|\mathtt{II}_2\| &\le (C_0+1)(t_2-t_1)^{1-\epsilon}L\left(1 + \sup_{0\le s\le T}\|u(s)\|\right) \\ &\le C_{\mathtt{II}}(t_2-t_1)^{1-\epsilon}\left(1 + \|u_0\|\right),\end{aligned}$$

for $C_{\mathtt{II}} := (C_0 + 1)L(1 + K_T)$ where K_T is given by (3.42). Finally, $\|u(t_2) - u(t_1)\| \le \|\mathtt{I}\| + \|\mathtt{II}\|$ and the estimates on the norms of $\mathtt{I}$ and $\mathtt{II}$ complete the proof. □

3.4 Method of lines and finite differences for semilinear PDEs

The method of lines applies to a broad class of PDEs, including semilinear evolution equations like (3.39), and gives a system of ODEs that approximates the PDE. The resulting ODEs can be approximated in time by the Euler methods to find a practical computational method. In §3.5, the method of lines is used with the Galerkin method. In the present section, we take a simpler approach and apply the method of lines to a reaction–diffusion equation using finite differences.

Specifically, consider the PDE

$$u_t = \varepsilon\, u_{xx} + f(u), \qquad u(0,x) = u_0(x), \tag{3.44}$$

on the domain $D = (0,a)$ for $\varepsilon > 0$, initial data $u_0 \colon [0,a] \to \mathbb{R}$, homogeneous Dirichlet boundary conditions, and a reaction term $f \colon \mathbb{R} \to \mathbb{R}$. Introduce a grid of $J+1$ uniformly spaced points $x_j = jh$ on $[0,a]$ for $j = 0,\dots,J$ and $h = a/J$. By Taylor's theorem (Theorem A.1) for $u \in \mathrm{C}^4(\mathbb{R})$, we have, for $j = 1,\dots,J-1$,

$$u_{xx}(t,x_j) = \frac{u(t,x_{j+1}) - 2u(t,x_j) + u(t,x_{j-1})}{h^2} + r_j(t), \tag{3.45}$$

where the remainder r_j is $\mathcal{O}(h^2)$. At each grid point x_j, (3.44) becomes

$$\frac{d}{dt}u(t,x_j) = \varepsilon\left(\frac{u(t,x_{j+1}) - 2u(t,x_j) + u(t,x_{j-1})}{h^2}\right) + f(u(t,x_j)) + r_j(t). \tag{3.46}$$

Let $\boldsymbol{u}(t) \in \mathbb{R}^{J-1}$ be the vector with components $u(t,x_j)$, $j = 1,\dots,J-1$. Then, (3.46) gives the system of ODEs

$$\frac{d\boldsymbol{u}}{dt} = -\varepsilon\, A^{\mathrm{D}}\boldsymbol{u} + \boldsymbol{f}(\boldsymbol{u}) + \boldsymbol{r}(t), \tag{3.47}$$

where $\boldsymbol{f}(\boldsymbol{u}) = [f(u_1),\dots,f(u_{J-1})]^{\mathsf{T}}$, A^{D} is the $(J-1)\times(J-1)$ matrix

$$A^{\mathrm{D}} := \frac{1}{h^2}\begin{pmatrix} 2 & -1 & & & \\ -1 & 2 & -1 & & \\ & -1 & 2 & -1 & \\ & & \ddots & \ddots & -1 \\ & & & -1 & 2 \end{pmatrix},$$

and the remainder term is $\boldsymbol{r}(t) = [r_1(t),\dots,r_{J-1}(t)]^{\mathsf{T}}$. The definition of A^{D} is specific to Dirichlet boundary conditions. The so-called *centred* finite difference approximation is obtained by neglecting the remainder term in (3.46). We denote by $u_j(t)$ the resulting approximation to $u(t,x_j)$ and set $\boldsymbol{u}_J(t) := [u_1(t),\dots,u_{J-1}(t)]^{\mathsf{T}}$. The method of lines for the finite difference approximation is the following semilinear system of ODEs:

$$\frac{d\boldsymbol{u}_J}{dt} = -\varepsilon\, A^{\mathrm{D}}\boldsymbol{u}_J + \boldsymbol{f}(\boldsymbol{u}_J) \tag{3.48}$$

with initial data $\boldsymbol{u}_J(0) = [u_1(0),\dots,u_{J-1}(0)]^{\mathsf{T}}$.

The method can be applied to problems with other boundary conditions. Periodic conditions are considered in Exercise 3.13 and non-homogeneous Dirichlet conditions are considered in Exercise 3.12. Next, we briefly consider Neumann boundary conditions.

Example 3.32 (Neumann boundary conditions) For a given function g on the boundary ∂D of the domain D, a Neumann boundary condition is of the form

$$\frac{\partial u(\boldsymbol{x})}{\partial \boldsymbol{n}} = \nabla u(\boldsymbol{x}) \cdot \boldsymbol{n}(\boldsymbol{x}) = g(\boldsymbol{x}), \qquad \boldsymbol{x} \in \partial D,$$

where $\boldsymbol{n}$ is the unit outward normal vector to the boundary. On $D = (0,a)$, this simply means that we specify u_x at $x = 0$ and $x = a$. In contrast to the Dirichlet case, $u(t,0)$ and $u(t,a)$ are now unknown and we solve for $\boldsymbol{u}_J(t) = [u_0(t), u_1(t), \ldots, u_J(t)]^{\mathsf{T}}$ to approximate $[u(t,x_0), u(t,x_1), \ldots, u(t,x_J)]^{\mathsf{T}}$. To implement the boundary condition, we use the second-order approximation

$$u_x(t,x) = \frac{u(t,x+h) - u(t,x-h)}{2h} + \mathcal{O}(h^2), \tag{3.49}$$

which may be proved by Taylor's theorem for $u \in \mathrm{C}^3(\mathbb{R})$. To interpret (3.49) at the boundary, introduce two fictitious points, $x_{-1} = -h$ and $x_{J+1} = a + h$, which are then eliminated using the Neumann condition. For example, suppose $g = 0$ at $x = 0$ and $x = a$. By approximating $u_x(t,0)$ and $u_x(t,a)$ by (3.49), neglecting the $\mathcal{O}(h^2)$ term, and applying the boundary condition, we find

$$\frac{u(t,h) - u(t,-h)}{2h} \approx 0 \quad \text{and} \quad \frac{u(t,(J+1)h) - u(t,(J-1)h)}{2h} \approx 0,$$

and we assert that $u_{-1}(t) = u_1(t)$ and $u_{J+1}(t) = u_{J-1}(t)$. The approximation $u_0(t)$ to $u(t,0)$ satisfies

$$\frac{du_0(t)}{dt} = \varepsilon \frac{2u_1(t) - 2u_0(t)}{h^2} + f(u_0(t)).$$

Similarly, the approximation $u_J(t)$ to $u(t,a)$ satisfies

$$\frac{du_J(t)}{dt} = \varepsilon \frac{2u_{J-1}(t) - 2u_J(t)}{h^2} + f(u_J(t)).$$

Importantly, the second-order approximation to u_x in (3.49) preserves the overall error of $\mathcal{O}(h^2)$. We obtain the following system in $\mathbb{R}^{J+1}$

$$\frac{d\boldsymbol{u}_J}{dt} = -\varepsilon\, A^{\mathrm{N}} \boldsymbol{u}_J + \boldsymbol{f}(\boldsymbol{u}_J), \tag{3.50}$$

where A^{N} is now the $(J+1) \times (J+1)$ matrix

$$A^{\mathrm{N}} := \frac{1}{h^2} \begin{pmatrix} 2 & -2 & & & \\ -1 & 2 & -1 & & \\ & -1 & 2 & -1 & \\ & & \ddots & \ddots & -1 \\ & & & -2 & 2 \end{pmatrix}. \tag{3.51}$$

Once a semilinear system of ODEs (3.50) has been obtained, it can be discretised in time. We choose the semi-implicit time-stepping method (3.19). The region of absolute stability (see Figure 3.3) of this method includes $\{\lambda \Delta t \in \mathbb{C} : \operatorname{Re} \lambda < 0\}$, which contains all eigenvalues of $-A = \partial_{xx}$, and the method does not introduce an awkward restriction on the time step. In Algorithm 3.4, we have implemented this method and we may set homogeneous Dirichlet or Neumann, as well as periodic boundary conditions.

Algorithm 3.4 Code to compute the semi-implicit Euler and centred finite difference approximation to (3.44). The inputs are the initial vector `u0`$= [u_0(x_0), \dots, u_0(x_J)]^{\mathsf{T}}$, the final time `T`, the domain size a, number of steps `N` (so that $\Delta t = T/N$), the dimension `J` (so that $h = a/J$), the parameter ε, and a function handle `fhandle` to evaluate $\boldsymbol{f}(\boldsymbol{u})$. The input `bctype='D'`, `'N'`, or `'P'` indicates homogeneous Dirichlet, homogeneous Neumann, or periodic boundary conditions, respectively. The outputs are `t`$= [t_0, \dots, t_N]^{\mathsf{T}}$ and a matrix `ut`, with columns approximating $[u(t_n, x_0), \dots, u(t_n, x_J)]^{\mathsf{T}}$, $n = 0, \dots, N$.

```
function [t,ut]=pde_fd(u0,T,a,N,J,epsilon,fhandle,bctype)
Dt=T/N;  t=[0:Dt:T]'; h=a/J;
% set matrix A according to boundary conditions
e=ones(J+1,1);   A=spdiags([-e 2*e -e], -1:1, J+1, J+1);
switch lower(bctype)
  case {'dirichlet','d'}
    ind=2:J;  A=A(ind,ind);
  case {'periodic','p'}
    ind=1:J; A=A(ind,ind); A(1,end)=-1; A(end,1)=-1;
  case {'neumann','n'}
    ind=1:J+1; A(1,2)=-2; A(end,end-1)=-2;
end
EE=speye(length(ind))+Dt*epsilon*A/h^2;
ut=zeros(J+1,length(t)); % initialize vectors
ut(:,1)=u0; u_n=u0(ind); % set initial condition
for k=1:N, % time loop
  fu=fhandle(u_n); % evaluate f(u_n)
  u_new=EE\(u_n+Dt*fu); % linear solve for (1+epsilon A)
  ut(ind,k+1)=u_new; u_n=u_new;
end
if lower(bctype)=='p' | lower(bctype)=='periodic'
    ut(end,:)=ut(1,:); % correct for periodic case
end
```

Example 3.33 (Nagumo equation) Consider the Nagumo equation

$$u_t = \varepsilon u_{xx} + u(1-u)(u-\alpha), \tag{3.52}$$

on $D = (0, a)$ with homogeneous Neumann boundary conditions. Choose parameters $a = 20$, $\varepsilon = 1$, $\alpha = -1/2$, and initial data $u_0(x) = \left(1 + \exp(-(2-x)/\sqrt{2})\right)^{-1}$. We call Algorithm 3.4 (and set `bctype='N'`) to solve the PDE numerically using the semi-implicit Euler method in time and centred finite differences in space. For example, to compute an approximation on $[0, T] = [0, 10]$ with $h = a/512$ and $\Delta t = T/N = 0.001$, we use the following MATLAB commands:

```
>> T=10; N=10^4; epsilon=1;
>> a=20; J=512; h=a/J; x=(0:h:a)';
>> u0=1./(1+exp(-(2-x)/sqrt(2)));
>> [t,ut]=pde_fd(u0,T,a,N,J,epsilon,@(u) u.*(1-u).*(u+0.5),'N');
```

A plot of the resulting numerical solution is given in Figure 3.4(b).

We discuss the error in the finite difference approximation (3.48) to the solution of (3.44) for homogeneous Dirichlet conditions. To measure the error $\boldsymbol{e}(t) = \boldsymbol{u}(t) - \boldsymbol{u}_J(t)$, we use the

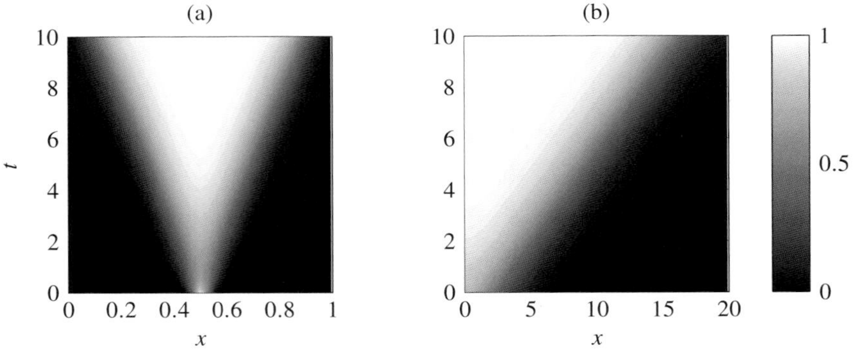

Figure 3.4 (a) A finite element approximation of the Nagumo equation for $t \in [0, 10]$ on $D = (0, 1)$ with homogeneous Dirichlet boundary conditions (see Example 3.44). (b) A finite difference solution of the Nagumo equation on $D = (0, 20)$ with homogeneous Neumann boundary conditions for $t \in [0, 10]$ (see Example 3.33).

norm on $\mathbb{R}^{J-1}$ defined by

$$\|\boldsymbol{e}(t)\|_{2,h} := \left(\sum_{j=1}^{J-1} e_j^2(t)\, h \right)^{1/2}. \tag{3.53}$$

This is equivalent to taking piecewise constant interpolants $\tilde{u}(t), u_J(t) \colon [0, a] \to \mathbb{R}$ of $\boldsymbol{u}(t)$ and $\boldsymbol{u}_J(t)$ and measuring the error by the $L^2(0, a)$ norm of $\tilde{u}(t) - u_J(t)$.

Theorem 3.34 *For a time interval* $[0, T]$*, consider* (3.44) *with homogeneous Dirichlet boundary conditions on* $D = (0, a)$*. Suppose that* $f \colon \mathbb{R} \to \mathbb{R}$ *is Lipschitz continuous and the solution* $u(t, x)$ *has four uniformly bounded spatial derivatives, so that* $u \in \mathrm{C}([0, T], \mathrm{C}^4(\bar{D}))$*. Then, for* $h = a/J$ *and some* $K > 0$ *independent of* J*,*

$$\sup_{0 \le t \le T} \|\boldsymbol{u}(t) - \boldsymbol{u}_J(t)\|_{2,h} \le K h^2$$

where $\boldsymbol{u}(t) = [u(t, x_1), \ldots, u(t, x_{J-1})]^{\mathsf{T}}$ *and* $\boldsymbol{u}_J(t)$ *is the solution to* (3.48).

Proof See Exercise 3.14. □

See also Theorem 5.13. The assumption that $u(t) \in C^4(\bar{D})$ in Theorem 3.34 is a strong requirement. For many stochastic PDEs, the forcing is too irregular and this level of regularity does not hold.

3.5 Galerkin methods for semilinear PDEs

Galerkin methods, which we first saw in Chapter 2, allow greater flexibility in spatial discretisation than finite differences and their application to semilinear PDEs fits well with our theoretical framework of ODEs on a Hilbert space H.

First, recall the semilinear evolution equation (3.39)

$$\frac{du}{dt} = -Au + f(u), \qquad u(0) = u_0 \in H, \tag{3.54}$$

where A satisfies Assumption 3.19 and generates an analytic semigroup e^{-At} and $f\colon H \to H$. The Galerkin method is based on the weak formulation. Before writing this down, observe that, in general, the mild solution u and its derivative du/dt belong to the following fractional power spaces.

Lemma 3.35 *Let the assumptions of Theorem 3.29 hold and $u(t)$ be the mild solution of* (3.54). *Then,*

$$u(t) \in \mathcal{D}(A^{1/2}) \quad \text{and} \quad \frac{du(t)}{dt} \in \mathcal{D}(A^{-1/2}), \qquad \forall t > 0.$$

Proof Using the definition of mild solution (Definition 3.28),

$$A^{1/2}u(t) = A^{1/2}\mathrm{e}^{-At}u_0 + \int_0^t A^{1/2}\mathrm{e}^{-A(t-s)} f(u(s))\,ds.$$

Using Lemma 3.22, we can show $\|A^{1/2}u(t)\| < \infty$. For example, for some $K > 0$,

$$\begin{aligned}\left\|\int_0^t A^{1/2}\mathrm{e}^{-A(t-s)} f(u(s))\,ds\right\| &\le \int_0^t \|A^{1/2}\mathrm{e}^{-A(t-s)}\|_{\mathcal{L}(H)} \|f(u(s))\|\,ds \\ &\le \int_0^t \frac{K}{(t-s)^{1/2}} \sup_{0\le s\le t} \|f(u(s))\|\,ds < \infty\end{aligned}$$

using (3.41) and (3.42). This gives $u(t) \in \mathcal{D}(A^{1/2})$. Similarly, by differentiating the variation of constants formula (3.40), we have

$$\frac{du(t)}{dt} = -A\mathrm{e}^{-At}u_0 - \int_0^t A\mathrm{e}^{-A(t-s)} f(u(s))\,ds + f(u(t)),$$

and this can be used to show $\frac{du}{dt} \in \mathcal{D}(A^{-1/2})$. □

This lemma does not provide enough smoothness of $u(t)$ to interpret (3.54) directly, as we do not know whether $u(t) \in \mathcal{D}(A)$. We can, however, develop the weak form by using test functions $v \in \mathcal{D}(A^{1/2})$. Taking the inner product of (3.54) with v gives

$$\left\langle \frac{du}{dt}, v\right\rangle = -\langle Au, v\rangle + \langle f(u), v\rangle. \tag{3.55}$$

Lemma 1.89 gives

$$\left\langle \frac{du}{dt}, v\right\rangle = \left\langle A^{-1/2}\frac{du}{dt}, A^{1/2}v\right\rangle \quad \text{and} \quad \langle Au, v\rangle = \langle A^{1/2}u, A^{1/2}v\rangle$$

for $\frac{du}{dt} \in H$, $u \in \mathcal{D}(A)$, and $v \in \mathcal{D}(A^{1/2})$. Even if the conditions on $\frac{du}{dt}$ and u are not met, the two expressions on the right-hand side are well defined for $t > 0$ due to Lemma 3.35. This leads to the following definition of weak solution. In the following, V^* is often called the dual space of $D(A^{1/2})$ (it can be identified with the dual space given in Definition 1.56).

Definition 3.36 (weak solution) Let $V := \mathcal{D}(A^{1/2})$ and $V^* := \mathcal{D}(A^{-1/2})$. We say that $u\colon [0,T] \to V$ is a weak solution of (3.54) if, for almost all $s \in [0,T]$, we have $du(s)/dt \in V^*$ and

$$\left\langle \frac{du(s)}{dt}, v \right\rangle = -a(u(s),v) + \big\langle f(u(s)), v \big\rangle, \qquad \forall v \in V, \tag{3.56}$$

where $a(u,v) := \big\langle A^{1/2}u, A^{1/2}v \big\rangle \equiv \langle u,v \rangle_{1/2}$ for $u,v \in V$.

This weak formulation leads to the Galerkin approximation. First, introduce a finite-dimensional subspace $\widetilde{V} = \mathrm{span}\{\psi_1,\psi_2,\dots,\psi_J\} \subset V$. We then seek $\tilde{u}(t) = \sum_{j=1}^{J} \hat{u}_j(t)\psi_j$ such that

$$\left\langle \frac{d\tilde{u}}{dt}, v \right\rangle = -a(\tilde{u},v) + \big\langle f(\tilde{u}), v \big\rangle, \qquad \forall v \in \widetilde{V}, \tag{3.57}$$

which is (3.56) with V replaced by $\widetilde{V}$ and we do not write the (s). Equivalently,

$$\left\langle \frac{d\tilde{u}}{dt}, \psi_j \right\rangle = -a(\tilde{u},\psi_j) + \big\langle f(\tilde{u}), \psi_j \big\rangle, \qquad j = 1,\dots,J. \tag{3.58}$$

It is convenient to introduce the orthogonal projection $\widetilde{P}\colon H \to \widetilde{V}$ (see Definition 1.38 and which should not be confused with the Galerkin projection given in Definition 2.19). In this notation, (3.58) is written as

$$\frac{d\tilde{u}}{dt} = -\widetilde{A}\,\tilde{u} + \widetilde{P} f(\tilde{u}) \tag{3.59}$$

for $\widetilde{A}\colon \widetilde{V} \to \widetilde{V}$ defined by

$$\big\langle \widetilde{A}w, v \big\rangle = a(w,v), \qquad \forall w,v \in \widetilde{V}. \tag{3.60}$$

We normally choose initial data $\tilde{u}(0) = \widetilde{P}u_0$. This is an example of the method of lines, where the ODE (3.54) is approximated by a finite-dimensional ODE (3.59). Euler methods apply to (3.59) to give a time discretisation.

Spectral Galerkin approximation

Let A satisfy Assumption 3.19, so that the eigenfunctions ϕ_j form a basis for H. We use the eigenfunctions ϕ_j for ψ_j to define the Galerkin subspace $\widetilde{V} = V_J := \mathrm{span}\{\phi_1,\dots,\phi_J\}$ and the orthogonal projection $P_J\colon H \to V_J$ defined by

$$P_J u := \sum_{j=1}^{J} \hat{u}_j \phi_j, \qquad \text{where } \hat{u}_j := \frac{1}{\|\phi_j\|^2}\langle u, \phi_j \rangle \text{ for all } u \in H. \tag{3.61}$$

We write $A_J = \widetilde{A}$ and $u_J = \tilde{u}$ and the resulting approximation (3.59) is

$$\frac{du_J}{dt} = -A_J u_J + P_J f(u_J), \qquad u_J(0) = P_J u_0. \tag{3.62}$$

In this case, $A_J = P_J A$, as ϕ_j are eigenfunctions of A. The solution $u_J(t)$ is known as the spectral Galerkin approximation to the solution $u(t)$ of (3.54). For $j = 1,\dots,J$,

$$\frac{d\hat{u}_j}{dt} = -\lambda_j \hat{u}_j + \hat{f}_j(u_J), \qquad \hat{u}_j(0) = \hat{u}_{0,j}, \tag{3.63}$$

where $\hat{u}_{0,j}$, $\hat{u}_j(t)$, $\hat{f}_j(u)$ are the coefficients of u_0, $u_J(t)$ and $f(u)$ in the basis ϕ_j, respectively. If we let $\hat{u}_J(t) := [\hat{u}_1(t), \hat{u}_2(t), \dots, \hat{u}_J(t)]^{\mathsf{T}}$, then (3.63) gives the semilinear system of ODEs

$$\frac{d\hat{\boldsymbol{u}}_J}{dt} = -M\hat{\boldsymbol{u}}_J + \hat{\boldsymbol{f}}_J(\hat{\boldsymbol{u}}_J) \tag{3.64}$$

where $\hat{\boldsymbol{f}}_J(\hat{\boldsymbol{u}}_J) := [\hat{f}_1(u_J), \hat{f}_2(u_J), \dots, \hat{f}_J(u_J)]^{\mathsf{T}}$ and M is a diagonal matrix with entries $m_{jj} = \lambda_j$ (see also Examples 3.39 and 3.40). We stress that (3.62) describes the evolution of the function $u_J(t)$ whereas, equivalently, the system of ODEs (3.64) describes the evolution of the vector of coefficients $\hat{\boldsymbol{u}}_J(t)$.

At this point, we analyse the error in the Galerkin approximation (3.62) of the linear evolution equation (3.54) with $f = 0$ in the norm $\|\cdot\|$ on H.

Lemma 3.37 (spatial error, linear equation) *Suppose that Assumption 3.19 holds. Let $u(t)$ satisfy* (3.54) *and $u_J(t)$ satisfy* (3.62) *with $f = 0$. If $u_0 \in \mathcal{D}(A^\gamma)$ for some $\gamma \ge 0$, then*

$$\|u(t) - u_J(t)\| \le \frac{1}{\lambda_{J+1}^\gamma}\|u_0\|_\gamma, \qquad t \ge 0. \tag{3.65}$$

Proof A_J is the infinitesimal generator of a semigroup e^{-tA_J} and it is straightforward to see $e^{-tA_J} = e^{-tA}P_J$. We then have $u(t) = e^{-tA}u_0$ and $u_J(t) = e^{-tA_J}P_J\, u_0$. The error $u(t) - u_J(t) = (e^{-tA} - e^{-tA_J}P_J)u_0 = e^{-tA}(I - P_J)u_0$, so that

$$\|u(t) - u_J(t)\| \le \|(I - P_J)A^{-\gamma}\|_{\mathcal{L}(H)}\|A^\gamma u_0\| \le \frac{1}{\lambda_{J+1}^\gamma}\|u_0\|_\gamma,$$

where we have used Lemma 1.94. □

An error analysis for the semilinear evolution equation is covered in §3.7.

Example 3.38 (heat equation) Consider the heat equation in Example 3.17 where the eigenvalues of A are given by $\lambda_j = j^2$. Then, $H = L^2(0,\pi)$ and Lemma 3.37 gives convergence of the spectral Galerkin approximation in space for initial data $u_0 \in \mathcal{D}(A) = H^2(0,\pi) \cap H_0^1(0,\pi)$. Specifically,

$$\sup_{t\ge 0}\|u(t) - u_J(t)\|_{L^2(0,\pi)} = \mathcal{O}(\lambda_{J+1}^{-1}) = \mathcal{O}(J^{-2}).$$

Spectral Galerkin with the semi-implicit Euler method

To obtain a practical computational method, we discretise the ODE (3.62) for $u_J(t)$ in time by the semi-implicit Euler method. We denote the approximation to $u_J(t_n)$ by $u_{J,n}$ for $n = 0, 1, 2, \dots$, where $t_n = n\Delta t$ and Δt is the time step. Equivalently, we may discretise the ODEs (3.64) and (3.63) for the vector $\hat{\boldsymbol{u}}_J$ with components $\hat{u}_j$, $j = 1, \dots, J$, and we denote our approximation to $\hat{\boldsymbol{u}}_J(t_n)$ by $\hat{\boldsymbol{u}}_{J,n}$ and our approximation to $\hat{u}_j(t_n)$ by $\hat{u}_{j,n}$. These are related through

$$u_{J,n} = \sum_{j=1}^{J}\hat{u}_{j,n}\phi_j. \tag{3.66}$$

By applying the semi-implicit Euler method (3.19) to (3.62), $u_{J,n}$ is given by

$$u_{J,n+1} = u_{J,n} - \Delta t\, A_J\, u_{J,n+1} + \Delta t\, P_J\, f(u_{J,n}). \tag{3.67}$$

Equivalently, $\hat{\boldsymbol{u}}_{J,n}$ is defined by

$$\hat{\boldsymbol{u}}_{J,n+1} = \hat{\boldsymbol{u}}_{J,n} - \Delta t\, M\, \hat{\boldsymbol{u}}_{J,n+1} + \Delta t\, \hat{\boldsymbol{f}}_J(\hat{\boldsymbol{u}}_{J,n}),$$

which arises by applying (3.19) to (3.64). Rearranging gives

$$(I + \Delta t\, M)\, \hat{\boldsymbol{u}}_{J,n+1} = \hat{\boldsymbol{u}}_{J,n} + \Delta t\, \hat{\boldsymbol{f}}_J(\hat{\boldsymbol{u}}_{J,n}), \tag{3.68}$$

and so, for each coefficient in (3.63), we have

$$(1 + \Delta t\, \lambda_j)\hat{u}_{j,n+1} = \hat{u}_{j,n} + \Delta t\, \hat{f}_j(u_{J,n}), \qquad j = 1, \dots, J.$$

Equation (3.67) is useful for proving convergence and examining theoretical properties of the numerical method. The system (3.68) is used for the implementation in Algorithm 3.5.

Algorithm 3.5 Code to find the semi-implicit Euler and spectral Galerkin approximation for (3.44). The inputs and outputs are similar to those in Algorithm 3.4. Here J is the dimension of the Galerkin subspace.

```
function [t,ut]=pde_oned_Gal(u0,T,a,N,J,epsilon,fhandle)
Dt=T/N; t=[0:Dt:T]'; ut=zeros(J+1,N+1);
% set linear operator
lambda=2*pi*[0:J/2 -J/2+1:-1]'/a; M= epsilon*lambda.^2;
EE=1./(1+Dt*M); % diagonal of (1+ Dt M)^{-1}
ut(:,1)=u0; u=u0(1:J); uh=fft(u); % set initial condition
for n=1:N, % time loop
    fhu=fft(fhandle(u)); % evaluate fhat(u)
    uh_new=EE.*(uh+Dt*fhu); % semi-implicit Euler step
    u=real(ifft(uh_new)); ut(1:J,n+1)=u; uh=uh_new;
end
ut(J+1,:)=ut(1,:); % make periodic
```

Before we examine convergence, we give an example for the Allen–Cahn equation

$$u_t = \varepsilon\, \Delta u + u - u^3, \qquad u(0) = u_0 \in L^2(D), \tag{3.69}$$

on $D = (0, a)$ and $D = (0, a_1) \times (0, a_2)$ with periodic boundary conditions (see (3.3)) .

Example 3.39 (Allen–Cahn equation in one dimension) To derive the spectral Galerkin method for (3.69) on $(0, a) = (0, 2\pi)$ with periodic boundary conditions, let $A = -\varepsilon\, \partial_{xx}$ with $\mathcal{D}(A) = H^2_{\text{per}}(0, 2\pi)$ and $f(u) = u - u^3$, similar to Example 3.25. In complex notation (see Example 1.84), the eigenfunctions of $-\varepsilon\, \partial_{xx}$ are $\phi_j(x) = e^{ijx}$ and the eigenvalues are $\lambda_j = \varepsilon j^2$ for $j \in \mathbb{Z}$. Then, (3.63) becomes

$$\frac{d\hat{u}_j}{dt} = -\varepsilon j^2 \hat{u}_j + \hat{f}_j(u_J), \tag{3.70}$$

where $\hat{u}_j$, $\hat{f}_j(u_J)$ are the coefficients of u_J and $f(u_J)$ in the basis ϕ_j. This gives a system for $\hat{\boldsymbol{u}}_J(t)$ of the form (3.64) where the diagonal matrix M has entries $m_{jj} = \varepsilon j^2$. We choose J even and order the J elements of $\hat{\boldsymbol{u}}_J$ so that

$$\hat{\boldsymbol{u}}_J(t) = \left[\hat{u}_0(t), \hat{u}_1(t), \dots, \hat{u}_{J/2}(t), \hat{u}_{-J/2+1}(t), \dots, \hat{u}_{-1}(t)\right] \in \mathbb{R}^J.$$

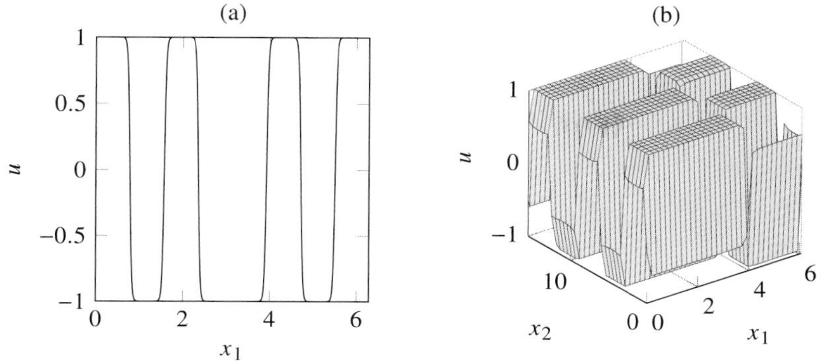

Figure 3.5 Plots of the numerical solution to (3.69) obtained with the semi-implicit Euler method in time and spectral Galerkin method in space at $T = 10$ with $\Delta t = 0.01$ in (a) one dimension (using Algorithm 3.5) with $J = 512$ and (b) two dimensions using Algorithm 3.6 with $J_1 = 128$ and $J_2 = 256$.

This ordering of the $\hat{u}_j$, like the scaling chosen for ϕ_j, is convenient for using the fast Fourier transform (FFT). The semi-implicit Euler method in time applied to (3.70) gives

$$\hat{u}_{j,n+1} = \big(1 + \Delta t\, \varepsilon \lambda_j\big)^{-1}\big(\hat{u}_{j,n} + \Delta t\, \hat{f}_j(u_{J,n})\big) \tag{3.71}$$

for $j = 0, \pm 1, \ldots, \pm(J/2 - 1), J/2$.

For the implementation, the coefficient functions $\hat{f}_j$, and hence $P_J f$, are not easy to compute exactly and we replace P_J with I_J, the trigonometric interpolant defined in (1.40). This approximation is justified when f is sufficiently smooth (see Exercise 1.23). We actually compute $U_{J,n} \approx u_{J,n}$, where $U_{J,n}$ has coefficients $\hat{U}_{j,n}$ in the basis ϕ_j and (3.71) becomes

$$\hat{U}_{j,n+1} = \big(1 + \Delta t\, \varepsilon \lambda_j\big)^{-1}\big(\hat{U}_{j,n} + \Delta t\, \hat{F}_j(U_{J,n})\big), \tag{3.72}$$

where $\hat{F}_j(U)$ are the coefficients of $I_J f(U)$ for $U \in H$ given by (1.41). Initial data are given by the coefficients $\hat{U}_{j,0}$ of $I_J u_0$. We may easily convert from the coefficients $\hat{U}_{j,n}$ to the function $U_{J,n}$ using (1.40) and computationally this is the MATLAB function `fft`. In Algorithm 3.5, we implement this method and store (in the variable `M`) only the non-zero elements of the linear term in (3.70).

With $\varepsilon = 10^{-3}$ and initial data $u_0(x) = \sin(4x)\sin(x)$, we approximate the solution to (3.69) for $t \in [0, 10]$ using $\Delta t = 0.01$ and $J = 512$ by calling Algorithm 3.5 with the following MATLAB commands:

```
>> T=10; N=1000; a=2*pi; J=512; epsilon=1e-3;
>> x=(0:a/J:a)'; u0=sin(4*x).*sin(x);
>> [t,ut]=pde_oned_Gal(u0,T,a,N,J,epsilon,@(u) u-u.^3);
```

A plot of the solution at time $T = 10$ is shown in Figure 3.5(a).

The next example is in two spatial dimensions and uses the two-dimensional discrete Fourier transform (DFT) and inverse discrete Fourier transform (IDFT); see Definition 7.27.

Example 3.40 (Allen–Cahn equation in two dimensions) Consider (3.69) again but now on the domain $D = (0, a_1) \times (0, a_2) = (0, 2\pi) \times (0, 16)$ (see also Example 3.18). To derive the spectral Galerkin method for (3.69), let $A = -\varepsilon\,\Delta$ and write the eigenfunctions $\phi_{j_1,j_2}(\boldsymbol{x}) = e^{2\pi i j_1 x_1/a_1} e^{2\pi i j_2 x_2/a_2}$ with corresponding eigenvalues

$$\lambda_{j_1,j_2} = \varepsilon\left(\left(\frac{2\pi j_1}{a_1}\right)^2 + \left(\frac{2\pi j_2}{a_2}\right)^2\right) = \varepsilon\left(j_1^2 + \left(\frac{\pi j_2}{8}\right)^2\right).$$

For even numbers $J_1, J_2 \in \mathbb{N}$, let $j_i \in \mathfrak{I}_{J_i} := \{-J_i/2 + 1, \ldots, J_i/2\}$ for $i = 1, 2$. Then, the spectral Galerkin approximation takes the form

$$u_{J_1,J_2}(t, \boldsymbol{x}) = \sum_{j_1 \in \mathfrak{I}_{J_1}, j_2 \in \mathfrak{I}_{J_2}} \hat{u}_{j_1,j_2}(t) \phi_{j_1,j_2}(\boldsymbol{x})$$

for $\hat{u}_{j_1,j_2}(t)$ to be determined. Note that this is a double sum. Substituting into (3.57) and taking $v = \phi_{j_1,j_2}$, we obtain for each coefficient $\hat{u}_{j_1,j_2}(t)$ an ODE

$$\frac{d\hat{u}_{j_1,j_2}}{dt} = -\varepsilon\lambda_{j_1,j_2}\hat{u}_{j_1,j_2} + \hat{f}_{j_1,j_2}(u_{J_1,J_2}), \tag{3.73}$$

where $\hat{f}_{j_1,j_2}(u_{J_1,J_2})$ are the coefficients of $f(u_{J_1,J_2})$ in the basis ϕ_{j_1,j_2}. The semi-implicit Euler method is now applied.

Algorithm 3.6 Code to apply the semi-implicit Euler method to a spectral Galerkin approximation of (3.35) with periodic boundary conditions on $D = (0, a_1) \times (0, a_2)$. The inputs are as in Algorithm 3.5 with `a`=[a_1,a_2] and `J`=[J_1,J_2]. The outputs are a vector `t` of times $t_0, \ldots, t_N$, and a three-dimensional array `ut` where `ut(j,k,n)` approximates $u_{J_1,J_2}(t_n, \boldsymbol{x}_{j-1,k-1})$ for $\boldsymbol{x}_{j,k} = [ja_1/J_1, ka_2/J_2]$.

```
function [t,ut]=pde_twod_Gal(u0,T,a,N,J,epsilon,fhandle)
Dt=T/N; t=[0:Dt:T]'; ut=zeros(J(1)+1,J(2)+1,N);
% set linear operators
lambdax=2*pi*[0:J(1)/2 -J(1)/2+1:-1]'/a(1);
lambday=2*pi*[0:J(2)/2 -J(2)/2+1:-1]'/a(2);
[lambdaxx lambdayy]=meshgrid(lambday,lambdax);
M=epsilon*(lambdaxx.^2+lambdayy.^2); EE=1./(1+Dt*M);
ut(:,:,1)=u0; u=u0(1:J(1),1:J(2)); uh=fft2(u); % set initial data
for n=1:N, % time loop
  fhu=fft2(fhandle(u)); % compute fhat
  uh_new=EE.*(uh+Dt*fhu);
  u=real(ifft2(uh_new)); ut(1:J(1), 1:J(2), n+1)=u; uh=uh_new;
end
ut(J(1)+1,:,:)=ut(1,:,:); ut(:,J(2)+1,:)=ut(:,1,:); % make periodic
```

To exploit the DFT in two dimensions, we write the system of ODEs in (3.73) as a $J_1 \times J_2$ matrix problem and use component-wise operations in MATLAB. For each n, let $U_{J_1,J_2,n}$ be a matrix with entries $\hat{u}_{j_1,j_2,n}$ that approximate $\hat{u}_{j_1,j_2}(t_n)$ and, in Algorithm 3.6, `M` is a dense $J_1 \times J_2$ matrix with entries $m_{i,j} = \varepsilon\lambda_{j_1,j_2}$. The MATLAB commands `fft2` and `ifft2` apply the Fourier transforms. The implementation is similar to the one-dimensional case; see Example 3.39.

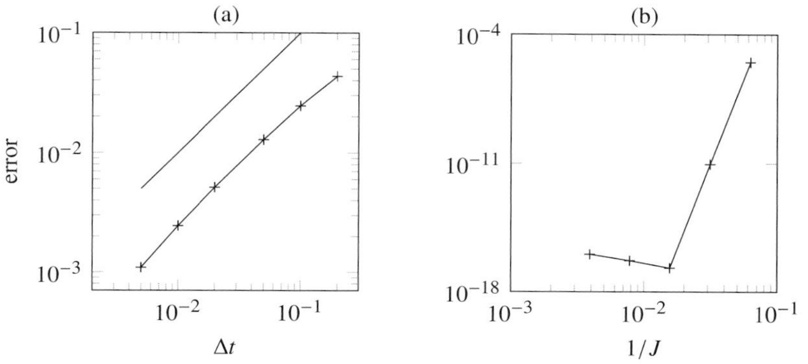

Figure 3.6 Convergence of the semi-implicit Euler and spectral Galerkin approximation for the Allen–Cahn equation (3.69) on $(0, 2\pi)$. We see (a) errors of order Δt (reference line has slope 1) computed with a fixed spatial discretisation $J = 1024$ and, in (b), convergence in space as $J \to \infty$ with a fixed time discretisation $\Delta t = 10^{-3}$. The convergence in (b) exceeds the predicted rate and reaches machine precision at $J = 16$.

We numerically solve (3.69) over $t \in [0, 10]$ for $\varepsilon = 10^{-3}$ and $u_0(\boldsymbol{x}) = \sin(x_1)\cos(\pi x_2/8)$ with discretisation parameters $J_1 = 128$, $J_2 = 256$, $\Delta t = 0.01$ by calling Algorithm 3.6 with the following commands:

```
>> T=10; N=1000; a=[2*pi 16]; J=[128 256]; epsilon=1e-3;
>> x=(0:a(1)/J(1):a(1))'; y=(0:a(2)/J(2):a(2))';
>> [xx yy]=meshgrid(y,x); u0=sin(xx).*cos(pi*yy/8);
>> [t,ut]=pde_twod_Gal(u0,T,a,N,J,epsilon,@(u) u-u.^3);
```

A plot of the approximation obtained at time $T = 10$ is shown in Figure 3.5(b).

Error analysis

We show theoretically for the linear problem (i.e., (3.54) with $f = 0$) and numerically for the Allen–Cahn equation that the approximations $u_{J,n}$ in (3.67) converge to the exact solution $u(t_n)$ in the limit of $\Delta t \to 0$ and $J \to \infty$. We first perform numerical experiments for the Allen–Cahn equation on $D = (0, a)$ and estimate the error $\|u(t_n) - u_{J,n}\|_{L^2(0,a)}$.

Example 3.41 (numerical convergence) Consider the Allen–Cahn equation with $\varepsilon = 1$ on the domain $D = (0, 2\pi)$ with periodic boundary conditions and initial data $u_0(x) = \sin(4x)\sin(x)$. We investigate the approximation given by the spectral Galerkin and semi-implicit Euler methods. Since we do not have an analytic solution, we use a reference solution $u_{J,N_{\text{ref}}}$ (computed with a small time step $\Delta t_{\text{ref}} = T/N_{\text{ref}}$) to examine convergence in time and a second reference solution $u_{J_{\text{ref}},N}$ (computed with a large value of J_{ref}) to examine convergence in space. The error $\|u(t_N) - u_{J,N}\|_{L^2(0,a)}$ is then approximated by $\|u_{J_{\text{ref}},N} - u_{J,N}\|_{L^2(0,a)}$ or $\|u_{J,N_{\text{ref}}} - u_{J,N}\|_{L^2(0,a)}$ as appropriate. Both these norms are computed exactly by the trapezium rule with $h_{\text{ref}} = a/J_{\text{ref}}$ (see Exercise 1.24).

Below, we take $T = 10$, $N = N_{\text{ref}} = 10^4$ (so $\Delta t_{\text{ref}} = 10^{-3}$) and $J = J_{\text{ref}} = 1024$. Setting $\Delta t = \kappa \Delta t_{\text{ref}}$, for $\kappa = 5, 10, 20, 50, 100, 200$ and using the following Matlab commands, we call Algorithm 3.8 (which calls Algorithm 3.7) to approximate the error associated with the time discretisation.

```
>> epsilon=1; a=2*pi; J=1024;
>> x=(0:a/J:a)'; u0=sin(4*x).*sin(x);
>> kappa=[5,10,20,50,100,200]'; T=10; Nref=10^4;
>> [dt,errT]=pde_oned_Gal_convDt(u0,T,a,Nref,kappa,J,epsilon,@(u) u-u.^3);
```

The error $\|u_{J,N_{\text{ref}}} - u_{J,N}\|_{L^2(0,a)}$ is illustrated numerically to be $\mathcal{O}(\Delta t)$ in Figure 3.6(a).

To examine the spatial error, we compute a sequence of approximations $u_{J,N}$, where each $J < J_{\text{ref}}$. Then, at each step of the computation, we systematically set to zero $\hat{u}_j$ and $\hat{f}_j$ in (3.68) for j such that $\lambda_j > \lambda_J$. The variable IJJ in Algorithm 3.7 is used to set these coefficients to zero. We then estimate the size of the error by $\|u_{J_{\text{ref}},N} - u_{J,N}\|_{L^2(0,a)}$. Convergence is illustrated numerically in Figure 3.6(b). The following commands call Algorithm 3.9 to determine the error in the spatial discretisation with J coefficients for $J = 512, 256, 128, 64, 32$ and 16, solving to time $T = 10$ with $\Delta t = 0.001$:

```
>> epsilon=1; a=2*pi; Jref=1024; T=10; N=10^4;
>> x=(0:a/Jref:a)'; u0=sin(4*x).*sin(x);
>> J=Jref*[1/2 1/4 1/8 1/16 1/32 1/64]';
>> errJ=pde_oned_Gal_convJ(u0,T,a,N,Jref,J,epsilon,@(u) u-u.^3);
```

The error $\|u_{J\text{ref},N} - u_{J,N}\|_{L^2(0,a)}$ is plotted against J in Figure 3.6(b) and shows rapid convergence, as suggested by Lemma 3.37 or Theorem 3.43 with γ large.

For the rigorous convergence analysis of the linear case, we cannot use Theorem 3.7 to establish convergence as the Lipschitz constant of the function $\boldsymbol{u} \mapsto M\boldsymbol{u}$ that appears in (3.64) grows like λ_J. Instead we exploit the smoothing properties of e^{-At}. We examine errors in the H norm $\|\cdot\|$.

Lemma 3.42 (time-stepping error, linear equation) *Suppose that Assumption 3.19 holds. Let $u_{J,n}$ satisfy (3.67) and $u_J(t)$ satisfy (3.62) with $f = 0$. For $\gamma \in [0,1]$ and $u_0 \in \mathcal{D}(A^\gamma)$,*

$$\left\|u_J(t_n) - u_{J,n}\right\| \le \Delta t^\gamma \left\|u_0\right\|_\gamma, \qquad n \in \mathbb{N}.$$

Proof Let $T_n := e^{-t_n A_J} - (I + \Delta t\, A_J)^{-n}$. As $u_J(t_n) - u_{J,n} = T_n P_J u_0$, it suffices to show the following quantity is order Δt^γ:

$$\left\|T_n P_J A^{-\gamma}\right\|_{\mathcal{L}(H)} = \Delta t^\gamma \left\|\left(e^{-n\Delta t\, A} - (I + \Delta t\, A)^{-n}\right) P_J \left(\Delta t\, A\right)^{-\gamma}\right\|_{\mathcal{L}(H)}.$$

The operator norm on a finite-dimensional space is characterised by the spectral radius (Example 1.71). Thus,

$$\left\|T_n P_J A^{-\gamma}\right\|_{\mathcal{L}(H)} = \Delta t^\gamma \sup_{j=1,\dots,J} \left|\left(e^{-n\Delta t\, \lambda_j} - (1 + \Delta t\, \lambda_j)^{-n}\right)(\Delta t\, \lambda_j)^{-\gamma}\right|.$$

Using elementary calculus, we may show that $0 \le (1+x)^{-1} - e^{-x} \le \frac{1}{2}\min\{1, x^2\}$ for $x \ge 0$.

Algorithm 3.7 Code to approximate solutions of (3.44) with periodic boundary conditions on $D = (0,a)$ using the semi-implicit Euler and spectral Galerkin methods. The inputs and outputs are similar to those in Algorithm 3.5 with `Nref` given in place of `N`. Additional input parameters `Jref=` J_{ref} and `kappa=` κ are given and the approximation computed with $\Delta t = T\kappa/N_{\text{ref}}$ and J modes. The solution `ut` is returned on the grid with spacing $h_{\text{ref}} = 1/J_{\text{ref}}$.

```
function [t,ut]=pde_oned_Gal_JDt(u0,T,a,Nref,kappa,Jref,J,epsilon,...
                                  fhandle)
N=Nref/kappa;
Dt=T/N; t=[0:Dt:T]';
% initialise
ut=zeros(Jref+1,N+1);
% use IJJ to set unwanted modes to zero.
IJJ=J/2+1:Jref-J/2-1;
% set linear operator
lambda=2*pi*[0:Jref/2 -Jref/2+1:-1]'/a;
M=epsilon*lambda.^2; EE=1./(1+Dt*M); EE(IJJ)=0;
% set initial condition
ut(:,1)=u0; u=u0(1:Jref); uh=fft(u); uh(IJJ)=0;
for n=1:N, % time loop
    fhu=fft(fhandle(u)); fhu(IJJ)=0;
    uh_new=EE.*(uh+Dt*fhu); % semi-implicit Euler step
    u=real(ifft(uh)); ut(1:Jref,n+1)=u;
    uh=uh_new; uh(IJJ)=0;
end
ut(Jref+1,:)=ut(1,:); % make periodic
```

Algorithm 3.8 Code to examine convergence as $\Delta t \to 0$ of the semi-implicit Euler and spectral Galerkin approximation of solutions to (3.44). The inputs `u0`, `T`, `J`, `epsilon`, and `fhandle` are similar to those in Algorithm 3.5. In addition, the number of reference steps `Nref` (so that $\Delta t_{\text{ref}} = T/N_{\text{ref}}$) and a vector `kappa` with entries κ_i are given to determine a vector `dt` of time steps $\Delta t_i = \kappa_i \Delta t_{\text{ref}}$. The algorithm computes a vector of the corresponding errors `errT` and returns `dt` and `errT`.

```
function [dt,errT]=pde_oned_Gal_convDt(u0,T,a,Nref,kappa,J,epsilon,...
                                        fhandle)
% reference soln
[t,ureft]=pde_oned_Gal_JDt(u0,T,a,Nref,1,J,J,epsilon,fhandle);
for i=1:length(kappa) % approximations
    [t,ut]=pde_oned_Gal_JDt(u0,T,a,Nref,kappa(i),J,J,epsilon,fhandle);
  S(i)=sum((ureft(1:end-1,end)-ut(1:end-1,end)).^2)*(a/J);
end
errT=sqrt(S); dtref=T/Nref; dt=kappa*dtref;
```

Hence, for $\mu := (1+x)^{-1}$,

$$\left|\left(e^{-nx} - (1+x)^{-n}\right)x^{-\gamma}\right| = \left(x^{-\gamma}\left(e^{-x} - (1+x)^{-1}\right)\sum_{j=0}^{n-1}(1+x)^{-(n-1-j)}e^{-jx}\right)$$
$$\le x^{-\gamma}\frac{1}{2}\min\{1,x^2\}\frac{1}{1-\mu} \le x^{-\gamma}\min\{1,x^2\}\left(\frac{1}{2x}+\frac{1}{2}\right).$$

Hence, the supremum is bounded by 1 and $\|T_n P_J A^{-\gamma}\|_{\mathcal{L}(H)} \le \Delta t^{\gamma}$. □

Algorithm 3.9 Code to examine convergence as $J \to \infty$ of the semi-implicit Euler and spectral Galerkin approximation of solutions to (3.44). The inputs `u0`, `T`, `N`, `epsilon`, and `fhandle` are similar to those in Algorithms 3.5. In addition, `Jref` gives the dimension of the reference solution. A vector `J` of spatial dimensions is given and the algorithm returns the corresponding errors `errJ`.

```
function [errJ]=pde_oned_Gal_convJ(u0,T,a,N,Jref,J,epsilon, ...
                                    fhandle)
% reference soln
[tref,ureft]=pde_oned_Gal_JDt(u0,T,a,N,1,Jref,Jref,epsilon,fhandle);
for i=1:length(J) % approximation with J=J(i)
  [t,ut]=pde_oned_Gal_JDt(u0,T,a,N,1,Jref,J(i),epsilon,fhandle);
  S(i)=sum((ureft(1:end-1,end)-ut(1:end-1,end)).^2)*a/Jref;
end
errJ=sqrt(S);
```

We combine the spatial and time-stepping errors to describe the total error in the following theorem. For this theorem, we give the additional step necessary to allow zero eigenvalues in Assumption 3.19.

Theorem 3.43 (total error, linear equation) *Suppose that A has a complete orthonormal set of eigenfunctions $\{\phi_j : j \in \mathbb{N}\}$ and eigenvalues $\lambda_j \geq 0$, ordered so that $\lambda_{j+1} \geq \lambda_j$. Suppose that the initial data $u_0 \in \mathcal{D}(A^\gamma)$, for some $\gamma > 0$. For $f = 0$, the solution $u(t)$ of* (3.54) *and $u_{J,n}$ from* (3.67) *satisfy*

$$\left\|u(t_n) - u_{J,n}\right\| \leq \left(\frac{1}{\lambda_{J+1}^\gamma} + \Delta t^{\min\{\gamma,1\}}\right)\left\|u_0\right\|_\gamma, \qquad n \in \mathbb{N}. \tag{3.74}$$

Proof Split the error into the spatial and time-stepping errors,

$$\left\|u(t_n) - u_{J,n}\right\| \leq \left\|u(t_n) - u_J(t_n)\right\| + \left\|u_J(t_n) - u_{J,n}\right\|.$$

If $\lambda_j > 0$ for all j, (3.74) follows by applying Lemmas 3.37 and 3.42. Notice that Lemma 3.42 must be applied with $\gamma \in [0,1]$ and hence the restriction to $\Delta t^{\min\{\gamma,1\}}$. For j such that $\lambda_j = 0$, it easy to see that $\hat{u}_j(t_n) = \hat{u}_{j,n}$ for all n and this mode does not contribute to the total error. □

For $\gamma \leq 1$, the spatial and time-stepping errors are balanced in the limit $J \to \infty$ and $\Delta t \to 0$ if $\lambda_J \Delta t$ is held constant. This is known as the Courant–Friedrichs–Lewy (CFL) condition.

Convergence for the semilinear evolution equation (3.54) is studied in §3.7.

3.6 Finite elements for reaction–diffusion equations

We focus now on the Galerkin finite element method on a domain $D = (0, a)$. We partition D into elements of width h (as in §2.1) and choose $V^h \subset V$ to be the associated set of piecewise linear polynomials. We apply the Galerkin finite element method to the reaction–diffusion equation (3.44) with homogeneous Dirichlet boundary conditions. In this case, we can formulate the PDE as an ODE on $H = L^2(0, a)$ (see (3.36)) and the operator $-A = \varepsilon \frac{d^2}{dx^2}$

with $\mathcal{D}(A) = H^2(0,a) \cap H_0^1(0,a)$. Substituting u_h for $\tilde{u}$ in (3.57) gives

$$\left\langle \frac{du_h}{dt}, v \right\rangle_{L^2(0,a)} = -a(u_h, v) + \left\langle f(u_h), v \right\rangle_{L^2(0,a)}, \qquad \forall v \in V^h, \tag{3.75}$$

and $a(u,v) := \int_0^a \varepsilon\, u'(x)v'(x)\,dx$ as in (2.9) (with $p = \epsilon$ and $q = 0$).

Let P_{h,L^2} be the orthogonal projection $P_{h,L^2}\colon L^2(0,a) \to V^h$ with respect to the $L^2(0,a)$ inner product (see Definition 1.38). Note that P_{h,L^2} is different to the Galerkin projection P_h in Definition 2.19, which was defined with respect to the energy norm. As initial data, we take $u_{h,0} = P_{h,L^2}u_0$ and define the finite-dimensional operator $A_h\colon V^h \to V^h$ by

$$\left\langle A_h w, v \right\rangle_{L^2(0,a)} = a(w,v), \qquad \forall w, v \in V^h. \tag{3.76}$$

In the spectral Galerkin method, the operators A_J and A have the same first J eigenvalues; this is not the case for A_h and A. Following (3.59), we use A_h to write an evolution equation for the approximation $u_h(t) \in V^h$ for $t > 0$; that is,

$$\frac{du_h}{dt} = -A_h u_h + P_{h,L^2} f(u_h), \tag{3.77}$$

with initial condition $u_{h,0} = P_{h,L^2}u_0$.

We recall the basis $\{\phi_j : j = 1, \dots, J\}$ for V^h given by the global basis functions (see Definition 2.23). Here $J = n_e - 1$ is the number of interior vertices, which are chosen due to the homogeneous Dirichlet boundary conditions, n_e is the number of elements, and $h = a/n_e$. We write

$$u_h(t,x) = \sum_{j=1}^{J} u_j(t)\phi_j(x). \tag{3.78}$$

Let $\boldsymbol{u}_h(t) := [u_1(t), u_2(t), \dots, u_J(t)]^{\mathsf{T}}$. This satisfies the system of ODEs

$$M\frac{d\boldsymbol{u}_h}{dt} = -K\boldsymbol{u}_h + \boldsymbol{f}(\boldsymbol{u}_h). \tag{3.79}$$

The vector $\boldsymbol{f}(\boldsymbol{u}_h) \in \mathbb{R}^J$ has elements $f_j = \langle f(u_h), \phi_j \rangle_{L^2(0,a)}$. M is the mass matrix with elements $m_{ij} = \langle \phi_i, \phi_j \rangle_{L^2(0,a)}$ and K is the diffusion matrix with elements $k_{ij} = a(\phi_i, \phi_j)$. We may also rewrite (3.79) as

$$\frac{d\boldsymbol{u}_h}{dt} = -M^{-1}K\boldsymbol{u}_h + M^{-1}\boldsymbol{f}(\boldsymbol{u}_h). \tag{3.80}$$

This system of semilinear ODEs can be discretised in time by the methods considered in §3.1 and we focus on the semi-implicit Euler method (3.19). Introduce a time step Δt and denote by $\boldsymbol{u}_{h,n}$ the solution of

$$\left(M + \Delta t\, K\right)\boldsymbol{u}_{h,n+1} = M\boldsymbol{u}_{h,n} + \Delta t\, \boldsymbol{f}(\boldsymbol{u}_{h,n}) \tag{3.81}$$

with $\boldsymbol{u}_{h,0} = \boldsymbol{u}_h(0)$. Then, $\boldsymbol{u}_{h,n}$ approximates $\boldsymbol{u}_h(t_n)$ for $t_n = n\Delta t$ and $n \in \mathbb{N}$.

Algorithm 3.11 implements (3.81) for (3.44). In this case, the matrices M and K in (3.81) correspond to the matrices found in Algorithms 2.1 and 2.2 for $p = \epsilon$ and $q = 1$ (i.e., we take (2.3) and match $-u_{xx}$ to the p term and split the Euler approximation to u_t between the q term and the right-hand side). The vector $\boldsymbol{f}$ is evaluated using quadrature in Algorithm 3.10. Algorithm 3.11 uses $\boldsymbol{f}$, M, and K to find the approximation $\boldsymbol{u}_{h,n}$.

Algorithm 3.10 Code to approximate $\boldsymbol{f}$, where $f_j = \langle f(u_h), \phi_j \rangle_{L^2(0,a)}$. Inputs are the number of finite elements `ne`, the uniform mesh width `h`, and the nodal values `f` of $f(u_h)$. The output `b` is the required vector $\boldsymbol{f}$.

```
function b=oned_linear_FEM_b(ne,h,f)
nvtx=ne+1; elt2vert=[1:ne; 2:(ne+1)]';
bks=zeros(ne,2); b=zeros(nvtx,1);
bks(:,1) = f(1:end-1,:).*(h/3)+f(2:end,:).*(h/6);
bks(:,2) = f(1:end-1,:).*(h/6)+f(2:end,:).*(h/3);
for row_no=1:2
 nrow=elt2vert(:,row_no);
 b=b+sparse(nrow,1,bks(:,row_no),nvtx,1);
end
b(1,:)=[];b(end,:)=[];
```

Algorithm 3.11 Code to find the semi-implicit Euler and finite element approximation of solutions to (3.44) with homogeneous Dirichlet boundary conditions on $D = (0,a)$. The inputs and outputs are similar to those in Algorithm 3.4 and, in place of J, we pass the number of elements `ne`=n_e.

```
function [t,ut]=pde_fem(u0,T,a,N,ne,epsilon,fhandle)
h=a/ne; nvtx=ne+1; Dt=T/N; t=[0:Dt:T]';
p=epsilon*ones(ne,1);q=ones(ne,1);f=ones(ne,1);
[uh,A,b,KK,MM]=oned_linear_FEM(ne,p,q,f);
EE=(MM+Dt*KK); ZM=0;
ut=zeros(nvtx,N+1); ut(:,1)=u0; u=u0; % set initial condition
for k=1:N, % time loop
  fu=fhandle(u);   b=oned_linear_FEM_b(ne,h,fu);
  u_new=EE\(MM*u(2:end-1,:)+Dt*b);
  u=[ZM;u_new;ZM]; ut(:,k+1)=u;
end
```

Example 3.44 (Nagumo equation) Consider the Nagumo equation (3.52) with $a = 1$, $\varepsilon = 10^{-3}$, $\alpha = -0.5$, and $u_0(x) = \exp\big(-(x-1/2)^2/\varepsilon\big)$. We generate a numerical approximation by the semi-implicit Euler method in time and piecewise linear finite elements in space on a uniform mesh of $n_e = 512$ elements with 513 vertices. This means there are $J = 511$ interior vertices and two boundary vertices at $x = 0$ and $x = 1$. Choose $\Delta t = 0.001$ and integrate up to $T = 10$. The resulting numerical solution is obtained by calling Algorithm 3.11 with the following commands:

```
>> T=10; N=1e4; a=1; ne=512; h=a/ne; epsilon=1e-3;
>> x=(0:h:a)'; u0=exp(-(x-0.5).^2/epsilon);
>> [t,ut]=pde_fem(u0,T,a,N,ne,epsilon,@(u) u.*(1-u).*(u+0.5));
```

The solution is plotted in Figure 3.4(a) for $t \in [0,10]$. In Algorithm 3.11, we use Algorithm 2.1 to obtain the diffusion matrix K and mass matrix M only.

Rather than prove convergence results here, we illustrate numerically the theoretical rates of convergence that show errors of order Δt in time and order h^2 in space in the $L^2(0,a)$ norm. In Chapter 2, we saw errors in the energy norm of order h in space (see Example 2.33). In the weaker $L^2(0,a)$ norm, we see a faster rate of convergence. Again, we approximate the norm using (3.53).

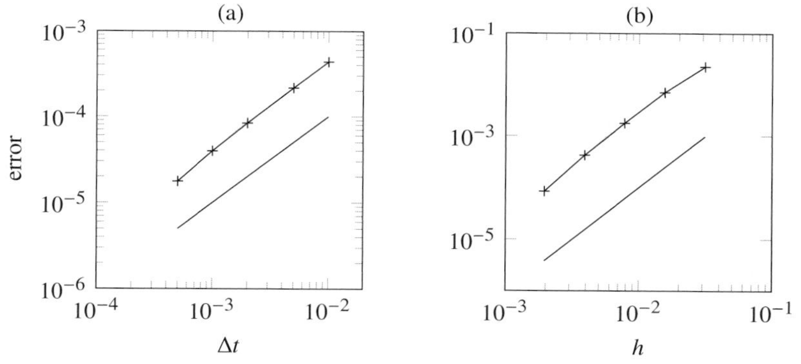

Figure 3.7 Convergence of the semi-implicit Euler and finite element discretisation of (3.52). See Example 3.45. In (a), convergence in time shows errors of order Δt (reference line has slope 1) and in (b) convergence in space shows errors of order h^2 (reference line has slope 2).

Example 3.45 (numerical convergence) Continuing with the Nagumo equation from Example 3.44, we estimate the error by computing two reference solutions $u_{h_{\text{ref}},N}$ with $\Delta t = T/N$ and $h_{\text{ref}} = a/n_{e,\text{ref}}$ and $u_{h,N_{\text{ref}}}$ with $\Delta t_{\text{ref}} = T/N_{\text{ref}}$ and $h = a/n_e$, for a large reference number of elements $n_{e,\text{ref}}$ and time steps N_{ref}. The errors are then estimated by comparing the reference solutions to solutions $u_{h,N}$ computed with $h = Lh_{\text{ref}}$, $N = N_{\text{ref}}/\kappa$, and $\Delta t = \kappa \Delta t_{\text{ref}}$ for different values of L and κ. The solutions are computed by Algorithm 3.12.

Algorithm 3.12 Code to approximate solutions of (3.44) using (3.81) with different h and Δt. The inputs and outputs are similar to those in Algorithm 3.11 with `neref`$= n_{e,\text{ref}}$ and `Nref`$= N_{\text{ref}}$ passed in place of `ne` and `N`. The additional inputs `kappa` and `L` determine the time step $\Delta t = \kappa \Delta t_{\text{ref}}$ for $\Delta t_{\text{ref}} = T/N_{\text{ref}}$ and mesh width $h = Lh_{\text{ref}}$ for $h_{\text{ref}} = a/n_{e,\text{ref}}$ used for the computation of `t` and `ut`.

```
function [t,ut]=pde_fem_hDt(u0,T,a,Nref,kappa,neref,L,epsilon,...
                              fhandle)
ne=neref/L; assert(mod(ne,1)==0); h=a/ne; nvtx=ne+1;
dtref=T/Nref; Dt=kappa*dtref; t=[0:Dt:T]';
p=epsilon*ones(ne,1);q=ones(ne,1);f=ones(ne,1);
[uh,A,b,K,M]=oned_linear_FEM(ne,p,q,f); EE=M+Dt*K; ZM=0;
ut=zeros(nvtx,Nref/kappa+1); ut(:,1)=u0; u=u0;% set initial condition
for k=1:Nref/kappa,% time loop
  fu=fhandle(u);  b=oned_linear_FEM_b(ne,h,fu);
  u_new=EE\(M*u(2:end-1,:)+Dt*b);
  u=[ZM;u_new;ZM]; ut(:,k+1)=u;
end
```

First, we examine convergence of the time discretisation as we did in Example 3.41 and, in Figure 3.7(a), observe errors of order Δt. For convergence of the spatial discretisation, we take $L = 2, 4, 8, 16$ and 32 and compute approximate solutions with $n_e = n_{e,\text{ref}}/L$ elements and their errors relative to the approximation with $n_{e,\text{ref}} = 512$ elements. We also fix $T = 1$ and $N = 10^4$ so that $\Delta t = 0.001$ and $\kappa = 1$.

The following MATLAB commands compute the errors for this set of L values by calling Algorithm 3.13:

```
>> T=1; N=1e4; kappa=1; epsilon=1e-3;
>> neref=512; a=1; L=[2,4,8,16,32]';
>> href=a/neref; x=(0:href:a)'; u0=exp(-(x-0.5).^2/epsilon);
>> [h,errh]=pde_fem_convh(u0,T,a,N,neref,L,epsilon,@(u) u.*(1-u).*(u+0.5));
```

In Figure 3.7(c), we see errors of order h^2. The plot is on a log-log scale and the reference line has slope 2.

Algorithm 3.13 Code to examine convergence of the finite element approximation as $h \to 0$. Inputs are nodal values `u00` of the initial data on the reference mesh, final time `T`, size of domain `a`, `N` so that $\Delta t = T/N$, `neref=` $n_{e,\text{ref}}$ to determine the reference mesh of width $h_{\text{ref}} = a/n_{e,\text{ref}}$, a vector `L` of values L_j so that $h_j = L_j h_{\text{ref}}$, `epsilon=` ε and a handle `fhandle` to evaluate $\boldsymbol{f}(\boldsymbol{u})$. The outputs are a vector `h` of mesh widths h_j and a vector `err` of the errors computed relative to the reference solution.

```
function [h,err]=pde_fem_convh(u00,T,a,N,neref,L,epsilon,fhandle)
href=a/neref; xref=[0:href:a]';kappa=1;
[t,uref]=pde_fem_hDt(u00,T,a,N,kappa,neref,1,epsilon,fhandle);
for j=1:length(L)
    h(j)=href*L(j); u0=u00(1:L(j):end);
    [t,u]=pde_fem_hDt(u0,T,a,N,kappa,neref,L(j),epsilon,fhandle);
    x=[0:h(j):a]'; uinterp=interp1(x,u,xref);
    S(j)=sum(uref(:,end)-uinterp(:,end)).^2*href;
end
err=sqrt(S);
```

3.7 Non-smooth error analysis

To finish this chapter, we study convergence of the Galerkin approximation with semi-implicit Euler time stepping under an abstract assumption on the Galerkin subspace $\widetilde{V}$. We build up to Theorem 3.55, which gives convergence for the semilinear evolution equation (3.54). Our starting point is the linear evolution equation,

$$\frac{du}{dt} = -A\,u, \qquad \text{given } u(0) = u_0, \tag{3.82}$$

where A satisfies Assumption 3.19. Let $\widetilde{V} \subset V := \mathcal{D}(A^{1/2})$ be a finite-dimensional subspace of V and seek an approximation $\tilde{u}(t) \in \widetilde{V}$ to $u(t)$. In particular, we consider $\tilde{u}(t)$ satisfying the semi-discrete system

$$\frac{d\tilde{u}}{dt} = -\widetilde{A}\,\tilde{u}, \qquad \text{given } \tilde{u}(0) = \widetilde{P}u_0, \tag{3.83}$$

where $\widetilde{A} \colon \widetilde{V} \to \widetilde{V}$ is defined by (3.60). Each of the spatial discretisations discussed in §3.5 and §3.6 leads to an equation of the form (3.83).

Example 3.46 (spectral Galerkin) Applying the spectral Galerkin method to (3.82) leads to (3.62) with $f = 0$. In that case, $\widetilde{V} = V_J := \text{span}\{\phi_1, \dots, \phi_J\}$ where ϕ_j is the jth eigenfunction of A. Moreover, the projection $\widetilde{P} = P_J$ where $P_J : H \to V_J$ is given in (3.61) and $\widetilde{A} = A_J := P_J A$.

Example 3.47 (finite element) Consider the case of the linear heat equation $u_t = \Delta u$ with homogeneous Dirichlet boundary conditions (see Example 3.17). Then $H = L^2(D)$, $A = -\Delta$ with domain $\mathcal{D}(A) = H^2(D) \cap H_0^1(D)$, and $V = H_0^1(D)$. The semi-discrete system (3.83) associated with a finite element discretisation is given by (3.77) with $f = 0$. In this case, $\widetilde{V} = V^h \subset H_0^1(D)$ is a set of piecewise (e.g., linear) polynomials associated with a mesh $\mathcal{T}_h$ on D. The discrete operator $\widetilde{A} = A_h : V^h \to V^h$ is defined by (3.76) and $\widetilde{P} = P_{h,L^2}$. We consider the finite element discretisation further in Example 3.49.

Regardless of the specific choice of spatial discretisation, we arrive at the following full space and time discretisation of (3.82) by using the implicit Euler method:

$$(I + \Delta t\, \widetilde{A})\tilde{u}_{n+1} = \tilde{u}_n$$

where $\tilde{u}_0 = \widetilde{P} u_0$. By introducing $\widetilde{S}_{\Delta t} := \left(I + \Delta t\, \widetilde{A}\right)^{-1}$, we write

$$\tilde{u}_{n+1} = \widetilde{S}_{\Delta t} \tilde{u}_n. \tag{3.84}$$

Our goal is to bound the error $\|\tilde{u}_n - u(t_n)\|$ at time $t_n = n\Delta t$, where $\|\cdot\|$ is the norm on H. To perform the convergence analysis, we extend the definition of $\widetilde{A}^{-1}$ to H and make the following assumption on the Galerkin subspace $\widetilde{V}$.

Assumption 3.48 Suppose that $\widetilde{A}^{-1} \in \mathcal{L}(H)$ satisfies $\widetilde{A}^{-1}\widetilde{A} = I$ on $\widetilde{V}$ and $\widetilde{A}^{-1}(I - \widetilde{P}) = 0$ and is non-negative definite. Further, for some $C, \delta > 0$, suppose that

$$\|(\widetilde{A}^{-1} - A^{-1})f\| \le C\,\delta^2 \|f\|, \qquad \forall f \in H \tag{3.85}$$

and

$$\|v\| \le C\delta^{-2}\|A^{-1}v\|, \qquad \forall v \in \widetilde{V}. \tag{3.86}$$

The constant C should be independent of the choice of $\widetilde{V}$ taken from a specific family of subspaces. For example, if $\widetilde{V} = V_J$ is the spectral Galerkin subspace of Example 3.46, the assumption holds with $\delta = 1/\sqrt{\lambda_J}$ and $C = 1$ (see Lemma 1.94). If V^h is a space of piecewise linear functions, we think of δ as the mesh width h and we want C to be independent of h. We demonstrate this in the next example.

Example 3.49 (finite element approximation) Continuing from Example 3.47, we show that Assumption 3.48 holds with $\delta = h$ for $\widetilde{V} = V^h$, the space of piecewise linear polynomials on a shape-regular mesh. Note that $u = A^{-1}f$ is the weak solution of the elliptic problem $-\Delta u = -\nabla^2 u = f$ with homogeneous Dirichlet boundary conditions and (see (2.49) and (2.50))

$$a(u,\phi) = \int_D \nabla u(\boldsymbol{x}) \cdot \nabla \phi(\boldsymbol{x})\, d\boldsymbol{x} = \int_D f(\boldsymbol{x})\, \phi(\boldsymbol{x})\, d\boldsymbol{x} = \ell(\phi), \qquad \forall \phi \in V. \tag{3.87}$$

Similarly, $u_h = A_h^{-1} f$ is the solution of

$$a(u_h, \phi) = \ell(\phi), \qquad \forall \phi \in V^h. \tag{3.88}$$

Let $e := u_h - u$ and we aim to show that $\|e\|_{L^2(D)} \le Ch^2\|f\|_{L^2(D)}$. Let $\psi \in V$ be the solution of the auxiliary problem $A\psi = e$. Then, ψ obeys the weak form

$$a(\psi,\phi) = \ell_{\text{aux}}(\phi) := \int_D e(\boldsymbol{x})\,\phi(\boldsymbol{x})\,d\boldsymbol{x}, \qquad \forall \phi \in V.$$

For $\psi_h \in V^h$, by Galerkin orthogonality (2.21),

$$\ell_{\text{aux}}(e) = a(\psi,e) = a(e,\psi - \psi_h) \tag{3.89}$$

and the Cauchy–Schwarz inequality yields

$$\|e\|^2_{L^2(D)} = \langle e,e\rangle_{L^2(D)} = \ell_{\text{aux}}(e) \le \|e\|_{H^1_0(D)}\|\psi - \psi_h\|_{H^1_0(D)}.$$

By Theorem 2.67, for a constant C,

$$\|e\|_{H^1_0(D)} \le C\,h\,\|f\|_{L^2(D)}.$$

The constant C is independent of the choice of shape-regular mesh. For the same reason, if $\psi_h \in V^h$ is the FEM approximation to (3.89), Theorem 2.67 again gives

$$\|\psi - \psi_h\|_{H^1_0(D)} \le C\,h\,\|e\|_{L^2(D)}.$$

Thus,

$$\|e\|^2_{L^2(D)} \le C^2\,h^2\,\|f\|_{L^2(D)}\|e\|_{L^2(D)}.$$

We have shown that $\|e\|_{L^2(D)} \le Ch^2\|f\|_{L^2(D)}$. This is known as the *Aubin–Nitsche* duality argument and, because $e = u - u_h = A^{-1}f - A_h^{-1}f$, it implies (3.85).

For $f \in L^2(D)$, $\langle A_h^{-1}f,f\rangle_{L^2(D)} = a(u_h,u_h) \ge 0$. This shows A_h^{-1} is non-negative definite. (3.86) is left as an exercise.

Linear error analysis

Using the Assumption 3.48 on $\widetilde{V}$, we now complete the error analysis for the linear evolution equation (3.82). Using the operator A^{-1}, rewrite (3.82) as

$$A^{-1}\frac{du}{dt} + u = 0 \tag{3.90}$$

and similarly, for the semi-discrete equation (3.83),

$$\widetilde{A}^{-1}\frac{d\tilde{u}}{dt} + \tilde{u} = 0. \tag{3.91}$$

To obtain an equation for the error $e(t) = u(t) - \tilde{u}(t)$, first note that

$$\widetilde{A}^{-1}\frac{de}{dt} + e = \widetilde{A}^{-1}\left(\frac{du}{dt} - \frac{d\tilde{u}}{dt}\right) + (u - \tilde{u}).$$

Rearranging and using (3.91), we find

$$\widetilde{A}^{-1}\frac{de}{dt} + e = \widetilde{A}^{-1}\frac{du}{dt} + u - \left(\widetilde{A}^{-1}\frac{d\tilde{u}}{dt} + \tilde{u}\right) = \left(A^{-1}\frac{du}{dt} + u\right) - \left(A^{-1}\frac{du}{dt} - \widetilde{A}^{-1}\frac{du}{dt}\right),$$

where we have added in and subtracted out $A^{-1}\frac{du}{dt}$. By (3.90),

$$\widetilde{A}^{-1}\frac{de}{dt} + e = \left(A^{-1} - \widetilde{A}^{-1}\right) A\, u =: \rho. \tag{3.92}$$

Our first goal is to estimate $\|e(t_n)\| = \|u(t_n) - \tilde{u}(t_n)\|$ and we start with two preliminary lemmas.

Lemma 3.50 *Let Assumptions 3.19 and 3.48 hold. Let $u(t)$, $\tilde{u}(t)$ obey* (3.90), (3.91) *respectively and denote $e(t) = u(t) - \tilde{u}(t)$. Define $\rho(t)$ as in* (3.92). *Then,*

$$\int_0^t \|e(s)\|^2\, ds \le \int_0^t \|\rho(s)\|^2\, ds.$$

Proof See Exercise 3.21. □

Lemma 3.51 *Let the assumptions of Lemma 3.50 hold. Then,*

$$\|e(t)\| \le \sqrt{10} \sup_{0\le s\le t} \left(\|\rho(s)\| + s\left\|\frac{d\rho(s)}{ds}\right\|\right).$$

Proof The proof is by an energy argument. Take the inner product of (3.92) with $2\frac{de(t)}{dt}$:

$$2\left\langle \widetilde{A}^{-1}\frac{de(t)}{dt}, \frac{de(t)}{dt}\right\rangle + \frac{d}{dt}\|e(t)\|^2 = 2\left\langle \rho(t), \frac{de(t)}{dt}\right\rangle,$$

where we use that $\frac{d}{dt}\|e(t)\|^2 = 2\langle e, \frac{de}{dt}\rangle$. Since $\widetilde{A}^{-1}$ is non-negative definite and rewriting the right-hand side,

$$\frac{d}{dt}\|e(t)\|^2 \le 2\left\langle \rho(t), \frac{de(t)}{dt}\right\rangle = 2\frac{d}{dt}\langle \rho(t), e(t)\rangle - 2\left\langle \frac{d\rho(t)}{dt}, e(t)\right\rangle.$$

Multiplying by t and noting that $\frac{d}{dt}(t\,\langle \rho, e\rangle) = t\,\frac{d}{dt}\langle \rho, e\rangle + \langle \rho, e\rangle$, we find

$$\frac{d}{dt}\left(t\|e(t)\|^2\right) \le 2\frac{d}{dt}\left(t\langle \rho(t), e(t)\rangle\right) - 2\langle \rho(t), e(t)\rangle - 2t\left\langle \frac{d\rho(t)}{dt}, e(t)\right\rangle.$$

Integrating the above and using the Cauchy–Schwarz inequality gives

$$\begin{aligned} t\|e(t)\|^2 &\le 2\,t\,\|\rho(t)\|\,\|e(t)\| \\ &\quad + \int_0^t \left(2\|\rho(s)\|\,\|e(s)\| + 2\,s\left\|\frac{d\rho(s)}{ds}\right\|\,\|e(s)\|\right) ds. \end{aligned} \tag{3.93}$$

We use the inequality $2ab \le \alpha a^2 + \frac{1}{\alpha}b^2$ for $a, b, \alpha > 0$ with $\alpha = 2$, to see

$$2\,t\,\|\rho(t)\|\,\|e(t)\| \le 2\,t\,\|\rho(t)\|^2 + \frac{t}{2}\|e(t)\|^2.$$

With $\alpha = 1$, the same argument gives $2\|\rho(s)\|\,\|e(s)\| \le \|\rho(s)\|^2 + \|e(s)\|^2$ and also $2s\left\|\frac{d\rho(s)}{ds}\right\|\,\|e(s)\| \le s^2\|\rho(t)\|^2 + \|e(t)\|^2$. Hence, from (3.93),

$$\frac{1}{2}\|e(t)\|^2 \le 2\|\rho(t)\|^2 + \frac{1}{t}\int_0^t \left(2\|e(s)\|^2 + \|\rho(s)\|^2 + s^2\left\|\frac{d\rho(s)}{ds}\right\|^2\right) ds.$$

By Lemma 3.50,

$$\|e(t)\|^2 \le 4\|\rho(t)\|^2 + \frac{2}{t}\int_0^t \left(3\|\rho(s)\|^2 + s^2\left\|\frac{d\rho(s)}{ds}\right\|^2\right) ds.$$

The proof is completed by replacing the functions of s by their supremum. □

In (3.83), we work on the subspace $\widetilde{V}$ and replace the linear operator A of (3.82) by $\widetilde{A}$, to define an approximation $\tilde{u}(t)$ to $u(t)$. The resulting difference in the solution $e(t) = u(t) - \tilde{u}(t)$ is similar to the spatial error studied in Lemma 3.37 for the Galerkin approximation. In the following, we describe the spatial error introduced by working on a subspace $\widetilde{V}$ satisfying Assumption 3.48. Again, the estimate depends on the smoothness of the initial data, as described by γ.

Proposition 3.52 (spatial error) *Let Assumptions 3.19 and 3.48 hold and suppose that $u(t)$, $\tilde{u}(t)$ satisfy* (3.90), (3.91) *respectively. For $\gamma \le 1$, there exists $K > 0$ independent of δ such that the error $e(t) = u(t) - \tilde{u}(t)$ satisfies*

$$\|e(t)\| \le K\frac{\delta^2}{t^{1-\gamma}}\|u_0\|_\gamma, \qquad \forall u_0 \in \mathcal{D}(A^\gamma), \quad t > 0. \tag{3.94}$$

Proof The error $e(t)$ satisfies (3.92) and $\rho(t) = \left(A^{-1} - \widetilde{A}^{-1}\right) A\, u(t)$. Using $u(t) = \mathrm{e}^{-At}u_0$ and Assumption 3.48, we have $\|\rho(t)\| \le C\delta^2\|Ae^{-At}u_0\|$. Write $A = A^{1-\gamma}A^\gamma$ and apply Lemma 3.22(ii), to find $K_1 > 0$ such that

$$\|\rho(t)\| \le CK_1\frac{\delta^2}{t^{1-\gamma}}\|A^\gamma u_0\|.$$

Similarly, we can show for a constant $K_2 > 0$ that

$$t\left\|\frac{d\rho(t)}{dt}\right\| \le C\delta^2 t\left\|A^2\mathrm{e}^{-At}u_0\right\| \le CK_2\frac{\delta^2}{t^{1-\gamma}}\|A^\gamma u_0\|.$$

These estimates, together with Lemma 3.51, complete the proof of (3.94). □

The error estimate for the time discretisation $\tilde{u}(t_n) - \tilde{u}_n$ comes in two flavours: for smooth initial data $u_0 \in \mathcal{D}(A)$, the error bound (3.95) is uniform over $t_n > 0$; otherwise for $u_0 \in \mathcal{D}(A^\gamma)$ and $\gamma \in [-1, 1)$, the bounds (3.95) and (3.96) are not uniform as $t_n \to 0$. We include initial data $u_0 \in \mathcal{D}(A^{-\alpha})$ for $\alpha > 0$, which may not belong to H. The case $\alpha > 0$ is necessary to study space–time white noise forcing in §10.6 and is the only place that (3.86) is employed in this book.

Proposition 3.53 (time-stepping error) *Let Assumptions 3.19 and 3.48 hold. Let $\tilde{u}(t)$ solve* (3.83) *and $\tilde{u}_n$ solve* (3.84). *For $\gamma \in [0, 1]$, there exists $K_2 > 0$ independent of δ and Δt such that*

$$\|\tilde{u}(t_n) - u_n\| \le K_2\frac{\Delta t + \delta^{2\gamma}\Delta t^{1-\gamma}}{t_n^{1-\gamma}}\|u_0\|_\gamma, \qquad \forall u_0 \in \mathcal{D}(A^\gamma), \quad n \in \mathbb{N}. \tag{3.95}$$

For $\alpha \in [0, 1]$, there exists $K_1 > 0$ independent of δ and Δt such that

$$\|\tilde{u}(t_n) - u_n\| \le K_1\frac{\Delta t\, \delta^{-2\alpha}}{t_n}\|u_0\|_{-\alpha}, \qquad \forall u_0 \in \mathcal{D}(A^{-\alpha}), \quad n \in \mathbb{N}, \tag{3.96}$$

Proof Define the operator T_n by $T_n\tilde{u}_0 := (\mathrm{e}^{-t_n\tilde{A}} - \tilde{S}^n_{\Delta t})\tilde{u}_0$. Then, $T_n\tilde{P}u_0 = \tilde{u}(t_n) - \tilde{u}_n$. Since $\tilde{S}^n_{\Delta t} = (I + \Delta t\,\tilde{A})^{-n}$, we obtain

$$\|T_n\tilde{P}\|_{\mathcal{L}(H)} = \|(\mathrm{e}^{-n\Delta t\,\tilde{A}} - (I + \Delta t\,\tilde{A})^{-n})\tilde{P}\|_{\mathcal{L}(H)}.$$

Since $\tilde{P}$ projects onto a finite-dimensional space, as in Example 1.71, this norm is characterised by the spectral radius. If $\tilde{\lambda}_j$, $j = 1, 2, \dots, J$, are the eigenvalues of $\tilde{A}$,

$$\|T_n\tilde{P}\|_{\mathcal{L}(H)} = \sup_{j=1,\dots,J}\left|\mathrm{e}^{-n\Delta t\,\tilde{\lambda}_j} - (1 + \Delta t\,\tilde{\lambda}_j)^{-n}\right|.$$

By Exercise 3.19, $\|T_n\tilde{P}\|_{\mathcal{L}(H)} \le 1/n$. This gives (3.96) in the case $\alpha = 0$.

For the case $\alpha \in (0,1]$, using (3.86) and Exercise 3.20, we see $\|\tilde{u}\| \le C^\alpha\delta^{-2\alpha}\|A^{-\alpha}\tilde{u}\|$ for $\tilde{u} \in \tilde{V}$. Thus, $\|\tilde{P}u\| \le C^\alpha\delta^{-2\alpha}\|u\|_{-\alpha}$ for any $u \in V$ and

$$\begin{aligned}\|T_n\tilde{P}u\| &\le \|T_n\tilde{P}\|_{\mathcal{L}(H)}\|\tilde{P}u\| \\ &\le \frac{1}{n}C^\alpha\delta^{-2\alpha}\|u\|_{-\alpha}.\end{aligned}$$

The proof of (3.96) is done.

By following the arguments in Lemma 3.42, we find $\|\tilde{u}(t_n) - u_n\| \le \Delta t\,\|\tilde{A}u_0\|$ or equivalently $\|T_n\tilde{A}^{-1}\|_{\mathcal{L}(H)} \le \Delta t$. Then, using also (3.85),

$$\begin{aligned}\|T_nu_0\| &\le \|T_n\tilde{A}^{-1}Au_0\| + \|T_n(A^{-1} - \tilde{A}^{-1})Au_0\| \\ &\le \|T_n\tilde{A}^{-1}\|_{\mathcal{L}(H)}\|Au_0\| + \|T_n\|_{\mathcal{L}(H)}\|(A^{-1} - \tilde{A}^{-1})Au_0\| \\ &\le \Delta t\,\|Au_0\| + \|T_n\|_{\mathcal{L}(H)}C\delta^2\|Au_0\|.\end{aligned}$$

This gives (3.95) in the case $\gamma = 1$. The extension to fractional power spaces follows from Exercise 3.20 because $\|T_n\tilde{P}A^{-\gamma}u_0\| \le \|T_n\tilde{P}A^{-1}u_0\|^\gamma\|T_n\tilde{P}u_0\|^{1-\gamma}$. □

For the total error in our approximation, we develop the analogue of Theorem 3.43 for the spectral Galerkin method (where $\delta = 1/\sqrt{\lambda_J}$). In this case, the two errors are balanced if $\Delta t/\delta^2$ is fixed in the limit $\delta \to 0$ and $\Delta t \to 0$ (the CFL condition as discussed after Theorem 3.43). Define $\tilde{T}_n$ by

$$\tilde{T}_nu_0 := (\mathrm{e}^{-t_nA} - \tilde{S}^n_{\Delta t}\tilde{P})u_0 = u(t_n) - \tilde{u}_n. \tag{3.97}$$

Theorem 3.54 (total error) *Suppose that Assumptions 3.19 and 3.48 hold and denote by $u(t)$ the solution of* (3.82) *and by $\tilde{u}_n$ the solution of* (3.84).

For $\alpha \in [0,1]$, there exists $K_1 > 0$ independent of δ and Δt such that

$$\|\tilde{T}_nu_0\| \le K_1\left(\frac{\Delta t\,\delta^{-2\alpha}}{t_n} + \frac{\delta^2}{t_n^{1+\alpha}}\right)\|u_0\|_{-\alpha}, \qquad \forall u_0 \in \mathcal{D}(A^\alpha), \quad n > 0. \tag{3.98}$$

For $\gamma \in (0,1]$, there exists $K_2 > 0$ independent of δ and Δt such that

$$\|\tilde{T}_nu_0\| \le K_2\frac{\Delta t + \delta^{2\gamma}\Delta t^{1-\gamma} + \delta^2}{t_n^{1-\gamma}}\|u_0\|_\gamma, \qquad \forall u_0 \in \mathcal{D}(A^\gamma), \quad n > 0. \tag{3.99}$$

Proof Simply sum the estimates in Propositions 3.52 and 3.53. □

Error analysis for the semilinear evolution equation

The linear error analysis is important for proving convergence of the *semilinear* evolution equation (3.54) of the approximation $\tilde{u}_n$ to $u(t_n)$ defined by

$$\tilde{u}_{n+1} = \tilde{u}_n - \Delta t\, \widetilde{A}\, \tilde{u}_{n+1} + \widetilde{P} f(\tilde{u}_n)\Delta t, \qquad \tilde{u}_0 = \widetilde{P} u_0. \tag{3.100}$$

For simplicity, we assume the initial data is smooth, specifically $u_0 \in \mathcal{D}(A)$.

Theorem 3.55 (convergence) *Suppose that A satisfies Assumption 3.19, $f\colon H \to H$ satisfies* (3.41), *and $u_0 \in \mathcal{D}(A)$. Let $\tilde{u}_n$ be given by* (3.100) *under Assumption 3.48. For all $\epsilon > 0$, there exists $K > 0$ independent of δ and Δt such that*

$$\max_{0 \le t_n \le T} \left\| u(t_n) - \tilde{u}_n \right\| \le K\big(\Delta t + \delta^2\big)\Delta t^{-\epsilon}.$$

If the CFL condition $\Delta t/\delta^2 = C$ holds, we see that

$$\max_{0 \le t_n \le T} \left\| u(t_n) - \tilde{u}_n \right\| \le \Delta t^{1-\epsilon}\big(K + K/C\big).$$

Proof Let $\widetilde{S}_{\Delta t} := (1 + \Delta t\, \widetilde{A})^{-1}$. Then, (3.100) becomes

$$\tilde{u}_{n+1} = \widetilde{S}_{\Delta t} \tilde{u}_n + \widetilde{S}_{\Delta t} \widetilde{P} f(\tilde{u}_n)\, \Delta t.$$

Over n steps, this gives

$$\tilde{u}_n = \widetilde{S}_{\Delta t}^n\, \widetilde{P}\, u_0 + \sum_{k=0}^{n-1} \widetilde{S}_{\Delta t}^{n-k}\, \widetilde{P} f(\tilde{u}_k)\, \Delta t. \tag{3.101}$$

Subtracting (3.101) from the variation of constants formula (3.40), the error $u(t_n) - \tilde{u}_n = \mathtt{I} + \mathtt{II}$ for

$$\begin{aligned}
\mathtt{I} &:= \mathrm{e}^{-t_n A} u_0 - \widetilde{S}_{\Delta t}^n\, \widetilde{P}\, u_0,\\
\mathtt{II} &:= \sum_{k=0}^{n-1} \left(\int_{t_k}^{t_{k+1}} \mathrm{e}^{-(t_n - s)A} f(u(s))\, ds - \widetilde{S}_{\Delta t}^{n-k}\, \widetilde{P}\, f(\tilde{u}_k) \Delta t \right).
\end{aligned}$$

For the first term, (3.99) with $\gamma = 1$ provides a $C_{\mathtt{I}} > 0$ such that

$$\|\mathtt{I}\| \le C_{\mathtt{I}}\big(\Delta t + \delta^2\big).$$

For the second term, write $\mathrm{e}^{-(t_n - s)A} f(u(s)) - \widetilde{S}_{\Delta t}^{n-k} \widetilde{P} f(\tilde{u}_k) = X_1 + X_2 + X_3 + X_4$, where

$$\begin{aligned}
X_1 &:= \big(\mathrm{e}^{-(t_n - s)A} - \mathrm{e}^{-(t_n - t_k)A}\big) f(u(s)), & X_2 &:= \big(\mathrm{e}^{-(t_n - t_k)A} - \widetilde{S}_{\Delta t}^{n-k} \widetilde{P}\big) f(u(s))),\\
X_3 &:= \widetilde{S}_{\Delta t}^{n-k} \widetilde{P}\big(f(u(s)) - f(u(t_k))\big), & X_4 &:= \widetilde{S}_{\Delta t}^{n-k} \widetilde{P}\big(f(u(t_k)) - f(\tilde{u}_k)\big).
\end{aligned}$$

For the first term,

$$\begin{aligned}\|\mathtt{II}_1\| &:= \left\|\sum_{k=0}^{n-1}\int_{t_k}^{t_{k+1}} X_1\,ds\right\| \\ &\le \sum_{k=0}^{n-1}\int_{t_k}^{t_{k+1}} \|(\mathrm{e}^{-(t_n-s)A} - \mathrm{e}^{-(t_n-t_k)A})f(u(s))\|\,ds \\ &\le \sum_{k=0}^{n-1}\int_{t_k}^{t_{k+1}} \|\mathrm{e}^{-(t_n-s)A} - \mathrm{e}^{-(t_n-t_k)A}\|_{\mathcal{L}(H)}\|f(u(s))\|\,ds.\end{aligned}$$

By Lemma 3.22 for $0 < s \le t$, there exists $K_1 > 0$ such that

$$\|\mathrm{e}^{-sA} - \mathrm{e}^{-tA}\|_{\mathcal{L}(H)} = \|A\mathrm{e}^{-sA}A^{-1}(I - \mathrm{e}^{-(t-s)A})\|_{\mathcal{L}(H)} \le K_1\frac{(t-s)}{s}.$$

Then, with (3.41) and (3.42),

$$\begin{aligned}\|\mathtt{II}_1\| &\le \left(\Delta t + \sum_{k=0}^{n-2}\int_{t_k}^{t_{k+1}} K_1\frac{s-t_k}{t_n-s}\,ds\right)L\sup_{0\le s\le t_n}(1+\|u(s)\|) \\ &\le \left(\Delta t + \sum_{k=0}^{n-2}\frac{\Delta t}{t_n-t_{k+1}}\int_{t_k}^{t_{k+1}} K_1\,ds\right)L\,(1+K_T)(1+\|u_0\|).\end{aligned}$$

The sum $\sum_{k=0}^{n-2}\frac{\Delta t}{t_n-t_{k+1}} = \sum_{k=1}^{n-1}1/k$ grows logarithmically in n. As $n \le T/\Delta t$, we can find $C_1 > 0$ such that $\|\mathtt{II}_1\| \le C_1\Delta t^{1-\epsilon}$.

Recall the definition of $\widetilde{T}_n$ from (3.97) and note that $X_2 = \widetilde{T}_{n-k}f(u(s))$. Then, by Theorem 3.54 with $\alpha = 0$, there exists $K_2 > 0$ such that

$$\begin{aligned}\|\mathtt{II}_2\| &= \left\|\sum_{k=0}^{n-1}\int_{t_k}^{t_{k+1}} X_2\,ds\right\| \le \sum_{k=0}^{n-1}\int_{t_k}^{t_{k+1}} \|\widetilde{T}_{n-k}f(u(s))\|\,ds \\ &\le \sum_{k=0}^{n-1}\int_{t_k}^{t_{k+1}} K_2\frac{\Delta t+\delta^2}{t_n-t_k}\|f(u(s))\|\,ds.\end{aligned}$$

Arguing as for $\mathtt{II}_1$, we can find $C_2 > 0$ such that $\|\mathtt{II}_2\| \le C_2(\Delta t+\delta^2)\,\Delta t^{-\epsilon}$.

For $\mathtt{II}_3$, apply the Lipschitz assumption (3.41) on f and $\|\widetilde{S}_{\Delta t}^{n-k}\|_{\mathcal{L}(H)}, \|\widetilde{P}\|_{\mathcal{L}(H)} \le 1$, to find

$$\begin{aligned}\|\mathtt{II}_3\| &= \left\|\sum_{k=0}^{n-1}\int_{t_k}^{t_{k+1}} X_3\,ds\right\| \le \sum_{k=0}^{n-1}\int_{t_k}^{t_{k+1}} \|f(u(s)) - f(u(t_k))\|\,ds \\ &\le L\sum_{k=0}^{n-1}\int_{t_k}^{t_{k+1}} \|u(s)-u(t_k)\|\,ds.\end{aligned}$$

The regularity of the solution in time provided by Proposition 3.31 with $u_0 \in \mathcal{D}(A)$ gives $C_3 > 0$ such that

$$\|\mathtt{II}_3\| \le K_{RT}\sum_{k=0}^{n-1}\int_{t_k}^{t_{k+1}} (s-t_k)^{1-\epsilon}\,ds \le C_3\Delta t^{1-\epsilon}.$$

For the final term, we use the Lipschitz condition (3.41) on f, to find

$$\|\mathtt{II}_4\| \le L \sum_{k=0}^{n-1} \int_{t_k}^{t_{k+1}} \|u(t_k) - u_{J,k}\| \, ds.$$

Combining all the estimates of $\|\mathtt{I}\|$ and $\|\mathtt{II}_i\|$, we have

$$\|u(t_n) - \tilde{u}_n\| \le E + L \sum_{k=0}^{n-1} \|u(t_k) - u_{J,k}\| \Delta t,$$

for $E := C_{\mathtt{I}}(\Delta t + \delta^2) + C_{\mathtt{II}}(\Delta t + \delta^2)\Delta t^{-\epsilon}$ and $C_{\mathtt{II}} = C_1 + C_2 + C_3$. The constants $C_{\mathtt{I}}, C_{\mathtt{II}}$ can be chosen independent of Δt and δ. Assuming that $L\Delta t \neq 1$, the discrete Gronwall lemma (Lemma A.14) provides

$$\|u(t_n) - \tilde{u}_n\| \le E \frac{1 - L\Delta t}{1 - L^n \Delta t^n} \le E. \qquad \square$$

3.8 Notes

There is a wealth of good references on the numerical solution of ODEs. See, for example, (Hairer and Wanner, 1996; Hairer, Nørsett, and Wanner, 1993; Iserles, 1996; Lambert, 1991). The book Stuart and Humphries (1996) takes a semigroup approach to time-stepping methods and examines dynamical properties of numerical approximations.

Classic texts on the properties of semigroups are (Henry, 1981; Pazy, 1983). This approach is also covered more recently, for example, in (Evans, 2010; Renardy and Rogers, 2004; Robinson, 2001; Temam, 1988). We examined semigroups e^{-At} for A subject to Assumption 3.19 and these are examples of *analytic* semigroups. That is, for the C_0-semigroup $S(t)\colon X \to X$ of Definition 3.15, the mapping $t \mapsto S(t)u$ is analytic on $[0, \infty)$ for all $u \in X$. It can be shown that if A is a so-called sectorial operator then $-A$ is the infinitesimal generator of an analytic semigroup. A sectorial operator has eigenvalues lying in a well-defined sector of the complex plane. Assumption 3.19 on the eigenvalues ensures that A is sectorial. In fact, it can be shown (Fujita and Suzuki, 1991) that if

$$Au = -\sum_{i,j=1}^{d} \frac{\partial}{\partial x_i}\left(a_{ij}(x)\frac{\partial u}{\partial x_j}\right) + \sum_{i=1}^{d} a_i(x)\frac{\partial u}{\partial x_i} + a_0(x)u,$$

where a_{ij}, a_j are continuously differentiable on $\bar{D}$ and the $d \times d$ matrix with entries a_{ij} is symmetric and positive definite, then both $-A$ and $-A_h$ are infinitesimal generators of an analytic semigroup, where A_h arises from a finite element discretisation based on a piecewise linear triangulation of D.

To make Example 3.14 and the infinitesimal generator precise, it is necessary to introduce the notion of uniform continuity. A function $u\colon \mathbb{R} \to \mathbb{R}$ is uniformly continuous if, for all $\epsilon > 0$, there exists $\delta > 0$ such that $|u(x) - u(y)| < \epsilon$ for all $x, y \in \mathbb{R}$ whenever $|x - y| < \delta$. Then, the set X of functions $u\colon \mathbb{R} \to \mathbb{R}$ that are bounded and uniformly continuous form a Banach space with the norm $\|u\|_\infty = \sup_{x \in \mathbb{R}} |u(x)|$. For this choice of Banach space X, the limit in (3.27) is well defined.

As in §2.1, the semilinear evolution equation has different forms of solutions and we can talk about 'classical' and 'strong' solutions, in addition to the mild and weak solutions found in Definitions 3.28 and 3.36. Roughly, a classical solution requires that u has continuous space and time derivatives. For a strong solution, we reduce the continuity requirement and interpret the derivatives in a weak sense and seek solutions in a Sobolev space. In general, a strong or weak solution is a mild solution. The reverse implication is more difficult and the mild solution is a classical solution if $u_0 \in \mathcal{D}(A)$ and some additional assumptions on f hold. For further discussion, see for example (Evans, 2010; Renardy and Rogers, 2004; Robinson, 2001).

We established existence and uniqueness of a mild solution in §3.3 for Lipschitz continuous functions $f\colon H \to H$ in (3.39) and the examples focused on Nemytskii operators. Examples where f is non-local, such as in Exercise 3.16, lead to f that cannot be written as a Nemytskii operator. Our definition of Lipschitz continuous in (3.41) is the *global Lipschitz condition*, as the constant is uniform for all $u_1, u_2 \in H$, and is too restrictive for most PDEs of interest. See, for example, the Allen–Cahn equation (3.3) and the Nagumo equation (3.52). Results on existence and uniqueness of mild and weak solutions can be proved for locally Lipschitz nonlinearities $f(u)$ with conditions on the growth as $\|u\| \to \infty$. These are often based on a spectral Galerkin approximation and an energy argument is used to obtain uniform estimates and pass to the limit. See, for example, (Evans, 2010; Robinson, 2001; Temam, 1988).

The numerical solution of time-dependent PDEs is a vast topic. We only considered in any detail a semi-implicit Euler time-stepping method best applied to parabolic semilinear PDEs and we touched on three types of spatial approximation. We considered finite difference methods in §3.4 and these methods are widely used to discretise PDEs because of their simplicity. More details on their analysis can be found, for example, in (Morton and Mayers, 2005; Strikwerda, 2004).

The spectral Galerkin methods are particularly well suited to solve PDEs such as (3.39) on regular domains for Neumann, Dirichlet, or periodic boundary conditions where the eigenfunctions of $-A$ are known. In cases where the solutions are regular, exponential rates of convergence can be observed as in Figure 3.6. The books (Canuto et al., 1988; Gottlieb and Orszag, 1977) are comprehensive texts on spectral methods and their analysis. More recently, Trefethen (2000) gives an introduction to a broad class of spectral methods. In our implementation of the spectral Galerkin method, we compute $P_J f$ by using the trigonometric interpolant $I_J f$ and ignored aliasing errors in computing the nonlinearities. These errors arise from the fact that $P_J u\, P_J v \neq P_J uv$ for $u, v \in H$. The approximation error $\|P_J f - I_J f\|$ is examined in Exercise 1.23(a) and is small if f is smooth. A standard technique to control this type of error is to compute any nonlinear terms using the *2/3rds rule* (Canuto et al., 1988) and to perform computations with $3J/2$ coefficients combined with an orthogonal projection onto J coefficients.

Finite element methods are commonly used to solve a wide range of time-dependent and stationary problems. In general, they are better able to approximate solutions of PDEs on more complicated domains and with more complicated boundary conditions than spectral Galerkin methods. See, for example, (Hundsdorfer and Verwer, 2003; Quarteroni and Valli, 2008; Thomée, 2006) on time-dependent problems and the references given in the Notes of Chapter 2.

One important family of spatial discretisations that we have not considered is finite volume methods; see, for example, Eymard, Gallouët, and Herbin (2000). These are often preferred for numerical simulations that are subject to conservation laws. They have similar features to finite element methods.

The non-smooth error analysis presented in §3.7 follows that in Thomée (2006). However, Thomée (2006) considers more general approximations in time in a finite element context. The exponents in (3.85) and (3.86) need not be 2, which is suitable for the elliptic operators that we consider, and Assumption 3.48 can be extended to include more general problems. In Theorem 3.55, the assumption on $f\colon H \to H$ is weak and results in an ill-behaved $\Delta t^{-\epsilon}$ term in the error bound. This does not show up in our numerical experiments as $f(u)$ has more regularity.

Exercises

3.1 Consider the differential equation

$$\frac{d^n u}{dt^n} + a_{n-1}\frac{d^{n-1}u}{dt^{n-1}} + \cdots + a_0 u = f(u)$$

for $a_0,\ldots,a_{n-1} \in \mathbb{R}$, subject to $\frac{d^k u(0)}{dt^k} = u_0^k \in \mathbb{R}$ for $k = 0,\ldots,n-1$. Rewrite the equation for u as an IVP for a first-order system of ODEs in terms of $\boldsymbol{u}(t) = \left[u(t),\ldots,\frac{d^{n-1}u(t)}{dt^{n-1}}\right]^{\mathsf{T}}$.

3.2 Show that the solution of the IVP

$$\frac{du}{dt} = u^2, \qquad u(0) = 1$$

is not well defined for all $t > 0$. Why does Theorem 3.2 not apply?

3.3 Show that there are at least two solutions to the IVP

$$\frac{du}{dt} = 2|u|^{1/2}, \qquad u(0) = 0.$$

Why does Theorem 3.2 not apply?

3.4 Suppose that

$$z(t) \le z_0 + \int_0^t \big(a + bz(s)\big)\,ds \tag{3.102}$$

for some constants $a, b \in \mathbb{R}$. Prove that, if $b \neq 0$,

$$z(t) \le \mathrm{e}^{bt} z_0 + \frac{a}{b}(\mathrm{e}^{bt} - 1), \qquad t \ge 0.$$

This is known as *Gronwall's inequality*.

3.5 Determine the region of absolute stability for the trapezoidal method given in (3.20).

3.6 Prove that the trapezoidal method (3.20) is of second order. State any assumptions you make on the nonlinearity $\boldsymbol{f}$ and on the solution $\boldsymbol{u}(t)$.

3.7 a. Implement Heun's method (3.21) in Matlab for the population model (3.14) and approximate $\boldsymbol{u}(t)$ for $t \in [0,10]$ starting from $\boldsymbol{u}_0 = [0.5, 0.1]^{\mathsf{T}}$ with $\Delta t = 0.01$.

b. Test the convergence of Heun's method numerically using method A, following Example 3.8. Estimate the order of convergence by fitting a linear polynomial (use the Matlab command `polyfit`) to the estimated errors versus Δt on a log-log plot.

3.8 Use Heun's method (3.21) to estimate by method C the error in the explicit Euler approximation to the solution of (3.14) at time $T = 1$. Take the reference time step $\Delta t_{\text{ref}} = 10^{-6}$ and consider $\Delta t_\kappa = 10^{-5}, 10^{-4}, 10^{-3}, 10^{-2}, 10^{-1}$. Recall methods A and B are considered in Example 3.8.

3.9 Suppose that $\boldsymbol{u}(T)$ is approximated by $\boldsymbol{u}_N$ with time step Δt such that

$$\boldsymbol{u}(T) = \boldsymbol{u}_N + C_1 \Delta t + \mathcal{O}(\Delta t^2), \qquad N\Delta t = T,$$

for C_1 independent of Δt. Show that if $\boldsymbol{u}_{N_0}$ is computed with Δt_0 and $\boldsymbol{u}_{N_1}$ with $\Delta t_1 = \kappa \Delta t_0$, for some $\kappa \in \mathbb{N}$, then

$$\boldsymbol{u}_R := \frac{\kappa \boldsymbol{u}_{N_1} - \boldsymbol{u}_{N_0}}{\kappa - 1} = \boldsymbol{u}(T) + \mathcal{O}(\Delta t_0^2).$$

This is known as *Richardson extrapolation.*

3.10 Use the explicit Euler method to solve (3.14) with time steps $\Delta t_1 = 0.01$ and $\Delta t_0 = 0.1$, taking $u_1(0) = 0.5$ and $u_2(0) = 0.1$. Estimate (by method A) the error at $T = 1$ for each approximation using a reference solution computed with time step $\Delta t_{\text{ref}} = 10^{-6}$. Estimate the error in $\boldsymbol{u}_R$ from Exercise 3.9.

3.11 Consider the wave equation

$$u_{tt} = \Delta u, \qquad \text{given } u(0,x) = u_0(x) \text{ and } u_t(0,x) = v(x)$$

with homogeneous Dirichlet boundary conditions on $D = (0,1)$. Write this as a first-order system in time for $\boldsymbol{u}(t) = [u_1(t), u_2(t)]^{\mathsf{T}}$ to obtain a semilinear equation of the form $\frac{d\boldsymbol{u}}{dt} = -A\boldsymbol{u}$ where $\mathcal{D}(A) = H_0^1(D) \times L^2(D)$.

3.12 Consider the finite difference approximation to (3.44) with Dirichlet boundary conditions on $D = (0,a)$ of the form $u(t,0) = g_0(t)$ and $u(t,a) = g_a(t)$. Introduce a vector $\boldsymbol{b}(t) = [g_0(t), 0, \dots, 0, g_a(t)]^{\mathsf{T}}$, and write down the semilinear system of ODEs to be solved.

3.13 Consider the finite difference approximation to (3.44) with periodic boundary conditions on $D = (0,a)$. With $\boldsymbol{u}_J = [u_1, \dots, u_J]^{\mathsf{T}}$, write down the semilinear system of ODEs to be solved.

3.14 Use the variation of constants formula (3.40) to prove Theorem 3.34.

3.15 Show that the Langmuir reaction term in (3.38) is globally Lipschitz continuous with $\alpha = 1$.

3.16 Consider the PDE

$$u_t = \varepsilon\, \Delta u - (\boldsymbol{v} \cdot \nabla) u \tag{3.103}$$

where $\varepsilon > 0$, $\boldsymbol{v} = (\psi_y, -\psi_x)^{\mathsf{T}}$, and $\Delta\psi = -u$ (see Example 10.3). Use a spectral Galerkin approximation in space and semi-implicit Euler method in time to solve (3.103) in Matlab. At each time step, the Poisson equation $\Delta\psi = -u$ needs to be solved for ψ. Take $\boldsymbol{x} \in D = (0,2\pi) \times (0,2\pi)$ with periodic boundary conditions, $\varepsilon = 0.001$, and initial data $u(0,\boldsymbol{x}) = \sin(x_1)\cos(4x_2)$. Solve until $T = 100$ with $\Delta t = 0.01$ and $J_1 = J_2 = 64$. Hint: modify Algorithm 3.6. Owing to the nature of (3.103), it is simpler to explicitly code the nonlinearity rather than use `fhandle`.

3.17 Consider the semilinear IVP (3.18) and use the variation of constants formula (3.34) to derive the numerical method

$$u_{n+1} = e^{-\Delta t\, M} u_n + \phi_1(-\Delta t\, M) f(u_n)\, \Delta t \qquad (3.104)$$

where $\phi_1(z) := z^{-1}(e^z - 1)$. This is an example of an *exponential integrator*.

3.18 a. Implement the method (3.104) in MATLAB for the spectral Galerkin approximation of the PDE

$$u_t = \varepsilon\, \Delta u + u - u^3$$

on $D = (0, 2\pi) \times (0, 2\pi)$ with periodic boundary conditions, $\varepsilon = 0.001$, and initial data $u(0, \boldsymbol{x}) = \sin(x_1)\cos(x_2)$. Test your code with $J_1 = J_2 = 64$ and $\Delta t = 10^{-4}$.

b. By numerically estimating the error using method A for a fixed spatial resolution, show that the time discretisation error is order Δt. Take $\Delta t_{\text{ref}} = 10^{-4}$ and $\Delta t_\kappa = 5 \times 10^{-4}, 10^{-3}, 2 \times 10^{-3}, 10^{-2}, 2 \times 10^{-2}$ and $J_1 = J_2 = 64$.

3.19 Show that

$$0 \le (1 + x)^{-n} - e^{-nx} \le \frac{1}{n}, \qquad x \ge 0, \quad n \in \mathbb{N}.$$

3.20 Let A satisfy Assumption 1.85. By using Hölder's inequality (1.7), show that for $0 \le \alpha \le 1$

$$\|A^\alpha u\| \le \|Au\|^\alpha \|u\|^{1-\alpha}, \qquad u \in \mathcal{D}(A).$$

This is known as an *interpolation inequality*. For $B \in \mathcal{L}(H)$, show that

$$\|BA^\alpha u\| \le \|BAu\|^\alpha \|Bu\|^{1-\alpha}, \qquad u \in \mathcal{D}(A).$$

3.21 Use an energy argument to prove Lemma 3.50. Hint: see the proof of Lemma 3.51.

4
Probability Theory

Solutions to stochastic differential equations are random variables and this chapter develops the concept of random variables. §4.1 reviews probability theory and interprets the notions of measure space and measurable function from Chapter 1 in the probabilistic setting. §4.2 explores the relationship between expectation and least-squares approximation and develops the concept of conditional expectation. §4.3 outlines modes of convergence of random variables and includes the law of large numbers and the central limit theorem. Finally, in preparation for solving stochastic differential equations numerically, §4.4 discusses random number generation and sampling methods for random variables, as well as the ubiquitous Monte Carlo method.

4.1 Probability spaces and random variables

Think of an experiment, such as rolling a die or tossing a coin, with an outcome that changes randomly with each repetition. As the experiment is repeated, the frequencies of the outcomes vary and statistical and probabilistic tools are needed to analyse the results. In particular, we assign probabilities to each outcome as a limit of the frequency of occurrence relative to the total number of trials. These ideas are simple and intuitive in an experiment with a *finite* number of possibilities such as tossing a coin or a die. To express these concepts in an *uncountable* setting, such as the stochastic processes and random fields arising in the study of stochastic PDEs, we use an abstract measure space $(\Omega, \mathcal{F}, \mathbb{P})$, called a probability space. We review first the concepts of sample space Ω, σ-algebra $\mathcal{F}$, and probability measure $\mathbb{P}$.

Probability space $(\Omega, \mathcal{F}, \mathbb{P})$

In an experiment, the sample space Ω is the set of all possible outcomes. For a single coin toss, there are only two possible outcomes and $\Omega = \{H, T\}$ for head or tail. In a sequence of M coin tosses, there are 2^M possible outcomes and $\Omega = \{H, T\}^M$. For a single roll of a die, there are six possible outcomes and $\Omega = \{1, 2, 3, 4, 5, 6\}$ and, in a sequence of M rolls, $\Omega = \{1, 2, 3, 4, 5, 6\}^M$. In general, the sample space is simply an abstract set, one large enough to parameterize all possible outcomes of the experiment.

$\mathcal{F}$ is a collection of subsets of Ω (a subset of the power set of Ω) and comprises those subsets to which we can assign probabilities. Formally, it is a σ-algebra (see Definition 1.15).

Any measurable set $F \in \mathcal{F}$ is known as an *event* and each event corresponds to a set of outcomes in the experiment. For example, in a model of three coin tosses, the probability of obtaining *all heads or all tails* is of interest and so the set $F = \{(H,H,H),(T,T,T)\}$ is included as an event in the σ-algebra $\mathcal{F}$.

Recall that measures were introduced in Definition 1.17. For a probability space, the probability measure $\mathbb{P}$ assigns a probability, a number between 0 and 1, to every measurable set in $\mathcal{F}$; sets with greater probabilities represent events that are more likely to occur in an experiment. For the single coin toss experiment, both $\{H\}$ and $\{T\}$ are elements of the σ-algebra $\mathcal{F} = \{\{\},\{H\},\{T\},\{H,T\}\}$ of subsets of $\Omega = \{H,T\}$ and $\mathbb{P}(\{H\}) = \mathbb{P}(\{T\}) = 1/2$ if the coin is fair.

Definition 4.1 (probability measure) A measure $\mathbb{P}$ on $(\Omega,\mathcal{F})$ is a *probability measure* if it has unit total mass, $\mathbb{P}(\Omega) = 1$.

Definition 4.2 (almost surely) An event F happens *almost surely (a.s. or* $\mathbb{P}$*-a.s.)* with respect to a probability measure $\mathbb{P}$ if $\mathbb{P}(F) = 1$.

In the single coin toss experiment, $\{H,T\} = \Omega$ and the only set of measure zero is the empty set so that throwing *a head or a tail* happens surely. For an example of almost surely, see Exercise 4.1.

Random variables and expectations

In an experiment, measurements or observations are modelled mathematically as *random variables*. These are functions that assign values to each outcome $\omega \in \Omega$. For example, in a sequence of M coin tosses, the number of observed heads is a random variable and is modelled as a function X from the sample space $\Omega = \{H,T\}^M$ to the set $\{0,1,\dots,M\}$. The probability measure $\mathbb{P}$ on Ω describes the statistical behaviour of X, the number of heads observed. As a concrete example, consider the probability of obtaining one head, which is written $\mathbb{P}(F)$ where $F = \{\omega \in \Omega\colon X(\omega) = 1\} \in \mathcal{F}$ or $\mathbb{P}(X = 1)$ for short. If $M = 3$, this is $\mathbb{P}\big(\{(H,T,T),(T,H,T),(T,T,H)\}\big) = 3/8$ if the coin is fair.

Recalling the definition of a *measurable function* (see Definition 1.19), we rigorously define random variables.

Definition 4.3 (random variables, realisation) Let $(\Omega,\mathcal{F},\mathbb{P})$ be a probability space and let $(\Psi,\mathcal{G})$ be a measurable space. Then, X is a *Ψ-valued random variable* if X is a measurable function from $(\Omega,\mathcal{F})$ to $(\Psi,\mathcal{G})$. To emphasise the σ-algebra on Ω, we may write that X is an $\mathcal{F}$-measurable random variable. The observed value of $X(\omega)$ for a given $\omega \in \Omega$ belongs to Ψ and is called a *realisation* of X.

We often consider *real-valued* random variables X with $(\Psi,\mathcal{G}) = (\mathbb{R},\mathcal{B}(\mathbb{R}))$, where $\mathcal{B}(\mathbb{R})$ is the Borel σ-algebra (see Definition 1.16).

Every random variable X from $(\Omega,\mathcal{F})$ to $(\Psi,\mathcal{G})$ has an associated probability distribution $\mathbb{P}_X$ defined as follows.

Definition 4.4 (probability distribution) Let X be a Ψ-valued random variable where $(\Psi,\mathcal{G})$ is a measurable space and $(\Omega,\mathcal{F},\mathbb{P})$ is the underlying probability space. The *probability distribution* $\mathbb{P}_X$ *of* X (also called the law of X) is the probability measure on $(\Psi,\mathcal{G})$ defined by $\mathbb{P}_X(G) := \mathbb{P}\big(X^{-1}(G)\big)$ for pullback sets $X^{-1}(G) := \{\omega \in \Omega\colon X(\omega) \in G\}$ and $G \in \mathcal{G}$.

By establishing that $\mathbb{P}_X$ is a measure on $(\Psi, \mathcal{G})$ (see Exercise 4.2), it follows immediately that $\mathbb{P}_X$ is a valid probability measure as $\mathbb{P}_X(\Psi) = \mathbb{P}(X^{-1}(\Psi)) = \mathbb{P}(\Omega) = 1$. This means that $(\Psi, \mathcal{G}, \mathbb{P}_X)$ is also a probability space.

Example 4.5 (single coin toss) Consider the single coin toss experiment with $\Omega = \{H, T\}$. If X denotes the number of heads, $X(H) = 1$ and $X(T) = 0$ and X maps from Ω to the measurable space $(\Psi, \mathcal{G})$, with $\Psi = \{0, 1\}$ and with $\mathcal{G} = \{\{\}, \{0\}, \{1\}, \{0, 1\}\}$. The pullback sets are $X^{-1}(\{0\}) = \{T\}$, $X^{-1}(\{1\}) = \{H\}$, $X^{-1}(\{0, 1\}) = \Omega$ and $X^{-1}(\{\}) = \{\}$. The probability distribution $\mathbb{P}_X$ on $(\Psi, \mathcal{G})$ satisfies $\mathbb{P}_X(\{0\}) = \mathbb{P}(\{T\}) = 1/2$ and $\mathbb{P}_X(\{1\}) = \mathbb{P}(\{H\}) = 1/2$.

Statistical quantities, such as the mean, variance and higher-order moments, are defined as expectations of random variables. Recall that integrals of Banach space-valued functions were described in Definition 1.21.

Definition 4.6 (expectation) Let X be a Banach space-valued random variable on the probability space $(\Omega, \mathcal{F}, \mathbb{P})$. If X is integrable, the *expectation* of X is

$$\mathbb{E}[X] := \int_\Omega X(\omega)\, d\mathbb{P}(\omega), \tag{4.1}$$

the integral of X with respect to the probability measure $\mathbb{P}$. If X is an integer-valued random variable,

$$\mathbb{E}[X] := \sum_{j \in \mathbb{Z}} j\, \mathbb{P}(X = j). \tag{4.2}$$

Definition 4.7 (moments) The kth order moment of a real-valued random variable X is $\mathbb{E}[X^k]$. The first moment $\mu := \mathbb{E}[X]$ is the expectation or *mean* value. The moments of the deviation from the mean $\mathbb{E}[(X - \mu)^k]$, known as central moments or moments about the mean, are important. The second central moment $\mathbb{E}[(X - \mu)^2]$ measures the spread of the distribution around the mean and is known as the *variance*,

$$\operatorname{Var} X := \mathbb{E}[(X - \mu)^2] = \mathbb{E}[X^2] - \mu^2.$$

The quantity $\sigma := \sqrt{\operatorname{Var} X}$ is known as the *standard deviation*.

It is not usually possible to compute the integral in (4.1) directly. However, we can perform a change of variables or, equivalently, change the measure. To see this, consider a real-valued random variable X from $(\Omega, \mathcal{F})$ to $(D, \mathcal{B}(D))$ where $D \subset \mathbb{R}$. Choose any $B \in \mathcal{B}(D)$ and define $F = X^{-1}(B) \in \mathcal{F}$. Denote the probability distribution of X by $\mathbb{P}_X$. Definition 4.4 with $\Psi = D$ gives

$$\int_\Omega 1_F(\omega)\, d\mathbb{P}(\omega) = \mathbb{P}(F) = \mathbb{P}_X(B) = \int_D 1_B(x)\, d\mathbb{P}_X(x).$$

For integrable functions $g \colon D \to \mathbb{R}$, it follows from the construction in Definition 1.21 that

$$\int_\Omega g(X(\omega))\, d\mathbb{P}(\omega) = \int_D g(x)\, d\mathbb{P}_X(x). \tag{4.3}$$

In particular,

$$\mathbb{E}[X] = \int_\Omega X(\omega)\, d\mathbb{P}(\omega) = \int_D x\, d\mathbb{P}_X(x). \tag{4.4}$$

Observe that the expectation is now expressed as an integral over the observation space D rather than the abstract sample space Ω. In many cases, integrals with respect to $\mathbb{P}_X$ can be expressed as integrals with respect to a so-called probability density function (pdf) and then computed using standard calculus.

Definition 4.8 (pdf) Let $\mathbb{P}$ be a probability measure on $(D, \mathcal{B}(D))$ for some $D \subset \mathbb{R}$. If there exists a function $p\colon D \to [0, \infty)$ such that $\mathbb{P}(B) = \int_B p(x)\,dx$ for any $B \in \mathcal{B}(D)$, we say $\mathbb{P}$ has a density p with respect to Lebesgue measure and we call p the *probability density function* (pdf). If X is a D-valued random variable on $(\Omega, \mathcal{F}, \mathbb{P})$, the pdf p_X of X (if it exists) is the pdf of the probability distribution $\mathbb{P}_X$.

When it is clear that we are speaking about the random variable X, we omit the subscript and simply write $p(x)$ for the pdf of X. Let X be a real-valued random variable from $(\Omega, \mathcal{F})$ to $(D, \mathcal{B}(D))$ with $D \subset \mathbb{R}$. Using Definition 4.8 and (4.4), it follows that

$$\mathbb{E}[X] = \int_\Omega X(\omega)\,d\mathbb{P}(\omega) = \int_D x\,d\mathbb{P}_X(x) = \int_D x p(x)\,dx. \tag{4.5}$$

If the pdf $p(x)$ of X is known, calculating probabilities of events is also straightforward. For example,

$$\mathbb{P}\big(X \in (a,b)\big) = \mathbb{P}\big(\{\omega \in \Omega\colon a < X(\omega) < b\}\big) = \mathbb{P}_X\big((a,b)\big) = \int_a^b p(x)\,dx.$$

Examples of real-valued random variables

We give some of the distributions and corresponding pdfs commonly encountered in practice. A function $p\colon D \to \mathbb{R}$ is a well-defined pdf of a D-valued random variable if $p(x) \geq 0$ for all $x \in D$ and

$$\int_D p(x)\,dx = 1. \tag{4.6}$$

Example 4.9 (uniform distribution) A random variable X is *uniformly distributed on* $D = [a,b]$ if its pdf $p(x) = 1/(b-a)$ for $a \leq x \leq b$. We write $X \sim \mathrm{U}(a,b)$. Using (4.5), we have

$$\mathbb{E}[X] = \int_a^b x \frac{1}{b-a}\,dx = \frac{a+b}{2}, \qquad \mathbb{E}[X^2] = \int_a^b x^2 \frac{1}{b-a}\,dx = \frac{b^3 - a^3}{3(b-a)},$$

so that $\operatorname{Var} X = \mathbb{E}[X^2] - \mathbb{E}[X]^2 = (b-a)^2/12$. Notice that $\int_a^b (b-a)^{-1}\,dx = 1$ and that the pdf $p(x)$ defines a probability measure.

Example 4.10 (Gaussian distribution) A random variable X is said to follow the *Gaussian or normal distribution* on $D = \mathbb{R}$ if, for some $\mu \in \mathbb{R}$ and $\sigma > 0$, its pdf is

$$p(x) = \frac{1}{\sqrt{2\pi\sigma^2}} \exp\left(\frac{-(x-\mu)^2}{2\sigma^2}\right). \tag{4.7}$$

We write $X \sim \mathrm{N}(\mu, \sigma^2)$. By Exercise 4.3, $p(x)$ is a valid pdf and the random variable X has mean μ and variance σ^2.

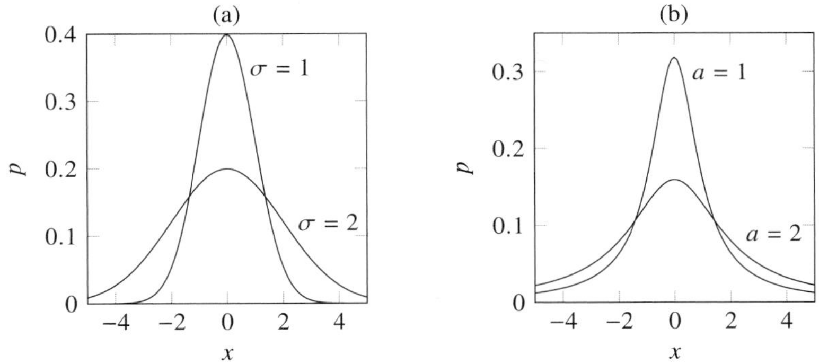

Figure 4.1 (a) pdf of $X \sim \mathrm{N}(0, \sigma^2)$ with $\sigma = 1$ and $\sigma = 2$; (b) pdf of $X \sim \mathrm{Cauchy}(0, a)$ with $a = 1$ and $a = 2$. Although the two pdfs look similar, the Cauchy distribution decays so slowly at infinity that its moments are infinite.

The probability that X is within R of the mean, $\mathbb{P}(|X - \mu| \le R)$, is given in terms of the error function $\operatorname{erf}(x) := (2/\sqrt{\pi}) \int_0^x \mathrm{e}^{-t^2}\,dt$ (see Exercise 4.4) by

$$\mathbb{P}\Big(|X - \mu| \le R\Big) = \operatorname{erf}\left(\frac{R}{\sqrt{2\sigma^2}}\right). \tag{4.8}$$

Example 4.11 (Cauchy distribution) A random variable X follows the *Cauchy distribution* on $D = \mathbb{R}$, written $X \sim \mathrm{Cauchy}(x_0, a)$, if its pdf is

$$p(x) = \frac{a}{\pi(a^2 + (x - x_0)^2)}.$$

The parameters x_0 and a control the location of the peak of the distribution and the spread, respectively. Neither the mean nor variance of X is defined, however, as the moments are infinite; see Exercise 4.5. Figure 4.1 shows the pdf of the Cauchy distribution in comparison to the Gaussian distribution.

Lemma 4.12 (change of variable) *Suppose $Y : \Omega \to \mathbb{R}$ is a real-valued random variable and $g : (a, b) \to \mathbb{R}$ is continuously differentiable with inverse function g^{-1}. If p_Y is the pdf of Y, the pdf of the random variable $X : \Omega \to (a, b)$ defined via $X = g^{-1}(Y)$ is*

$$p_X(x) = p_Y(g(x))|g'(x)|, \qquad \textit{for } a < x < b.$$

Proof Let $D = (a, b)$. X is a random variable from $(\Omega, \mathcal{F})$ to $(D, \mathcal{B}(D))$ and $Y = g(X)$ is a random variable from $(\Omega, \mathcal{F})$ to $(\mathbb{R}, \mathcal{B}(\mathbb{R}))$. Let $B \in \mathcal{B}(D)$ and $g(B) := \{g(x) : x \in B\} \in \mathcal{B}(\mathbb{R})$. If $g(x)$ is monotonically increasing, $y = g(x)$ and $dy = g'(x)dx$ gives

$$\mathbb{P}_X(B) = \mathbb{P}_Y(g(B)) = \int_{g(B)} p_Y(y)\,dy = \int_B p_Y(g(x))|g'(x)|\,dx.$$

The same formula holds if $g(x)$ is monotonically decreasing because $g(B)$ is not oriented. From Definition 4.8, we see that the pdf of X is indeed $p_Y(g(x))|g'(x)|$. □

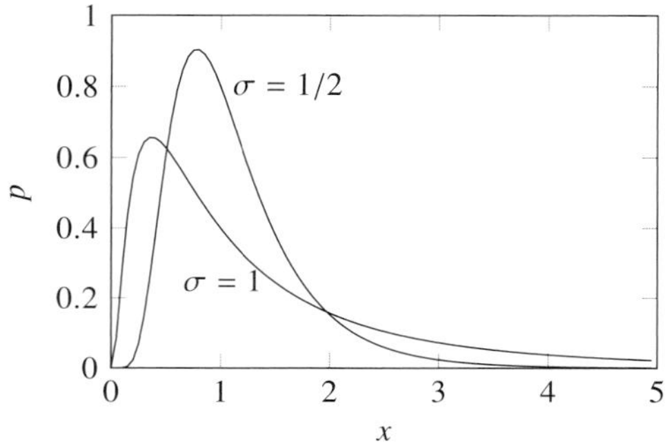

Figure 4.2 The pdf of a log-normal random variable $X = e^Y$, where $Y \sim N(0, \sigma^2)$ with $\sigma = 1/2$, 1.

Example 4.13 (log-normal distribution) If $Y \sim N(0, \sigma^2)$, the random variable $X = e^Y$ is known as a *log-normal* random variable and is an important example of a random variable taking values in $D = (0, \infty)$. With $g(x) = \log x$, Lemma 4.12 gives the pdf of X for $0 < x < \infty$ as

$$p_X(x) = p_Y(g(x))|g'(x)| = \frac{p_Y(\log x)}{x} = \frac{1}{\sqrt{2\pi\sigma^2 x^2}} e^{-(\log x)^2/2\sigma^2}.$$

This is illustrated in Figure 4.2. Using the pdf, it can then be verified that

$$\mathbb{E}\big[X\big] = e^{\sigma^2/2}, \qquad \operatorname{Var} X = (e^{\sigma^2} - 1)e^{\sigma^2}. \tag{4.9}$$

Correlation and independence

Let X and Y be two real-valued random variables on $(\Omega, \mathcal{F}, \mathbb{P})$ so that X is measurable from $(\Omega, \mathcal{F})$ to $(D_X, \mathcal{B}(D_X))$ and Y is measurable from $(\Omega, \mathcal{F})$ to $(D_Y, \mathcal{B}(D_Y))$, with $D_X \subset \mathbb{R}$ and $D_Y \subset \mathbb{R}$. To describe the relationship between the random variables, we introduce the concept of covariance.

Definition 4.14 (covariance) The *covariance* of two real-valued random variables X and Y is

$$\operatorname{Cov}(X, Y) := \mathbb{E}\Big[(X - \mu_X)(Y - \mu_Y)\Big] = \mathbb{E}\big[XY\big] - \mu_X \mu_Y,$$

where $\mu_X = \mathbb{E}[X]$ and $\mu_Y = \mathbb{E}[Y]$. In particular, $\operatorname{Cov}(X, X) = \operatorname{Var} X$.

To obtain $\operatorname{Cov}(X, Y)$, we need to compute $\mathbb{E}[XY]$. The product XY is a function of the bivariate random variable $\boldsymbol{X} = [X, Y]^{\mathsf{T}} : \Omega \to D_X \times D_Y$ so

$$\mathbb{E}\big[XY\big] = \int_\Omega X(\omega)Y(\omega)\, d\mathbb{P}(\omega) = \int_{D_X \times D_Y} xy\, d\mathbb{P}_{X,Y}(x, y), \tag{4.10}$$

where $\mathbb{P}_{X,Y}$ is the so-called joint probability distribution of $\boldsymbol{X}$.

Definition 4.15 (joint probability distribution) The *joint probability distribution* of X and Y is the probability distribution $\mathbb{P}_{X,Y}$ of the bivariate random variable $\boldsymbol{X} = [X, Y]^{\mathsf{T}}$. That is,

$\mathbb{P}_{X,Y}(B) := \mathbb{P}(\{\omega \in \Omega \colon X(\omega) \in B\})$ for $B \in \mathcal{B}(D_X \times D_Y)$ (see Definition 1.23). If it exists, the density $p_{X,Y}$ of $\mathbb{P}_{X,Y}$ is known as the *joint pdf* and

$$\mathbb{P}_{X,Y}(B) = \int_B p_{X,Y}(x,y)\,dx\,dy. \tag{4.11}$$

Often the covariance is rescaled so it lies in the range $[-1,1]$ leading to the so-called *correlation coefficient*

$$\rho(X,Y) := \frac{\mathrm{Cov}(X,Y)}{\sigma_X \sigma_Y}, \tag{4.12}$$

where σ_X, σ_Y are the standard deviations of X and Y respectively (see Exercise 4.6). We do not use the correlation coefficient in this book. If $\mathrm{Cov}(X,Y) > 0$, then simultaneous realisations of $X - \mu_X$ and $Y - \mu_Y$ tend to have the same sign, as in the case $X = Y$ where $\mathrm{Cov}(X,X) = \mathrm{Var}\,X \geq 0$. If $\mathrm{Cov}(X,Y) < 0$ then simultaneous realisations of $X - \mu_X$ and $Y - \mu_Y$ tend to have the opposite sign. We see examples of random variables with positive and negative correlation when we discuss stochastic processes and random fields.

Uncorrelated random variables are also important.

Definition 4.16 (uncorrelated random variables) If $\mathrm{Cov}(X,Y) = 0$, the random variables X,Y are said to be *uncorrelated*. We say a family of random variables $\{X_j\}$ are *pairwise uncorrelated* if X_i and X_j are uncorrelated for $i \neq j$.

Random variables that are uncorrelated may actually be strongly related to each other. For example, if

$$X \sim \mathrm{N}(0,1) \quad \text{and} \quad Y = \cos(X),$$

then $\mu_X = 0$ and $\mathrm{Cov}(X,Y) = \mathbb{E}[X\cos(X)]$. Using (4.3) with $D = \mathbb{R}$ and (4.7) with $\mu = 0$ and $\sigma^2 = 1$ gives

$$\mathbb{E}\big[X\cos(X)\big] = \int_{\mathbb{R}} x\cos(x)\,d\mathbb{P}_X(x) = \frac{1}{\sqrt{2\pi}}\int_{\mathbb{R}} x\cos(x)\exp\left(\frac{-x^2}{2}\right)dx = 0.$$

Hence, X and Y are uncorrelated. To express the stronger idea of *independent* random variables, we first introduce sub σ-algebras.

Definition 4.17 (sub σ-algebra) A σ-algebra $\mathcal{G}$ is a *sub σ-algebra* of $\mathcal{F}$ if $F \in \mathcal{F}$ for every $F \in \mathcal{G}$.

Every random variable X on $(\Omega, \mathcal{F}, \mathbb{P})$ generates a sub σ-algebra of $\mathcal{F}$.

Definition 4.18 (σ-algebra generated by X) Let X be a Ψ-valued random variable where $(\Psi, \mathcal{G})$ is a measurable space and $(\Omega, \mathcal{F}, \mathbb{P})$ is the underlying probability space. The *σ-algebra generated by X*, denoted $\sigma(X)$, is the set of pullback sets $\{X^{-1}(G) \colon G \in \mathcal{G}\}$.

Using Definition 1.15, it is easy to prove that $\sigma(X)$ is a σ-algebra on Ω (see Exercise 4.2). Since X is $\mathcal{F}$-measurable, $X^{-1}(G) \in \mathcal{F}$ for each $G \in \mathcal{G}$ and so $\sigma(X) \subset \mathcal{F}$. Note that $(\Omega, \sigma(X), \mathbb{P})$ is also a probability space and X is $\sigma(X)$-measurable. In fact, $\sigma(X)$ is the smallest σ-algebra such that X is measurable. In Example 4.5, $\sigma(X) = \mathcal{F}$ but in general $\sigma(X)$ may be considerably smaller than $\mathcal{F}$.

Example 4.19 ($\sigma(X)$ for two throws of a die) Consider two throws of a fair die and define

$$X := \begin{cases} 1, & \text{if a six is thrown on the first throw,} \\ 0, & \text{otherwise.} \end{cases}$$

Consider the measurable space $(\Omega, \mathcal{F})$ defined by $\Omega = \{(i,j)\colon i,j = 1,2,\dots,6\}$ and the associated power set $\mathcal{F}$. Now, X is a measurable function from $(\Omega, \mathcal{F})$ to $(\Psi, \mathcal{G})$ where $\Psi = \{0,1\}$ and $\mathcal{G} = \{\{\}, \{0\}, \{1\}, \{0,1\}\}$. The σ-algebra generated by X is

$$\sigma(X) = \{\{\}, X^{-1}(\{0\}), X^{-1}(\{1\}), \Omega\},$$

where

$$X^{-1}(\{0\}) = \{(i,j)\colon i = 1,2,\dots,5,\ j = 1,2,\dots,6\},$$
$$X^{-1}(\{1\}) = \{(6,j)\colon j = 1,2,\dots,6\}.$$

It is easy to see that $\sigma(X)$ is a σ-algebra and, here, a strict subset of $\mathcal{F}$.

Definition 4.20 (independence of σ-algebras)

(i) Two events $F, G \in \mathcal{F}$ are *independent* if $\mathbb{P}(F \cap G) = \mathbb{P}(F)\mathbb{P}(G)$.
(ii) Two sub σ-algebras $\mathcal{F}_1, \mathcal{F}_2$ of the σ-algebra $\mathcal{F}$ are *independent* if events F_1, F_2 are independent for all $F_1 \in \mathcal{F}_1$ and $F_2 \in \mathcal{F}_2$.

Independent random variables are defined as follows.

Definition 4.21 (independence of random variables) Two random variables X, Y on a probability space $(\Omega, \mathcal{F}, \mathbb{P})$ are *independent* if the σ-algebras $\sigma(X)$ and $\sigma(Y)$ generated by X and Y are independent. We say a family of random variables $\{X_i\}$ on $(\Omega, \mathcal{F}, \mathbb{P})$ is *pairwise independent* if X_i and X_j are independent for $i \neq j$.

Example 4.22 (independence for die throws) Continuing Example 4.19, we define

$$X_i := \begin{cases} 1, & \text{if a six is thrown on the } i\text{th throw,} \\ 0, & \text{otherwise.} \end{cases}$$

The σ-algebras $\sigma(X_1)$ and $\sigma(X_2)$ are independent and X_1 and X_2 are independent random variables. See Exercise 4.8.

Example 4.23 (constant random variables) Let X and Y be real-valued random variables on $(\Omega, \mathcal{F}, \mathbb{P})$ and suppose that Y is constant. Then $Y(\omega) = a$ for all $\omega \in \Omega$ for some $a \in \mathbb{R}$ and $\sigma(Y) = \{\{\}, \Omega\}$. For any $F_2 \in \sigma(X)$, we have $\{\} \cap F_2 = \{\}$ and $\Omega \cap F_2 = F_2$. Hence $\mathbb{P}(\{\} \cap F_2) = \mathbb{P}(\{\}) = 0 = \mathbb{P}(\{\})\,\mathbb{P}(F_2)$ and $\mathbb{P}(\Omega \cap F_2) = \mathbb{P}(F_2) = 1\,\mathbb{P}(F_2) = \mathbb{P}(\Omega)\,\mathbb{P}(F_2)$. Therefore, $\sigma(X)$ and $\sigma(Y)$ are independent σ-algebras and X and Y are independent random variables by Definition 4.21.

Independence of *real-valued* random variables can be expressed conveniently using the joint distribution $\mathbb{P}_{X,Y}$. Specifically, X and Y are independent if and only if $\mathbb{P}_{X,Y}$ equals the product measure $\mathbb{P}_X \times \mathbb{P}_Y$ (see Definition 1.23). Further, if X and Y are real-valued with densities p_X and p_Y, they are independent if and only if the *joint pdf* $p_{X,Y}$ is

$$p_{X,Y}(x,y) = p_X(x)p_Y(y), \qquad x, y \in \mathbb{R}.$$

We show that independent real-valued random variables are uncorrelated.

Lemma 4.24 *If X and Y are independent real-valued random variables and $\mathbb{E}[|X|]$, $\mathbb{E}[|Y|] < \infty$, then X and Y are uncorrelated.*

Proof Using (4.10) with $\mathbb{P}_{X,Y} = \mathbb{P}_X \times \mathbb{P}_Y$ and $D_X = \mathbb{R} = D_Y$ gives

$$\mathbb{E}[XY] = \int_{\mathbb{R}^2} xy \, d\mathbb{P}_{X,Y}(x,y) = \int_{\mathbb{R}^2} xy \, d(\mathbb{P}_X \times \mathbb{P}_Y)(x,y).$$

Since $|xy| = |x||y|$, Fubini's theorem applies because

$$\int_{\mathbb{R}} \left(\int_{\mathbb{R}} |xy| \, d\mathbb{P}_X(x) \right) d\mathbb{P}_Y(y) = \int_{\mathbb{R}} \mathbb{E}[|X|]|y| \, d\mathbb{P}_Y(y) = \mathbb{E}[|X|] \, \mathbb{E}[|Y|] < \infty.$$

Hence,

$$\mathbb{E}[XY] = \int_{\mathbb{R}} \left(\int_{\mathbb{R}} yx \, d\mathbb{P}_X(x) \right) d\mathbb{P}_Y(y) = \mu_X \mu_Y.$$

By Definition 4.16, X and Y are uncorrelated. □

Uncorrelated random variables are not independent in general, except in the special case of Gaussian random variables (see Lemma 4.33).

Examples of $\mathbb{R}^d$-valued random variables

Random variables $\boldsymbol{X} = [X_1, \dots, X_d]^{\mathsf{T}}$ from $(\Omega, \mathcal{F}, \mathbb{P})$ to $(D, \mathcal{B}(D))$, where $D \subset \mathbb{R}^d$ with $d > 1$, are referred to as *multivariate* random variables (bivariate if $d = 2$). Each component X_j is a real-valued random variable and the mean is a vector $\boldsymbol{\mu} \in \mathbb{R}^d$ with entries $\mu_j = \mathbb{E}[X_j]$; that is,

$$\boldsymbol{\mu} = \mathbb{E}[\boldsymbol{X}] = \int_{\Omega} \boldsymbol{X}(\omega) \, d\mathbb{P}(\omega) = \left[\mathbb{E}[X_1], \dots, \mathbb{E}[X_d]\right]^{\mathsf{T}}.$$

The distribution $\mathbb{P}_{\boldsymbol{X}}$ and the pdf $p(\boldsymbol{x})$ are defined analogously to the bivariate case in Definition 4.15. If the pdf of $\boldsymbol{X}$ is known, the probability

$$\mathbb{P}(\boldsymbol{X} \in B) = \mathbb{P}\big(\{\omega \in \Omega \colon \boldsymbol{X}(\omega) \in B\}\big) = \mathbb{P}_{\boldsymbol{X}}(B) = \int_B p(\boldsymbol{x}) \, d\boldsymbol{x}, \qquad B \in \mathcal{B}(D).$$

The components of $\boldsymbol{X}$ are *independent* if and only if $\mathbb{P}_{\boldsymbol{X}}$ is the product measure $\mathbb{P}_{X_1} \times \mathbb{P}_{X_2} \times \cdots \times \mathbb{P}_{X_d}$ (see Definition 1.23 for the bivariate case) or, equivalently, if and only if

$$p(\boldsymbol{x}) = p_{X_1}(x_1) p_{X_2}(x_2) \cdots p_{X_d}(x_d),$$

where p_{X_j} is the pdf of X_j.

Definition 4.25 (multivariate uniform) Let $D \in \mathcal{B}(\mathbb{R}^d)$ with $0 < \mathrm{Leb}(D) < \infty$. A D-valued random variable $\boldsymbol{X}$ has the *uniform distribution* on D, written $\boldsymbol{X} \sim \mathrm{U}(D)$, if its pdf is $p(\boldsymbol{x}) = 1/\mathrm{Leb}(D)$ for $\boldsymbol{x} \in D$.

The density of $\boldsymbol{X} \sim \mathrm{U}([0,a] \times [0,b])$ is $p(\boldsymbol{x}) = 1/ab$, which is the product of the densities p_{X_1} and p_{X_2} of $X_1 \sim \mathrm{U}(0,a)$ and $X_2 \sim \mathrm{U}(0,b)$. Hence, the components of a $\mathrm{U}([0,a] \times [0,b])$ random variable are independent.

In general, the relationship between the components of a multivariate random variable is more tangled and is described by a covariance matrix.

Definition 4.26 (covariance for multivariate random variables) The *covariance* of $\mathbb{R}^d$-valued random variables $\boldsymbol{X} = [X_1, \dots, X_d]^\mathsf{T}$, $\boldsymbol{Y} = [Y_1, \dots, Y_d]^\mathsf{T}$ is the $d \times d$ matrix

$$\operatorname{Cov}(\boldsymbol{X}, \boldsymbol{Y}) := \mathbb{E}\Big[(\boldsymbol{X} - \mathbb{E}[\boldsymbol{X}])(\boldsymbol{Y} - \mathbb{E}[\boldsymbol{Y}])^\mathsf{T}\Big],$$

whose (i, j)th entry is $c_{ij} = \operatorname{Cov}(X_i, Y_j)$. We say $\boldsymbol{X}, \boldsymbol{Y}$ are *uncorrelated* if $\operatorname{Cov}(\boldsymbol{X}, \boldsymbol{Y}) = 0$ (the $d \times d$ zero matrix). The matrix $\operatorname{Cov}(\boldsymbol{X}, \boldsymbol{X})$ is known as the covariance matrix of $\boldsymbol{X}$.

Lemma 4.27 *Suppose that $\boldsymbol{X}$ is an $\mathbb{R}^d$-valued random variable with mean vector $\boldsymbol{\mu}$ and covariance matrix C. Then, C is a symmetric non-negative definite matrix and* $\operatorname{Tr} C = \mathbb{E}\big[\|\boldsymbol{X} - \boldsymbol{\mu}\|_2^2\big]$.

Proof C is symmetric because $\operatorname{Cov}(X_i, X_j) = \operatorname{Cov}(X_j, X_i)$ and is non-negative definite because, for any $\boldsymbol{\xi} \in \mathbb{R}^d$,

$$\boldsymbol{\xi}^\mathsf{T} C \boldsymbol{\xi} = \boldsymbol{\xi}^\mathsf{T} \mathbb{E}\Big[(\boldsymbol{X} - \boldsymbol{\mu})(\boldsymbol{X} - \boldsymbol{\mu})^\mathsf{T}\Big] \boldsymbol{\xi} = \mathbb{E}\Big[\big(\boldsymbol{\xi}^\mathsf{T}(\boldsymbol{X} - \boldsymbol{\mu})\big)^2\Big] \geq 0.$$

Write $\boldsymbol{X} = [X_1, \dots, X_d]^\mathsf{T}$; then $c_{jj} = \operatorname{Cov}(X_j, X_j) = \operatorname{Var} X_j$ and

$$\operatorname{Tr} C := \sum_{j=1}^{d} c_{jj} = \sum_{j=1}^{d} \mathbb{E}\big[(X_j - \mu_j)^2\big] = \mathbb{E}\Big[\|\boldsymbol{X} - \boldsymbol{\mu}\|_2^2\Big]. \qquad \square$$

When the covariance matrix C is positive definite, we say $\boldsymbol{X}$ is a *multivariate Gaussian* random variable with mean $\boldsymbol{\mu}$ and covariance matrix C, written $\boldsymbol{X} \sim \mathrm{N}(\boldsymbol{\mu}, C)$, if its pdf is given by

$$p(\boldsymbol{x}) = \frac{1}{\sqrt{(2\pi)^d \det C}} \exp\left(\frac{-(\boldsymbol{x} - \boldsymbol{\mu})^\mathsf{T} C^{-1} (\boldsymbol{x} - \boldsymbol{\mu})}{2}\right). \tag{4.13}$$

Note that (4.13) agrees with the pdf (4.7) of the univariate Gaussian when $d = 1$ since then $C = \sigma^2$. Further, if $\boldsymbol{\mu} = [\mu_1, \dots, \mu_d]^\mathsf{T}$, the components X_j of $\boldsymbol{X}$ have distribution $\mathrm{N}(\mu_j, c_{jj})$. It is not true, however, that any family of Gaussian random variables X_j form a multivariate Gaussian random variable $\boldsymbol{X}$ (see Exercise 4.7).

We now extend this definition to symmetric and non-negative definite matrices C, in which case C^{-1} may not exist and (4.13) is not well defined. To include this case, we use the characteristic function, the name given to the Fourier transform in probability theory.

Definition 4.28 (characteristic function) The *characteristic function* of an $\mathbb{R}^d$-valued random variable $\boldsymbol{X}$ is $\mathbb{E}\Big[\mathrm{e}^{\mathrm{i}\boldsymbol{\lambda}^\mathsf{T}\boldsymbol{X}}\Big]$ for $\boldsymbol{\lambda} \in \mathbb{R}^d$. If $\boldsymbol{X}$ has pdf $p(\boldsymbol{x})$, the characteristic function is $\mathbb{E}\Big[\mathrm{e}^{\mathrm{i}\boldsymbol{\lambda}^\mathsf{T}\boldsymbol{X}}\Big] = (2\pi)^{d/2} \hat{p}(-\boldsymbol{\lambda})$,where $\hat{p}(\boldsymbol{\lambda})$ is the Fourier transform of $p(\boldsymbol{x})$, introduced in Definition 1.102.

We follow the convention in probability theory that $\hat{p}(-\boldsymbol{\lambda})$ and not $\hat{p}(\boldsymbol{\lambda})$ defines the characteristic function.

Proposition 4.29 *Suppose that $\boldsymbol{\mu} \in \mathbb{R}^d$ and C is a $d \times d$ symmetric positive-definite matrix. A random variable $\boldsymbol{X}$ has a density $p(\boldsymbol{x})$ given by* (4.13) *if and only if its characteristic function is*

$$\mathbb{E}\Big[\mathrm{e}^{\mathrm{i}\boldsymbol{\lambda}^\mathsf{T}\boldsymbol{X}}\Big] = \mathrm{e}^{\mathrm{i}\boldsymbol{\lambda}^\mathsf{T}\boldsymbol{\mu} - \boldsymbol{\lambda}^\mathsf{T} C \boldsymbol{\lambda}/2}. \tag{4.14}$$

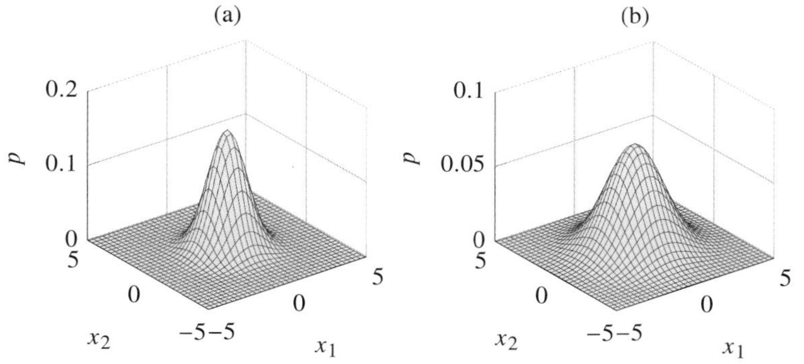

Figure 4.3 The pdf of $X \sim \mathrm{N}(\mathbf{0}, \sigma^2 I)$ with $d = 2$ and (a) $\sigma = 1$, (b) $\sigma = 3/2$.

Proof Suppose X has density $p(x)$ given by (4.13) and let $Z = C^{-1/2}(X - \mu)$. A change of variables (as in Lemma 4.12) shows that Z has density

$$p_Z(z) = p(z)\det(C^{1/2}) = \frac{1}{(2\pi)^{d/2}} e^{-z^{\mathsf{T}} z/2}.$$

Applying (1.46) with $C = I/2$ and $\det C = 2^{-d}$ gives

$$\mathbb{E}\left[e^{i\lambda^{\mathsf{T}} Z}\right] = (2\pi)^{d/2} \hat{p}_z(-\lambda) = e^{-\lambda^{\mathsf{T}}\lambda/2}.$$

Substituting for Z, we find

$$\mathbb{E}\left[e^{i\lambda^{\mathsf{T}} X}\right] = \mathbb{E}\left[e^{i\lambda^{\mathsf{T}}(\mu + C^{1/2} Z)}\right] = e^{i\lambda^{\mathsf{T}}\mu}\,\mathbb{E}\left[e^{i(C^{1/2}\lambda)^{\mathsf{T}} Z}\right] = e^{i\lambda^{\mathsf{T}}\mu} e^{-\lambda^{\mathsf{T}} C\lambda/2}.$$

That is, the characteristic function of X is (4.14). The converse is established by applying the inverse Fourier transform. □

The characteristic function (4.14) is well defined even when C is singular and provides a definition of the multivariate Gaussian distribution for all covariance matrices.

Definition 4.30 (multivariate Gaussian) An $\mathbb{R}^d$-valued random variable X follows the Gaussian distribution $\mathrm{N}(\mu, C)$, where $\mu \in \mathbb{R}^d$ and C is a symmetric non-negative definite matrix, if the characteristic function of X is (4.14).

Example 4.31 (pdf of a multivariate Gaussian) If $X \sim \mathrm{N}(\mathbf{0}, C)$ with covariance matrix $C = \sigma^2 I_d$, where I_d is the $d \times d$ identity matrix, then X is a multivariate random variable with pdf and characteristic function

$$p(x) = \frac{1}{\sqrt{(2\pi)^d \sigma^{2d}}} \exp\left(-\frac{x^{\mathsf{T}} x}{2\sigma^2}\right), \qquad \mathbb{E}\left[e^{i\lambda^{\mathsf{T}} X}\right] = e^{-\sigma^2\lambda^{\mathsf{T}}\lambda/2},$$

respectively. The pdf is plotted in Figure 4.3 for the case $d = 2$ and should be compared with Figure 4.1(a). Since C is diagonal, the components X_1 and X_2 of X are uncorrelated.

We give two further properties of Gaussian random variables. The first says that affine transformations of multivariate Gaussian random variables are also Gaussian and the second that uncorrelated components X_j of a multivariate Gaussian X are independent.

Proposition 4.32 *Consider an $\mathbb{R}^{d_1}$ random variable $\boldsymbol{X} \sim \mathrm{N}(\boldsymbol{\mu}, C)$. For a $d_2 \times d_1$ matrix A and $\boldsymbol{b} \in \mathbb{R}^{d_2}$, the $\mathbb{R}^{d_2}$ random variable $A\boldsymbol{X} + \boldsymbol{b}$ has distribution $\mathrm{N}(A\boldsymbol{\mu} + \boldsymbol{b}, ACA^{\mathsf{T}})$. In addition, if $X \sim \mathrm{N}(\mu_X, \sigma_X^2)$ and $Y \sim \mathrm{N}(\mu_Y, \sigma_Y^2)$ are independent, then $X + Y \sim \mathrm{N}(\mu_X + \mu_Y, \sigma_X^2 + \sigma_Y^2)$.*

Proof Using (4.14), the characteristic function of $A\boldsymbol{X} + \boldsymbol{b}$ is

$$\mathbb{E}\left[\mathrm{e}^{\mathrm{i}\boldsymbol{\lambda}^{\mathsf{T}}(A\boldsymbol{X}+\boldsymbol{b})}\right] = \mathrm{e}^{\mathrm{i}\boldsymbol{\lambda}^{\mathsf{T}}\boldsymbol{b}}\,\mathbb{E}\left[\mathrm{e}^{\mathrm{i}(A^{\mathsf{T}}\boldsymbol{\lambda})^{\mathsf{T}}\boldsymbol{X}}\right]$$
$$= \mathrm{e}^{\mathrm{i}\boldsymbol{\lambda}^{\mathsf{T}}\boldsymbol{b}}\left(\mathrm{e}^{\mathrm{i}(A^{\mathsf{T}}\boldsymbol{\lambda})^{\mathsf{T}}\boldsymbol{\mu}-\boldsymbol{\lambda}^{\mathsf{T}}ACA^{\mathsf{T}}\boldsymbol{\lambda}/2}\right) = \mathrm{e}^{\mathrm{i}\boldsymbol{\lambda}^{\mathsf{T}}(A\boldsymbol{\mu}+\boldsymbol{b})-\boldsymbol{\lambda}^{\mathsf{T}}ACA^{\mathsf{T}}\boldsymbol{\lambda}/2}.$$

Using Definition 4.30, we conclude that $A\boldsymbol{X} + \boldsymbol{b} \sim \mathrm{N}(A\boldsymbol{\mu} + \boldsymbol{b}, ACA^{\mathsf{T}})$. For the second result, $\mathbb{E}\big[\mathrm{e}^{\mathrm{i}\lambda(X+Y)}\big] = \mathbb{E}\big[\mathrm{e}^{\mathrm{i}\lambda X}\big]\mathbb{E}\big[\mathrm{e}^{\mathrm{i}\lambda Y}\big]$ by independence. Hence,

$$\mathbb{E}\big[\mathrm{e}^{\mathrm{i}\lambda(X+Y)}\big] = \mathrm{e}^{(\mu_X+\mu_Y)\mathrm{i}\lambda-(\sigma_X^2+\sigma_Y^2)\lambda^2/2}$$

and $X + Y$ is therefore Gaussian with distribution $\mathrm{N}(\mu_X + \mu_Y, \sigma_X^2 + \sigma_Y^2)$. □

Lemma 4.33 *Suppose $\boldsymbol{X} = [X_1, \ldots, X_d]^{\mathsf{T}}$ is a multivariate Gaussian random variable. If the covariance matrix C is diagonal, the individual random variables X_j are pairwise independent.*

Proof As C is diagonal, the determinant of C is $\det C = \prod_{j=1}^{d} c_{jj} = \prod_{j=1}^{d} \sigma_j^2$ with $\sigma_j^2 = \operatorname{Var} X_j$. Suppose first that $\sigma_j^2 > 0$ for each j so that $\det C \neq 0$ and the density $p(\boldsymbol{x})$ is well defined. Then,

$$(\boldsymbol{x} - \boldsymbol{\mu})^{\mathsf{T}} C^{-1} (\boldsymbol{x} - \boldsymbol{\mu}) = \sum_{j=1}^{d} \frac{(x_j - \mu_j)^2}{\sigma_j^2}.$$

If p_{X_j} denotes the pdf of X_j, the pdf of $\boldsymbol{X}$ is

$$p(\boldsymbol{x}) = \prod_{j=1}^{d} \frac{1}{\sqrt{2\pi\sigma_j^2}} \mathrm{e}^{-(x_j-\mu_j)^2/2\sigma_j^2} = \prod_{j=1}^{d} p_{X_j}(x_j),$$

which is a product of the densities of each X_j. Hence, any two components of $\boldsymbol{X}$ are independent.

If $\sigma_j^2 = 0$ for some j, then X_j is a.s. constant and is independent of all other components (see Example 4.23). Using Proposition 4.32 (with $\boldsymbol{b} = \boldsymbol{0}$ and a clever choice of A), the subvector of $\boldsymbol{X}$ containing the components with non-zero variance is a multivariate Gaussian with a diagonal covariance matrix. Repeating the above argument completes the proof. □

Hilbert space-valued random variables

Random variables can take values in any set Ψ equipped with a σ-algebra. We frequently consider random variables taking values in a Hilbert space H and always consider H with the Borel σ-algebra $\mathcal{B}(H)$ (see Definition 1.16). Families of such random variables, with well-defined moments, form Banach and Hilbert spaces. Recall Definition 1.26 of the L^p spaces.

Definition 4.34 ($L^p(\Omega, H)$ spaces) Let $(\Omega, \mathcal{F}, \mathbb{P})$ be a probability space and let H be a Hilbert space with norm $\|\cdot\|$. Then, $L^p(\Omega, H)$ with $1 \le p < \infty$ is the space of H-valued $\mathcal{F}$-measurable random variables $X\colon \Omega \to H$ with $\mathbb{E}[\|X\|^p] < \infty$ and is a Banach space with the norm

$$\|X\|_{L^p(\Omega,H)} := \left(\int_\Omega \|X(\omega)\|^p \, d\mathbb{P}(\omega)\right)^{1/p} = \mathbb{E}\left[\|X\|^p\right]^{1/p}.$$

$L^\infty(\Omega, H)$ is the Banach space of H-valued random variables $X\colon \Omega \to H$ where

$$\|X\|_{L^\infty(\Omega,H)} := \operatorname*{ess\,sup}_{\omega \in \Omega} \|X(\omega)\| < \infty.$$

We frequently use $L^2(\Omega, H)$ (the case $p = 2$) in later chapters. It is the set of H-valued random variables X with $\mathbb{E}\left[\|X\|^2\right] < \infty$, known as the *mean-square integrable random variables*, and is a Hilbert space with the inner product

$$\langle X, Y \rangle_{L^2(\Omega,H)} := \int_\Omega \langle X(\omega), Y(\omega) \rangle \, d\mathbb{P}(\omega) = \mathbb{E}\left[\langle X, Y \rangle\right].$$

The norms $\|\cdot\|_{L^p(\Omega,H)}$ and the inner product $\langle \cdot, \cdot \rangle_{L^2(\Omega,H)}$ are defined in terms of the norm $\|\cdot\|$ and inner product $\langle \cdot, \cdot \rangle$ on H. For random variables $X, Y \in L^2(\Omega, H)$, the *Cauchy–Schwarz* inequality (Lemma 1.32) takes the form

$$\left|\langle X, Y \rangle_{L^2(\Omega,H)}\right| \le \|X\|_{L^2(\Omega,H)} \|Y\|_{L^2(\Omega,H)}$$

or equivalently

$$\mathbb{E}\left[\langle X, Y \rangle\right]^2 \le \mathbb{E}\left[\|X\|^2\right] \mathbb{E}\left[\|Y\|^2\right].$$

The notion of independence described in Definition 4.21 includes Hilbert space-valued random variables. However, the notions of uncorrelated and covariance depend on the Hilbert space inner product. For now, we consider a real-valued Hilbert space H so that $\langle X, \phi \rangle$ is a real-valued random variable for any $\phi \in H$. We meet complex-valued random variables in Chapter 6.

Definition 4.35 (uncorrelated, covariance operator) Let H be a Hilbert space. A linear operator $\mathcal{C}\colon H \to H$ is the *covariance* of H-valued random variables X and Y if

$$\langle \mathcal{C}\phi, \psi \rangle = \mathrm{Cov}\big(\langle X, \phi \rangle, \langle Y, \psi \rangle\big), \qquad \forall \phi, \psi \in H.$$

If $\mathcal{C}$ is the zero operator, we say that X and Y are *uncorrelated*. In the case $X = Y$, we say $\mathcal{C}$ is the covariance of X.

It is straightforward to generalise Lemma 4.24 to show that independent random variables $X, Y \in L^1(\Omega, H)$ are uncorrelated.

Example 4.36 (covariance operator for $H = \mathbb{R}^d$) Consider the Hilbert space $H = \mathbb{R}^d$ and the multivariate random variable $\boldsymbol{X} \in L^2(\Omega, H)$. We show that, in this finite dimensional case, the covariance matrix coincides with the covariance operator.

Since $\boldsymbol{X}$ has finite second moments, the covariance matrix C is well defined. Denote $\mathbb{E}[\boldsymbol{X}] = \boldsymbol{\mu}_X, \mathbb{E}[\boldsymbol{Y}] = \boldsymbol{\mu}_Y$. For $\boldsymbol{\phi}, \boldsymbol{\psi} \in \mathbb{R}^d$,

$$\begin{aligned}\operatorname{Cov}\big(\langle \boldsymbol{X}, \boldsymbol{\phi}\rangle, \langle \boldsymbol{X}, \boldsymbol{\psi}\rangle\big) &= \operatorname{Cov}\big(\boldsymbol{\phi}^{\mathsf{T}}\boldsymbol{X}, \boldsymbol{\psi}^{\mathsf{T}}\boldsymbol{X}\big) = \mathbb{E}\big[\boldsymbol{\phi}^{\mathsf{T}}(\boldsymbol{X} - \boldsymbol{\mu}_X)(\boldsymbol{X} - \boldsymbol{\mu}_X)^{\mathsf{T}}\boldsymbol{\psi}\big] \\ &= \boldsymbol{\phi}^{\mathsf{T}}\mathbb{E}\big[(\boldsymbol{X} - \boldsymbol{\mu}_X)(\boldsymbol{X} - \boldsymbol{\mu}_X)^{\mathsf{T}}\big]\boldsymbol{\psi} = \boldsymbol{\phi}^{\mathsf{T}} C \boldsymbol{\psi} = (C\boldsymbol{\phi})^{\mathsf{T}}\boldsymbol{\psi} = \langle C\boldsymbol{\phi}, \boldsymbol{\psi}\rangle.\end{aligned}$$

The operator $\mathcal{C}$ is defined by $\mathcal{C}\boldsymbol{\phi} = C\boldsymbol{\phi}$ and is the usual covariance matrix.

The following version of Lemma 4.27 applies to H-valued random variables and gives properties of the covariance operator $\mathcal{C}$.

Lemma 4.37 *Let H be a separable Hilbert space and let $X \in L^2(\Omega, H)$ with mean μ and covariance operator $\mathcal{C}$. The covariance operator $\mathcal{C}$ is symmetric non-negative definite and is of trace class with* $\operatorname{Tr}\mathcal{C} = \mathbb{E}\big[\|X - \mu\|^2\big]$.

Proof For any $\phi, \psi \in H$, $\langle \mathcal{C}\phi, \psi\rangle = \operatorname{Cov}(\langle X, \phi\rangle, \langle X, \psi\rangle)$ is symmetric in ϕ and ψ. Moreover, $\langle \mathcal{C}\phi, \phi\rangle = \operatorname{Cov}(\langle X, \phi\rangle, \langle X, \phi\rangle) = \operatorname{Var}\langle X, \phi\rangle \geq 0$ for any $\phi \in H$. Hence, the operator $\mathcal{C}$ is non-negative definite.

Since H is a separable Hilbert space, it has an orthonormal basis $\{\phi_j : j = 1, \ldots, \infty\}$. Then, $\langle \mathcal{C}\phi_j, \phi_j\rangle = \operatorname{Var}\langle X, \phi_j\rangle$ and, from Definition 1.79 of the trace,

$$\operatorname{Tr}\mathcal{C} := \sum_{j=1}^{\infty} \operatorname{Var}\langle X, \phi_j\rangle = \mathbb{E}\left[\sum_{j=1}^{\infty} \left|\langle X - \mu, \phi_j\rangle\right|^2\right] = \mathbb{E}\big[\|X - \mu\|^2\big].$$

As $X - \mu \in L^2(\Omega, H)$, $\operatorname{Tr}\mathcal{C} < \infty$ and $\mathcal{C}$ is of trace class. □

There are no translation invariant measures on *infinite-dimensional* Hilbert spaces (see Exercise 1.18) and uniformly distributed random variables are not well defined. However, Gaussian random variables taking values in $H = L^2(D)$ with $D \subset \mathbb{R}^d$ are used in Chapter 7 in our study of Gaussian random fields.

Definition 4.38 (H-valued Gaussian) Let H be a real Hilbert space. An H-valued random variable X is *Gaussian* if $\langle X, \phi\rangle$ is a real-valued Gaussian random variable for all $\phi \in H$.

In contrast to multivariate Gaussian random variables, no covariance is introduced in Definition 4.38. We show in Corollary 4.41, however, that the covariance operator of a real H-valued Gaussian random variable is well defined. First we need two preliminary lemmas.

Lemma 4.39 *If X is an H-valued Gaussian random variable, then, for any $\phi_1, \ldots, \phi_d \in H$, $\boldsymbol{X} = \big[\langle X, \phi_1\rangle, \ldots, \langle X, \phi_d\rangle\big]^{\mathsf{T}}$ is an $\mathbb{R}^d$-valued Gaussian.*

Proof We assume for simplicity that $\mathbb{E}[X] = 0$. Consider $\boldsymbol{\lambda} = [\lambda_1, \ldots, \lambda_d]^{\mathsf{T}} \in \mathbb{R}^d$ and $\phi := \sum_{j=1}^{d} \lambda_j \phi_j$. Then, $\phi \in H$ and $\langle \phi, X\rangle = \boldsymbol{\lambda}^{\mathsf{T}}\boldsymbol{X}$ is a real-valued Gaussian random variable with mean zero and variance

$$\operatorname{Var}\langle \phi, X\rangle = \operatorname{Var}(\boldsymbol{\lambda}^{\mathsf{T}}\boldsymbol{X}) = \mathbb{E}[\boldsymbol{\lambda}^{\mathsf{T}}\boldsymbol{X}\boldsymbol{X}^{\mathsf{T}}\boldsymbol{\lambda}] = \boldsymbol{\lambda}^{\mathsf{T}}\mathbb{E}[\boldsymbol{X}\boldsymbol{X}^{\mathsf{T}}]\boldsymbol{\lambda} = \boldsymbol{\lambda}^{\mathsf{T}}C\boldsymbol{\lambda},$$

where C is the covariance matrix of $\boldsymbol{X}$. That is, $\langle \phi, X\rangle \sim \mathrm{N}(0, \boldsymbol{\lambda}^{\mathsf{T}}C\boldsymbol{\lambda})$ and the characteristic function $\mathbb{E}\big[e^{i\langle \phi, X\rangle}\big] = e^{-\boldsymbol{\lambda}^{\mathsf{T}}C\boldsymbol{\lambda}/2}$. As $\langle \phi, X\rangle = \boldsymbol{\lambda}^{\mathsf{T}}\boldsymbol{X}$, we see that $\mathbb{E}\big[e^{i\boldsymbol{\lambda}^{\mathsf{T}}\boldsymbol{X}}\big] = e^{-\boldsymbol{\lambda}^{\mathsf{T}}C\boldsymbol{\lambda}/2}$ and $\boldsymbol{X}$ is Gaussian by Definition 4.30. □

Lemma 4.40 *If $\boldsymbol{X}$ is an $\mathbb{R}^d$-valued Gaussian with mean $\mathbb{E}[\boldsymbol{X}] = \boldsymbol{0}$, then*

$$\mathbb{E}\left[e^{-\|\boldsymbol{X}\|_2^2}\right] \le \left(1 + 2\mathbb{E}\left[\|\boldsymbol{X}\|_2^2\right]\right)^{-1/2}.$$

Proof Let C denote the covariance matrix of $\boldsymbol{X}$ and assume it is non-singular. By (4.13),

$$\begin{aligned}\mathbb{E}\left[e^{-\|\boldsymbol{X}\|_2^2}\right] &= \frac{1}{\sqrt{(2\pi)^d \det C}} \int_{\mathbb{R}^d} e^{-\|\boldsymbol{x}\|_2^2} e^{-\boldsymbol{x}^{\mathsf{T}} C^{-1}\boldsymbol{x}/2}\, d\boldsymbol{x} \\ &= \frac{1}{\sqrt{(2\pi)^d \det C}} \int_{\mathbb{R}^d} e^{-\boldsymbol{x}^{\mathsf{T}} Q^{-1}\boldsymbol{x}/2}\, d\boldsymbol{x},\end{aligned}$$

with $Q^{-1} = C^{-1}(2C + I)$. We have $\det Q = \det C/\det(2C + I)$ and, since the probability density of $\mathrm{N}(\boldsymbol{0}, Q)$ has unit mass,

$$\frac{1}{\sqrt{(2\pi)^d \det Q}} \int_{\mathbb{R}} e^{-\boldsymbol{x}^{\mathsf{T}} Q^{-1}\boldsymbol{x}/2}\, dx = 1.$$

Therefore,

$$\mathbb{E}\left[e^{-\|\boldsymbol{X}\|_2^2}\right] = \frac{\sqrt{\det Q}}{\sqrt{\det C}} = \frac{1}{\sqrt{\det(2C + I)}}. \tag{4.15}$$

In fact, (4.15) holds even when C is singular. Now, as C is non-negative definite, $\det(I + 2C) \ge 1 + \operatorname{Tr} 2C$ (see Exercise 4.9) and

$$\mathbb{E}\left[e^{-\|\boldsymbol{X}\|_2^2}\right] \le (1 + 2\operatorname{Tr} C)^{-1/2}.$$

Finally, by Lemma 4.27 with $\boldsymbol{\mu} = \boldsymbol{0}$, we have $\operatorname{Tr} C = \mathbb{E}\left[\|\boldsymbol{X}\|_2^2\right]$. □

Corollary 4.41 *Let H be a separable Hilbert space and X be an H-valued Gaussian with $\mu = \mathbb{E}[X]$. Then $X \in L^2(\Omega, H)$ and the covariance operator $\mathcal{C}$ of X is a well-defined trace class operator on X. We write $X \sim \mathrm{N}(\mu, \mathcal{C})$.*

Proof For simplicity, we assume $\mu = \mathbb{E}[X] = 0$. As H is separable, there exists an orthonormal basis $\{\phi_j : j = 1, \ldots, \infty\}$ of H. Write $X = \sum_{j=1}^{\infty} x_j \phi_j$ with $x_j = \langle X, \phi_j \rangle$ and let $X_N = P_N X = \sum_{j=1}^{N} x_j \phi_j$. Lemma 4.39 says that $\boldsymbol{X}_N = [x_1, \ldots, x_N]^{\mathsf{T}}$ is a finite-dimensional Gaussian random variable and Lemma 4.40 gives

$$\mathbb{E}\left[e^{-\|\boldsymbol{X}_N\|_2^2}\right] \le \left(1 + 2\mathbb{E}\left[\|\boldsymbol{X}_N\|_2^2\right]\right)^{-1/2}.$$

Furthermore, the $\|\cdot\|_2$ norm of the vector of components $\boldsymbol{X}_N \in \mathbb{R}^N$ and the H norm of $X_N \in H$ are equal (see (1.9)). Taking the limits as $N \to \infty$, we achieve

$$\mathbb{E}\left[e^{-\|X\|^2}\right] \le \left(1 + 2\mathbb{E}\left[\|X\|^2\right]\right)^{-1/2}. \tag{4.16}$$

As X is H-valued, $\|X\| < \infty$ a.s. and $\mathbb{E}\left[e^{-\|X\|^2}\right] > 0$ so (4.16) implies that $\mathbb{E}\left[\|X\|^2\right] < \infty$. In particular, $X \in L^2(\Omega, H)$ and, by Lemma 4.37 with $\mu = 0$, the trace of the covariance operator satisfies $\operatorname{Tr} \mathcal{C} = \mathbb{E}\left[\|X\|^2\right] < \infty$. □

Example 4.42 (the space $L^2(\Omega, H^1(D))$) For a domain $D \subset \mathbb{R}^d$, $H^1(D)$ is a separable Hilbert space and so Gaussian random variables X taking values in $H^1(D)$ belong to $L^2(\Omega, H^1(D))$.

4.2 Least-squares approximation and conditional expectation

The *expectation* $\mathbb{E}[X]$ (see (4.1)) of a real-valued random variable X summarises all the possible outcomes of X in single real number. In other words, we are approximating X by a real number. To make this precise, suppose $X \in L^2(\Omega)$, so that X is a *real-valued* random variable on $(\Omega, \mathcal{F}, \mathbb{P})$ with $\mathbb{E}\big[X^2\big] < \infty$ and seek the constant $a \in \mathbb{R}$ that minimises the error $V(\alpha) := \mathbb{E}\big[(X-\alpha)^2\big]$ over $\alpha \in \mathbb{R}$. Solving $\frac{dV}{d\alpha}(a) = 0$ gives $\mathbb{E}[2(X-a)] = 0$ and $a = \mathbb{E}[X]$. The least-squares constant approximation of X is indeed its expectation $\mathbb{E}[X]$.

Instead of approximating X by a constant, we could approximate it by a random variable from an appropriate set G and solve

$$\min_{Y \in G} \mathbb{E}\big[(X-Y)^2\big].$$

This is useful when we have some partial information or observations of X that we can feed into our approximation. We quantify the extra observations by a sub σ-algebra $\mathcal{A}$ of $\mathcal{F}$ and choose G equal to the set of $\mathcal{A}$-measurable random variables. We illustrate the idea using a simple example with a finite sample space.

Example 4.43 (two throws of a die) Following Example 4.22, consider the random variables X_1, X_2 on the probability space $(\Omega, \mathcal{F}, \mathbb{P})$ taking values in $\{0,1\}$, with $X_i = 1$ indicating a six is thrown and $X_i = 0$ indicating a six is not thrown on the ith throw of a fair die. Let $X = X_1 + X_2$, the total number of sixes obtained, and consider the following three situations.

Blind: If we are 'blind' to the throws (i.e., there are no observations available), all we can do is approximate X by a constant and we know the least-squares constant approximation is $\mathbb{E}[X]$. Intuition says that $\mathbb{E}[X] = 1/3$ since the probability that $X_1 = 1$ is $1/6$ and the probability that $X_2 = 1$ is also $1/6$. To be pedantic, $\mathbb{E}[X]$ in this discrete situation is given by the sum

$$\mathbb{E}[X] = \sum_{n \in \{0,1\}} \sum_{m \in \{0,1\}} \mathbb{P}(X_1 = m, X_2 = n)(n+m) = \frac{1}{3}.$$

In terms of σ-algebras, a blind approximation to X is one chosen from the $\mathcal{A}$-measurable functions for $\mathcal{A} = \big\{\{\}, \Omega\big\}$. $\mathcal{A}$ contains the least information of any σ-algebra and captures the idea of blind. From Example 4.23, we know the $\mathcal{A}$ measurable random variables are the constant functions and hence $G = \big\{Y : \Omega \to \{a\} \text{ for any } a \in \mathbb{R}\big\}$.

Partially observed: Suppose we observe the first throw but not the second; that is, we know that $X_1 = 1$ or $X_1 = 0$ depending on the outcome of the first throw. Our intuition says that the best approximation to X is $X_1 + \mathbb{E}[X_2] = X_1 + 1/6$, the observation X_1 plus an average due to the hidden throw $(1/6)$. In terms of σ-algebras, we capture the idea that X_1 is known by looking at $\sigma(X_1)$-measurable random variables (where $\sigma(X_1)$ is derived in Example 4.19). The Doob–Dynkin lemma (Lemma 4.46) shows that any $\sigma(X_1)$-measurable random variable can be written as a measurable function of X_1 and hence

$$G := \big\{Y : \Omega \to \mathbb{R} : Y = \phi(X_1) \text{ for any function } \phi\colon \{0,1\} \to \mathbb{R}\big\}.$$

Intuitively, if we have sight of the first throw, the outcome of any random variable in G is known.

To find the least-squares approximation to X in G, let $V(Y) = \mathbb{E}\big[(X - Y)^2\big]$ and put $Y = \phi(X_1)$, so that

$$V(Y) = \mathbb{E}\big[(X_1 + X_2 - \phi(X_1))^2\big] = \sum_{n\in\{0,1\}} \sum_{m\in\{0,1\}} \mathbb{P}(X_1 = n, X_2 = m)\big(n + m - \phi(n)\big)^2.$$

Minimising $V(Y)$ with respect to Y amounts to finding $\phi(n)$, for $n = 0, 1$, that minimise the above quantity. Setting $\frac{\partial V(Y)}{\partial \phi(n)} = 0$ gives

$$\begin{aligned} 0 &= \sum_{m\in\{0,1\}} \mathbb{P}(X_1 = n,\, X_2 = m)\big(n + m - \phi(n)\big) \\ &= \mathbb{P}(X_1 = n)\big(n - \phi(n)\big) + \sum_{m\in\{0,1\}} \mathbb{P}(X_1 = n,\, X_2 = m)\, m \\ &= \mathbb{P}(X_1 = n)\big(n - \phi(n)\big) + \mathbb{P}(X_1 = n,\, X_2 = 1). \end{aligned}$$

Since X_1 and X_2 are independent, $\mathbb{P}(X_1 = n,\, X_2 = 1) = \mathbb{P}(X_1 = n)\mathbb{P}(X_2 = 1) = \mathbb{P}(X_1 = n)/6$. Cancelling $\mathbb{P}(X_1 = n)$, we find $\phi(n) = n + 1/6$. The least-squares approximation to X in G is therefore $\phi(X_1) = X_1 + 1/6$, as expected.

Fully observed: If we observe the value of X_1 *and* the value of X_2, G is the set of random variables that depend on both X_1 and X_2. In other words, G is the set of $\mathcal{A}$-measurable functions where $\mathcal{A}$ is the σ-algebra $\mathcal{F}$ containing all 2^{36} events in the probability space. Since X is itself an $\mathcal{F}$-measurable random variable, the least-squares estimator is exact and equals X, the total number of sixes.

Below, we define conditional expectation for square-integrable H-valued random variables. First, we introduce the following Hilbert space.

Definition 4.44 (the Hilbert space $L^2(\Omega, \mathcal{F}, H)$) For a Hilbert space H, the space of $\mathcal{F}$-measurable random variables from $(\Omega, \mathcal{F}, \mathbb{P})$ to $(H, \mathcal{B}(H))$ with $\mathbb{E}\big[\|X\|^2\big] < \infty$ is denoted $L^2(\Omega, \mathcal{F}, H)$ and forms a Hilbert space with the inner product

$$\langle X, Y\rangle_{L^2(\Omega,\mathcal{F},H)} := \mathbb{E}\big[\langle X, Y\rangle\big], \qquad \text{for } X, Y \in L^2(\Omega, \mathcal{F}, H).$$

This definition agrees with the one given for the space $L^2(\Omega, H)$ in Definition 4.34, except that, here, we make the underlying measurable space explicit. We may now consider the spaces $L^2(\Omega, \mathcal{A}, H)$ of square-integrable H-valued random variables that are $\mathcal{A}$-measurable for sub σ-algebras $\mathcal{A}$ of $\mathcal{F}$. We assume that $L^2(\Omega, \mathcal{A}, H)$ is a closed subspace of $L^2(\Omega, \mathcal{F}, H)$. Recall the orthogonal projection P of Definition 1.36 from a Hilbert space to a closed subspace G. If the Hilbert space in question is $L^2(\Omega, \mathcal{F}, H)$ and we take $G = L^2(\Omega, \mathcal{A}, H)$, Theorem 1.37 says that PX for $X \in L^2(\Omega, \mathcal{F}, H)$ is the element in G that minimises $\|X - Y\|^2_{L^2(\Omega,\mathcal{F},H)} = \mathbb{E}\big[\|X - Y\|^2\big]$, over all choices of $Y \in G$. In other words, PX is the least-squares approximation of X in G. This leads to our definition of conditional expectation.

Definition 4.45 (conditional expectation given a σ-algebra) Let $X \in L^2(\Omega, \mathcal{F}, H)$. If $\mathcal{A}$ is a sub σ-algebra of $\mathcal{F}$, the *conditional expectation* of X given $\mathcal{A}$, denoted $\mathbb{E}[X \,|\, \mathcal{A}]$, is defined as $\mathbb{E}[X \,|\, \mathcal{A}] := PX$, where P is the orthogonal projection from $L^2(\Omega, \mathcal{F}, H)$ to $L^2(\Omega, \mathcal{A}, H)$.

The conditional expectation $\mathbb{E}[X \mid \mathcal{A}]$ is, by definition, an H-valued $\mathcal{A}$-measurable random variable. For the two extreme cases in Example 4.43 (blind and fully observed), we have two simple expressions for the conditional expectation:

$$\text{if } \mathcal{A} = \{\Omega, \{\}\} \quad \text{then } \mathbb{E}\big[X \mid \mathcal{A}\big] = \mathbb{E}\big[X\big], \tag{4.17}$$

$$\text{if } \mathcal{A} = \mathcal{F} \quad \text{then } \mathbb{E}\big[X \mid \mathcal{A}\big] = X. \tag{4.18}$$

However, we are often interested in the case where Y is a second random variable and $\mathcal{A} = \sigma(Y)$. Since $\mathbb{E}[X \mid \sigma(Y)]$ is $\sigma(Y)$-measurable, the following important result says that $\mathbb{E}[X \mid \sigma(Y)]$ is a function of Y.

Lemma 4.46 (Doob–Dynkin) *Let H_1, H_2 be separable Hilbert spaces and let Z, Y be H_1, H_2-valued random variables, respectively. If Z is $\sigma(Y)$-measurable, then $Z = \phi(Y)$ for some Borel measurable function $\phi\colon H_2 \to H_1$.*

Definition 4.47 (conditional expectation given a random variable) Let H_1, H_2 be separable Hilbert spaces with $X \in L^2(\Omega, \mathcal{F}, H_1)$. Let Y be an H_2-valued random variable. The *conditional expectation* of X given Y is $\mathbb{E}[X \mid Y] := \mathbb{E}[X \mid \sigma(Y)]$. Furthermore, $\mathbb{E}[X \mid Y = y] = \phi(y)$, where $\phi\colon H_2 \to H_1$ is a measurable function such that $\mathbb{E}[X \mid \sigma(Y)] = \phi(Y)$ (which exists by Lemma 4.46).

For the partially observed case in Example 4.43, we explicitly derived $\phi\colon \mathbb{R} \to \mathbb{R}$. We have $\mathbb{E}[X \mid X_1] = \mathbb{E}[X \mid \sigma(X_1)] = \phi(X_1) = X_1 + 1/6$. If the first throw is a six, $\mathbb{E}[X \mid X_1 = 1] = \phi(1) = 7/6$ and we expect $7/6$ sixes. Otherwise, we obtain $\mathbb{E}[X \mid X_1 = 0] = \phi(0) = 1/6$. Conditional probabilities are also explored in Exercise 4.11. When the joint pdf of multivariate random variables $\boldsymbol{X}$ and $\boldsymbol{Y}$ is known, an explicit formula for the conditional expectation of $\boldsymbol{X}$ given $\boldsymbol{Y}$ is available.

Lemma 4.48 *Let $\boldsymbol{X}, \boldsymbol{Y}$ be $\mathbb{R}^n$ and $\mathbb{R}^m$-valued random variables, respectively. Suppose that $\boldsymbol{X}, \boldsymbol{Y}$ has joint pdf $p_{\boldsymbol{X},\boldsymbol{Y}}$ and $\boldsymbol{Y}$ has pdf $p_{\boldsymbol{Y}}$. Let $F\colon \mathbb{R}^n \times \mathbb{R}^m \to \mathbb{R}$ be a real-valued measurable function such that $F(\boldsymbol{X}, \boldsymbol{Y}) \in L^2(\Omega, \mathcal{F}, \mathbb{R})$. Then, almost surely, $\mathbb{E}\big[F(\boldsymbol{X}, \boldsymbol{Y}) \,\big|\, \boldsymbol{Y} = \boldsymbol{y}\big]$ is a minimiser of*

$$V(c; \boldsymbol{y}) := \int_{\mathbb{R}^n} \big(F(\boldsymbol{x}, \boldsymbol{y}) - c\big)^2 p_{\boldsymbol{X},\boldsymbol{Y}}(\boldsymbol{x}, \boldsymbol{y})\, d\boldsymbol{x} \quad \text{over } c \in \mathbb{R}.$$

Furthermore, if $p_{\boldsymbol{Y}}(\boldsymbol{y}) \neq 0$,

$$\mathbb{E}\big[F(\boldsymbol{X}, \boldsymbol{Y}) \,\big|\, \boldsymbol{Y} = \boldsymbol{y}\big] = \frac{1}{p_{\boldsymbol{Y}}(\boldsymbol{y})} \int_{\mathbb{R}^n} F(\boldsymbol{x}, \boldsymbol{y}) p_{\boldsymbol{X},\boldsymbol{Y}}(\boldsymbol{x}, \boldsymbol{y})\, d\boldsymbol{x}.$$

Proof We can write $\mathbb{E}[F(\boldsymbol{X}, \boldsymbol{Y}) \mid \boldsymbol{Y}] := \mathbb{E}[F(\boldsymbol{X}, \boldsymbol{Y}) \mid \sigma(\boldsymbol{Y})] = \phi(\boldsymbol{Y})$ for a measurable function $\phi\colon \mathbb{R}^m \to \mathbb{R}$ (by Lemma 4.46 with $H_2 = \mathbb{R}^m$, $H_1 = \mathbb{R}$). From Definition 4.45, ϕ is a minimiser of

$$V(\psi) := \int_{\mathbb{R}^m} \int_{\mathbb{R}^n} \big(F(\boldsymbol{x}, \boldsymbol{y}) - \psi(\boldsymbol{y})\big)^2 p_{\boldsymbol{X},\boldsymbol{Y}}(\boldsymbol{x}, \boldsymbol{y})\, d\boldsymbol{x}\, d\boldsymbol{y}$$

over measurable functions $\psi\colon \mathbb{R}^m \to \mathbb{R}$. As $p_{\boldsymbol{X},\boldsymbol{Y}} \geq 0$, ϕ is the minimiser of the inner integrand for almost all $\boldsymbol{y}$; that is, $\phi(\boldsymbol{y})$ is the minimiser of

$$V(c; \boldsymbol{y}) = \int_{\mathbb{R}^n} \big(F(\boldsymbol{x}, \boldsymbol{y}) - c\big)^2 p_{\boldsymbol{X},\boldsymbol{Y}}(\boldsymbol{x}, \boldsymbol{y})\, d\boldsymbol{x} \quad \text{over } c \in \mathbb{R}.$$

By differentiating and setting $V'(\phi(\boldsymbol{y}); \boldsymbol{y}) = 0$, we have

$$\phi(\boldsymbol{y}) \int_{\mathbb{R}^n} p_{\boldsymbol{X},\boldsymbol{Y}}(\boldsymbol{x},\boldsymbol{y})\, d\boldsymbol{x} = \phi(\boldsymbol{y}) p_{\boldsymbol{Y}}(\boldsymbol{y}) = \int_{\mathbb{R}^n} F(\boldsymbol{x},\boldsymbol{y}) p_{\boldsymbol{X},\boldsymbol{Y}}(\boldsymbol{x},\boldsymbol{y})\, d\boldsymbol{x}.$$

As $\phi(\boldsymbol{y}) = \mathbb{E}[F(\boldsymbol{X},\boldsymbol{Y}) \,|\, \boldsymbol{Y} = \boldsymbol{y}]$, the proof is complete. □

Example 4.49 (taking out what is known) Let X, Y be real-valued random variables with joint pdf $p_{X,Y}$. Suppose that $XY \in L^2(\Omega, \mathcal{F}, \mathbb{R})$. By Lemma 4.48 with $n = 1, m = 1$ and $F(X,Y) = XY$, we see $\mathbb{E}[XY \,|\, Y = y]$ is the minimiser of

$$V(c; y) = \int_{\mathbb{R}} (xy - c)^2 p_{X,Y}(x,y)\, dx \qquad \text{for almost all } y \in Y.$$

For $y \neq 0$,

$$V(c; y) = y^2 \int_{\mathbb{R}} \big(x - (c/y)\big)^2 p_{X,Y}(x,y)\, dx.$$

Then $c/y = \mathbb{E}[X \,|\, Y = y]$ and $\mathbb{E}[XY \,|\, Y = y] = c = y\,\mathbb{E}[X \,|\, Y = y]$. In other words, $\mathbb{E}[XY \,|\, Y] = Y\mathbb{E}[X \,|\, Y]$; we can take Y out of the expectation because it is observed when we condition on Y.

Consider a sample space Ψ equipped with a σ-algebra $\mathcal{G}$. The probability distribution $\mathbb{P}_X$ of a Ψ-valued random variable X is given by $\mathbb{P}_X(G) = \mathbb{P}(\{\omega \in \Omega \colon X(\omega) \in G\})$ for $G \in \mathcal{G}$ (see Definition 4.4). Written as an integral, we have $\mathbb{P}_X(G) = \mathbb{E}[1_G(X)]$. Because $1_G(X)$ is a square-integrable real-valued random variable, our definition of conditional expectation (Definition 4.45) also provides a definition for the conditional probability that $X \in G$ given $\mathcal{A}$.

Definition 4.50 (conditional probability) For a Ψ-valued random variable X on $(\Omega, \mathcal{F}, \mathbb{P})$ and a sub σ-algebra $\mathcal{A}$ of $\mathcal{F}$, define

$$\mathbb{P}_X(G \,|\, \mathcal{A}) := \mathbb{E}[1_G(X) \,|\, \mathcal{A}], \qquad G \in \mathcal{G}.$$

Then, $\mathbb{P}_X(G \,|\, \mathcal{A})$ is a $[0,1]$-valued random variable and the map $G \mapsto \mathbb{P}_X(G \,|\, \mathcal{A})$ is a measure-valued random variable.

For an H-valued random variable Y and $y \in H$, define

$$\mathbb{P}_X(G \,|\, Y = y) := \mathbb{P}(X \in G \,|\, Y = y) = \mathbb{E}\big[1_G(X) \,\big|\, Y = y\big]. \tag{4.19}$$

The map $G \mapsto \mathbb{P}_X(G \,|\, Y = y)$ is a measure on $(\Psi, \mathcal{G})$. We say $\mathbb{P}_X(G \,|\, Y = y)$ is the *conditional probability* that $X \in G$ given $Y = y$.

In many cases, $\mathbb{P}_X(\cdot \,|\, Y = y)$ defines a probability distribution, called *the conditional probability distribution of X given Y*. As an example, we derive the conditional probability distribution of a multivariate Gaussian random variable $\boldsymbol{X} = [X_1, \ldots, X_{n+m}]^{\mathsf{T}}$ subject to conditions on $\boldsymbol{Y} = [X_1, \ldots, X_n]^{\mathsf{T}}$. The calculation assumes that the covariance matrix of $\boldsymbol{Y}$ is non-singular.

Example 4.51 (conditional Gaussian) Let X be a $\mathbb{R}^{n+m}$-valued Gaussian random variable with distribution $\mathrm{N}(\boldsymbol{\mu}, Q)$, mean $\boldsymbol{\mu} \in \mathbb{R}^{n+m}$, and covariance matrix Q. Write

$$X = \begin{pmatrix} X_1 \\ X_2 \end{pmatrix}, \qquad Q = \begin{pmatrix} Q_{11} & Q_{12} \\ Q_{12}^{\mathsf{T}} & Q_{22} \end{pmatrix}$$

for $X_1 \in \mathbb{R}^n$, $X_2 \in \mathbb{R}^m$. Here, Q_{11}, Q_{12} and Q_{22} are $n \times n$, $n \times m$, and $m \times m$ matrices respectively. Q_{11} is the covariance matrix of X_1 and we assume it is non-singular. We compute the pdf of the conditional distribution of X_2 given X_1. We have

$$LQL^{\mathsf{T}} = \begin{pmatrix} Q_{11} & 0 \\ 0 & Q_{22} - Q_{12}^{\mathsf{T}} Q_{11}^{-1} Q_{12} \end{pmatrix} = \bar{Q} = \begin{pmatrix} Q_{11} & 0 \\ 0 & \bar{Q}_{22} \end{pmatrix},$$

where

$$L = \begin{pmatrix} I_n & 0 \\ -Q_{12}^{\mathsf{T}} Q_{11}^{-1} & I_m \end{pmatrix}.$$

Using this factorisation, we obtain $Q^{-1} = L^{\mathsf{T}} \bar{Q}^{-1} L$ and, for any $\boldsymbol{x} \in \mathbb{R}^{n+m}$,

$$\boldsymbol{x}^{\mathsf{T}} Q^{-1} \boldsymbol{x} = \boldsymbol{x}^{\mathsf{T}} L^{\mathsf{T}} \bar{Q}^{-1} L \boldsymbol{x} = (L\boldsymbol{x})^{\mathsf{T}} \bar{Q}^{-1} (L\boldsymbol{x}).$$

The pdf of X can then be written as

$$\begin{aligned} p_X(\boldsymbol{x}_1, \boldsymbol{x}_2) &= \frac{1}{(2\pi \det Q)^{(n+m)/2}} \mathrm{e}^{-(\boldsymbol{x}-\boldsymbol{\mu})^{\mathsf{T}} Q^{-1} (\boldsymbol{x}-\boldsymbol{\mu})} \\ &= \frac{1}{(2\pi \det Q)^{(n+m)/2}} \mathrm{e}^{-(\boldsymbol{x}_1-\boldsymbol{\mu}_1)^{\mathsf{T}} Q_{11}^{-1} (\boldsymbol{x}_1-\boldsymbol{\mu}_1)} \mathrm{e}^{-(\bar{\boldsymbol{x}}_2-\bar{\boldsymbol{\mu}}_2)^{\mathsf{T}} \bar{Q}_{22}^{-1} (\bar{\boldsymbol{x}}_2-\bar{\boldsymbol{\mu}}_2)}, \end{aligned}$$

where

$$\begin{pmatrix} \bar{\boldsymbol{x}}_1 \\ \bar{\boldsymbol{x}}_2 \end{pmatrix} = L\boldsymbol{x} = \begin{pmatrix} \boldsymbol{x}_1 \\ \boldsymbol{x}_2 - Q_{12}^{\mathsf{T}} Q_{11}^{-1} \boldsymbol{x}_1 \end{pmatrix}, \qquad \begin{pmatrix} \bar{\boldsymbol{\mu}}_1 \\ \bar{\boldsymbol{\mu}}_2 \end{pmatrix} = L\boldsymbol{\mu} = \begin{pmatrix} \boldsymbol{\mu}_1 \\ \boldsymbol{\mu}_2 - Q_{12}^{\mathsf{T}} Q_{11}^{-1} \boldsymbol{\mu}_1 \end{pmatrix}.$$

Lemma 4.48 and (4.19) give

$$\mathbb{P}(X \in G \,|\, X_1 = \boldsymbol{y}) = \frac{1}{p_{X_1}(\boldsymbol{y})} \int_G p_X(\boldsymbol{y}, \boldsymbol{x}_2)\, d\boldsymbol{x}_2.$$

The density of X given $X_1 = \boldsymbol{y}$ is then (using $\bar{\boldsymbol{x}}_2 = \boldsymbol{x}_2 - Q_{12}^{\mathsf{T}} Q_{11}^{-1} \boldsymbol{y}$)

$$\frac{1}{p_{X_1}(\boldsymbol{y})} \frac{1}{(2\pi \det Q)^{(n+m)/2}} \mathrm{e}^{-(\boldsymbol{y}-\boldsymbol{\mu}_1)^{\mathsf{T}} Q_{11}^{-1} (\boldsymbol{y}-\boldsymbol{\mu}_1)} \mathrm{e}^{-(\boldsymbol{x}_2-\hat{\boldsymbol{\mu}}_2)^{\mathsf{T}} \bar{Q}_{22}^{-1} (\boldsymbol{x}_2-\hat{\boldsymbol{\mu}}_2)},$$

for $\hat{\boldsymbol{\mu}}_2 := \boldsymbol{\mu}_2 - Q_{12}^{\mathsf{T}} Q_{11}^{-1} (\boldsymbol{\mu}_1 - \boldsymbol{y})$. Using $\det Q = \det Q_{11} \det \bar{Q}_{22}$ and cancelling out $p_{X_1}(\boldsymbol{y})$, we find the conditional distribution of X_2 given $X_1 = \boldsymbol{y}$ is $\mathrm{N}(\hat{\boldsymbol{\mu}}_2, \bar{Q}_{22})$.

We have defined conditional expectation for square-integrable random variables. We end this section with the following general definition, which allows for integrable random variables (and possibly $\mathbb{E}\big[\|X\|^2\big] = \infty$).

Theorem 4.52 *Let H be a separable Hilbert space and X be a H-valued random variable on the probability space $(\Omega, \mathcal{F}, \mathbb{P})$ with $\mathbb{E}[\|X\|] < \infty$. For any sub σ-algebra $\mathcal{A}$ of $\mathcal{F}$,* *there exists an $\mathcal{A}$-measurable random variable $\mathbb{E}[X \mid \mathcal{A}]$, unique almost surely, such that $\mathbb{E}\big[\|\mathbb{E}[X \mid \mathcal{A}]\|\big] < \infty$ and*

$$\int_G \mathbb{E}[X \mid \mathcal{A}](\omega)\, d\mathbb{P}(\omega) = \int_G X(\omega)\, d\mathbb{P}(\omega), \qquad \forall G \in \mathcal{A}.$$

Furthermore the following properties hold:

Linearity: *If X, Y are $\mathcal{F}$-measurable random variables, then for $a, b \in \mathbb{R}$*

$$\mathbb{E}\big[aX + bY \,\big|\, \mathcal{A}\big] = a\mathbb{E}\big[X \,\big|\, \mathcal{A}\big] + b\mathbb{E}\big[Y \,\big|\, \mathcal{A}\big] \qquad a.s.$$

Independence: $\mathbb{E}\big[X \,\big|\, \mathcal{A}\big] = \mathbb{E}[X]$ *a.s. if $\mathcal{A}$ and $\sigma(X)$ are independent σ-algebras. In particular,* $\mathbb{E}\big[X \,\big|\, \{\Omega, \{\}\}\big] = \mathbb{E}[X]$.

Taking out what is known: *If Y is $\mathcal{A}$-measurable, then $\mathbb{E}\big[Y \,\big|\, \mathcal{A}\big] = Y$ and $\mathbb{E}\big[XY \,\big|\, \mathcal{A}\big] = Y\mathbb{E}\big[X \,\big|\, \mathcal{A}\big]$ a.s.*

Tower property: *If $\mathcal{A}_1$ is a sub σ-algebra of $\mathcal{A}_2$, $\mathbb{E}\big[\mathbb{E}[X \,|\, \mathcal{A}_2] \,\big|\, \mathcal{A}_1\big] = \mathbb{E}\big[X \,\big|\, \mathcal{A}_1\big]$ a.s.*

Example 4.53 (martingale) Let X_n be a sequence of real-valued random variables with $\mathbb{E}[|X_n|] < \infty$ and let $\mathcal{F}_n = \sigma(X_1, \ldots, X_n)$, the smallest σ-algebra such that $X_1, \ldots, X_n$ are measurable. We say X_n is a martingale if $\mathbb{E}[X_{n+1} \,|\, \mathcal{F}_n] = X_n$. A martingale is the mathematical abstraction of a fair game: if X_n denotes the winnings on the nth round of the game, the game is fair if the winnings $X_{n+1} - X_n$ over the next round equal zero on average; that is, the expectation of X_{n+1} given that we have played the first n rounds $X_1, \ldots, X_n$ is equal to X_n. In this context, $\mathcal{F}_{n-1}$ is a sub σ-algebra of $\mathcal{F}_n$ and the tower property says that $\mathbb{E}\big[\mathbb{E}[X_n \,|\, \mathcal{F}_n] \,\big|\, \mathcal{F}_{n-1}\big] = X_{n-1}$.

4.3 Convergence of random variables

To understand numerical approximations to stochastic differential equations, we examine their convergence to the exact solution. For stochastic problems, the approximations X_n and the exact solution X are random variables, and it is necessary to precisely define limits of sequences of random variables. There are two basic types of approximation and a number of associated modes of convergence: (1) Approximation of each realisation $X(\omega)$ as a function of the sample $\omega \in \Omega$. In Definition 4.54, this is made precise using the concepts of convergence almost surely, convergence in probability, and convergence in pth mean. (2) Approximation of averages $\mathbb{E}[\phi(X)]$ for a test function ϕ. This is described in Definition 4.54 using the concept of convergence in distribution.

Definition 4.54 (convergence of random variables) Let H be a Hilbert space and X_n be a sequence of H-valued random variables. We say X_n converges to $X \in H$

almost surely if $X_n(\omega) \to X(\omega)$ for almost all $\omega \in \Omega$. That is, if

$$\mathbb{P}\Big(\|X_n - X\| \to 0 \quad \text{as } n \to \infty\Big) = 1.$$

By Lemma 1.20, this implies X is also a random variable.

in probability if $\mathbb{P}(\|X_n - X\| > \epsilon) \to 0$ as $n \to \infty$ for any $\epsilon > 0$.

in pth mean or in $L^p(\Omega, H)$ if $\mathbb{E}\left[\|X_n - X\|^p\right] \to 0$ as $n \to \infty$. When $p = 2$, this is known as *convergence in mean square.*

in distribution if $\mathbb{E}[\phi(X_n)] \to \mathbb{E}[\phi(X)]$ as $n \to \infty$ for any bounded continuous function $\phi \colon H \to \mathbb{R}$. Convergence in distribution is also known as convergence of laws or weak convergence. We often study convergence of $\mathbb{E}[\phi(X_n)]$ for ϕ in a strict subset (such as Lipschitz continuous functions) of the bounded continuous functions; while this is often sufficient for practical application, it does not itself imply convergence in distribution.

The basic relationships between the types of convergence listed above are summarised in the following diagram. For $q > p$,

in qth mean	$\Rightarrow$	in pth mean	$\Rightarrow$			
				in probability	$\Rightarrow$	in distribution
		almost surely	$\Rightarrow$			

In addition, convergence almost surely implies convergence in pth mean if the random variables X_n are uniformly bounded; see Exercise 4.13.

To describe the effectiveness of a particular approximation scheme, we look for specific rates of convergence. Theorem 4.58 relates the rates of different types of convergence of $X_n \to X$, starting from a given rate of pth mean convergence, which is often the easiest to prove in the context of stochastic PDEs. We start with three important lemmas.

Lemma 4.55 (Chebyshev's inequality) *If $p \geq 1$ and $X \in L^p(\Omega, H)$, then for any $R > 0$*

$$\mathbb{P}\left(\|X\| \geq R\right) \leq R^{-p}\, \mathbb{E}\left[\|X\|^p\right]. \tag{4.20}$$

Proof See Exercise 4.12. □

Lemma 4.56 (Jensen's inequality) *If $\phi \colon \mathbb{R} \to \mathbb{R}$ is a convex function and X is a real-valued random variable with $\mathbb{E}[X] < \infty$, then*

$$\phi(\mathbb{E}[X]) \leq \mathbb{E}[\phi(X)]. \tag{4.21}$$

In particular, for $p > 1$, $\left(\mathbb{E}[|X|]\right)^p \leq \mathbb{E}[|X|^p]$.

Proof See Exercise 1.9. □

Lemma 4.57 (Borel–Cantelli) *Let $F_n \in \mathcal{F}$ be events associated with a probability space $(\Omega, \mathcal{F}, \mathbb{P})$. If $\sum_{n=1}^{\infty} \mathbb{P}(F_n) < \infty$, then*

$$\mathbb{P}(F_n \textit{ occurs for infinitely many } n) = 0.$$

In particular, if $\sum_{n=1}^{\infty} \mathbb{P}(|X_n| > 1) < \infty$ for a sequence X_n of real-valued random variables, there exists an $\mathbb{N}$-valued random variable n^* such that

$$\mathbb{P}\left(\omega \in \Omega \colon |X_n(\omega)| \leq 1 \textit{ for } n \geq n^*(\omega)\right) = 1. \tag{4.22}$$

Proof Let E be the set of $\omega \in \Omega$ that belong to infinitely many F_n and $G_k = \cup_{n=k}^{\infty} F_n$. If $\omega \in E$, then ω belongs to infinitely many F_n and hence $\omega \in G_k$ for all k. This implies $\mathbb{P}(E) \leq \mathbb{P}(G_k) \leq \sum_{n=k}^{\infty} \mathbb{P}(F_n)$ for all k. As $\sum_{n=1}^{\infty} \mathbb{P}(F_n) < \infty$, $\mathbb{P}(G_k) \to 0$ as $k \to \infty$ and we conclude that $\mathbb{P}(E) = 0$.

Define the $\mathbb{N}$-valued random variable

$$n^*(\omega) = \begin{cases} k, & \text{if } |X_{k-1}(\omega)| > 1 \text{ and } |X_n(\omega)| \le 1 \text{ for } n \ge k, \quad k = 2,3,\dots, \\ 1, & \text{otherwise.} \end{cases}$$

Consider $F_n := \{\omega \in \Omega \colon |X_n(\omega)| > 1\}$. The series $\sum_{n=1}^{\infty} \mathbb{P}(F_n)$ is finite by assumption and the probability that F_n occurs infinitely often is zero. Thus, $\mathbb{P}(n^* = 1) = 0$ and (4.22) holds. □

Theorem 4.58 *Suppose that $X_n \to X$ in pth mean. In particular, suppose that, for some $r > 0$ and constant $K(p) > 0$ depending on p,*

$$\|X_n - X\|_{L^p(\Omega,H)} = \mathbb{E}\left[\|X_n - X\|^p\right]^{1/p} \le K(p) n^{-r}. \tag{4.23}$$

Then we have the following convergence properties:

(i) *For any $\epsilon > 0$, $X_n \to X$ in probability and*

$$\mathbb{P}\left(\|X_n - X\| \ge n^{-r+\epsilon}\right) \le K(p)^p\, n^{-p\epsilon}. \tag{4.24}$$

(ii) *$\mathbb{E}[\phi(X_n)] \to \mathbb{E}[\phi(X)]$ for all globally Lipschitz continuous functions on H (see Definition 1.14). If L denotes the Lipschitz constant of ϕ,*

$$\left|\mathbb{E}[\phi(X_n)] - \mathbb{E}[\phi(X)]\right| \le L\, K(p)\, n^{-r}.$$

(iii) *If (4.23) holds for every p sufficiently large, then $X_n \to X$ a.s. Further, for each $\epsilon > 0$, there exists a non-negative random variable K such that $\|X_n(\omega) - X(\omega)\| \le K(\omega)\, n^{-r+\epsilon}$ for almost all ω.*

Proof (i) By the Chebyshev inequality in Lemma 4.55, for any $\epsilon > 0$,

$$\mathbb{P}\left(\|X_n - X\| \ge n^{-r+\epsilon}\right) \le \frac{\mathbb{E}[\|X_n - X\|^p]}{n^{-rp+p\epsilon}} \le K(p)^p\, n^{-p\epsilon}.$$

(ii) By Nensen's inequality, $\mathbb{E}[\|X_n - X\|] \le \mathbb{E}[\|X_n - X\|^p]^{1/p} \le K(p) n^{-r}$. Let L denote the Lipschitz constant of ϕ. Then

$$\left|\mathbb{E}[\phi(X_n)] - \mathbb{E}[\phi(X)]\right| \le \mathbb{E}\left[|\phi(X_n) - \phi(X)|\right] \le L\,\mathbb{E}\left[\|X_n - X\|\right] \le L K(p) n^{-r}.$$

(iii) Let F_n be the event that $\|X_n - X\| > n^{-r+\epsilon}$. By (4.24), $\mathbb{P}(F_n) \le K(p)^p n^{-p\epsilon}$ for any $\epsilon > 0$ and, with p sufficiently large, $\sum_{n=1}^{\infty} \mathbb{P}(F_n) < \infty$. The Borel–Cantelli lemma applies and there exists an $\mathbb{N}$-valued random variable n^* such that

$$\mathbb{P}\left(\|X_n(\omega) - X(\omega)\| \le n^{-r+\epsilon} \text{ for } n \ge n^*(\omega)\right) = 1.$$

Define $K(\omega) := \max\{1, \|X_n(\omega) - X(\omega)\| n^{r-\epsilon} \colon n = 1, \dots, n^*(\omega)\}$. Clearly then, we have $\|X_n(\omega) - X(\omega)\| \le K(\omega) n^{-r+\epsilon}$ for all n a.s. □

As examples of convergence of random variables, we study sums of random variables and the approximation of Gaussian random variables.

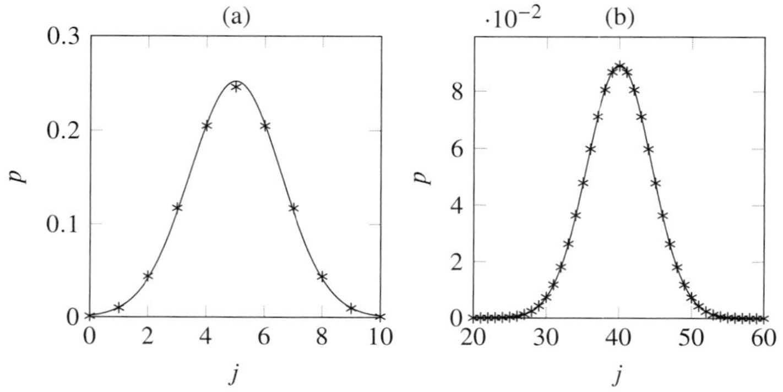

Figure 4.4 The $*$ marks the probabilities for the binomial distribution $\mathrm{B}(M,p)$ and the line is the pdf of the Gaussian distribution $\mathrm{N}(Mp, Mp(1-p))$ for (a) $M = 10$, $p = 1/2$ and (b) $M = 80$, $p = 1/2$. The probabilities come closer to the pdf as M increases.

Sums of random variables

Definition 4.59 (*iid*) A set of random variables $\{X_j : j = 1,2,\dots\}$ is said to be *independent and identically distributed* (*iid*) if each X_j has identical distribution and the distinct random variables X_i, X_j, for $i \neq j$, are independent.

Sums of *iid* random variables arise frequently in statistical analysis and in sampling algorithms. Consider, for example, the binomial distribution, which models the number of successes of M identical and independent experiments, each with probability of success p.

Definition 4.60 (binomial distribution) Let $Y_M := \sum_{j=1}^{M} X_j$ where X_j are *iid* random variables with $\mathbb{P}(X_j = 1) = p$ and $\mathbb{P}(X_j = 0) = 1 - p$. Then

$$\mathbb{E}\big[X_j\big] = p \quad \text{and} \quad \operatorname{Var} X_j = p(1-p)$$

as in Exercise 4.14. We say Y_M has the *binomial distribution* and write $Y_M \sim \mathrm{B}(M,p)$, and

$$\mathbb{P}(Y_M = j) = \binom{M}{j} p^j (1-p)^{M-j}, \qquad j = 0,\dots,M$$

with $\mathbb{E}[Y_M] = Mp$ and $\operatorname{Var} Y_M = Mp(1-p)$.

The binomial distribution 'looks like' the Gaussian distribution when the number of samples M is large, as illustrated by the plots of $\mathbb{P}(Y_M = j)$ against j in Figure 4.4. This illustrates a general result known as the *central limit theorem*, which says the Gaussian distribution arises generically from sums of *iid* random variables from *any* distribution with finite second moments.

Before stating the rigorous result in Theorem 4.62 for $\mathbb{R}^d$-valued random variables, we discuss first the case of *iid real-valued* random variables X_j, each with mean μ and variance σ^2. Define

$$\overline{X}_M := \frac{1}{M}(X_1 + \cdots + X_M), \qquad X_M^* := \sqrt{M}\big(\overline{X}_M - \mu\big). \tag{4.25}$$

Here $\overline{X}_M$ can be thought of as the *sample mean* of M independent samples $X_1,\dots,X_M$ of a random variable X of interest. The statistical properties of the random variables $\overline{X}_M$ and X^*_M are described by the following important results.

Strong law of large numbers: If $\mathbb{E}[|X_j|] < \infty$, the sample mean $\overline{X}_M$ of $X_1,\dots,X_M$ converges to μ almost surely.

Central limit theorem: If the random variables X_j each have finite second moments with $\operatorname{Var} X_j = \sigma^2$, the distribution of $\overline{X}_M$ converges to that of a Gaussian random variable with mean μ and variance $\frac{\sigma^2}{M}$. Thus, X^*_M converges in distribution to $Z \sim \mathrm{N}(0,\sigma^2)$.

Berry–Esséen inequality: If $\mathbb{E}\big[|X_j|^3\big] < \infty$, the rate of convergence of X^*_M to $\mathrm{N}(0,\sigma^2)$ is $\mathcal{O}(M^{-1/2})$. In particular, if $Z \sim \mathrm{N}(0,\sigma^2)$,

$$\sup_{z\in\mathbb{R}}\left|\mathbb{P}(X^*_M \le z) - \mathbb{P}(Z \le z)\right| \le \frac{\mathbb{E}\big[|X_1-\mu|^3\big]}{\sigma^3\sqrt{M}}. \tag{4.26}$$

Example 4.61 (binomial distribution) For the binomial distribution in Definition 4.60 with $\mu = p$ and $\sigma^2 = p(1-p)$, we have

$$\overline{X}_M = \frac{1}{M}Y_M, \qquad X^*_M = \frac{1}{\sqrt{M}}(Y_M - Mp).$$

The strong law of large numbers says that $Y_M \to Mp$ as $M \to \infty$ almost surely. The central limit theorem says that the distribution $\mathrm{B}(M,p)$ converges to the Gaussian distribution $\mathrm{N}(Mp, Mp(1-p))$ as $M \to \infty$. Recall that if $Z \sim \mathrm{N}(0,\sigma^2)$, we have $\mathbb{P}(Z \le z) = \frac{1}{2}\big(1 + \operatorname{erf}\big(z/\sqrt{2\sigma^2}\big)\big)$ (see Exercise 4.4). Choosing $z = \sigma R$, the Berry–Esséen inequality (4.26) says

$$\left|\mathbb{P}\left(\frac{Y_M - Mp}{\sqrt{Mp(1-p)}} \le R\right) - \frac{1}{2}\left(1 + \operatorname{erf}\left(\frac{R}{\sqrt{2}}\right)\right)\right| \le \frac{1-2p+2p^2}{\sqrt{Mp(1-p)}}, \tag{4.27}$$

or, equivalently,

$$\left|\mathbb{P}\left(|\overline{X}_M - p| \le \frac{R\sqrt{p(1-p)}}{\sqrt{M}}\right) - \operatorname{erf}\left(\frac{R}{\sqrt{2}}\right)\right| \le \frac{2(1-2p+2p^2)}{\sqrt{Mp(1-p)}}. \tag{4.28}$$

See Exercise 4.14.

We state the multi-dimensional versions of the law of large numbers and the central limit theorem.

Theorem 4.62 (sums of *iid* $\mathbb{R}^d$-valued random variables) *Let $\boldsymbol{X}_j$ be iid $\mathbb{R}^d$-valued random variables with mean $\boldsymbol{\mu}$.*

(i) *If $\mathbb{E}[\|\boldsymbol{X}_j\|_2] < \infty$, then $\overline{\boldsymbol{X}}_M = \frac{1}{M}\sum_{j=1}^M \boldsymbol{X}_j \to \boldsymbol{\mu}$ almost surely as $M \to \infty$.*

(ii) *If $\mathbb{E}\big[\|\boldsymbol{X}_j\|_2^2\big] < \infty$ and each $\boldsymbol{X}_j$ has covariance matrix C, then*

$$\boldsymbol{X}^*_M = \sqrt{M}\big(\overline{\boldsymbol{X}}_M - \boldsymbol{\mu}\big)$$

converges in distribution to $\mathrm{N}(\boldsymbol{0},C)$.

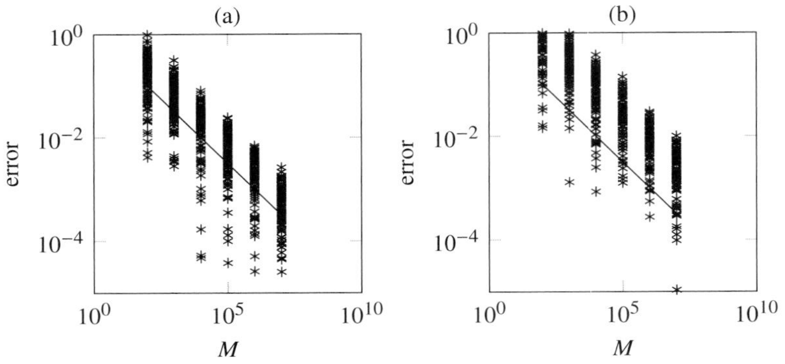

Figure 4.5 Plots of (a) $|\overline{X}_M - \mu|$ and (b) $|\sigma_M^2 - \sigma^2|$ against the number of samples M. 100 realisations are plotted for each value of M. The underlying distribution is $\mathrm{N}(\mu, \sigma^2)$ with $\mu = 1$ and $\sigma = 3$. The errors run parallel to the straight line which has slope $-1/2$.

Estimating the mean and variance of a random variable

We often have access to M samples of a random variable X without knowing its distribution and wish to compute one of the moments, say θ, of X. To do this, we use an estimator θ_M that converges to the true moment θ in the limit $M \to \infty$. If the estimator θ_M is a random variable, we say it is *unbiased* when $\mathbb{E}[\theta_M] = \theta$. For example, suppose that $X_1, \ldots, X_M$ are independent samples of X (i.e., X_j are *iid* random variables with the same distribution as X) and let $\overline{X}_M$ be the sample average as in (4.25). Then, $\overline{X}_M$ is an estimator for the true mean $\mu = \mathbb{E}[X]$. If $\mathbb{E}[|X|] < \infty$, we know $\overline{X}_M \to \mu$ almost surely as $M \to \infty$ by the strong law of large numbers. Notice that $\overline{X}_M$ is a random variable as it depends on the particular realisations of $X_1, \ldots, X_M$.

It is easy to show that $\mathbb{E}\left[\overline{X}_M\right] = \mu$ and hence $\overline{X}_M$ is an unbiased estimator of μ. Further information about the accuracy of $\overline{X}_M$ can be gained from the central limit theorem, an idea we take up for Monte Carlo methods in §4.4. Similarly, consider

$$\sigma_M^2 = \frac{1}{M-1} \sum_{j=1}^{M} \left(X_j - \overline{X}_M\right)^2. \tag{4.29}$$

As shown in Exercise 4.15, $\mathbb{E}[\sigma_M^2] = \sigma^2 = \operatorname{Var} X$ and $\sigma_M^2 \to \sigma^2$ almost surely as $M \to \infty$. Thus, σ_M^2 is an unbiased estimator for the true variance of X. The choice of scaling $1/(M-1)$, needed to ensure the estimator is unbiased, differs from that used for $\overline{X}_M$ (although in the large M limit this difference is negligible).

Given a set of M observations of a random variable of interest, the Matlab commands `mean` and `var` compute the above estimators for the mean and variance. We demonstrate this in Algorithm 4.1. In the case of $X \sim \mathrm{N}(\mu, \sigma^2)$ with $\mu = 1$ and $\sigma = 3$, Figure 4.5(a) shows $|\mu - \overline{X}_M|$ for 100 realisations of $\overline{X}_M$, plotted against M, and Figure 4.5(b) shows a similar plot for $|\sigma^2 - \sigma_M^2|$. The decrease in the error is observed to be proportional to $M^{-1/2}$.

Algorithm 4.1 Code to estimate the mean and variance of $X \sim \mathrm{N}(\mu, \sigma^2)$. The inputs are mu=$\mu$, sigma=$\sigma$, and the number of samples M. The outputs mu_M and sigma_sq_M are estimators for μ and σ^2.

```
function [mu_M, sig_sq_M]=est_mean_var_func(mu,sigma,M)
X=randn(M,1);
X=mu+sigma*X; % generate M samples from N(mu, sigma^2)
mu_M=mean(X);   % estimate mean
sig_sq_M=var(X); % estimate variance
```

Approximating multivariate Gaussian random variables

Suppose we wish to approximate a mean-zero Gaussian random variable $\boldsymbol{X} \sim \mathrm{N}(\boldsymbol{0}, C_X)$ by a second mean-zero Gaussian random variable $\boldsymbol{Y} \sim \mathrm{N}(\boldsymbol{0}, C_Y)$. The following result describes the convergence of $\boldsymbol{Y}$ to $\boldsymbol{X}$ with respect to $L^2(\mathbb{R}^d)$ test functions in terms of the $C_X - C_Y$.

Theorem 4.63 *Let C_X be a $d \times d$ symmetric positive-definite matrix. For each $\epsilon > 0$, there exists a constant $L > 0$ such that*

$$\left|\mathbb{E}\left[\phi(\boldsymbol{X})\right] - \mathbb{E}\left[\phi(\boldsymbol{Y})\right]\right| \le L\, \|\phi\|_{L^2(\mathbb{R}^d)} \|C_X - C_Y\|_{\mathrm{F}}, \qquad \forall \phi \in L^2(\mathbb{R}^d),$$

for any $\boldsymbol{X} \sim \mathrm{N}(\boldsymbol{0}, C_X)$ and $\boldsymbol{Y} \sim \mathrm{N}(\boldsymbol{0}, C_Y)$ where C_Y is a symmetric positive-definite matrix such that $\|C_X - C_Y\|_{\mathrm{F}} \le \epsilon$.

Proof Using the Parseval identity (1.43), we have that

$$\mathbb{E}\left[\phi(\boldsymbol{X})\right] = \int_{\mathbb{R}^d} \phi(\boldsymbol{x}) p(\boldsymbol{x})\, d\boldsymbol{x} = \int_{\mathbb{R}^d} \hat{\phi}(\boldsymbol{\lambda}) \hat{p}(\boldsymbol{\lambda})\, d\boldsymbol{\lambda}$$

where $p(\boldsymbol{x})$ is the pdf of the Gaussian distribution $\mathrm{N}(\boldsymbol{0}, C_X)$. From Definition 1.102 and (4.14),

$$\hat{p}(\boldsymbol{\lambda}) = \frac{1}{(2\pi)^{d/2}} \mathbb{E}\left[\mathrm{e}^{-\mathrm{i}\boldsymbol{\lambda}^{\mathsf{T}}\boldsymbol{X}}\right] = \frac{1}{(2\pi)^{d/2}} \mathrm{e}^{-\boldsymbol{\lambda}^{\mathsf{T}} C_X \boldsymbol{\lambda}/2}$$

and so

$$\mathbb{E}\left[\phi(\boldsymbol{X})\right] - \mathbb{E}\left[\phi(\boldsymbol{Y})\right] = \frac{1}{(2\pi)^{d/2}} \int_{\mathbb{R}^d} \hat{\phi}(\boldsymbol{\lambda}) \left(\mathrm{e}^{-\boldsymbol{\lambda}^{\mathsf{T}} C_X \boldsymbol{\lambda}/2} - \mathrm{e}^{-\boldsymbol{\lambda}^{\mathsf{T}} C_Y \boldsymbol{\lambda}/2}\right) d\boldsymbol{\lambda}.$$

By the Cauchy–Schwarz inequality and (1.43),

$$\left|\mathbb{E}\left[\phi(\boldsymbol{X})\right] - \mathbb{E}\left[\phi(\boldsymbol{Y})\right]\right| \le \frac{1}{(2\pi)^{d/2}} \|\phi\|_{L^2(\mathbb{R}^d)} \left\|\mathrm{e}^{-\boldsymbol{\lambda}^{\mathsf{T}} C_X \boldsymbol{\lambda}/2} - \mathrm{e}^{-\boldsymbol{\lambda}^{\mathsf{T}} C_Y \boldsymbol{\lambda}/2}\right\|_{L^2(\mathbb{R}^d)}. \tag{4.30}$$

The Gaussian integral for symmetric positive-definite matrices Q is

$$\int_{\mathbb{R}^d} \mathrm{e}^{-\boldsymbol{x}^{\mathsf{T}} Q \boldsymbol{x}}\, d\boldsymbol{x} = \pi^{d/2} \det Q^{-1/2}.$$

As C_X, C_Y, and $(C_X + C_Y)/2$ are positive definite, this gives

$$\begin{aligned}\left\|\mathrm{e}^{-\boldsymbol{\lambda}^{\mathsf{T}} C_X \boldsymbol{\lambda}/2} - \mathrm{e}^{-\boldsymbol{\lambda}^{\mathsf{T}} C_Y \boldsymbol{\lambda}/2}\right\|_{L^2(\mathbb{R}^d)}^2 &= \int_{\mathbb{R}^d} \left(\mathrm{e}^{-\boldsymbol{\lambda}^{\mathsf{T}} C_X \boldsymbol{\lambda}/2} - \mathrm{e}^{-\boldsymbol{\lambda}^{\mathsf{T}} C_Y \boldsymbol{\lambda}/2}\right)^2 d\boldsymbol{\lambda} \\ &= \pi^{d/2} \left(\det C_X^{-1/2} + \det C_Y^{-1/2} - 2 \det \left(\frac{C_X + C_Y}{2}\right)^{-1/2}\right).\end{aligned}$$

With $\Gamma := C_Y - C_X$,

$$\det C_X^{-1/2} + \det C_Y^{-1/2} - 2\det\left(\frac{C_X + C_Y}{2}\right)^{-1/2}$$
$$= \det C_X^{-1/2}\left(\det I + \det(I + C_X^{-1}\Gamma)^{-1/2} - 2\det(I + C_X^{-1}\Gamma/2)^{-1/2}\right).$$

For any square matrix A, $\det(I + A) = 1 + \operatorname{Tr} A + \mathcal{O}(\|A\|_{\mathrm{F}}^2)$ so that

$$\det C_X^{-1/2} + \det C_Y^{-1/2} - 2\det\left(\frac{C_X + C_Y}{2}\right)^{-1/2}$$
$$= \det C_X^{-1/2}\left(1 + (1 - \tfrac{1}{2}\operatorname{Tr} C_X^{-1}\Gamma) - 2\left(1 - \frac{1}{4}\operatorname{Tr}(C_X^{-1}\Gamma)\right) + \mathcal{O}\left(\|C_X^{-1}\Gamma\|_{\mathrm{F}}^2\right)\right).$$

The first-order terms cancel and we can choose $L > 0$ (which depends on C_X and ϵ) such that

$$\left\|\mathrm{e}^{-\lambda^{\mathsf{T}} C_X \lambda/2} - \mathrm{e}^{-\lambda^{\mathsf{T}} C_Y \lambda/2}\right\|_{L^2(\mathbb{R}^d)}^2 \le L^2 \|\Gamma\|_{\mathrm{F}}^2$$

if $\|\Gamma\|_{\mathrm{F}} = \|C_X - C_Y\|_{\mathrm{F}} \le \epsilon$. With (4.30), this completes the proof. □

We use this result to analyse approximations of correlated Gaussian random fields in Chapter 7. It can be extended to broader classes of test functions, including polynomially growing test functions; see Exercises 4.16 and 4.17.

4.4 Random number generation

True random number generators use measurements of a truly random event and are not widely used in computational experiments. Pseudo-random numbers are generated from a seed by a deterministic algorithm and are random in the sense that they pass statistical tests: their means, variances and distributions may be computed and agree with the underlying distribution to a high degree of accuracy. Further, correlations between pairs of pseudo-random numbers may be computed and indicate that successive instances are uncorrelated to a high degree of accuracy. Matlab and other scientific computing environments make pseudo-random number generators available. In Matlab, the command `RandStream.list` gives a list of available algorithms. For example, `mt19937ar` refers to the Mersenne Twister algorithm. Pseudo-random numbers are random enough for most purposes, though you can certainly find non-random effects, such as periodicity, whereby the sequence of random numbers is repeated after a large number of calls. The size of the period, however, is often so large (the period of the Mersenne Twister is approximately 4.3×10^{6001}) that periodic effects are never encountered.

Matlab R2008b introduced pseudo-random number streams and each stream gives a different way to generate a sequence of pseudo-random values. For each stream, Matlab generates random numbers starting from an integer seed value using a specific algorithm. Since default values for the seed and algorithm are set when an instance of Matlab is started, Matlab will produce the *same* sequence of random numbers each time it is started. Try it! In practice, we often wish to specify the seed to get different random numbers and/or to reproduce exactly the same pseudo-random numbers.

You can see the properties of the default stream as follows:

```
>> RandStream.getGlobalStream
ans =
mt19937ar random stream (current global stream)
            Seed: 0
  NormalTransform: Ziggurat
```

This says the seed is 0 and lists two algorithms for generating random numbers: `mt19937ar` for sampling the uniform distribution and `Ziggurat` for sampling the Gaussian distribution. Using these algorithms, MATLAB generates an $n \times m$ matrix of independent samples from $\mathrm{U}(0,1)$ given the command `rand(n,m)` and an $n \times m$ matrix of independent samples from $\mathrm{N}(0,1)$ given the command `randn(n,m)`.

Algorithm 4.2 Code to create a stream (equivalent to the default start up values in MATLAB) and set it to be the default. By using `reset`, the same M random numbers are produced in `r0` and `r00`.

```
function [r0,r00] = setseed0(M);
% create a stream s0 (here equivalent to restarting matlab)
s0 = RandStream('mt19937ar','Seed',0)
% set the default stream to be s0
RandStream.setGlobalStream(s0);
% draw M N(0,1) numbers from s0
r0=randn(M,1);
% Return to the start of s0
reset(s0);
% draw the same M N(0,1) numbers from s0
r00=randn(M,1);
```

In Algorithm 4.2, we create a stream `s0`, make `s0` the default stream, and draw M values from $\mathrm{N}(0,1)$ using the default stream. Next, the command `reset` resets the stream and we redraw the same numbers. Using streams with different seeds, we obtain different sets of pseudo-random numbers that are (typically) uncorrelated.

Algorithm 4.3 Code to return the covariance matrix formed from two streams of random numbers for sample size M.

```
function [QM]= setseed1(M);
s1 = RandStream('mt19937ar','Seed',1); % create a new stream s1
r1=randn(s1,M,1); % draw M N(0,1) numbers from s1
r0=randn(M,1); % draw M N(0,1) numbers from default stream
QM=cov([r1,r0]); % covariance matrix
```

In Algorithm 4.3, we define a stream `s1` and draw M samples from $\mathrm{N}(0,1)$ using `s1` and also M samples using the default stream. We use the MATLAB command `cov` to compute the sample covariance matrix Q_M of the two sets of samples. In other words, the two sets of samples are treated as realisations of two random variables R_0, R_1 (corresponding to the respective streams) and Q_M is an estimator of the covariance matrix of the bivariate random variable $(R_0, R_1)^{\mathsf{T}}$ based on M samples.

With $M = 1000$, we find

```
>> QM=setseed1(1000)
QM =
    0.9970    0.0056
    0.0056    0.9979
```

The output matrix QM is close to diagonal and in particular $\mathrm{Cov}(R_0, R_1) \approx 0$ and R_0, R_1 are close to being uncorrelated. Lemma 4.33 suggests the streams give independent samples from $\mathrm{N}(0, 1)$.

Some of MATLAB's random number generators (e.g., mrg32k3a) support multiple streams that have the same generator algorithm and seed value but are statistically independent. These algorithms allow substreams to be created from the same seed and the substream contains a pointer to the current position of the algorithm in the stream. Substreams are used in Algorithm 4.4 to draw random numbers and redraw the same numbers using the 4th substream. If the MATLAB parallel computing toolbox is available, this technique is valuable for drawing the same numbers in a parallel computation and makes such a computation reproducible. See Exercise 4.25.

Algorithm 4.4 Code to use a substream to reproduce the 4th draw of numbers in a serial case. In parallel, if the MATLAB parallel computing toolbox is available, substreams will reproduce random numbers in a repeatable manner.

```
function [sr,sr4,pr]=parseed
stream=RandStream.create('mrg32k3a','Seed',0);
N=5; M=6; sr=zeros(N,M);
for j=1:N, % draw numbers in serial using substreams indexed by loop
  s=stream; s.Substream=j; sr(j,1:M)=randn(s,1,M);
end
s.Substream=4; % draw from substream 4 to recreate 4th row
sr4=randn(s,1,M); % draw the same numbers in parallel
matlabpool open % start a worker pool of default size
pr=zeros(N,M); reset(stream); % reset the stream
parfor j=1:N, % do a loop in parallel
  s=stream; s.Substream=j; pr(j,:)=randn(s,1,M);
end
matlabpool close % close the worker pool
```

Algorithms for generating random numbers are fast. In MATLAB, the computation time for the generation of a single random number is comparable to a couple of multiplications or an evaluation of a trigonometric function, as we see in Exercise 4.18. Hence, where random numbers are required in the simulation of stochastic PDEs, the cost of generating the random numbers is an insignificant part of the whole computation. In this book, we assume a high quality supply of uniform and Gaussian samples are cheaply available for use in other algorithms.

We describe two methods for sampling other distributions using uniform and Gaussian samples: change of variables and so-called rejection sampling.

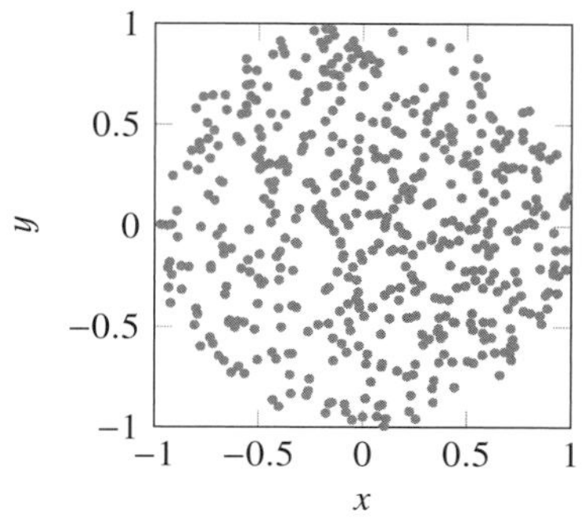

Figure 4.6 Using Algorithm 4.5, a plot of 500 independent samples from $\mathrm{U}(B)$.

Sampling by change of variables

Example 4.64 (uniform distribution on unit ball) Denote the unit ball in $\mathbb{R}^2$ by $B := \{\boldsymbol{x} \in \mathbb{R}^2 : \|\boldsymbol{x}\|_2 \le 1\}$ and suppose we want to sample the random variable $\boldsymbol{X} \sim \mathrm{U}(B)$; see Definition 4.25. First, write $\boldsymbol{X}$ in polar coordinates so that $\boldsymbol{X} = [x, y]^{\mathsf{T}} = [R\cos\theta, R\sin\theta]^{\mathsf{T}}$. Since the distribution is uniform in the angular variable, we have $\theta \sim \mathrm{U}(0, 2\pi)$. The density function of the radial variable R is proportional to the circumference of the corresponding circle, so that $p_R(r) = 2r$ for $0 \le r \le 1$. Using Lemma 4.12, if $S \sim \mathrm{U}(0,1)$ and $g(r) = r^2$, then $R = g^{-1}(S)$ has precisely the pdf $p_R(r) = g'(r) = 2r$. Consequently, we may generate R as $R = S^{1/2}$ for $S \sim \mathrm{U}(0,1)$. This technique is implemented in Algorithm 4.5 and a sample of 500 points is shown in Figure 4.6.

Algorithm 4.5 Code to generate a sample X from the distribution $\mathrm{U}(B)$ via the change of variables method.

```
function X=uniform_ball
theta=rand*2*pi; S=rand;
X=sqrt(S)*[cos(theta), sin(theta)];
```

Example 4.65 (uniform distribution on S^2) To sample $\boldsymbol{X} \sim \mathrm{U}(S^2)$, the uniform distribution on S^2 (the surface of the unit sphere), let $\boldsymbol{X} = [x, y, z]^{\mathsf{T}}$. The cross-section of S^2 at a constant z is a circle with radius $f(z) = \sqrt{1 - z^2}$. The area of S^2 bounded between $z = z_1$ and $z = z_2$ is given by

$$2\pi \int_{z_1}^{z_2} f(z)\,\sqrt{1 + f'(z)^2}\,dz = 2\pi \int_{z_1}^{z_2} 1\,dz = 2\pi(z_2 - z_1).$$

Hence, the density is uniform in the z direction. Given $z \sim \mathrm{U}(-1, 1)$, each pair $[x, y]^{\mathsf{T}}$ is uniformly distributed on a circle of radius $\sqrt{1 - z^2}$. This observation provides the following sampling algorithm based on the change of coordinates $[x, y]^{\mathsf{T}} = [r\cos\theta, r\sin\theta]^{\mathsf{T}}$. Sample $z \sim \mathrm{U}(-1, 1)$ and $\theta \sim \mathrm{U}(0, 2\pi)$ independently and then compute $r = \sqrt{1 - z^2}$. This simple method is implemented in Algorithm 4.6 and a sample of 100 points is plotted in Figure 4.7.

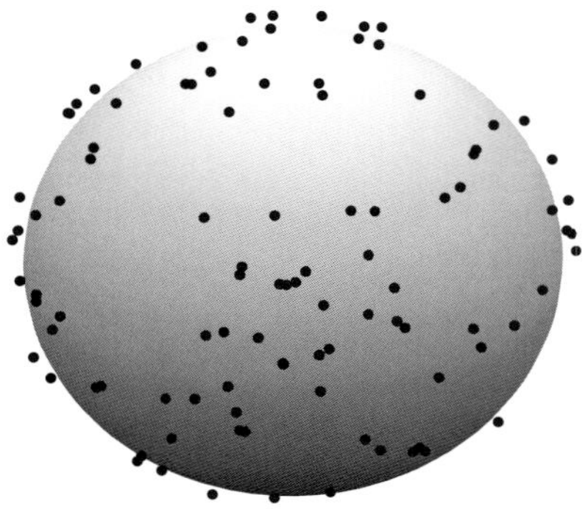

Figure 4.7 Using Algorithm 4.6, a plot of 100 independent samples from $\mathrm{U}(S^2)$.

Algorithm 4.6 Code to generate a sample `X` from the distribution $\mathrm{U}(S^2)$ via the change of variables method.

```
function X=uniform_sphere
z=-1+2*rand; theta=2*pi*rand; r=sqrt(1-z*z);
X=[r*cos(theta); r*sin(theta); z];
```

Example 4.66 (Cauchy distribution) Let $X \sim \mathrm{Cauchy}(0,1)$, a random variable with the Cauchy distribution and pdf $p_X(x) = 1/\pi(1+x^2)$. To sample X, we use the change of variable $X = \tan\theta$ for some $\theta \in (0,\pi)$. By Lemma 4.12 with $g(\theta) = \arctan\theta$ and $g'(\theta) = 1/(1+\theta^2)$, X is Cauchy if $\theta \sim \mathrm{U}(0,\pi)$. Thus, a Cauchy random variable is generated from one sample of $\theta \sim \mathrm{U}(0,\pi)$ followed by the transformation $X = \tan\theta$. See also Exercise 4.19.

Rejection sampling

In rejection sampling, we repeatedly sample from a trial distribution until some acceptance criterion is satisfied; by choosing the trial distribution and acceptance criterion carefully, a wide range of distributions can be sampled.

Example 4.67 (uniform distribution on unit ball) Consider a sequence $\boldsymbol{y}_j$ of *iid* samples from $\mathrm{U}([-1,1]^2)$, the uniform distribution on the *square* $[-1,1]^2$ that circumscribes the ball B. Let $\boldsymbol{y} = \boldsymbol{y}_n$ where $n = \min\{j : \|\boldsymbol{y}_j\|_2 \le 1\}$. The accepted sample $\boldsymbol{y}_n$ is uniformly distributed across B because the trial samples are uniform on $[-1,1]^2$; rejecting samples outside the ball B does not change that. See Algorithm 4.7 for a MATLAB implementation.

Algorithm 4.7 Code to generate one sample from the distribution $\mathrm{U}(B)$ via rejection sampling. The output `X` is the sample; `M` is the number of rejections required to obtain `X`.

```
function [X, M]=reject_uniform
M=0; X=[1;1]; % make sure initial X is rejected
while norm(X)>1, % reject
    M=M+1; X=2*rand(1,2)-1; % generate sample
end;
```

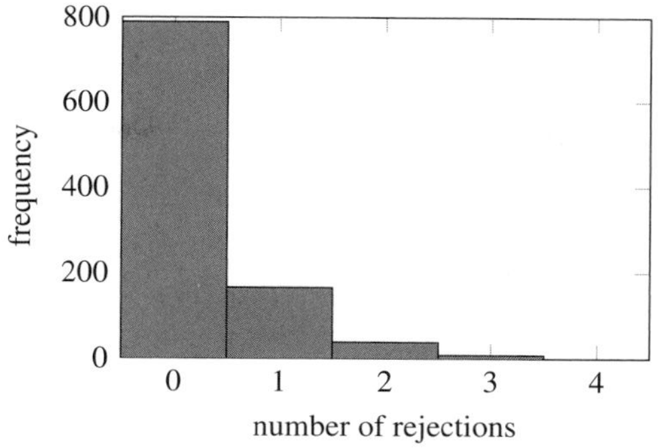

Figure 4.8 Histogram of the number of rejected trials in applying Algorithm 4.7 to generate 1000 samples.

The effectiveness of the method depends on how many samples are rejected before the acceptance criterion is satisfied and on the speed of sampling the trial distribution. The number of rejected trials varies randomly. If the probability of acceptance is p, the probability of acceptance at the jth trial is $(1-p)^{j-1}p$ and the expected number of trials (using (4.2)) is

$$\sum_{j=1}^{\infty} jp(1-p)^{j-1} = -p\frac{d}{dp}\sum_{j=0}^{\infty}(1-p)^j = \frac{1}{p}.$$

It is therefore beneficial to choose a sampling distribution where the probability of acceptance p is close to 1.

Example 4.68 In Example 4.67, a sample is accepted with probability $p = \pi/4$ and so the expected number of trials is $4/\pi \approx 1.27$. On average, we reject 0.27 trials. Figure 4.8 shows a histogram of the number of rejected trials in 1000 runs of Algorithm 4.7.

Sampling from the multivariate Gaussian $\mathrm{N}(\mu, C)$

The Gaussian distribution $\mathrm{N}(\mu, C)$ is parameterized by its mean μ and covariance matrix C. If the covariance matrix C is positive definite, it has a Cholesky factorisation $C = R^{\mathsf{T}}R$, where R is an upper triangular matrix. The Cholesky factorisation provides a simple way to sample from the multivariate Gaussian distribution as follows. Compute d independent samples $Z_1, \dots, Z_d \sim \mathrm{N}(0,1)$ and let $\boldsymbol{X} = \boldsymbol{\mu} + R^{\mathsf{T}}\boldsymbol{Z}$, where $\boldsymbol{Z} = [Z_1, \dots, Z_d]^{\mathsf{T}} \sim \mathrm{N}(\boldsymbol{0}, I_d)$. $\boldsymbol{X}$ is a linear combination of Gaussians and is therefore itself Gaussian (using Proposition 4.32). $\mathbb{E}[\boldsymbol{X}] = \boldsymbol{\mu} + \boldsymbol{0} = \boldsymbol{\mu}$ and the covariance matrix of $\boldsymbol{X}$ is

$$\mathbb{E}\left[(\boldsymbol{X}-\boldsymbol{\mu})(\boldsymbol{X}-\boldsymbol{\mu})^{\mathsf{T}}\right] = \mathbb{E}\left[(R^{\mathsf{T}}\boldsymbol{Z})(R^{\mathsf{T}}\boldsymbol{Z})^{\mathsf{T}}\right] = R^{\mathsf{T}}\mathbb{E}\left[\boldsymbol{Z}\boldsymbol{Z}^{\mathsf{T}}\right]R = R^{\mathsf{T}}\, I_d\, R = C.$$

Hence $\boldsymbol{X}$ has the desired distribution $\mathrm{N}(\boldsymbol{\mu}, C)$. The method is easy to implement (see Algorithm 4.8) but the computational cost is dominated by the Cholesky factorisation, which is $\mathcal{O}(d^3)$ and so is not recommended when d is large.

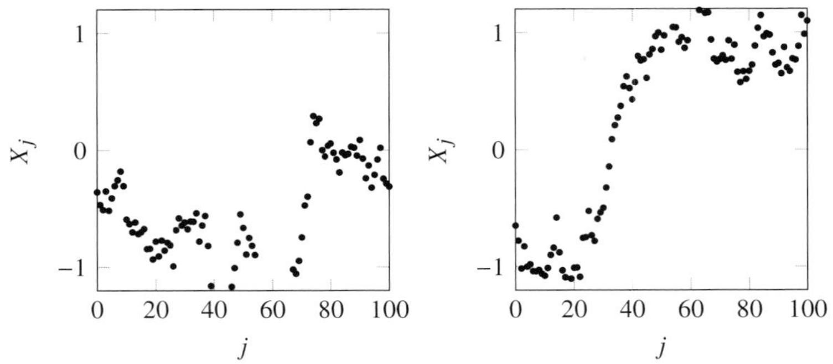

Figure 4.9 Two samples of $\boldsymbol{X} \sim \mathrm{N}(\mathbf{0}, C)$ with $d = 101$ where C is the matrix defined in Example 4.69. The components X_j of $\boldsymbol{X}$ are plotted against j for $j = 1, \dots, 101$.

Example 4.69 Let C be the 101×101 matrix with entries

$$c_{ij} = \exp(-|t_i - t_j|), \qquad t_i = \frac{i-1}{100}, \; t_j = \frac{j-1}{100}, \qquad \text{for } i, j = 1, \dots 101.$$

Two samples of $\boldsymbol{X} \sim \mathrm{N}(\mathbf{0}, C)$ generated using Algorithm 4.8 are shown in Figure 4.9. The covariance C is positive definite here, but its minimum eigenvalue is 5×10^{-3}. Cholesky factorisation fails if a singular matrix C is supplied to Algorithm 4.8.

Algorithm 4.8 Code to generate a sample `X` from the distribution $\mathrm{N}(\boldsymbol{\mu}, C)$ using Cholesky factorisation. The input `mu` is a vector of length d and `C` is a symmetric positive-definite matrix of size $d \times d$.

```
function X = gauss_chol(mu,C)
R=chol(C); Z=randn(size(mu)); X=mu+R'*Z;
```

To compute N samples from the same Gaussian distribution, the Cholesky factorisation needs to be computed only once. When C is dense, the cost of a standard matrix-vector product is $\mathcal{O}(d^2)$. The work for the matrix-vector multiplications is $\mathcal{O}(Nd^2)$ and is the most costly part of the computation when $N \gg d$. When d is large, as in the case of sampling Gaussian random fields (see Chapter 7), the cost of even one matrix-vector product can be unacceptably high. We develop more sophisticated algorithms that are suitable for large d, as well as for general non-negative definite covariance matrices, in Chapters 5, 6 and 7.

Monte Carlo

The name of Monte Carlo, a European tourist resort famous for its casinos and gambling, is used for a class of methods developed to solve large problems in physics, which are based like casino games on random variables. The idea of using random variables with a given distribution to solve deterministic problems goes back much further, with methods for computing π dating back to the eighteenth century. One such classic method is Buffon's needle, a way of computing π from the position of a needle dropped randomly on a plane. Today, Monte Carlo methods are an important and widely used tool in scientific computing.

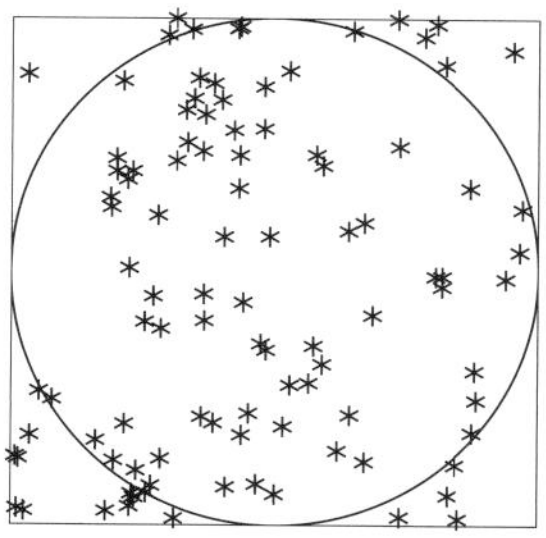

Figure 4.10 Representative experiment for Example 4.70. 76 out of 100 samples lie inside the circle, giving an approximation for π of $4 \times 76/100 = 3.04$.

Example 4.70 (Monte Carlo estimate of π) A simple method for estimating π is as follows. Drop M points onto a square, independently and uniformly distributed, and count the number $M_\circ$ of points that lie in the circle inscribed in that square, as in Figure 4.10. The area of the inscribed circle divided by the area of the square is $\pi/4$. In the limit $M \to \infty$, the fraction $M_\circ/M$ converges to $\pi/4$, the probability of landing in the circle (a consequence of the law of large numbers). To estimate π, we simply compute $4M_\circ/M$ for large M.

Generally speaking, Monte Carlo methods repeat an experiment using different input data and then average the separate outputs X_j to produce a result $\overline{X}_M = (X_1 + \cdots + X_M)/M$. Usually, the 'experiment' consists of calling a computer program with inputs that are produced by a random number generator. If the experiment is carefully designed, $\overline{X}_M$ converges to a quantity of interest as the number of experiments M increases. In Example 4.70, X_j equals 1 if the jth randomly generated point belongs to the circle and equals 0 otherwise. When solving stochastic differential equations via Monte Carlo methods, the outputs X_j of the experiment are approximations to the solution of a deterministic differential equation with different samples of the random inputs (e.g., boundary and initial conditions or material coefficients).

Monte Carlo methods are beautifully simple but notoriously slow. The convergence rate of the approximation is $\mathcal{O}(M^{-1/2})$— 'to reduce the error tenfold calls for a hundredfold increase in the observations' (Hammersley and Handscomb, 1965). In modern numerical analysis, we are familiar with methods of high order for a broad range of problems, yet Monte Carlo remains popular despite its $\mathcal{O}(M^{-1/2})$ convergence rate. It offers a number of important advantages: (1) The Monte Carlo method applies in any dimension and the convergence rate, although slow, is dimension independent. (2) The Monte Carlo method allows the reuse of existing deterministic algorithms. For stochastic differential equations, each experiment requires the numerical solution of a deterministic differential equation. State-of-the-art solvers for these intermediary problems are often available. Users want to take advantage of familiar tools and Monte Carlo allows us to repeat deterministic calculations many times with different input data and perform statistical analysis later. Alternative methods are available as we shall see, but involve a fundamental rethink of the solution process. (3) The Monte Carlo method is iterative. Performing just one simulation provides an estimate of a quantity of interest, so for problems where the experiment is very expensive to perform, the Monte Carlo method produces approximations, no matter how crude, with one experiment and produces a new approximation with each new experiment. (4) The Monte Carlo method is straightforward to parallelise. Since it relies on repeated simulations of independent numerical experiments, these experiments can be spread across different processors or cores.

Analysis of Monte Carlo

To analyse the Monte Carlo method, let X_j be a sequence of *iid* real-valued random variables on the probability space $(\Omega, \mathcal{F}, \mathbb{P})$ each with mean μ and variance σ^2. The strong law of large numbers ensures the sample mean $\overline{X}_M$ converges to the true mean μ almost surely. In the example illustrated in Figure 4.10, $M_\circ/M$, where $M_\circ = \sum_{j=1}^{M} X_j$ and $\mathbb{E}[X_j] = \pi/4$, converges to $\pi/4$. To understand the *rate* of convergence of $\overline{X}_M$ to μ, we compute the variance of $\overline{X}_M$:

$$\operatorname{Var} \overline{X}_M = \operatorname{Var}\left(\frac{X_1 + \cdots + X_M}{M}\right) = \frac{1}{M^2} \sum_{j=1}^{M} \operatorname{Var} X_j = \frac{\sigma^2}{M}, \tag{4.31}$$

as the X_j are uncorrelated. Hence $\mathbb{E}\left[(\overline{X}_M - \mu)^2\right] \to 0$ as $M \to \infty$. Theorem 4.58(i) with $p = 2$ and $r = 1/2$ gives, for any $\epsilon > 0$,

$$\mathbb{P}\left(\left|\overline{X}_M - \mu\right| \geq M^{-1/2+\epsilon}\right) \leq \frac{\sigma^2}{M^{2\epsilon}}. \tag{4.32}$$

Hence, as the number of samples M increases, the probability of the error being larger than $\mathcal{O}(M^{-1/2+\epsilon})$ converges to zero for any $\epsilon > 0$.

More detailed information about the convergence of $\overline{X}_M$ to μ is provided by the central limit theorem and the Berry–Esséen inequality under the assumption that $\mathbb{E}\left[|X_j|^3\right] < \infty$. Consider the random variable X_M^* defined in (4.25) and let $Z \sim \mathrm{N}(0, \sigma^2)$. By (4.26), we have $\mathbb{P}(X_M^* \leq z) = \mathbb{P}(Z \leq z) + \mathcal{O}(M^{-1/2})$ and so

$$\begin{aligned}
\mathbb{P}\left(|X_M^*| \leq R\right) &= \mathbb{P}\left(X_M^* \leq R\right) - \mathbb{P}\left(X_M^* < -R\right) \\
&= \mathbb{P}\left(Z \leq R\right) - \mathbb{P}\left(Z < -R\right) + \mathcal{O}(M^{-1/2}) \\
&= \mathbb{P}\left(|Z| \leq R\right) + \mathcal{O}(M^{-1/2}).
\end{aligned}$$

Using the error function as in (4.8), we obtain

$$\mathbb{P}\left(|X_M^*| \leq R\right) = \operatorname{erf}\left(\frac{R}{\sqrt{2\sigma^2}}\right) + \mathcal{O}(M^{-1/2}). \tag{4.33}$$

For $R = 2\sigma$, we have $\operatorname{erf}(\sqrt{2}) \approx 0.9545$ and we obtain

$$\mathbb{P}\left(\left|\overline{X}_M - \mu\right| < \frac{2\sigma}{\sqrt{M}}\right) \approx 0.9545 + \mathcal{O}(M^{-1/2}). \tag{4.34}$$

The interval $\left[\overline{X}_M - \frac{2\sigma}{\sqrt{M}}, \overline{X}_M + \frac{2\sigma}{\sqrt{M}}\right]$ is known as a *95% confidence interval*, as it contains the true mean μ with at least probability 0.95. If σ is unknown, it can be estimated using the unbiased estimator σ_M^2 in (4.29). In that case, the radius of the 95% confidence interval can be estimated via

$$\frac{2\sigma}{\sqrt{M}} \approx \frac{2\sigma_M}{\sqrt{M}}. \tag{4.35}$$

We can be more specific about the error term $\mathcal{O}(M^{-1/2})$ in (4.34) by applying the Berry–Esséen inequality. We illustrate this now for the case of the binomial distribution.

Example 4.71 (Monte Carlo estimate of π) In estimating π in Example 4.70, we have M random variables X_j defined by $\mathbb{P}(X_j = 1) = \pi/4$ and $\mathbb{P}(X_j = 0) = 1 - \pi/4$ with mean $\mu = p = \pi/4$ and variance $\sigma^2 = p(1-p)$. Thus, the number $M_\circ = M\overline{X}_M$ of points lying in the circle follows a binomial distribution $\mathrm{B}(M,p)$.

From the Berry–Esséen inequality for the binomial distribution (4.28) with $R = 2$, we have

$$\mathbb{P}\left(\left|\overline{X}_M - p\right| \le \frac{2\sqrt{p(1-p)}}{\sqrt{M}}\right) > 0.95 - \frac{2(1-2p+2p^2)}{\sqrt{Mp(1-p)}},$$

and, with $p = \pi/4$, we obtain

$$\mathbb{P}\left(\left|\overline{X}_M - \frac{\pi}{4}\right| \le \frac{0.8211}{\sqrt{M}}\right) > 0.95 - \frac{3.23}{\sqrt{M}}.$$

We show how the Monte Carlo method is used to solve an initial-value problem with random initial data.

Example 4.72 (ODE with random initial data) Consider the following simple model of two interacting populations u_1 and u_2:

$$\frac{d}{dt}\begin{pmatrix} u_1 \\ u_2 \end{pmatrix} = \begin{pmatrix} u_1(1-u_2) \\ u_2(u_1-1) \end{pmatrix}, \qquad \boldsymbol{u}(0) = \boldsymbol{u}_0. \tag{4.36}$$

Often ecologists collect data to estimate current populations (with some accuracy ϵ) and use this information to predict future (mean) populations. It is also possible to estimate the variance in these mean populations. Here, we concentrate on the means and the errors in the predictions from sampling over many different initial conditions by a Monte Carlo method. We model the uncertainty in the initial population by assuming $\boldsymbol{u}_0 \sim \mathrm{U}(D)$; that is, the initial condition is uniformly distributed on

$$D = \left[\bar{u}_{0,1} - \epsilon, \bar{u}_{0,1} + \epsilon\right] \times \left[\bar{u}_{0,2} - \epsilon, \bar{u}_{0,2} + \epsilon\right].$$

We aim to estimate the average of the first component, $\mathbb{E}[u_1(T)]$, at time T.

As analytic solutions are unavailable, we apply the explicit Euler approximation (3.12) with time step Δt. Let $\boldsymbol{u}_n$ denote the explicit Euler approximation to (4.36) at time $t = n\Delta t$ with random initial data $\boldsymbol{u}_0$. At each time step, the numerical solution $\boldsymbol{u}_n$ is a random variable. To approximate $\mathbb{E}[u_1(T)]$, we define $\phi(\boldsymbol{u}) = u_1$ and estimate $\mathbb{E}[\phi(\boldsymbol{u}_N)]$ for $N\Delta t = T$ by the Monte Carlo method. Let $\overline{X}_M$ denote the sample mean of M independent samples of $\phi(\boldsymbol{u}_N)$ (the first component of the explicit Euler approximation to $\phi(\boldsymbol{u}(T)) = u_1(T)$ at time $t = T$). To obtain this quantity, we apply the explicit Euler method to the deterministic ODE with multiple independent realisations of the initial conditions. Algorithm 4.9 generates $\overline{X}_M$ and the estimated radius of the 95% confidence interval using (4.35).

Clearly, we expect $\overline{X}_M$ to approximate $\mathbb{E}[\phi(\boldsymbol{u}(T))]$ well when Δt is small and M is large. The error has two components: the Monte Carlo sample error and the time-stepping error of

Algorithm 4.9 Code to perform Monte Carlo for Example 4.72. The inputs are the number of trials, M, the value T (where the estimate of $\mathbb{E}[u_1(T)]$ is required), the Euler time step Δt, the mean initial data $\boldsymbol{u}_0$, (a vector of length two) and the uncertainty ϵ. The outputs are `bar_x`, the Monte Carlo estimate for $\mathbb{E}[u_1(T)]$ and `sig95`, the estimate (4.35).

```
function [bar_x, sig95]=pop_monte(M,T,Dt,baru0,epsilon)
u=[];
for j=1:M,
    u0 = baru0+epsilon*(2*rand(1,2)-1); % sample initial data
    u(j,:)=pop_solve(u0,T,Dt); % solve ODE
end;
[bar_x, sig95]=monte(u(:,1))% analyse first component
```

```
function [sample_av, conf95]=monte(samples)
M=length(samples);
conf95 = 2*sqrt(var(samples)/M); sample_av = mean(samples);
```

the explicit Euler method. Let $\mu = \mathbb{E}[\phi(\boldsymbol{u}(T))]$ be the exact answer. The error with N time steps and M samples is

$$\epsilon_M = \left|\mu - \bar{X}_M\right| \le \underbrace{\left|\mu - \mathbb{E}\left[\phi(\boldsymbol{u}_N)\right]\right|}_{\text{explicit Euler error}} + \underbrace{\left|\mathbb{E}\left[\phi(\boldsymbol{u}_N)\right] - \bar{X}_M\right|}_{\text{Monte Carlo error}}.$$

For the first error, the Euler error estimate of Theorem 3.7 says that $\|\boldsymbol{u}(T) - \boldsymbol{u}_N\|_2 \le K\Delta t$ for some constant K. Hence, $|\phi(\boldsymbol{u}(T)) - \phi(\boldsymbol{u}_N)| \le K\,L\,\Delta t$ as ϕ is Lipschitz with global Lipschitz constant $L = 1$. Hence,

$$\left|\mu - \mathbb{E}\left[\phi(\boldsymbol{u}_N)\right]\right| = \left|\mathbb{E}\left[\phi(\boldsymbol{u}(T))\right] - \mathbb{E}\left[\phi(\boldsymbol{u}_N)\right]\right| \le K\,L\,\Delta t. \tag{4.37}$$

For the Monte Carlo error, we apply the central limit theory and, in particular, use (4.34), which describes a 95% confidence interval. If $\operatorname{Var}\phi(\boldsymbol{u}_N) = \sigma^2$,

$$\mathbb{P}\left(\left|\mathbb{E}[\phi(\boldsymbol{u}_N)] - \bar{X}_N\right| \le \frac{2\sigma}{\sqrt{M}}\right) > 0.95 + \mathcal{O}(M^{-1/2}).$$

Substituting (4.37), we conclude that

$$\mathbb{P}\left(\epsilon_M \le KL\Delta t + \frac{2\sigma}{\sqrt{M}}\right) > 0.95 + \mathcal{O}(M^{-1/2}).$$

We look to achieve a fixed level of accuracy δ by balancing the two types of errors with $LK\Delta t = \delta/2$ and $2\sigma/\sqrt{M} = \delta/2$. This suggests choosing $\Delta t \approx \delta/2KL$ and $M \approx 16\sigma^2/\delta^2$. This calculation provides a guideline on the correct relationship between the time step Δt and number of Monte Carlo samples M as we reduce δ. In many situations, however, K and σ^2 are unknown and the best we can do is keep $\Delta t^2 M$ fixed as we reduce the time step Δt and increase the number of samples M.

Figure 4.11 shows two sample trajectories of the solution $\boldsymbol{u}$ of (4.36) with initial data $\bar{\boldsymbol{u}}_0 = [0.5, 2]^{\mathsf{T}}$ and $\epsilon = 0.2$. The Euler–Monte Carlo approximation to $\mathbb{E}[u_1(T)]$ is also plotted, together with the 95% confidence intervals for the estimate. Observe that the confidence intervals shrink as $M \to \infty$.

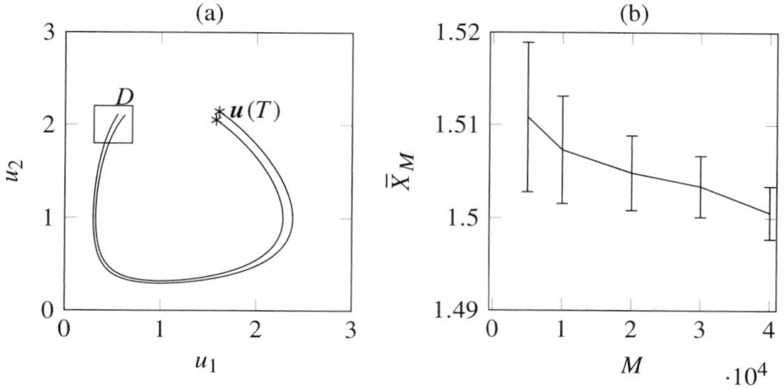

Figure 4.11 For $\bar{\boldsymbol{u}}_0 = [0.5, 2]^\mathsf{T}$ and $\epsilon = 0.2$, (a) shows trajectories for two random initial conditions drawn from $\mathrm{U}(D)$ and $*$ marks the final position $\boldsymbol{u}(T)$ with $T = 6$; (b) shows the approximation $\overline{X}_M$ to $\mathbb{E}[u_1(T)]$ with 95% confidence intervals using the Euler method with $\Delta t = T/10\sqrt{M}$ and M samples.

Improving the Monte Carlo method

We briefly introduce two approaches to improving the Monte Carlo convergence rate: quasi-Monte Carlo methods and variance reduction techniques. *Quasi-Monte Carlo* methods, unlike the standard Monte Carlo method, do not generate samples using random number generators (or pseudo-random number generators). Instead, they use a sequence of quasi-random numbers, also known as a low discrepancy sequence. Roughly speaking, the smaller the discrepancy of a sequence of M numbers, the more equidistributed they are. Quasi-random numbers do not try to be independent. They are deterministic sequences of numbers that cover an interval of interest uniformly so that the convergence rate of the sample average increases.

We give one example for the uniform distribution; quasi-Monte Carlo can be applied to other distributions by a change of variable.

Example 4.73 (van der Corput sequence with base 3) Let x_n denote the sequence

$$0, \frac{1}{3}, \frac{2}{3}, \frac{1}{9}, \frac{4}{9}, \frac{7}{9}, \frac{2}{9}, \frac{5}{9}, \frac{8}{9}, \frac{1}{27}, \frac{10}{27}, \dots \tag{4.38}$$

Each number can be written as an irreducible fraction $k/3^j$ and the sequence is ordered with increasing j. The order in k is understood by writing k in base 3 (e.g., $4_3 = 11$, $7_3 = 21$, $2_3 = 02$) and ordering by the reverse of the base 3 representation (hence the ordering $4/9, 7/9, 2/9$ because $11 < 12 < 20$). The sequence is elementary to compute and uniformly fills out the unit interval.

By replacing *iid* random numbers from $\mathrm{U}(0, 1)$ in a standard Monte Carlo calculation for $\mathbb{E}[\phi(X)]$, for some $\phi \in \mathrm{C}_\mathrm{c}^\infty(0, 1)$ and $X \sim \mathrm{U}(0, 1)$, by the van der Corput sequence of length M, we obtain a quasi-Monte Carlo method. The error is reduced from $\mathcal{O}(M^{-1/2})$ to $\mathcal{O}(M^{-2})$ as shown in Exercise 4.24. Although this improvement is impressive, the method does not generalise easily and the rate of convergence depends on the problem; in general, the rate of convergence for a quasi-Monte Carlo method depends on the dimension d.

The second approach to improving the Monte Carlo method is *variance reduction.*

The constant in (4.32) is the variance σ^2 of the random sample X_j. If we can design an experiment with a smaller variance that achieves the same computation, then the Monte Carlo error is reduced. In the Monte Carlo estimate of π in Example 4.70, the variance can be reduced by replacing the square with a region that contains and more closely matches the circle. This is investigated in Exercise 4.23. In Chapter 8, we examine the multilevel Monte Carlo method as applied to SDEs as a method of variance reduction. Here we outline another approach to achieving variance reduction for general problems, known as *antithetic sampling*.

To compute $\mathbb{E}[X]$ by the standard Monte Carlo method, we generate independent samples X_j of X for $j = 1,\dots,M$ and compute the sample average $\overline{X}_M$. Suppose now that X_j^{a} are also samples of X and denote by $\overline{X}_M^{\mathrm{a}}$ the sample average of $X_1^{\mathrm{a}},\dots,X_M^{\mathrm{a}}$. Both $\overline{X}_M$ and $\overline{X}_M^{\mathrm{a}}$ converge to $\mathbb{E}[X]$ as $M \to \infty$, as does the mean $\big(\overline{X}_M + \overline{X}_M^{\mathrm{a}}\big)/2$. When X_j^{a} and X_j are negatively correlated, X_j^{a} are known as *antithetic* samples and the approximation $\big(\overline{X}_M + \overline{X}_M^{\mathrm{a}}\big)/2$ is a more reliable approximation of $\mathbb{E}[X]$ than $\overline{X}_{2M}$.

Theorem 4.74 *Suppose that X_j, X_j^{a} are identically distributed real-valued random variables such that $\operatorname{Cov}(X_j, X_k) = \operatorname{Cov}(X_j^{\mathrm{a}}, X_k^{\mathrm{a}}) = 0$ for $j \neq k$. Let $\overline{X}_M$ and $\overline{X}_M^{\mathrm{a}}$ denote the sample averages of X_j and X_j^{a} for $j = 1,\dots,M$ respectively. Then*

$$\operatorname{Var}\left(\frac{\overline{X}_M + \overline{X}_M^{\mathrm{a}}}{2}\right) = \operatorname{Var}(\overline{X}_{2M}) + \frac{1}{2}\operatorname{Cov}\big(\overline{X}_M, \overline{X}_M^{\mathrm{a}}\big) \le \operatorname{Var}(\overline{X}_M). \tag{4.39}$$

Proof If X, Y are uncorrelated, $\operatorname{Var}((X+Y)/2) = (\operatorname{Var}(X) + \operatorname{Var}(Y))/4$ by Exercise 4.6. As both $\overline{X}_M$ and $\overline{X}_M^{\mathrm{a}}$ are means of uncorrelated samples from the same distribution, $\operatorname{Var}\big(\overline{X}_{2M}\big) = \operatorname{Var}(\overline{X}_M)/2 = \operatorname{Var}(\overline{X}_M^{\mathrm{a}})/2$ and

$$\begin{aligned}\operatorname{Var}\left(\frac{\overline{X}_M + \overline{X}_M^{\mathrm{a}}}{2}\right) &= \frac{1}{4}\Big(\operatorname{Var}\overline{X}_M + \operatorname{Var}\overline{X}_M^{\mathrm{a}} + 2\operatorname{Cov}\big(\overline{X}_M, \overline{X}_M^{\mathrm{a}}\big)\Big)\\ &= \operatorname{Var}(\overline{X}_{2M}) + \frac{1}{2}\operatorname{Cov}\big(\overline{X}_M, \overline{X}_M^{\mathrm{a}}\big).\end{aligned} \tag{4.40}$$

By Exercise 4.6,

$$\operatorname{Var}\left(\frac{\overline{X}_M + \overline{X}_M^{\mathrm{a}}}{2}\right) \le 2\left(\operatorname{Var}\left(\frac{\overline{X}_M}{2}\right) + \operatorname{Var}\left(\frac{\overline{X}_M^{\mathrm{a}}}{2}\right)\right) = \operatorname{Var}(\overline{X}_M). \qquad \square$$

In the worst case, (4.39) says that the variance of the sample average of M independent samples and M antithetic samples is less than the variance of the sample average of M independent samples. The best case is that the antithetic samples are negatively correlated, $\operatorname{Cov}\big(\overline{X}_M, \overline{X}_M^{\mathrm{a}}\big) < 0$, in which case (4.40) says that the variance of the sample average of M independent samples and M antithetic samples is less than the variance of the mean of $2M$ independent samples.

Example 4.75 (antithetic) Continuing from Example 4.72, let $\boldsymbol{A}(\boldsymbol{u}) = 2\bar{\boldsymbol{u}}_0 - \boldsymbol{u}$ so that $\boldsymbol{A}(\boldsymbol{u}) \sim \mathrm{U}(D)$ if $\boldsymbol{u} \sim \mathrm{U}(D)$. The approximation $\boldsymbol{u}_n^{\mathrm{a}}$ generated by the explicit Euler method with initial data $\boldsymbol{A}(\boldsymbol{u}_0)$ has the same distribution as the approximation $\boldsymbol{u}_n$ generated with initial data $\boldsymbol{u}_0 \sim \mathrm{U}(D)$. We apply the Monte Carlo method with antithetic samples to approximate $\mathbb{E}[\phi(\boldsymbol{u}(T))]$ (where ϕ is as in Example 4.72) by computing the sample

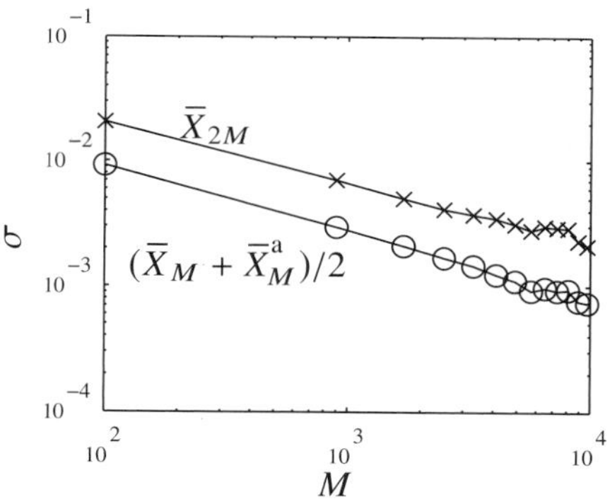

Figure 4.12 Standard deviation σ of $\overline{X}_{2M}$ and of $\big(\overline{X}_M + \overline{X}^{\mathrm{a}}_M\big)/2$. The method of antithetic samples gives a smaller standard deviation, though in both cases the error decreases like $\mathcal{O}(M^{-1/2})$.

average $\overline{X}_M$ of M independent samples $\phi(\boldsymbol{u}_{N,j})$ of $\phi(\boldsymbol{u}_N)$ and the sample average $\overline{X}^{\mathrm{a}}_M$ of the corresponding antithetic samples $\phi(\boldsymbol{u}^{\mathrm{a}}_{N,j})$, with $j = 1, \dots, M$, and then computing $\big(\overline{X}_M + \overline{X}^{\mathrm{a}}_M\big)/2$. This is implemented in Algorithm 4.10. We do not verify the correlation conditions of Theorem 4.74. Figure 4.12 shows the dependence of the standard deviation of the approximation on the number of samples. A reduction of the standard deviation and hence of the width of the confidence intervals for the estimate of $\mathbb{E}[\phi(\boldsymbol{u}(T))]$ is observed when using $\big(\overline{X}_M + \overline{X}^{\mathrm{a}}_M\big)/2$ in place of $\overline{X}_{2M}$. This method does not improve the accuracy of the explicit Euler time stepping or change the overall rate of convergence of the approximation.

Algorithm 4.10 Code to perform Monte Carlo with antithetic sampling for Example 4.72. The inputs and outputs are the same as for Algorithm 4.9.

```
function[bar_x,sig95]=pop_monte_anti(M,T,Dt,baru0,epsilon)
u=[];
for j=1:M,
    % two solutions of DE with correlated initial condition
    u0 = baru0+epsilon*(2*rand(1,2)-1); u(j, :)=pop_solve(u0,T,Dt);
    u0 = 2*baru0-u0;    u(j+M, :)=pop_solve(u0,T,Dt);
end;
[bar_x, sig95]=monte(u(:,1))% analyse first component
```

4.5 Notes

In this chapter, we have touched on the main points of probability theory that are relevant to our study of stochastic PDEs. Many books are available that cover probability theory thoroughly, including (Billingsley, 1995; Breiman, 1992; Brzeźniak and Zastawniak, 1999; Fristedt and Gray, 1997; Grimmett and Stirzaker, 2001). For further details of central limit theorems, see (Bhattacharya and Ranga Rao, 1986; Kuelbs and Kurtz, 1974; Serfling, 1980).

In particular, the Berry–Esséen inequality (4.26) can be sharpened to

$$\sup_{z\in\mathbb{R}}\left|\mathbb{P}(X_M^* \le z) - \mathbb{P}(Z \le z)\right| \le \frac{C\mathbb{E}\left[|X_1 - \mu|^3\right]}{\sigma^3\sqrt{M}},$$

where C is known to satisfy $0.4097 \le C \le 0.7056$ (Shevtsova, 2007).

We omitted the proof of the Doob–Dynkin lemma (Lemma 4.46). As H is a separable Hilbert space, we may write the random variables X and Y using a countable set of real-valued coordinates. Using this observation, the proof is completed by applying Bogachev (2007, Theorem 2.12.3).

Sampling techniques for random variables are discussed in Asmussen and Glynn (2007) and Ross (1997). See Matsumoto and Nishimura (1998) for the Mersenne Twister algorithm and, for further details on the algorithms used in MATLAB, see (Marsaglia and Tsang, 2000; Moler, 1995, 2001). See also (Caflisch, 1998; Fishman, 1996) for further information on the Monte Carlo method. We have omitted many important methods for sampling random variables. For example, the best known of the quasi-random number generators are the Sobol and Niederreiter sequences (Bratley and Fox, 1988; Joe and Kuo, 2003; Niederreiter, 1992). Further, a large class of important methods is Markov chain Monte Carlo, which samples a distribution as a long-time simulation of a Markov chain (Diaconis, 2009).

Exercises

4.1 Show that if $X \sim \mathrm{N}(0,1)$ (see Example 4.10), then the random variable X takes values in the irrational numbers almost surely.

4.2 Prove that $\mathbb{P}_X$ in Definition 4.4 is a measure on $(\Psi, \mathcal{G})$ and that $\sigma(X)$ in Definition 4.18 is a σ-algebra on Ω.

4.3 Show that $p(x)$ given by (4.7) defines a probability density function and that $X \sim \mathrm{N}(\mu,\sigma^2)$ has mean μ and variance σ^2. Hint: use (1.49).

4.4 Let $X \sim \mathrm{N}(\mu,\sigma^2)$ and prove (4.8). Show that if $Z \sim \mathrm{N}(0,\sigma^2)$, then $\mathbb{P}(Z \le z) = \frac{1}{2}\left(1 + \mathrm{erf}\left(z/\sqrt{2\sigma^2}\right)\right)$.

4.5 Show that all moments $\mathbb{E}\left[X^k\right] = \infty$, for $X \sim \mathrm{Cauchy}(0,1)$ and $k \ge 1$.

4.6 Let X, Y be real-valued random variables with finite second moments. Prove that

a. $-1 \le \rho(X,Y) \le 1$ where ρ is defined in (4.12).

b. $\mathrm{Var}(X+Y) \le 2\,\mathrm{Var}(X) + 2\,\mathrm{Var}(Y)$.

c. X, Y are uncorrelated if and only if $\mathbb{E}[XY] = \mathbb{E}[X]\,\mathbb{E}[Y]$ and if and only if $\mathrm{Var}(X + Y) = \mathrm{Var}(X) + \mathrm{Var}(Y)$.

4.7 Give an example where X, Y are Gaussian random variables but $\boldsymbol{X} = [X,Y]^{\mathsf{T}}$ is *not* a bivariate Gaussian random variable. Hint: for some $R > 0$, let

$$Y := \begin{cases} -X, & |X| > R, \\ X, & |X| \le R. \end{cases}$$

4.8 Using Definition 4.21, show that the random variables X_1 and X_2 in Example 4.22 are independent.

4.9 Suppose that C is a $d \times d$ symmetric non-negative definite matrix.

a. Prove $\det(I + C) \ge 1 + \mathrm{Tr}\, C$.

b. Let $C^{1/2}$ denote the symmetric non-negative definite matrix such that $C^{1/2}C^{1/2} = C$. Show that $C^{1/2}$ is well defined.

c. If $\boldsymbol{Z} \sim \mathrm{N}(\boldsymbol{0}, I)$, prove that $\boldsymbol{X} = C^{1/2}\boldsymbol{Z} + \boldsymbol{\mu} \sim \mathrm{N}(\boldsymbol{\mu}, C)$.

4.10 Prove that the moments of $X \sim \mathrm{N}(0, \sigma^2)$ are

$$\mathbb{E}\big[X^{2k}\big] = \frac{(2k)!}{2^k k!}\,\sigma^{2k}, \qquad \mathbb{E}\big[X^{2k+1}\big] = 0. \tag{4.41}$$

Hint: use (4.14) and apply Taylor expansions to $\mathrm{e}^{\mathrm{i}\lambda X}$ and $\mathrm{e}^{-\sigma^2\lambda^2/2}$.

4.11 Let X, Y be discrete random variables with $\mathbb{P}(X = x_k) = p_X(k)$ and $\mathbb{P}(Y = y_j) = p_Y(j)$ and $\mathbb{P}(X = x_k, Y = y_j) = p_{X,Y}(j,k)$. Assume that any $\sigma(Y)$-measurable random variable Z may be written $Z = \phi(Y)$ for a function ϕ. Now, let $\phi_i = \mathbb{E}[X \,|\, Y = y_i]$ and prove, starting from Definition 4.45, that

$$\phi_i = \frac{\sum_k x_k p_{X,Y}(k,i)}{p_Y(i)}.$$

By recalling Baye's theorem $\mathbb{P}(A \cap B) = \mathbb{P}(A|B)\,\mathbb{P}(B)$, show that

$$\mathbb{E}\big[X \,|\, Y = y_i\big] = \sum_k x_k \mathbb{P}(X = X_k \,|\, Y = y_i).$$

4.12 Prove Chebyshev's inequality (4.20) given in Lemma 4.55.

4.13 Let X_j, X be real-valued random variables.

a. Suppose that $X_j, X \in L^2(\Omega)$ and $X_j \to X$ in mean square. Prove that $X_j \to X$ in probability.

b. Show that if $X_j \to X$ almost surely and $|X_j| \le \overline{X}$ for some $\overline{X} \in L^2(\Omega)$, then $X_j \to X$ in mean square.

4.14 If $\mathbb{P}(X = 1) = p$ and $\mathbb{P}(X = 0) = 1 - p$, calculate $\mu = \mathbb{E}[X]$ and $\mathbb{E}\big[(X - \mu)^k\big]$ for $k = 2, 3$. Derive the mean and variance of the binomial distribution $\mathrm{B}(M, p)$ and derive (4.27) and (4.28) from the Berry–Esséen inequality (4.26).

4.15 Consider an *iid* sequence of real-valued random variables X_j each with mean μ and variance σ^2 and let $\overline{X}_M := (X_1 + \cdots + X_M)/M$ and $\sigma_M^2 := \sum_{j=1}^M (X_j - \overline{X}_M)^2/(M-1)$.

a. Prove that σ_M^2 is an unbiased estimator of σ^2.

b. Prove that $\sigma_M^2 \to \sigma^2$ as $M \to \infty$ almost surely.

4.16 Let $\phi(\boldsymbol{x}) = \boldsymbol{x}^{\mathsf{T}} A \boldsymbol{x}$ where A is $d \times d$ matrix and $\boldsymbol{x} \in \mathbb{R}^d$. Show that if $\boldsymbol{X} \sim \mathrm{N}(\boldsymbol{0}, C_X)$, $\boldsymbol{Y} \sim \mathrm{N}(\boldsymbol{0}, C_Y)$, where C_X, C_Y are $d \times d$ positive-definite matrices, then there exists a constant $K > 0$ such that

$$\Big|\mathbb{E}\big[\phi(\boldsymbol{X})\big] - \mathbb{E}\big[\phi(\boldsymbol{Y})\big]\Big| \le K \big\|C_X - C_Y\big\|_{\mathrm{F}}.$$

4.17 Suppose that $\sigma_X^2 > 0$ and consider a function $\phi\colon \mathbb{R} \to \mathbb{R}$ and polynomial $q(x)$ such that $\phi/q \in L^2(\mathbb{R})$. Show that, for each $\epsilon > 0$, there exists a constant $L > 0$ such that

$$\Big|\mathbb{E}\big[\phi(X)\big] - \mathbb{E}\big[\phi(Y)\big]\Big| \le L\,|\sigma_X^2 - \sigma_Y^2|,$$

for any $X \sim \mathrm{N}(0, \sigma_X^2)$, $Y \sim \mathrm{N}(0, \sigma_Y^2)$, and $\sigma_Y^2 > 0$ with $|\sigma_X^2 - \sigma_Y^2| \le \epsilon$. Hint: first generalise Example 1.105 to show $q(\mathrm{i}\frac{d}{d\lambda})\hat{p}(\lambda) = \widehat{qp}(\lambda)$.

4.18 Experiment with the MATLAB random number generators `rand` and `randn`. Use `tic` and `toc` to verify that the cost of sampling from the uniform and Gaussian distributions is comparable to the cost of evaluating one trigonometric function or performing a few scalar multiplications.

4.19 Let $Z = X/Y$, where $X, Y \sim \mathrm{N}(0,1)$ *iid*. Prove that $Z \sim \mathrm{Cauchy}(0,1)$. Develop this observation into a sampling algorithm for the distribution Cauchy(0, 1). By performing simulations in MATLAB, compare the efficiency of the algorithm to the change of variable method discussed in Example 4.66.

4.20 For a stream `s`, the variable `s.State` contains the current state of the random number generator. Write MATLAB code to store the state of the random number generator and then reproduce the same numbers. Experiment with choosing the seed `sum(100*clock)` so that it depends on the system clock.

4.21 Develop an algorithm based on rejection sampling to generate samples from the uniform distribution on S^d (the surface of the unit ball in $\mathbb{R}^{d+1}$). Compute the mean of the number of trials that are rejected as a function of d.

4.22 Implement Algorithms 4.5 and 4.7 and compare their efficiency for sampling points from the unit ball in $\mathbb{R}^2$.

4.23 By inscribing a circle inside a regular hexagon, show how uniform sampling on the hexagon can be used to approximate π. Compute the variance of the sample and compare the Monte Carlo error to the results obtained by using the strategy in Example 4.70.

4.24 Suppose that $\phi \in \mathrm{C}_\mathrm{c}^\infty(0,1)$. Let $\bar{\phi}_M := \frac{1}{M}\sum_{j=1}^{M}\phi(x_j)$, where $M = 3^p$ and x_j denotes the van der Corput sequence in (4.38). For any $n \geq 2$, prove that $\left|\bar{\phi}_M - \mathbb{E}[\phi(X)]\right| = \mathcal{O}(M^{-n})$ when $X \sim \mathrm{U}(0,1)$. Use Theorem A.6.

4.25 Modify the code in Algorithm 4.9 to perform the Monte Carlo simulations in parallel and show how to make the computation reproducible using substreams. You will need the MATLAB parallel computing toolbox and the commands

```
parfor,  matlabpool open,  matlabpool close.
```

Compare timings of the serial and parallel code as the number of samples M increases. Plot the serial and parallel times against M. If you include the time to call `matlabpool`, how large is M before you see an advantage in the parallelization?

5

Stochastic Processes

Random variables in stochastic PDEs usually occur in the form of *stochastic processes* and *random fields*. Stochastic processes are families of random variables parameterized by a scalar variable $t \in \mathbb{R}$ denoting time and random fields are families of random variables parameterized by $\boldsymbol{x} \in \mathbb{R}^d$ for $d > 1$ often denoting space. Random fields are developed in Chapter 7 and, in this chapter, we focus on stochastic processes. Starting from a simple random walk, §5.1 introduces stochastic processes by developing Brownian motion. §5.2 develops the theory of second-order and particularly Gaussian processes. §5.3 gives some examples of Gaussian processes, including the Brownian bridge, fractional Brownian motion, and white noise. §5.4 develops the Karhunen–Loève expansion for coloured noise and, finally, §5.5 analyses the continuity and differentiability of stochastic processes as functions of t.

5.1 Introduction and Brownian motion

Our main focus is *continuous* time stochastic processes. In this section, we develop *Brownian motion*, the prime example of such a process, by studying the following discrete time process.

Example 5.1 (random walk) Consider the simple random walk X_n on the integers defined by

$$X_0 := 0, \qquad X_{n+1} := X_n + \xi_n, \qquad \text{for } n = 0, 1, 2, \ldots, \tag{5.1}$$

where ξ_n are *iid* random variables with $\mathbb{P}(\xi_n = \pm 1) = 1/2$. Then $X_0, X_1, X_2, \ldots$ is a sequence of integer-valued random variables. Since $X_0 = 0$, X_n takes even integer values when n is even and odd integer values when n is odd. Figure 5.1 shows an example of 30 steps of the random walk. The mean and covariance of the random walk are easily computed. As $\mathbb{E}[\xi_n] = 0$, we see the mean $\mathbb{E}[X_n] = 0$ for each n. Furthermore, since $\mathbb{E}[\xi_j \xi_k] = 0$ for $j \neq k$ and $\mathbb{E}[\xi_j^2] = 1$, the covariance of X_n and X_m is

$$\operatorname{Cov}(X_n, X_m) = \mathbb{E}\big[X_n X_m\big] = \mathbb{E}\left[\sum_{j=1}^{n} \xi_{j-1} \sum_{k=1}^{m} \xi_{k-1}\right] = \min\{n, m\}.$$

The random walk X_n defined by (5.1) is a set of random variables indexed by the *non-negative integers* $n = 0, 1, 2, \ldots$. Setting $X(t)$ to be the piecewise linear interpolant of X_n gives a set of random variables indexed by a *real-valued* parameter $t \in \mathbb{R}^+$. This is our first example of a *stochastic process*.

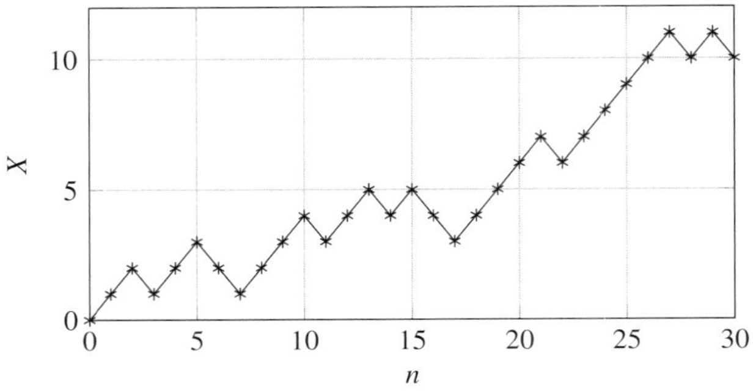

Figure 5.1 An example of 30 steps of the random walk X_n (denoted by $*$) and the associated piecewise linear interpolant $X(t)$ (solid line), which is continuous on $[0,30]$.

Definition 5.2 (stochastic process) Given a set $\mathcal{T} \subset \mathbb{R}$, a measurable space $(H, \mathcal{H})$, and a probability space $(\Omega, \mathcal{F}, \mathbb{P})$, a H-valued *stochastic process* is a set of H-valued random variables $\{X(t) \colon t \in \mathcal{T}\}$. We sometimes drop the set $\mathcal{T}$ and simply write $X(t)$ to denote the process. This should not be interpreted as a simple function of t and, to emphasise the dependence on ω and that $X \colon \mathcal{T} \times \Omega \to H$, we may write $X(t,\omega)$.

Definition 5.3 (sample path) For a fixed $\omega \in \Omega$, the *sample path* of an H-valued stochastic process $\{X(t) \colon t \in \mathcal{T}\}$ is the deterministic function $f \colon \mathcal{T} \to H$ defined by $f(t) \coloneqq X(t,\omega)$ for $t \in \mathcal{T}$. We often denote the function f by $X(\cdot,\omega)$. We say a process X has *continuous sample paths* if all samples paths of X are continuous.

The interpolated random walk in Example 5.1 is a real-valued stochastic process with continuous sample paths, an example of which is shown in Figure 5.1.

Convergence of the random walk to a Gaussian distribution

By using the central limit theorem, we show that the random walk X_n converges to a Gaussian distribution in the limit of large n. More importantly, after applying the *diffusion scaling* to the random walk, we find a stochastic process that leads to Brownian motion.

Example 5.4 (distribution of random walk) Let $p_{nj} = \mathbb{P}(X_n = j)$, the probability that X_n equals $j \in \mathbb{Z}$ at step $n = 0, 1, 2 \ldots$. The update rule (5.1) gives

$$p_{n+1,j} = \frac{1}{2}\big(p_{n,j-1} + p_{n,j+1}\big), \qquad p_{0j} = \begin{cases} 1, & j = 0, \\ 0, & j \neq 0. \end{cases} \tag{5.2}$$

The probabilities $p_{50,j}$ are plotted in Figure 5.2 and a Gaussian profile is clearly visible for even values of j (the odd values are zero since $n = 50$ is even).

By Definition 4.54, X_n converges to a real-valued random variable X *in distribution* if $\mathbb{E}[\phi(X_n)] \to \mathbb{E}[\phi(X)]$ for all bounded continuous functions $\phi \colon \mathbb{R} \to \mathbb{R}$. Averages of continuous functions of the random walk X_n blur the distinction between even and odd grid points and we have the following description of the limit of the random walk.

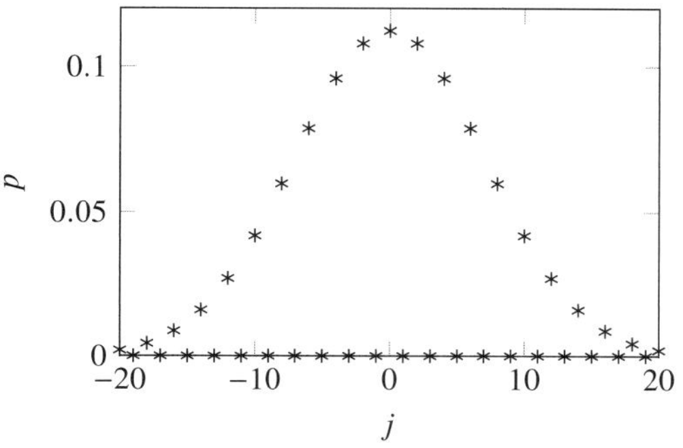

Figure 5.2 Probabilities $p_{50,j} = \mathbb{P}(X_{50} = j)$ for the random walk from Example 5.1. Since $n = 50$ is even, X_{50} takes only even integer values and $p_{50,j} = 0$ if j is odd.

Lemma 5.5 (i) $\frac{1}{n}X_n \to 0$ *a.s. and*

(ii) $\frac{1}{\sqrt{n}}X_n \to \mathrm{N}(0,1)$ *in distribution as* $n \to \infty$.

Proof Recall that $X_n = \sum_{j=1}^{n} \xi_{j-1}$, with $\mathbb{E}[\xi_j] = 0$ and $\operatorname{Var}\xi_j = 1$. As the ξ_j are *iid*, the law of large numbers tells us that $\frac{1}{n}X_n \to 0$ a.s. and the central limit theorem says that $\frac{1}{\sqrt{n}}X_n \to \mathrm{N}(0,1)$ in distribution as $n \to \infty$. □

The random walk is Gaussian in the large time limit; we now derive a family of stochastic processes $\{Y_N(t)\colon t \in \mathbb{R}^+\}$ for $N \in \mathbb{N}$, where $Y_N(t)$ is Gaussian in the large N limit for any fixed time t. This is achieved by applying the so-called *diffusion scaling* $t \mapsto t/N$ and $X \mapsto X/\sqrt{N}$ to the linear interpolant $X(t)$ of the random walk X_n for $N = 1,2,\dots$. Define

$$Y_N(t) := \frac{1}{\sqrt{N}}X(tN), \qquad t \in \mathbb{R}^+, \tag{5.3}$$

which is also written as

$$Y_N(t) = \frac{1}{\sqrt{N}}X_{\lfloor tN \rfloor} + \frac{tN - \lfloor tN \rfloor}{\sqrt{N}}\xi_{\lfloor tN \rfloor}, \tag{5.4}$$

where $\lfloor tN \rfloor$ is the largest integer not greater than tN. $\{Y_N(t)\colon t \in \mathbb{R}^+\}$ is a real-valued stochastic process for each N and has continuous sample paths. Figure 5.3 shows examples for $N = 100$ and $N = 400$.

Lemma 5.6 *Let $Y_N(t)$ be the rescaled linear interpolant of the random walk defined in* (5.3). *Then, $Y_N(t) \to \mathrm{N}(0,t)$ in distribution as $N \to \infty$ for any $t \in \mathbb{R}^+$.*

Proof By Lemma 5.5, the first term in (5.4), $\frac{1}{\sqrt{N}}X_{\lfloor tN \rfloor}$, converges in distribution to $\mathrm{N}(0,t)$ as $N \to \infty$. The second term in (5.4) converges to zero in probability; that is, for any $\epsilon > 0$,

$$\mathbb{P}\left(\left|\frac{tN - \lfloor tN \rfloor}{\sqrt{N}}\xi_{\lfloor tN \rfloor}\right| > \epsilon\right) \to 0 \quad \text{as} \quad N \to \infty$$

(see Exercise 5.1). As convergence to zero in probability implies convergence to zero in distribution, we conclude $Y_N(t) \to \mathrm{N}(0,t)$ in distribution as $N \to \infty$ for each $t \in \mathbb{R}^+$. □

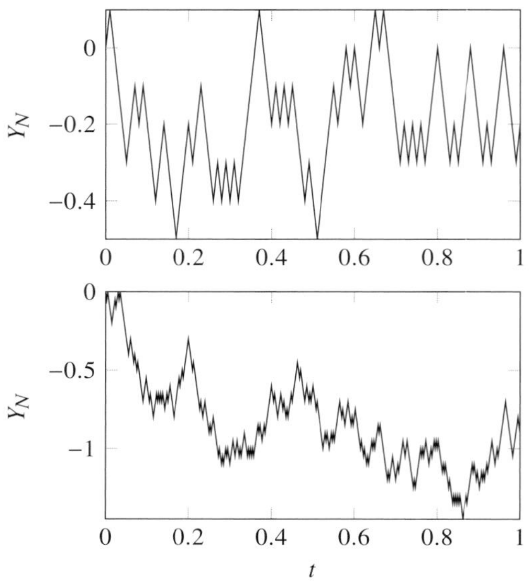

Figure 5.3 Sample paths of $Y_N(t)$ on $[0,1]$ for $N = 100$ (top) and $N = 400$ (bottom).

For a fixed $t \in \mathbb{R}^+$, we can use (5.4) to show that $Y_N(t) \in L^2(\Omega)$ for each N. However, the sequence $Y_N(t)$ for $N \in \mathbb{N}$ is not a Cauchy sequence and has no limit in $L^2(\Omega)$ (see Exercise 5.2).

We are often interested in the distributions of the multivariate random variables $\boldsymbol{X}$ given by $X(t)$ at a finite set of t-values.

Definition 5.7 (finite-dimensional distributions) Let $\{X(t)\colon t \in \mathcal{T}\}$ be a real-valued stochastic process. For $t_1, \dots, t_M \in \mathcal{T}$, define $\boldsymbol{X} := [X(t_1), \dots, X(t_M)]^\mathsf{T}$. Then, $\boldsymbol{X}$ is an $\mathbb{R}^M$-valued random variable and the probability distributions $\mathbb{P}_{\boldsymbol{X}}$ on $(\mathbb{R}^M, \mathcal{B}(\mathbb{R}^M))$ (see Definition 4.4) are known as the *finite-dimensional distributions* of $X(t)$.

Extending Lemma 5.6, we see the finite-dimensional distributions $\mathbb{P}_{\boldsymbol{Y}_N}$ of the rescaled random walk $Y_N(t)$ are Gaussian in the limit as $N \to \infty$.

Lemma 5.8 *Let $Y_N(t)$ be the real-valued stochastic process defined in* (5.3). *For any $t_1, \dots, t_M \in \mathbb{R}^+$, let $\boldsymbol{Y}_N = [Y_N(t_1), \dots, Y_N(t_M)]^\mathsf{T}$. Then $\boldsymbol{Y}_N \to \mathrm{N}(\boldsymbol{0}, C)$ in distribution as $N \to \infty$ where $C \in \mathbb{R}^{M \times M}$ has entries $c_{ij} = \min\{t_i, t_j\}$.*

Proof We assume without loss of generality that $t_1 < t_2 < \cdots < t_M$ and consider the vector of increments

$$\frac{1}{\sqrt{N}}\Big(X_{\lfloor t_2 N\rfloor} - X_{\lfloor t_1 N\rfloor}, X_{\lfloor t_3 N\rfloor} - X_{\lfloor t_2 N\rfloor}, \dots, X_{\lfloor t_M N\rfloor} - X_{\lfloor t_{M-1} N\rfloor}\Big)^\mathsf{T}. \tag{5.5}$$

Note that $X_{\lfloor t_{i+1}N\rfloor} - X_{\lfloor t_i N\rfloor} = \sum_{j=\lfloor t_i N\rfloor+1}^{\lfloor t_{i+1}N\rfloor} \xi_j$ and, under the assumption $t_i < t_{i+1}$, each ξ_j appears in only one increment. Hence, components of the increment vector (5.5) are pairwise independent and, by Lemma 5.5, converge in distribution to a Gaussian random variable. This implies the increment vector converges to a multivariate Gaussian. Now,

$$\boldsymbol{Y}_N = \begin{pmatrix} Y_N(t_1) \\ \vdots \\ Y_N(t_M) \end{pmatrix} = \frac{1}{\sqrt{N}} \begin{pmatrix} X_{\lfloor t_1 N \rfloor} \\ \vdots \\ X_{\lfloor t_M N \rfloor} \end{pmatrix} + \frac{1}{\sqrt{N}} \begin{pmatrix} (t_1 N - \lfloor t_1 N \rfloor)\xi_{\lfloor t_1 N \rfloor} \\ \vdots \\ (t_M N - \lfloor t_M N \rfloor)\xi_{\lfloor t_M N \rfloor} \end{pmatrix}. \tag{5.6}$$

The first term is an affine transformation of (5.5) and hence its limit is a multivariate Gaussian by Proposition 4.32 (with $\boldsymbol{b} = \boldsymbol{0}$ and $A \in \mathbb{R}^{M\times(M-1)}$). To determine the mean and covariance matrix, recall that for the random walk X_n, $\mathbb{E}[X_n] = 0$ and $\mathbb{E}[X_n X_m] = \min\{n,m\}$. Hence $\mathbb{E}\left[\left(1/\sqrt{N}\right)X_{\lfloor t_i N \rfloor}\right] = 0$ and

$$\mathbb{E}\left[\frac{1}{\sqrt{N}}X_{\lfloor t_i N \rfloor}, \frac{1}{\sqrt{N}}X_{\lfloor t_j N \rfloor}\right] = \frac{1}{N}\min\{\lfloor t_i N \rfloor, \lfloor t_j N \rfloor\} \to \min\{t_i, t_j\},$$

as $N \to \infty$, for $i, j = 1, \ldots, M$. We conclude that

$$\frac{1}{\sqrt{N}} \begin{pmatrix} X_{\lfloor t_1 N \rfloor} \\ \vdots \\ X_{\lfloor t_M N \rfloor} \end{pmatrix} \to \mathrm{N}(\boldsymbol{0}, C)$$

in distribution where $c_{ij} = \min\{t_i, t_j\}$. Arguing as for $Y_N(t)$ in Lemma 5.6, the second term in (5.6) converges to zero in probability, and hence in distribution, as $N \to \infty$. Summing, we see that $\boldsymbol{Y}_N \to \mathrm{N}(\boldsymbol{0}, C)$ in distribution. □

Lemma 5.8 agrees with Lemma 5.6 when $M = 1$.

Gaussian processes and Brownian motion

Before giving a rigorous definition of Brownian motion, we introduce second-order and Gaussian processes.

Definition 5.9 (second order) A real-valued stochastic process $\{X(t) : t \in \mathcal{T}\}$ is *second order* if $X(t) \in L^2(\Omega)$ for each $t \in \mathcal{T}$. The mean function is defined by $\mu(t) := \mathbb{E}[X(t)]$ and the covariance function is defined by $C(s,t) := \mathrm{Cov}(X(s), X(t))$ for all $s, t \in \mathcal{T}$.

The key point about second-order processes is that their mean and covariance functions are well defined.

Definition 5.10 (real-valued Gaussian process) A real-valued second-order stochastic process $\{X(t) : t \in \mathcal{T}\}$ is *Gaussian* if $\boldsymbol{X} = [X(t_1), \ldots, X(t_M)]^{\mathsf{T}}$ follows a multivariate Gaussian distribution for any $t_1, \ldots, t_M \in \mathcal{T}$ and any $M \in \mathbb{N}$.

The mean and covariance functions are key to the following definition of Brownian motion.

Definition 5.11 (Brownian motion) We say $\{W(t) : t \in \mathbb{R}^+\}$ is a *Brownian motion* if it is a real-valued Gaussian process with continuous sample paths, mean function $\mu(t) = 0$, and covariance function $C(s,t) = \min\{s,t\}$.

The name *Brownian* refers to Robert Brown, who identified Brownian motion in the movement of pollen particles. Brownian motion is frequently called the Wiener process (and hence the choice of notation $W(t)$), after Norbert Wiener, who made a significant contribution to the mathematical theory. In this chapter, we focus on real-valued processes and use the name Brownian motion. We consider H-valued Wiener processes in Chapter 10.

Notice that the argument so far does not prove the existence of Brownian motion. The process $Y_N(t)$ in (5.3) has continuous sample paths and $Y_N(0) = 0$ for each $N \in \mathbb{N}$. Furthermore, Lemma 5.8 says $Y_N(t)$ is a Gaussian process in the limit $N \to \infty$ with mean $\mu(t) = 0$ and covariance $C(s,t) = \min\{s,t\}$. We can therefore think of Brownian motion $W(t)$ as the limiting process associated with the rescaled linear interpolant $Y_N(t)$. However, we do not make precise the notion of $Y_N(t)$ converging to $W(t)$ as $N \to \infty$ beyond Lemma 5.8. We prove that a stochastic process satisfying Definition 5.11 exists in Corollary 5.42.

Let us take a look at some properties of Brownian motion. As with the random walk, increments $W(t) - W(s)$ over disjoint intervals are independent. Indeed, for $p,r,s,t \geq 0$,

$$\begin{aligned}\mathbb{E}\big[(W(r) - W(p))\,(W(t) - W(s))\big] &= \mathrm{Cov}(W(r), W(t)) - \mathrm{Cov}(W(r), W(s)) \\ &\quad - \mathrm{Cov}(W(p), W(t)) + \mathrm{Cov}(W(p), W(s)) \\ &= r \wedge t - r \wedge s - p \wedge t + p \wedge s,\end{aligned}$$

where $s \wedge t := \min\{s,t\}$. When $p \leq r \leq s \leq t$, $\mathrm{Cov}(W(r) - W(p), W(t) - W(s)) = 0$ and so the increments are uncorrelated. As two increments have a joint Gaussian distribution, the increments are independent. With $r = t$ and $p = s$, we find that $\mathrm{Var}(W(t) - W(s)) = |t - s|$ and hence $W(t) - W(s) \sim \mathrm{N}(0, |t - s|)$.

As a result of these observations, we give an equivalent definition of Brownian motion (see also §8.2). *Brownian motion* $W(t)$ is a real-valued stochastic process on $\mathcal{T} = [0,\infty)$ such that

(i) $W(0) = 0$ a.s.,
(ii) the increments $W(t) - W(s)$ have distribution $\mathrm{N}(0, |t - s|)$ and increments over disjoint intervals are independent, and
(iii) $W(t)$ is continuous as a function of t.

Algorithm 5.1 Code to sample Brownian motion on $[0,T]$. The input `t` is a vector of sample points $t_j < t_{j+1}$ and we assume $t_1 = 0$. The output `X` is a vector whose jth component is a sample of $W(t_j)$.

```
function X=bmotion(t)
X(1)=0; % start at 0
for n=2:length(t),
    dt=t(n)-t(n-1); X(n)=X(n-1)+sqrt(dt)*randn;
end;
```

From this definition, Brownian motion is easily sampled at discrete points $0 = t_1 < t_2 < \cdots < t_N = T$ by drawing increments from a normal distribution. First, $t_1 = 0$ and $W(t_1) = 0$. Next, $W(t_2) - W(t_1) = W(t_2) = \xi_1$, where $\xi_1 \sim \mathrm{N}(0, |t_2 - t_1|)$. Continuing iteratively, we

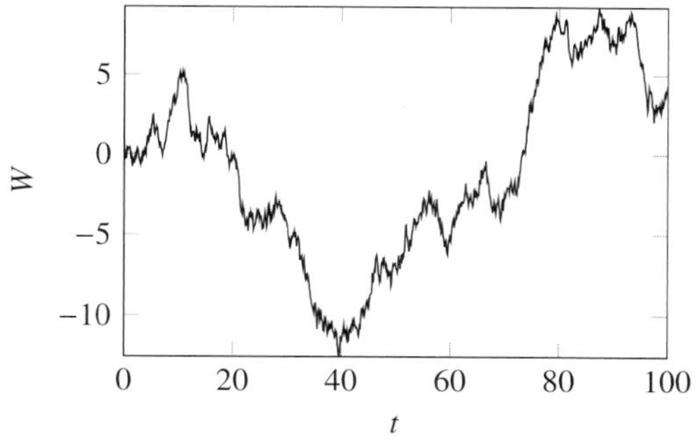

Figure 5.4 Approximate sample path of $W(t)$ on $[0,T]$ with $T = 100$, generated using 1001 points with uniform spacing $\Delta t = 0.1$ and linear interpolation. The graph was produced with Algorithm 5.1 using the commands `t=[0:0.1:100]`, `W=bmotion(t)`, `plot(t,W)`.

define a discrete Brownian path by

$$W(t_{j+1}) = W(t_j) + \xi_j, \tag{5.7}$$

for $j = 1, \dots, N-1$, where $\xi_j \sim \mathrm{N}(0, |t_{j+1} - t_j|)$ are independent random variables. See Algorithm 5.1. Figure 5.4 shows a sample path of $W(t)$ on the interval $[0, 100]$ generated by the `plot` command in MATLAB. `plot` performs linear interpolation and the figure actually shows a *continuous* approximation to the sample path.

It is usual to consider Brownian motion on $\mathcal{T} = \mathbb{R}^+$ (as in Definition 5.11) but we also have use for a two-sided Brownian motion on $\mathcal{T} = \mathbb{R}$.

Definition 5.12 (two-sided Brownian motion) A process $\{W(t): t \in \mathbb{R}\}$ is a *two-sided Brownian motion* if it has continuous sample paths and $\{W(t): t \in \mathbb{R}^+\}$ and $\{W(-t): t \in \mathbb{R}^+\}$ are independent Brownian motions.

A random walk, Brownian motion, and the heat equation

To end this section, we explore the interesting relationship between the random walk and a finite difference approximation to the heat equation (see §3.4) and learn that the limiting distribution of the stochastic process $Y_N(t)$ (i.e., Brownian motion) is controlled by the heat equation. Recall the random walk X_n of Example 5.1. We rescale X_n to define the random walk $Y_n = X_n / \sqrt{N}$, for $N \in \mathbb{N}$, on the lattice

$$(t_n, x_j) = (n\Delta t, j\Delta x), \qquad j \in \mathbb{Z}, \quad n = 0, 1, 2, \dots$$

with spacing $\Delta t = \Delta x^2 = 1/N$. From (5.2), we see that $p_{nj} = \mathbb{P}(Y_n = x_j)$ satisfies

$$p_{n+1,j} = \frac{1}{2}\big(p_{n,j-1} + p_{n,j+1}\big). \tag{5.8}$$

Rearranging and dividing by $\Delta t = \Delta x^2$, we have

$$\frac{p_{n+1,j} - p_{nj}}{\Delta t} = \frac{1}{2}\frac{p_{n,j+1} - 2p_{nj} + p_{n,j-1}}{\Delta x^2}.$$

This we recognise as a finite difference approximation (see §3.4) of the heat equation

$$\frac{\partial u}{\partial t} = \frac{1}{2}\frac{\partial^2 u}{\partial x^2}, \qquad x \in (-\infty, \infty), \quad t \ge 0. \tag{5.9}$$

To make the connection with an initial condition $u(0,x) = \phi(x)$ for a given $\phi\colon \mathbb{R} \to \mathbb{R}$, we work with

$$U_{nj} := \mathbb{E}\big[\phi(x_j + Y_n)\big] = \sum_{k\in\mathbb{Z}} \phi(x_j + x_k) p_{nk}.$$

From (5.8),

$$\begin{aligned} U_{n+1,j} &= \sum_{k\in\mathbb{Z}} \phi(x_j + x_k) p_{n+1,k} = \sum_{k\in\mathbb{Z}} \phi(x_j + x_k)\frac{1}{2}\big(p_{n,k-1} + p_{n,k+1}\big) \\ &= \frac{1}{2}\sum_{k\in\mathbb{Z}} \big(\phi(x_j + x_{k+1}) + \phi(x_j + x_{k-1})\big) p_{nk} = \frac{1}{2}\big(U_{n,j+1} + U_{n,j-1}\big), \end{aligned}$$

where we have relabelled the k's and used that $x_j + x_{k\pm 1} = x_{j\pm 1} + x_k$. Since $\Delta t = \Delta x^2$ and $Y_0 = 0$, we see that the U_{nj} satisfy

$$\frac{U_{n+1,j} - U_{nj}}{\Delta t} = \frac{U_{n,j-1} - 2U_{nj} + U_{n,j+1}}{2\Delta x^2}, \qquad U_{0j} = \phi(x_j). \tag{5.10}$$

This difference equation is again recognisable as a finite difference approximation to the heat equation. In particular, the finite difference approximation u_{nj} (with centred differencing in space and explicit Euler in time) to the solution $u(t_n, x_j)$ of (5.9) with initial condition $u(0,x) = \phi(x)$ is defined by

$$\frac{u_{n+1,j} - u_{nj}}{\Delta t} = \frac{u_{n,j-1} - 2u_{nj} + u_{n,j+1}}{2\Delta x^2}, \qquad u_{0j} = \phi(x_j). \tag{5.11}$$

As the update rules and initial conditions are the same in (5.10) and (5.11), the finite difference approximation u_{nj} and the averages U_{nj} of the random walk are identical. By exploiting this connection, we prove that the finite difference approximation converges.

Theorem 5.13 *If the initial data ϕ is bounded and continuous, the finite difference approximation u_{nj} defined in* (5.11) *converges to the solution $u(t,x)$ of* (5.9) *with $u(0,x) = \phi(x)$: that is, $u_{nj} \to u(t_n, x_j)$ for (t_n, x_j) fixed as $\Delta t = \Delta x^2 \to 0$.*

Proof As $Y_n = Y_N(t_n)$, Lemma 5.6 tells us that $Y_n \to \mathrm{N}(0, t_n)$ and hence $x_j + Y_n \to \mathrm{N}(x_j, t_n)$ in distribution as $N \to \infty$ with (t_n, x_j) fixed. In terms of the pdf (4.7), the convergence in distribution says that

$$U_{nj} = \mathbb{E}\big[\phi(x_j + Y_n)\big] \to \int_{\mathbb{R}} \phi(x) \frac{1}{\sqrt{2\pi t_n}} \mathrm{e}^{(x-x_j)^2/2t_n}\, dx \tag{5.12}$$

for all bounded continuous functions ϕ. The exact solution $u(t,x)$ of the heat equation (5.9) is given by

$$u(t,x) = \frac{1}{\sqrt{2\pi t}} \int_{\mathbb{R}} \mathrm{e}^{-(x-y)^2/2t} \phi(y)\, dy \tag{5.13}$$

(see Exercise 5.4). Because $U_{nj} = u_{nj}$, (5.12) and (5.13) give that $u_{nj} \to u(t_n, x_j)$ as $N \to \infty$ with (t_n, x_j) fixed. □

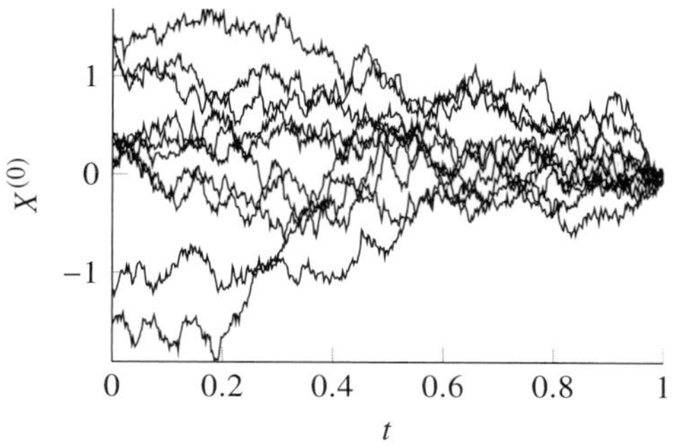

Figure 5.5 Sample paths of the stochastic process $X^x(t) = x + W(T - t)$, with $0 \le t \le T$, in the case $T = 1$ and $x = 0$. At $t = T$, the solution of the heat equation (5.9) with initial condition $u(0, x) = \phi(x)$, can be written as $u(T, x) = \mathbb{E}[\phi(X^x(0))]$.

Example 5.14 (backward Kolmogorov equation) For $T > 0$, define the stochastic process

$$X^x(t) = x + W(T - t), \qquad x \in \mathbb{R},$$

where $W(t)$ is a Brownian motion. Several sample paths are shown in Figure 5.5 for $T = 1$ and $x = 0$. The process $X^x(t)$ is like Brownian motion, but evolving backwards in time and translated by x. Now, define

$$u(T, x) := \mathbb{E}\big[\phi(X^x(0))\big], \tag{5.14}$$

so that $u(T, x)$ is the expected value of ϕ evaluated at $X^x(t)$ at $t = 0$. According to Definition 5.11, $W(t) \sim \mathrm{N}(0, t)$ for each $t \in \mathbb{R}^+$ and so $X^x(0) \sim \mathrm{N}(x, T)$ and it is clear that $u(T, x)$ is the exact solution of the heat equation (5.9) at $t = T$. With this interpretation of $u(T, x)$, the PDE (5.9) is known as *Kolmogorov's backward equation*, which we discuss again in Proposition 8.43 for stochastic ODEs.

5.2 Gaussian processes and the covariance function

In Definition 4.30, we saw that the distribution of a real-valued multivariate Gaussian random variable $\boldsymbol{X}$ is determined by a mean vector $\boldsymbol{\mu}$ and a covariance matrix C, which may be any symmetric, non-negative definite matrix. We are now interested in real-valued stochastic processes $\{X(t) : t \in \mathcal{T}\}$ and a similar pattern holds. The set of covariance functions $C : \mathcal{T} \times \mathcal{T} \to \mathbb{R}$ is exactly the set of symmetric, non-negative definite functions. We show in Theorem 5.18 that every symmetric, non-negative definite function is the covariance of some stochastic process. If the mean is also given and the process is Gaussian, we show in Corollary 5.19 that the distribution of the process is then uniquely defined. First we interpret a stochastic process as a random variable. Denote by $\mathbb{R}^{\mathcal{T}}$ the set of all real-valued deterministic functions $f : \mathcal{T} \to \mathbb{R}$, and let $\mathcal{B}(\mathbb{R}^{\mathcal{T}})$ denote the smallest σ-algebra that contains all sets, for $t_1, \dots, t_N \in \mathcal{T}$ and $F \in \mathcal{B}(\mathbb{R}^N)$,

$$B = \big\{ f \in \mathbb{R}^{\mathcal{T}} : [f(t_1), \dots, f(t_N)]^{\mathsf{T}} \in F \big\}. \tag{5.15}$$

Definition 5.3 says that the sample paths $X(\cdot,\omega)$ of a real-valued stochastic process $\{X(t)\colon t \in \mathcal{T}\}$ belong to $\mathbb{R}^{\mathcal{T}}$. It is also true that the sample paths define a $\mathbb{R}^{\mathcal{T}}$-valued random variable.

Lemma 5.15 *Let $(\Omega,\mathcal{F},\mathbb{P})$ denote the underlying probability space. The map from $\omega \mapsto X(\cdot,\omega)$ from $(\Omega,\mathcal{F})$ to the measurable space $(\mathbb{R}^{\mathcal{T}},\mathcal{B}(\mathbb{R}^{\mathcal{T}}))$ is measurable. Thus, the sample path is a $\mathbb{R}^{\mathcal{T}}$-valued random variable.*

This allows us to define what it means for processes and sample paths to be independent.

Definition 5.16 (independent processes, sample paths)

(i) Two real-valued processes $\{X(t)\colon t \in \mathcal{T}\}$, $\{Y(s)\colon s \in \mathcal{T}\}$ are *independent processes* if the associated $(\mathbb{R}^{\mathcal{T}},\mathcal{B}(\mathbb{R}^{\mathcal{T}}))$ random variables are independent. This means

$$\boldsymbol{X} = [X(t_1),\dots,X(t_M)]^{\mathsf{T}}, \qquad \boldsymbol{Y} = [Y(s_1),\dots,Y(s_M)]^{\mathsf{T}},$$

are independent multivariate random variables for any $t_i, s_i \in \mathcal{T}$ and $i = 1,\dots,M$.

(ii) We say $f_i\colon \mathcal{T} \to \mathbb{R}$, $i = 1,2$, are *independent sample paths* of a real-valued process $\{X(t)\colon t \in \mathcal{T}\}$ if $f_i(t) = X_i(t,\omega)$ for some $\omega \in \Omega$, where $X_i(t)$ are *iid* processes with the same distribution as $X(t)$.

We state the Daniel–Kolmogorov theorem. It says that if a family of distributions, $\nu_{t_1,\dots,t_N}$ for any $t_1,\dots,t_n \in \mathcal{T}$, satisfies two natural consistency conditions, there exists a unique stochastic process $\{X(t)\colon t \in \mathcal{T}\}$ with finite-dimensional distributions (see Definition 5.7) given by $\nu_{t_1,\dots,t_N}$. Denote the probability distribution of a process $X(t)$ on the measurable space $(\mathbb{R}^{\mathcal{T}},\mathcal{B}(\mathbb{R}^{\mathcal{T}}))$ by $\mathbb{P}_X$ (see Definition 4.4).

Theorem 5.17 (Daniel–Kolmogorov) *Suppose that for each set $\{t_1,\dots,t_N\} \subset \mathcal{T}$, there exists a probability measure $\nu_{t_1,\dots,t_N}$ on $\mathbb{R}^N$ such that*

(i) *for any permutation σ of $\{1,\dots,N\}$ and any $F \in \mathcal{B}(\mathbb{R}^N)$,*

$$\nu_{t_{\sigma(1)},\dots,t_{\sigma(N)}}(\sigma(F)) = \nu_{t_1,\dots,t_N}(F)$$

where $\sigma(F) := \{[X(t_{\sigma(1)}),\dots,X(t_{\sigma(N)})]^{\mathsf{T}}$ for $[X(t_1),\dots,X(t_N)]^{\mathsf{T}} \in F\}$, and

(ii) *for $M < N$ and any $F \in \mathcal{B}(\mathbb{R}^M)$,*

$$\nu_{t_1,\dots,t_N}(F \times \mathbb{R}^{N-M}) = \nu_{t_1,\dots,t_M}(F).$$

Then there exists a stochastic process $\{X(t)\colon t \in \mathcal{T}\}$ with finite-dimensional distributions $\nu_{t_1,\dots,t_N}$. If $X(t), Y(t)$ are two such processes then $\mathbb{P}_X(B) = \mathbb{P}_Y(B)$ for any $B \in \mathcal{B}(\mathbb{R}^{\mathcal{T}})$.

Recall from Definition 1.75 that a function $C\colon \mathcal{T} \times \mathcal{T} \to \mathbb{R}$ is *non-negative definite* if for all $t_1,\dots,t_N \in \mathcal{T}$ and $a_1,\dots,a_N \in \mathbb{R}$, $\sum_{j,k=1}^{N} a_j C(t_j,t_k) a_k \geq 0$. We show the set of covariance functions is exactly the set of symmetric, non-negative definite functions.

Theorem 5.18 *Let $\mathcal{T} \subset \mathbb{R}$. The following statements are equivalent.*

(i) *There exists a real-valued second-order stochastic process $\{X(t)\colon t \in \mathcal{T}\}$ with mean function $\mu(t)$ and covariance function $C(s,t)$.*

(ii) *The function μ maps from $\mathcal{T} \to \mathbb{R}$ and the function C maps from $\mathcal{T} \times \mathcal{T} \to \mathbb{R}$. Further, C is symmetric and non-negative definite.*

Proof Assume that (i) is true. As the process is second-order, the mean $\mu(t)$ and covariance $C(s,t)$ are well defined in $\mathbb{R}$. Then, $C(s,t)$ is non-negative definite, because for any $a_1,\ldots,a_N \in \mathbb{R}$ and $t_1,\ldots,t_N \in \mathcal{T}$

$$\sum_{j,k=1}^{N} a_j C(t_j,t_k) a_k = \mathbb{E}\left[\sum_{j,k=1}^{N} (X(t_j)-\mu(t_j))(X(t_k)-\mu(t_k)) a_j a_k\right]$$
$$= \mathbb{E}\left[\left|\sum_{j=1}^{N} a_j\big(X(t_j)-\mu(t_j)\big)\right|^2\right] \geq 0.$$

$C(s,t)$ is symmetric as $\operatorname{Cov}(X(s),X(t)) = \operatorname{Cov}(X(t),X(s))$ for all $s,t \in \mathcal{T}$.

Now assume that (ii) is true. Consider any $t_1,\ldots,t_N \in \mathcal{T}$ and let $C_N \in \mathbb{R}^{N\times N}$ be the matrix with entries $c_{jk} = C(t_j,t_k)$ for $j,k = 1,\ldots,N$. C_N is a valid covariance matrix; it is clearly symmetric and is non-negative definite as

$$\boldsymbol{a}^{\mathsf{T}} C_N \boldsymbol{a} = \sum_{j,k=1}^{N} a_j C(t_j,t_k) a_k \geq 0, \qquad \forall\, \boldsymbol{a} \in \mathbb{R}^N.$$

Next, choose $\nu_{t_1,\ldots,t_N} = \mathrm{N}(\mathbf{0},C_N)$ and let $\boldsymbol{Y} = [Y_1,\ldots,Y_N]^{\mathsf{T}}$ be a random variable with distribution $\nu_{t_1,\ldots,t_N}$. The Daniel–Kolmogorov theorem applies because $[Y_{\sigma(1)},\ldots,Y_{\sigma(N)}]^{\mathsf{T}}$ has mean zero and covariance matrix with (j,k) entry $C(t_{\sigma(j)},t_{\sigma(k)})$, and hence distribution $\nu_{t_{\sigma(1)},\ldots,t_{\sigma(N)}}$. Further,

$$\mathbb{P}\big([Y_1,\ldots,Y_N]^{\mathsf{T}} \in F \times \mathbb{R}^{N-M}\big) = \mathbb{P}\big([Y_1,\ldots,Y_M]^{\mathsf{T}} \in F\big), \qquad \forall F \in \mathcal{B}(\mathbb{R}^M).$$

We conclude that there exists a real-valued stochastic process $Y(t)$ with the finite-dimensional distributions $\nu_{t_1,\ldots,t_N}$. In particular, the distribution of $[Y(s),Y(t)]^{\mathsf{T}}$ is

$$\nu_{s,t} = \mathrm{N}\left(\begin{pmatrix}0\\0\end{pmatrix}, \begin{pmatrix}C(s,s) & C(s,t)\\ C(s,t) & C(t,t)\end{pmatrix}\right)$$

and the process $Y(t)$ has mean zero and covariance function $C(s,t)$. Finally, $X(t) = \mu(t)+Y(t)$ has the desired mean and covariance. □

For Gaussian processes, we can go further and observe a one-to-one correspondence between mean and covariance functions and their distributions.

Corollary 5.19 *The probability distribution $\mathbb{P}_X$ on $(\mathbb{R}^{\mathcal{T}},\mathcal{B}(\mathbb{R}^{\mathcal{T}}))$ of a real-valued Gaussian process $X(t)$ is uniquely determined by its mean $\mu\colon \mathcal{T} \to \mathbb{R}$ and covariance function $C\colon \mathcal{T}\times\mathcal{T} \to \mathbb{R}$.*

Proof Let $X(t)$ and $Y(t)$ be real-valued Gaussian processes with the same mean and covariance function. Because the multivariate Gaussian distribution is determined by its mean and covariance matrix, the processes $X(t)$ and $Y(t)$ have the same finite-dimensional distributions. That is, $\boldsymbol{X} = [X(t_1),\ldots,X(t_N)]^{\mathsf{T}}$ and $\boldsymbol{Y} = [Y(t_1),\ldots,Y(t_N)]^{\mathsf{T}}$ have the same distribution for any $t_1,\ldots,t_N \in \mathcal{T}$. The Daniel–Kolmogorov theorem says the associated probability distributions $\mathbb{P}_X, \mathbb{P}_Y$ on $(\mathbb{R}^{\mathcal{T}},\mathcal{B}(\mathcal{T}))$ are equal. □

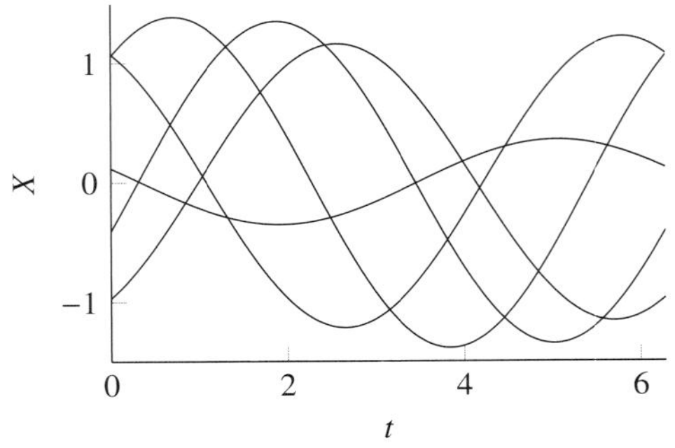

Figure 5.6 Five sample paths of the Gaussian process $X(t)$ from (5.16) on $[0, 2\pi]$, which has mean $\mu(t) = 0$ and covariance $C(s,t) = \cos(s - t)$.

Example 5.20 (cosine covariance function) Let $C(s,t) = \cos(s - t)$. Clearly $C(s,t) = C(t,s)$. Consider any $t_1, \ldots, t_N \in \mathcal{T} = \mathbb{R}$ and $a_1, \ldots, a_N \in \mathbb{R}$. Then,

$$\begin{aligned}\sum_{j,k=1}^{N} a_j \cos(t_j - t_k) a_k &= \sum_{j,k=1}^{N} a_j \big(\cos t_j \cos t_k + \sin t_j \sin t_k\big) a_k \\ &= \sum_{j=1}^{N} a_j \cos t_j \sum_{k=1}^{N} a_k \cos t_k + \sum_{j=1}^{N} a_j \sin t_j \sum_{k=1}^{N} a_k \sin t_k \\ &= \left|\sum_{j=1}^{N} a_j \cos t_j\right|^2 + \left|\sum_{j=1}^{N} a_j \sin t_j\right|^2 \geq 0.\end{aligned}$$

Thus, $C \colon \mathbb{R} \times \mathbb{R} \to \mathbb{R}$ is a non-negative definite function and, by Theorem 5.18, there exists a second-order process $\{X(t) \colon t \in \mathbb{R}\}$ with covariance function $C(s,t) = \cos(t - s)$ and, say, mean $\mu(t) = 0$.

As in Exercise 5.5,

$$X(t) = \xi_1 \cos(t) + \xi_2 \sin(t), \qquad \xi_i \sim \mathrm{N}(0,1) \; iid, \tag{5.16}$$

is a Gaussian process with mean $\mu(t) = 0$ and covariance $\cos(s - t)$. Five sample paths are plotted in Figure 5.6 on the interval $[0, 2\pi]$. Notice how the wave of positive and negative correlation described by the covariance function appears in the sample paths.

The next result shows that the function $C(s,t) = \min\{s,t\}$ is a valid covariance function for stochastic processes associated with $\mathcal{T} = \mathbb{R}^+$. When we combine this result with our study of sample path continuity in Corollary 5.42, this establishes existence of Brownian motion $\{W(t) \colon t \in \mathbb{R}^+\}$.

Lemma 5.21 *The function $C \colon \mathbb{R}^+ \times \mathbb{R}^+ \to \mathbb{R}$ defined by $C(s,t) = \min\{s,t\}$ is symmetric and non-negative definite.*

Proof Symmetry is clear. We need to show that, for $t_1, \ldots, t_N \in \mathbb{R}^+$ and $a_1, \ldots, a_N \in \mathbb{R}$,

$$\sum_{i=1}^{N} \sum_{j=1}^{N} a_i a_j C(t_i, t_j) \geq 0. \tag{5.17}$$

Let C_N be the $N \times N$ matrix with entries $C(t_i,t_j) = \min\{t_i,t_j\}$ for $i,j = 1,2,\dots N$. Then, the left-hand side of (5.17) is $\boldsymbol{a}^\mathsf{T} C_N \boldsymbol{a}$ and we must show that C_N is a non-negative definite matrix. Without loss of generality, we may assume $0 \le t_1 < t_2 < \cdots < t_N$. Let E_J be the matrix whose entries are all zero, except in the bottom right $J \times J$ corner, where all the entries are equal to 1. Then,

$$C_N = \begin{pmatrix} t_1 & t_1 & t_1 & \dots & t_1 \\ t_1 & t_2 & t_2 & \dots & t_2 \\ t_1 & t_2 & t_3 & & t_3 \\ \vdots & \vdots & & \ddots & \\ t_1 & t_2 & t_3 & & t_N \end{pmatrix} = \sum_{J=1}^{N} c_J E_J, \qquad E_J := \begin{pmatrix} 0 & 0 & & \dots & & & 0 \\ \vdots & \ddots & & & & & \vdots \\ \vdots & \cdots & 0 & & & \cdots & 0 \\ 0 & \cdots & 0 & 1 & & \cdots & 1 \\ \vdots & & \vdots & \vdots & & & \vdots \\ 0 & \cdots & 0 & 1 & & \cdots & 1 \end{pmatrix}.$$

where $c_N = t_1$, and $c_J = t_{N-J+1} - t_{N-J}$ for $J = 1,\dots,N-1$. Each matrix E_J is non-negative definite, as is easily verified by showing that $\boldsymbol{x}^\mathsf{T} E_J \boldsymbol{x} = (\sum_{j=N+1-J}^{N} x_j)^2$. Since $0 \le t_1 < t_2 < \cdots < t_N$, the coefficients c_J are positive and the matrix C_N is non-negative definite. □

5.3 Brownian bridge, fractional Brownian motion, and white noise

In this section, we present two additional examples of Gaussian processes, the Brownian bridge and fractional Brownian motion (fBm), and compare white and coloured noise.

Brownian bridge

When sampling a Brownian motion $W(t)$ numerically, we determine sample paths at specific points, say at $t = 0$ and $t = T$. The values of the Brownian motion in $(0,T)$ are usually unknown and it is interesting to consider their distribution. This leads to the *Brownian bridge*, which is the stochastic process $B(t)$ on $[0,T]$, whose distribution is obtained by conditioning the distribution of $W(t)$ by boundary conditions at $t = 0$ and $t = T$. Brownian motion starts at the origin, $W(0) = 0$ a.s., and we further specify the boundary condition $W(T) = b$, for some b. For $0 \le t_1 < \cdots < t_N \le T$, the finite-dimensional distributions of a Brownian bridge $B(t)$ are thus given by (compare with (4.19))

$$\mathbb{P}\left(\begin{pmatrix} B(t_1) \\ \vdots \\ B(t_N) \end{pmatrix} \in F\right) = \mathbb{E}\left[1_F\begin{pmatrix} W(t_1) \\ \vdots \\ W(t_N) \end{pmatrix} \middle| W(T) = b\right], \qquad F \in \mathcal{B}(\mathbb{R}^N).$$

Then, for measurable functions $\phi\colon \mathbb{R}^N \to \mathbb{R}$,

$$\mathbb{E}\big[\phi(B(t_1),\dots,B(t_N))\big] = \mathbb{E}\Big[\phi(W(t_1),\dots,W(t_N)) \,\Big|\, W(T) = b\Big]. \tag{5.18}$$

For $T > 0$, the variance of $W(T)$ is positive and the finite-dimensional distributions of the Brownian bridge are Gaussian as in Example 5.18. The Brownian bridge is a Gaussian process and its mean function is $\mu(t) = \mathbb{E}[W(t) \,|\, W(T) = b]$ (using (5.18) with $\phi(x) = x$).

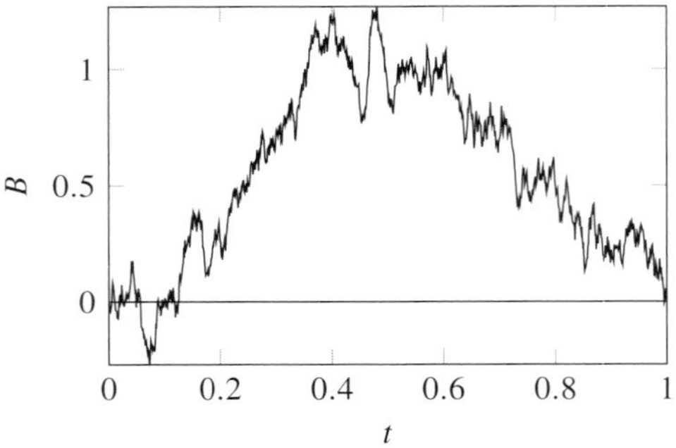

Figure 5.7 Approximate sample path of the Brownian bridge $B(t)$ on $[0,1]$ with boundary conditions $B(0) = B(1) = 0$, generated using 1001 points with uniform spacing $\Delta t = 0.001$. The graph was produced with Algorithm 5.2 using the commands `t=[0:0.001:1], X=bb(t), plot(t,X)`.

The covariance is

$$\begin{aligned} C(s,t) &= \mathbb{E}\big[\big(B(s) - \mu(s)\big)\big(B(t) - \mu(t)\big)\big] \\ &= \mathbb{E}\Big[\big(W(s) - \mu(s)\big)\big(W(t) - \mu(t)\big)\,\Big|\, W(T) = b\Big], \end{aligned}$$

using (5.18) with $\phi(x,y) = (x - \mu(s))(y - \mu(t))$. Explicit expressions for $\mu(t)$ and $C(s,t)$ can be found using the formula in Example 4.51 (see also Exercise 5.8). In the following lemma, we compute the mean and covariance for the standard Brownian bridge (the case $T = 1$ and $b = 0$) by a different method.

Lemma 5.22 *The Brownian bridge $\{B(t) : t \in [0,1]\}$ with conditions $B(0) = B(1) = 0$ is the Gaussian process with mean $\mu(t) = 0$ and covariance function $C(s,t) = \min\{s,t\} - st$.*

Proof For $0 \le t \le 1$,

$$\operatorname{Cov}\big(W(t) - tW(1),\, W(1)\big) = \mathbb{E}\Big[\big(W(t) - tW(1)\big)\, W(1)\Big] = \min\{t,1\} - t1 = 0.$$

As their covariance is zero and their joint distribution Gaussian, Lemma 4.33 implies that

$$W(t) - tW(1) \text{ and } W(1) \text{ are independent.} \tag{5.19}$$

With Theorem 4.52 (independence), this gives

$$\mathbb{E}\Big[W(t) - tW(1)\,\Big|\, W(1)\Big] = \mathbb{E}\big[W(t) - tW(1)\big] = 0, \qquad \text{a.s.}$$

and, from Theorem 4.52 ('taking out what is known' and linearity),

$$\mathbb{E}\big[W(t) \,|\, W(1)\big] = \mathbb{E}\big[t\, W(1) \,|\, W(1)\big] = t\, W(1), \qquad \text{a.s.} \tag{5.20}$$

Given $W(1)$, the average of $W(t)$ increases linearly from 0 to the value of $W(1)$. Hence, the mean of $B(t)$ is $\mu(t) = \mathbb{E}[W(t) \,|\, W(1) = 0] = 0$. The covariance of $B(t)$ is

$$C(s,t) = \operatorname{Cov}\big(B(s), B(t)\big) = \mathbb{E}\Big[\big(W(s) - \mu(s)\big)\big(W(t) - \mu(t)\big)\,\Big|\, W(1) = 0\Big]$$

and, as $\mu(t) = 0$,

$$\begin{aligned} C(s,t) &= \mathbb{E}\Big[W(s)W(t)\,\Big|\, W(1) = 0\Big] \\ &= \mathbb{E}\Big[\big(W(s) - sW(1)\big)\big(W(t) - t\, W(1)\big)\,\Big|\, W(1) = 0\Big]. \end{aligned}$$

The random variable $(W(s) - sW(1))(W(t) - tW(1))$ is independent of $W(1)$ and hence by the independence property of Theorem 4.52, we have

$$\begin{aligned} C(s,t) &= \mathbb{E}\big[(W(s) - sW(1))(W(t) - tW(1))\big] \\ &= \mathbb{E}\big[W(t)W(s)\big] - t\,\mathbb{E}\big[W(1)W(s)\big] - s\,\mathbb{E}\big[W(1)W(t)\big] + s\,t\,\mathbb{E}\big[W(1)^2\big] \\ &= \min\{s,t\} - ts - st + ts = \min\{s,t\} - st. \end{aligned}$$

□

A sample path of the Brownian bridge is shown in Figure 5.7. It was generated by Algorithm 5.2, which exploits the fact that $W(t) - tW(1)$ has the same distribution as the standard Brownian bridge $B(t)$; see Exercise 5.7. Given Algorithm 5.1 for generating samples of $W(t)$ at fixed values of t, this provides an easy way to approximate sample paths of $B(t)$.

Algorithm 5.2 Code to approximate sample paths of the standard Brownian bridge $B(t)$ on $[0,T]$. The input `t` is a vector of sample times $t_j < t_{j+1}$ and output `X` is a vector whose jth component is a sample of $B(t_j)$.

```
function X=bb(t)
W=bmotion(t); X=W-W(end)*(t-t(1))/(t(end)-t(1));
```

Fractional Brownian motion

Brownian motion has two properties that we have not yet mentioned and that define a whole family of stochastic processes known as *fractional Brownian motion* (fBm).

Self-similarity: Given a Brownian motion $\{W(t) \colon t \in \mathbb{R}^+\}$ and a scaling factor $\alpha > 0$, let $\widetilde{W}(t) := \alpha^{1/2} W(t/\alpha)$. $\widetilde{W}(t)$ is a Gaussian process with continuous sample paths, mean function $\mu(t) = 0$ and covariance function

$$C(s,t) = \mathbb{E}\left[\widetilde{W}(s)\widetilde{W}(t)\right] = \alpha \min\{s/\alpha, t/\alpha\} = \min\{s,t\}.$$

By Definition 5.11, $\widetilde{W}(t)$ is also a Brownian motion and, because of this, we say Brownian motion is *self-similar*.

Stationary increments: The increment $W(t) - W(s)$ of a Brownian motion over the interval $[s,t]$ has distribution $\mathrm{N}(0, |t-s|)$ and hence the distribution of $W(t) - W(s)$ is the same as that of $W(t+h) - W(s+h)$ for any $h > 0$. That is, the distribution of the increments is independent of translations and we say that Brownian motion has *stationary* increments.

Fractional Brownian motion (fBm) is a family of mean-zero Gaussian processes $B^H(t)$, parameterized by $H \in (0,1)$ (the Hurst parameter), which are self-similar and have stationary increments. The self-similar property says that for any $\alpha > 0$,

$$\widetilde{B}^H(t) := \alpha^H B^H(t/\alpha) \tag{5.21}$$

is a fBm with the same Hurst parameter H as the original fBm $B^H(t)$. If we normalise the process and require that $\mathbb{E}\big[B^H(t)\big] = 0$ and $\mathbb{E}\big[B^H(1)^2\big] = 1$ then we arrive at the definition of fBm. That is, these conditions specify the mean and covariance function of $B^H(t)$ and, using Corollary 5.19, the distribution of a Gaussian process is defined by those functions.

To derive the covariance function, suppose that $\mathbb{E}\big[B^H(t)^2\big] = q(t)$. For self-similarity, we require $q(t) = \mathbb{E}\big[\widetilde{B}^H(t)^2\big]$ and so $q(t) = \alpha^{2H} q(t/\alpha)$. This implies $q(t)$ is proportional to t^{2H} by substituting $\alpha = t$. Since $\mathbb{E}\big[B^H(1)^2\big] = 1$, we must have $q(t) = t^{2H}$ and, from the stationary increment property, $\mathbb{E}\big[(B^H(t) - B^H(s))^2\big] = q(t-s)$. Noting that $x^2 + y^2 - (x-y)^2 = 2xy$, we then obtain

$$\begin{aligned}\mathbb{E}\big[B^H(t)B^H(s)\big] &= \frac{1}{2}\Big(\mathbb{E}\big[B^H(t)^2\big] + \mathbb{E}\big[B^H(s)^2\big] - \mathbb{E}\Big[\big(B^H(t) - B^H(s)\big)^2\Big]\Big)\\ &= \frac{1}{2}\big(t^{2H} + s^{2H} - |t-s|^{2H}\big).\end{aligned}$$

Definition 5.23 (fractional Brownian motion) A stochastic process $B^H(t)$ is a fractional Brownian motion (fBm) with Hurst parameter $H \in (0,1)$ if the process is Gaussian with mean zero and covariance function

$$\mathbb{E}\big[B^H(t)B^H(s)\big] = \frac{1}{2}\big(|t|^{2H} + |s|^{2H} - |t-s|^{2H}\big).$$

The existence of fractional Brownian motion for $H \in (0,1)$ follows from Corollary 5.19 as the covariance function is non-negative definite (see Exercise 5.11). For $H = 1/2$, the covariance function is $(|t| + |s| - |t-s|)/2 = \min\{s,t\}$, which is the covariance of Brownian motion. Hence, an fBm with Hurst parameter $H = 1/2$ and continuous sample paths is a Brownian motion.

Example 5.24 (correlation of increments) Brownian motion $W(t)$ is easy to simulate via (5.7) because the increments are independent. Increments of $B^H(t)$ are correlated and it is much harder to simulate on a computer. To see this, consider the increments over the intervals $[0,1]$ and $[1,2]$. We have

$$\begin{aligned}&\mathbb{E}\Big[\big(B^H(2) - B^H(1)\big)\big(B^H(1) - B^H(0)\big)\Big]\\ &\quad= \mathbb{E}\big[B^H(2)B^H(1)\big] - \mathbb{E}\big[B^H(2)B^H(0)\big] - \mathbb{E}\big[B^H(1)^2\big] + \mathbb{E}\big[B^H(1)B^H(0)\big]\\ &\quad= \frac{1}{2}\big(2^{2H} + 1 - 1\big) - 0 - \frac{1}{2}(2) + 0 = \frac{1}{2}\big(4^H - 2\big).\end{aligned}$$

That is, $\mathrm{Cov}\big(B^H(2) - B^H(1), B^H(1) - B^H(0)\big) = (4^H - 2)/2$ and the increments $B^H(2) - B^H(1)$ and $B^H(1) - B^H(0)$ are positively correlated if $H > 1/2$ and negatively correlated if $H < 1/2$. In general, for $H > 1/2$, all increments of $B^H(t)$ are positively correlated, meaning that trends are likely to continue over long periods. For $H < 1/2$, increments are negatively correlated and we expect strong variations in the process. Figure 5.8 shows approximations to sample paths of fBm for $H = 1/4$ and $H = 3/4$, generated using Algorithm 5.3 (which will be introduced in §5.4). Observe that the sample path of the $H = 1/4$ process has a more oscillatory trajectory.

White and coloured noise

White and coloured noise derive their names from light, which consists of a linear combination of waves of different wavelengths (seen in the colours of the rainbow). A *homogeneous* mix of wavelengths results in white light, whereas a *heterogeneous* mix of wavelengths gives coloured light. Similarly, stochastic processes are classified as white or coloured noise

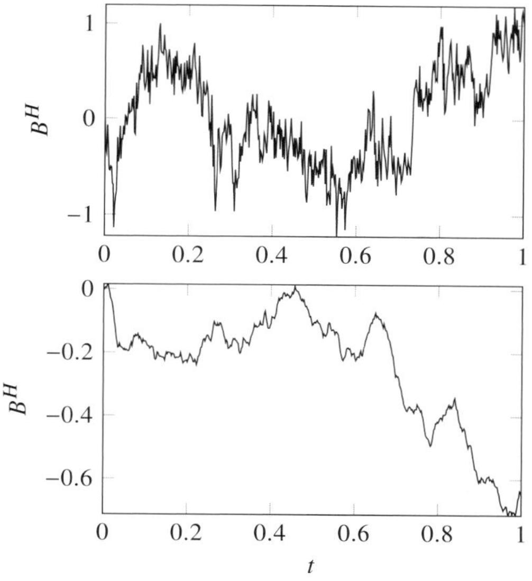

Figure 5.8 Approximate sample paths of fBm with $H = 1/4$ (top) and $H = 3/4$ (bottom) on $[0,1]$ generated using 501 sample points with uniform spacing $\Delta t = 0.002$. The graph was produced with Algorithm 5.3 using the commands `t=[0:0.002:1],X=fbm(t,H),plot(t,X)`.

depending on their representation in an orthonormal basis of functions (with basis functions corresponding to elementary waves of the colours of light). First we consider white noise.

White noise contains a homogeneous mix of all the different basis functions. To fix ideas, consider $\mathcal{T} = [0,1]$ and let $\{\phi_j : j = 1,2,\dots\}$ be an orthonormal basis of $L^2(0,1)$. If, for example, $\phi_j(t) = \sqrt{2}\sin(j\pi t)$ then each ϕ_j represents a wave and j is known as the wave number. White noise is the stochastic process consisting of equal random amounts of each basis function. Again to fix ideas, we consider

$$\zeta(t) = \sum_{j=1}^{\infty} \xi_j \phi_j(t), \qquad \text{where } \xi_j \sim \mathrm{N}(0,1) \textit{ iid}. \tag{5.22}$$

Clearly, $\zeta(t)$ is a mean-zero stochastic process with covariance function

$$\mathrm{Cov}(\zeta(s),\zeta(t)) = \sum_{j,k=1}^{\infty} \mathrm{Cov}(\xi_j,\xi_k)\phi_j(s)\phi_k(t) = \sum_{j=1}^{\infty} \phi_j(s)\phi_j(t). \tag{5.23}$$

Following the argument in Example 1.92 for the chosen basis $\{\phi_j\}$, we see that the covariance function of $\zeta(t)$ is in fact $C(s,t) = \delta(s-t)$. This has two implications. First, for $s \neq t$, $\mathrm{Cov}(\zeta(s),\zeta(t)) = 0$ and white noise at two distinct points $s,t \in \mathcal{T}$ is uncorrelated. In addition, $C(t,t) = \mathrm{Var}(\zeta(t)) = \delta(0)$ and, since the delta function is infinite at zero, $\mathbb{E}\big[\zeta(t)^2\big] = \infty$. That is, white noise is not well defined in $L^2(\Omega)$ as the sum (5.22) does not converge. To understand white noise rigorously, stochastic integrals are introduced in §8.2.

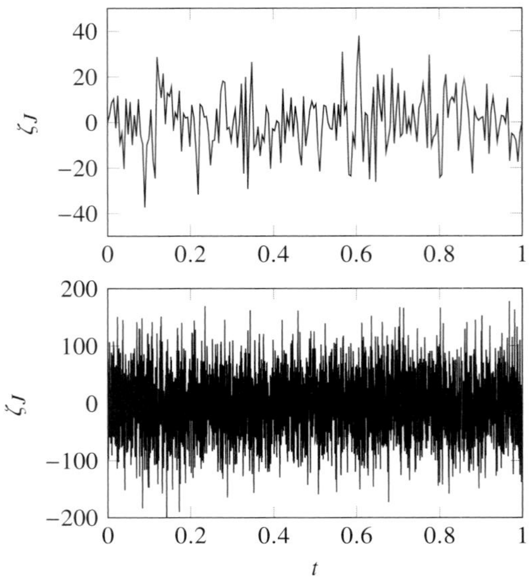

Figure 5.9 A sample path of ζ_J for $J = 200$ (top) and $J = 3200$ (bottom). From a 16-fold increase in the number of modes, observe that ζ_J increases in size approximately 4-fold. Unlike the Brownian bridge, sample paths of $\zeta_J(t)$ do not converge to $L^2(0,1)$ functions as $J \to \infty$.

The divergence of the sum in (5.22) is illustrated in Figure 5.9, where we plot a sample path of

$$\zeta_J(t) = \sum_{j=1}^{J} \xi_j \phi_j(t), \tag{5.24}$$

for $J = 200$ and $J = 3200$. Notice that ζ_J is becoming larger as we increase J, indicating its divergence. For coloured noise, our analogy to light suggests a heterogeneous mix of wave numbers. To fix ideas, consider a stochastic process defined for $t \in \mathcal{T}$ by

$$X(t) = \sum_{j=1}^{\infty} \sqrt{\nu_j}\, \xi_j \phi_j(t), \qquad \text{where } \xi_j \sim \mathrm{N}(0,1) \text{ } iid \text{ and } \nu_j \geq 0. \tag{5.25}$$

When the ν_j vary with j, the process $X(t)$ is said to be a coloured noise and, in contrast to white noise, the random variables $X(t), X(s)$ are correlated. Most of the processes we have encountered so far take this form and this includes the Brownian bridge. We derive rigorously an expansion of the form (5.25) for the Brownian bridge in Example 5.30 and find that ν_j behaves like $1/j^2$. The expansion (5.25) was suggested purely from the analogy with coloured light and, in fact, it is a useful way to examine many stochastic processes. When the basis functions are chosen as the eigenfunctions of the covariance function, this is termed the *Karhunen–Loève* expansion.

5.4 The Karhunen–Loève expansion

The mean-zero Gaussian process with covariance $\cos(s-t)$ in (5.16) is written as a linear combination of two functions, $\cos(t)$ and $\sin(t)$, with coefficients ξ_1, ξ_2 that are uncorrelated (independent in this case) random variables. Looking closely, we see that $\cos(t)$ and $\sin(t)$ are eigenfunctions of the integral operator on $L^2(0,2\pi)$ defined by the covariance $\cos(s-t)$. That is,

$$\int_0^{2\pi} \cos(s-t)\cos(t)\,dt = \frac{1}{2}\int_0^{2\pi} \cos(s-2t) + \cos(s)\,dt = \pi\cos(s)$$

(and similarly for $\sin(t)$). This construction, writing a process as a linear combination of the eigenfunctions of its covariance with uncorrelated random coefficients, is called the *Karhunen–Loève expansion*. Before developing the general theory for real-valued processes, we consider the spectral decomposition of real-valued symmetric matrices, which leads to a discrete version of the Karhunen–Loève expansion in finite dimensions.

Matrix spectral decomposition

Let $\{X(t)\colon t \in \mathcal{T}\}$ be a real-valued Gaussian process with mean function $\mu(t)$ and covariance function $C(s,t)$. For $t_1, \ldots, t_N \in \mathcal{T}$, define

$$\boldsymbol{X} = \left[X(t_1), \ldots, X(t_N)\right]^{\mathsf{T}} \sim \mathrm{N}(\boldsymbol{\mu}, C_N), \tag{5.26}$$

where $\boldsymbol{\mu} = [\mu(t_1), \ldots, \mu(t_N)]^{\mathsf{T}}$ and $C_N \in \mathbb{R}^{N\times N}$ has entries $c_{ij} = C(t_i, t_j)$. We saw in §4.4 that if C_N is written as

$$C_N = V^{\mathsf{T}}V, \tag{5.27}$$

then samples of $\boldsymbol{X} \sim \mathrm{N}(\boldsymbol{\mu}, C_N)$ can be generated via

$$\boldsymbol{X} = \boldsymbol{\mu} + V^{\mathsf{T}}\boldsymbol{\xi}, \tag{5.28}$$

where $\boldsymbol{\xi} := [\xi_1, \ldots, \xi_N]^{\mathsf{T}}$ with *iid* components $\xi_j \sim \mathrm{N}(0,1)$. When C_N is positive definite, Cholesky factorisation can be used to find a matrix V satisfying (5.27). An alternative method is the spectral decomposition.

Theorem 5.25 (spectral decomposition) *Every $N \times N$ real-valued symmetric matrix A may be written as $A = U\Sigma U^{\mathsf{T}}$, where U is an orthonormal matrix whose columns $\boldsymbol{u}_j$ are eigenvectors of A and Σ is an $N \times N$ diagonal matrix whose entries ν_j are the real eigenvalues of A. The form $U\Sigma U^{\mathsf{T}}$ is known as the* spectral decomposition *of A.*

The covariance matrix C_N of $\boldsymbol{X}$ in (5.26) is square, real valued and symmetric, so it has a spectral decomposition $C_N = U\Sigma U^{\mathsf{T}}$. Further, since C_N is non-negative definite, we can order the eigenvalues so that $\nu_1 \ge \nu_2 \ge \cdots \ge \nu_N \ge 0$. Defining $V^{\mathsf{T}} := U\Sigma^{1/2}$, where $\Sigma^{1/2}$ is the diagonal matrix with entries $\sqrt{\nu_j}$, we see that $V^{\mathsf{T}}V = C_N$. With this choice of V, (5.28) becomes

$$\boldsymbol{X} = \boldsymbol{\mu} + \sum_{j=1}^{N} \sqrt{\nu_j}\,\boldsymbol{u}_j\,\xi_j, \qquad \xi_j \sim \mathrm{N}(0,1)\ iid, \tag{5.29}$$

and we can generate samples from $\mathrm{N}(\boldsymbol{\mu}, C_N)$ after computing the eigenvectors and eigenvalues of C_N. Note that (5.29) is well defined even if one of the eigenvalues is zero; the spectral decomposition is defined when C_N is singular and so is more robust than Cholesky factorisation.

Example 5.26 (sampling fBm) Consider an fBm $B^H(t)$ on the interval $[0,T]$ and the set of points $t_j = (j-1)\Delta t$ for $\Delta t = T/(N-1)$ and $j = 1,\dots,N$. Here, the covariance matrix C_N associated with $C(s,t) = |t|^{2H} + |s|^{2H} - |t-s|^{2H}$ is singular because $C(t_1,t) = C(0,t) = 0$ for all t. Cholesky factorisation fails. Algorithm 5.3 demonstrates how to sample fBm with Hurst parameter $H \in (0,1)$ at a set of grid points on $[0,T]$ using the spectral decomposition. Approximate sample paths generated with this method are shown in Figure 5.8.

Algorithm 5.3 Code to sample fBm $B^H(t)$ on $[0,T]$. The input `H` is the Hurst parameter and `t` is a vector of sample times $t_j < t_{j+1}$. The jth element of the output `X` is a sample of $B^H(t_j)$.

```
function X=fbm(t,H)
N=length(t); C_N=[];
for i=1:N, % compute covariance matrix
    for j=1:N,
        ti=t(i); tj=t(j);
        C_N(i,j)=0.5*(ti^(2*H)+tj^(2*H)-abs(ti-tj)^(2*H));
    end;
end;
[U,S]=eig(C_N); xsi=randn(N, 1); X=U*(S^0.5)*xsi;
```

Truncated spectral decomposition

It is often appropriate to truncate the spectral decomposition of an $N \times N$ symmetric matrix and compute only a few of the most important eigenvalues and eigenvectors. This can yield significant computational savings. Written in terms of the columns $\boldsymbol{u}_1,\dots,\boldsymbol{u}_N$ of U, the decomposition of C_N is

$$C_N = U\Sigma U^{\mathsf{T}} = \sum_{j=1}^{N} \nu_j \boldsymbol{u}_j \boldsymbol{u}_j^{\mathsf{T}}, \tag{5.30}$$

a weighted sum of N rank one $N \times N$ matrices. Since $\nu_j \geq \nu_{j+1}$, terms corresponding to large j contribute less to the sum in (5.30). Let Σ_n denote the $n \times n$ diagonal matrix with entries $\nu_1,\dots,\nu_n$ (here, the largest eigenvalues of C_N) and let U_n be the $N \times n$ matrix $U_n = [\boldsymbol{u}_1,\dots,\boldsymbol{u}_n]$ whose columns are the associated eigenvectors. The truncated spectral decomposition of C_N is

$$C_n = U_n \Sigma_n U_n^{\mathsf{T}} = \sum_{j=1}^{n} \nu_j \boldsymbol{u}_j \boldsymbol{u}_j^{\mathsf{T}}. \tag{5.31}$$

The error in approximating C_N by C_n in the operator norm (recall Example 1.71) is

$$\|C_N - C_n\|_{\mathcal{L}(\mathbb{R}^N)} = \left\| \sum_{j=n+1}^{N} \nu_j \boldsymbol{u}_j \boldsymbol{u}_j^{\mathsf{T}} \right\|_{\mathcal{L}(\mathbb{R}^N)} = \nu_{n+1}.$$

Suppose that we truncate (5.29) after $n < N$ terms, and form

$$\widehat{\boldsymbol{X}} = \boldsymbol{\mu} + \sum_{j=1}^{n} \sqrt{\nu_j}\, \boldsymbol{u}_j\, \xi_j, \qquad \xi_j \sim \mathrm{N}(0,1) \text{ } iid, \tag{5.32}$$

which is otherwise written as $\widehat{\boldsymbol{X}} = \boldsymbol{\mu} + U_n \Sigma_n^{1/2} \boldsymbol{\xi}$ for $\boldsymbol{\xi} \sim \mathrm{N}(\boldsymbol{0}, I_n)$ in terms of the truncated spectral decomposition of C_N. We see that $\widehat{\boldsymbol{X}} \sim \mathrm{N}(\boldsymbol{\mu}, C_n)$, where C_n is given by (5.31).

The mean-square error in approximating $\boldsymbol{X}$ by $\widehat{\boldsymbol{X}}$ is

$$\mathbb{E}\left[\left\|\boldsymbol{X} - \widehat{\boldsymbol{X}}\right\|_2^2\right] = \mathbb{E}\left[\sum_{j=n+1}^{N} \sum_{k=n+1}^{N} \sqrt{\nu_j}\, \sqrt{\nu_k}\, \boldsymbol{u}_j^{\mathsf{T}} \boldsymbol{u}_k \xi_j \xi_k\right] = \sum_{j=n+1}^{N} \nu_j,$$

and the error in approximating averages (see Theorem 4.63) satisfies, for some $K > 0$,

$$\begin{aligned} \left|\mathbb{E}[\phi(\boldsymbol{X})] - \mathbb{E}\left[\phi(\widehat{\boldsymbol{X}})\right]\right| &\le K \,\|\phi\|_{L^2(\mathbb{R}^N)} \|C_n - C_N\|_{\mathrm{F}}^2 \\ &= K\, \|\phi\|_{L^2(\mathbb{R}^N)} \sum_{j=n+1}^{N} \nu_j^2, \qquad \forall \phi \in L^2(\mathbb{R}^N). \end{aligned}$$

Both errors are minimised by retaining terms corresponding to the n largest eigenvalues of C_N and, if the eigenvalues of C_N decay rapidly, we obtain a good approximation to $\boldsymbol{X}$ using few eigenvectors. Viewed another way, (5.32) says that we can approximate N-variate Gaussian random variables $\boldsymbol{X}$ with $n < N$ uncorrelated $\mathrm{N}(0,1)$ random variables ξ_j. See also Theorem 4.63. The truncated spectral decomposition therefore provides a form of model order reduction.

Karhunen–Loève expansion

The generalisation of the spectral decomposition and the representation (5.29) to stochastic processes $X(t)$ is called the *Karhunen–Loève expansion*. Let $\mu(t) = \mathbb{E}[X(t)]$ and consider the mean-zero process $X(t) - \mu(t)$. We are interested in writing sample paths $X(t,\omega) - \mu(t)$ as a series in an orthonormal basis $\{\phi_j : j \in \mathbb{N}\}$ of $L^2(\mathcal{T})$. That is,

$$X(t,\omega) - \mu(t) = \sum_{j=1}^{\infty} \gamma_j(\omega) \phi_j(t) \tag{5.33}$$

where the coefficients γ_j are random variables given by

$$\gamma_j(\omega) := \left\langle X(t,\omega) - \mu(t), \phi_j(t) \right\rangle_{L^2(\mathcal{T})}.$$

Let $C(s,t)$ denote the covariance function of the process $X(t)$ and define the integral operator $\mathcal{C}$ by

$$(\mathcal{C}f)(t) := \int_{\mathcal{T}} C(s,t) f(s)\, ds \qquad \text{for } f \in L^2(\mathcal{T}). \tag{5.34}$$

The *Karhunen–Loève expansion* is the name given to (5.33) when the functions ϕ_j are chosen to be the eigenfunctions of the operator $\mathcal{C}$. The similarities with the spectral decomposition and the expansion (5.29) of a multivariate random variable are clear and, in fact, (5.29) is often called the discrete Karhunen–Loève expansion.

The expansion (5.33) makes sense for a given ω if $X(t,\omega) - \mu(t)$ belongs to $L^2(\mathcal{T})$. We show that this is true for almost all $\omega \in \Omega$ if $X \in L^2(\Omega, L^2(\mathcal{T}))$.

Lemma 5.27 *If $X \in L^2(\Omega, L^2(\mathcal{T}))$, the mean function $\mu \in L^2(\mathcal{T})$ and the sample path $X(t,\omega) \in L^2(\mathcal{T})$ for almost all $\omega \in \Omega$.*

Proof By assumption, $\|X\|_{L^2(\Omega,L^2(\mathcal{T}))} = \mathbb{E}\big[\|X\|^2_{L^2(\mathcal{T})}\big]^{1/2} < \infty$ and $\|X(t,\omega)\|_{L^2(\mathcal{T})} < \infty$ for almost all $\omega \in \Omega$ and the sample paths $X(t,\omega)$ belong to $L^2(\mathcal{T})$ almost surely. Jensen's inequality gives $\mu(t)^2 = (\mathbb{E}[X(t)])^2 \le \mathbb{E}\big[X(t)^2\big]$ and so

$$\|\mu\|^2_{L^2(\mathcal{T})} = \int_{\mathcal{T}} \mu(t)^2\,dt \le \int_{\mathcal{T}} \mathbb{E}\big[X(t)^2\big]\,dt.$$

The right-hand side

$$\begin{aligned}\int_{\mathcal{T}} \mathbb{E}\big[X(t)^2\big]\,dt &= \int_{\mathcal{T}}\int_{\Omega} |X(t,\omega)|^2\,d\mathbb{P}(\omega)\,dt \\ &= \int_{\Omega}\int_{\mathcal{T}} |X(t,\omega)|^2\,dt\,d\mathbb{P}(\omega) = \|X\|^2_{L^2(\Omega,L^2(\mathcal{T}))}\end{aligned}$$

where we apply Fubini's theorem (Theorem 1.24). Clearly then, $\|\mu\|_{L^2(\mathcal{T})} < \infty$ and the mean function $\mu \in L^2(\mathcal{T})$. □

Theorem 5.28 (L^2 convergence of Karhunen–Loève) *Consider a process $\{X(t)\colon t \in \mathcal{T}\}$ and suppose that $X \in L^2(\Omega, L^2(\mathcal{T}))$. Then,*

$$X(t,\omega) = \mu(t) + \sum_{j=1}^{\infty} \sqrt{\nu_j}\,\phi_j(t)\,\xi_j(\omega) \tag{5.35}$$

where the sum converges in $L^2(\Omega, L^2(\mathcal{T}))$,

$$\xi_j(\omega) := \frac{1}{\sqrt{\nu_j}}\big\langle X(t,\omega) - \mu(t), \phi_j(t)\big\rangle_{L^2(\mathcal{T})},$$

and $\{\nu_j, \phi_j\}$ denote the eigenvalues and eigenfunctions of the covariance operator in (5.34) with $\nu_1 \ge \nu_2 \ge \cdots \ge 0$. The random variables ξ_j have mean zero, unit variance and are pairwise uncorrelated. If the process is Gaussian, then $\xi_j \sim \mathrm{N}(0,1)$ iid.

Proof The Cauchy–Schwarz inequality gives

$$\begin{aligned}\int_{\mathcal{T}\times\mathcal{T}} C(s,t)^2\,ds\,dt &= \int_{\mathcal{T}\times\mathcal{T}} \Big(\mathbb{E}\big[(X(s)-\mu(s))(X(t)-\mu(t))\big]\Big)^2\,ds\,dt \\ &\le \int_{\mathcal{T}} \mathbb{E}\big[(X(s)-\mu(s))^2\big]\,ds \int_{\mathcal{T}} \mathbb{E}\big[(X(t)-\mu(t))^2\big]\,dt.\end{aligned}$$

This is finite as $X - \mu \in L^2(\Omega, L^2(\mathcal{T}))$ and we see that $C \in L^2(\mathcal{T}\times\mathcal{T})$. The operator $\mathcal{C}$ is symmetric and compact (see Theorem 1.68 and Lemma 1.72) and the Hilbert–Schmidt spectral theorem (Theorem 1.73) gives an orthonormal basis $\{\phi_j\}$ of eigenfunctions for the range of $\mathcal{C}$, which we can supplement if necessary to form an orthonormal basis for $L^2(\mathcal{T})$. As C is also non-negative definite, the corresponding eigenvalues ν_j are non-negative. Now (5.35) is simply (5.33) with the basis $\{\phi_j\}$ of eigenfunctions and $\xi_j(\omega) = \gamma_j(\omega)/\sqrt{\nu_j}$.

Define the truncated expansion

$$X_J(t,\omega) = \mu(t) + \sum_{j=1}^{J} \sqrt{\nu_j}\,\phi_j(t)\,\xi_j(\omega). \tag{5.36}$$

As ϕ_j is a basis for $L^2(\mathcal{T})$, $X_J(t,\omega) \to X(t,\omega)$ in $L^2(\mathcal{T})$ almost surely. By Lemma 1.41, $\|X_J(t,\omega)\|_{L^2(\mathcal{T})} \le \|X(t,\omega)\|_{L^2(\mathcal{T})}$ and hence $X_J \to X$ in $L^2(\Omega, L^2(\mathcal{T}))$ (see Exercise 4.13).

Finally, the random variables ξ_j defined by $\xi_j(\omega) = \gamma_j(\omega)/\sqrt{\nu_j}$ satisfy $\mathbb{E}[\gamma_j] = 0$ and Fubini's theorem gives

$$\begin{aligned}\operatorname{Cov}(\gamma_j,\gamma_k) &= \mathbb{E}\left[\int_{\mathcal{T}}\int_{\mathcal{T}} (X(s)-\mu(s))\phi_j(s)\big(X(t)-\mu(t)\big)\phi_k(t)\,ds\,dt\right]\\ &= \int_{\mathcal{T}}\int_{\mathcal{T}} C(s,t)\phi_j(t)\phi_k(s)\,ds\,dt = \big\langle \phi_j, \mathcal{C}\phi_k\big\rangle_{L^2(\mathcal{T})}.\end{aligned}$$

Since the eigenfunctions ϕ_j are orthonormal,

$$\big\langle \phi_j, \mathcal{C}\phi_k\big\rangle_{L^2(\mathcal{T})} = \big\langle \phi_j, \nu_k\phi_k\big\rangle_{L^2(\mathcal{T})} = \begin{cases}\nu_j, & j = k,\\ 0, & j \ne k,\end{cases}$$

and $\operatorname{Cov}(\gamma_j,\gamma_k) = 0$ when $j \ne k$. Hence, the γ_j are pairwise uncorrelated. Since $\xi_j = \gamma_j/\sqrt{\nu_j}$, we also see that ξ_j are pairwise uncorrelated and satisfy $\mathbb{E}[\xi_j] = 0$ and $\operatorname{Var}(\xi_j) = 1$. If the process X is Gaussian, the coefficients γ_j, and hence the ξ_j, have a joint Gaussian distribution, and so by Lemma 4.33 are pairwise independent. □

The truncated Karhunen–Loève expansion $X_J(t,\omega)$ introduced in (5.36) defines a stochastic process $X_J(t)$ and its covariance function is

$$C_J(s,t) = \sum_{j=1}^{J} \nu_j\phi_j(s)\phi_j(t) \tag{5.37}$$

(see Exercise 5.6). Since $\phi_j(s)\phi_k(t)$ is an orthonormal basis for $L^2(\mathcal{T}\times\mathcal{T})$, it follows that the covariance C_J converges to C in $L^2(\mathcal{T}\times\mathcal{T})$ (as in Theorem 1.65). If in addition $\mathcal{T}$ is bounded and C is continuous, then $C_J(s,t) \to C(s,t)$ and $\mathbb{E}\big[(X(t)-X_J(t))^2\big] \to 0$ uniformly for $s,t \in \mathcal{T}$.

Theorem 5.29 (uniform convergence of Karhunen–Loève) *Consider a real-valued stochastic process $X \in L^2(\Omega, L^2(\mathcal{T}))$ and let $X_J(t,\omega)$ be the stochastic process defined in* (5.36)*. If $\mathcal{T} \subset \mathbb{R}$ is a closed and bounded set and the covariance function $C \in \mathrm{C}(\mathcal{T}\times\mathcal{T})$, then $\phi_j \in \mathrm{C}(\mathcal{T})$ and the series expansion of C converges uniformly. In particular,*

$$\sup_{s,t\in\mathcal{T}} \big|C(s,t) - C_J(s,t)\big| \le \sup_{t\in\mathcal{T}} \sum_{j=J+1}^{\infty} \nu_j\phi_j(t)^2 \to 0, \qquad \text{as } J \to \infty, \tag{5.38}$$

where C_J is defined in (5.37)*, and* $\sup_{t\in\mathcal{T}} \mathbb{E}\big[(X(t)-X_J(t))^2\big] \to 0$ *as $J \to \infty$.*

Proof Mercer's theorem (Theorem 1.80) tells us that $\phi_j \in \mathrm{C}(\mathcal{T})$, the estimate (5.38) holds and that the series $C(s,t) = \sum_{j=1}^{\infty} \nu_j \phi_j(s)\phi_j(t)$ converges in $\mathrm{C}(\mathcal{T} \times \mathcal{T})$. Note that

$$\sup_{t\in\mathcal{T}} \mathbb{E}\left[\left(X(t) - X_J(t)\right)^2\right] = \sup_{t\in\mathcal{T}} \sum_{j=J+1}^{\infty} \nu_j \phi_j(t)^2. \tag{5.39}$$

Using (5.38) with $s = t$, we see that

$$\sup_{t\in\mathcal{T}} \mathbb{E}\left[\left(X(t) - X_J(t)\right)^2\right] \to 0 \qquad \text{as } J \to \infty.$$ □

The truncated expansion $X_J(t)$ provides an approximation to the stochastic process $X(t)$. For all $t \in \mathcal{T}$,

$$\mathrm{Var}(X(t)) - \mathrm{Var}(X_J(t)) = \mathbb{E}\left[\left(X(t) - X_J(t)\right)^2\right] = \sum_{j=J+1}^{\infty} \nu_j \phi_j(t)^2 \geq 0 \tag{5.40}$$

and the random variable $X_J(t)$ always underestimates the variance. Observe that (5.39) gives an error estimate for the approximation of $X(t)$ as a real-valued random variable for each $t \in \mathcal{T}$. If we treat $X(t)$ as an $L^2(\mathcal{T})$-valued random variable, a simple calculation gives

$$\|X - X_J\|^2_{L^2(\Omega, L^2(\mathcal{T}))} = \mathbb{E}\left[\sum_{j=J+1}^{\infty} \nu_j \|\phi_j\|^2_{L^2(\mathcal{T})} \xi_j^2\right] = \sum_{j=J+1}^{\infty} \nu_j \tag{5.41}$$

and

$$\|X - X_J\|^2_{L^2(\Omega, L^2(\mathcal{T}))} = \mathbb{E}\left[\|X - \mu\|^2_{L^2(\mathcal{T})} - \|X_J - \mu\|^2_{L^2(\mathcal{T})}\right] = \int_{\mathcal{T}} \mathrm{Var}\, X(t)\, dt - \sum_{j=1}^{J} \nu_j.$$

When available, the first J eigenvalues ν_j may be summed and used to determine the correct truncation parameter J for a given error tolerance. For example, in many cases $\mathrm{Var}\, X(t) = \sigma^2$ is constant on $\mathcal{T}$ and

$$\mathbb{E}\left[\|X - X_J\|^2_{L^2(\mathcal{T})}\right] = \sigma^2 \mathrm{Leb}(\mathcal{T}) - \sum_{j=1}^{J} \nu_j,$$

where σ^2 and $\mathrm{Leb}(\mathcal{T})$ are known and we may evaluate the sum to determine the size of the error.

In general, the eigenvalues and eigenfunctions are not readily available and must be approximated by numerical methods. We give an example in our treatment of random fields in §7.4. We show how to find the Karhunen–Loève expansion of the standard Brownian bridge $B(t)$, for which explicit formulae for the eigenvalues and eigenfunctions are available.

Example 5.30 (Karhunen–Loève expansion of the Brownian bridge) The covariance function of the standard Brownian bridge $B(t)$ on $\mathcal{T} = [0,1]$ is $C(s,t) = \min\{s,t\} - st$ (see Lemma 5.22). To compute the Karhunen–Loève expansion of $B(t)$, we need the eigenfunctions of

$$(\mathcal{C}f)(t) = \int_0^1 (\min\{s,t\} - st) f(s)\, ds.$$

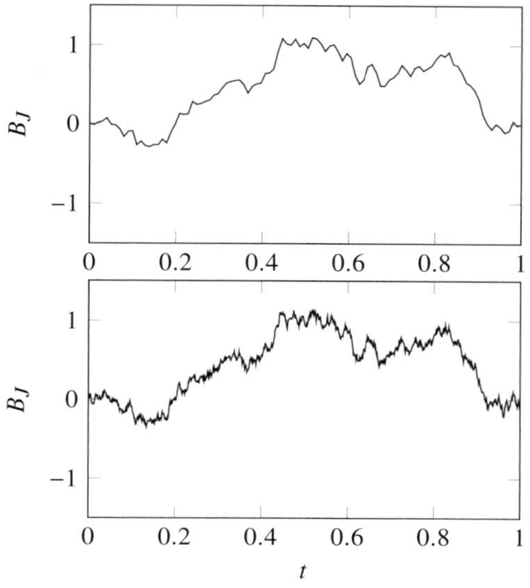

Figure 5.10 Sample path of the truncated Karhunen–Loève expansion $B_J(t)$ in (5.44) with $J = 100$ (top) and $J = 1000$ (bottom). As the index j increases, the eigenfunctions ϕ_j become more oscillatory and the eigenvalues ν_j decay. Adding more terms in the Karhunen–Loève expansion corresponds to adding oscillations of increasing frequency and decreasing amplitude. The sample path $B_J(t,\omega) \to B(t,\omega)$ in $L^2(0,1)$ as $J \to \infty$.

The eigenvalue problem for $\mathcal{C}$ is to find $\phi \in L^2(0,1)$ and $\nu \in \mathbb{R}$ such that

$$(1-t)\int_0^t s\phi(s)\,ds + t\int_t^1 (1-s)\phi(s)\,ds = \nu\phi(t).$$

By evaluating the expression at $t = 0, 1$ and differentiating twice, this is reduced to

$$-\phi(t) = \nu\phi''(t), \qquad \phi(0) = 0 = \phi(1),$$

or, equivalently with $\lambda = \nu^{-1}$,

$$\phi''(t) + \lambda\phi(t) = 0, \qquad \phi(0) = 0 = \phi(1),$$

Non-zero solutions exist only for $\lambda > 0$, in which case the general solution is

$$\phi(t) = A\cos\left(\sqrt{\lambda}t\right) + B\sin\left(\sqrt{\lambda}t\right)$$

for $A, B \in \mathbb{R}$. Imposing the boundary conditions gives $A = 0$ and $B\sin\left(\sqrt{\lambda}\right) = 0$ and non-zero solutions exist when $\sin\left(\sqrt{\lambda}\right) = 0$. Hence, the eigenvalues and (normalised) eigenfunctions are

$$\nu_j = \frac{1}{\pi^2 j^2}, \quad \phi_j(t) = \sqrt{2}\sin(j\pi t), \qquad j = 1, 2, \dots \tag{5.42}$$

The Brownian bridge can then be represented as

$$B(t) = \sum_{j=1}^{\infty} \frac{\sqrt{2}}{\pi j}\sin(j\pi t)\,\xi_j, \qquad \xi_j \sim \mathrm{N}(0,1)\ iid. \tag{5.43}$$

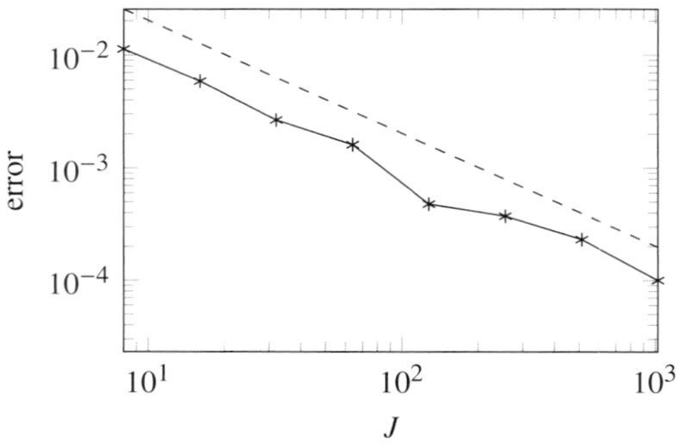

Figure 5.11 Plot of the error $\mathrm{Var}(B(t)) - \mathrm{Var}(B_J(t))$ for $t = 1/2$ (solid line) in comparison to the error bound $2/\pi^2 J$ in (5.45) (dashed line). The true variance is $\mathrm{Var}(B(t)) = C(t,t) = t - t^2$, which at $t = 1/2$ is $\mathrm{Var}(B(1/2)) = 1/4$. The variance of $B_J(1/2)$ is computed by sampling (5.44) 10^8 times.

We can truncate the Karhunen–Loève expansion (5.43) and approximate sample paths of $B(t)$ on $\mathcal{T} = [0,1]$ via

$$B_J(t) = \sum_{j=1}^{J} \frac{\sqrt{2}}{\pi j} \sin(j\pi t)\, \xi_j, \qquad \xi_j \sim \mathrm{N}(0,1)\ iid. \tag{5.44}$$

Examples are plotted in Figure 5.10. Finite series of this type are important in Chapter 6 for sampling stationary processes.

From (5.40), the error in approximating $B(t)$ by $B_J(t)$ in the mean-square sense is

$$\mathrm{Var}(B(t)) - \mathrm{Var}(B_J(t)) = \sum_{j=J+1}^{\infty} \frac{2}{\pi^2 j^2} \sin^2(j\pi t) \le \frac{2}{\pi^2 J} \tag{5.45}$$

(because $|\sin(j\pi t)| \le 1$ and $\sum_{j=J+1}^{\infty} j^{-2} \le \int_J^\infty j^{-2}\, dj = J^{-1}$). This is illustrated in Figure 5.11. We also have from (5.41) that

$$\mathbb{E}\left[\|B - B_J\|^2_{L^2(0,1)}\right] = \sum_{j=J+1}^{\infty} \frac{1}{\pi^2 j^2} \le \frac{1}{\pi^2 J},$$

which says the $L^2(\Omega, L^2(0,1))$ norm of the error behaves like $\mathcal{O}(J^{-1/2})$. The rate $J^{-1/2}$ is typical for pathwise approximations of Brownian motion. It is notable again that the error in the approximation of averages, $\mathbb{E}[\phi(B_J(t))] - \mathbb{E}[\phi(B(t))] = \mathcal{O}(1/J)$ for $\phi \in L^2(\mathbb{R})$ as given by Theorem 4.63, converges at double the rate.

5.5 Regularity of stochastic processes

Regularity is key to understanding rates of convergence of many numerical algorithms and this is true also for the numerical approximation of stochastic processes. In defining continuity for processes, we make precise the meaning of $X(s) \to X(t)$ as $s \to t$. As we saw in Definition 4.54, there are many notions of limit for random variables and this leads to different notions of regularity for stochastic processes. We study first *mean-square continuity and differentiability* (Theorems 5.32 and 5.35) for real-valued processes, which, like the Karhunen–Loève theorem, is strongly related to the covariance function. We then also study the continuity of *sample paths* (Theorem 5.40).

Definition 5.31 (mean-square continuity) A stochastic process $\{X(t)\colon t \in \mathcal{T}\}$ is *mean-square continuous* if, for all $t \in \mathcal{T}$,

$$\|X(t+h) - X(t)\|_{L^2(\Omega)} = \mathbb{E}\left[|X(t+h) - X(t)|^2\right] \to 0 \quad \text{as } h \to 0.$$

We assume in the rest of this section that $X(t)$ is a stochastic process with mean zero. In that case, the regularity of the process can be gauged solely from the regularity of the covariance function $C(s,t)$.

Theorem 5.32 *Let $\{X(t)\colon t \in \mathcal{T}\}$ be a mean-zero process. The covariance function C is continuous at (t,t) if and only if $\mathbb{E}\left[(X(t+h) - X(t))^2\right] \to 0$ as $h \to 0$. In particular, if $C \in \mathrm{C}(\mathcal{T} \times \mathcal{T})$, then $\{X(t)\colon t \in \mathcal{T}\}$ is mean-square continuous.*

Proof A straightforward calculation gives

$$\begin{aligned}\mathbb{E}\left[(X(t+h) - X(t))^2\right] &= \mathbb{E}\left[X(t+h)^2\right] - 2\mathbb{E}\left[X(t+h)X(t)\right] + \mathbb{E}\left[X(t)^2\right]\\ &= C(t+h,t+h) - 2C(t+h,t) + C(t,t).\end{aligned}$$

Clearly then, the continuity of C implies mean-square continuity. For $h,k > 0$,

$$\begin{aligned}C(t+h,t+k) - C(t,t) &= \mathbb{E}\left[\big(X(t+h) - X(t)\big)\big(X(t+k) - X(t)\big)\right]\\ &\quad + \mathbb{E}\left[\big(X(t+h) - X(t)\big)X(t)\right] + \mathbb{E}\left[\big(X(t+k) - X(t)\big)X(t)\right].\end{aligned}$$

By the Cauchy–Schwarz inequality, if $X(t)$ is mean-square continuous, the right-hand side converges to 0 as $h,k \to 0$ and hence C is continuous at (t,t). □

Apart from white noise, all the processes we have encountered so far have continuous covariance functions and are mean-square continuous. The mean-square derivative of a stochastic process, denoted $\frac{dX(t)}{dt}$, is the stochastic process defined as follows.

Definition 5.33 (mean-square derivative) A process $\{X(t)\colon t \in \mathcal{T}\}$ is *mean-square differentiable* with mean-square derivative $\frac{dX(t)}{dt}$ if, for all $t \in \mathcal{T}$, we have as $h \to 0$

$$\left\|\frac{X(t+h) - X(t)}{h} - \frac{dX(t)}{dt}\right\|_{L^2(\Omega)} = \mathbb{E}\left[\left|\frac{X(t+h) - X(t)}{h} - \frac{dX(t)}{dt}\right|^2\right]^{1/2} \to 0.$$

It can be shown that $X(t)$ is mean-square differentiable when the second derivatives of the covariance function are well defined. To prove this, we need the following preliminary result.

Lemma 5.34 *Suppose that X_h for $h \in \mathbb{R}$ is a family of real-valued random variables. Then $\mathbb{E}[X_h X_k]$ converges to a limit L as $h,k \to 0$ if and only if $\mathbb{E}\left[|X_h - X|^2\right] \to 0$ as $h \to 0$ for some random variable X.*

Proof If $\mathbb{E}[X_h X_k] \to L$ as $h,k \to 0$ then $\mathbb{E}\big[(X_h - X_k)^2\big] = \mathbb{E}\big[X_h^2 - 2X_h X_k + X_k^2\big] \to L - 2L + L = 0$ as $h,k \to 0$. As $L^2(\Omega)$ is complete, X_h has a limit, say X. Conversely, if $X_h \to X$ in mean-square, then

$$\mathbb{E}\Big[X_h X_k - XX\Big] = \mathbb{E}\Big[(X_h - X)X_k + X(X_k - X)\Big] \to 0$$

as $h,k \to 0$ by Cauchy–Schwarz. Hence, $\mathbb{E}[X_h X_k] \to \mathbb{E}\big[X^2\big] = L$ as $h,k \to 0$. □

Theorem 5.35 *Let $\{X(t)\colon t \in \mathcal{T}\}$ be a stochastic process with mean zero. Suppose that the covariance function $C \in \mathrm{C}^2(\mathcal{T} \times \mathcal{T})$. Then $X(t)$ is mean-square differentiable and the derivative $\frac{dX(t)}{dt}$ has covariance function $\frac{\partial^2 C(s,t)}{\partial s \partial t}$.*

Proof For any $s,t \in \mathcal{T}$ and real constants $h,k > 0$,

$$\begin{aligned}\mathrm{Cov}\left(\frac{X(s+h) - X(s)}{h}, \frac{X(t+k) - X(t)}{k}\right) &= \frac{1}{hk}\mathbb{E}\Big[\big(X(s+h) - X(s)\big)\big(X(t+k) - X(t)\big)\Big]\\ &= \frac{1}{hk}\big(C(s+h,t+k) - C(s+h,t) - C(s,t+k) + C(s,t)\big).\end{aligned}$$

If $C \in \mathrm{C}^2(\mathcal{T} \times \mathcal{T})$, a simple calculation with Taylor series shows that the right-hand side converges to $\frac{\partial^2 C(s,t)}{\partial s \partial t}$ as $h,k \to 0$. Let $X_h = (X(s+h) - X(s))/h$ and $X_k = (X(t+k) - X(t))/k$. Then,

$$\mathbb{E}\big[X_h X_k\big] \to \frac{\partial^2 C(s,t)}{\partial s \partial t} \qquad \text{as } h,k \to 0.$$

Lemma 5.34 now says that the limit in mean square of X_h exists for each $t \in \mathcal{T}$ and $X(t)$ is therefore mean-square differentiable and the derivative $\frac{dX(t)}{dt}$ has covariance $\frac{\partial^2 C(s,t)}{\partial s \partial t}$. □

By analogy, higher-order mean-square derivatives of a stochastic process are defined and it can be shown that, if it exists, the rth mean-square derivative $\frac{d^r X(t)}{dt^r}$ has covariance function $\frac{\partial^{2r} C(s,t)}{\partial s^r \partial t^r}$.

Example 5.36 (exponential and Gaussian covariance) For a time scale $\ell > 0$, consider

$$C_1(s,t) = \mathrm{e}^{-|s-t|/\ell}, \qquad C_2(s,t) = \mathrm{e}^{-|s-t|^2/\ell^2},$$

which are known as the exponential and Gaussian covariance functions respectively. We verify that C_1 and C_2 are valid covariance functions by using Fourier techniques in Example 6.7. To understand the mean-square differentiability of the corresponding stochastic processes, observe that

$$\frac{\partial C_1}{\partial s} = -\frac{1}{\ell}\,\mathrm{sgn}(s-t)\mathrm{e}^{-|s-t|/\ell}, \qquad \frac{\partial C_2}{\partial s} = -2\frac{1}{\ell^2}(s-t)\mathrm{e}^{-(s-t)^2/\ell^2}.$$

The first derivative of C_1 is not defined at $s = t$ and a stochastic process with exponential covariance is therefore not mean-square differentiable. However, the first derivative of C_2 is well defined for all values of s and t. We also have

$$\frac{\partial^2 C_2}{\partial s \partial t} = -4\frac{(s-t)^2}{\ell^4}\mathrm{e}^{-(s-t)^2/\ell^2} + 2\frac{1}{\ell^2}\mathrm{e}^{-(s-t)^2/\ell^2},$$

which is finite for all values of s and t. The second derivative of C_2 is well defined and the Gaussian covariance gives a mean-square differentiable process. Sample paths of Gaussian

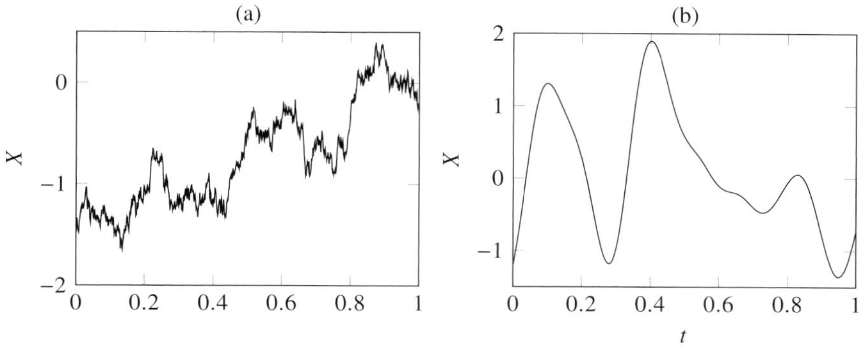

Figure 5.12 Example sample paths of the mean-zero Gaussian process with covariance (a) $C_1(s,t) = \mathrm{e}^{-|s-t|}$ and (b) $C_2(s,t) = \mathrm{e}^{-(s-t)^2/\ell^2}$ with $\ell = 1/10$. Samples of the process with exponential covariance are rougher, due to the non-differentiability of C_1 at $s = t$.

stochastic processes with covariance functions $C_1(s,t)$ and $C_2(s,t)$ are plotted in Figure 5.12 and the sample path for the Gaussian covariance is much smoother than the one for the exponential covariance.

Linear interpolation of Gaussian processes

Numerical methods sample stochastic processes $\{X(t)\colon t \in \mathcal{T}\}$ at a finite number of points $t_1,\ldots,t_N \in \mathcal{T}$ and it is common to approximate $X(t)$ between the sample points by linear interpolation. The piecewise linear interpolant $I(t)$ of $X(t)$ is defined via

$$I(t) := \frac{t - t_j}{t_{j+1} - t_j} X(t_{j+1}) + \frac{t - t_{j+1}}{t_j - t_{j+1}} X(t_j), \qquad t_j \le t < t_{j+1}. \tag{5.46}$$

$I(t)$ is a stochastic process and $I(t,\omega)$ is the piecewise linear function in t that agrees with $X(t,\omega)$ at $t = t_j$, $j = 1,\ldots,N$. It is easy to see that $\mathbb{E}[I(t)] = \mathbb{E}[X(t)]$ if $X(t)$ has a constant mean and that $I(t)$ is Gaussian if $X(t)$ is Gaussian. However, the covariance functions of the two processes are different. If $X(t)$ is mean-square differentiable, the following result shows that the covariance of $I(t)$ converges to the covariance of $X(t)$ as the grid spacing $\Delta t \to 0$. The proof makes use of the mean-square derivative of $X(t)$ from Definition 5.33.

Proposition 5.37 *Let $\{X(t)\colon t \in \mathcal{T}\}$ be a second-order mean-zero process. Let $\{I(t)\colon t \in \mathcal{T}\}$ be the process* (5.46)*, constructed using uniformly spaced points t_j with spacing Δt. Suppose that $X(t)$ is differentiable in mean square. Then, for every $s,t \in \mathcal{T}$, there exists $K, \Delta t^* > 0$ such that*

$$\left|C_X(s,t) - C_I(s,t)\right| \le K \Delta t^2, \qquad \text{for } \Delta t < \Delta t^*, \tag{5.47}$$

where C_X, C_I denote the covariance functions of $X(t)$ and $I(t)$.

Proof As $I(t)$ is linear on $t_j \le t \le t_{j+1}$ and agrees with $X(t)$ at $t = t_j, t_{j+1}$,

$$\frac{X(t) - I(t)}{t - t_j} = \frac{X(t) - X(t_j)}{t - t_j} - \frac{I(t) - I(t_j)}{t - t_j} = \frac{X(t) - X(t_j)}{t - t_j} - \frac{X(t_{j+1}) - X(t_j)}{t_{j+1} - t_j}.$$

From Definition 5.33, as $\Delta t \to 0$ with $t_j \le t \le t_{j+1}$,

$$\frac{X(t) - X(t_j)}{t - t_j} - \frac{X(t_{j+1}) - X(t)}{t_{j+1} - t} \to \frac{dX(t)}{dt} - \frac{dX(t)}{dt} = 0 \qquad \text{in } L^2(\Omega).$$

Hence, $\frac{X(t)-I(t)}{t-t_j} \to 0$ in $L^2(\Omega)$ as $\Delta t \to 0$. Thus, for all $t \in \mathcal{T}$, we can find $K, \Delta t^* > 0$ such that

$$\left\|X(t) - I(t)\right\|_{L^2(\Omega)} \le K\Delta t, \qquad \text{for } \Delta t < \Delta t^*. \tag{5.48}$$

As $X(t)$ and $I(t)$ have zero mean, the Cauchy–Schwarz inequality using $C_X(s,t) = \langle X(s), X(t)\rangle_{L^2(\Omega)}$ gives

$$\begin{aligned}\left|C_X(s,t) - C_I(s,t)\right| &= \left|\left\langle X(s) - I(s), X(t)\right\rangle_{L^2(\Omega)} + \left\langle I(s), X(t) - I(t)\right\rangle_{L^2(\Omega)}\right| \\ &\le \left\|X(s) - I(s)\right\|_{L^2(\Omega)}\left\|X(t)\right\|_{L^2(\Omega)} + \left\|I(s)\right\|_{L^2(\Omega)}\left\|X(t) - I(t)\right\|_{L^2(\Omega)}.\end{aligned}$$

The estimate (5.47) follows from (5.48). □

Corollary 5.38 (weak convergence of interpolants) *Suppose that $\{X(t) \colon t \in \mathcal{T}\}$ is a mean-zero Gaussian process and its covariance function $C_X \in \mathrm{C}^2(\mathcal{T} \times \mathcal{T})$ and is positive definite. Let $\{I(t) \colon t \in \mathcal{T}\}$ be the piecewise linear interpolant* (5.46), *constructed using uniformly spaced points t_j with spacing Δt. Let $s_1, \ldots, s_m \in \mathcal{T}$ (not necessarily equal to the t_j) and define*

$$\boldsymbol{X} = [X(s_1), \ldots, X(s_m)]^{\mathsf{T}} \qquad \boldsymbol{I} = [I(s_1), \ldots, I(s_m)]^{\mathsf{T}}.$$

Then, we can find $K, \Delta t^* > 0$ such that for $\Delta t \le \Delta t^*$

$$\left|\mathbb{E}\left[\phi(\boldsymbol{X})\right] - \mathbb{E}\left[\phi(\boldsymbol{I})\right]\right| \le K\|\phi\|_{L^2(\mathbb{R}^m)}\Delta t^2, \qquad \forall \phi \in L^2(\mathbb{R}^m). \tag{5.49}$$

Proof As $C_X \in \mathrm{C}^2(\mathcal{T} \times \mathcal{T})$, $X(t)$ is mean-square differentiable and Proposition 5.37 applies. The vectors $\boldsymbol{X}$ and $\boldsymbol{I}$ are multivariate Gaussian random variables and their covariance matrices satisfy $\|C_X - C_I\|_{\mathrm{F}} = \mathcal{O}(\Delta t^2)$. Finally, Theorem 4.63 gives (5.49). □

Sample path continuity

Mean-square continuity and differentiability is measured using the expectation; that is, by an average over all sample paths. We now consider *sample path continuity* of a stochastic process, which addresses the continuity of each sample path individually. This is a delicate issue and, for example, it is not possible to show that every sample path of a Gaussian process is continuous. Even though the distribution of a Gaussian process is uniquely defined on $\mathcal{B}(\mathbb{R}^{\mathcal{T}})$, sample path continuity cannot hold in general. $\mathcal{B}(\mathbb{R}^{\mathcal{T}})$ is constructed from a countable set of conditions and cannot determine continuity, which depends on a continuous parameter (and therefore an uncountable set of conditions). Instead we introduce, under a condition on the moments of the increments $X(t) - X(s)$, a 'version of $X(t)$', which does have continuous sample paths (see Theorem 5.40).

Definition 5.39 (version) A stochastic process $\{Y(t) \colon t \in \mathcal{T}\}$ is a *version* or *modification* of a process $\{X(t) \colon t \in \mathcal{T}\}$ if $\mathbb{P}(X(t) = Y(t)) = 1$ for all $t \in \mathcal{T}$. We say $\{X(t) \colon t \in \mathcal{T}\}$ has a *continuous version* if there exists a version of $X(t)$ with continuous sample paths.

Theorem 5.40 *Let $\mathcal{T} \subset \mathbb{R}$ be a closed interval and $\{X(t)\colon t \in \mathcal{T}\}$ be a stochastic process such that, for some $p,r,K > 0$,*

$$\mathbb{E}\Big[\big|X(t) - X(s)\big|^p\Big] \le K|t-s|^{1+r}, \qquad \forall s,t \in \mathcal{T}. \tag{5.50}$$

Then, there exists a continuous version of $X(t)$.

Before giving the proof, we show how to apply the theorem to fBm.

Example 5.41 (regularity of fBm) We have seen in §5.3 that fBm $B^H(t)$ has the covariance function $C(s,t) = \big(|t|^{2H} + |s|^{2H} - |t-s|^{2H}\big)/2$ and so

$$\frac{\partial^2 C}{\partial s \partial t} = -H(2H-1)|t-s|^{2H-2}.$$

The second derivative is infinite at $t = s$ for $H < 1$ and so none of the corresponding fractional Brownian motions has a mean-square derivative. The increments of fBm are Gaussian random variables with mean zero and variance $|t-s|^{2H}$ and Exercise 4.10 gives

$$\mathbb{E}\Big[\big|B^H(t) - B^H(s)\big|^{2n}\Big] = \frac{(2n)!}{2^n n!}|t-s|^{2nH}. \tag{5.51}$$

Hence if $n > 1/2H$, Theorem 5.40 applies with $p = \lceil 1/H \rceil$ and $r = pH - 1 > 0$. We conclude that fBm has a continuous version for any $H \in (0,1)$.

Theorem 5.40 also finishes the proof of existence of Brownian motion.

Corollary 5.42 (existence of Brownian motion) *The Gaussian process $\{X(t)\colon t \in \mathbb{R}^+\}$ with mean zero and covariance function $\min\{s,t\}$ has a continuous version $W(t)$. Hence, Brownian motion exists.*

Proof $X(t)$ is a fBm with $H = 1/2$. From the previous example, we conclude there exists a version $W(t)$ of $X(t)$ that is continuous on $\mathbb{R}^+$. The version $W(t)$ has the same finite-dimensional distributions as $X(t)$ and so satisfies the conditions to be a Brownian motion (Definition 5.11). □

The following technical lemma is required for the proof of Theorem 5.40.

Lemma 5.43 *Suppose that $\{X(t)\colon t \in \mathcal{T}\}$ and $\{Y(t)\colon t \in \mathcal{T}\}$ are stochastic processes such that*

(i) *$X(t) = Y(t)$ almost surely for t in a dense subset of $\mathcal{T}$,*
(ii) *all sample paths of $Y(t)$ are continuous, and*
(iii) *for some $p,r,K > 0$,*

$$\mathbb{E}\Big[\big|X(t) - X(s)\big|^p\Big] \le K|t-s|^r, \qquad s,t \in \mathcal{T}.$$

Then, $Y(t)$ is a continuous version of $X(t)$.

Proof Fix $t \in \mathcal{T}$. Because $X(s) = Y(s)$ for s in a dense subset of $\mathcal{T}$, we may choose t_n for $n \in \mathbb{N}$ such that $X(t_n) = Y(t_n)$ a.s. and such that $|t - t_n| \le 2^{-n}$. Using Chebyshev's inequality (Lemma 4.55), we obtain

$$\mathbb{P}\Big(|Y(t_n) - X(t)| \ge \epsilon\Big) = \mathbb{P}\Big(|X(t_n) - X(t)| \ge \epsilon\Big) \le \frac{K2^{-nr}}{\epsilon^p}.$$

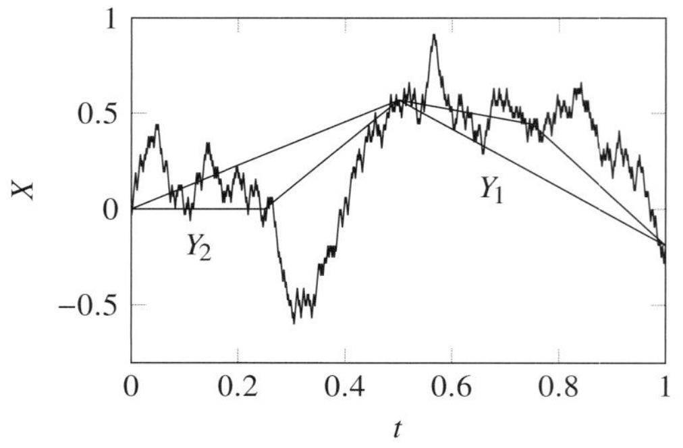

Figure 5.13 A sample path of a stochastic process $X(t)$ and the corresponding piecewise-linear interpolating processes $Y_1(t)$ and $Y_2(t)$. $Y_1(t)$ interpolates $X(t)$ at $t_0 = 0, t_1 = 1/2$ and $t_2 = 1$, while $Y_2(t)$ interpolates $X(t)$ at $t_0 = 0$, $t_1 = 1/4$, $t_2 = 1/2$, $t_3 = 3/4$, and $t_4 = 1$.

Let $\epsilon_n = 2^{-nr/2p}$ and $F_n = \{\omega \in \Omega\colon |Y(t_n) - X(t)| \geq \epsilon_n\}$. Then $\mathbb{P}(F_n) \leq K2^{-nr/2}$ and $\sum_{n=1}^{\infty} \mathbb{P}(F_n) < \infty$. By the Borel–Cantelli lemma (Lemma 4.57), there exists an $\mathbb{N}$-valued random variable n^* such that $|Y(t_n) - X(t)| < \epsilon_n$ almost surely for $n > n^*(\omega)$. By taking $\epsilon_n \to 0$ and noting that sample paths of Y are continuous, we see $X(t) = Y(t)$ a.s. and the proof is complete. □

Proof of Theorem 5.40 We assume that $\mathcal{T} = [0, 1]$; the extension to general $\mathcal{T}$ is discussed at the end of the proof. For $n \in \mathbb{N}$, divide the interval $[0, 1]$ into 2^n pieces and let $t_j = j2^{-n}$ for $j = 0, 1, \ldots, 2^n$. The set of points $\{t_j\}$ is known as a dyadic grid and the grids are nested for increasing values of n. Let $Y_n(t)$ be the piecewise linear interpolant of the process $X(t)$ defined via (5.46) on the grid $\{t_j\}$. As the dyadic grids are nested, the grid points used to construct $Y_n(t)$ are a subset of those used to construct $Y_{n+1}(t)$. See Figure 5.13.

For a fixed n, the process $Y_n(t)$ can be viewed as a $\mathrm{C}([0, 1])$-valued random variable, which we denote Y_n. The main part of the proof is to show that the random variables Y_n converge almost surely as $n \to \infty$ to a $\mathrm{C}([0, 1])$-valued random variable Y. The limit Y defines a stochastic process $Y(t)$ with continuous sample paths. Because $Y_m(j2^{-n}) = Y_n(j2^{-n})$ for $m \geq n$, the process $Y(t)$ satisfies $Y(j2^{-n}) = Y_n(j2^{-n}) = X(j2^{-n})$ almost surely. Then, Lemma 5.43 implies $Y(t)$ is a continuous version of $X(t)$.

We show that Y_n converges almost surely to a limit point Y in $\mathrm{C}([0, 1])$ as follows. As sample paths of $Y_n(t)$ and $Y_{n+1}(t)$ are linear on the interval $[j2^{-n}, (j + 1)2^{-n}]$ and agree at the end points, the largest difference occurs at the midpoint, $\hat{t}$. That is,

$$\max_{j2^{-n} \leq t \leq (j+1)2^{-n}} \left|Y_{n+1}(t) - Y_n(t)\right| = \left|Y_{n+1}(\hat{t}) - Y_n(\hat{t})\right|.$$

Replace $Y_n(\hat{t})$ by the average of the endpoints:

$$\begin{aligned}\max_{j2^{-n} \leq t \leq (j+1)2^{-n}} \left|Y_{n+1}(t) - Y_n(t)\right| &= \left|Y_{n+1}(\hat{t}) - \tfrac{1}{2}\big(Y_n(j2^{-n}) + Y_n((j + 1)2^{-n})\big)\right| \\ &\leq \tfrac{1}{2}\Big(\left|Y_{n+1}(\hat{t}) - Y_n(j2^{-n})\right| + \left|Y_{n+1}(\hat{t}) - Y_n((j + 1)2^{-n})\right|\Big).\end{aligned}$$

As $Y_n(t)$ agrees with $X(t)$ at $t = j2^{-n}$ and $t = (j + 1)2^{-n}$, and $Y_{n+1}(t)$ agrees with $X(t)$ at $\hat{t}$, we see that

$$\mathbb{P}\Big(\max_{j2^{-n}\le t\le (j+1)2^{-n}} |Y_n(t)-Y_{n+1}(t)|\ge \epsilon\Big) \le \mathbb{P}\big(|X(\hat t)-X(j2^{-n})|\ge\epsilon\big) + \mathbb{P}\big(|X(\hat t)-X((j+1)2^{-n})|\ge\epsilon\big).$$

By Chebyshev's inequality and (5.50), for any $s\in[0,1]$,

$$\mathbb{P}\Big(|X(\hat t)-X(s)|\ge\epsilon\Big) \le \frac{\mathbb{E}\big[|X(\hat t)-X(s)|^p\big]}{\epsilon^p} \le K\frac{|\hat t-s|^{1+r}}{\epsilon^p}.$$

Therefore,

$$\mathbb{P}\Big(\max_{j2^{-n}\le t\le (j+1)2^{-n}} |Y_n(t)-Y_{n+1}(t)|\ge \epsilon\Big) \le 2K\frac{2^{-(n+1)(1+r)}}{\epsilon^p}.$$

There are 2^n intervals of width 2^{-n} in $[0,1]$ and hence

$$\begin{aligned}\mathbb{P}\Big(\|Y_n-Y_{n+1}\|_\infty\ge\epsilon\Big) &\le \sum_{j=0}^{2^n-1}\mathbb{P}\Big(\max_{j2^{-n}\le t\le (j+1)2^{-n}} |Y_n(t)-Y_{n+1}(t)|\ge \epsilon\Big)\\ &\le 2^{n+1}K\frac{2^{-(n+1)(1+r)}}{\epsilon^p} = \frac{K}{\epsilon^p}2^{-(n+1)r}.\end{aligned}$$

With $\epsilon_n = 2^{-(n+1)r/2p}$, $\mathbb{P}\big(\|Y_n-Y_{n+1}\|_\infty\ge\epsilon_n\big)\le K2^{-(n+1)r/2}$. Let F_n be the event that $\|Y_n-Y_{n+1}\|_\infty\ge\epsilon_n$. Then $\sum_{n=1}^\infty \mathbb{P}(F_n)<\infty$ and the Borel–Cantelli lemma (Lemma 4.57) applies. Thus, there exists an $\mathbb{N}$-valued random variable n^* such that for $n>n^*(\omega)$

$$\big\|Y_n(\omega)-Y_{n+1}(\omega)\big\|_\infty\le\epsilon_n,\quad \text{almost surely.}$$

Note that $\sum_n\epsilon_n<\infty$ as it is a geometric series for $r,p>0$. Consequently, the series $Y_n = Y_0+\sum_{k=1}^n(Y_k-Y_{k-1})$ converges absolutely in $\mathrm{C}([0,1])$ for almost all ω. Let

$$Y = \begin{cases}\lim_{n\to\infty} Y_0+\sum_{k=1}^n(Y_k-Y_{k-1}), & \text{the sum converges,}\\ X(0)+t\,X(1), & \text{otherwise.}\end{cases}$$

Note that $Y(t)=X(t)$ for $t=0,1$. Then Y is a well defined $\mathrm{C}([0,1])$ random variable such that $Y_n\to Y$ in $\mathrm{C}([0,1])$ almost surely.

It remains only to consider other possibilities for $\mathcal{T}$. To treat bounded intervals, we can rescale time to map $\{X(t)\colon t\in\mathcal{T}\}$ onto a process on $[0,1]$. The condition (5.50) is still satisfied. For unbounded intervals, divide the interval into a countable union of bounded subintervals and construct the continuous version on each subinterval. The version constructed agrees with X at the end points of the subintervals and hence the version is continuous on the whole interval. □

5.6 Notes

Stochastic processes are described in depth in (Cramér and Leadbetter, 2004; Grimmett and Stirzaker, 2001; Loève, 1977, 1978; Yaglom, 1962) and are discussed from a more applied viewpoint in (Chorin and Hald, 2006; Gardiner, 2009; Öttinger, 1996; van Kampen, 1997). Nelson (1967) describes Brownian motion and its historical development.

The stochastic process $Y_N(t)$ introduced in (5.3) has continuous sample paths and defines a family of probability measures μ_N on $\mathrm{C}(\mathbb{R}^+)$. The measures μ_N converge to the measure on $\mathrm{C}(\mathbb{R}^+)$ defined by Brownian motion (known as the *Donsker invariance principle*). More details can be found in (Karatzas and Shreve, 1991, p. 70; Fristedt and Gray, 1997, p. 374; Rogers and Williams, 2000, p. 16). The backward Kolmogorov equation mentioned in Example 5.14 is developed in more generality in Øksendal (2003) and Brownian motion $W(t)$ is studied in detail in (Karatzas and Shreve, 1991; Mörters and Peres, 2010).

The theory of second-order stochastic processes is developed in Cramér and Leadbetter (2004) and as a special case of the theory of second-order random fields in (Abrahamsen, 1997; Schlather, 1999). The Daniel–Kolmogorov theorem (often called the Kolmogorov extension theorem) is described in (Breiman, 1992, p. 252; Cramér and Leadbetter, 2004, p. 33; Rogers and Williams, 2000, p. 123).

Fractional Brownian motion was introduced by Kolmogorov (1940) and popularised in Mandelbrot and van Ness (1968); see also Sottinen (2003). An efficient method for sampling fBm is the circulant embedding method (see Chapter 6 for details) applied to the increment process; see Exercise 6.23.

In Definition 5.25, we introduced the terminology *spectral decomposition*. This result is often also called the *(symmetric) Schur decomposition* or the *eigendecomposition*. For square $N \times N$ matrices A, we always have $AV = VD$ where D is the diagonal matrix of eigenvalues and V is an $N \times N$ matrix whose columns are eigenvectors. For real-valued symmetric matrices A (such as the covariance matrices studied in this chapter), the eigenvalues are real and there are N linearly independent eigenvectors which are orthogonal and can be chosen to have norm one, so that V^{-1} exists and, moreover, $V^{-1} = V^{\mathsf{T}}$. In that case, $A = VDV^{-1} = VDV^{\mathsf{T}}$ and the eigendecomposition coincides with our definition of the spectral decomposition. For more details about matrix decompositions, see Golub and Van Loan (2013).

The Karhunen–Loève theorem can be found in Loève (1978) under the heading proper orthogonal decomposition theorem for the case of bounded domains and mean-square continuous processes (or equivalently mean-zero processes with continuous covariance functions). It is widely used in data analysis and signal processing.

The theory of mean-square and sample path regularity is described in Cramér and Leadbetter (2004). Theorem 5.40 is given as the Kolmogorov–Centsov theorem in Karatzas and Shreve (1991). We revisit pathwise regularity in the context of Gaussian random fields in Theorem 7.68.

Exercises

5.1 In the proof of Lemma 5.6, show that $\frac{1}{\sqrt{N}} X_{\lfloor tN \rfloor} \to \mathrm{N}(0,t)$ in distribution and $\frac{tN - \lfloor tN \rfloor}{\sqrt{N}} \xi_{\lfloor tN \rfloor} \to 0$ in probability as $N \to \infty$.

5.2 Using (5.4) for fixed $t \in \mathbb{R}^+$, show that $Y_N(t) \in L^2(\Omega)$ for each $N \in \mathbb{N}$, but that $Y_N(t)$ is not a Cauchy sequence and has no limit in $L^2(\Omega)$.

5.3 Let $\xi_{-1} \sim \mathrm{N}(0,1)$ and $Y_0(t)$ be the linear function on $[0,1]$ with $Y_0(0) = 0$ and $Y_0(1) = \xi_{-1}$. For $j \in \mathbb{N}$ and $N = 0,1,2,\ldots$, let $t_j^N = 2^{-N} j$ and let $Y_N(t)$ be the piecewise linear function such that

$$Y_{N+1}\left(t_j^N\right) = Y_N\left(t_j^N\right), \qquad j = 0,\ldots,2^N,$$
$$Y_{N+1}\left(t_{2j+1}^{N+1}\right) = \frac{1}{2}\left(Y_N\left(t_j^N\right) + Y_N\left(t_{j+1}^N\right)\right) + \xi_{j,N},$$

where $\xi_{j,N} \sim \mathrm{N}(0, 2^{-N-2})$ *iid* (and independent of ξ_{-1}).

a. Prove that $\mathbb{E}\left[Y_N\left(t_j^N\right)\right] = 0$ and

$$\mathrm{Cov}\left(Y_N\left(t_j^N\right), Y_N\left(t_k^N\right)\right) = \min\left\{t_j^N, t_k^N\right\}, \qquad j,k = 0,\ldots,2^N.$$

b. For any $s_1,\ldots,s_M \in [0,1]$, prove that $[Y_N(s_1),\ldots,Y_N(s_M)]^\mathsf{T}$ convergence in distribution to $\mathrm{N}(0,C)$, where C has entries $c_{ij} = \min\{s_i, s_j\}$.

c. Prove that $Y_N(t)$ has a limit in $L^2(\Omega)$ as $N \to \infty$.

This is known as the *Lévy construction* of Brownian motion.

5.4 Show that the solution of the free space heat equation (5.9) can be written as (5.13).

5.5 Show that the stochastic process $X(t) = \xi_1 \cos t + \xi_2 \sin t$ from (5.16), where $\xi_1, \xi_2 \sim \mathrm{N}(0,1)$ *iid*, is a Gaussian process. Derive the covariance function $C(s,t)$ of $X(t)$.

5.6 Show that $C_J(s,t)$ in (5.37) is the covariance function of the stochastic process $X_J(t)$ defined in (5.36).

5.7 Let $W(t)$ be a Brownian motion and define $B(t) = W(t) - tW(1)$ for $0 \le t \le 1$. Prove that $B(t)$ is a Brownian bridge subject to $B(0) = B(1) = 0$.

5.8 By following Example 4.51 with $\boldsymbol{X} = [W(T), W(s), W(t)]^\mathsf{T}$ subject to $W(T) = b$, derive the mean and covariance of the Brownian bridge $B(t)$ on $[0,T]$ subject to conditions $B(0) = 0$ and $B(T) = b$.

5.9 Repeat the previous exercise, this time following the proof of Lemma 5.22. Further prove that $B(t) = bt/T + (1 - t/T)W(t)$ is a Brownian bridge subject to $B(0) = 0$ and $B(T) = b$.

5.10 If $X(t)$ is a stochastic process, we say it has *long-range dependence* if

$$\sum_{n=1}^{\infty} \mathbb{E}\left[\Delta(1)\Delta(n)\right] = \infty$$

where $\Delta(n) = X(n) - X(n-1)$. Prove that the fBm $B^H(t)$ has long-range dependence for $H < 1/2$.

5.11 Prove that the covariance function $C(s,t) = \frac{1}{2}\left(t^{2H} + s^{2H} - |t-s|^{2H}\right)$ is non-negative definite. Hint: expand $\phi(\lambda) = \mathrm{e}^{-\epsilon|\lambda|^\alpha}$ in ϵ up to $\mathcal{O}(\epsilon^2)$. You may assume that $\phi(\lambda)$ is a positive-definite function for any $\epsilon > 0$ and $\alpha \in (0,2]$.

5.12 Show that the Karhunen–Loève expansion for Brownian motion on the interval $[0, 1]$ is

$$W(t) = \sum_{j=0}^{\infty} \frac{\sqrt{2}}{(j + 1/2)\pi} \sin\bigl(t(j + 1/2)\pi\bigr)\, \xi_j,$$

with *iid* $\xi_j \sim \mathrm{N}(0, 1)$. Write a MATLAB code to sample Brownian motion on $[0, 1]$ by truncating its Karhunen–Loève expansion.

5.13 Let $\{W(t) \colon t \in \mathbb{R}^+\}$ be a Brownian motion and define $Y_N(t)$ to be the piecewise linear interpolant of $W(t)$ on the grid points $t_k = k2^{-N}$ for $k \in \mathbb{N}$. For $t \geq 0$, prove that $Y_n(t) \to W(t)$ almost surely as $n \to \infty$ and, for any $\epsilon > 0$, that there exists a random variable K such that $|Y_n(t) - W(t)| \leq K\Delta t^{1/2-\epsilon}$. Hint: use Theorem 4.58.

6
Stationary Gaussian Processes

Equilibria such as fixed points and periodic orbits are key to understanding dynamical systems. For example, the solution $u(t) = \mathrm{e}^{-\lambda t} u_0$ of the deterministic initial-value problem

$$\frac{du}{dt} = -\lambda u, \qquad u(0) = u_0 \in \mathbb{R},$$

with parameter $\lambda > 0$, converges to zero as $t \to \infty$. The fixed point $u = 0$ describes the long-time behaviour of the system. For stochastic processes, equilibrium behaviours are described by so-called *stationary* processes. For example, the stochastic process $u(t)$ that solves the initial-value problem

$$\frac{du}{dt} = -\lambda u + \sqrt{2}\zeta(t), \qquad u(0) = u_0,$$

with white noise forcing $\zeta(t)$ (see Example 8.1) converges as $t \to \infty$ to a Gaussian process $X(t)$ with mean zero and exponential covariance $C(s,t) = \mathrm{e}^{-|s-t|}$. $X(t)$ is an example of a stationary process.

The aim of this chapter is to outline the theory of stationary Gaussian processes and to develop numerical methods for generating sample paths. In §6.1, we study real-valued stationary processes and show that, for a wide class of such processes, the covariance may be written as a Fourier integral. Since the Fourier transform is the main technique for studying stationary processes, we introduce complex-valued random variables and complex-valued Gaussian stochastic processes in §6.2. We develop stochastic integrals with respect to Brownian motion and the stochastic Fourier integral representation of complex-valued stationary Gaussian processes in §6.3. In §6.4, we apply quadrature to the Fourier integral and find an approximate sampling method. Finally, in §6.5, we describe the circulant embedding method, which provides an efficient way to sample stationary Gaussian processes exactly.

6.1 Real-valued stationary processes

We begin with a formal definition of stationary processes. Our main interest is in stationary processes that are *real valued*, though we will see advantages in working with complex-valued processes in §6.3. Recall, first, Definition 5.2 of a stochastic process.

Definition 6.1 (stationary process) A stochastic process $\{X(t) : t \in \mathbb{R}\}$ is *stationary* if the mean $\mu(t)$ is independent of t and the covariance function $C(s,t)$ depends only on the difference $s - t$. We then write $C(s,t) = C(s - t, 0) = c(s - t)$, for some function $c(t)$ known as the *stationary covariance function*.

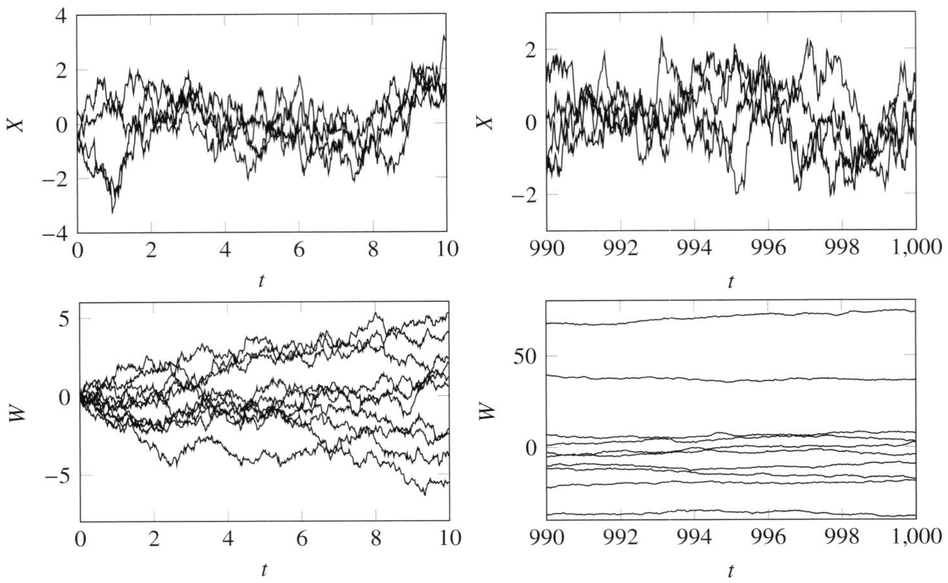

Figure 6.1 Sample paths of the Gaussian process $X(t)$ with $\mu(t) = 0$ and $C(s,t) = e^{-|s-t|}$ (top) and sample paths of Brownian motion $W(t)$ (bottom).

Note that $c(t)$ is an even function as $C(s,t) = C(t,s)$ and stationary processes are defined for $t \in \mathcal{T} = \mathbb{R}$ to allow for translations $t + h$ of t for any $h \in \mathbb{R}$. We may also show (see Exercise 6.1) that $c(t)$ is non-negative definite in the following sense.

Definition 6.2 (non-negative definite) A function $c\colon \mathbb{R} \to \mathbb{C}$ is *non-negative definite* if, for any $a_1,\dots,a_n \in \mathbb{C}$ and $t_1,\dots,t_n \in \mathbb{R}$,

$$\sum_{i,j=1}^{n} a_i c(t_i - t_j)\bar{a}_j \geq 0. \tag{6.1}$$

It is *positive definite* if the inequality is strict for $a_1,\dots,a_n \neq 0$.

For Gaussian processes, Definition 6.1 also implies that the distribution of the process is invariant to translations in t. Of the examples studied in Chapter 5, the mean-zero Gaussian processes with covariance $C(s,t) = \delta(s-t)$ (white noise), the cosine covariance $\cos(s-t)$, and the exponential covariance $e^{-|s-t|}$, all have distributions that do not change when t is translated and provide important examples of real-valued Gaussian stationary processes.

Brownian motion $W(t)$ is Gaussian, but not stationary. Figure 6.1 illustrates this well. The figure shows sample paths of both $W(t)$ and the stationary Gaussian process $X(t)$ with mean $\mu(t) = 0$ and covariance $C(s,t) = e^{-|s-t|}$. The behaviour of $X(t)$ on the intervals $[0,10]$ and $[990,1000]$ is similar. In contrast, Brownian motion, which has covariance $C(s,t) = \min\{s,t\}$, looks very different on the two intervals. The variance of $W(t)$ increases linearly as t increases (see Definition 5.11) and so the distribution changes when t is shifted.

Theorem 5.18 says that the set of symmetric, non-negative definite functions coincides with the set of covariance functions of second-order stochastic processes. A similar classification holds for covariance functions of stationary processes that are mean-square continuous. In particular, the Wiener–Khintchine theorem (Theorem 6.5) says that stationary covariance functions $c(t)$ of such processes may be written as Fourier integrals. The proof depends on the following result from analysis.

Theorem 6.3 (Bochner) *A function $c\colon \mathbb{R} \to \mathbb{C}$ is non-negative definite and continuous if and only if*

$$c(t) = \int_{\mathbb{R}} e^{itv}\, dF(v),$$

for some measure F on $\mathbb{R}$ that is finite (i.e., $F(\mathbb{R}) < \infty$).

See Exercise 6.3 for a partial proof. Theorem 6.3 includes the case of real-valued stationary covariance functions $c(t)$.

Example 6.4 Let $c(t) = \int_{\mathbb{R}} e^{itv}\, dF(v)$ where $dF(v) = f(v)v$ and $f(v) = \frac{1}{2}(\delta(1-v) + \delta(1+v))$ and δ is the Dirac delta function. Since $f(v)$ is even,

$$c(t) = \int_{\mathbb{R}} \cos(tv)\frac{1}{2}(\delta(1-v) + \delta(1+v))\, dv = \frac{1}{2}(\cos t + \cos(-t)) = \cos t.$$

This is the cosine covariance function from Example 5.20, which we proved is non-negative definite. Note also that $F(\mathbb{R}) = \int_{\mathbb{R}} 1 dF(v) = c(0) = 1 < \infty$.

Theorem 6.5 (Wiener–Khintchine) *The following statements are equivalent:*

(i) *There exists a mean-square continuous, real-valued stationary process $X(t)$ with stationary covariance function $c(t)$.*
(ii) *The function $c\colon \mathbb{R} \to \mathbb{R}$ is such that*

$$c(t) = \int_{\mathbb{R}} e^{ivt}\, dF(v), \tag{6.2}$$

for some measure F on $\mathbb{R}$ with $F(\mathbb{R}) < \infty$.

Proof Assume that (ii) holds. By Bochner's theorem, $c(t)$ and hence $C(s,t) = c(s-t)$ are non-negative definite and continuous. Further, $C(s,t)$ is symmetric and real valued because $c(t)$ is even (see Exercise 6.2) and real valued. By Theorem 5.18, there exists a real-valued mean-zero stochastic process $X(t)$ with covariance $C(s,t)$. As $C(s,t)$ is continuous, $X(t)$ is mean-square continuous by Theorem 5.32. We have constructed a mean-square continuous, real-valued process with stationary covariance $c(t)$.

Assume that (i) holds. As $X(t)$ is mean-square continuous, the covariance $C(s,t)$ and $c(s-t) = C(s,t)$ are continuous. As $c(t)$ is also non-negative definite, Bochner's theorem applies and this gives the representation (6.2) of the stationary covariance function. □

In the Wiener–Khintchine theorem, F is known as the *spectral distribution* and, if it exists, the density f of F is the *spectral density function*. From (6.2),

$$c(t) = \int_{\mathbb{R}} e^{ivt} f(v)\, dv. \tag{6.3}$$

or $c(t) = \sqrt{2\pi}\check{f}(t)$ in terms of the inverse Fourier transform (1.42).

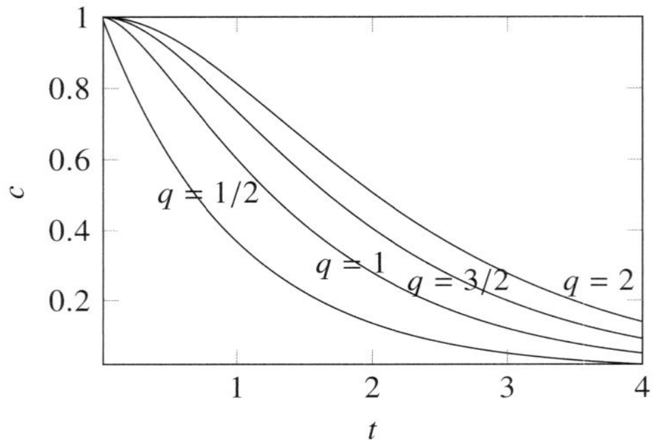

Figure 6.2 The Whittle–Matérn covariance $c_q(t) = |t|^q K_q(|t|)/2^{q-1}\Gamma(q)$ for $q = 1/2, 1, 3/2, 2$.

If $f : \mathbb{R} \to \mathbb{R}^+$ is integrable (i.e., $\int_{\mathbb{R}} f(\nu)\,d\nu < \infty$), then f is the density of a measure F with $F(\mathbb{R}) < \infty$. By the Wiener–Khintchine theorem, there exists a stationary process $X(t)$ with covariance $c(t)$ given by (6.3). For $c(t)$ to be real valued, f must be an *even* function since (6.3) gives

$$c(t) = \int_{\mathbb{R}} \cos(\nu t) f(\nu)\,d\nu + \mathrm{i} \int_{\mathbb{R}} \sin(\nu t) f(\nu)\,d\nu.$$

Alternatively, given a function $c : \mathbb{R} \to \mathbb{R}$, we may compute $f(\nu) = \hat{c}(\nu)/\sqrt{2\pi}$ and, if f is non-negative and integrable, the Wiener–Khintchine theorem says that $c(t)$ is a valid covariance function.

Example 6.6 (exponential covariance) Let $c(t) = \mathrm{e}^{-|t|/\ell}$, where $\ell > 0$ is a constant. Using (1.44), the spectral density $f(\nu)$ is

$$f(\nu) = \frac{\hat{c}(\nu)}{\sqrt{2\pi}} = \frac{1}{2\pi} \int_{-\infty}^{\infty} \mathrm{e}^{-\mathrm{i}t\nu} c(t)\,dt = \frac{\ell}{\pi(1 + \ell^2 \nu^2)}. \tag{6.4}$$

Here, f is non-negative and, in fact, is the density of Cauchy$(0, 1/\ell)$ (see Example 4.11), which defines the spectral distribution F. Clearly, $F(\mathbb{R}) = c(0) = 1 < \infty$. The Wiener–Khintchine theorem implies that $c(t)$ is a valid covariance function and is associated with a well-defined stationary process.

Example 6.7 (Gaussian covariance) Let $c(t) = \mathrm{e}^{-(t/\ell)^2}$, for some $\ell > 0$. Using (1.45),

$$f(\nu) = \frac{\hat{c}(\nu)}{\sqrt{2\pi}} = \frac{1}{2\pi} \int_{-\infty}^{\infty} \mathrm{e}^{-\mathrm{i}t\nu} \mathrm{e}^{-(t/\ell)^2}\,dt = \frac{\ell}{\sqrt{4\pi}} e^{-\ell^2 \nu^2/4}.$$

This time, we see f is the density of the Gaussian distribution $\mathrm{N}(0, 2/\ell^2)$ (hence the name Gaussian covariance) and again $F(\mathbb{R}) = c(0) = 1 < \infty$ and the spectral distribution F is a probability measure. The covariance $C(s,t) = \mathrm{e}^{-|s-t|^2/\ell^2} = c(s - t)$ was introduced in Example 5.36 without checking it is non-negative definite. By the Wiener–Khintchine theorem, we now know that the Gaussian covariance is well defined.

Example 6.8 (Whittle–Matérn covariance) For $q > 0$, let

$$c_q(t) := \frac{|t|^q}{2^{q-1}\Gamma(q)} K_q(|t|), \tag{6.5}$$

where $\Gamma(q)$ is the gamma function and $K_q(t)$ is the modified Bessel function of order q. The function $c_q(t)$ is known as the *Whittle–Matérn covariance* and is plotted for $q = 1/2$, 1, 3/2, 2 in Figure 6.2. To show it is well defined, note that $K_q(t) := \check{g}(t)/t^q$ for

$$g(\nu) = \frac{2^{q-1/2}\Gamma(q+1/2)}{(1+\nu^2)^{q+1/2}}$$

(see Definition A.5). From (6.5), the spectral density is

$$f(\nu) = \frac{\hat{c}_q(\nu)}{\sqrt{2\pi}} = \frac{g(\nu)}{2^{q-1}\Gamma(q)\sqrt{2\pi}} = \frac{\Gamma(q+1/2)}{\Gamma(q)\Gamma(1/2)} \frac{1}{(1+\nu^2)^{q+1/2}}, \tag{6.6}$$

(since $\Gamma(1/2) = \sqrt{\pi}$). Observe that $f(\nu)$ is non-negative and integrable for $q > 0$ and the scaling constants are chosen so that $c_q(0) = \int_{\mathbb{R}} f(\nu)\,d\nu = 1$. By the Wiener–Khintchine theorem, $c_q(t)$ is a well-defined covariance function. In the special case $q = 1/2$, $f(\nu) = 1/\pi(1+\nu^2)$ and $c_{1/2}(t) = \mathrm{e}^{-t}$, which is the usual exponential covariance with $\ell = 1$.

We may also show (Exercise 6.6) that

$$f(\nu) = \lim_{T\to\infty} \mathbb{E}\left[\frac{1}{2\pi T}\left|\int_0^T X(t)\mathrm{e}^{-\mathrm{i}\nu t}\,dt\right|^2\right]. \tag{6.7}$$

This relationship can be used to estimate the spectral density from a data set.

Example 6.9 (approximation of the spectral density) We show how to approximate the spectral density $f(\nu)$ using (6.7) and a set of sample paths. We first approximate

$$f_T(\nu) := \frac{1}{2\pi T}\left|\int_0^T X(t,\omega)\mathrm{e}^{-\mathrm{i}\nu t}\,dt\right|^2, \qquad \nu \in \mathbb{R}, \tag{6.8}$$

for a sample path $X(\cdot,\omega)$. For $\nu_k = 2\pi k/T$, we have $f_T(\nu_k) = (T/2\pi)u_k^2$, where u_k are the Fourier coefficients of $X(\cdot,\omega)$ on $(0,T)$ (see (1.35) with $a = 0$ and $b = T$). Hence, we may approximate

$$f_T(\nu_k) \approx \frac{1}{2\pi}U_k^2, \qquad \text{for } k = -J/2+1,\dots,J/2,$$

where U_k is the trapezium rule approximation (1.39). To gain an approximation for $f(\nu)$, we use the piecewise linear interpolant of the sample average of $f_T(\nu_k)$ for independent samples paths $X(\cdot,\omega_j)$ for $j = 1,\dots,M$ of the process $X(t)$. When the number of samples is large and T is large, we expect a good approximation to the spectral density. In the case of white noise and the mean-zero Gaussian process with stationary covariance $\mathrm{e}^{-|t|}$, we see in Figures 6.3 and 6.4 good agreement with 10 samples and $T = 1$. See also Exercise 6.22. This technique to estimate an unknown spectral density may be applied directly to measured data from experiments as well as numerical simulations.

Algorithm 6.1 Code to approximate $f_T(\nu)$ at $\nu_k = 2\pi k/T$ for $k = -J/2+1,\dots,J/2$ for a sample path $X(\cdot,\omega) \in L^2(0,T)$. The inputs are a vector `X` containing $X(j\Delta t,\omega)$ for $j = 0,1\dots,J$ and $\Delta t = T/J$, and the length of the time interval `T`. The outputs are `f`, a vector of approximations to $f(\nu_k)$, and `nu`, a vector of $\nu_0,\dots,\nu_{J/2},\nu_{-J/2+1},\dots,\nu_{-1}$.

```
function [f, nu]=spectral_density(X, T)
J=length(X)-1;[Uk,nu]=get_coeffs(X, 0,T);
f=abs(Uk).^2*T/(2*pi);
```

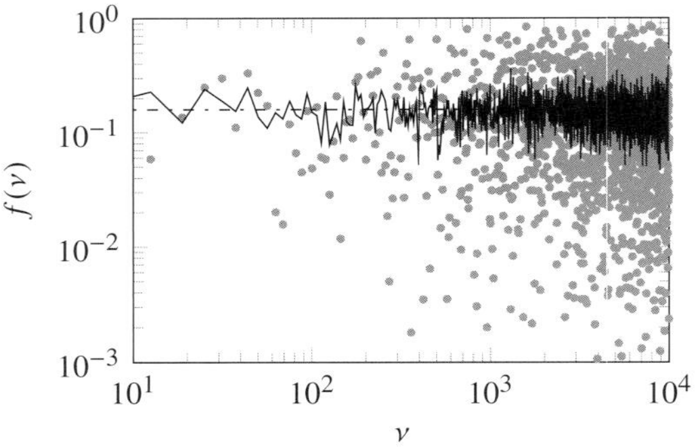

Figure 6.3 Spectral density of white noise $\zeta_J(t)$ on $[0,1]$ with $J = 3200$. The • show $f_T(\nu_k)$ for one sample path computed by Algorithm 6.1; the solid line shows the piecewise linear interpolant of an average of $f_T(\nu_k)$ over 10 independent sample paths; the dash-dot line shows the exact spectral density $f(\nu) = 1/2\pi$.

Example 6.10 (Karhunen–Loève) The Karhunen–Loève expansion of §5.4 represents a stochastic process in a basis of eigenfunctions of the covariance. In general, these eigenfunctions are difficult to compute and must be approximated numerically (as in Examples 7.62 and 7.63). We show now how to use (6.7) to define a stochastic process on $[0,T]$ using a Karhunen–Loève expansion with a known set of eigenfunctions (the Fourier basis) that has a given covariance in the limit $T \to \infty$.

As an example, we consider the stationary covariance

$$c(t) = \frac{1}{2\ell}\exp\left(\frac{-\pi t^2}{4\ell^2}\right) \tag{6.9}$$

for a correlation length $\ell > 0$. Note that $\int_{\mathbb{R}} c(t)\,dt = 1$ and that $c(t) \to 0$ as $\ell \to 0$ for $t \neq 0$. Hence, $c(t)$ is a rescaling of the Gaussian covariance of Example 6.7 that approximates the delta function. This covariance has spectral density

$$f(\nu) = \frac{1}{2\pi}\exp\left(\frac{-\ell^2\nu^2}{\pi}\right).$$

Let $Y(t)$ be a mean-zero stationary process with covariance $c(t)$ and write the sample paths of $Y(t)$ as Fourier series on $[0,2T]$ with Fourier coefficient y_k,

$$Y(t,\omega) = \sum_{k\in\mathbb{Z}} y_k(\omega)e^{i\nu_k t} \qquad \text{for } \nu_k = \frac{2\pi k}{2T}.$$

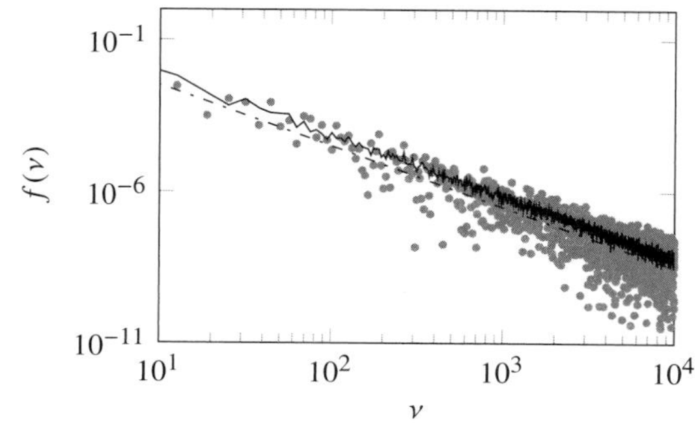

Figure 6.4 Spectral density of the mean-zero Gaussian process with stationary covariance $e^{-|t|}$. The • shows $f_T(\nu_k)$ for one sample path computed by Algorithm 6.1; the solid line shows the piecewise linear interpolant of an average of $f_T(\nu_k)$ over 10 independent sample paths; the dash-dot line shows the exact spectral density (6.4).

Then, $\mathbb{E}[y_k] = 0$ and $\operatorname{Var} y_k = 2\pi f_T(\nu_k)/2T$ as in Example 6.9. Assume that y_k are real and independent. Writing the sample paths of the real part of $Y(t)$ as a Fourier cosine series on $[0,T]$ with Fourier coefficients y_k,

$$\operatorname{Re} Y(t,\omega) = y_0(\omega) + \sum_{k=1}^{\infty} y_k'(\omega)\cos(\nu_k t) \qquad \text{for } \nu_k = \frac{\pi k}{T},$$

where $\operatorname{Var} y_0 = \pi f_T(\nu_0)/T$ and $\operatorname{Var} y_k' = \operatorname{Var}(y_k + y_{-k}) = 2\pi f_T(\nu_k)/T$.

Though f_T is not known explicitly, (6.7) implies that $f_T \to f$ as $T \to \infty$. Hence, we achieve a real-valued process $X(t)$ on $[0,T]$ that has similar statistical properties to $Y(t)$ and is represented in the Fourier basis, by replacing f_T with the known density f. In other words, we work with the process $X(t)$ defined by the Karhunen–Loève expansion

$$X(t) = \sqrt{\eta_0}\frac{1}{\sqrt{T}}\xi_0 + \sum_{k=1}^{\infty} \sqrt{\eta_k}\frac{\cos(\nu_k t)}{\sqrt{T/2}}\xi_k, \qquad t \in [0,T],$$

where ξ_k for $k = 0,1,2,\ldots$ are mean-zero *iid* random variables with unit variance and

$$\begin{aligned} \operatorname{Var} y_0 &= \frac{\pi f_T(\nu_0)}{T} &\longmapsto&\quad \frac{\pi f(\nu_0)}{T} = \frac{1}{2T} &=: \frac{\eta_0}{T}, \\ \operatorname{Var} y_k' &= \frac{2\pi f_T(\nu_k)}{T} &\longmapsto&\quad \frac{2\pi f(\nu_k)}{T} = \frac{1}{T}\exp\left(\frac{-\pi k^2\ell^2}{T^2}\right) &=: \frac{\eta_k}{T/2}. \end{aligned}$$

We further truncate $X(t)$ and consider

$$X_J(t) := \sqrt{\eta_0}\frac{1}{\sqrt{T}}\xi_0 + \sum_{k=1}^{2J-1} \sqrt{\eta_k}\frac{\cos(\nu_k t)}{\sqrt{T/2}}\xi_k = \operatorname{Re}\sum_{k=0}^{2J-1} a_k e^{i\nu_k t}, \tag{6.10}$$

where

$$a_k^2 := \frac{\exp(-\pi k^2\ell^2/T^2)\xi_k^2}{T} \times \begin{cases} 1/2, & k = 0, \\ 1, & k = 1,\ldots,2J-1. \end{cases}$$

To sample $X(t_j)$ for $t_j = T(j-1)/J$ and $j = 1,\ldots,J$, we may use the fast Fourier transform (FFT) and set $X_J(t_j) = \hat{a}_j$ for $j = 1,\ldots,J$ (note that $\nu_k t_j = 2\pi k(j-1)/2J$).

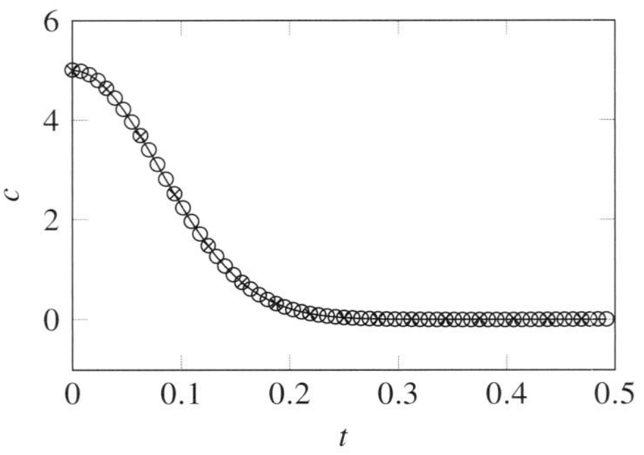

Figure 6.5 Plot of the Gaussian covariance (6.9) for $\ell = 0.1$, alongside the sample covariance for $\text{Cov}(X_J(t), X_J(0))$ with $J = 64$ (o) and $J = 16$ (x) generated using 10^5 samples.

See Algorithm 6.2. The covariance $\text{Cov}(u_J(t), u_J(0))$ is compared to $c(t)$ in Figure 6.5 for $\ell = 0.1$ and good agreement is found even for $J = 16$.

Algorithm 6.2 Code to sample the truncated Karhunen–Loève expansion (6.10) on $t_j = T(j-1)/J$ for $j = 1, \dots, J$ in the case $\xi_k \sim \text{U}(-\sqrt{3}, \sqrt{3})$ *iid*. The inputs are `T`, `J`, and `ell`$=\ell$. The output is a column vector of $X_J(t_1), \dots, X_J(t_J)$.

```
function XJ=fkl_1d(J,T,ell)
kk=[0:2*J-1]'; % col vector for range of k
b=sqrt(1/T*exp(-pi*(kk*ell/T).^2));
b(1)=sqrt(0.5)*b(1);
xi=rand(2*J,1)*sqrt(12)-sqrt(3);
a=b.*xi; XJ=real(fft(a)); XJ=XJ(1:J);
```

Mean-square differentiability

The spectral density provides a way to understand the regularity of a stationary process $X(t)$, as measured by its mean-square differentiability (see Definition 5.33). The key is the decay of the spectral density $f(\nu)$ as $|\nu| \to \infty$.

Lemma 6.11 *Consider a mean-zero stationary process $X(t)$ with spectral density $f(\nu)$. If there exists $q, L_q > 0$ such that*

$$f(\nu) \le \frac{L_q}{|\nu|^{2q+1}}, \tag{6.11}$$

then $X(t)$ is n times mean-square differentiable for $n < q$.

Proof By Theorem 5.35, a process is mean-square differentiable if its covariance function is twice differentiable. From Exercise 6.13, the covariance is twice differentiable if (6.11) holds with $q > 1$. The general case is similar. A process is n times mean-square differentiable if the covariance is $2n$ times mean-square differentiable. This happens if $\int_{\mathbb{R}} \nu^{2n} f(\nu)\, d\nu < \infty$, which holds under the decay condition for $q > n$. □

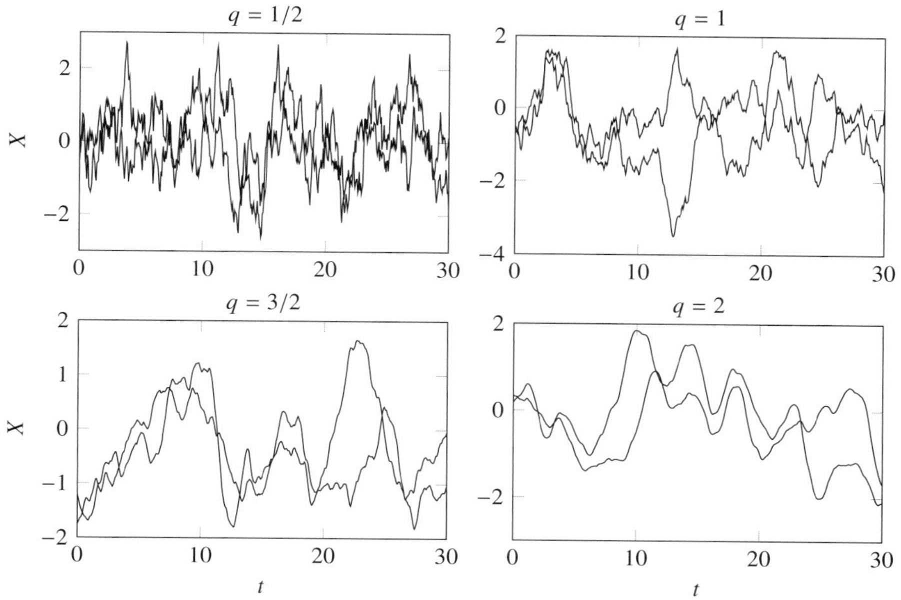

Figure 6.6 Sample paths of a mean-zero Gaussian process with the Whittle–Matérn covariance $c_q(t)$ for $q = 1/2$, $q = 1$, $q = 3/2$, and $q = 2$.

For the Gaussian covariance, the decay condition is satisfied for any $q > 0$ and mean-zero processes with the Gaussian covariance are mean-square differentiable to any order. Indeed, looking at the sample paths in Figure 5.12, we see they are very smooth. For the Whittle–Matérn covariance $c_q(t)$, the decay condition (6.11) holds and any mean-zero stationary process with covariance $c_q(t)$ is n times mean-square differentiable for $n < q$. This is useful in applications, where q may be chosen to construct processes with a particular regularity. Example sample paths for the Whittle–Matérn covariance are plotted in Figure 6.6 and we see the paths are smoother for larger values of q.

6.2 Complex-valued random variables and stochastic processes

Since the Fourier transform is the key tool for analysing stationary processes, complex numbers and complex-valued random variables are important for our study. We begin by extending some ideas from Chapter 5 to the complex setting. Consider a probability space $(\Omega, \mathcal{F}, \mathbb{P})$ and the measure space $(\mathbb{C}, \mathcal{B}(\mathbb{C}))$ comprising the complex numbers with the Borel σ-algebra (see Definition 1.16). We are interested in complex-valued random variables, also known as $\mathbb{C}$-valued random variables in Definition 4.3.

Definition 6.12 (complex-valued random variable) A *complex-valued random variable* Z is a measurable function from $(\Omega, \mathcal{F})$ to $(\mathbb{C}, \mathcal{B}(\mathbb{C}))$. We write $Z = X + \mathrm{i}Y$ where X and Y are real-valued random variables. The *mean* is $\mu_Z = \mathbb{E}[Z] = \mathbb{E}[X] + \mathrm{i}\mathbb{E}[Y]$ and the *variance* is $\mathrm{Var}(Z) = \mathrm{Cov}(Z, Z) = \mathbb{E}\big[|Z - \mu_Z|^2\big] = \mathrm{Var}(X) + \mathrm{Var}(Y)$. If Z_1, Z_2 are a pair of complex-valued random variables, the *covariance* is

$$\operatorname{Cov}(Z_1, Z_2) = \mathbb{E}\Big[(Z_1 - \mu_{Z_1})\overline{(Z_2 - \mu_{Z_2})}\Big],$$

where, as usual, $\bar{}$ denotes complex conjugation, $\mu_{Z_1} = \mathbb{E}[Z_1]$ and $\mu_{Z_2} = \mathbb{E}[Z_2]$. In general, $\operatorname{Cov}(Z_1, Z_2)$ is complex valued. If $\boldsymbol{Z}$ is a $\mathbb{C}^d$-valued random variable, the mean is the vector $\boldsymbol{\mu}_Z := \mathbb{E}[\boldsymbol{Z}] \in \mathbb{C}^d$ and the covariance matrix $C_Z := \mathbb{E}[(\boldsymbol{Z} - \boldsymbol{\mu}_Z)(\boldsymbol{Z} - \boldsymbol{\mu}_Z)^*]$, where * denotes the conjugate or Hermitian transpose.

If the real and imaginary parts of a complex-valued random variable $\boldsymbol{Z}$ are uncorrelated, then the covariance matrix is always real valued.

Lemma 6.13 *Let $\boldsymbol{Z} = \boldsymbol{X} + \mathrm{i}\boldsymbol{Y}$, where $\boldsymbol{X}, \boldsymbol{Y}$ are $\mathbb{R}^d$-valued random variables. If $\boldsymbol{X}$ and $\boldsymbol{Y}$ are uncorrelated, the covariance matrix C_Z of $\boldsymbol{Z}$ is a $d \times d$ real-valued matrix. Further, $C_Z = C_X + C_Y$, where C_X and C_Y are the covariance matrices of $\boldsymbol{X}$ and $\boldsymbol{Y}$, respectively.*

Proof By substituting $\boldsymbol{Z} \mapsto \boldsymbol{Z} - \mathbb{E}[\boldsymbol{Z}]$, which does not change the covariance matrix, we may assume $\boldsymbol{Z}$ and hence $\boldsymbol{X}, \boldsymbol{Y}$ have mean zero. Definition 6.12 gives

$$\begin{aligned} C_Z &= \mathbb{E}\big[(\boldsymbol{X} + \mathrm{i}\boldsymbol{Y})(\boldsymbol{X} + \mathrm{i}\boldsymbol{Y})^*\big] = \mathbb{E}\big[(\boldsymbol{X} + \mathrm{i}\boldsymbol{Y})(\boldsymbol{X} - \mathrm{i}\boldsymbol{Y})^{\mathsf{T}}\big] \\ &= \mathbb{E}\big[\boldsymbol{X}\boldsymbol{X}^{\mathsf{T}} + \boldsymbol{Y}\boldsymbol{Y}^{\mathsf{T}}\big] + \mathrm{i}\mathbb{E}\big[-\boldsymbol{X}\boldsymbol{Y}^{\mathsf{T}} + \boldsymbol{Y}\boldsymbol{X}^{\mathsf{T}}\big]. \end{aligned}$$

As $\boldsymbol{X}, \boldsymbol{Y}$ are uncorrelated (Definition 4.26), $\mathbb{E}\big[\boldsymbol{X}\boldsymbol{Y}^{\mathsf{T}}\big] = \mathbb{E}\big[\boldsymbol{Y}\boldsymbol{X}^{\mathsf{T}}\big] = 0$ and $C_Z = C_X + C_Y$. □

If $d = 1$, we always have $C_Z = C_X + C_Y$, where $C_Z = \operatorname{Var}(Z)$, $C_X = \operatorname{Var}(X)$ and $C_Y = \operatorname{Var}(Y)$.

The next result gives conditions under which the real and imaginary parts $\boldsymbol{X}$ and $\boldsymbol{Y}$ are uncorrelated, and provides formulae for the covariance matrices C_X and C_Y. For convenience, we work with mean-zero random variables.

Lemma 6.14 *Let $\boldsymbol{Z} = \boldsymbol{X} + \mathrm{i}\boldsymbol{Y}$ be a $\mathbb{C}^d$-valued mean-zero random variable. If $\mathbb{E}[\boldsymbol{Z}\boldsymbol{Z}^*]$ and $\mathbb{E}\big[\boldsymbol{Z}\boldsymbol{Z}^{\mathsf{T}}\big]$ are $d \times d$ real-valued matrices, then $\boldsymbol{X}$ and $\boldsymbol{Y}$ are uncorrelated and*

$$C_X = \tfrac{1}{2}\big[\mathbb{E}[\boldsymbol{Z}\boldsymbol{Z}^*] + \mathbb{E}[\boldsymbol{Z}\boldsymbol{Z}^{\mathsf{T}}]\big], \qquad C_Y = \tfrac{1}{2}\big[\mathbb{E}[\boldsymbol{Z}\boldsymbol{Z}^*] - \mathbb{E}[\boldsymbol{Z}\boldsymbol{Z}^{\mathsf{T}}]\big].$$

Proof Using $\boldsymbol{X} = \big(\boldsymbol{Z} + \bar{\boldsymbol{Z}}\big)/2$ and $\boldsymbol{Y} = \big(\boldsymbol{Z} - \bar{\boldsymbol{Z}}\big)/2\mathrm{i}$, we have

$$\operatorname{Cov}(\boldsymbol{X}, \boldsymbol{Y}) = \mathbb{E}\big[\boldsymbol{X}\boldsymbol{Y}^{\mathsf{T}}\big] = \frac{1}{4\mathrm{i}}\Big(\mathbb{E}\big[\boldsymbol{Z}\boldsymbol{Z}^{\mathsf{T}}\big] - \mathbb{E}[\boldsymbol{Z}\boldsymbol{Z}^*] + \mathbb{E}\big[\bar{\boldsymbol{Z}}\boldsymbol{Z}^{\mathsf{T}}\big] - \mathbb{E}\big[\bar{\boldsymbol{Z}}\boldsymbol{Z}^*\big]\Big).$$

Because $\mathbb{E}[\boldsymbol{Z}\boldsymbol{Z}^*]$ and $\mathbb{E}\big[\boldsymbol{Z}\boldsymbol{Z}^{\mathsf{T}}\big]$ are real valued, $\mathbb{E}[\boldsymbol{Z}\boldsymbol{Z}^*] = \mathbb{E}\big[\bar{\boldsymbol{Z}}\boldsymbol{Z}^{\mathsf{T}}\big]$ and $\mathbb{E}\big[\boldsymbol{Z}\boldsymbol{Z}^{\mathsf{T}}\big] = \mathbb{E}\big[\bar{\boldsymbol{Z}}\boldsymbol{Z}^*\big]$. Hence, $\operatorname{Cov}(\boldsymbol{X}, \boldsymbol{Y}) = 0$ and $\boldsymbol{X}$ and $\boldsymbol{Y}$ are uncorrelated. Further, since $\boldsymbol{Z}$ has mean zero, $\boldsymbol{X}$ and $\boldsymbol{Y}$ have mean zero. The covariance matrix of $\boldsymbol{X}$ is

$$\begin{aligned} C_X &= \mathbb{E}\big[\boldsymbol{X}\boldsymbol{X}^{\mathsf{T}}\big] = \mathbb{E}\left[\left(\frac{\boldsymbol{Z} + \bar{\boldsymbol{Z}}}{2}\right)\left(\frac{\boldsymbol{Z} + \bar{\boldsymbol{Z}}}{2}\right)^{\mathsf{T}}\right] \\ &= \frac{1}{4}\Big[\mathbb{E}\big[\boldsymbol{Z}\boldsymbol{Z}^{\mathsf{T}}\big] + \mathbb{E}[\boldsymbol{Z}\boldsymbol{Z}^*] + \mathbb{E}\big[\bar{\boldsymbol{Z}}\boldsymbol{Z}^{\mathsf{T}}\big] + \mathbb{E}\big[\bar{\boldsymbol{Z}}\boldsymbol{Z}^*\big]\Big] = \frac{1}{2}\big[\mathbb{E}[\boldsymbol{Z}\boldsymbol{Z}^*] + \mathbb{E}[\boldsymbol{Z}\boldsymbol{Z}^{\mathsf{T}}]\big], \end{aligned}$$

as required. The calculation for C_Y is similar. □

The Gaussian and multivariate Gaussian distributions are defined in Example 4.10 and Definition 4.30. We now define a special kind of complex-valued Gaussian distribution, which is valuable for studying stationary processes.

Definition 6.15 (complex Gaussian $\mathrm{CN}(\boldsymbol{\mu}_Z, C_Z)$) A $\mathbb{C}^d$-valued random variable $\boldsymbol{Z} = \boldsymbol{X} + \mathrm{i}\boldsymbol{Y}$ follows the *complex Gaussian* distribution $\mathrm{CN}(\boldsymbol{\mu}_Z, C_Z)$ for $\boldsymbol{\mu}_Z \in \mathbb{C}^d$ and $C_Z \in \mathbb{R}^{d\times d}$ if the real and imaginary parts of $\boldsymbol{Z} - \boldsymbol{\mu}_Z$ are independent and each has distribution $\mathrm{N}(\boldsymbol{0}, C_Z/2)$. Then, $\boldsymbol{X} \sim \mathrm{N}(\mathrm{Re}\,\boldsymbol{\mu}_Z, C_Z/2)$ and $\boldsymbol{Y} \sim \mathrm{N}(\mathrm{Im}\,\boldsymbol{\mu}_Z, C_Z/2)$ and

$$\begin{pmatrix}\boldsymbol{X}\\ \boldsymbol{Y}\end{pmatrix} \sim \mathrm{N}\left(\begin{pmatrix}\mathrm{Re}\,\boldsymbol{\mu}_Z\\ \mathrm{Im}\,\boldsymbol{\mu}_Z\end{pmatrix}, \begin{pmatrix}C_Z/2 & 0\\ 0 & C_Z/2\end{pmatrix}\right).$$

Note that C_Z in Definition 6.15 is *real valued* and, using Lemma 6.13, it is easy to show that C_Z is the covariance matrix of $\boldsymbol{Z}$.

Example 6.16 (complex Gaussian random variable) In the case $d = 1$, we write $Z \sim \mathrm{CN}(\mu_Z, \sigma_Z^2)$ for $\sigma_Z^2 := \mathrm{Var}\, Z = \mathbb{E}\left[|Z - \mu_Z|^2\right]$. If $Z = X + \mathrm{i}Y$ then $X \sim \mathrm{N}(\mathrm{Re}\,\mu_Z, \sigma_Z^2/2)$ and $Y \sim \mathrm{N}(\mathrm{Im}\,\mu_Z, \sigma_Z^2/2)$. We can use the MATLAB command `randn` to draw two independent samples X and Y from the distribution $\mathrm{N}(0, 1/2)$ to obtain a sample from $\mathrm{CN}(0, 1)$ as follows:

```
>> Z=sqrt(1/2)*randn(1,2)*[1;sqrt(-1)]
Z =
   0.3802 + 1.2968i
```

If it is possible to draw a sample $Z \in \mathbb{C}$ from $\mathrm{CN}(0, 1)$ directly, then the real and imaginary parts of Z provide independent $\mathrm{N}(0, 1/2)$ samples.

The definitions of *stochastic process* and *Gaussian process* from Chapter 5 are now extended to complex-valued processes.

Definition 6.17 (complex-valued process) A *complex-valued stochastic process* is a set of complex-valued random variables $\{Z(t) \colon t \in \mathcal{T}\}$ on a probability space $(\Omega, \mathcal{F}, \mathbb{P})$ with $\mathcal{T} \subset \mathbb{R}$. We say the process is *second order* if $Z(t) \in L^2(\Omega, \mathbb{C})$ for each $t \in \mathcal{T}$.

Definition 6.18 (complex-valued Gaussian process) A complex-valued stochastic process $\{Z(t) \colon t \in \mathcal{T}\}$ is *Gaussian* if $\boldsymbol{Z} = [Z(t_1), \dots, Z(t_M)]^{\mathsf{T}}$ follows the complex Gaussian distribution for all $t_1, \dots, t_M \in \mathcal{T}$ and $M \in \mathbb{N}$. If $\mu_Z \colon \mathcal{T} \to \mathbb{C}$ and $C_Z \colon \mathcal{T} \times \mathcal{T} \to \mathbb{R}$ are the mean and covariance of $Z(t)$, then $\boldsymbol{Z} \sim \mathrm{CN}(\boldsymbol{\mu}_Z, C_Z)$ with $[\boldsymbol{\mu}_Z]_i = \mathbb{E}[Z(t_i)] = \mu_Z(t_i)$ and the (i, j) entry of C_Z is $\mathrm{Cov}(Z(t_i), Z(t_j)) = C_Z(t_i, t_j)$.

Example 6.19 (harmonic oscillation) Let

$$Z(t) = \mathrm{e}^{\mathrm{i}t\nu}\xi,$$

where the angular frequency is $\nu \in \mathbb{R}$ and $\xi \sim \mathrm{CN}(0, 1)$. Then $x := \mathrm{Re}\,\xi$ and $y := \mathrm{Im}\,\xi$ are *iid* $\mathrm{N}(0, 1/2)$ random variables. $\{Z(t) \colon t \in \mathbb{R}\}$ is certainly a family of mean-zero complex-valued Gaussian random variables. However, $Z(t)$ is not a complex Gaussian process in the sense of Definition 6.18. To see this, consider

$$\boldsymbol{Z} = \begin{pmatrix}Z(0)\\ Z(t)\end{pmatrix} = \begin{pmatrix}x + \mathrm{i}y\\ x\cos(t\nu) - y\sin(t\nu) + \mathrm{i}(y\cos(t\nu) + x\sin(t\nu))\end{pmatrix}.$$

The real and imaginary parts,

$$\boldsymbol{X} = \begin{pmatrix} x \\ x\cos(t\nu) - y\sin(t\nu) \end{pmatrix}, \qquad \boldsymbol{Y} = \begin{pmatrix} y \\ y\cos(t\nu) + x\sin(t\nu) \end{pmatrix},$$

are mean-zero Gaussian random variables. However, $\mathbb{E}\big[\boldsymbol{X}\boldsymbol{Y}^\mathsf{T}\big] \neq 0$ and they are correlated. For example, since x, y are uncorrelated and $\operatorname{Var} x = 1/2$,

$$\mathbb{E}\big[\boldsymbol{X}\boldsymbol{Y}^\mathsf{T}\big]_{1,2} = \mathbb{E}\big[\operatorname{Re} Z(0)\ \operatorname{Im} Z(t)\big] = \mathbb{E}\big[xy\cos(t\nu) + x^2\sin(t\nu)\big] = \tfrac{1}{2}\sin(t\nu),$$

which is non-zero for some t and ν. The real and imaginary parts of $\boldsymbol{Z}$ are therefore not independent and so $Z(t)$ is not a complex Gaussian process.

Example 6.20 (harmonic oscillation as complex Gaussian) Let

$$Z(t) = \mathrm{e}^{-\mathrm{i}t\nu}\xi_- + \mathrm{e}^{+\mathrm{i}t\nu}\xi_+$$

for $\nu \in \mathbb{R}$ and $\xi_\pm \sim \mathrm{CN}(0,1)$ *iid*. $Z(t)$ is an example of a random Fourier series (see Exercise 6.7). In this case, $Z(t)$ does define a complex Gaussian process. Consider $\boldsymbol{Z} = [Z(t_1), \dots, Z(t_M)]^\mathsf{T}$, for any $M \in \mathbb{N}$. It is easy to show that the real and imaginary parts of $\boldsymbol{Z}$ are multivariate, mean-zero Gaussian random variables. Further, as $\xi_\pm$ are uncorrelated,

$$\begin{aligned} \mathbb{E}\big[\boldsymbol{Z}\boldsymbol{Z}^*\big]_{i,j} &= \mathrm{e}^{-\mathrm{i}(t_i - t_j)\nu}\mathbb{E}\big[\xi_-\bar{\xi}_-\big] + \mathrm{e}^{\mathrm{i}(t_i - t_j)\nu}\mathbb{E}\big[\xi_+\bar{\xi}_+\big], \\ \mathbb{E}\Big[\boldsymbol{Z}\boldsymbol{Z}^\mathsf{T}\Big]_{i,j} &= \mathrm{e}^{-\mathrm{i}(t_i + t_j)\nu}\mathbb{E}\big[\xi_-\xi_-\big] + \mathrm{e}^{\mathrm{i}(t_i + t_j)\nu}\mathbb{E}\big[\xi_+\xi_+\big]. \end{aligned}$$

As $\xi_\pm \sim \mathrm{CN}(0,1)$, $\mathbb{E}\big[\xi_-\bar{\xi}_-\big] = \mathbb{E}\big[\xi_+\bar{\xi}_+\big] = \operatorname{Var}(\xi_\pm) = 1$ and, writing $\xi_\pm = x + \mathrm{i}y$, we see $\mathbb{E}\big[\xi_\pm^2\big] = \mathbb{E}\big[x^2 - y^2\big] + \mathrm{i}2\mathbb{E}[xy] = 0$. Hence, for $i, j = 1, \dots, M$,

$$\mathbb{E}\big[\boldsymbol{Z}\boldsymbol{Z}^*\big]_{i,j} = 2\cos\big((t_i - t_j)\nu\big), \qquad \mathbb{E}\Big[\boldsymbol{Z}\boldsymbol{Z}^\mathsf{T}\Big]_{i,j} = 0.$$

Lemma 6.14 implies that the real and imaginary parts of $\boldsymbol{Z}$ are uncorrelated and have the same covariance matrix C with entries $c_{ij} = \cos((t_i - t_j)\nu)$. Hence, the real and imaginary parts of $\boldsymbol{Z}$ are *iid* Gaussian random variables, and $Z(t)$ is a complex Gaussian process. Furthermore, $Z(t)$ is stationary with $\mu(t) = 0$ and stationary covariance function $c_Z(t) = 2\cos(t\nu)$.

6.3 Stochastic integrals

Example 6.20 considers a sum of two elementary stochastic processes $\mathrm{e}^{\pm\mathrm{i}t\nu}\xi_\pm$ and shows that their sum defines a complex-valued stationary process. We generalise this result from sums to integrals and find that a large class of stationary processes can be written as stochastic integrals. In particular, we look at stationary processes defined by integrals of the form $Z(t) = \int_{\mathbb{R}} \mathrm{e}^{\mathrm{i}t\nu}\sqrt{f(\nu)}\,d\mathcal{W}(\nu)$ (see Definition 6.28), where $f(\nu)$ is a density function, $\mathrm{e}^{\mathrm{i}t\nu}$ is the harmonic oscillation studied in Example 6.19, and $\mathcal{W}(\nu)$ is a complex version of Brownian motion.

First, we develop the real-valued stochastic integral $\int_{\mathbb{R}} g(\nu)\,dW(\nu)$, for a two-sided Brownian motion $\{W(\nu) \colon \nu \in \mathbb{R}\}$ (see Definition 5.12) and $g \in L^2(\mathbb{R})$ (for now, g is independent of t.) We would like to define

$$\int_{\mathbb{R}} g(\nu)\,dW(\nu) = \int_{\mathbb{R}} g(\nu)\frac{dW}{d\nu}\,d\nu, \tag{6.12}$$

and use standard integration theory to interpret the integral on the right-hand side. The difficulty lies in interpreting the derivative $\frac{dW}{dv}$, as we know from Example 5.41 that Brownian motion is not differentiable in the classical sense. In fact, the derivative of Brownian motion is white noise. The numerical derivative $D_{\Delta v}(v) = (W(v+\Delta v)-W(v))/\Delta v$, for $\Delta v > 0$, is investigated in Exercise 6.11. It diverges as $\Delta v \to 0$ and the spectral density is flat, just like for white noise $\zeta(t)$. Indeed, because Brownian motion has stationary increments, the process $\{D_{\Delta v}(v)\colon v \in \mathbb{R}\}$ is stationary with mean zero and stationary covariance function

$$c_{\Delta v}(v) = \mathbb{E}\Big[D_{\Delta v}(v)D_{\Delta v}(0)\Big] = \frac{1}{\Delta v^2}\mathbb{E}\Big[\big(W(v+\Delta v) - W(v)\big)W(\Delta v)\Big].$$

For $v > 0$, we obtain

$$c_{\Delta v}(v) = \frac{1}{\Delta v^2}\Big(\min\{v+\Delta v, \Delta v\} - \min\{v, \Delta v\}\Big).$$

Using $c_{\Delta v}(v) = c_{\Delta v}(-v)$, it is then elementary to check that

$$c_{\Delta v}(v) = 0 \text{ for } |v| > \Delta v \qquad \text{and} \qquad \int_{\mathbb{R}} c_{\Delta v}(v)\,dv = 1.$$

Hence, $c_{\Delta v}(v) \to \delta(v)$ as $\Delta v \to 0$ and the derivative of Brownian motion is white noise.

Formally, we can work with the derivative $dW(v)/dv$ of Brownian motion using the Karhunen–Loève expansion of white noise. From (5.22),

$$\frac{dW(v)}{dv} = \zeta(v) = \sum_{j=1}^{\infty} \xi_j \phi_j(v),$$

for $\xi_j \sim \mathrm{N}(0,1)$ *iid* and an orthonormal basis $\{\phi_j\colon j = 1,2,\ldots\}$ of $L^2(\mathbb{R})$. If we substitute this expansion into (6.12) and integrate term by term, we arrive at the definition of the stochastic integral. To show this makes sense, let

$$\zeta_J(v) := \sum_{j=1}^{J} \xi_j \phi_j(v) \tag{6.13}$$

denote the truncated white noise expansion of dW/dv. This is a well-defined function in $L^2(\mathbb{R})$ and we may consider $\int_{\mathbb{R}} g(v)\zeta_J(v)\,dv$ and its limit as $J \to \infty$.

Lemma 6.21 *For $g \in L^2(\mathbb{R})$, the sequence $\int_{\mathbb{R}} g(v)\zeta_J(v)\,dv$, for $J = 1,2,\ldots$ and ζ_J defined in* (6.13)*, has a well-defined limit in $L^2(\Omega)$.*

Proof As $g, \zeta_J \in L^2(\mathbb{R})$, the integral $\int_{\mathbb{R}} g(v)\zeta_J(v)\,dv = \langle g, \zeta_J\rangle_{L^2(\mathbb{R})}$ is well defined. The random variables ξ_j are *iid* with unit variance and hence

$$\mathbb{E}\left[\left|\int_{\mathbb{R}} g(v)\zeta_J(v)\,dv\right|^2\right] = \sum_{j=1}^{J} \mathbb{E}\big[\xi_j^2\big]\big\langle g, \phi_j\big\rangle_{L^2(\mathbb{R})}^2 = \sum_{j=1}^{J} \big\langle g, \phi_j\big\rangle_{L^2(\mathbb{R})}^2. \tag{6.14}$$

When $g \in L^2(\mathbb{R})$, we have by (1.9) that $\sum_{j=1}^{\infty} \langle g, \phi_j\rangle_{L^2(\mathbb{R})}^2 = \|g\|_{L^2(\mathbb{R})}^2 < \infty$. The sum (6.14) converges as $J \to \infty$ and $\int_{\mathbb{R}} g(v)\zeta_J(v)\,dv$ has a well-defined limit in $L^2(\Omega)$. □

Definition 6.22 (stochastic integral) The *stochastic integral of* $g \in L^2(\mathbb{R})$ *with respect to* $W(\nu)$ is the $L^2(\Omega)$ random variable defined by, where ζ_J is given by (6.13),

$$\int_{\mathbb{R}} g(\nu)\, dW(\nu) := \lim_{J\to\infty} \int_{\mathbb{R}} g(\nu)\zeta_J(\nu)\, d\nu. \tag{6.15}$$

For $g \in L^2(a,b)$,

$$\int_a^b g(\nu)\, dW(\nu) := \int_{\mathbb{R}} 1_{(a,b)}(\nu) g(\nu)\, dW(\nu). \tag{6.16}$$

The stochastic integral presented above is for non-random integrands $g(\nu)$. When we consider stochastic ODEs in Chapter 8, stochastic integrals are defined for a class of *random* integrands correlated to $W(\nu)$.

Proposition 6.23 *The following properties hold for* $a,b \in \mathbb{R} \cup \{\pm\infty\}$.

(i) *The integral is linear: if* $g,h \in L^2(a,b)$, *then*

$$\int_a^b g(\nu)\, dW(\nu) + \int_a^b h(\nu)\, dW(\nu) = \int_a^b \big(g(\nu) + h(\nu)\big)\, dW(\nu).$$

(ii) *If* $g \in L^2(a,b)$, *the stochastic integral* (6.16) *is a Gaussian random variable with distribution* $\mathrm{N}\big(0, \|g\|^2_{L^2(a,b)}\big)$.

(iii) *If* $g^i \in L^2(a,b)$ *and* $X^i = \int_a^b g^i(\nu)\, dW(\nu)$ *for* $i = 1,\ldots,N$, *then* $\boldsymbol{X} = [X^1,\ldots,X^N]^{\mathsf{T}}$ *is a multivariate Gaussian with distribution* $\mathrm{N}(\boldsymbol{0}, C)$, *where*

$$c_{ij} = \mathrm{Cov}(X^i, X^j) = \int_a^b g^i(\nu)\, g^j(\nu)\, d\nu = \langle g^i, g^j \rangle_{L^2(a,b)}. \tag{6.17}$$

Proof (i) Linearity follows easily from the definition.

(ii) The stochastic integral (6.15), with $g(\nu)$ replaced by $1_{(a,b)}(\nu)g(\nu)$, is a limit of a linear combination of Gaussian random variables ξ_j and is therefore also Gaussian (see Exercise 6.8). The integral has mean zero as $\mathbb{E}[\xi_j] = 0$ for each j. Taking the limit as $J \to \infty$ in (6.14) reveals that the variance is $\|g\|^2_{L^2(a,b)}$.

(iii) The vector $\boldsymbol{X}$ is also a limit of a linear transformation of a vector of Gaussian random variables ξ_j and hence is Gaussian. Each component of $\boldsymbol{X}$ has mean zero and so $\boldsymbol{\mu} = \boldsymbol{0}$. To find the covariance, write $g^i = \sum_{k=1}^{\infty} g^i_k \phi_k$, where $g^i_k = \langle g^i, \phi_k \rangle_{L^2(\mathbb{R})}$, so that

$$\mathbb{E}\left[\int_{\mathbb{R}} g^i(\nu)\, dW(\nu) \int_{\mathbb{R}} g^j(\nu)\, dW(\nu)\right] = \mathbb{E}\left[\sum_{k=1}^{\infty} g^i_k \xi_k \sum_{\ell=1}^{\infty} g^j_\ell \xi_\ell\right]$$
$$= \sum_{k=1}^{\infty} g^i_k g^j_k = \langle g^i, g^j \rangle_{L^2(\mathbb{R})}.$$

To obtain (6.17) for integrals over (a,b), replace $g^i(\nu)$ by $1_{(a,b)}(\nu) g^i(\nu)$ and observe that $\langle 1_{(a,b)} g^i, 1_{(a,b)} g^j \rangle_{L^2(\mathbb{R})} = \langle g^i, g^j \rangle_{L^2(a,b)}$. □

If the interval (a,b) or the integrand $g(\nu)$ depends on t, the stochastic integrals in (6.15) and (6.16) return Gaussian stochastic processes. For example, $X(t) = \int_0^t dW(\nu)$ is a Gaussian random variable for each t. In Lemma 6.24, we show the mean and covariance of $X(t)$ match those of Brownian motion. However, $X(t)$ is not strictly speaking a Brownian motion because the stochastic integral is defined in $L^2(\Omega)$ and pathwise continuity cannot be verified.

Lemma 6.24 *For $t \geq 0$, let $X(t) = \int_0^t dW(\nu)$. $X(t)$ is a Gaussian process with mean zero and covariance* $\min\{s,t\}$. *Furthermore, $X(t)$ has a continuous version, which is a Brownian motion.*

Proof Choosing $t_1,\dots,t_M \in \mathbb{R}$ and defining $X^i = X(t_i) = \int_{\mathbb{R}} 1_{(0,t_i)}(\nu)\, dW(\nu)$, Proposition 6.23(iii) tells us that the finite-dimensional distributions of $X(t)$ are Gaussian with mean zero, and so $X(t)$ is a mean-zero Gaussian stochastic process. To compute the covariance,

$$\mathbb{E}\left[\int_0^s dW(\nu) \int_0^t dW(\nu)\right] = \mathbb{E}\left[\int_{\mathbb{R}} 1_{(0,s)}(\nu)\, dW(\nu) \int_{\mathbb{R}} 1_{(0,t)}(\nu)\, dW(\nu)\right]$$
$$= \left\langle 1_{(0,s)}, 1_{(0,t)}\right\rangle_{L^2(\mathbb{R})} = \min\{s,t\}.$$

This means that $X(t)$ has the correct mean and covariance function for Brownian motion. Corollary 5.42 tells us that a continuous version of $X(t)$ exists. The resulting process is a Brownian motion. □

Complex-valued stochastic integrals

The Wiener–Khintchine theorem says that covariance functions of mean-square continuous stationary processes can be expressed as Fourier integrals. We introduce a stochastic Fourier integral and define a complex-valued stationary process, whose real and imaginary parts have a target Gaussian distribution. When quadrature is introduced in §6.4, this provides a method for approximating sample paths of real-valued Gaussian stationary processes.

Our first task is to introduce a complex-valued version of the stochastic integral (6.16); that is, an integral of the form $\int_{\mathbb{R}} g(\nu)\, d\mathcal{W}(\nu)$, where $g \in L^2(\mathbb{R},\mathbb{C})$ and $\mathcal{W}(\nu)$ is the complex version of Brownian motion defined by

$$\mathcal{W}(\nu) := W_1(\nu) + \mathrm{i}W_2(\nu), \tag{6.18}$$

for independent two-sided Brownian motions $W_1(\nu)$, $W_2(\nu)$.

Definition 6.25 (complex-valued stochastic integral) For $g \in L^2(\mathbb{R},\mathbb{C})$ and $\mathcal{W}(\nu)$ given by (6.18), the stochastic integral

$$Z = \int_{\mathbb{R}} g(\nu)\, d\mathcal{W}(\nu) \tag{6.19}$$

is the complex-valued random variable $Z = X + \mathrm{i}Y$ where

$$X := \int_{\mathbb{R}} g_1(\nu)\, dW_1(\nu) - \int_{\mathbb{R}} g_2(\nu)\, dW_2(\nu),$$
$$Y := \int_{\mathbb{R}} g_2(\nu)\, dW_1(\nu) + \int_{\mathbb{R}} g_1(\nu)\, dW_2(\nu),$$

for $g_1(\nu) := \operatorname{Re} g(\nu)$ and $g_2(\nu) := \operatorname{Im} g(\nu)$.

Clearly, Z is a complex-valued random variable. To prove that Z follows the complex Gaussian distribution, we need the following results.

Lemma 6.26 *For any $g, h \in L^2(\mathbb{R}, \mathbb{C})$,*

$$\mathbb{E}\left[\int_{\mathbb{R}} g(\nu)\, d\mathcal{W}(\nu) \overline{\int_{\mathbb{R}} h(\nu)\, d\mathcal{W}(\nu)}\right] = 2\langle g, h\rangle_{L^2(\mathbb{R},\mathbb{C})}, \tag{6.20}$$

$$\mathbb{E}\left[\int_{\mathbb{R}} g(\nu)\, d\mathcal{W}(\nu) \int_{\mathbb{R}} h(\nu)\, d\mathcal{W}(\nu)\right] = 0. \tag{6.21}$$

Proof Apply Proposition 6.23 and see Exercise 6.9. □

Corollary 6.27 *Let $f_i \in L^2(\mathbb{R}, \mathbb{C})$ and consider the stochastic integrals*

$$Z_i = \int_{\mathbb{R}} f_i(\nu)\, d\mathcal{W}(\nu), \qquad i = 1, \dots, N.$$

If $\langle f_i, f_j\rangle_{L^2(\mathbb{R},\mathbb{C})}$ is real for each $i, j = 1, \dots, N$, then $\boldsymbol{Z} = [Z_1, \dots, Z_N]^{\mathsf{T}}$ is a mean-zero complex Gaussian random variable.

Proof The real part $\boldsymbol{X}$ and imaginary part $\boldsymbol{Y}$ of $\boldsymbol{Z}$ are both mean-zero real-valued Gaussian random variables, because of Proposition 6.23(ii). It remains to check that $\boldsymbol{Z}$ is a complex Gaussian. By Lemma 6.26,

$$\mathbb{E}\big[Z_i \bar{Z}_j\big] = 2\langle f_i, f_j\rangle_{L^2(\mathbb{R},\mathbb{C})} \qquad \text{and} \qquad \mathbb{E}\big[Z_i Z_j\big] = 0.$$

Hence, $\mathbb{E}[\boldsymbol{Z}\boldsymbol{Z}^*]$ and $\mathbb{E}\big[\boldsymbol{Z}\boldsymbol{Z}^{\mathsf{T}}\big]$ are $N \times N$ real-valued matrices and Lemma 6.14 implies that $\boldsymbol{X}$ and $\boldsymbol{Y}$ are uncorrelated. Further, since $\mathbb{E}\big[\boldsymbol{Z}\boldsymbol{Z}^{\mathsf{T}}\big] = 0$, the covariance matrices of $\boldsymbol{X}$ and $\boldsymbol{Y}$ are identical and equal to $\frac{1}{2}\mathbb{E}[\boldsymbol{Z}\boldsymbol{Z}^*]$. Hence, $\boldsymbol{Z}$ is a mean-zero complex Gaussian random variable. □

We are most interested in complex-valued stochastic integrals of Fourier type, for which the integrand $g(\nu)$ depends on t.

Definition 6.28 (stochastic Fourier integral) We say $\{Z(t)\colon t \in \mathbb{R}\}$ is a *stochastic Fourier integral process* if

$$Z(t) = \int_{\mathbb{R}} e^{it\nu} \sqrt{f(\nu)}\, d\mathcal{W}(\nu), \tag{6.22}$$

for some integrable function $f\colon \mathbb{R} \to \mathbb{R}^+$ and $\mathcal{W}(\nu)$ defined by (6.18).

It is clear from the preceding discussion that $Z(t)$ defines a mean-zero complex-valued stochastic process with covariance function $C_Z(s,t) = \mathbb{E}\Big[Z(s)\overline{Z(t)}\Big]$.

Example 6.29 (exponential covariance) Let

$$Z(t) = \int_{\mathbb{R}} e^{it\nu} \sqrt{\frac{\ell}{\pi(1 + \ell^2\nu^2)}}\, d\mathcal{W}(\nu).$$

From (6.22) and Lemma 6.26, the covariance of $Z(t)$ is

$$C_Z(s,t) = \mathbb{E}\Big[Z(s)\overline{Z(t)}\Big] = 2\int_{\mathbb{R}} e^{i(s-t)\nu} \frac{\ell}{\pi(1 + \ell^2\nu^2)}\, d\nu.$$

This equals $2c(s-t)$ for $c(t) = e^{-|t|/\ell}$ by Example 6.6. $Z(t)$ is therefore a stationary complex-valued process with mean zero and stationary exponential covariance $2c(t)$.

Example 6.30 (Gaussian covariance) Let

$$Z(t) = \int_{\mathbb{R}} e^{itv} \sqrt{\frac{\ell}{\sqrt{4\pi}} e^{-v^2\ell^2/4}}\, d\mathcal{W}(v).$$

This time, the covariance is

$$C_Z(s,t) = 2\mathbb{E}\left[Z(s)\overline{Z(t)}\right] = 2\int_{\mathbb{R}} e^{i(s-t)v} \frac{\ell}{\sqrt{4\pi}} e^{-v^2\ell^2/4}\, dv = 2c(s-t)$$

for $c(t) = e^{-t^2/\ell^2}$ by Example 6.7. Again, $Z(t)$ is a stationary complex-valued process with mean zero and the stationary Gaussian covariance $2c(t)$.

Complex Gaussian process

When f is even, as in Examples 6.29 and 6.30, the stochastic Fourier integral $Z(t)$ in (6.22), in addition to being stationary, is a complex Gaussian process (see Proposition 6.32). Significantly, this means that the real and imaginary parts of $Z(t)$ are independent copies of a real-valued stationary Gaussian process. Recall Definition 5.16 of independent sample paths.

Lemma 6.31 *$Z(t)$ is a complex-valued Gaussian process with $\mu(t) = 0$ and real-valued covariance $C_Z(s,t) = 2C(s,t)$ if and only if the real and imaginary parts of $Z(t)$ are independent Gaussian processes, each with mean zero and covariance $C(s,t)$.*

Proof Write $Z(t) = X(t)+iY(t)$, for mean-zero real-valued processes $X(t), Y(t)$. Then, $Z(t)$ is a complex Gaussian process if and only if $\boldsymbol{Z} = [Z(t_1),\dots,Z(t_M)]^{\mathsf{T}}$ is a complex Gaussian for any $t_1,\dots,t_M \in \mathcal{T}$, if and only if $\boldsymbol{X} = [X(t_1),\dots,X(t_M)]^{\mathsf{T}}, \boldsymbol{Y} = [Y(t_1),\dots,Y(t_M)]^{\mathsf{T}}$ are *iid* multivariate Gaussian random variables, if and only if the real and imaginary parts of $Z(t)$ are *iid* real-valued Gaussian processes. With $t_1 = s$ and $t_2 = t$,

$$\boldsymbol{Z} = \begin{pmatrix} Z(s) \\ Z(t) \end{pmatrix} \sim \mathrm{CN}(\boldsymbol{0}, 2C) \quad \text{with} \quad C = \begin{pmatrix} C(s,s) & C(s,t) \\ C(s,t) & C(t,t) \end{pmatrix}, \qquad \forall s,t \in \mathcal{T}$$

if and only if

$$\begin{pmatrix} X(s) \\ X(t) \end{pmatrix}, \begin{pmatrix} Y(s) \\ Y(t) \end{pmatrix} \sim \mathrm{N}(\boldsymbol{0}, C) \quad iid.$$

This establishes the relationship between the covariance $C(s,t)$ of $X(t)$ and $Y(t)$ and the covariance $2C(s,t)$ of the complex-valued process $Z(t)$. □

The next result gives conditions for $Z(t)$ to be a stationary complex Gaussian process.

Proposition 6.32 *Consider an integrable and even function $f\colon \mathbb{R} \to \mathbb{R}^+$. The process $Z(t)$ defined by* (6.22) *is a stationary complex Gaussian process with mean zero and real-valued covariance $2c(t)$, where $c(t)$ is given by* (6.3).

Proof Because f is even,

$$c(t) = \int_{\mathbb{R}} e^{itv} f(v)\, dv = \int_{\mathbb{R}} \cos(tv) f(v)\, dv.$$

Crucially, this means that $c(t)$ is *real valued* and Corollary 6.27 implies that $\mathbf{Z} := [Z(t_1), \dots, Z(t_N)]^{\mathsf{T}}$ follows a complex Gaussian distribution.

Hence, $Z(t)$ is a complex Gaussian process and, by Lemma 6.31, the real and imaginary parts of $Z(t)$ are *iid* with stationary covariance $c(t)$. The process $Z(t)$ is stationary because the mean is zero and the covariance $C_Z(s,t)$ is a function of $s-t$. Using (6.20),

$$\operatorname{Cov}(Z(s), Z(t)) = 2\left\langle e^{isv}\sqrt{f(v)}, e^{itv}\sqrt{f(v)}\right\rangle_{L^2(\mathbb{R},\mathbb{C})} = 2\int_{\mathbb{R}} e^{i(s-t)v} f(v)\,dv,$$

which equals $2c(s-t)$, as required. □

Example 6.33 (sinc covariance) Let $Z(t)$ be defined as in (6.22), where

$$f(v) = \begin{cases} \ell/2\pi, & |v| < \pi/\ell, \\ 0, & \pi/\ell \le |v|, \end{cases} \tag{6.23}$$

for a given constant $\ell > 0$. From the above results, we know $Z(t)$ is a mean-zero complex Gaussian process. Using (6.20), the covariance is

$$C_Z(s,t) = \mathbb{E}\left[Z(s)\overline{Z(t)}\right] = 2\int_{\mathbb{R}} e^{i(s-t)v} f(v)\,dv = 2\int_{-\pi/\ell}^{\pi/\ell} e^{i(s-t)v}\frac{\ell}{2\pi}\,dv,$$

and straightforward integration gives

$$C_Z(s,t) = \begin{cases} 2, & s = t, \\ 2\dfrac{\sin((s-t)\pi/\ell)}{(s-t)\pi/\ell}, & s \ne t. \end{cases} \tag{6.24}$$

That is, $C_Z(s,t) = 2c(s-t)$, where $c(t) = \operatorname{sinc}(t/\ell)$ and $\operatorname{sinc}(x) := \sin(\pi x)/\pi x$. The real and imaginary parts of $Z(t)$ are independent real-valued stationary Gaussian processes and can be used to generate independent sample paths of a Gaussian process with mean zero and stationary covariance $c(t)$. Figure 6.7 shows two independent sample paths of the process with $\ell = 2$.

6.4 Sampling by quadrature

The stochastic Fourier integral (6.22) provides a natural route to the approximation of $Z(t)$. For a given set of N sample times t_n, the idea is to approximate the complex random variables $Z(t_n)$ via quadrature. We use a generalisation of the trapezium rule for stochastic integrals. As the integral (6.22) is over the real line, we first approximate $Z(t)$ by

$$Z_R(t) := \int_{-R}^{R} e^{itv}\sqrt{f(v)}\,dW(v)$$

for a finite interval $[-R, R]$. To approximate $Z_R(t)$ for a fixed t, we introduce a grid $v_j = -R + j\Delta v$, $j = 0, \dots, J-1$, of J uniformly spaced quadrature points on the interval $[-R, R]$ with spacing $\Delta v = 2R/(J-1)$. Consider the quadrature rule $\widetilde{Z}_R(t)$ defined by

$$\widetilde{Z}_R(t) := \sum_{j=0}^{J-1} e^{itv_j}\sqrt{f(v_j)}\,\Delta W_j, \tag{6.25}$$

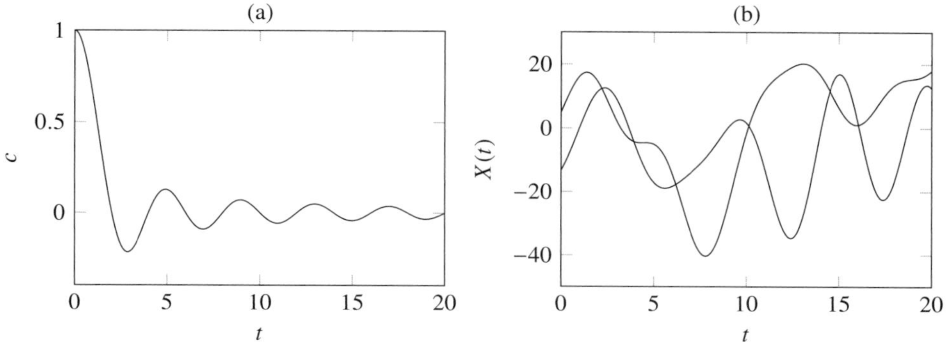

Figure 6.7 Let $Z(t)$ be a mean-zero complex Gaussian process with stationary covariance function $2c(t)$ for $c(t) = \mathrm{sinc}(t/2)$ as shown in (a). In (b), we plot approximate sample paths of the real and imaginary parts of $Z(t)$, computed using Algorithm 6.3 with the commands `t=[0:0.1:20]';` `Z=quad_sinc(t, 100, 2); plot(t,[real(Z),imag(Z)]);`

where

$$\Delta \mathcal{W}_j := \begin{cases} \mathcal{W}(\nu_0 + \Delta\nu/2) - \mathcal{W}(\nu_0), & j = 0, \\ \mathcal{W}(\nu_j + \Delta\nu/2) - \mathcal{W}(\nu_j - \Delta\nu/2), & j = 1, \dots, J-2, \\ \mathcal{W}(\nu_{J-1}) - \mathcal{W}(\nu_{J-1} - \Delta\nu/2), & j = J-1. \end{cases} \tag{6.26}$$

The increments $\Delta\mathcal{W}_j \sim \mathrm{CN}(0, 2\Delta\nu)$ for $j = 1, \dots, J-2$ and $\Delta\mathcal{W}_j \sim \mathrm{CN}(0, \Delta\nu)$ for $j = 0, J-1$. They are pairwise independent and can be easily sampled (as in Example 6.16), so that $\widetilde{Z}_R(t)$ is easy to sample for a given t.

Example 6.34 (sinc covariance) The spectral density of the covariance function $c(t) = \mathrm{sinc}(t/\ell)$ of Example 6.33 has support $[-\pi/\ell, \pi/\ell]$ (see (6.23)). This is convenient for the quadrature method as we may choose the range $R = \pi/\ell$ to exactly capture the support. Algorithm 6.3 approximates the mean-zero complex Gaussian process $Z(t)$ with stationary covariance $c_Z(t) = 2\,\mathrm{sinc}(t/\ell)$. The algorithm returns $\widetilde{Z}_R(t)$ sampled at a given set of sample times $t = t_n$ and Figure 6.7 plots the real and imaginary parts of $\widetilde{Z}_R(t_n)$ for $\ell = 2$.

Algorithm 6.3 Code to sample $\widetilde{Z}_R(t)$ for the sinc covariance $2\,\mathrm{sinc}(t/\ell)$ in Example 6.33. The inputs are a vector of sample times `t`, the number of quadrature points `J`, and the parameter `ell` $= \ell$ for the covariance function. The output `Z` is a vector containing $\widetilde{Z}_R(t)$ evaluated at the sample times in `t`.

```
function Z=quad_sinc(t, J, ell)
R=pi/ell;  nustep=2*R/J;
Z=exp(-sqrt(-1)*t*R)*randn(1,2)*[sqrt(-1);1]/sqrt(2);
for j=1:J-2,
    nu=-R+j*nustep;
    Z=Z+exp(sqrt(-1)*t*nu)*randn(1,2)*[sqrt(-1);1];
end;
Z=Z+exp(sqrt(-1)*t*R)*randn(1,2)*[sqrt(-1);1]/sqrt(2)
Z=Z*sqrt(ell/(2*pi));
```

Error analysis

The accuracy of the approximation $\widetilde{Z}_R(t)$ to $Z(t)$ depends on the spacing $\Delta\nu$ and the range parameter R. In the following lemmas, we find upper bounds on the root-mean-square error $\|Z(t)-\widetilde{Z}_R(t)\|_{L^2(\Omega,\mathbb{C})} = \mathbb{E}\big[|Z(t)-\widetilde{Z}_R(t)|^2\big]^{1/2}$ (Lemma 6.35) and the error in the covariance function (Lemma 6.36). The latter has important consequences for the approximation of averages in Corollary 6.37. To obtain these error bounds, we assume the spectral density function $f(\nu)$ decays as $|\nu|\to\infty$ according to (6.11). As we saw in Lemma 6.11, the decay rate determines the regularity of the process and, of course, this influences the quality of the approximation. We begin with the root-mean-square error.

Introducing the variable $\hat\nu$ defined via $\hat\nu := \nu_j$ if $\nu \in [\nu_j - \Delta\nu/2, \nu_j + \Delta\nu/2)$ for $j = 0,\dots,J-1$, we define a piecewise constant function $\mathrm{e}^{\mathrm{i}t\hat\nu}\sqrt{f(\hat\nu)}$ of ν. Then, the quadrature rule (6.25) may be conveniently written as the integral

$$\widetilde{Z}_R(t) = \int_{-R}^{R} \mathrm{e}^{\mathrm{i}t\hat\nu}\sqrt{f(\hat\nu)}\,dW(\nu).$$

Lemma 6.35 (root-mean-square error) *Suppose that $\sqrt{f(\nu)}$ is globally Lipschitz continuous with Lipschitz constant L and that the decay condition* (6.11) *holds for some constants $L_q, q > 0$. For $t > 0$ and $K_t := 2\big(t\sqrt{\|f\|_\infty} + L\big)$,*

$$\|Z(t)-\widetilde{Z}_R(t)\|_{L^2(\Omega,\mathbb{C})} \le K_t\sqrt{R}\Delta\nu + \frac{\sqrt{2L_q}}{\sqrt{q}}\frac{1}{R^q}. \tag{6.27}$$

Proof First,

$$\|Z(t)-\widetilde{Z}_R(t)\|_{L^2(\Omega,\mathbb{C})} \le \|Z(t)-Z_R(t)\|_{L^2(\Omega,\mathbb{C})} + \|Z_R(t)-\widetilde{Z}_R(t)\|_{L^2(\Omega,\mathbb{C})},$$

where $\|Z\|_{L^2(\Omega,\mathbb{C})} := \mathbb{E}\big[|Z|^2\big]^{1/2}$. The error due to the quadrature spacing $\Delta\nu$ is

$$Z_R(t)-\widetilde{Z}_R(t) = \int_{-R}^{R}\Big(\mathrm{e}^{\mathrm{i}t\nu}\sqrt{f(\nu)} - \mathrm{e}^{\mathrm{i}t\hat\nu}\sqrt{f(\hat\nu)}\Big)\,dW(\nu)$$

and, from (6.20),

$$\mathbb{E}\Big[\big|Z_R(t)-\widetilde{Z}_R(t)\big|^2\Big] = 2\int_{-R}^{R}\Big|\mathrm{e}^{\mathrm{i}t\nu}\sqrt{f(\nu)} - \mathrm{e}^{\mathrm{i}t\hat\nu}\sqrt{f(\hat\nu)}\Big|^2\,d\nu.$$

As the Lipschitz constant of $\sqrt{f(\nu)}$ is L, the function $\mathrm{e}^{\mathrm{i}t\nu}\sqrt{f(\nu)}$ has Lipschitz constant $t\sqrt{\|f\|_\infty}+L$. Therefore,

$$\mathbb{E}\Big[\big|Z_R(t)-\widetilde{Z}_R(t)\big|^2\Big] \le 4R\Delta\nu^2\Big(t\sqrt{\|f\|_\infty}+L\Big)^2.$$

Using the decay condition (6.11), the error due to the finite range is

$$\begin{aligned}\mathbb{E}\Big[\big|Z(t)-Z_R(t)\big|^2\Big] &= 2\int_{-\infty}^{-R}|f(\nu)|\,d\nu + 2\int_{R}^{\infty}|f(\nu)|\,d\nu\\ &\le 4\int_R^\infty \frac{L_q}{\nu^{2q+1}}\,d\nu = \frac{4L_q}{2qR^{2q}}.\end{aligned}$$

Combining the two errors, we obtain (6.27). □

Now, let

$$c_R(t) := \int_{-R}^{R} e^{itv} f(v)\, dv, \tag{6.28}$$

$$\tilde{c}_R(t) := \left(\frac{e^{itv_0} f(v_0)}{2} + \sum_{j=1}^{J-2} e^{itv_j} f(v_j) + \frac{e^{itv_{J-1}} f(v_{J-1})}{2} \right) \Delta v. \tag{6.29}$$

The approximations $Z_R(t)$ and $\widetilde{Z}_R(t)$ of $Z(t)$ have stationary covariance functions $2c_R(t)$ and $2\tilde{c}_R(t)$ respectively. We look at the error $2|c(t) - \tilde{c}_R(t)|$ in approximating the stationary covariance $2c(t)$ of $Z(t)$ by that of $\widetilde{Z}_R(t)$, making use of $c_R(t)$ in the proof.

Lemma 6.36 (error in covariance) *Suppose that $f \in C^4(\mathbb{R})$ is even and satisfies the decay condition* (6.11). *If $K_t := \sup_{v \in \mathbb{R}} \left| \frac{d^4}{dv^4} e^{itv} f(v) \right|$,*

$$\left| c(t) - \tilde{c}_R(t) \right| \le \frac{1}{6} \Delta v^2 \left[\left| t f(R) \right| + \left| f'(R) \right| \right] + \frac{L_q}{q} \frac{1}{R^{2q}} + \frac{K_t}{360} \Delta v^4 R. \tag{6.30}$$

Proof $\tilde{c}_R(t)$ is the trapezium rule approximation to $\int_{-R}^{R} \psi(v)\, dv = c_R(t)$ with $\psi(v) = e^{itv} f(v)$. Now,

$$\psi'(v) = (it f(v) + f'(v)) e^{itv},$$

and $\psi'(R) - \psi'(-R) = -2t f(R) \sin tR + 2 f'(R) \cos tR$. Using Theorem A.6, the trapezium rule error is

$$\left| c_R(t) - \tilde{c}_R(t) \right| \le \frac{\Delta v^2}{6} \left[\left| t f(R) \right| + \left| f'(R) \right| \right] + \Delta v^4 \frac{2R}{720} K_t. \tag{6.31}$$

This is the error in the covariance due to the grid space Δv. The error due to the finite range R is similar to the mean-square case. From (6.28),

$$\left| c(t) - c_R(t) \right| \le \int_{-\infty}^{-R} f(v)\, dv + \int_{R}^{\infty} f(v)\, dv \le \frac{2L_q}{2q R^{2q}}.$$

Since $|c(t) - \tilde{c}_R(t)| \le |c(t) - c_R(t)| + |c_R(t) - \tilde{c}_R(t)|$, we obtain (6.30). □

Both upper bounds for the approximation error converge to zero as $R \to \infty$ and $\Delta v \to 0$. The constant K_t in the upper bound for the error grows with t in both lemmas. The predicted growth of the covariance error with respect to t and Δv can be clearly seen in Figure 6.8 for the sinc covariance. However, the spectral density (6.23) of the sinc covariance is not continuous and does not satisfy the conditions of Lemma 6.36.

Evaluating $\widetilde{Z}_R(t)$ using the fast Fourier transform

To approximate a sample path $Z(t,\omega)$ for $t \in [0,T]$ and a fixed $\omega \in \Omega$, we evaluate $\widetilde{Z}_R(t,\omega)$ at a set of N uniformly spaced sample times $t_n = n\Delta t$ for $\Delta t = T/(N-1)$ and $n = 0,\ldots,N-1$. $\widetilde{Z}_R(t_n,\omega)$ forms a vector in $\mathbb{C}^N$ and can be written in terms of the discrete Fourier transform (DFT), which allows efficient evaluation by the FFT. To do this, however, we must choose the grid spacing Δt, the quadrature spacing Δv, and range parameter R very carefully. Details are given below.

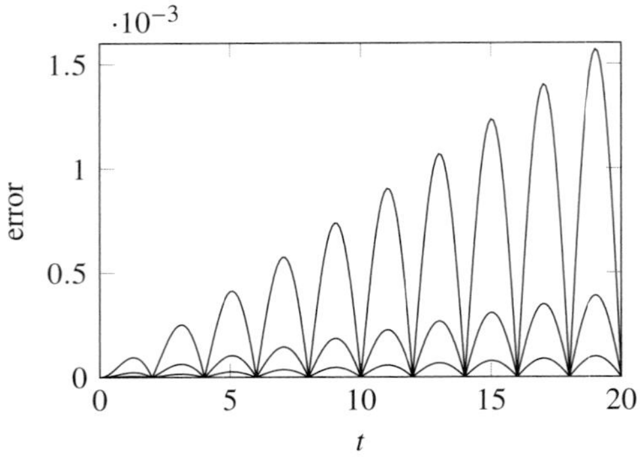

Figure 6.8 The error $|c(t) - \tilde{c}_R(t)|$ between the sinc covariance $c(t) = \text{sinc}(t/\ell)$ with $\ell = 2$ and the approximation $\tilde{c}_R(t)$ from (6.28), with $R = \pi/\ell$ and three choices of quadrature spacing $\Delta\nu = R/40$, $R/80$, $R/160$. The error grows as t and $\Delta\nu$ increase.

For quadrature points $\nu_j = -R + j\Delta\nu$, we have

$$\tilde{Z}_R(t_n) = \sum_{j=0}^{J-1} e^{it_n\nu_j} \sqrt{f(\nu_j)}\, \Delta W_j. \tag{6.32}$$

To turn this into a DFT, we need to convert the sum in

$$e^{it_n\nu_j} = e^{it_n(-R+j\Delta\nu)} = e^{-iRt_n} e^{in\Delta t j\Delta\nu}$$

with $t_n = n\Delta t$ for $n = 0, \dots, N-1$, into a sum in $e^{i2\pi nj/N}$, with $j = 0, \dots, J-1$. We can achieve this by simply setting $J = N$ and $\Delta t\Delta\nu = 2\pi/N$. However, this is too restrictive as it gives $\Delta\nu = 2\pi/N\Delta t$, which does not become small as N increases with $\Delta t = T/(N-1)$. Instead, we choose $\Delta t(M\Delta\nu) = 2\pi/N$ and $J = NM$, which gives $\Delta\nu = 2\pi/NM\Delta t$. Now, as $2R = (J-1)\Delta\nu$, the three discretisation parameters are

$$\Delta t = \frac{T}{N-1}, \qquad \Delta\nu = \frac{2\pi}{NM\Delta t}, \qquad R = \frac{(NM-1)}{NM}\frac{\pi}{\Delta t}. \tag{6.33}$$

From (6.27) and (6.30), for the root-mean-square error and the covariance error to converge to zero, we need $\Delta\nu \to 0$ and $R \to \infty$. Hence, we must let $M \to \infty$ and $N \to \infty$ (where N is the number of sample times and $J = NM$ is the number of quadrature points).

With the above choices, the quadrature points can be written as $\nu_j = -R + j\Delta\nu$ for $j = 0, 1 \dots, NM-1$ with $j = kM + m$ for $m = 0, \dots, M-1$ and $k = 0, \dots, N-1$. Then, with the increments (6.26),

$$\begin{aligned}
\tilde{Z}_R(t_n) &= e^{-iRt_n} \sum_{j=0}^{NM-1} e^{ij\Delta\nu t_n} \sqrt{f(\nu_j)}\, \Delta W_j \\
&= e^{-iRt_n} \sum_{m=0}^{M-1} \sum_{k=0}^{N-1} e^{i(kM+m)\Delta\nu t_n} \sqrt{f(\nu_{kM+m})}\, \Delta W_{kM+m} \\
&= \sum_{m=0}^{M-1} e^{i(-Rt_n + m\Delta\nu t_n)} \sum_{k=0}^{N-1} e^{ikM\Delta\nu n\Delta t} \sqrt{f(\nu_{kM+m})}\, \Delta W_{kM+m}.
\end{aligned}$$

Finally, using $2\pi/N = M\Delta\nu\,\Delta t$ from (6.33),

$$\widetilde{Z}_R(t_n) = \sum_{m=0}^{M-1} \mathrm{e}^{\mathrm{i}(-R+m\Delta\nu)t_n} \sqrt{f(\nu_{kM+m})}\,\Delta W_{kM+m}. \tag{6.34}$$

Algorithm 6.4 Code to sample $\widetilde{Z}_R(t)$ at N uniformly spaced sample times in $[0,T]$. The inputs are the length of the sample interval `T`, the number of sample times `N`, a discretisation parameter `M` from (6.33), and a function handle `fhandle` for the spectral density. The outputs are the vector of values `t` at which $\widetilde{Z}_R$ is sampled and a vector `Z` of length N, whose nth entry is a sample of $\widetilde{Z}_R(t_n)$.

```
function [t, Z]=squad(T, N, M, fhandle)
dt=T/(N-1); t=dt*[0:N-1]';
R=pi/dt; dnu=2*pi/(N*dt*M);
Z=zeros(N, 1); coeff=zeros(N,1);
for m=1:M,
    for k=1:N,
        nu=-R+((k-1)*M+(m-1))*dnu;
        xi=randn(1,2)*[1;sqrt(-1)];
        coeff(k)=sqrt(fhandle(nu)*dnu)*xi;
        if ((m==1 && k==1) || (m==M && k==N))
            coeff(k)=coeff(k)/sqrt(2);
        end;
    end;
    Zi=N*ifft(coeff);
    Z=Z+exp(sqrt(-1)*(-R+(m-1)*dnu)*t).*Zi;
end;
```

To evaluate this expression simultaneously for $n = 0,\ldots,N-1$, we evaluate the inner sum with M FFTs of size N and then sum M vectors of size N. This is implemented in Algorithm 6.4 and the total amount of work is $\mathcal{O}(NM + NM\log N)$. The naive method of evaluating $\widetilde{Z}_R(t_n)$ for each $n = 0,\ldots,N-1$ using (6.33) with $J = NM$ quadrature points has complexity $\mathcal{O}(N^2M)$.

Non-uniformly spaced sample times

Often, a numerical method calls for samples of a real-valued stationary Gaussian process at a set of sample times t that are not uniformly spaced. We encounter such a situation in §7.3 when studying the turning bands method for sampling a class of random fields. To take advantage of the FFT in approximating $Z(t) = X(t) + \mathrm{i}Y(t)$ via quadrature, we require N uniformly spaced sample times $t_n = n\Delta t$, $n = 0,1,\ldots N-1$. However, we can approximate $X(t)$ (and $Y(t)$ is similar) at a general set of sample times $t = s_1,\ldots,s_P$ using interpolation as follows:

(i) For given N, M, define Δt, R, $\Delta\nu$ by (6.33) where (assuming the s_p are ordered) $T = s_P - s_1$ is the length of the sample interval.
(ii) Generate $\widetilde{Z}_R(t_n)$ at $t_n = n\Delta t$ for $n = 0,\ldots,N-1$ by (6.34) using Algorithm 6.4.
(iii) Denote the linear interpolant of $\widetilde{X}_R(t) = \mathrm{Re}\,\widetilde{Z}_R(t)$ at $t = t_n$, for $n = 0,\ldots,N-1$, by $I(t)$ and evaluate $X_p = I(s_p)$ at $p = 1,\ldots,P$.

Algorithm 6.5 Code to sample the real-valued stationary Gaussian process with mean zero and spectral density $f(\nu)$, where $f(\nu)$ is defined by `fhandle`. Two independent sample paths evaluated at the vector of times `s` (which may be non-uniformly spaced) are returned in `X`, `Y`, which are calculated as the real and imaginary parts of $I(t)$, with `N` and `M` giving the discretisation parameters (see (6.33)).

```
function [X,Y]=interp_quad(s, N, M, fhandle)
T=max(s)-min(s);
[t, Z]=squad(T, N, M, fhandle);
I=interp1(t+min(s), Z, s);
X=real(I); Y=imag(I);
```

This is implemented in Algorithm 6.5. The complexity is $\mathcal{O}(P + NM\log N)$ and is efficient when $P > N$, which is the case in §7.3.

When approximating $[X(s_1),\dots,X(s_P)]^{\mathsf{T}}$ by $[X_1,\dots,X_P]^{\mathsf{T}}$, the error is controlled by the size of N and M. The following corollary gives the error in approximating averages of the form $\mathbb{E}[\phi(X(s_1),\dots,X(s_P))]$ for a test function $\phi \in L^2(\mathbb{R}^P)$ by $\mathbb{E}[\phi(X_1,\dots,X_P)]$.

Corollary 6.37 (weak error) *Suppose that $X(t)$ is a real-valued stationary process with spectral density $f \in \mathrm{C}^4(\mathbb{R})$ satisfying the decay condition* (6.11) *for some $q > 1$. Further suppose that $|f'(R)| \le K_q/R^{2q+2}$ for some $K_q > 0$. For any $s_1,\dots,s_P \in [0,T]$, let*

$$\boldsymbol{X}_{\text{true}} := \big[X(s_1),\dots,X(s_P)\big]^{\mathsf{T}}, \qquad \boldsymbol{X}_{\text{approx}} := [X_1,\dots,X_P]^{\mathsf{T}}.$$

There exists $K,\bar{N} > 0$ such that

$$\big|\mathbb{E}\big[\phi(\boldsymbol{X}_{\text{true}})\big] - \mathbb{E}\big[\phi(\boldsymbol{X}_{\text{approx}})\big]\big| \le K\|\phi\|_{L^2(\mathbb{R}^P)}\Big(\frac{N}{M^4} + \frac{1}{N^2}\Big) \tag{6.35}$$

for $N > \bar{N}$ and $M \in \mathbb{N}$ and any $\phi \in L^2(\mathbb{R}^P)$.

Proof $\boldsymbol{X}_{\text{true}}$ and $\boldsymbol{X}_{\text{approx}}$ are multivariate Gaussian random variables with mean zero. Denote their covariance matrices by C_{true}, C_{approx} respectively. The covariance matrix C_{true} has entries $c_Z(s_j - s_k)/2$. Denote the covariance function of the linear interpolant $I(t)$ of $\tilde{X}_R(t) = \operatorname{Re}\tilde{Z}_R(t)$ by $C_I(s,t)$, so that the covariance matrix C_{approx} has entries $C_I(s_j,s_k)$. With $R,\Delta\nu$ given by (6.33), Lemma 6.36 provides $K_1 > 0$ such that

$$|c_Z(s-t) - \tilde{c}_R(s-t)| \le K_1\Big(\frac{1}{M^2}\frac{1}{N^{2q}} + \frac{N}{M^4} + \frac{1}{N^{2q}}\Big), \qquad \forall\, s,t \in [0,T].$$

As $q > 1$, Lemma 6.11 gives that $\tilde{Z}_R(t)$ is mean-square differentiable. $I(t)$ is the linear interpolant of $\operatorname{Re}\tilde{Z}_R(t)$, which has covariance $\tilde{c}_R(t)/2$. By Proposition 5.37, for constants $K_2,\bar{N} > 0$, we have for $N > \bar{N}$

$$\big|\tilde{c}_R(s-t) - C_I(s,t)\big| \le \frac{K_2}{N^2}, \qquad s,t \in [0,T].$$

Hence, using again $q > 1$,

$$\max_{j,k=0,\dots,P-1}\big|c_Z(s_j - s_k) - C_I(s_j,s_k)\big| \le K_1\frac{N}{M^4} + \frac{K_1 + K_2}{N^2}. \tag{6.36}$$

This implies that $\|C_{\text{true}} - C_{\text{approx}}\|_{\mathrm{F}} = \mathcal{O}(NM^{-4} + N^{-2})$. C_{true} is positive definite because the covariance of $X(t)$ is a positive-definite function due to Exercise 6.12. To complete the proof, apply Theorem 4.63. □

Algorithm 6.6 Code to sample the mean-zero stationary Gaussian process with the Whittle–Matérn covariance $c_q(t)$. The outputs and the inputs `s`, `N`, `M` are as in Algorithm 6.5. The last input `q` defines the Whittle–Matérn parameter.

```
function [X,Y]=quad_wm(s, N, M, q)
[X,Y]=interp_quad(s, N, M, @(nu) f_wm(nu,q));

function f=f_wm(nu, q) % spectral density
const=gamma(q+0.5)/(gamma(q)*gamma(0.5));
f=const/((1+nu*nu)^(q+0.5));
```

Example 6.38 (Whittle–Matérn covariance) The spectral density of the Whittle–Matérn covariance is given by (6.6) and the quadrature sampling method is easy to implement; see Algorithm 6.6. The spectral density belongs to $\mathrm{C}^4(\mathbb{R})$ and satisfies the decay condition on $f'(R)$ given in Corollary 6.37. Hence, for $q > 1$, the weak error is $\mathcal{O}(NM^{-4} + N^{-2})$ and this suggests keeping N^3/M^4 constant as we increase accuracy in the limit $N \to \infty$. For $q \le 1$, the process is not mean-square differentiable and Corollary 6.37 does not apply. In that case, it can be shown that the weak error is $\mathcal{O}(NM^{-4} + N^{-2q})$ and it is advantageous to keep N^{1+2q}/M^4 constant as $N \to \infty$.

6.5 Sampling by circulant embedding

Let $X(t)$ be a mean-zero stationary Gaussian process. The quadrature method described above provides samples of $X(t)$ on a time interval $[0,T]$ that are approximate. We develop an *exact* method for sampling $X(t)$ at uniformly spaced sample times $t_n = n\Delta t$ for $n = 0,\dots,N-1$ known as *the circulant embedding method.* If it is required to sample $X(t)$ at non-uniformly spaced times $t = s_1,\dots,s_P$, linear interpolation can be applied as in §6.4. This introduces a weak approximation error, as in (6.35), of size $\mathcal{O}(\Delta t^2)$ for mean-square differentiable processes. Because the interpolation is based on samples from the correct distribution, the method is more accurate than the quadrature method.

Denote the vector of samples by $\boldsymbol{X}$,

$$\boldsymbol{X} := \left[X(t_0), X(t_2), \dots, X(t_{N-1})\right]^{\mathsf{T}}, \qquad t_n = n\Delta t, \quad \Delta t = \frac{T}{N-1}.$$

The distribution of $\boldsymbol{X}$ is $\mathrm{N}(\mathbf{0}, C)$ for an $N \times N$ covariance matrix C and we already have exact sampling methods for this distribution: for example, we may calculate the spectral decomposition of C as in (5.27) and then form (5.28) (with $\boldsymbol{\mu} = \mathbf{0}$). If C is circulant (see Definition 6.41), the spectral decomposition takes a special form in terms of the DFT and the calculations can be done efficiently using the FFT. Covariance matrices are not in general circulant. However, symmetric Toeplitz matrices, which arise when sampling stationary processes at uniformly spaced times, can be embedded inside circulant matrices and the DFT factorisation applied. This leads to a very efficient sampling method for stationary processes.

Real-valued Toeplitz and circulant matrices

We begin with some basic definitions and results from linear algebra.

Definition 6.39 (Toeplitz matrix) An $N \times N$ real-valued matrix $C = (c_{ij})$ is *Toeplitz* if $c_{ij} = c_{i-j}$ for some real numbers $c_{1-N}, \dots, c_{N-1}$. We have

$$C = \begin{pmatrix} c_0 & c_{-1} & \dots & c_{2-N} & c_{1-N} \\ c_1 & c_0 & c_{-1} & \ddots & c_{2-N} \\ \vdots & \ddots & \ddots & \ddots & \vdots \\ c_{N-2} & \ddots & c_1 & c_0 & c_{-1} \\ c_{N-1} & c_{N-2} & \dots & c_1 & c_0 \end{pmatrix},$$

and the entries of C are constant on each diagonal. An $N \times N$ Toeplitz matrix is uniquely defined by the vector $\boldsymbol{c} = [c_{1-N}, \dots c_{-1}, c_0, c_1, \dots c_{N-1}]^{\mathsf{T}} \in \mathbb{R}^{2N-1}$, containing the entries of the first row and column. A *symmetric Toeplitz* matrix has $c_{i-j} = c_{j-i}$ and is defined by its first column $\boldsymbol{c}_1 = [c_0, c_1, \dots, c_{N-1}]^{\mathsf{T}} \in \mathbb{R}^N$.

Example 6.40 Consider the following 3×3 Toeplitz matrices:

$$A = \begin{pmatrix} 1 & 2 & 3 \\ 4 & 1 & 2 \\ 5 & 4 & 1 \end{pmatrix}, \qquad B = \begin{pmatrix} 1 & 2 & 3 \\ 2 & 1 & 2 \\ 3 & 2 & 1 \end{pmatrix}. \tag{6.37}$$

Note that A can be generated from the vector $\boldsymbol{c} = [3, 2, 1, 4, 5]^{\mathsf{T}}$ and B, which is symmetric, can be generated from $\boldsymbol{c}_1 = [1, 2, 3]^{\mathsf{T}}$.

Definition 6.41 (circulant matrix) An $N \times N$ real-valued Toeplitz matrix $C = (c_{ij})$ is *circulant* if each column is a circular shift of the elements of the preceding column (so that the last entry becomes the first entry). That is,

$$C = \begin{pmatrix} c_0 & c_{N-1} & \dots & c_2 & c_1 \\ c_1 & c_0 & c_{N-1} & \dots & c_2 \\ \vdots & \ddots & \ddots & \ddots & \vdots \\ c_{N-2} & \ddots & c_1 & c_0 & c_{N-1} \\ c_{N-1} & c_{N-2} & \dots & c_1 & c_0 \end{pmatrix},$$

which is uniquely determined by the first column $\boldsymbol{c}_1 = [c_0, c_1, \dots, c_{N-1}] \in \mathbb{R}^N$. We have $c_{ij} = c_{i-j}$ for $1 \le j \le i$ and $c_{ij} = c_{i-j+N}$ for $i + 1 \le j \le N$. *Symmetric circulant* matrices also have

$$c_{N-j} = c_j, \qquad \text{for } j = 1, 2, \dots, N-1, \tag{6.38}$$

and hence at most $\lfloor N/2 \rfloor + 1$ distinct entries.

Definition 6.42 (Hermitian vector) A vector $\boldsymbol{v} = [v_1, v_2, \dots, v_N]^{\mathsf{T}} \in \mathbb{C}^N$ is a *Hermitian vector* if $v_1 \in \mathbb{R}$ and

$$v_j = \bar{v}_{N-j+2}, \qquad \text{for } j = 2, \dots, N. \tag{6.39}$$

The first column $\boldsymbol{c}_1 = [c_0, \dots, c_{N-1}]^{\mathsf{T}}$ of a symmetric circulant matrix $C \in \mathbb{R}^{N \times N}$ is always a Hermitian vector as $c_0 \in \mathbb{R}$ and (6.38) coincides with (6.39).

Example 6.43 Consider the following symmetric circulant matrices:

$$C = \begin{pmatrix} 3 & 2 & 1 & 2 \\ 2 & 3 & 2 & 1 \\ 1 & 2 & 3 & 2 \\ 2 & 1 & 2 & 3 \end{pmatrix}, \qquad D = \begin{pmatrix} 1 & 2 & 3 & 3 & 2 \\ 2 & 1 & 2 & 3 & 3 \\ 3 & 2 & 1 & 2 & 3 \\ 3 & 3 & 2 & 1 & 2 \\ 2 & 3 & 3 & 2 & 1 \end{pmatrix}. \tag{6.40}$$

Here, C can be generated from $\boldsymbol{c}_1 = [3, 2, 1, 2]^{\mathsf{T}}$, for which $c_3 = c_1$ and D can be generated from $\boldsymbol{c}_1 = [1, 2, 3, 3, 2]^{\mathsf{T}}$, for which $c_5 = c_1$ and $c_4 = c_2$. In both cases, $\boldsymbol{c}_1$ is a Hermitian vector and both matrices have only three distinct entries.

Hermitian vectors, and hence symmetric circulant matrices, have a special relationship with the DFT.

Lemma 6.44 *Let W be the $N \times N$ Fourier matrix (see Definition 1.95) with entries $w_{ij} = \frac{1}{\sqrt{N}} \omega^{(i-1)(j-1)}$ for $\omega := \mathrm{e}^{-2\pi \mathrm{i}/N}$. For $\boldsymbol{c} \in \mathbb{R}^N$, $W\boldsymbol{c}$ and $W^*\boldsymbol{c}$ are Hermitian. For a Hermitian vector $\boldsymbol{c} \in \mathbb{C}^N$, $W\boldsymbol{c}$ and $W^*\boldsymbol{c}$ are real.*

Proof See Exercise 6.16. □

The eigenvalues of a circulant matrix are given by N times the inverse DFT of the first column and the eigenvectors are the columns of the Fourier matrix W. This leads to the following factorisation of a circulant matrix.

Proposition 6.45 (Fourier representation of circulant matrices) *Let C be an $N \times N$ real-valued circulant matrix with first column $\boldsymbol{c}_1$. Then $C = WDW^*$, where W is the $N \times N$ Fourier matrix and D is a diagonal matrix with diagonal $\boldsymbol{d} = \sqrt{N} W^* \boldsymbol{c}_1$.*

If C is symmetric, $\boldsymbol{d} = [d_1, \dots, d_N]^{\mathsf{T}} \in \mathbb{R}^N$ and there are at most $\lfloor N/2 \rfloor + 1$ distinct eigenvalues: d_1 and $d_j = d_{N-j+2}$ for $j = 2, \dots, \lfloor N/2 \rfloor + 1$.

Proof The ijth entry of CW is

$$\begin{aligned}
(CW)_{ij} &= \sum_{k=1}^{N} (C)_{ik} (W)_{kj} = \frac{1}{\sqrt{N}} \sum_{k=1}^{N} (C)_{ik} \omega^{(k-1)(j-1)} \\
&= \frac{1}{\sqrt{N}} \sum_{l=0}^{i-1} c_l \omega^{(i-l-1)(j-1)} + \frac{1}{\sqrt{N}} \sum_{l=i}^{N-1} c_l \omega^{(N+i-l-1)(j-1)} \\
&= \frac{1}{\sqrt{N}} \sum_{l=0}^{N-1} c_l \omega^{(i-l-1)(j-1)} = \frac{\omega^{(i-1)(j-1)}}{\sqrt{N}} \sum_{l=0}^{N-1} c_l \omega^{-l(j-1)} = (W)_{ij} d_j
\end{aligned}$$

where $\boldsymbol{d} = \sqrt{N} W^* \boldsymbol{c}_1$. Hence, $CW = WD$ where D is the diagonal matrix with diagonal $\boldsymbol{d}$. The columns $\boldsymbol{w}_j$ of W are eigenvectors and $\boldsymbol{d}$ contains the eigenvalues. Since W is unitary (see Definition 1.95), $C = WDW^*$.

If C is symmetric, $\boldsymbol{c}_1$ is Hermitian and $\boldsymbol{d}$ is real and Hermitian by Lemma 6.44. Hence $d_j = d_{N-j+2}$ for $j = 2, \dots, N$, giving at most $\lfloor N/2 \rfloor + 1$ distinct eigenvalues. □

Example 6.46 The symmetric circulant matrix C in Example 6.43 has eigenvalues $d_1 = 8, d_2 = 2, d_3 = 0$ and $d_4 = 2$. The vector $\boldsymbol{d} = [8,2,0,2]^{\mathsf{T}}$ is a real, Hermitian vector with 3 distinct entries.

Circulant embedding

An $N \times N$ symmetric Toeplitz matrix C can always be embedded inside a larger symmetric circulant matrix $\widetilde{C}$. This is known as *circulant embedding*. We give a simple example to illustrate the idea.

Example 6.47 (minimal circulant embedding) Consider the 4×4 symmetric Toeplitz matrix C with first column $\boldsymbol{c}_1 = [c_0, c_1, c_2, c_3]^{\mathsf{T}} = [5, 2, 3, 4]^{\mathsf{T}}$. We can add rows and columns to C to form a larger matrix $\widetilde{C}$ as follows.

$$C = \begin{pmatrix} 5 & 2 & 3 & 4 \\ 2 & 5 & 2 & 3 \\ 3 & 2 & 5 & 2 \\ 4 & 3 & 2 & 5 \end{pmatrix}, \qquad \widetilde{C} = \left(\begin{array}{cccc|c} 5 & 2 & 3 & 4 & \\ 2 & 5 & 2 & 3 & \\ 3 & 2 & 5 & 2 & \\ 4 & 3 & 2 & 5 & \\ \hline & & & & \\ & & & & \end{array}\right).$$

To make $\widetilde{C}$ circulant, the last entry of the first column must equal the first entry in the second column, which by symmetry of C is $c_1 = 2$. The penultimate entry must equal the first entry in the third column, which is $c_2 = 3$, etc. If we want $\widetilde{C}$ to be circulant, but as small as possible, the first entry to be added in the first column after $c_3 = 4$ should match the leading entry in the 3rd column, which is $c_2 = 3$. Here, we only need to add two rows and columns to C to obtain the circulant matrix

$$\widetilde{C} = \left(\begin{array}{cccc|cc} 5 & 2 & 3 & 4 & 3 & 2 \\ 2 & 5 & 2 & 3 & 4 & 3 \\ 3 & 2 & 5 & 2 & 3 & 4 \\ 4 & 3 & 2 & 5 & 2 & 3 \\ \hline 3 & 4 & 3 & 2 & 5 & 2 \\ 2 & 3 & 4 & 3 & 2 & 5 \end{array}\right), \tag{6.41}$$

with first column $\tilde{\boldsymbol{c}}_1 = [5, 2, 3, 4, 3, 2]^{\mathsf{T}} = [c_0, c_1, c_2, c_3, c_2, c_1]^{\mathsf{T}}$. We call the 6×6 matrix $\widetilde{C}$ in (6.41) the *minimal circulant extension* of C.

Definition 6.48 (minimal circulant extension) Given a symmetric Toeplitz matrix $C \in \mathbb{R}^{N \times N}$ with first column $\boldsymbol{c}_1 = [c_0, c_1, \dots, c_{N-1}]^{\mathsf{T}}$, the *minimal circulant extension* is the circulant matrix $\widetilde{C} \in \mathbb{R}^{2N' \times 2N'}$ with first column $\tilde{\boldsymbol{c}}_1 = [c_0, c_1, \dots, c_{N'}, c_{N'-1}, \dots, c_1]^{\mathsf{T}} \in \mathbb{R}^{2N'}$, where $N' = N - 1$.

Note that $[\tilde{\boldsymbol{c}}_1]_j = [\tilde{\boldsymbol{c}}_1]_{2N'-j+2}$ for $j = 2, \dots, N'$ so $\tilde{\boldsymbol{c}}_1$ is a Hermitian vector and $\widetilde{C}$ is symmetric. By construction, the leading principal $N \times N$ submatrix of $\widetilde{C}$ is C.

The minimal circulant extension is the *smallest* symmetric circulant matrix inside which we can embed C. By introducing a symmetric Toeplitz matrix $C^* \in \mathbb{R}^{(N+M)\times(N+M)}$ that has C as its leading principal $N \times N$ submatrix, we can embed C into a larger symmetric circulant matrix. This is known as circulant embedding with padding and is important for finding circulant extensions that are non-negative definite and hence covariance matrices.

Example 6.49 (circulant extension with padding) Let C be the 4×4 matrix given in Example 6.47 and let $M = 1$. For any $x \in \mathbb{R}$, the 8×8 matrix

$$\widetilde{C} = \left(\begin{array}{cccc|cccc} 5 & 2 & 3 & 4 & x & 4 & 3 & 2 \\ 2 & 5 & 2 & 3 & 4 & x & 4 & 3 \\ 3 & 2 & 5 & 2 & 3 & 4 & x & 4 \\ 4 & 3 & 2 & 5 & 2 & 3 & 4 & x \\ \hline x & 4 & 3 & 2 & 5 & 2 & 3 & 4 \\ 4 & x & 4 & 3 & 2 & 5 & 2 & 3 \\ 3 & 4 & x & 4 & 3 & 2 & 5 & 2 \\ 2 & 3 & 4 & x & 4 & 3 & 2 & 5 \end{array}\right)$$

is also a circulant extension of C, but has two more rows and columns than the minimal circulant extension (6.41). Here, $\widetilde{C}$ is the minimal circulant extension of the 5×5 symmetric Toeplitz matrix C^* with first column $\boldsymbol{c}_1^* = [5, 2, 3, 4, x]^{\mathsf{T}}$.

Definition 6.50 (circulant extension with padding) Given a symmetric Toeplitz matrix $C \in \mathbb{R}^{N \times N}$ with first column $\boldsymbol{c}_1 = [c_0, c_1, \ldots, c_{N-1}]^{\mathsf{T}}$ and $\boldsymbol{x} = [x_1, \ldots, x_M]^{\mathsf{T}} \in \mathbb{R}^M$, let C^* be the symmetric Toeplitz matrix with first column

$$\boldsymbol{c}_1^* = [c_0, c_1, \ldots, c_{N-1}, x_1, \ldots, x_M]^{\mathsf{T}}.$$

Denoting $N' = N + M - 1$, the *circulant extension* of C *with padding* $\boldsymbol{x}$ is the minimal circulant extension $\widetilde{C} \in \mathbb{R}^{2N' \times 2N'}$ of C^*.

When $M = 0$, $C^* = C$ and $\widetilde{C}$ coincides with the minimal circulant extension. For any choice of M, the leading principal $N \times N$ submatrix of $\widetilde{C}$ is C.

Sampling N(0, *C*) *with circulant covariance*

Suppose we want to draw samples from $\mathrm{N}(\mathbf{0}, C)$, where $C \in \mathbb{R}^{N \times N}$ is a circulant covariance matrix. As C is symmetric and non-negative definite, the eigenvalues d_j are real and non-negative and we define $D^{1/2}$ as the real-valued diagonal matrix with entries $\sqrt{d_j}$. Consider

$$\boldsymbol{Z} = W D^{1/2} \boldsymbol{\xi} = \sum_{j=1}^{N} \sqrt{d_j} \boldsymbol{w}_j \xi_j, \tag{6.42}$$

where $\boldsymbol{\xi} = [\xi_1, \ldots, \xi_N]^{\mathsf{T}} \sim \mathrm{N}(\mathbf{0}, I_N)$. Note that (6.42) is a discrete Karhunen–Loève expansion with eigenvectors $\boldsymbol{u}_j = \boldsymbol{w}_j$ and eigenvalues $\nu_j = d_j$ (compare with (5.29)). Clearly, $\boldsymbol{Z}$ is a complex-valued random variable and has covariance $\mathbb{E}[\boldsymbol{Z}\boldsymbol{Z}^*] = W D^{1/2} D^{1/2} W^* = C$. The next result says that if $\boldsymbol{\xi}$ is an appropriate *complex-valued* Gaussian vector then $\boldsymbol{Z} \sim \mathrm{CN}(\mathbf{0}, 2C)$, and so the real and imaginary parts provide independent samples from $\mathrm{N}(\mathbf{0}, C)$.

Lemma 6.51 *Let* $\boldsymbol{Z} := W D^{1/2} \boldsymbol{\xi}$*, where* $\boldsymbol{\xi} \sim \mathrm{CN}(\mathbf{0}, 2I_N)$ *and* $D \in \mathbb{R}^{N\times N}$ *is a diagonal matrix with non-negative entries. Then,* $\boldsymbol{Z} \sim \mathrm{CN}(\mathbf{0}, 2C)$ *for* $C = WDW^*$.

Proof Since $\boldsymbol{\xi}$ has mean zero, so does $\boldsymbol{Z}$. Notice that

$$\mathbb{E}[\boldsymbol{Z}\boldsymbol{Z}^*] = \mathbb{E}\left[W D^{1/2} \boldsymbol{\xi}\, \boldsymbol{\xi}^* D^{1/2} W^*\right] = W D^{1/2} 2 I_N D^{1/2} W^* = 2C,$$

and

$$\mathbb{E}\left[\boldsymbol{Z}\boldsymbol{Z}^{\mathsf{T}}\right] = \mathbb{E}\left[W D^{1/2} \boldsymbol{\xi}\, \boldsymbol{\xi}^{\mathsf{T}} D^{1/2} W^{\mathsf{T}}\right] = 0,$$

because $\mathbb{E}[\boldsymbol{\xi}\boldsymbol{\xi}^{\mathsf{T}}] = 0$ (as in Example 6.20) and $D^{1/2}$ is real valued. Using Lemma 6.14, $\boldsymbol{X} = \operatorname{Re}\boldsymbol{Z}$ and $\boldsymbol{Y} = \operatorname{Im}\boldsymbol{Z}$ are uncorrelated and each has covariance matrix C. Then $\boldsymbol{X}, \boldsymbol{Y} \sim \mathrm{N}(\mathbf{0}, C)$ *iid* and $\boldsymbol{Z} \sim \mathrm{CN}(\mathbf{0}, 2C)$. □

Algorithm 6.7 Code to generate a pair of samples from $\mathrm{N}(\mathbf{0}, C)$ where $C \in \mathbb{R}^{N\times N}$ is a circulant covariance matrix. The input `c` is the first column of C and is a column vector of length N. The outputs `X` and `Y` are independent samples from $\mathrm{N}(\mathbf{0}, C)$.

```
function [X,Y] = circ_cov_sample(c)
N=length(c);   d=ifft(c,'symmetric')*N;
xi=randn(N,2)*[1; sqrt(-1)];
Z=fft((d.^0.5).*xi)/sqrt(N);
X=real(Z); Y=imag(Z);
```

Recall from (1.31) that the DFT of $\boldsymbol{u} \in \mathbb{C}^N$ can be computed in Matlab using the command `fft(`$\boldsymbol{u}$`)` and is expressed in terms of the Fourier matrix as $\sqrt{N} W \boldsymbol{u}$. Similarly, the inverse DFT (1.32) of $\boldsymbol{u} \in \mathbb{C}^N$ is $W^* \boldsymbol{u}/\sqrt{N}$ and is computed in Matlab via `ifft(`$\boldsymbol{u}$`)`. Then $\boldsymbol{Z} = \texttt{fft}(\boldsymbol{y})/\sqrt{N}$ where $\boldsymbol{y} = D^{1/2}\boldsymbol{\xi}$ and $\boldsymbol{d} = N\,\texttt{ifft}(\boldsymbol{c}_1)$. This is implemented in Algorithm 6.7. When C is non-negative definite, the algorithm generates two samples from $\mathrm{N}(\mathbf{0}, C)$ using one DFT and one inverse DFT on vectors of length N. The computational cost using the FFT is $\mathcal{O}(N \log N)$ and this is significantly less than for sampling methods based on the Cholesky or spectral decomposition. Each additional pair of samples requires only one more FFT, since we do not need to recompute $\boldsymbol{d}$.

Example 6.52 The symmetric circulant matrix C in Example 6.43 is a valid covariance matrix. Using Algorithm 6.7, we generate two independent samples from $\mathrm{N}(\mathbf{0}, C)$ in Matlab as follows:

```
>> [X,Y] = circ_cov_sample([3;2;1;2])
  X =                                Y =
       2.6668                           -0.2316
       1.9273                            1.1379
      -1.1460                            1.1332
      -0.4066                           -0.2363
```

Sampling N(0, *C*) *with Toeplitz covariance*

Consider a real-valued Gaussian process $X(t)$ with mean zero and stationary covariance $c(t)$. To sample $X(t)$ at N uniformly spaced points on $[0, T]$, we sample $\boldsymbol{X} = [X(t_0), \ldots,$

$X(t_{N-1})]^{\mathsf{T}}$ for $t_n = n\Delta t$ and $\Delta t = T/(N-1)$. The covariance matrix of $\boldsymbol{X}$ has entries

$$c_{ij} = \operatorname{Cov}(X(t_{i-1}), X(t_{j-1})) = c(t_{i-1} - t_{j-1}) = c((i-j)\Delta t).$$

Since c is even, c_{ij} depends only on $|i-j|$ and C is a symmetric Toeplitz matrix by Definition 6.39 with first column $\boldsymbol{c}_1 = [c_0, \dots, c_{N-1}]^{\mathsf{T}}$ and $c_n = c(t_n)$. Unfortunately, C is not in general circulant.

Example 6.53 (exponential covariance) Consider specifically the exponential covariance $c(t) = \mathrm{e}^{-|t|/\ell}$. The covariance matrix of $\boldsymbol{X}$ has entries $c_{ij} = \mathrm{e}^{-|t_{i-1}-t_{j-1}|/\ell} = \mathrm{e}^{-|i-j|\Delta t/\ell}$. It is clear that $c_{ij} = c_{i-j} = c_{j-i}$ so C is symmetric and Toeplitz. However, $c_{21} = c_1 = \mathrm{e}^{-\Delta t/\ell} \neq c_{1N} = \mathrm{e}^{-(N-1)\Delta t/\ell}$ and hence C is not circulant. See also Exercise 6.20.

If a covariance matrix $C \in \mathbb{R}^{N\times N}$ is Toeplitz but not circulant (as in Example 6.53), we cannot apply Algorithm 6.7 to generate samples from $\mathrm{N}(\boldsymbol{0}, C)$. However, if the minimal circulant extension $\widetilde{C} \in \mathbb{R}^{2N'\times 2N'}$, where $N' = N-1$, of C is non-negative definite then $\mathrm{N}(\boldsymbol{0}, \widetilde{C})$ is a valid Gaussian distribution. Since

$$\widetilde{C} = \begin{pmatrix} C & B^{\mathsf{T}} \\ B & D \end{pmatrix}, \tag{6.43}$$

for some matrices $B \in \mathbb{R}^{(2N'-N)\times N}$ and $D \in \mathbb{R}^{(2N'-N)\times(2N'-N)}$, it is easy to show that if $\widetilde{\boldsymbol{X}} = [X_1, \dots, X_{2N'}]^{\mathsf{T}} \sim \mathrm{N}(\boldsymbol{0}, \widetilde{C})$, then $\boldsymbol{X} = [X_1, \dots, X_N]^{\mathsf{T}} \sim \mathrm{N}(\boldsymbol{0}, C)$ (see Exercise 6.17). That is, we can generate samples from $\mathrm{N}(\boldsymbol{0}, C)$ by generating $\mathrm{N}(\boldsymbol{0}, \widetilde{C})$ samples and reading off the first N components. This is known as the *minimal circulant embedding method* and is implemented in Algorithm 6.8. The cost of generating two independent samples from $\mathrm{N}(\boldsymbol{0}, \widetilde{C})$ is $\mathcal{O}(N \log N)$. Although we have to factorise a matrix of size $2N' \times 2N'$, the efficiency of the FFT means that (when N is large enough) minimal circulant embedding is far cheaper than standard factorisation methods for the original $N \times N$ matrix C.

Algorithm 6.8 Code to generate two independent samples from $\mathrm{N}(\boldsymbol{0}, C)$ where $C \in \mathbb{R}^{N\times N}$ is a Toeplitz covariance matrix. The input `c` is the first column of C and is a column vector of length N. The outputs `X` and `Y` are independent samples from $\mathrm{N}(\boldsymbol{0}, C)$.

```
function [X,Y]=circulant_embed_sample(c)
% create first column of C_tilde
tilde_c=[c; c(end-1:-1:2)];
% obtain 2 samples from N(0,C_tilde)
[X,Y]=circ_cov_sample(tilde_c);
% extract samples from N(0,C)
N=length(c); X=X(1:N); Y=Y(1:N);
```

Example 6.54 The matrix C in Example 6.47 is symmetric and Toeplitz and is a valid covariance matrix. The minimal circulant extension $\widetilde{C}$ in (6.41) is also non-negative definite and we generate two independent samples from $\mathrm{N}(\mathbf{0},C)$ in Matlab using Algorithm 6.8 as follows:

```
>> [X,Y]=circulant_embed_sample([5;2;3;4])
X =                          Y =
      -0.0176                      3.0063
       4.6240                     -1.8170
      -1.1264                     -1.5457
      -1.2369                     -0.9103
```

The next result gives an explicit formula for the (possibly negative) eigenvalues of the minimal circulant extension that is useful in analysis.

Lemma 6.55 *Let $\widetilde{C} \in \mathbb{R}^{2N' \times 2N'}$ be the minimal circulant extension of a symmetric Toeplitz matrix C with first column $\boldsymbol{c}_1 = [c_0, c_1, \dots, c_{N'}]^{\mathsf{T}}$. Let $\omega := \mathrm{e}^{-\pi\mathrm{i}/N'}$ and $c_{-j} := c_j$ for $j = 1, \dots, N'$. The eigenvalues of $\widetilde{C}$ are real and given by*

$$d_k = \sum_{j=-N'+1}^{N'} c_j \omega^{j(k-1)} = c_0 + c_{N'}(-1)^{(k-1)} + 2\sum_{j=1}^{N'-1} \mathrm{Re}\big(c_j \omega^{j(k-1)}\big), \tag{6.44}$$

for $k = 1, \dots, 2N'$, of which at most $N' + 1$ are distinct.

Proof The circulant extension $\widetilde{C}$ is the $2N' \times 2N'$ circulant matrix with first column $\tilde{\boldsymbol{c}}_1 = [c_0, c_1, \dots, c_{N'-1}, c_{N'}, c_{N'-1}, \dots, c_1]^{\mathsf{T}}$. $\widetilde{C}$ is symmetric and Proposition 6.45 implies that the eigenvalues of $\widetilde{C}$ are real and at most $\lfloor 2N'/2 \rfloor + 1 = N' + 1$ are distinct. (6.44) follows from the expression $\boldsymbol{d} = \sqrt{2N'}W^*\tilde{\boldsymbol{c}}_1$, where W is the $2N' \times 2N'$ Fourier matrix. See Exercise 6.18 with $P = N'$. □

For the exponential covariance, the minimal circulant extension is always non-negative definite.

Algorithm 6.9 Code to generate two independent samples of the Gaussian process $X(t)$ in Example 6.56 at $t_n = n\Delta t$ for $n = 0, \dots, N-1$. The inputs are the number of sample points `N`, the spacing `dt` $= \Delta t$, and the covariance parameter `ell` $= \ell$. The outputs are `t`, a vector containing the sample times, `X` and `Y`, the required samples, and `c`, the first column of C.

```
function [t,X,Y,c]=circulant_exp(N,dt,ell)
t=[0:(N-1)]'*dt;
c=exp(-abs(t)/ell);
[X,Y]=circulant_embed_sample(c);
```

Example 6.56 (exponential covariance) Continuing from Example 6.53, let C be the covariance matrix of the vector $\boldsymbol{X}$ of samples of the mean-zero Gaussian process $X(t)$ with covariance $c(t) = \mathrm{e}^{-|t|/\ell}$ at N uniformly spaced points on $[0,T]$. The minimum eigenvalue of the minimal circulant extension $\widetilde{C}$ of C decays to zero as $N \to \infty$ (see Exercise 6.20) but remains non-negative. Hence, the circulant embedding method gives exact samples from $\mathrm{N}(\mathbf{0},C)$. Two sample paths of $X(t)$ generated using Algorithm 6.9 are plotted in Figure 6.9 for the case $T = 1$ and $\ell = 1$.

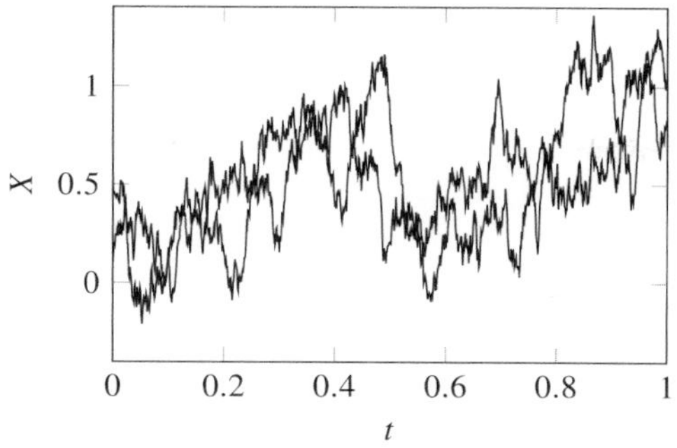

Figure 6.9 Two sample paths of the stationary Gaussian process $X(t)$ with $\mu(t) = 0$ and $c(t) = e^{-|t|}$ on the interval $[0, 1]$. The samples were generated using Algorithm 6.9 with the commands `N` $= 10^3$; `dt=1/(N-1); [t,X,Y,c]=circulant_exp(N,dt,1).`

We prove $\widetilde{C}$ is positive definite. The first column of the covariance matrix C is $\boldsymbol{c}_1 = [c_0, c_1, \dots, c_{N-1}]^{\mathsf{T}}$ where $c_j = e^{-j\Delta t/\ell}$. Fix k and let $\mu_k = e^{-\Delta t/\ell}\omega^{(k-1)}$ where $\omega = e^{-i\pi/N'}$, so that $\mu_k^j = c_j\omega^{j(k-1)}$. From (6.44), the eigenvalues of $\widetilde{C}$ are

$$d_k = c_0 + c_{N'}(-1)^{(k-1)} + 2\sum_{j=1}^{N'-1} \mathrm{Re}\big(c_j\omega^{j(k-1)}\big), \qquad k = 1, \dots, 2N',$$

(of which at most N values are distinct). Now,

$$d_k = 1 + \mu_k^{N'} + 2\,\mathrm{Re}\sum_{j=1}^{N'-1} \mu_k^j.$$

Using the summation formula for the geometric series and $2\,\mathrm{Re}\, Z = Z + \bar{Z}$,

$$d_k = 1 + \mu_k^{N'} + \frac{\mu_k - \mu_k^{N'}}{1 - \mu_k} + \frac{\bar{\mu}_k - \mu_k^{N'}}{1 - \bar{\mu}_k}.$$

Since $(1 - \mu_k)(1 - \bar{\mu}_k) = 1 - 2\,\mathrm{Re}\,\mu_k + |\mu_k|^2$, this simplifies to

$$d_k = \frac{(1 - |\mu_k|^2)(1 - \mu_k^{N'})}{1 - 2\,\mathrm{Re}\,\mu_k + |\mu_k|^2}.$$

Note that $|\mu_k| < 1$ and $\mu_k^{N'} = \pm c_{N'}$ (depending on whether k is odd or even). However, $0 < c_{N'} < 1$ and the numerator is strictly positive in both cases. Also,

$$1 - 2\,\mathrm{Re}\,\mu_k + |\mu_k|^2 \geq 1 - 2|\mu_k| + |\mu_k|^2 = \big(1 - |\mu_k|\big)^2 > 0.$$

The numerator and denominator are strictly positive and so $d_k > 0$. Hence, the minimal circulant extension $\widetilde{C}$ is always positive definite.

Padding and approximate circulant embedding

When the circulant extension $\widetilde{C}$ is indefinite, the diagonal matrix D in the factorisation $\widetilde{C} = WDW^*$ has both positive and negative entries. If D has diagonal $[d_1, \dots, d_{2N'}]^{\mathsf{T}}$, we

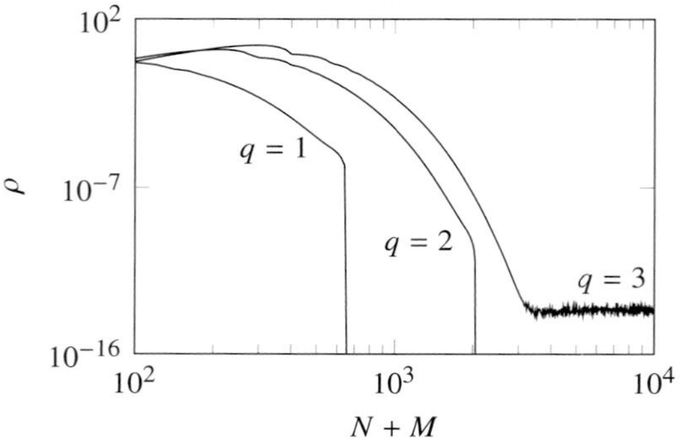

Figure 6.10 Plot of $\rho(D_-)$, the largest negative eigenvalue of the minimal circulant extension of C^*, against $N + M$ for the Whittle–Matérn covariance function $c_q(t)$ with $q = 1, 2, 3$ for $\Delta t = T/(N-1)$, $T = 1$, and $N = 100$.

may write

$$D = D_+ - D_-, \tag{6.45}$$

where $D_\pm$ is the diagonal matrix with kth diagonal entry $\max\{0, \pm d_k\}$. The natural definition of the square root is now $D^{1/2} := D_+^{1/2} + \mathrm{i} D_-^{1/2}$, which is a complex-valued matrix. For $\boldsymbol{\xi} \sim \mathrm{CN}(\mathbf{0}, 2I_{2N'})$, the random variable $\widetilde{\boldsymbol{Z}} = W D^{1/2} \boldsymbol{\xi}$ has covariance matrix

$$\begin{aligned} \mathbb{E}\left[\widetilde{\boldsymbol{Z}} \widetilde{\boldsymbol{Z}}^*\right] &= W D^{1/2} \mathbb{E}[\boldsymbol{\xi}\,\boldsymbol{\xi}^*] \bar{D}^{1/2} W^* \\ &= 2W\left(D_+^{1/2} + \mathrm{i} D_-^{1/2}\right)\left(D_+^{1/2} - \mathrm{i} D_-^{1/2}\right) W^* = 2(\widetilde{C}_+ + \widetilde{C}_-), \end{aligned} \tag{6.46}$$

where $C_\pm := W D_\pm W^*$. As $\widetilde{C} = \widetilde{C}_+ - \widetilde{C}_-$, we see $\widetilde{\boldsymbol{Z}}$ does not have the desired distribution.

The existence of negative eigenvalues introduces an error into the circulant embedding method. In many cases, we can avoid the negative eigenvalues and the resulting error by embedding C into a larger circulant matrix by introducing a padding vector $\boldsymbol{x} \in \mathbb{R}^M$ as in Definition 6.50. The covariance matrices C of interest have first column

$$\boldsymbol{c}_1 = [c_0, \dots, c_{N-1}]^{\mathsf{T}}, \qquad c_n = c(t_n), \quad t_n = n\Delta t,$$

where $c(t)$ is the stationary covariance function. We define a symmetric Toeplitz matrix C^* of size $(N + M) \times (N + M)$ with first column

$$\boldsymbol{c}_1^* = \begin{pmatrix} \boldsymbol{c}_1 \\ \boldsymbol{x} \end{pmatrix}, \qquad \boldsymbol{x} = \left[c_N, \dots, c_{N+M-1}\right]^{\mathsf{T}}. \tag{6.47}$$

C^* coincides with the covariance matrix associated with the sample times $t_0, \dots, t_{N+M-1}$. We denote the minimal circulant extension of C^* by $\widetilde{C}$ and note it has dimension $2N' \times 2N'$ for $N' = N + M - 1$. Often, $\widetilde{C}$ is non-negative when M is chosen large enough and this is effective in the following example.

Example 6.57 (Whittle–Matérn $q = 1$) Let $X(t)$ be a mean-zero Gaussian process with the Whittle–Matérn covariance $c_q(t)$ with $q = 1$. We consider sampling $X(t)$ on $[0, T]$ by the circulant embedding method for $N = 100$ uniformly spaced times $t_n = n\Delta t$ for $\Delta t = T/(N-1)$ and $n = 0, \dots, N-1$. The circulant embedding method based on the minimal circulant extension of C, implemented as `circulant_wm` in Algorithm 6.11, fails to give an exact sample because of negative eigenvalues.

```
>> N=100; M=0; dt=1/(N-1);
>> [t,X,Y,c]=circulant_wm(N, M, dt, 1);
 rho(D_minus)= 0.92134
```

Here $\rho(D_{-})$ denotes the spectral radius of D_{-} (see (6.45)), which is the size of the most negative eigenvalue of $\widetilde{C}$. Now consider padding and work with the minimal circulant extension $\widetilde{C}$ of the $(N+M)\times(N+M)$ matrix C^{*} given by (6.47). Figure 6.10 shows how the size of the most negative eigenvalue $\rho(D_{-})$ of the padded circulant extension $\widetilde{C}$ depends on $N+M$. In particular, for $N+M=7N$, the embedding is non-negative definite and the following calls give exact samples from $\mathrm{N}(\mathbf{0},C)$:

```
>> N=100; M=6*N; dt=1/(N-1);
>> [t,X,Y,c]=circulant_wm(N, M,dt,1);
```

The padding in the previous example significantly increases the dimension of the vectors used in the FFT calculations (by a factor of 7 in this case) and hence the amount of work involved. In some cases, we cannot make the circulant embedding non-negative definite and we turn to approximate circulant embedding instead.

First, we compute the distribution of $\widetilde{\boldsymbol{Z}}_{\mathbf{1}}=WD_{+}^{1/2}\boldsymbol{\xi}$.

Lemma 6.58 *Consider a circulant matrix* $\widetilde{C}\in\mathbb{R}^{2N'\times 2N'}$ *with Fourier representation* $\widetilde{C}=WDW^{*}$ *and write* $D=D_{+}-D_{-}$ *as in* (6.45). *If* $\widetilde{\boldsymbol{Z}}_1:=WD_{+}^{1/2}\boldsymbol{\xi}$, *where* $\boldsymbol{\xi}\sim\mathrm{CN}(\mathbf{0},2I_{2N'})$, *then* $\widetilde{\boldsymbol{Z}}_1\sim\mathrm{CN}(\mathbf{0},2(\widetilde{C}+\widetilde{C}_{-}))$ *for* $\widetilde{C}_{-}=WD_{-}W^{*}$.

Proof See Exercise 6.19. □

The real and imaginary parts of $\widetilde{\boldsymbol{Z}}_1$ have covariance $\widetilde{C}+\widetilde{C}_{-}$, which differs from the target covariance matrix $\widetilde{C}$. When the norm of $\widetilde{C}_{-}$ is small, the first N components $\boldsymbol{X}$ of $\operatorname{Re}\widetilde{\boldsymbol{Z}}_1$ can be used to gain a good approximation to $\mathrm{N}(\mathbf{0},C)$ and this is implemented in Algorithm 6.10. It is possible to show that $\boldsymbol{X}\sim\mathrm{N}(\mathbf{0},C+E)$, where E is an $N\times N$ matrix such that $\|E\|_2\le\rho(D_{-})$. See Exercise 6.19. Note that $D_{-}=0$ when $\widetilde{C}$ is non-negative definite. Because it computes the eigenvalues, the algorithm is able to gauge its own accuracy using $\rho(D_{-})$ as an error estimate.

Algorithm 6.10 Code to approximate $\mathrm{N}(\mathbf{0},C)$ with circulant embedding. The syntax is the same as Algorithm 6.8. When negative eigenvalues arise, a warning message with the value of $\rho(D_{-})$ is given.

```
function [X,Y]=circulant_embed_approx(c)
tilde_c=[c; c(end-1:-1:2)]; tilde_N=length(tilde_c);
d=ifft(tilde_c,'symmetric')*tilde_N;
d_minus=max(-d,0); d_pos=max(d,0);
if (max(d_minus)>0)
    disp(sprintf('rho(D_minus)=%0.5g', max(d_minus)));
end;
xi=randn(tilde_N,2)*[1; sqrt(-1)];
Z=fft((d_pos.^0.5).*xi)/sqrt(tilde_N);
N=length(c); X=real(Z(1:N)); Y=imag(Z(1:N));
```

Example 6.59 (Whittle–Matérn $q = 3$) Following Example 6.57, we consider sampling a mean-zero Gaussian process $X(t)$ with the Whittle–Matérn covariance $c_q(t)$, this time with $q = 3$. Again we are interested in sampling on $[0,T] = [0,1]$ with $N = 100$ uniformly spaced points. As we see from Figure 6.10, the circulant extension with padding has negative eigenvalues up to dimension $N + M < 10^4$ and, though decreasing as M increases, $\rho(D_-) > 10^{-14}$ over this range (note the computation of $\rho(D_-)$ for $N + M \approx 10^4$ is limited by machine precision). With $N = 100$ and $M = 9900$, we run Algorithm 6.11 as follows:

```
>> N=100; M=9900; dt=1/(N-1);
>> [t,X,Y,c]= circulant_wm(N, M, dt,3);
rho(D_minus)=5.6843e-14
```

If $\mathrm{N}(\mathbf{0}, C)$ denotes the target distribution, the approximate circulant embedding method with $N+M = 10^4$ samples from the distribution $\mathrm{N}(\mathbf{0}, C+E)$, where $\|E\|_2 \le \rho(D_-) \approx 5.7\times 10^{-14}$.

Algorithm 6.11 Code to generate two independent samples for the Gaussian process with the Whittle–Matérn covariance. The outputs and first two inputs are as for Algorithm 6.9 and `q` is the Whittle–Matérn parameter.

```
function [t, X, Y, c]=circulant_wm( N, M, dt, q)
Ndash=N+M-1; c=zeros(Ndash+1, 1); t=dt*[0:Ndash]';
c(1)=1; % t=0 is special, due to singularity in Bessel fn
const=2^(q-1)*gamma(q);
for i=2:Ndash+1,
    c(i)=(t(i)^q)*besselk(q, t(i))/const;
end;
[X,Y]=circulant_embed_approx(c); X=X(1:N); Y=Y(1:N); t=t(1:N);
```

The following theorem gives an upper bound on $\rho(D_-)$ proportional to $\int_{t_{N'-1}}^{\infty} |c(t)|\,dt$ for a range of covariance functions. When the covariance decays rapidly (e.g., the Gaussian and Whittle–Matérn covariances in Figure 6.11), $\int_{t_{N'-1}}^{\infty} |c(t)|\,dt$ decays to zero rapidly and the theorem implies that that $\rho(D_-)$ and the error in approximate circulant embedding converge to zero as $N' \to \infty$.

Theorem 6.60 *Let $t_n = n\Delta t$ for $n \in \mathbb{Z}$ and $\Delta t > 0$. Suppose that $c(t)$ is a stationary covariance function such that $\sum_{n=0}^{\infty} |c(t_n)| < \infty$. For $N' = N + M - 1$, let $\widetilde{C}$ be the minimal circulant extension of the matrix C^* with first column $\mathbf{c}_1^*$ defined by (6.47). Then, any eigenvalue d of $\widetilde{C}$ satisfies $d \ge -2\sum_{n=N'}^{\infty} |c(t_n)|$ and $d \ge (-2/\Delta t)\int_{t_{N'-1}}^{\infty} |c(t)|\,dt$ if $c(t)$ decays monotonically for $t > t_{N'-1}$.*

Proof By Lemma 6.55, any eigenvalue d of $\widetilde{C}$ can be written for a $2N'$th root of unity ω as $d = \sum_{j=-N'+1}^{N'} c_j \omega^j$ for $c_j := c(t_j)$. Then, for any $k \in \mathbb{Z}$,

$$d = \sum_{j=k-N'+1}^{k+N'} c_{j-k}\,\omega^{j-k}.$$

For $m \in \mathbb{N}$, define $\mathcal{J} = \{-mN' - N', \dots, mN' + N'\}$ and note that

$$\sum_{j,k \in \mathcal{J}} c_{j-k}\,\omega^{j-k} = \sum_{|k| \le mN'} \sum_{j=k-N'+1}^{k+N'} c_{j-k}\,\omega^{j-k} + A_1 + A_2 = \sum_{|k| \le mN'} d + A_1 + A_2,$$

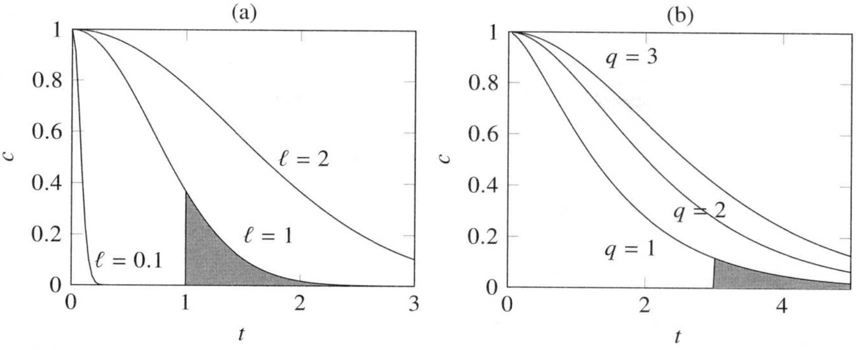

Figure 6.11 Stationary covariance functions (a) $c(t) = \mathrm{e}^{-(|t|/\ell)^2}$ for $\ell = 0.1, 1, 2$ and (b) the Whittle–Matérn covariance for $c(t) = c_q(t)$ for $q = 1, 2, 3$. The areas of the shaded regions are $\int_{t_{N'-1}}^{\infty} |c(t)|\, dt$ for (a) $t_{N'-1} = 1$ and $\ell = 1$ and (b) $t_{N'-1} = 3$ and $q = 1$ and the areas rapidly become smaller as N' is increased.

where

$$A_1 := \sum_{|k|>mN',\ k\in\mathfrak{I}} \omega^{-k} \sum_{j\in\mathfrak{I}} c_{j-k}\omega^j,$$

$$A_2 := \sum_{|k|\le mN'} \omega^{-k} \left[\sum_{j=-mN'-N'}^{k-N'} + \sum_{j=k+N'+1}^{mN'+N'} \right] c_{j-k}\omega^j.$$

As $c(t)$ is a non-negative definite function, $\sum_{j,k\in\mathfrak{I}} \omega^{j-k} c_{j-k} \ge 0$. Hence,

$$0 \le (2mN' + 1)d + A_1 + A_2.$$

Because $|\omega| = 1$ and the sum A_1 has at most $2N'$ terms in k,

$$A_1 \le 4N' \sum_{n=0}^{\infty} |c_n|.$$

Further, with $n = j - k$,

$$A_2 = (2mN' + 1)\left[\sum_{n=-\infty}^{-N'} + \sum_{n=N'+1}^{\infty} \right] |c_n| \le 2(2mN + 1) \sum_{n=N'}^{\infty} |c_n|.$$

Using the bounds on A_1 and A_2,

$$0 \le (2mN' + 1)d + 4N' \sum_{j=0}^{\infty} |c_j| + 2(2mN' + 1) \sum_{n=N'}^{\infty} |c_n|.$$

Divide by $2mN' + 1$ and take limits as $m \to \infty$, to find $0 \le d + 2\sum_{n=N'}^{\infty} |c_n|$. When $c(t)$ is monotonically decreasing, it is easy to see that $\sum_{n=N'}^{\infty} |c_n| \Delta t \le \int_{t_{N'-1}}^{\infty} |c(t)|\, dt$. □

6.6 Notes

References for stationary processes and related topics include (Cramér and Leadbetter, 2004; Yaglom, 1962). Definition 6.1 should not be confused with the notion of *strictly stationary*, which asserts that all finite dimensional distributions of $X(t_1 + h), \dots, X(t_N + h)$ are independent of h. For Gaussian processes, the two definitions are equivalent (as the finite dimensional distributions are defined by the first two moments). In general, stationary processes in the sense of Definition 6.1 (also called *weakly* or *second-order* or *wide-sense* stationary by other authors) are not expected to be strictly stationary. The Wiener–Khintchine theorem is discussed in (Brown and Hwang, 1992; Gardiner, 2009; Peebles, 1993).

The complex Gaussian distribution introduced in Definition 6.15 is narrow, though suitable for our purposes, and is a particular case of the circular symmetric Gaussian distribution with *real* covariance. See (Goodman, 1963; Picinbono, 1996). In general, a random variable $\boldsymbol{Z} = \boldsymbol{X} + \mathrm{i}\boldsymbol{Y} \in \mathbb{C}^n$ is complex Gaussian if $[\boldsymbol{X}^\mathsf{T}, \boldsymbol{Y}^\mathsf{T}]^\mathsf{T} \in \mathbb{R}^{2n}$ is Gaussian, and this requires three parameters: the mean $\boldsymbol{\mu} = \mathbb{E}[\boldsymbol{Z}]$, the covariance $\mathrm{Cov}(\boldsymbol{Z}, \boldsymbol{Z}) \in \mathbb{C}^{n\times n}$, and a relation matrix $\mathrm{Cov}(\boldsymbol{Z}, \bar{\boldsymbol{Z}}) \in \mathbb{C}^{n\times n}$.

We have shown that particular stochastic integrals yield Gaussian processes with particular covariances. More powerful versions of this result show any stationary process can be written as a stochastic Fourier integral with respect to some stochastic process (particular ones with orthogonal increments). See Cramér and Leadbetter (2004).

The quadrature methods discussed in §6.4 are described in more detail in (Shinozuka, 1971; Shinozuka and Jan, 1972). The circulant embedding method outlined in §6.5 is developed in (Dietrich and Newsam, 1993, 1997; Gneiting et al., 2006; Newsam and Dietrich, 1994; Wood and Chan, 1994). One interesting result not covered says that, if $c(t)$ is positive, monotone decreasing and convex, then the minimal embedding is non-negative definite. Unfortunately, when $c'(0) = 0$, which is the case for all smooth stochastic processes, the convexity condition fails to hold. The circulant embedding method can be used to simulate fBm as in Exercise 6.23; see (Davies and Harte, 1987; Dieker and Mandjes, 2003).

Exercises

6.1 Show that if $\{Z(t) : t \in \mathbb{R}\}$ is a mean-zero complex-valued stationary process then $c(t) = \mathbb{E}\left[Z(t)\overline{Z(0)}\right]$ is a non-negative definite function.

6.2 Show that if $c : \mathbb{R} \to \mathbb{R}$ is non-negative definite then $c(t)$ is an even function of t.

6.3 Show that if $c(t) = \int_{\mathbb{R}} \mathrm{e}^{\mathrm{i}t\nu} f(\nu)\, d\nu$ for a non-negative and integrable function f, then $c(t)$ is non-negative definite.

6.4 Let $c(t)$ be a stationary covariance function of a real-valued stationary process $X(t)$ with mean zero. Prove that $c(t)$ is uniformly continuous for all $t \in \mathbb{R}$ if and only if $c(t)$ is continuous at $t = 0$.

6.5 Let $Z(t) = \xi \mathrm{e}^{\mathrm{i}\nu t}$ where ξ is a complex-valued random variable and $\nu \in \mathbb{C}$. If $Z(t)$ is stationary, prove that $\nu \in \mathbb{R}$ and $\mathbb{E}[\nu\xi] = 0$.

6.6 Let $X(t)$ be a mean-square continuous stationary process with mean μ and covariance $c(t)$. Suppose that $c(t)$ is integrable. Prove that the spectral density $f(\nu)$ is well defined and that (6.7) holds. Use Theorems 1.22 and 1.24.

6.7 Suppose that $\nu_j \in \mathbb{R}$ and $\xi_j \sim \mathrm{CN}(0,1)$ *iid*. For $t \in \mathbb{R}$, consider the random Fourier series

$$Z_J(t) := \sum_{j=-J}^{J} \mathrm{e}^{\mathrm{i}t\nu_j} \sqrt{f_j}\, \xi_j, \qquad f_j \geq 0.$$

a. Show that $Z_J(t)$ is a stationary process.
b. If $\nu_j = -\nu_{-j}$ and $f_j = f_{|j|}$, show that $Z_J(t)$ is a complex Gaussian process.

6.8 Suppose that X_n, for $n \in \mathbb{N}$, are Gaussian random variables such that $X_n \to X$ in $L^2(\Omega)$ as $n \to \infty$. By use of the characteristic function, show that X is a Gaussian random variable.

6.9 Prove (6.20) and (6.21).

6.10 Consider $f\colon [-R,R] \to \mathbb{R}$ and let

$$X(t) = \sqrt{2}\left[\int_0^R \sqrt{f(\nu)}\cos(t\nu)\, dW_1(\nu) + \int_0^R \sqrt{f(\nu)}\sin(t\nu)\, dW_2(\nu)\right],$$

for *iid* Brownian motions $W_1(\nu)$ and $W_2(\nu)$. Find the spectral density function of $X(t)$.

6.11 For $J \in \mathbb{N}$, consider the truncated white noise $\zeta_J(t)$ defined in (5.24). For $\Delta t > 0$ and a Brownian motion $W(t)$, consider the numerical derivative $D_{\Delta t} W(t)$ defined by

$$D_{\Delta t} W(t) := \frac{W(t+\Delta t) - W(t)}{\Delta t}, \qquad t > 0.$$

Plot sample paths and spectral densities for $D_{\Delta t} W(t)$ and $\zeta_J(t)$ and compute the best linear fit to the spectral density. In both cases, observe the spectral density is nearly flat.

6.12 Suppose that $f(\nu)$ is a non-negative and continuous function that is not identically zero. If $c(t)$ is defined by (6.3), show that $c(t)$ is positive definite.

6.13 Consider an integrable function $f\colon \mathbb{R} \to \mathbb{R}^+$ such that (6.11) holds for $q > 1$. Show that $\int_{\mathbb{R}} \nu^2 f(\nu)\, d\nu < \infty$. By considering (6.3), show that $c \in \mathrm{C}^2(\mathbb{R})$.

6.14 Let F be a probability measure on $\mathbb{R}$ and let Λ be a real-valued random variable with probability distribution $\mathbb{P}_\Lambda = F$. Show that $Z(t) := \mathrm{e}^{\mathrm{i}\Lambda t}$ is a complex-valued random variable with

$$\mathbb{E}\left[Z(s)\overline{Z(t)}\right] = \int_{\mathbb{R}} \mathrm{e}^{\mathrm{i}\nu(s-t)}\, dF(\nu).$$

6.15 Consider independent real-valued random variables Λ_j, Φ_j and define

$$X_J(t) := \sqrt{\frac{2}{J}} \sum_{j=1}^{J} \cos(\Lambda_j t + \Phi_j).$$

a. If $\Lambda_j \sim \mathrm{N}(0, 2/\ell^2)$ and $\Phi_j \sim \mathrm{U}(0,\pi)$, show that $X_J(t)$ is a stationary process with covariance $c(t) = \mathrm{e}^{-(t/\ell)^2}$.
b. If $\Lambda_j \sim \mathrm{Cauchy}(0, 1/\ell)$ and $\Phi_j \sim \mathrm{U}(0,\pi)$, show that $X_J(t)$ is a stationary process with covariance $c(t) = \mathrm{e}^{-|t|/\ell}$.

Prove that $X_J(t)$ is not Gaussian, but $X_J(t)$ is Gaussian in the limit $J \to \infty$.

6.16 Prove Lemma 6.44.

6.17 Suppose that $\widetilde{\boldsymbol{X}} \sim \mathrm{N}(\mathbf{0}, \widetilde{C})$, where $\widetilde{\boldsymbol{X}} \in \mathbb{R}^{2N'}$. Let $\boldsymbol{X}$ be the vector containing the first N components of $\widetilde{\boldsymbol{X}}$. If the covariance matrix $\widetilde{C}$ has the form (6.43), prove that $\boldsymbol{X} \sim \mathrm{N}(\mathbf{0}, C)$.

6.18 Let $C \in \mathbb{R}^{2P \times 2P}$ be a symmetric, circulant matrix with first column equal to

$$[c_0, c_1, \ldots, c_P, c_{P-1}, \ldots, c_1]^{\mathsf{T}}.$$

Denote $c_j = c_{-j}$ for $j = 1, \ldots, P$. Show that the eigenvalues of C are, for $k = 1, \ldots, P$,

$$d_k = \sum_{j=-P+1}^{P} c_j \omega^{j(k-1)} = c_0 + c_{P/2}(-1)^{(k-1)} + 2 \sum_{j=1}^{P/2-1} c_j \cos\left(\frac{\pi j(k-1)}{P}\right).$$

6.19 In the notation of Lemma 6.58, let

$$\widetilde{\boldsymbol{Z}} := W D^{1/2} \boldsymbol{\xi} \quad \text{and} \quad \widetilde{\boldsymbol{Z}}_\alpha := \alpha W D_+^{1/2} \boldsymbol{\xi} \quad \text{for } \alpha > 0.$$

a. Show that $\widetilde{\boldsymbol{Z}} \sim \mathrm{CN}(\mathbf{0}, 2(\widetilde{C} + 2\widetilde{C}_-))$ and $\widetilde{\boldsymbol{Z}}_\alpha \sim \mathrm{CN}(\mathbf{0}, 2\alpha^2(\widetilde{C} + \widetilde{C}_-))$.

b. Let C_X, C_α denote the covariance matrices of the first N components of $\operatorname{Re} \widetilde{\boldsymbol{Z}}$, $\operatorname{Re} \widetilde{\boldsymbol{Z}}_\alpha$ respectively. Show that $\|C - C_X\|_2 \le 2\rho(D_-)$ and

$$\left\|C - C_\alpha\right\|_2 \le \max\left\{\rho(D_-), |1 - \alpha^2| \rho(D_+)\right\}.$$

c. Show that $\operatorname{Tr} \alpha^2(\widetilde{C} + \widetilde{C}_-) = \operatorname{Tr} \widetilde{C}$ for $\alpha^2 = \operatorname{Tr}(D)/\operatorname{Tr}(D_+)$ and hence that each component of $\widetilde{\boldsymbol{Z}}_\alpha$ has the correct one-dimensional distribution.

6.20 Consider the stationary Gaussian process $X(t)$ with mean zero and exponential covariance $c(t) = \mathrm{e}^{-|t|/\ell}$.

a. Using the MATLAB command `imagesc`, plot the covariance matrix $C \in \mathbb{R}^{N \times N}$ of $\boldsymbol{X} = [X(t_0), \ldots, X(t_{N-1})]^{\mathsf{T}}$, where $t_n = n\Delta t$ and $\Delta t = T/(N-1)$ for $\ell = 1, 0.1$.

b. Write a MATLAB routine to compute the minimum eigenvalue of the minimal circulant extension $\widetilde{C}$. Plot the minimum eigenvalue against N for increasing N and show that it is $\mathcal{O}(N^{-1})$ as $N \to \infty$.

6.21 Consider the stationary Gaussian process $X(t)$ with mean zero and Gaussian covariance $c(t) = \mathrm{e}^{-(t/\ell)^2}$.

a. Write a MATLAB routine that calls Algorithm 6.8 to generate sample paths of $X(t)$ at N points $t_n = n\Delta t$ with specified spacing Δt.

b. Suppose we want to sample $X(t)$ on $[0, 1]$ at $t_n = n\Delta t$ where $\Delta t = 1/(N-1)$ for $N = 100$. Consider the cases $\ell = 0.1$, 1, 10. Increase N and investigate whether the minimal circulant extension $\widetilde{C}$ is non-negative definite.

6.22 Choose a stationary Gaussian process $X(t)$ where the spectral density is known. Generate a set of samples of $X(t)$ on a uniformly spaced grid on a long time interval $[0, T]$ using a method from §6.4 or §6.5. Compute the spectral density (using Algorithm 6.1) for each sample and average to approximate $f(\nu)$ using (6.7). Compare your approximation to the exact spectral density.

6.23 Show that fBm is *not* a stationary process for any $H \in (0, 1)$. By considering the vector of increments $\boldsymbol{u} = [X(t_1) - X(t_0), \ldots, X(t_N) - X(t_{N-1})]^{\mathsf{T}}$, show how to sample $\boldsymbol{X} = [X(t_0), \ldots, X(t_N)]^{\mathsf{T}}$ for $t_n = n\Delta t$ for a fBm $X(t)$ by using the circulant embedding method.

7
Random Fields

We now turn from stochastic processes $\{u(t): t \geq 0\}$, which are families of random variables for a one-dimensional parameter t, to *random fields* $\{u(\boldsymbol{x}): \boldsymbol{x} \in D \subset \mathbb{R}^d\}$, which are families of random variables for a $d > 1$ dimensional parameter $\boldsymbol{x}$. Random fields are important in many applications and are used, for example, to model biological tissue, velocity fields in turbulent flows, permeability of rocks or other geological features, as well as temperature, rainfall and ocean heights in climate modelling. Depending on the application, random fields have different statistical characteristics, which we describe in terms of the mean $\mathbb{E}[u(\boldsymbol{x})]$ and covariance $\mathrm{Cov}(u(\boldsymbol{x}), u(\boldsymbol{y}))$. Important cases are stationary random fields (where the mean is constant and the covariance depends only on $\boldsymbol{x} - \boldsymbol{y}$), isotropic random fields (the covariance depends only on the distance $\|\boldsymbol{x} - \boldsymbol{y}\|_2$), or anisotropic random fields (the covariance is directionally dependent).

Random fields once constructed are typically used to obtain other quantities, such as cell movement, fluid pressures, vegetation patterns, temperatures, and flow rates, often through solving a PDE. A PDE is a differential equation with derivatives with respect to $\boldsymbol{x} \in \mathbb{R}^d$ for $d > 1$ and the solution $u(\boldsymbol{x})$ of a PDE with random coefficients is a family of random variables parameterized by $\boldsymbol{x}$. In other words, the solution $u(\boldsymbol{x})$ is also a random field. In Chapters 9 and 10, we develop solution methods for such PDEs and we show how to calculate statistics (e.g., mean and variance) of quantities derived from the PDE and hence quantify uncertainty in the particular PDE model.

As a specific example, consider the elliptic boundary-value problem

$$-\nabla \cdot a\nabla u = f \quad \text{in } D \subset \mathbb{R}^{2,3}, \qquad u = g \quad \text{on } \partial D, \tag{7.1}$$

which provides a simple model of fluid flow in a porous medium where u is the fluid pressure, $\boldsymbol{v} = -a\nabla u$ is the fluid velocity, and a is the permeability of the porous medium. If a, f, and g are functions of $\boldsymbol{x} \in D$ then the solution u is a deterministic function of $d = 2$ or 3 variables. Usually we do not know the permeability a everywhere and, in geophysics, it is common to model it as a log-normal random field; that is, $a(\boldsymbol{x}) = \mathrm{e}^{z(\boldsymbol{x})}$ where $\{z(\boldsymbol{x}): \boldsymbol{x} \in D\}$ is a Gaussian random field with a given mean and covariance. A popular choice for the covariance is

$$\mathrm{Cov}\big(z(\boldsymbol{x}), z(\boldsymbol{y})\big) = \sigma^2 \exp\big(-\|\boldsymbol{x} - \boldsymbol{y}\|/\ell\big), \qquad \boldsymbol{x}, \boldsymbol{y} \in D, \tag{7.2}$$

for some norm $\|\cdot\|$, where σ^2 is the variance and $\ell > 0$ is a correlation length. When $\boldsymbol{x}, \boldsymbol{y}$ are close together, $z(\boldsymbol{x}), z(\boldsymbol{y})$ are more highly correlated and are likely to have similar values.

As $\boldsymbol{x}$ and $\boldsymbol{y}$ become further apart, the covariance of $z(\boldsymbol{x})$ and $z(\boldsymbol{y})$ decreases exponentially, at a rate that depends on ℓ, and $z(\boldsymbol{x}), z(\boldsymbol{y})$ are effectively uncorrelated when $\|\boldsymbol{x}-\boldsymbol{y}\| \gg \ell$.

The simplest strategy for solving (7.1) is the Monte Carlo method and this is studied in Chapter 9. It requires samples of the random field $a(\boldsymbol{x})$ and we focus on methods for generating samples of Gaussian random fields that are stationary or isotropic in two or three dimensions. The chapter is organised as follows. §7.1 summarises basic definitions, introduces the Brownian sheet, and outlines the theory of Gaussian, stationary, and isotropic random fields. The next two sections introduce efficient sampling methods or so-called random field generators for Gaussian random fields. In §7.2, we extend the circulant embedding method from Chapter 6 to stationary Gaussian random fields on $D \subset \mathbb{R}^2$ and, in §7.3, we introduce the turning bands method for generating realisations of isotropic fields on $D \subset \mathbb{R}^{2,3}$. §7.4 outlines the Karhunen–Loève expansion for random fields and, finally, §7.5 analyses the continuity and differentiability of random fields as functions of $\boldsymbol{x} \in D$.

7.1 Second-order random fields

In the following definition, D is a subset of $\mathbb{R}^d$ and we do not restrict ourselves to the case where D is an open set (as in Definition 1.3).

Definition 7.1 (random field) For a set $D \subset \mathbb{R}^d$, a (real-valued) *random field* $\{u(\boldsymbol{x})\colon \boldsymbol{x} \in D\}$ is a set of real-valued random variables on a probability space $(\Omega, \mathcal{F}, \mathbb{P})$. Although we often drop $\omega \in \Omega$ and simply write $u(\boldsymbol{x})$, it should be noted that $u\colon D \times \Omega \to \mathbb{R}$.

Most of the terminology introduced for stochastic processes $\{X(t)\colon t \in \mathcal{T}\}$ in Chapters 5 and 6 carries over straightforwardly to random fields $\{u(\boldsymbol{x})\colon \boldsymbol{x} \in D\}$ when we replace $t \in \mathcal{T}$ with $\boldsymbol{x} \in D$. Rather than *sample paths* used for stochastic processes, we usually speak of *realisations* of a random field.

Definition 7.2 (realisation) For a fixed $\omega \in \Omega$, the associated *realisation* of a random field $\{u(\boldsymbol{x})\colon \boldsymbol{x} \in D\}$ is the deterministic function $f\colon D \to \mathbb{R}$ defined by $f(\boldsymbol{x}) := u(\boldsymbol{x}, \omega)$ for $\boldsymbol{x} \in D$. We often denote the function f by $u(\cdot, \omega)$.

In parallel to Definitions 5.9 and 5.31 and Theorems 5.32 and 5.35, we have the following.

Definition 7.3 (second-order) For a set $D \subset \mathbb{R}^d$, a random field $\{u(\boldsymbol{x})\colon \boldsymbol{x} \in D\}$ is *second-order* if $u(\boldsymbol{x}) \in L^2(\Omega)$ for every $\boldsymbol{x} \in D$. We say a second-order random field has *mean* function $\mu(\boldsymbol{x}) := \mathbb{E}[u(\boldsymbol{x})]$ and *covariance* function

$$C(\boldsymbol{x}, \boldsymbol{y}) = \operatorname{Cov}(u(\boldsymbol{x}), u(\boldsymbol{y})) := \mathbb{E}\big[(u(\boldsymbol{x}) - \mu(\boldsymbol{x}))(u(\boldsymbol{y}) - \mu(\boldsymbol{y}))\big], \qquad \boldsymbol{x}, \boldsymbol{y} \in D.$$

Theorem 7.4 (mean-square regularity) *Let $u(\boldsymbol{x})$ be a mean-zero second-order random field. If the covariance $C \in C(D \times D)$, then $u(\boldsymbol{x})$ is* mean-square continuous *so that $\|u(\boldsymbol{x}+\boldsymbol{h}) - u(\boldsymbol{x})\|_{L^2(\Omega)} \to 0$ as $\boldsymbol{h} \to \boldsymbol{0}$ for all $\boldsymbol{x} \in D$. If $C \in \mathrm{C}^2(D \times D)$, then $u(\boldsymbol{x})$ is mean-square differentiable. That is, a random field $\frac{\partial u(\boldsymbol{x})}{\partial x_i}$ exists such that*

$$\left\| \frac{u(\boldsymbol{x} + h\boldsymbol{e}_i) - u(\boldsymbol{x})}{h} - \frac{\partial u(\boldsymbol{x})}{\partial x_i} \right\|_{L^2(\Omega)} \to 0 \qquad \text{as } h \to 0$$

and $\frac{\partial u(\boldsymbol{x})}{\partial x_i}$ has covariance $C_i(\boldsymbol{x}, \boldsymbol{y}) = \frac{\partial^2 C(\boldsymbol{x}, \boldsymbol{y})}{\partial x_i \partial y_i}$ for $i = 1, \ldots, d$.

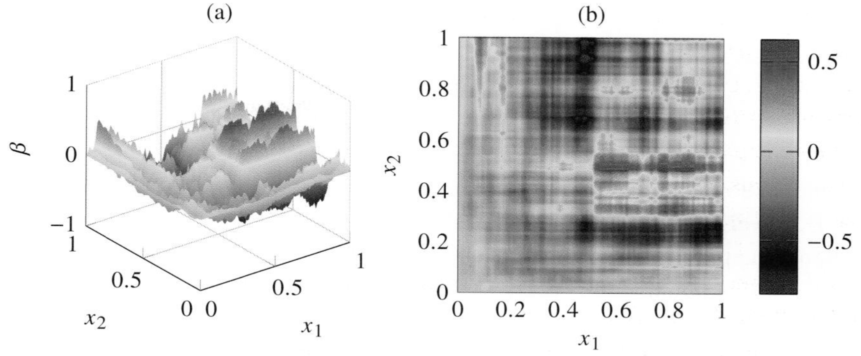

Figure 7.1 A realisation of the two-dimensional Brownian sheet $\beta(\boldsymbol{x})$ on $[0,1]\times[0,1]$. (a) shows a MATLAB `surf` plot and (b) shows a MATLAB `contourf` plot. We will usually prefer the latter format when plotting random fields.

In this book, we focus on Gaussian random fields.

Definition 7.5 (Gaussian random field) A *Gaussian random field* $\{u(\boldsymbol{x})\colon \boldsymbol{x} \in D\}$ is a second-order random field such that $\boldsymbol{u} = [u(\boldsymbol{x}_1), u(\boldsymbol{x}_2), \dots, u(\boldsymbol{x}_M)]^{\mathsf{T}}$ follows the multivariate Gaussian distribution for any $\boldsymbol{x}_1, \dots, \boldsymbol{x}_M \in D$ and any $M \in \mathbb{N}$. Specifically, $\boldsymbol{u} \sim \mathrm{N}(\boldsymbol{\mu}, C)$ where $\mu_i = \mu(\boldsymbol{x}_i)$ and $c_{ij} = C(\boldsymbol{x}_i, \boldsymbol{x}_j)$.

In analogue to Theorem 5.18 and Corollary 5.19 of §5.2, it can be shown that the distribution of a Gaussian random field is uniquely determined by its mean $\mu\colon D \to \mathbb{R}$ and covariance function $C\colon D \times D \to \mathbb{R}$. Realisations of Gaussian random fields with mean zero and a variety of covariance functions are plotted in Figures 7.1–7.5.

Our first example of a Gaussian random field, the *Brownian sheet* $\beta(\boldsymbol{x})$, is defined by its mean and covariance and is a multi-parameter version of Brownian motion (see Definition 5.11). Properties of $\beta(\boldsymbol{x})$ and the distribution of the increments $\beta(\boldsymbol{y}) - \beta(\boldsymbol{x})$ are explored in Exercises 7.1–7.3.

Definition 7.6 (Brownian sheet) A Brownian sheet $\beta(\boldsymbol{x})$ is a Gaussian random field on $D = (\mathbb{R}^+)^d$ with mean $\mu(\boldsymbol{x}) = 0$ and covariance

$$C(\boldsymbol{x}, \boldsymbol{y}) = \prod_{i=1}^{d} \min\{x_i, y_i\}, \qquad \boldsymbol{x}, \boldsymbol{y} \in D.$$

A realisation of $\beta(\boldsymbol{x})$ on $[0,1]\times[0,1]$ (with $d = 2$) is shown in Figure 7.1.

There are, however, many non-Gaussian random fields with a given covariance structure. For instance, the random field in Example 7.7 has the same mean and covariance as the two-dimensional Brownian sheet.

Example 7.7 (product of two Brownian motions) Consider $D = \mathbb{R}^+ \times \mathbb{R}^+$ and *iid* Brownian motions W_1, W_2. Define the random field $u(\boldsymbol{x})$ by

$$u(\boldsymbol{x}) := W_1(x_1) W_2(x_2), \qquad \boldsymbol{x} = \begin{pmatrix} x_1 \\ x_2 \end{pmatrix} \in D.$$

Then, $\mathbb{E}[u(\boldsymbol{x})] = \mathbb{E}[W_1(x_1)]\,\mathbb{E}[W_2(x_2)] = 0$ because $W_1(x_1)$ and $W_2(x_2)$ are uncorrelated and

$$\mathbb{E}\big[u(\boldsymbol{x})u(\boldsymbol{y})\big] = \mathbb{E}\big[W_1(x_1)W_1(y_1)\big]\,\mathbb{E}\big[W_2(x_2)W_2(y_2)\big]$$
$$= \min\{x_1, y_1\}\,\min\{x_2, y_2\}, \qquad \boldsymbol{x}, \boldsymbol{y} \in D.$$

Hence, the random field $\{u(\boldsymbol{x}) \colon \boldsymbol{x} \in D\}$ has the same mean and covariance as $\beta(\boldsymbol{x})$ (with $d = 2$). However, $u(\boldsymbol{x})$ is not a Brownian sheet because $u(\boldsymbol{x})$ is not Gaussian. To see this, we compute the probability density function (pdf) of the random variable $u(\boldsymbol{x})$ using the characteristic function as follows. By Lemma 4.48 and the independence of $W_1(x_1)$ and $W_2(x_2)$, we have

$$\mathbb{E}\big[e^{i\lambda u(x_1,x_2)}\big] = \mathbb{E}\Big[\mathbb{E}\Big[e^{i\lambda u(x_1,x_2)}\,\Big|\,W_1(x_1)\Big]\Big] = \mathbb{E}\left[\int_{\mathbb{R}} e^{i\lambda W_1(x_1)w_2}\frac{e^{-w_2^2/2x_2}}{\sqrt{2\pi x_2}}\,dw_2\right]$$
$$= \frac{1}{\sqrt{2\pi}}\int_{\mathbb{R}} \mathbb{E}\big[e^{i\lambda W_1(x_1)\sqrt{x_2}t}\big]e^{-t^2/2}\,dt, \qquad \text{for } t = w_2/\sqrt{x_2}.$$

Let $p(u)$ denote the pdf of $u(\boldsymbol{x}) = u(x_1, x_2)$. By Definition 4.28, $\hat{p}(-\lambda) = \mathbb{E}\big[e^{i\lambda u}\big]/\sqrt{2\pi}$. Using (4.14) with $\lambda \mapsto \lambda\sqrt{x_1x_2}\,t$ and $\mu = 0$,

$$\hat{p}(\lambda) = \frac{1}{2\pi}\int_{\mathbb{R}} e^{-\lambda^2x_1x_2t^2/2}e^{-t^2/2}\,dt = \frac{1}{2\pi}\int_{\mathbb{R}} e^{-(\lambda^2x_1x_2+1)t^2/2}\,dt = \frac{1}{\sqrt{2\pi(\lambda^2x_1x_2+1)}}.$$

Hence, from (A.5), $p(u) = (\pi x_1x_2)^{-1}K_0(|u|/x_1x_2)$ for any $\boldsymbol{x} = [x_1, x_2]^{\mathsf{T}} \in D$. Clearly, $u(\boldsymbol{x})$ is not a Gaussian random variable.

Recall that Hilbert space-valued random variables were defined in Chapter 4. $L^2(D)$-valued random variables can be viewed as second-order random fields.

Example 7.8 ($L^2(D)$-valued random variables) For $D \subset \mathbb{R}^d$, consider an $L^2(D)$-valued random variable u with mean $\mu \in L^2(D)$ and covariance operator $\mathcal{C}$; see Definitions 4.34 and 4.35. Then, $u(\boldsymbol{x})$ is a real-valued random variable for each $\boldsymbol{x}$ and the mean $\mathbb{E}[u(\boldsymbol{x})]$ and covariance $\mathrm{Cov}(u(\boldsymbol{x}), u(\boldsymbol{y}))$ are well defined (almost surely in $\boldsymbol{x}$). From the definition of the covariance operator (Definition 4.35), for $\phi, \psi \in L^2(D)$, Fubini's theorem (Theorem 1.24) gives

$$\langle \mathcal{C}\phi, \psi\rangle_{L^2(D)} = \mathrm{Cov}\Big(\langle\phi, u\rangle_{L^2(D)}, \langle\psi, u\rangle_{L^2(D)}\Big)$$
$$= \mathbb{E}\left[\int_D \phi(\boldsymbol{x})(u(\boldsymbol{x}) - \mu(\boldsymbol{x}))\,d\boldsymbol{x}\int_D \psi(\boldsymbol{y})(u(\boldsymbol{y}) - \mu(\boldsymbol{y}))\,d\boldsymbol{y}\right]$$
$$= \int_D\int_D \mathrm{Cov}\big(u(\boldsymbol{x}), u(\boldsymbol{y})\big)\,\phi(\boldsymbol{x})\,d\boldsymbol{x}\,\psi(\boldsymbol{y})\,d\boldsymbol{y},$$

so that

$$(\mathcal{C}\phi)(\boldsymbol{x}) = \int_D \mathrm{Cov}\big(u(\boldsymbol{x}), u(\boldsymbol{y})\big)\,\phi(\boldsymbol{y})\,d\boldsymbol{y},$$

and the kernel of the covariance operator $\mathcal{C}$ is $\mathrm{Cov}(u(\boldsymbol{x}), u(\boldsymbol{y}))$. Hence, any $L^2(D)$-valued random variable u defines a second-order random field $\{u(\boldsymbol{x}) \colon \boldsymbol{x} \in D\}$ with mean $\mu(\boldsymbol{x})$ and covariance $C(\boldsymbol{x}, \boldsymbol{y})$ equal to the kernel of $\mathcal{C}$.

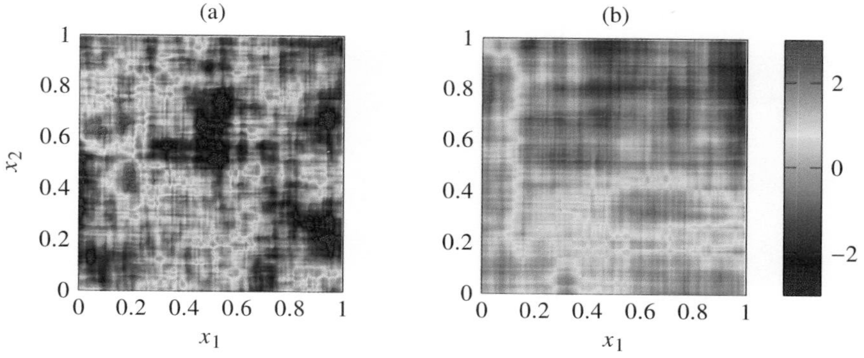

Figure 7.2 Realisations of the mean-zero Gaussian random field with stationary covariance $c(\boldsymbol{x}) = e^{-|x_1|/\ell_1 - |x_2|/\ell_2}$ on $[0,1] \times [0,1]$ for (a) $\ell_1 = \ell_2 = 1/10$ and (b) $\ell_1 = \ell_2 = 1/2$.

Stationary random fields

In Chapter 6, we introduced stationary processes and developed their analysis using Fourier theory. There are *two* classes of random fields that warrant special attention. We have *stationary* random fields, where the distribution is invariant to translations, and *isotropic* random fields, where the distribution is additionally invariant to rotations.

Definition 7.9 (stationary random field) We say a second-order random field $\{u(\boldsymbol{x}) : \boldsymbol{x} \in \mathbb{R}^d\}$ is *stationary* if the mean $\mu(\boldsymbol{x})$ is independent of $\boldsymbol{x}$ (i.e., constant) and the covariance has the form $C(\boldsymbol{x}, \boldsymbol{y}) = c(\boldsymbol{x} - \boldsymbol{y})$, for a function $c(\boldsymbol{x})$ known as the *stationary covariance*.

Example 7.10 (separable exponential) The function

$$c(\boldsymbol{x}) = \prod_{i=1}^{d} e^{-|x_i|/\ell_i}, \qquad \ell_i > 0, \tag{7.3}$$

is an example of a stationary covariance function. To verify $c(\boldsymbol{x})$ is a valid covariance function, see Exercise 7.5 and Example 7.12. The parameters ℓ_i are known as correlation lengths. A realisation of the mean-zero Gaussian random field with this covariance function is plotted in Figure 7.2 in the case $d = 2$ with $\ell_1 = \ell_2 = 1/10$ and $\ell_1 = \ell_2 = 1/2$.

Once again, the Wiener–Khintchine theorem (Theorem 6.5) tells us that stationary covariance functions $c(\boldsymbol{x})$ of mean-square continuous second-order random fields $u(\boldsymbol{x})$ may be written as Fourier integrals.

Theorem 7.11 (Wiener–Khintchine) *The following are equivalent:*

(i) *There exists a stationary random field $\{u(\boldsymbol{x}) : \boldsymbol{x} \in \mathbb{R}^d\}$ with stationary covariance function $c(\boldsymbol{x})$ that is mean-square continuous.*
(ii) *The function $c : \mathbb{R}^d \to \mathbb{R}$ is such that*

$$c(\boldsymbol{x}) = \int_{\mathbb{R}^d} e^{i\boldsymbol{\nu}^{\mathsf{T}}\boldsymbol{x}}\, dF(\boldsymbol{\nu}),$$

for some measure F on $\mathbb{R}^d$ with $F(\mathbb{R}^d) < \infty$.

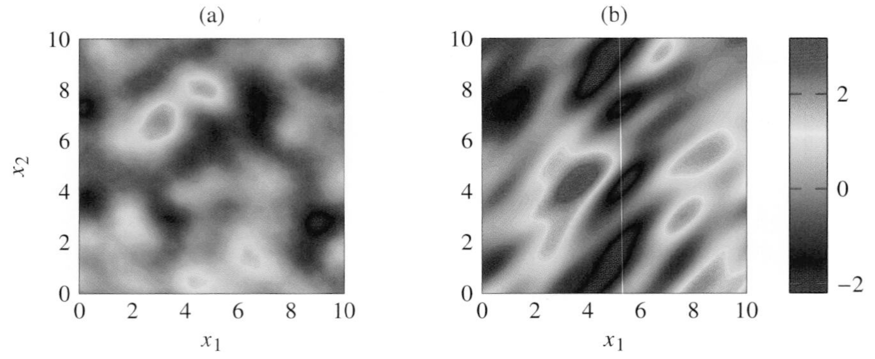

Figure 7.3 Realisations of the mean-zero Gaussian random field with stationary covariance $c(\boldsymbol{x}) = \mathrm{e}^{-\boldsymbol{x}^{\mathsf{T}} A \boldsymbol{x}}$ on $[0,10] \times [0,10]$ for (a) $A = I_2$ and (b) A given by (7.5). The covariance for (b) is not isotopic and, indeed, the realisation appears stretched from bottom left to top right.

The measure F is called the *spectral distribution*. If it exists, the density f of F is called the *spectral density* and $c(\boldsymbol{x}) = (2\pi)^{d/2} \check{f}(\boldsymbol{x})$. Alternatively, given $c\colon \mathbb{R}^d \to \mathbb{R}$, we may compute

$$f(\boldsymbol{\nu}) = \frac{1}{(2\pi)^d} \int_{\mathbb{R}^d} \mathrm{e}^{-\mathrm{i}\boldsymbol{\nu}^{\mathsf{T}}\boldsymbol{x}}\, c(\boldsymbol{x})\, d\boldsymbol{x} = \frac{1}{(2\pi)^{d/2}} \hat{c}(\boldsymbol{\nu}). \tag{7.4}$$

If f is non-negative and integrable then $c(\boldsymbol{x})$ is a valid covariance function.

Example 7.12 (separable exponential covariance) The Fourier transform of the separable exponential covariance (7.3) equals the product of the transforms of its individual factors $\mathrm{e}^{-|x_i|/\ell_i}$. That is,

$$f(\boldsymbol{\nu}) = \frac{1}{(2\pi)^d} \int_{\mathbb{R}^d} \mathrm{e}^{-\mathrm{i}\boldsymbol{x}^{\mathsf{T}}\boldsymbol{\nu}} c(\boldsymbol{x})\, d\boldsymbol{x} = \prod_{i=1}^{d} \frac{1}{2\pi} \int_{\mathbb{R}} \mathrm{e}^{-\mathrm{i}x_i \nu_i} \mathrm{e}^{-|x_i|/\ell_i}\, dx_i.$$

The terms in the product were computed in Example 6.6 and we conclude that

$$f(\boldsymbol{\nu}) = \prod_{i=1}^{d} \frac{\ell_i}{(\nu_i^2 + \ell_i^2)\pi}.$$

As $\ell_i > 0$, $f(\boldsymbol{\nu})$ is non-negative and is the density of a measure F with $F(\mathbb{R}) < \infty$. Theorem 7.11 then implies that $c(\boldsymbol{x})$ is a valid stationary covariance function for some mean-square continuous random field.

Example 7.13 (Gaussian covariance) For a symmetric positive-definite $d \times d$ matrix A, the function

$$c(\boldsymbol{x}) = \mathrm{e}^{-\boldsymbol{x}^{\mathsf{T}} A \boldsymbol{x}}, \qquad \boldsymbol{x} \in \mathbb{R}^d,$$

has Fourier transform (by (1.46))

$$f(\boldsymbol{\nu}) = \frac{1}{(2\pi)^d} \int_{\mathbb{R}^d} \mathrm{e}^{-\mathrm{i}\boldsymbol{x}^{\mathsf{T}}\boldsymbol{\nu}} \mathrm{e}^{-\boldsymbol{x}^{\mathsf{T}} A \boldsymbol{x}}\, d\boldsymbol{x} = \frac{1}{(2\pi)^{d/2} 2^{d/2} \sqrt{\det A}} \mathrm{e}^{-\boldsymbol{\nu}^{\mathsf{T}} A^{-1} \boldsymbol{\nu}/4}.$$

Note that $f(\boldsymbol{\nu})$ is non-negative and is the density of a measure F (in fact, the Gaussian distribution $\mathrm{N}(\mathbf{0}, 2A)$). Again, Theorem 7.11 implies $c(\boldsymbol{x})$ is a valid stationary covariance function. A realisation of the mean-zero Gaussian random field with this covariance is plotted in Figure 7.3 with $d = 2$ in the cases $A = I_2$ and

$$A = \begin{pmatrix} 1 & 0.8 \\ 0.8 & 1 \end{pmatrix}. \tag{7.5}$$

Isotropic random fields

Definition 7.14 (isotropic random field) A stationary random field $\{u(\boldsymbol{x}) : \boldsymbol{x} \in \mathbb{R}^d\}$ is *isotropic* if the stationary covariance $c(\boldsymbol{x})$ is invariant to rotations. Then μ is constant and $c(\boldsymbol{x}) = c^0(r)$, where $r := \|\boldsymbol{x}\|_2$, for a function $c^0 : \mathbb{R}^+ \to \mathbb{R}$ known as the *isotropic covariance*.

A simple example of an isotropic covariance is $c^0(r) = \mathrm{e}^{-r^2}$, given by $c(\boldsymbol{x}) = \mathrm{e}^{-\boldsymbol{x}^{\mathsf{T}} A \boldsymbol{x}}$ with $A = I_d$. In fact, $c(\boldsymbol{x}) = \mathrm{e}^{-\boldsymbol{x}^{\mathsf{T}} A \boldsymbol{x}}$ is isotropic whenever $A = \sigma I_d$ for some $\sigma > 0$. When A is not diagonal or is diagonal with distinct diagonal entries, $\mathrm{e}^{-\boldsymbol{x}^{\mathsf{T}} A \boldsymbol{x}}$ is not isotropic. Figure 7.3 shows a realisation of an isotropic and a non-isotropic (or anisotropic) field and the directional preference of the anisotropic example is clear.

For isotropic functions, the Fourier transform in the Wiener–Khintchine theorem becomes a Hankel transform (see Theorem 1.107) and

$$c^0(r) = \begin{cases} 2\displaystyle\int_0^\infty \cos(rs) f^0(s)\, ds, & d = 1, \\ 2\pi \displaystyle\int_0^\infty J_0(rs) f^0(s) s\, ds, & d = 2, \\ 4\pi \displaystyle\int_0^\infty \frac{1}{rs} \sin(rs) f^0(s) s^2\, ds, & d = 3, \end{cases} \tag{7.6}$$

where $f^0(s) = f(\boldsymbol{\nu})$ and $s = \|\boldsymbol{\nu}\|_2$. In the case $d > 3$, the covariance is given by the following result. The Bessel functions $J_p(r)$ are defined in §A.3 and the gamma function $\Gamma(x)$ is reviewed in (A.2).

Theorem 7.15 (isotropic covariance) *Let $\{u(\boldsymbol{x}) : \boldsymbol{x} \in \mathbb{R}^d\}$ be an isotropic random field with a mean-square continuous covariance function $c^0(r)$. There exists a finite measure F^0 on $\mathbb{R}^+$, known as the* radial spectral distribution, *such that*

$$c^0(r) = \Gamma(d/2) \int_0^\infty \frac{J_p(rs)}{(rs/2)^p}\, dF^0(s), \qquad p = \frac{d}{2} - 1. \tag{7.7}$$

If the spectral density $f(\boldsymbol{\nu})$ exists, $f^0(s) := f(\boldsymbol{\nu})$ for $s = \|\boldsymbol{\nu}\|_2$ is called the radial spectral density *function. Then $dF^0(s) = \omega_d s^{d-1} f^0(s)\, ds$ and*

$$f^0(s) = \frac{1}{(2\pi)^{d/2}} \int_0^\infty \frac{J_p(rs)}{(rs)^p} c^0(r) r^{d-1}\, dr. \tag{7.8}$$

Example 7.16 (isotropic exponential) We show that $c^0(r) = \mathrm{e}^{-r}$ defines a valid isotropic covariance function in $\mathbb{R}^d$ by showing its Hankel transform (7.8) is positive and therefore is the density of a radial spectral distribution function. After substituting $R = rs$,

$$f^0(s) = \frac{1}{(2\pi)^{d/2}} \int_0^\infty \frac{J_p(rs)}{(rs)^p} \mathrm{e}^{-r} r^{d-1}\, dr$$
$$= \frac{1}{(2\pi)^{d/2} s^d} \int_0^\infty \mathrm{e}^{-R/s} J_p(R) R^{d/2}\, dR. \tag{7.9}$$

We use the following identity for the Laplace transform (A.6),

$$\int_0^\infty \mathrm{e}^{-rs} r^p J_p(r)\, dr = \frac{2^p \Gamma(p + 1/2)}{\pi^{1/2}} \frac{1}{(1 + s^2)^{p+1/2}}.$$

Differentiating both sides with respect to s gives

$$\int_0^\infty \mathrm{e}^{-rs} J_p(r) r^{p+1}\, dr = \frac{2^p \Gamma(p + 1/2)}{\pi^{1/2}} \frac{2s(p + 1/2)}{(1 + s^2)^{p+3/2}}.$$

With (7.9) and $p = d/2 - 1$, we see

$$f^0(s) = \frac{1}{(2\pi)^{d/2} s^d} \frac{2^{(d-2)/2} \Gamma((d-1)/2)}{\pi^{1/2}} \frac{2((d-1)/2)/s}{(1 + (1/s^2))^{(d+1)/2}}$$
$$= \frac{\Gamma((d+1)/2)}{\pi^{(d+1)/2}} \frac{1}{(1 + s^2)^{(d+1)/2}}, \tag{7.10}$$

using $x\Gamma(x) = \Gamma(x + 1)$. Then $f^0(s)$ is positive and is a well-defined radial spectral density. Hence, e^{-r} is the isotropic covariance of a random field in $\mathbb{R}^d$ for any $d \in \mathbb{N}$. The first row of Figure 7.4 shows a realisation of the mean-zero Gaussian field with isotropic covariance $c^0(r) = \mathrm{e}^{-r/\ell}$ on $[0, 5] \times [0, 5]$ for $\ell = 1$ and $\ell = 1/10$.

Example 7.17 (Whittle–Matérn covariance) The Whittle–Matérn covariance, introduced for stochastic processes in Example 6.8, is a well-defined isotropic covariance function for random fields. We treat it here with a length scale $\ell > 0$ and consider

$$c_q^0(r) = \frac{1}{2^{q-1}\Gamma(q)} (r/\ell)^q K_q(r/\ell).$$

In this case, its radial spectral density is

$$f^0(s) = \frac{\Gamma(q + d/2)}{\Gamma(q)\pi^{d/2}} \frac{\ell^d}{(1 + \ell^2 s^2)^{q+d/2}}. \tag{7.11}$$

This can be derived from the integral identity (A.7). It agrees with (7.10) in the case $q = 1/2$, $\ell = 1$ and in fact $c_{1/2}^0(r) = \mathrm{e}^{-r/\ell}$. As in §6.1, the parameter q controls the regularity and the isotropic random field with the Whittle–Matérn covariance in $\mathbb{R}^d$ is n times mean-square differentiable for $n < q$. Figure 7.4 shows a realisation of a Gaussian random field with the Whittle–Matérn covariance function in the cases $q = 1/2, 1, 2$ and $\ell = 1, 1/10$. It is clear that the realisations become smoother as q is increased.

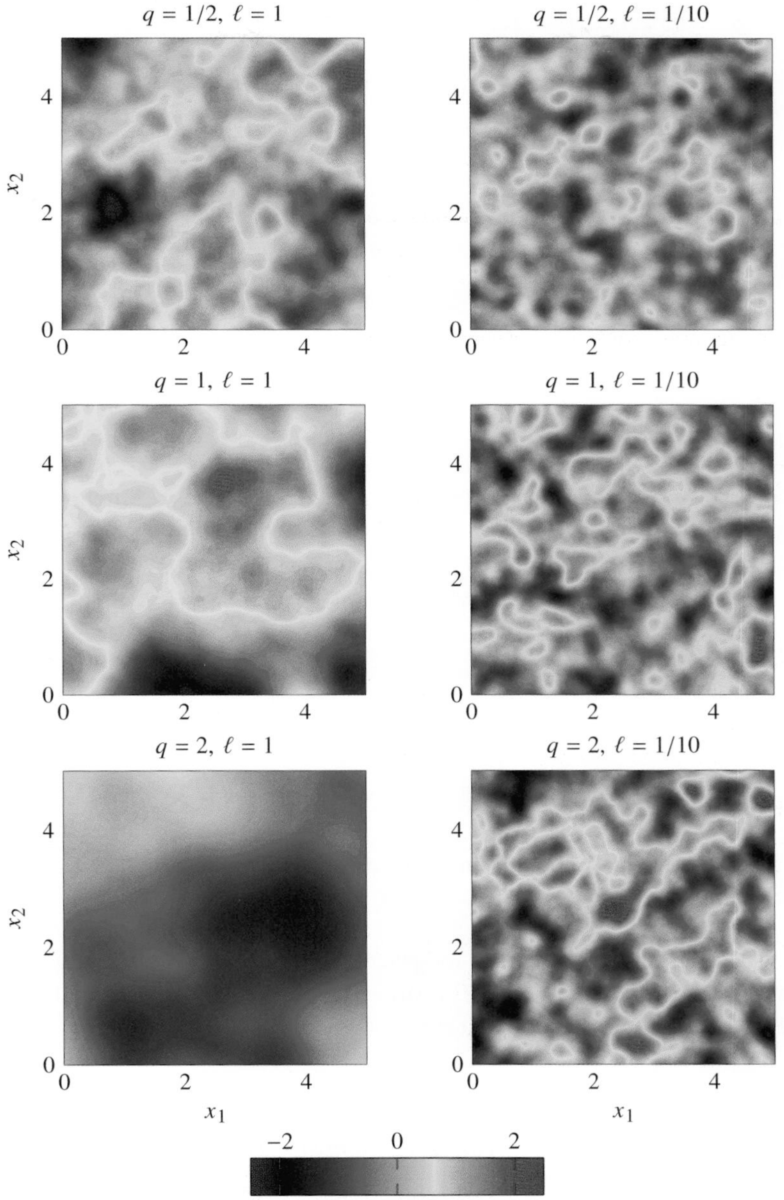

Figure 7.4 Realisations of the mean-zero Gaussian random field with the isotropic Whittle–Matérn covariance function $c_q^0(r)$: along rows, the correlation length ℓ is varied from $\ell = 1$ (first column) $\ell = 1/10$ (second column); along columns, q is varied from $q = 1/2$ (top row), $q = 1$ (middle row), $q = 2$ (bottom row). The case $q = 1/2$ is equivalent to the isotropic exponential covariance $c_{1/2}^0(r) = e^{-r/\ell}$.

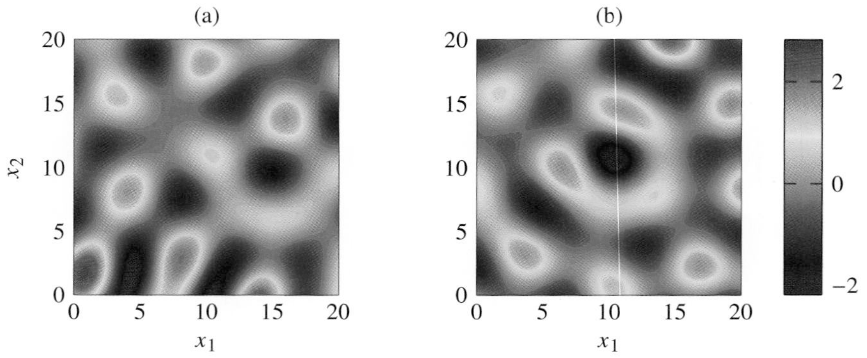

Figure 7.5 Two realisations of the mean-zero Gaussian random field with isotropic covariance $c^0(r) = J_0(r)$ on $[0,20] \times [0,20]$.

Example 7.18 (Bessel covariance) The Bessel covariance is the generalisation of the cosine covariance of Example 5.20 to dimension $d > 1$. It has a different character to the family of Whittle–Matérn covariance functions, which, like $c^0_{1/2}(r) = \mathrm{e}^{-r/\ell}$, decay monotonically. The Bessel functions $J_p(r)$ oscillate (see Figure A.2). By choosing $F^0(s) = \delta(s-1)$, (7.7) gives the covariance function $c^0(r) = \Gamma(d/2)J_p(r)/(r/2)^p$ for $p = (d-2)/2$, which is an oscillating function of r. In particular,

$$c^0(r) = \Gamma(d/2)\frac{J_p(r)}{(r/2)^p} = \begin{cases} \cos(r), & d = 1, \\ J_0(r), & d = 2, \\ \dfrac{\sin(r)}{r}, & d = 3. \end{cases}$$

Figure 7.5 shows two realisation of the mean-zero Gaussian field with covariance $c^0(r) = J_0(r)$ in two dimensions.

We now develop numerical methods for generating realisations of stationary and isotropic mean-zero Gaussian random fields. The strategy is to sample the Gaussian distribution obtained by sampling the random field at a finite set of points.

7.2 Circulant embedding in two dimensions

We know from Chapter 6 that if C is a symmetric non-negative definite *circulant* matrix, then one sample from $\mathrm{CN}(\mathbf{0}, 2C)$ provides two independent $\mathrm{N}(\mathbf{0}, C)$ samples. These are efficiently computed using the FFT. If C is Toeplitz but not circulant, it can be embedded inside a larger circulant matrix $\widetilde{C}$ and, if $\widetilde{C}$ is non-negative definite, then $\mathrm{N}(\mathbf{0}, C)$ samples can be recovered from $\mathrm{N}(\mathbf{0}, \widetilde{C})$ samples. The method can be extended to generate realisations of stationary Gaussian random fields $u(\boldsymbol{x})$ in two dimensions, if the sample points are uniformly spaced on a rectangular grid. It is more complicated to implement, however, due to the structure of the resulting covariance matrix. We begin with some definitions.

Real-valued BTTB and BCCB matrices

We will often need to organise the components of a matrix into a vector, and vice versa. To do this, we will make use of the following operators.

Definition 7.19 (vector operator) Consider a matrix $V \in \mathbb{C}^{n_1 \times n_2}$. The *vector operator* $\mathrm{vec}(\cdot)\colon \mathbb{C}^{n_1\times n_2} \to \mathbb{C}^{n_1 n_2}$ defined via

$$\mathrm{vec}(V) := \left[v_{11}, \dots, v_{n_1 1}, v_{12}, \dots, v_{n_1 2}, \dots, v_{1 n_2}, \dots, v_{n_1 n_2}\right]^{\mathsf{T}} = \boldsymbol{v} \in \mathbb{C}^{n_1 n_2},$$

corresponds to stacking the n_2 columns of V on top of each other.

Definition 7.20 (array operator) Consider a vector $\boldsymbol{v} \in \mathbb{C}^N$ where $N = n_1 n_2$ and n_1, n_2 are given. The *array operator* $\mathrm{array}(\cdot)\colon \mathbb{C}^N \to \mathbb{C}^{n_1\times n_2}$ defined via

$$\mathrm{array}(\boldsymbol{v}) := \begin{pmatrix} v_1 & v_{n_1+1} & \cdots & v_{(n_2-1)n_1+1} \\ v_2 & v_{n_1+2} & \cdots & v_{(n_2-1)n_1+2} \\ \vdots & \vdots & \cdots & \vdots \\ v_{n_1} & v_{2n_1} & \cdots & v_{n_1 n_2} \end{pmatrix} = V \in \mathbb{C}^{n_1\times n_2},$$

corresponds to making the n_2 consecutive blocks of $\boldsymbol{v}$ of length n_1 the columns of a matrix V of size $n_1 \times n_2$. If $V = \mathrm{array}(\boldsymbol{v})$, then $\boldsymbol{v} = \mathrm{vec}(V)$.

Recall Definitions 6.39 and 6.41. We now introduce block Toeplitz matrices with Toeplitz blocks and block circulant matrices with circulant blocks.

Definition 7.21 (block Toeplitz matrix with Toeplitz blocks) Let $N = n_1 n_2$. An $N \times N$ real-valued matrix C is said to be *block Toeplitz with Toeplitz blocks* (BTTB) if it has the form

$$C = \begin{pmatrix} C_0 & C_{-1} & \cdots & C_{2-n_2} & C_{1-n_2} \\ C_1 & C_0 & C_{-1} & \ddots & C_{2-n_2} \\ \vdots & \ddots & \ddots & \ddots & \vdots \\ C_{n_2-2} & \ddots & C_1 & C_0 & C_{-1} \\ C_{n_2-1} & C_{n_2-2} & \cdots & C_1 & C_0 \end{pmatrix},$$

and each of the blocks C_k is an $n_1 \times n_1$ Toeplitz matrix. Let $\boldsymbol{c}_k \in \mathbb{R}^{2n_1-1}$ be the vector containing the entries of the first row and column of C_k (as in Definition 6.39). Any BTTB matrix C can be generated from the reduced matrix

$$C_{\mathrm{red}} := \left[\boldsymbol{c}_{1-n_2}, \dots, \boldsymbol{c}_0, \dots, \boldsymbol{c}_{n_2-1}\right] \in \mathbb{R}^{(2n_1-1)\times(2n_2-1)}. \tag{7.12}$$

If C is *symmetric*, C_0 is symmetric and $C_{-j} = C_j^{\mathsf{T}}$ for $j = 1, \dots, n_2 - 1$. If C is *symmetric* and has *symmetric blocks* C_k, then C can be generated from

$$C_{\mathrm{red}} = \left[\boldsymbol{c}_0, \dots, \boldsymbol{c}_{n_2-1}\right] \in \mathbb{R}^{n_1\times n_2},$$

where $\boldsymbol{c}_k \in \mathbb{R}^{n_1}$ is the first *column* of C_k, or, equivalently, by $\boldsymbol{c}_{\mathrm{red}} = \mathrm{vec}(C_{\mathrm{red}})$, the first column of C.

Example 7.22 (symmetric BTTB matrices) Let $n_1 = 3$ and $n_2 = 2$ and consider the symmetric BTTB matrix

$$C = \begin{pmatrix} C_0 & C_1^{\mathsf{T}} \\ C_1 & C_0 \end{pmatrix}, \tag{7.13}$$

with 3×3 Toeplitz blocks

$$C_0 = \begin{pmatrix} 1 & 0.2 & 0.1 \\ 0.2 & 1 & 0.2 \\ 0.1 & 0.2 & 1 \end{pmatrix}, \qquad C_1 = \begin{pmatrix} 0.5 & 0.1 & 0.8 \\ 0.2 & 0.5 & 0.1 \\ 0.6 & 0.2 & 0.5 \end{pmatrix}. \tag{7.14}$$

C can be generated from the 5×3 reduced matrix

$$C_{\text{red}} = \begin{pmatrix} 0.6 & 0.1 & 0.8 \\ 0.2 & 0.2 & 0.1 \\ 0.5 & 1 & 0.5 \\ 0.1 & 0.2 & 0.2 \\ 0.8 & 0.1 & 0.6 \end{pmatrix}.$$

On the other hand, if we choose the symmetric matrix

$$C_1 = \begin{pmatrix} 0.5 & 0.1 & 0.8 \\ 0.1 & 0.5 & 0.1 \\ 0.8 & 0.1 & 0.5 \end{pmatrix}, \tag{7.15}$$

then C is a symmetric BTTB matrix with symmetric blocks, and can be generated from the 3×2 reduced matrix

$$C_{\text{red}} = \begin{pmatrix} 1 & 0.5 \\ 0.2 & 0.1 \\ 0.1 & 0.8 \end{pmatrix} \quad \text{or} \quad \boldsymbol{c}_{\text{red}} = \left[1, 0.2, 0.1, 0.5, 0.1, 0.8\right]^{\mathsf{T}}.$$

Definition 7.23 (block circulant matrix with circulant blocks) Let $N = n_1 n_2$. An $N \times N$ real-valued matrix C is *block circulant with circulant blocks* (BCCB) if it is a BTTB matrix of the form

$$C = \begin{pmatrix} C_0 & C_{n_2-1} & \cdots & C_2 & C_1 \\ C_1 & C_0 & C_{n_2-1} & \ddots & C_2 \\ \vdots & \ddots & \ddots & \ddots & \vdots \\ C_{n_2-2} & \ddots & C_1 & C_0 & C_{n_2-1} \\ C_{n_2-1} & C_{n_2-2} & \cdots & C_1 & C_0 \end{pmatrix}, \tag{7.16}$$

and each of the blocks $C_0, \ldots, C_{n_2-1}$ is an $n_1 \times n_1$ circulant matrix. Let $\boldsymbol{c}_k \in \mathbb{R}^{n_1}$ be the first column of C_k. Any BCCB matrix C is uniquely determined by the reduced $n_1 \times n_2$ matrix

$$C_{\text{red}} = [\boldsymbol{c}_0, \ldots, \boldsymbol{c}_{n_2-1}], \tag{7.17}$$

or the vector $\boldsymbol{c}_{\text{red}} = \text{vec}(C_{\text{red}}) \in \mathbb{R}^{n_1 n_2}$, which is the first column of C. If C is *symmetric*, C_0 is symmetric and $C_{n_2-j} = C_j^{\mathsf{T}}$ for $j = 1, \ldots, n_2 - 1$, so C_{red} has at most $\lfloor n_2/2 \rfloor + 1$ distinct columns. If C has *symmetric blocks*, each column of C_{red} is a Hermitian vector with at most $\lfloor n_1/2 \rfloor + 1$ distinct entries.

Example 7.24 (symmetric BCCB matrices) Let $n_1 = 3$ and $n_2 = 3$ and consider the symmetric BCCB matrix

$$C = \begin{pmatrix} C_0 & C_1^{\mathsf{T}} & C_1 \\ C_1 & C_0 & C_1^{\mathsf{T}} \\ C_1^{\mathsf{T}} & C_1 & C_0 \end{pmatrix} \tag{7.18}$$

with circulant blocks

$$C_0 = \begin{pmatrix} 1 & 0.2 & 0.2 \\ 0.2 & 1 & 0.2 \\ 0.2 & 0.2 & 1 \end{pmatrix}, \qquad C_1 = \begin{pmatrix} 0.5 & 0.2 & 0.1 \\ 0.1 & 0.5 & 0.2 \\ 0.2 & 0.1 & 0.5 \end{pmatrix}. \tag{7.19}$$

C can be generated from the 3×3 reduced matrix

$$C_{\text{red}} = \begin{pmatrix} 1 & 0.5 & 0.5 \\ 0.2 & 0.1 & 0.2 \\ 0.2 & 0.2 & 0.1 \end{pmatrix},$$

or $\boldsymbol{c}_{\text{red}} = [1, 0.2, 0.2, 0.5, 0.1, 0.2, 0.5, 0.2, 0.1]^{\mathsf{T}}$. If we choose

$$C_1 = \begin{pmatrix} 0.5 & 0.1 & 0.1 \\ 0.1 & 0.5 & 0.1 \\ 0.1 & 0.1 & 0.5 \end{pmatrix}, \tag{7.20}$$

which is symmetric, then C in (7.18) is a symmetric BCCB matrix with symmetric circulant blocks and can be generated from the 3×3 reduced matrix

$$C_{\text{red}} = \begin{pmatrix} 1 & 0.5 & 0.5 \\ 0.2 & 0.1 & 0.1 \\ 0.2 & 0.1 & 0.1 \end{pmatrix},$$

or $\boldsymbol{c}_{\text{red}} = [1, 0.2, 0.2, 0.5, 0.1, 0.1, 0.5, 0.1, 0.1]^{\mathsf{T}} = \text{vec}(C_{\text{red}})$. In this case, the reduced matrix C_{red} has two distinct Hermitian columns.

Fourier representation of BCCB matrices

BCCB matrices can be factorised efficiently using the *two-dimensional* DFT.

Definition 7.25 (matrix Kronecker product) Given two matrices $A \in \mathbb{R}^{n \times m}$ and $B \in \mathbb{R}^{r \times s}$, the *matrix Kronecker product* is the matrix

$$A \otimes B := \begin{pmatrix} a_{11}B & a_{12}B & \dots & a_{1m}B \\ a_{21}B & a_{22}B & \dots & a_{2m}B \\ \vdots & \vdots & \ddots & \vdots \\ a_{n1}B & a_{n2}B & \dots & a_{nm}B \end{pmatrix} \in \mathbb{R}^{nr \times ms}.$$

In particular, the Kronecker product satisfies $(A \otimes B)^* = A^* \otimes B^*$, where * denotes the Hermitian transpose.

Definition 7.26 (two-dimensional Fourier matrix) Let W_1 and W_2 be the $n_1 \times n_1$ and $n_2 \times n_2$ Fourier matrices respectively (see Definition 1.95). The *two-dimensional Fourier matrix* is the $N \times N$ matrix $W := W_2 \otimes W_1$ where $N = n_1 n_2$.

Definition 7.27 (two-dimensional DFT) The *two-dimensional discrete Fourier transform* of $V \in \mathbb{C}^{n_1 \times n_2}$ is the matrix $\hat{V} \in \mathbb{C}^{n_1 \times n_2}$ with entries

$$\hat{v}_{ij} := \sum_{m=1}^{n_2} \left(\sum_{l=1}^{n_1} v_{lm} \omega_1^{(i-1)(l-1)} \right) \omega_2^{(j-1)(m-1)}, \qquad \omega_1 := e^{-2\pi i/n_1},\ \omega_2 := e^{-2\pi i/n_2},$$

for $i = 1, \ldots, n_1$ and $j = 1, \ldots, n_2$. This amounts to taking one-dimensional DFTs of the columns of V, storing them in a matrix and then applying one-dimensional DFTs to each row. In MATLAB, this is computed using the command `fft2(V)` and, in matrix notation,

$$\hat{V} := \sqrt{n_1 n_2}\, \mathrm{array}\big((W_2 \otimes W_1)\boldsymbol{v}\big), \qquad \boldsymbol{v} := \mathrm{vec}(V).$$

Definition 7.28 (two-dimensional IDFT) The *two-dimensional inverse discrete Fourier transform* of $\hat{V} \in \mathbb{C}^{n_1 \times n_2}$ is the matrix $V \in \mathbb{C}^{n_1 \times n_2}$ with entries

$$v_{ij} := \frac{1}{n_1 n_2} \sum_{m=1}^{n_2} \left(\sum_{l=1}^{n_1} \hat{v}_{lm} \omega_1^{-(i-1)(l-1)} \right) \omega_2^{-(j-1)(m-1)},$$

for $i = 1, \ldots, n_1$ and $j = 1, \ldots, n_2$. In MATLAB, this is computed using the command `ifft2`($\hat{V}$) and in matrix notation,

$$V := \frac{1}{\sqrt{n_1 n_2}}\, \mathrm{array}\big((W_2 \otimes W_1)^* \hat{\boldsymbol{v}}\big), \qquad \hat{\boldsymbol{v}} := \mathrm{vec}(\hat{V}).$$

The next result is the analogue of Proposition 6.45. With our FFT scaling convention, it says the eigenvalues of an $N \times N$ BCCB matrix C with reduced matrix C_{red} are given by N times the two-dimensional inverse FFT of C_{red} and the eigenvectors are the columns of the Fourier matrix $W = W_2 \otimes W_1$. Hence, any BCCB matrix has a factorisation of the form $C = WDW^*$.

Proposition 7.29 (Fourier representation of BCCB matrices) *Let C be an $N \times N$ real-valued BCCB matrix of the form* (7.16), *where $N = n_1 n_2$. Then $C = WDW^*$, where W is the two-dimensional Fourier matrix and D is a diagonal matrix with diagonal $\boldsymbol{d} = \mathrm{vec}(\Lambda)$ where*

$$\Lambda := \sqrt{N}\, \mathrm{array}(W^* \boldsymbol{c}_{\mathrm{red}}), \qquad \boldsymbol{c}_{\mathrm{red}} := \mathrm{vec}(C_{\mathrm{red}}).$$

We do not need to assemble a BCCB matrix to compute its eigenvalues. We only need the first column, or equivalently, the reduced $n_1 \times n_2$ matrix, C_{red}.

Example 7.30 (eigenvalues of a symmetric BCCB matrix) For the symmetric BCCB matrix (7.18)–(7.19) in Example 7.24, we have $n_1 = 3$ and $n_2 = 3$ and we can compute the 9 real eigenvalues of C by applying a two-dimensional inverse DFT to C_{red} in MATLAB as follows:

```
>> C_red=[1 0.5 0.5; 0.2 0.1 0.2; 0.2 0.2 0.1];
>> Lambda=9*ifft2(C_red)
Lambda =
    3.0000    0.6000    0.6000
    1.5000    0.6000    0.3000
    1.5000    0.3000    0.6000
```

The vector $\boldsymbol{d}$ in Proposition 7.29 is obtained by stacking the columns of `Lambda`. In MATLAB, this is simply `d=Lambda(:)`. See also Exercise 7.7.

Proposition 6.45 says each block C_k of a BCCB matrix C has the decomposition $C_k = W_1 \Sigma_k W_1^*$, where Σ_k is a diagonal matrix with diagonal $\boldsymbol{\sigma}_k = \sqrt{n_1} W_1^* \boldsymbol{c}_k = n_1 \texttt{ifft}(\boldsymbol{c}_k)$ and $\boldsymbol{c}_k$ is the first column of C_k. We can also express the eigenvalues of a BCCB matrix C in terms of the n_2 sets of eigenvalues of its individual circulant blocks C_k. See Exercises 7.6 and 7.7.

Sampling $\mathrm{N}(\mathbf{0}, C)$ *with BCCB covariance*

Suppose we want to draw samples from $\mathrm{N}(\mathbf{0}, C)$ where $C \in \mathbb{R}^{N \times N}$ is a BCCB matrix, with $n_2 \times n_2$ blocks of size $n_1 \times n_1$ and $N = n_1 n_2$. For $C = W D W^*$ to be a covariance matrix, it must have real, non-negative eigenvalues. In that case, we can define the real-valued diagonal matrix $D^{1/2}$ and consider

$$\boldsymbol{Z} = W D^{1/2} \boldsymbol{\xi}, \qquad \boldsymbol{\xi} \sim \mathrm{CN}(\mathbf{0}, 2 I_N). \tag{7.21}$$

Lemma 6.51 tells us that the real and imaginary parts of $\boldsymbol{Z} = \boldsymbol{X} + \mathrm{i}\boldsymbol{Y}$ provide two independent $\mathrm{N}(\mathbf{0}, C)$ samples. We can compute samples of $\boldsymbol{Z}$ using the MATLAB command $\texttt{fft2}(V)$ where $V = \mathrm{array}(\boldsymbol{v})$ is the $n_1 \times n_2$ matrix obtained by rearranging the entries of $\boldsymbol{v} = D^{1/2} \boldsymbol{\xi}$. This is implemented in Algorithm 7.1.

Algorithm 7.1 Code to generate a pair of samples from $\mathrm{N}(\mathbf{0}, C)$ where $C \in \mathbb{R}^{N \times N}$ ($N = n_1 n_2$) is a BCCB covariance matrix. The input `C_red` is the reduced matrix which generates C, and `X,Y` are independent samples from $\mathrm{N}(\mathbf{0}, C)$, in matrix format.

```
function [X,Y]=circ_cov_sample_2d(C_red,n1,n2);
N=n1*n2;
Lambda=N*ifft2(C_red);
d=Lambda(:);
d_minus=max(-d,0);
if (max(d_minus)>0)
    disp(sprintf('Invalid covariance;rho(D_minus)=%0.5g',...
        max(d_minus)));
end;
xi=randn(n1,n2)+i.*randn(n1,n2);
V=(Lambda.^0.5).*xi;
Z=fft2(V)/sqrt(N); Z=Z(:);
X=real(Z); Y=imag(Z);
```

Example 7.31 The symmetric BCCB matrix C in (7.18) with blocks (7.19) is a valid covariance matrix. Using Algorithm 7.1, we can generate two independent samples from $\mathrm{N}(\mathbf{0}, C)$ as follows.

```
>> C_red=[1 0.5 0.5; 0.2 0.1 0.2; 0.2 0.2 0.1];
>> [X,Y]=circ_cov_sample_2d(C_red,3,3);
>> X=X(:); Y=Y(:);
```

The algorithm will return samples even if C is not non-negative definite. In that case, a warning message is returned and it should be noted that the samples do not have the correct covariance.

Circulant embedding

Definition 6.48 provides a way to embed a *symmetric* $n_1 \times n_1$ Toeplitz matrix into a *symmetric* $(2n_1 - 2) \times (2n_1 - 2)$ circulant matrix. For an $n_1 \times n_1$ non-symmetric Toeplitz matrix C, we need $2n_1 - 1$ entries to define that matrix and so the minimal circulant extension has size $(2n_1 - 1) \times (2n_1 - 1)$ and always has an odd number of rows and columns.

Example 7.32 (circulant extensions) Consider the 3×3 Toeplitz matrix C generated by $\boldsymbol{c} = [3, 2, 1, 4, 5]^{\mathsf{T}}$. Its minimal circulant extension is the 5×5 circulant matrix $\widetilde{C}$ generated by $\boldsymbol{c}_1 = [1, 4, 5, 3, 2]^{\mathsf{T}}$, which contains the entries of the first column of C, followed by the first row, in reverse order:

$$C = \begin{pmatrix} 1 & 2 & 3 \\ 4 & 1 & 2 \\ 5 & 4 & 1 \end{pmatrix}, \qquad \widetilde{C} = \begin{pmatrix} 1 & 2 & 3 & 5 & 4 \\ 4 & 1 & 2 & 3 & 5 \\ 5 & 4 & 1 & 2 & 3 \\ 3 & 5 & 4 & 1 & 2 \\ 2 & 3 & 5 & 4 & 1 \end{pmatrix}.$$

Neither C nor $\widetilde{C}$ is symmetric. Now let $x \in \mathbb{R}$. The 6×6 matrix

$$\widetilde{C} = \begin{pmatrix} 1 & 2 & 3 & x & 5 & 4 \\ 4 & 1 & 2 & 3 & x & 5 \\ 5 & 4 & 1 & 2 & 3 & x \\ x & 5 & 4 & 1 & 2 & 3 \\ 3 & x & 5 & 4 & 1 & 2 \\ 2 & 3 & x & 5 & 4 & 1 \end{pmatrix}$$

is also a circulant extension of C for any choice of x. It will be advantageous to have circulant extension matrices, which always have an *even* number of rows and columns for both symmetric and non-symmetric Toeplitz matrices. We will work with this so-called even extension and fix $x = 0$.

Definition 7.33 (even circulant extension) Let $C \in \mathbb{R}^{n_1 \times n_1}$ be a Toeplitz matrix generated by $\boldsymbol{c} = [c_{1-n_1}, \dots, c_{-1}, c_0, c_1, \dots, c_{n_1-1}]^{\mathsf{T}} \in \mathbb{R}^{2n_1-1}$ (see Definition 6.39). The *even circulant extension* is the circulant matrix $\widetilde{C} \in \mathbb{R}^{2n_1 \times 2n_1}$ with first column $\tilde{\boldsymbol{c}} = [c_0, c_1, \dots, c_{n_1-1}, 0, c_{1-n_1}, \dots, c_{-1}]^{\mathsf{T}} \in \mathbb{R}^{2n_1}$. Hence,

$$\widetilde{C} = \begin{pmatrix} C & B \\ B & C \end{pmatrix},$$

where C is the original Toeplitz matrix and $B \in \mathbb{R}^{n_1 \times n_1}$. Note that C is the leading principal $n_1 \times n_1$ submatrix of $\widetilde{C}$ and if C is symmetric, so is $\widetilde{C}$.

Example 7.34 (even circulant extension) Consider the 3×3 Toeplitz matrices C_0 and C_1 in (7.14). C_0 is symmetric but C_1 is not. Using Definition 7.33 we can embed both matrices into circulant 6×6 matrices as follows.

$$\widetilde{C}_0 = \left(\begin{array}{ccc|ccc} 1 & 0.2 & 0.1 & 0 & 0.1 & 0.2 \\ 0.2 & 1 & 0.2 & 0.1 & 0 & 0.1 \\ 0.1 & 0.2 & 1 & 0.2 & 0.1 & 0 \\ \hline 0 & 0.1 & 0.2 & 1 & 0.2 & 0.1 \\ 0.1 & 0 & 0.1 & 0.2 & 1 & 0.2 \\ 0.2 & 0.1 & 0 & 0.1 & 0.2 & 1 \end{array}\right) = \begin{pmatrix} C_0 & B_0 \\ B_0 & C_0 \end{pmatrix}, \tag{7.22}$$

$$\widetilde{C}_1 = \left(\begin{array}{ccc|ccc} 0.5 & 0.1 & 0.8 & 0 & 0.6 & 0.2 \\ 0.2 & 0.5 & 0.1 & 0.8 & 0 & 0.6 \\ 0.6 & 0.2 & 0.5 & 0.1 & 0.8 & 0 \\ \hline 0 & 0.6 & 0.2 & 0.5 & 0.1 & 0.8 \\ 0.8 & 0 & 0.6 & 0.2 & 0.5 & 0.1 \\ 0.1 & 0.8 & 0 & 0.6 & 0.2 & 0.5 \end{array}\right) = \begin{pmatrix} C_1 & B_1 \\ B_1 & C_1 \end{pmatrix}. \tag{7.23}$$

Now, any $N \times N$ BTTB matrix C, with $N = n_1 n_2$, can be embedded inside a larger BCCB matrix $\widetilde{C}$. The idea is to embed each $n_1 \times n_1$ Toeplitz block $C_k, k = 1 - n_2, \ldots, 0, \ldots, n_2 - 1$, into a larger circulant matrix $\widetilde{C}_k$ of size $2n_1 \times 2n_1$ and then assemble the $2n_2 - 1$ extension matrices $\widetilde{C}_k$ into a BCCB matrix.

Definition 7.35 (even BCCB extension of BTTB matrices) Given a BTTB matrix $C \in \mathbb{R}^{N \times N}$ where $N = n_1 n_2$, let $\widetilde{C}_k$ be the even circulant extension (see Definition 7.33) of the Toeplitz block $C_k, k = 1 - n_2, \ldots, 0, \ldots, n_2 - 1$. The *even BCCB extension* of C is the BCCB matrix

$$\widetilde{C} = \begin{pmatrix} \widetilde{C}_0 & \widetilde{C}_{-1} & \cdots & \widetilde{C}_{1-n_2} & \widetilde{0} & \widetilde{C}_{n_2-1} & \cdots & \widetilde{C}_1 \\ \widetilde{C}_1 & \widetilde{C}_0 & \ddots & \ddots & \ddots & \ddots & \ddots & \ddots \\ \vdots & \ddots & \ddots & \ddots & \ddots & \ddots & \ddots & \ddots \\ \widetilde{C}_{n_2-1} & \ddots & \ddots & \ddots & \ddots & \ddots & \ddots & \ddots \\ \widetilde{0} & \widetilde{C}_{n_2-1} & \ddots & \ddots & \ddots & \ddots & \ddots & \ddots \\ \widetilde{C}_{1-n_2} & \widetilde{0} & \ddots & \ddots & \ddots & \ddots & \ddots & \ddots \\ \vdots & \ddots & \ddots & \ddots & \ddots & \ddots & \ddots & \widetilde{C}_{-1} \\ \widetilde{C}_{-1} & \widetilde{C}_{-2} & \ddots & \ddots & \ddots & \ddots & \widetilde{C}_1 & \widetilde{C}_0 \end{pmatrix} \in \mathbb{R}^{4N \times 4N}, \tag{7.24}$$

where $\widetilde{0}$ is the $2n_1 \times 2n_1$ zero matrix. $\widetilde{C}$ has $2n_2 \times 2n_2$ blocks of size $2n_1 \times 2n_1$. If C is symmetric then C_0 is symmetric and $C_{-j} = C_j^{\mathrm{T}}$ for $j = 1, \ldots, n_2 - 1$. Hence $\widetilde{C}_0$ is symmetric, $\widetilde{C}_{-j} = \widetilde{C}_j^{\mathrm{T}}$ and $\widetilde{C}$ is also symmetric. Equivalently, if C is the BTTB matrix with reduced matrix $C_{\mathrm{red}} = [\boldsymbol{c}_{1-n_2}, \ldots, \boldsymbol{c}_{-1}, \boldsymbol{c}_0, \boldsymbol{c}_1, \ldots, \boldsymbol{c}_{n_2-1}] \in \mathbb{R}^{(2n_1-1)\times(2n_2-1)}$ as in (7.12), then $\widetilde{C} \in \mathbb{R}^{4N \times 4N}$ is the BCCB matrix with reduced matrix (see (7.17)) given by

$$\widetilde{C}_{\mathrm{red}} = \left[\tilde{\boldsymbol{c}}_0, \tilde{\boldsymbol{c}}_1, \ldots, \tilde{\boldsymbol{c}}_{n_2-1}, \tilde{\boldsymbol{0}}, \tilde{\boldsymbol{c}}_{1-n_2}, \ldots, \tilde{\boldsymbol{c}}_{-1}\right] \in \mathbb{R}^{2n_1 \times 2n_2},$$

where $\tilde{\boldsymbol{0}}$ is the zero vector of length $2n_1$ and $\tilde{\boldsymbol{c}}_k$ is the first column of the circulant extension $\widetilde{C}_k$ of the Toeplitz block C_k.

Example 7.36 (even BCCB extension) Consider the symmetric BTTB matrix

$$C = \left(\begin{array}{ccc|ccc} 1 & 0.2 & 0.1 & 0.5 & 0.2 & 0.6 \\ 0.2 & 1 & 0.2 & 0.1 & 0.5 & 0.2 \\ 0.1 & 0.2 & 1 & 0.8 & 0.1 & 0.5 \\ \hline 0.5 & 0.1 & 0.8 & 1 & 0.2 & 0.1 \\ 0.2 & 0.5 & 0.1 & 0.2 & 1 & 0.2 \\ 0.6 & 0.2 & 0.5 & 0.1 & 0.2 & 1 \end{array}\right) = \begin{pmatrix} C_0 & C_1^{\mathrm{T}} \\ C_1 & C_0 \end{pmatrix}$$

from Example 7.22 with 3×3 Toeplitz blocks C_0 and C_1. C can be generated from the 5×3 reduced matrix

$$C_{\mathrm{red}} = \begin{pmatrix} 0.6 & 0.1 & 0.8 \\ 0.2 & 0.2 & 0.1 \\ 0.5 & 1 & 0.5 \\ 0.1 & 0.2 & 0.2 \\ 0.8 & 0.1 & 0.6 \end{pmatrix} = \left[\boldsymbol{c}_{-1}, \boldsymbol{c}_0, \boldsymbol{c}_1\right]. \tag{7.25}$$

The BCCB extension $\widetilde{C}$ is the 24×24 symmetric BCCB matrix

$$\widetilde{C} = \begin{pmatrix} \widetilde{C}_0 & \widetilde{C}_1^{\mathrm{T}} & \widetilde{0} & \widetilde{C}_1 \\ \widetilde{C}_1 & \widetilde{C}_0 & \widetilde{C}_1^{\mathrm{T}} & \widetilde{0} \\ \widetilde{0} & \widetilde{C}_1 & \widetilde{C}_0 & \widetilde{C}_1^{\mathrm{T}} \\ \widetilde{C}_1^{\mathrm{T}} & \widetilde{0} & \widetilde{C}_1 & \widetilde{C}_0 \end{pmatrix}, \qquad \widetilde{C}_i := \begin{pmatrix} C_i & B_i \\ B_i & C_i \end{pmatrix}, \quad i = 1,2,$$

where $\widetilde{C}_0$ and $\widetilde{C}_1$ are the even circulant extensions of C_0 and C_1 defined in (7.22) and (7.23). To store $\widetilde{C}$, we need only the first column or equivalently, the 6×4 reduced matrix

$$\widetilde{C}_{\mathrm{red}} = \begin{pmatrix} 1 & 0.5 & 0 & 0.5 \\ 0.2 & 0.2 & 0 & 0.1 \\ 0.1 & 0.6 & 0 & 0.8 \\ 0 & 0 & 0 & 0 \\ 0.1 & 0.8 & 0 & 0.6 \\ 0.2 & 0.1 & 0 & 0.2 \end{pmatrix} = \left[\tilde{\boldsymbol{c}}_0, \tilde{\boldsymbol{c}}_1, \tilde{\boldsymbol{0}}, \tilde{\boldsymbol{c}}_{-1}\right]. \tag{7.26}$$

We will never assemble the extension matrix $\widetilde{C}$. Owing to the presence of the extra zero elements, it is easy to generate the reduced matrix $\widetilde{C}_{\mathrm{red}}$ from the reduced matrix of the underlying BTTB matrix. Compare (7.25) and (7.26). To obtain $\widetilde{C}_{\mathrm{red}}$, simply pad C_{red} with a row and column of zeros, split the resulting matrix into four blocks of dimension $n_1 \times n_2$, and permute the blocks as follows.

$$\begin{pmatrix} 0 & 0 & 0 & 0 \\ 0 & 0.6 & 0.1 & 0.8 \\ 0 & 0.2 & 0.2 & 0.1 \\ 0 & 0.5 & 1 & 0.5 \\ 0 & 0.1 & 0.2 & 0.2 \\ 0 & 0.8 & 0.1 & 0.6 \end{pmatrix} \longrightarrow \left(\begin{array}{cc|cc} 0 & 0 & 0 & 0 \\ 0 & 0.6 & 0.1 & 0.8 \\ 0 & 0.2 & 0.2 & 0.1 \\ \hline 0 & 0.5 & 1 & 0.5 \\ 0 & 0.1 & 0.2 & 0.2 \\ 0 & 0.8 & 0.1 & 0.6 \end{array}\right) \longrightarrow \left(\begin{array}{cc|cc} 1 & 0.5 & 0 & 0.5 \\ 0.2 & 0.2 & 0 & 0.1 \\ 0.1 & 0.6 & 0 & 0.8 \\ \hline 0 & 0 & 0 & 0 \\ 0.1 & 0.8 & 0 & 0.6 \\ 0.2 & 0.1 & 0 & 0.2 \end{array}\right).$$

This is done elegantly in MATLAB using the command `fftshift`. See Exercise 7.9.

Covariance matrices of stationary random fields

In two dimensions, covariance matrices associated with samples of stationary random fields $u(\boldsymbol{x})$ at uniformly spaced sample points $\boldsymbol{x} \in D$ lead to symmetric BTTB matrices. To see this, we first define a grid of sample points.

Definition 7.37 (uniformly spaced grid points) Let $D = [0, a_1] \times [0, a_2]$ and consider the $(n_1 - 1) \times (n_2 - 1)$ rectangular grid with $N = n_1 n_2$ vertices

$$\left\{ \boldsymbol{x} = \begin{pmatrix} x_{1,i} \\ x_{2,j} \end{pmatrix} : i = 0, \dots, n_1 - 1,\ j = 0, \dots, n_2 - 1 \right\},$$

where $x_{1,i} := i\Delta x_1$ and $x_{2,j} := j\Delta x_2$ for *uniform* spacings $\Delta x_1 := a_1/(n_1 - 1)$ and $\Delta x_2 := a_2/(n_2 - 1)$. We order these *grid points* from left to right and bottom to top so that, given $i = 0, \dots, n_1 - 1$ and $j = 0, \dots, n_2 - 1$, the kth one is

$$\boldsymbol{x}_k := \begin{pmatrix} x_{1,i} \\ x_{2,j} \end{pmatrix}, \qquad \text{where } k := i + jn_1 \text{ and } k = 0, \dots, N - 1.$$

Now, we want to develop an exact method for sampling the random variable

$$\boldsymbol{u} = \left[u(\boldsymbol{x}_0), u(\boldsymbol{x}_1), \dots, u(\boldsymbol{x}_{N-1})\right]^{\mathsf{T}} \sim \mathrm{N}(\boldsymbol{0}, C), \tag{7.27}$$

where $u(\boldsymbol{x})$ is a stationary, mean-zero, real-valued Gaussian process. With the prescribed ordering of sample points, the $(k + 1)$st entry of $\boldsymbol{u}$ is

$$u_{k+1} := u(\boldsymbol{x}_k) = u(\boldsymbol{x}_{i+jn_1}) = u\begin{pmatrix} x_{1,i} \\ x_{2,j} \end{pmatrix}, \qquad k = 0, \dots, N - 1.$$

Since $u(\boldsymbol{x})$ is stationary, the $N \times N$ covariance matrix C has entries

$$c_{k+1,l+1} = c(\boldsymbol{x}_k - \boldsymbol{x}_l) = c(\boldsymbol{x}_{i+jn_1} - \boldsymbol{x}_{r+sn_1}) = c\begin{pmatrix} (i - r)\Delta x_1 \\ (j - s)\Delta x_2 \end{pmatrix},$$

where $k := i + jn_1$ and $l := r + sn_1$ for $i, r = 0, \dots, n_1 - 1$ and $j, s = 0, \dots, n_2 - 1$. It is clear then that C has the block form

$$C = \begin{pmatrix} C_{0,0} & C_{0,1} & \cdots & C_{0,n_2-2} & C_{0,n_2-1} \\ C_{1,0} & C_{1,1} & C_{1,2} & \ddots & C_{1,n_2-1} \\ \vdots & \ddots & \ddots & \ddots & \vdots \\ C_{n_2-2,0} & \ddots & & C_{n_2-2,n_2-2} & C_{n_2-2,n_2-1} \\ C_{n_2-1,0} & C_{n_2-1,1} & \cdots & C_{n_2-1,n_2-2} & C_{n_2-1,n_2-1} \end{pmatrix}, \tag{7.28}$$

where $[C_{j,s}]_{i+1,r+1} := c((i - r)\Delta x_1, (j - s)\Delta x_2)$. On closer inspection, as $[C_{j,s}]_{i+1,r+1}$ is unchanged for constant $m := j - s$,

$$C = \begin{pmatrix} C_0 & C_{-1} & \cdots & C_{2-n_2} & C_{1-n_2} \\ C_1 & C_0 & C_{-1} & \ddots & C_{2-n_2} \\ \vdots & \ddots & \ddots & \ddots & \vdots \\ C_{n_2-2} & \ddots & C_1 & C_0 & C_{-1} \\ C_{n_2-1} & C_{n_2-2} & \cdots & C_1 & C_0 \end{pmatrix}, \tag{7.29}$$

where the $(i+1, r+1)$ element of the $n_1 \times n_1$ block C_m is

$$[C_m]_{i+1,r+1} := c\begin{pmatrix}(i-r)\Delta x_1\\ m\Delta x_2\end{pmatrix}. \tag{7.30}$$

As $[C_m]_{i+1,r+1}$ is also constant for a fixed value of $i-r$, each C_m is Toeplitz and so C is a BTTB matrix. Next, recall that $c(\boldsymbol{x}) = c(-\boldsymbol{x})$. It follows that

$$c\begin{pmatrix}(i-r)\Delta x_1\\ 0\end{pmatrix} = c\begin{pmatrix}(r-i)\Delta x_1\\ 0\end{pmatrix},$$

and so C_0 in (7.29) is symmetric. Furthermore, $C_{-m} = C_m^{\mathsf{T}}$ since

$$c\begin{pmatrix}(i-r)\Delta x_1\\ m\Delta x_2\end{pmatrix} = c\begin{pmatrix}(r-i)\Delta x_1\\ -m\Delta x_2\end{pmatrix}.$$

Hence, C is always a symmetric BTTB matrix of the form

$$C = \begin{pmatrix} C_0 & C_1^{\mathsf{T}} & \cdots & C_{n_2-2}^{\mathsf{T}} & C_{n_2-1}^{\mathsf{T}} \\ C_1 & C_0 & C_1^{\mathsf{T}} & \ddots & C_{n_2-1}^{\mathsf{T}} \\ \vdots & \ddots & \ddots & \ddots & \vdots \\ C_{n_2-2} & \ddots & C_1 & C_0 & C_1^{\mathsf{T}} \\ C_{n_2-1} & C_{n_2-2} & \cdots & C_1 & C_0 \end{pmatrix}. \tag{7.31}$$

If the stationary covariance is even in both coordinates, so that

$$c\begin{pmatrix}-x_1\\ x_2\end{pmatrix} = c\begin{pmatrix}x_1\\ x_2\end{pmatrix} = c\begin{pmatrix}x_1\\ -x_2\end{pmatrix},$$

the covariance matrix also has symmetric blocks. However, this is not necessarily the case.

Example 7.38 (BTTB covariance matrix) Let $D = [0,1]^2$ and set $n_1 = 3$ and $n_2 = 2$. Then $\Delta x_1 = 1/2$ and $\Delta x_2 = 1$ and the sample points are $\boldsymbol{x}_0 = [0,0]^{\mathsf{T}}$, $\boldsymbol{x}_1 = [1/2,0]^{\mathsf{T}}$, $\boldsymbol{x}_2 = [1,0]^{\mathsf{T}}$, $\boldsymbol{x}_3 = [0,1]^{\mathsf{T}}$, $\boldsymbol{x}_4 = [1/2,1]^{\mathsf{T}}$ and $\boldsymbol{x}_5 = [1,1]^{\mathsf{T}}$. The covariance matrix of $\boldsymbol{u} = [u(\boldsymbol{x}_0), u(\boldsymbol{x}_1), \dots, u(\boldsymbol{x}_5)]^{\mathsf{T}}$, where $u(\boldsymbol{x})$ is a stationary process with mean zero and stationary covariance $c(\boldsymbol{x})$, is

$$C = \left(\begin{array}{ccc|ccc}
c\begin{pmatrix}0\\0\end{pmatrix} & c\begin{pmatrix}-\frac{1}{2}\\0\end{pmatrix} & c\begin{pmatrix}-1\\0\end{pmatrix} & c\begin{pmatrix}0\\-1\end{pmatrix} & c\begin{pmatrix}-\frac{1}{2}\\-1\end{pmatrix} & c\begin{pmatrix}-1\\-1\end{pmatrix} \\
c\begin{pmatrix}\frac{1}{2}\\0\end{pmatrix} & c\begin{pmatrix}0\\0\end{pmatrix} & c\begin{pmatrix}-\frac{1}{2}\\0\end{pmatrix} & c\begin{pmatrix}\frac{1}{2}\\-1\end{pmatrix} & c\begin{pmatrix}0\\-1\end{pmatrix} & c\begin{pmatrix}-\frac{1}{2}\\-1\end{pmatrix} \\
c\begin{pmatrix}1\\0\end{pmatrix} & c\begin{pmatrix}\frac{1}{2}\\0\end{pmatrix} & c\begin{pmatrix}0\\0\end{pmatrix} & c\begin{pmatrix}1\\-1\end{pmatrix} & c\begin{pmatrix}\frac{1}{2}\\-1\end{pmatrix} & c\begin{pmatrix}0\\-1\end{pmatrix} \\
\hline
c\begin{pmatrix}0\\1\end{pmatrix} & c\begin{pmatrix}-\frac{1}{2}\\1\end{pmatrix} & c\begin{pmatrix}-1\\1\end{pmatrix} & c\begin{pmatrix}0\\0\end{pmatrix} & c\begin{pmatrix}-\frac{1}{2}\\0\end{pmatrix} & c\begin{pmatrix}-1\\0\end{pmatrix} \\
c\begin{pmatrix}\frac{1}{2}\\1\end{pmatrix} & c\begin{pmatrix}0\\1\end{pmatrix} & c\begin{pmatrix}-\frac{1}{2}\\1\end{pmatrix} & c\begin{pmatrix}\frac{1}{2}\\0\end{pmatrix} & c\begin{pmatrix}0\\0\end{pmatrix} & c\begin{pmatrix}-\frac{1}{2}\\0\end{pmatrix} \\
c\begin{pmatrix}1\\1\end{pmatrix} & c\begin{pmatrix}\frac{1}{2}\\1\end{pmatrix} & c\begin{pmatrix}0\\1\end{pmatrix} & c\begin{pmatrix}1\\0\end{pmatrix} & c\begin{pmatrix}\frac{1}{2}\\0\end{pmatrix} & c\begin{pmatrix}0\\0\end{pmatrix}
\end{array}\right).$$

Since $c(\boldsymbol{x}) = c(-\boldsymbol{x})$, C is a symmetric BTTB matrix. The off-diagonal blocks are only symmetric if $c\begin{pmatrix} x_1 \\ x_2 \end{pmatrix} = c\begin{pmatrix} -x_1 \\ x_2 \end{pmatrix}$. See Exercise 7.8. We can generate C from $C_{\text{red}} = [\boldsymbol{c}_{-1}, \boldsymbol{c}_0, \boldsymbol{c}_1]$ where

$$\boldsymbol{c}_{-1} := \left[c\begin{pmatrix} -1 \\ -1 \end{pmatrix}, c\begin{pmatrix} -\frac{1}{2} \\ -1 \end{pmatrix}, c\begin{pmatrix} 0 \\ -1 \end{pmatrix}, c\begin{pmatrix} \frac{1}{2} \\ -1 \end{pmatrix}, c\begin{pmatrix} 1 \\ -1 \end{pmatrix}\right]^{\mathsf{T}},$$
$$\boldsymbol{c}_{0} := \left[c\begin{pmatrix} -1 \\ 0 \end{pmatrix}, c\begin{pmatrix} -\frac{1}{2} \\ 0 \end{pmatrix}, c\begin{pmatrix} 0 \\ 0 \end{pmatrix}, c\begin{pmatrix} \frac{1}{2} \\ 0 \end{pmatrix}, c\begin{pmatrix} 1 \\ 0 \end{pmatrix}\right]^{\mathsf{T}},$$
$$\boldsymbol{c}_{1} := \left[c\begin{pmatrix} -1 \\ 1 \end{pmatrix}, c\begin{pmatrix} -\frac{1}{2} \\ 1 \end{pmatrix}, c\begin{pmatrix} 0 \\ 1 \end{pmatrix}, c\begin{pmatrix} \frac{1}{2} \\ 1 \end{pmatrix}, c\begin{pmatrix} 1 \\ 1 \end{pmatrix}\right]^{\mathsf{T}}.$$

When the sample points are uniformly spaced on a rectangular grid, as in the above example, it is simple to compute the reduced covariance matrix.

Algorithm 7.2 Code to generate the reduced symmetric BTTB covariance matrix `C_red` associated with the stationary covariance function `fhandle`. The inputs are `n1,n2`, the number of sample points in each direction, and `dx1,dx2`, the grid spacings Δx_1 and Δx_2.

```
function C_red=reduced_cov(n1,n2,dx1,dx2,fhandle);
C_red=zeros(2*n1-1, 2*n2-1);
for i=1:2*n1-1
    for j=1:2*n2-1
        C_red(i,j)=feval(fhandle, (i-n1)*dx1, (j-n2)*dx2);
    end
end
```

Definition 7.39 (reduced BTTB covariance matrix) The reduced covariance matrix $C_{\text{red}} \in \mathbb{R}^{(2n_1-1)\times(2n_2-1)}$ associated with the symmetric BTTB covariance matrix of a random field with stationary covariance $c(\boldsymbol{x})$, sampled at the $N = n_1 n_2$ uniformly spaced points in Definition 7.37, has entries

$$\left[C_{\text{red}}\right]_{ij} = c\begin{pmatrix} (i-n_1)\Delta x_1 \\ (j-n_2)\Delta x_2 \end{pmatrix}, \qquad i = 1,\dots,2n_1-1,\ j = 1,\dots,2n_2-1.$$

Algorithm 7.2 computes the reduced matrix associated with any stationary covariance function specified by `fhandle`.

Example 7.40 (reduced BTTB covariance matrix) Let $u(\boldsymbol{x})$ be a mean-zero random field with stationary covariance function

$$c(\boldsymbol{x}) = \mathrm{e}^{-\boldsymbol{x}^{\mathsf{T}} A \boldsymbol{x}}, \qquad \boldsymbol{x} = \begin{pmatrix} x_1 \\ x_2 \end{pmatrix}, \qquad A = \begin{pmatrix} a_{11} & a_{12} \\ a_{12} & a_{22} \end{pmatrix}, \tag{7.32}$$

where A is positive definite (as in Example 7.13). Now let $a_{11} = 1 = a_{22}$ and $a_{12} = 0.5$.

If we sample $u(\boldsymbol{x})$ at the grid points in Example 7.38, we can compute the 5×3 reduced covariance matrix using Algorithms 7.2 and 7.3 as follows:

```
>> fhandle =@(x1,x2)gaussA_exp(x1,x2,1,1,0.5);
>> C_red=reduced_cov(3,2,1/2,1,fhandle)
C_red =
    0.3679    0.3679    0.0498
    0.4724    0.7788    0.1738
    0.3679    1.0000    0.3679
    0.1738    0.7788    0.4724
    0.0498    0.3679    0.3679
```

In this example, $c(\boldsymbol{x})$ is uneven in both coordinates, so the first and third columns of C_{red} generate non-symmetric Toeplitz blocks $C_{-1} = C_1^{\mathsf{T}}$ and C_1, respectively. The second column generates a symmetric Toeplitz matrix C_0. The full 6×6 covariance matrix C is therefore a symmetric BTTB matrix with non-symmetric off-diagonal blocks. See also Exercise 7.10.

Algorithm 7.3 Code to evaluate the stationary covariance $c(\boldsymbol{x})$ in Example 7.40 at the point `(x1,x2)`. The inputs `a11,a22,a12` are parameters appearing in the covariance function and the output is the value of c.

```
function c=gaussA_exp(x1,x2,a11,a22,a12)
c=exp(-((x1^2*a11+x2^2*a22)-2*x1*x2*a12));
```

Sampling N(0, C) *with BTTB covariance*

When the covariance matrix $C \in \mathbb{R}^{N\times N}$ is BTTB but not BCCB, as in Examples 7.38 and 7.40, we cannot apply Algorithm 7.1 to generate samples from $\mathrm{N}(\mathbf{0}, C)$. However, if the even BCCB extension $\widetilde{C} \in \mathbb{R}^{4N\times 4N}$ is non-negative definite then $\mathrm{N}(\mathbf{0}, \widetilde{C})$ is a valid Gaussian distribution. To see how to recover $\boldsymbol{u} \sim \mathrm{N}(\mathbf{0}, C)$ samples from $\tilde{\boldsymbol{u}} \sim \mathrm{N}(\mathbf{0}, \widetilde{C})$ samples, we revisit the structure of $\widetilde{C}$.

When C is a covariance matrix, it is symmetric and so is the BCCB extension $\widetilde{C}$ in Definition 7.35. The leading principal submatrix $S \in \mathbb{R}^{2n_1n_2\times 2n_1n_2}$ of $\widetilde{C}$ is

$$S = \begin{pmatrix} \widetilde{C}_0 & \widetilde{C}_1^{\mathsf{T}} & \cdots & \widetilde{C}_{n_2-2}^{\mathsf{T}} & \widetilde{C}_{n_2-1}^{\mathsf{T}} \\ \widetilde{C}_1 & \widetilde{C}_0 & \widetilde{C}_1^{\mathsf{T}} & \ddots & \widetilde{C}_{n_2-2}^{\mathsf{T}} \\ \vdots & \ddots & \ddots & \ddots & \vdots \\ \widetilde{C}_{n_2-2} & \ddots & \widetilde{C}_1 & \widetilde{C}_0 & \widetilde{C}_1^{\mathsf{T}} \\ \widetilde{C}_{n_2-1} & \widetilde{C}_{n_2-2} & \cdots & \widetilde{C}_1 & \widetilde{C}_0 \end{pmatrix}, \qquad \widetilde{C}_i = \begin{pmatrix} C_i & B_i \\ B_i & C_i \end{pmatrix},$$

for $i = 0, \dots, n_2 - 1$ and $C_i, B_i \in \mathbb{R}^{n_1\times n_1}$. Now, given $\tilde{\boldsymbol{u}} \sim \mathrm{N}(\mathbf{0}, \widetilde{C})$, let

$$\boldsymbol{v} = \left[\tilde{u}_1, \tilde{u}_2, \dots, \tilde{u}_{2n_1n_2}\right]^{\mathsf{T}} \in \mathbb{R}^{2n_1n_2},$$

be the first $2n_1n_2$ components of $\tilde{\boldsymbol{u}}$. We know from Exercise 6.17 that $\boldsymbol{v} \sim \mathrm{N}(\mathbf{0}, S)$. Now compare S with the target $n_1n_2 \times n_1n_2$ covariance matrix C in (7.31). If we view S as an

$2n_2 \times 2n_2$ block matrix with blocks of size $n_1 \times n_1$, C is obtained from S by keeping only every other block column and block row, or, in other words, keeping only the leading principal $n_1 \times n_1$ submatrices of each of the blocks $\widetilde{C}_i$ and $\widetilde{C}_i^{\mathsf{T}}$. If we rearrange the components of $\boldsymbol{v}$ into an $n_1 \times 2n_2$ matrix and keep only every other column, then the resulting vector

$$\boldsymbol{u} = \operatorname{vec}(V) \in \mathbb{R}^{n_1 n_2}, \qquad V = \begin{pmatrix} \tilde{u}_1 & \tilde{u}_{2n_2+1} & \cdots & \tilde{u}_{(2n_2-2)n_1+1} \\ \tilde{u}_2 & \tilde{u}_{2n_2+2} & \cdots & \cdots \\ \vdots & \vdots & \vdots & \vdots \\ \tilde{u}_{n_1} & \tilde{u}_{3n_1} & \cdots & \tilde{u}_{(2n_2-1)n_1} \end{pmatrix},$$

has distribution $\mathrm{N}(\mathbf{0}, C)$. This can be proved in the same way as Exercise 6.17.

Example 7.41 Let $u(\boldsymbol{x})$ be a mean-zero Gaussian random field in two dimensions with stationary covariance

$$c(\boldsymbol{x}) = \exp(-|x_1|/\ell_1 - |x_2|/\ell_2). \tag{7.33}$$

If we sample $u(\boldsymbol{x})$ with $\ell_1 = 1/5$, $\ell_2 = 1/10$ on a uniformly spaced grid on $D = [0,1] \times [0,2]$ with $n_1 = 201$ and $n_2 = 401$ and grid spacings $\Delta x_1 = 1/200$, $\Delta x_2 = 1/200$, the resulting covariance matrix C has an even BCCB extension $\widetilde{C}$ which is non-negative definite. Algorithm 7.4 returns two exact samples $\boldsymbol{u}_1, \boldsymbol{u}_2$ from $\mathrm{N}(\mathbf{0}, C)$ and these are plotted in Figure 7.6. The reduced covariance matrix is generated using Algorithm 7.5, as follows:

```
>> fhandle1=@(x1,x2)sep_exp(x1,x2,1/5,1/10);
>> C_red=reduced_cov(201,401,1/200,1/200, fhandle1);
>> [u1,u2]=circ_embed_sample_2d(C_red,201,401);
```

Algorithm 7.4 Code to generate a pair of samples from $\mathrm{N}(\mathbf{0}, C)$ where $C \in \mathbb{R}^{N \times N}$ ($N = n_1 n_2$) is a BTTB covariance matrix. The input `C_red` is the reduced matrix which generates C. The outputs `u1,u2` are independent samples from $\mathrm{N}(\mathbf{0}, C)$, in matrix format.

```
function [u1,u2]=circ_embed_sample_2d(C_red,n1,n2);
N=n1*n2;
% form reduced matrix of BCCB extension of BTTB matrix C
tilde_C_red = zeros(2*n1,2*n2);
tilde_C_red(2:2*n1,2:2*n2) = C_red;
tilde_C_red = fftshift(tilde_C_red);
% sample from N(0, tilde_C)
[u1,u2]=circ_cov_sample_2d(tilde_C_red,2*n1,2*n2);
% recover samples from N(0,C)
u1=u1(:); u2=u2(:);
u1=u1(1:end/2);u1=reshape(u1,n1,2*n2);u1=u1(:,1:2:end);
u2=u2(1:end/2);u2=reshape(u2,n1,2*n2);u2=u2(:,1:2:end);
```

Algorithm 7.5 Code to evaluate the stationary covariance $c(\boldsymbol{x})$ in (7.33) at the point `(x1,x2)`. The inputs `ell1,ell2` are the parameters ℓ_1, ℓ_2 appearing in the covariance function and the output is the value of c.

```
function c=sep_exp(x1,x2,ell_1,ell_2)
c=exp(-abs(x1)/ell_1-abs(x2)/ell_2);
```

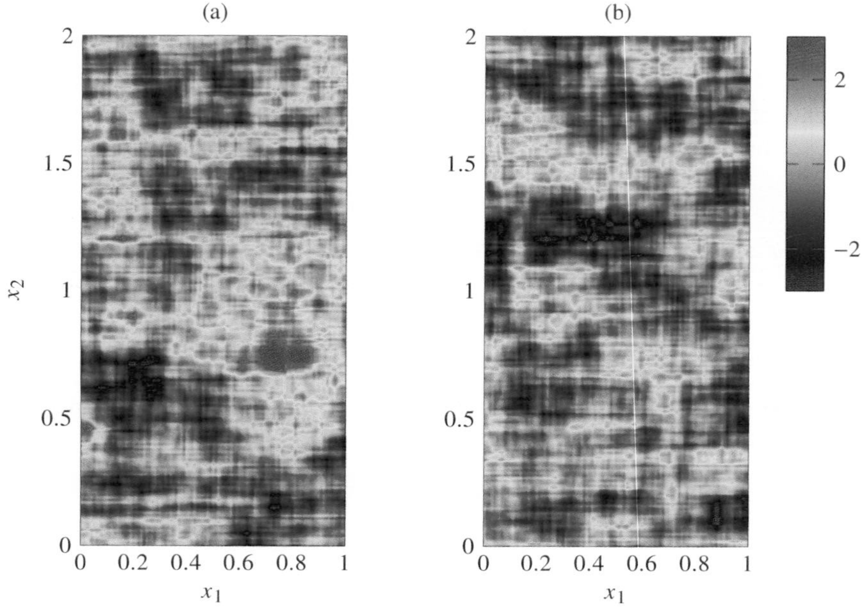

Figure 7.6 Two realisations of the Gaussian random field on $D = [0,1] \times [0,2]$ with mean zero and exponential covariance (7.33) with $\ell_1 = 1/5$ and $\ell_2 = 1/10$. The fields were generated using the circulant embedding method with $n_1 = 201$ and $n_2 = 401$ sample points in each direction, and grid spacings $\Delta x_1 = 1/200, \Delta x_2 = 1/200$.

The disadvantage is that, for a given $c(\boldsymbol{x})$, the $4n_1n_2 \times 4n_1n_2$ even BCCB circulant extension matrix $\widetilde{C}$ is not guaranteed to be non-negative definite (even for the separable exponential covariance). Larger extension matrices usually need to be considered. Indeed, Algorithm 7.4 (which calls Algorithm 7.1) returns a warning message and displays the size of the smallest eigenvalue, if the even BCCB extension corresponding to the user's choice of n_1, n_2 and $\Delta x_1, \Delta x_2$ is indefinite.

Padding and approximate circulant embedding

In cases where the even BCCB extension $\widetilde{C} = WDW^* \in \mathbb{R}^{4N \times 4N}$ is indefinite, the random variable defined by

$$\widetilde{\boldsymbol{Z}} := WD^{1/2}\boldsymbol{\xi}, \qquad \text{where } \boldsymbol{\xi} \sim \mathrm{CN}(\boldsymbol{0}, 2I_{4N}),$$

does not have the distribution $\mathrm{CN}(\boldsymbol{0}, 2\widetilde{C})$ and the real and imaginary parts do not have the distribution $\mathrm{N}(\boldsymbol{0}, \widetilde{C})$. This is due to the presence of negative eigenvalues and the fact that $D^{1/2}$ is a complex-valued matrix. From Chapter 6, we know that the real and imaginary parts of $\widetilde{\boldsymbol{Z}}$ actually have the distribution $\mathrm{N}(\boldsymbol{0}, \widetilde{C} + 2\widetilde{C}_-)$ where $\widetilde{C}_- = WD_-W^*, D = D_+ - D_-$ and $D_\pm$ is the diagonal matrix with kth diagonal entry $\max\{0, \pm d_k\}$. Furthermore, the error in the covariance of the resulting samples depends on the spectral radius of D_-, denoted $\rho(D_-)$, just as in the $d = 1$ case considered in Exercise 6.19.

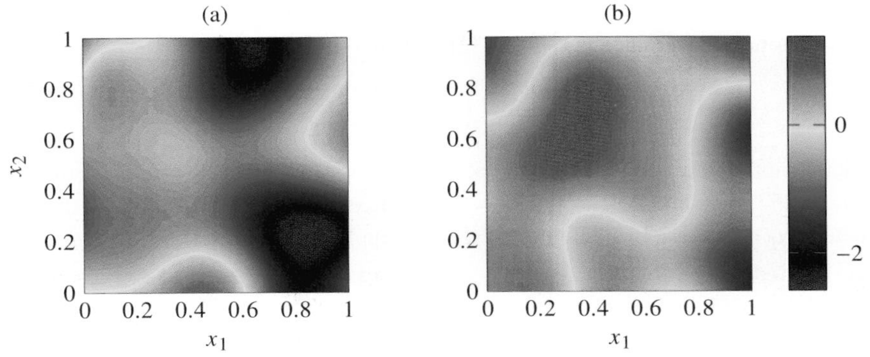

Figure 7.7 Two realisations of the Gaussian random field on $D = [0,1]^2$ with mean zero and Gaussian covariance (7.32) with $a_{11} = 10 = a_{22}$. The fields were generated using the circulant embedding method with $(n_1 + m_1)(n_2 + m_2)$ sample points with grid spacings $\Delta x_1 = 1/256 = \Delta x_2$ where $n_1 = 257 = n_2$ and $m_1 = 8n_1 = m_2$. The even BCCB extension is indefinite and so samples do not have the exact covariance.

We can avoid the negative eigenvalues and the resulting error in the covariance, by embedding the given symmetric BTTB matrix C into a larger BCCB matrix $\widetilde{C}$ as follows. Suppose we want to generate samples of a random field $u(\boldsymbol{x})$ on $[0,a_1] \times [0,a_2]$ where $a_1 = n_1\Delta x_1$, $a_2 = n_2\Delta x_2$, with n_1, n_2 and Δx_1, Δx_2 given. The covariance matrix $C \in \mathbb{R}^{n_1 n_2 \times n_1 n_2}$ of interest has $\boldsymbol{c}_{\text{red}} = \text{vec}(C_{\text{red}})$ for $C_{\text{red}} = [\boldsymbol{c}_{1-n_2}, \dots, \boldsymbol{c}_{-1}, \boldsymbol{c}_0, \boldsymbol{c}_1, \dots, \boldsymbol{c}_{n_2-1}]$, and for $i = 1, \dots, 2n_1 - 1$

$$[\boldsymbol{c}_k]_i = c\begin{pmatrix} (i - n_1)\Delta x_1 \\ k\Delta x_2 \end{pmatrix}, \qquad k = 1 - n_2, \dots, 0, \dots, n_2 - 1.$$

Now, suppose we form the symmetric BTTB matrix C^* with reduced matrix

$$C^*_{\text{red}} = \left[\boldsymbol{c}^*_{1-(n_2+m_2)}, \dots, \boldsymbol{c}^*_{-1}, \boldsymbol{c}^*_0, \boldsymbol{c}^*_1, \dots, \boldsymbol{c}^*_{(n_2+m_2)-1}\right],$$

where, for $i = 1, \dots, 2(n_1 + m_1) - 1$,

$$[\boldsymbol{c}^*_k]_i = c\begin{pmatrix} (i - (n_1 + m_1))\Delta x_1 \\ k\Delta x_2 \end{pmatrix}, \qquad k = 1 - (n_2 + m_2), \dots, 0, \dots, (n_2 + m_2) - 1,$$

for some $m_1, m_2 \in \mathbb{N}$. C^* coincides with the covariance matrix associated with an extended set of $(n_1 + m_1)(n_2 + m_2)$ grid points with the same mesh spacings $\Delta x_1, \Delta x_2$ on $[0, \hat{a}_1] \times [0, \hat{a}_2]$ where $\hat{a}_1 = (n_1 + m_1)\Delta x_1$, $\hat{a}_2 = (n_2 + m_2)\Delta x_2$. C can be recovered from C^* by taking the leading principal $n_1 \times n_1$ submatrices of the leading $n_2 \times n_2$ blocks. The even BCCB extension $\widetilde{C}$ of C^* has dimension $4\widehat{N} \times 4\widehat{N}$ where $\widehat{N} = (n_1 + m_1)(n_2 + m_2)$ and is often non-negative definite for some m_1, m_2 large enough. In some cases, the required values of m_1 and m_2 are so large that the computations become intractable. Approximate circulant embedding may be the only option.

Example 7.42 (Gaussian covariance) Let $u(\boldsymbol{x})$ be a mean-zero Gaussian random field with the stationary covariance $c(\boldsymbol{x})$ in (7.32) and choose

$$A = \begin{pmatrix} 10 & 0 \\ 0 & 10 \end{pmatrix}.$$

Consider sampling $u(\boldsymbol{x})$ on $[0,1] \times [0,1]$ at the vertices of a 256×256 uniform square grid. Then $n_1 = 257 = n_2$ and $\Delta x_1 = 1/256 = \Delta x_2$. The even BCCB extension, with padding parameters $m_1 = 0 = m_2$, is indefinite. To see this, apply Algorithms 7.2 and 7.6 as follows:

```
>> fhandle =@(x1,x2)gaussA_exp(x1,x2,10,10,0);
>> n1=257; n2=257; m1=0;m2=0; dx1=1/(n1-1); dx2=1/(n2-1);
>> C_red=reduced_cov(n1+m1,n2+m2,dx1,dx2,fhandle);
>> [u1,u2]=circ_embed_sample_2dB(C_red,n1,n2,m1,m2);
Invalid covariance;rho(D_minus)=0.064642
```

Algorithm 7.6 Code to generate a pair of samples from $\mathrm{N}(\mathbf{0}, C)$ where $C \in \mathbb{R}^{N \times N}$ ($N = n_1 n_2$) is symmetric, non-negative definite and BTTB. The input `C_red` is the reduced matrix which generates C. `n1,n2` provide the dimensions of C, and `m1,m2` are the padding parameters. The outputs `u1,u2` are independent samples from $\mathrm{N}(\mathbf{0}, C)$, in matrix format

```
function [u1,u2]=circ_embed_sample_2dB(C_red,n1,n2,m1,m2);
nn1=n1+m1;nn2=n2+m2;N=nn1*nn2;
% form reduced matrix of BCCB extension of BTTB matrix C*
tilde_C_red = zeros(2*nn1,2*nn2);
tilde_C_red(2:2*nn1,2:2*nn2) = C_red;
tilde_C_red = fftshift(tilde_C_red);
% sample from N(0, tilde_C)
[u1,u2]=circ_cov_sample_2d(tilde_C_red,2*nn1,2*nn2);
% recover samples from N(0,C)
u1=u1(:); u2=u2(:);
u1=u1(1:2*nn1*n2);u1=reshape(u1,nn1,2*n2);u1=u1(1:n1,1:2:end);
u2=u2(1:2*nn1*n2);u2=reshape(u2,nn1,2*n2);u2=u2(1:n1,1:2:end);
return
```

For the chosen discretisation parameters, $\rho(D_{-})$ is quite large. Now consider the $4\widehat{N} \times 4\widehat{N}$ BCCB extension $\widetilde{C}$ with $\widehat{N} = (n_1 + m_1)(n_2 + m_2)$ with increasing m_1 and m_2. The value of $\rho(D_{-})$ decays to zero, and so the extension matrix approaches a non-negative definite matrix, but stagnates at $\mathcal{O}(10^{-12})$. Try it for increasing values of m_1 and m_2. For example,

```
>> n1=257;n2=257;m1=8*n1;m2=8*n2;dx1=1/(n1-1);dx2=1/(n2-1);
>> C_red=reduced_cov(n1+m1,n2+m2,dx1,dx2,fhandle);
>> [u1,u2]=circ_embed_sample_2dB(C_red,n1,n2,m1,m2);
Invalid covariance;rho(D_minus)=4.1087e-12
```

Two realisations of the random field obtained when $m_1 = 8n_1$ and $m_2 = 8n_2$ are plotted in Figure 7.7.

If we choose $A = \sigma I_2$ to be a diagonal matrix with $\sigma \gg 1$, then the covariance $c(\boldsymbol{x}) = \mathrm{e}^{-\sigma \|\boldsymbol{x}\|_2^2}$ decays more rapidly than in the above example. In that case, a smaller error may be made with fewer sample points. For example, if

$$A = \begin{pmatrix} 10^3 & 0 \\ 0 & 10^3 \end{pmatrix},$$

then the even BCCB extension matrix associated with $m_1 = 0 = m_2$ (no padding) has a negative eigenvalue of size $\mathcal{O}(10^{-14})$. To see this, use the following MATLAB commands:

```
>> fhandle =@(x1,x2)gaussA_exp(x1,x2,1000,1000,0);
>> n1=257; n2=257; m1=0;m2=0; dx1=1/(n1-1); dx2=1/(n2-1);
>> C_red=reduced_cov(n1+m1,n2+m2,dx1,dx2,fhandle);
>> [u1,u2]=circ_embed_sample_2dB(C_red,n1,n2,m1,m2);
Invalid covariance;rho(D_minus)=5.0041e-14
```

Two realisations of the resulting Gaussian random field are plotted in Figure 7.8. In this case, $\rho(D\_)$ is close enough to machine precision that the resulting samples can be considered to have the exact covariance. In general, the user has to provide a threshold for $\rho(D\_)$ and experiment with padding until the resulting BCCB extension matrix is close enough to being non-negative definite.

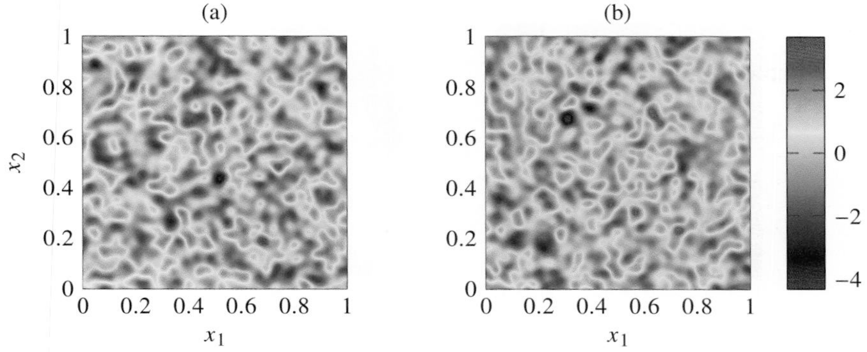

Figure 7.8 Two realisations of the Gaussian random field on $D = [0, 1]^2$ with mean zero and Gaussian covariance (7.32) with $a_{11} = 1000 = a_{22}$. The fields were generated using the circulant embedding method with $(n_1 + m_1)(n_2 + m_2)$ sample points with grid spacings $\Delta x_1 = 1/256 = \Delta x_2 = 1/256$ where $n_1 = 257 = n_2$ and $m_1 = 0 = m_2$. The even BCCB extension is indefinite and so the samples do not have the exact covariance.

7.3 Turning bands method

The circulant embedding method can be used to generate realisations of stationary Gaussian random fields $u(\boldsymbol{x})$. The turning bands method is a method for approximating realisations of *isotropic* Gaussian random fields (see Definition 7.14). The turning bands method uses a random vector $\boldsymbol{e} \sim \mathrm{U}(S^{d-1})$ (i.e., uniformly distributed on the surface of the unit ball $S^{d-1} := \{\boldsymbol{x} \in \mathbb{R}^d : \|\boldsymbol{x}\|_2 = 1\}$) and a stationary stochastic process $\{X(t) : t \in \mathbb{R}\}$ to define the so-called turning bands random field $v(\boldsymbol{x}) := X(\boldsymbol{x}^\mathsf{T}\boldsymbol{e})$ for $\boldsymbol{x} \in \mathbb{R}^d$. See Figure 7.9. Two realisations of $v(\boldsymbol{x})$ (for $X(t)$ given by (7.34)) are shown in Figure 7.10. We notice that if $\boldsymbol{x} - \boldsymbol{y}$ is orthogonal to $\boldsymbol{e}$, then $\boldsymbol{x}^\mathsf{T}\boldsymbol{e} = \boldsymbol{y}^\mathsf{T}\boldsymbol{e}$ and $v(\boldsymbol{x})$ and $v(\boldsymbol{y})$ are equal. Realisations of $v(\boldsymbol{x})$ therefore have a banded structure and this explains the name *turning bands*.

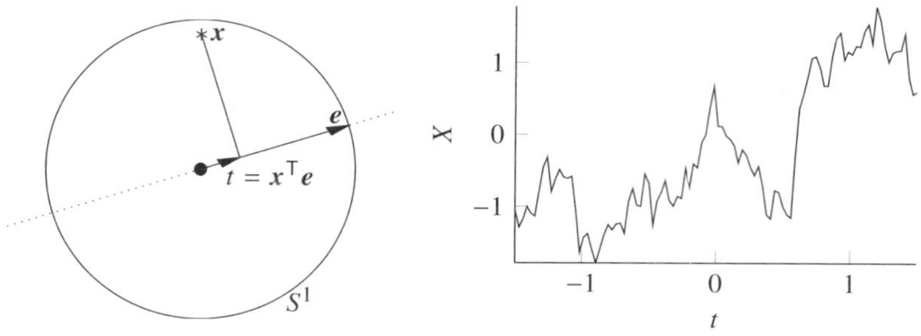

Figure 7.9 The turning bands field $v(\boldsymbol{x})$ is defined by evaluating a stationary process $X(t)$ at $t = \boldsymbol{x}^{\mathsf{T}}\boldsymbol{e}$, for a unit vector $\boldsymbol{e} \sim \mathrm{U}(S^{d-1})$. Here $d = 2$ and $\boldsymbol{e}$ is uniformly distributed on the unit circle S^1.

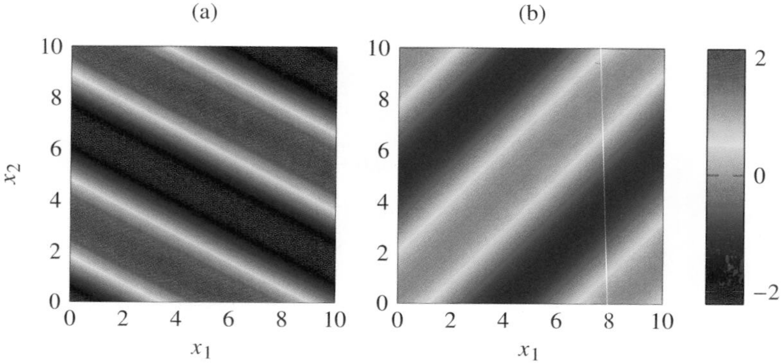

Figure 7.10 Two realisations of the random field $v(\boldsymbol{x})$ defined by (7.35). Notice the banded structure of the random field. The realisations are generated with the commands `grid=[0:0.05:10]';` `u=turn_band_simple(grid,grid);` using Algorithm 7.7.

The turning bands method approximates realisations of an isotropic random field $u(\boldsymbol{x})$ using easy to compute realisations of the turning bands random field $v(\boldsymbol{x})$. To do this, we define $v(\boldsymbol{x})$ in terms of a stationary process $X(t)$ with covariance function $c_X(t) = T_d c^0(t)$, where $c^0(r)$ is the isotropic covariance function of the target random field $u(\boldsymbol{x})$ and T_d is the turning bands operator.

Definition 7.43 (turning bands operator) Let $c^0(r)$ be an isotropic covariance function of a random field on $\mathbb{R}^d$ with radial spectral distribution $F^0(s)$. The *turning bands operator* T_d is defined by

$$T_d c^0(t) := \int_0^\infty \cos(st)\, dF^0(s).$$

$T_d c^0(t)$ is the covariance function of a stationary process with spectral distribution $F(s) = F^0(s)/2$. If $c^0(r)$ has radial spectral density function $f^0(s)$, then $T_d c^0(t)$ has spectral density $f^0(s)\omega_d s^{d-1}/2$ (with ω_d defined in Lemma A.7).

We show in Proposition 7.47 that if $X(t)$ has covariance $c_X(t) = T_d c^0(t)$ then the turning bands random field $v(\boldsymbol{x})$ in $\mathbb{R}^d$ has isotropic covariance $c^0(r)$. First, we give two examples.

Example 7.44 (three dimensions) The turning bands operator takes a particularly simple form in dimension $d = 3$. We have $J_{1/2}(r) = \sqrt{(2/\pi r)}\sin(r)$ from (1.47) and $\Gamma(3/2) = \sqrt{\pi}/2$. Hence,

$$\frac{d}{dt}\left(\Gamma(3/2)\, t\, \frac{J_{1/2}(ts)}{(ts/2)^{1/2}}\right) = \frac{d}{dt}\left(\frac{1}{s}\sin(ts)\right) = \cos(ts)$$

and, from (7.6) and (7.7), we see that

$$T_3 c^0(t) = \frac{d}{dt}\left(t c^0(t)\right).$$

For example, if we want to construct a three-dimensional random field with the isotropic exponential covariance e^{-r}, then we need to choose $X(t)$ with covariance $c_X(t) = T_3 \mathrm{e}^{-t} = (1-t)\mathrm{e}^{-t}$. Such a convenient expression for $T_d c^0$ is not available for $d = 2$.

Example 7.45 (Bessel covariance) The Bessel covariance of Example 7.18 has radial spectral distribution $F^0(s) = \delta(s-1)$ and hence for any $d \in \mathbb{N}$

$$T_d\left(\Gamma(d/2) J_p(t)/(t/2)^p\right) = \frac{1}{2}\cos(t), \qquad p = (d-2)/2.$$

We sample the mean-zero Gaussian process with covariance $c_X(t) = \frac{1}{2}\cos(t)$ following (5.16) by evaluating

$$X(t) = \frac{1}{\sqrt{2}}\cos(t)\xi_1 + \frac{1}{\sqrt{2}}\sin(t)\xi_2, \qquad \xi_i \sim \mathrm{N}(0,1)\ iid. \tag{7.34}$$

Then, the turning bands field

$$v(\boldsymbol{x}) := X(\boldsymbol{x}^{\mathsf{T}}\boldsymbol{e}) = \frac{1}{\sqrt{2}}\cos(\boldsymbol{x}^{\mathsf{T}}\boldsymbol{e})\xi_1 + \frac{1}{\sqrt{2}}\sin(\boldsymbol{x}^{\mathsf{T}}\boldsymbol{e})\xi_2, \qquad \boldsymbol{x} \in \mathbb{R}^d, \tag{7.35}$$

where $\boldsymbol{e} \sim \mathrm{U}(S^{d-1})$ and $\boldsymbol{e}, \xi_1, \xi_2$ are pairwise independent. In the case $d = 2$, Algorithm 7.7 evaluates (7.35) at a given set of points $\boldsymbol{x} \in \mathbb{R}^2$ and two realisations of $v(\boldsymbol{x})$ are plotted in Figure 7.10.

Algorithm 7.7 Code to sample $v(\boldsymbol{x})$ in (7.35) at a rectangular grid of points $[x_1, x_2]^{\mathsf{T}} \in \mathbb{R}^2$ with $\boldsymbol{e} \sim \mathrm{U}(S^1)$ and $X(t)$ given by (7.34). The inputs `grid1`, `grid2` are column vectors of coordinates of x_1 and x_2. The output `v` is a matrix with entries $v(x_1, x_2)$.

```
function v=turn_band_simple(grid1, grid2)
theta=2*pi*rand; e=[cos(theta); sin(theta)]; % sample e
[xx,yy]=ndgrid(grid1, grid2); tt=[xx(:), yy(:)] * e; % project
xi=randn(2,1); v=sqrt(1/2)*[cos(tt), sin(tt)]*xi; % sample v
v =reshape(v, length(grid1), length(grid2));
```

We show that the turning bands operator T_d has the correct form.

Lemma 7.46 *If the process $X(t)$ has mean zero and stationary covariance $c_X(t)$, the turning bands random field $v(\boldsymbol{x}) = X(\boldsymbol{x}^{\mathsf{T}}\boldsymbol{e})$ has mean zero and stationary covariance $\mathbb{E}\left[c_X(\boldsymbol{x}^{\mathsf{T}}\boldsymbol{e})\right]$ where $\boldsymbol{e} \sim \mathrm{U}(S^{d-1})$.*

Proof As $X(t)$ has zero mean and is independent of $\boldsymbol{e}$, $v(\boldsymbol{x}) = X(\boldsymbol{x}^\mathsf{T}\boldsymbol{e})$ has zero mean and

$$\operatorname{Cov}(v(\boldsymbol{x}), v(\boldsymbol{y})) = \mathbb{E}\big[v(\boldsymbol{x})v(\boldsymbol{y})\big] = \mathbb{E}\Big[X(\boldsymbol{x}^\mathsf{T}\boldsymbol{e})\,X(\boldsymbol{y}^\mathsf{T}\boldsymbol{e})\Big], \qquad \boldsymbol{x},\boldsymbol{y} \in \mathbb{R}^d.$$

Using conditional expectation on $\boldsymbol{e}$,

$$\operatorname{Cov}(v(\boldsymbol{x}), v(\boldsymbol{y})) = \mathbb{E}\Big[\mathbb{E}\Big[X(\boldsymbol{x}^\mathsf{T}\boldsymbol{e})\,X(\boldsymbol{y}^\mathsf{T}\boldsymbol{e})\,\Big|\,\boldsymbol{e}\Big]\Big].$$

As X and $\boldsymbol{e}$ are independent and c_X is the covariance of $X(t)$, Lemma 4.48 says that $\mathbb{E}\Big[X(\boldsymbol{x}^\mathsf{T}\boldsymbol{e})\,X(\boldsymbol{y}^\mathsf{T}\boldsymbol{e})\,\Big|\,\boldsymbol{e}\Big] = c_X((\boldsymbol{x}-\boldsymbol{y})^\mathsf{T}\boldsymbol{e})$ and

$$\operatorname{Cov}(v(\boldsymbol{x}), v(\boldsymbol{y})) = \mathbb{E}\big[c_X\big((\boldsymbol{x}-\boldsymbol{y})^\mathsf{T}\boldsymbol{e}\big)\big].$$

Finally, the right-hand side depends only on $\boldsymbol{x}-\boldsymbol{y}$, so that $v(\boldsymbol{x})$ has stationary covariance $\mathbb{E}\big[c_X(\boldsymbol{x}^\mathsf{T}\boldsymbol{e})\big]$. □

Proposition 7.47 (covariance of turning bands field) *Consider an isotropic covariance function $c^0(r)$ in $\mathbb{R}^d$ and a mean-zero process $\{X(t)\colon t \in \mathbb{R}\}$ with stationary covariance $T_d c^0(t)$. Then, the turning bands random field $v(\boldsymbol{x})$ has isotropic covariance $c^0(r)$.*

Proof The surface of the ball of radius r in $\mathbb{R}^d$ is denoted by $S^{d-1}(r)$ (and the unit ball by S^{d-1}) and has surface area $\omega_d r^{d-1}$ (see Lemma A.7). By Lemma 7.46, $\mathbb{E}\big[c_X(\boldsymbol{x}^\mathsf{T}\boldsymbol{e})\big]$ is the covariance of the turning bands field $v(\boldsymbol{x})$. Because $\boldsymbol{e} \sim \mathrm{U}(S^{d-1})$, we see that $\mathbb{E}\big[c_X(\boldsymbol{x}^\mathsf{T}\boldsymbol{e})\big]$ is isotropic and independent of $\boldsymbol{x} \in S^{d-1}(r)$. We choose $\boldsymbol{x} = [r, 0, \dots, 0]^\mathsf{T} \in \mathbb{R}^d$. Introduce coordinates $\theta, \mathbf{e}^*$ on S^{d-1} defined by $e_1 = \cos\theta$ for $\theta \in [0,\pi]$ and $\mathbf{e}^* := [e_2, \dots, e_d]^\mathsf{T} \in S^{d-2}(|\sin\theta|)$. Then $\boldsymbol{x}^\mathsf{T}\boldsymbol{e} = r e_1$ and

$$\mathbb{E}\big[c_X(\boldsymbol{x}^\mathsf{T}\boldsymbol{e})\big] = \frac{1}{\omega_d}\int_{S_{d-1}} c_X(\boldsymbol{x}^\mathsf{T}\boldsymbol{e})\,d\boldsymbol{e} = \frac{1}{\omega_d}\int_{S_{d-2}(|\sin\theta|)}\int_0^\pi c_X(r\cos\theta)\,d\theta\,d\mathbf{e}^*.$$

$S^{d-2}(|\sin\theta|)$ has surface area $\omega_{d-1}|\sin\theta|^{d-2}$ and

$$\mathbb{E}\big[c_X(\boldsymbol{x}^\mathsf{T}\boldsymbol{e})\big] = 2\frac{\omega_{d-1}}{\omega_d}\int_0^{\pi/2} c_X(r\cos\theta)\sin^{d-2}\theta\,d\theta.$$

Substitute $c_X(t) = T_d c^0(t) = \int_0^\infty \cos(st)\,dF^0(s)$ from Definition 7.43:

$$\begin{aligned}\mathbb{E}\big[c_X(\boldsymbol{x}^\mathsf{T}\boldsymbol{e})\big] &= 2\frac{\omega_{d-1}}{\omega_d}\int_0^{\pi/2}\int_0^\infty \cos(sr\cos\theta)\,dF^0(s)\,\sin^{d-2}\theta\,d\theta\\ &= 2\frac{\Gamma(d/2)}{\pi^{1/2}\Gamma((d-1)/2)}\int_0^\infty\int_0^{\pi/2}\cos(sr\cos\theta)\sin^{d-2}\theta\,d\theta\,dF^0(s),\end{aligned}$$

as $\omega_{d-1}/\omega_d = \pi^{-1/2}\Gamma(d/2)/\Gamma((d-1)/2)$. By (A.3) with $p = (d-2)/2$, we see $J_p(r) = \frac{2^{2-d/2}}{\pi^{(1/2}\Gamma((d-1)/2)}\int_0^{\pi/2}\cos(r\cos\theta)\sin^{d-2}\theta\,d\theta$ and hence

$$\mathbb{E}\big[c_X(\boldsymbol{x}^\mathsf{T}\boldsymbol{e})\big] = \Gamma(d/2)\int_0^\infty \frac{J_p(sr)}{(sr/2)^p}\,dF^0(s).$$

By Theorem 7.15, $\mathbb{E}\big[c_X(\boldsymbol{x}^\mathsf{T}\boldsymbol{e})\big] = c^0(r)$ and, by Lemma 7.46, the covariance function of the turning bands field $v(\boldsymbol{x})$ is $c^0(r)$. □

Turning bands method with multiple bands

Let $\{u(\boldsymbol{x})\colon \boldsymbol{x} \in \mathbb{R}^d\}$ be a random field with mean zero and isotropic covariance $c^0(r)$. We are interested in using the turning bands random field $\{v(\boldsymbol{x})\colon \boldsymbol{x} \in \mathbb{R}^d\}$ to generate approximate realisations of $u(\boldsymbol{x})$. If $\{X(t)\colon t \in \mathbb{R}\}$ is a mean-zero process with stationary covariance $T_d c^0(t)$, Proposition 7.47 shows that the mean and covariance of the random fields $v(\boldsymbol{x})$ and $u(\boldsymbol{x})$ are the same. However, correctness of the mean and covariance is insufficient and $v(\boldsymbol{x})$ does not necessarily approximate the distribution of $u(\boldsymbol{x})$. Even when $u(\boldsymbol{x})$ is a Gaussian random field, exactness of the mean and covariance is not enough, because the turning bands random field $v(\boldsymbol{x})$ is not Gaussian and the turning bands field has unnatural looking realisations. Compare, for example, the realisation of $v(\boldsymbol{x})$ shown in Figure 7.10 for $c^0(r) = J_0(r)$ with the well-resolved realisation shown in Figure 7.5 of a Gaussian random field with the same mean and covariance. The turning bands random field $v(\boldsymbol{x})$ is highly structured in space and this is not true of realisations of the target random field $u(\boldsymbol{x})$.

If $u(\boldsymbol{x})$ is Gaussian, however, we can use the turning band random field in combination with the central limit theorem as follows. Suppose that $X_j(t)$ for $j = 1, \dots, M$ are *iid* copies of $X(t)$ and that $\boldsymbol{e}_j \sim \mathrm{U}(S^{d-1})$. Further suppose that the random variables $X_j(t), \boldsymbol{e}_j$ are pairwise independent (where we treat $X_j(t)$ as an $\mathbb{R}^{\mathbb{R}}$ random variable as in Definition 5.16). Define

$$v_M(\boldsymbol{x}) := \frac{1}{\sqrt{M}} \sum_{j=1}^{M} X_j(\boldsymbol{x}^{\mathsf{T}} \boldsymbol{e}_j), \tag{7.36}$$

where we shall call the parameter M the *number of bands*. As $M \to \infty$, the central limit theorem implies convergence in distribution of $v_M(\boldsymbol{x})$ to the mean-zero Gaussian random field with isotropic covariance $c^0(r)$.

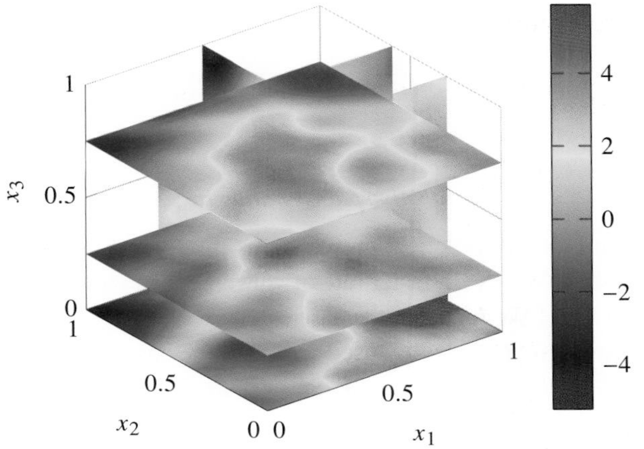

Figure 7.11 Approximate realisation of the mean-zero Gaussian random field in $\mathbb{R}^3$ with covariance $c(r) = \mathrm{e}^{-r}$. It is generated using Algorithm 7.8 and $M = 20$ bands by `grid=[0:0.04:1]';` `u=turn_band_exp_3d(grid, grid, grid, 20, 0, 1);`

Example 7.48 (isotropic exponential $d = 3$) Following Example 7.44, we use the turning bands method to approximate realisations of the mean-zero Gaussian random field $u(\boldsymbol{x})$ with isotropic covariance $c^0(r) = \mathrm{e}^{-r/\ell}$. Algorithm 7.8 implements the turning bands method in $\mathbb{R}^3$ using the circulant embedding algorithm and interpolation to approximate $X(t)$. In this case, $X(t)$ is the mean-zero Gaussian process with covariance $c_X(t) = (1 - t/\ell)\,\mathrm{e}^{-t/\ell}$. A realisation is shown in Figure 7.11. By sampling only $x_3 = 0$, Algorithm 7.8 is a convenient method for approximating realisations of the mean-zero Gaussian random field in $\mathbb{R}^2$ with the same covariance.

Algorithm 7.8 Code to sample $v_M(\boldsymbol{x})$ in (7.36) at a rectangular grid of points $[x_1, x_2, x_3]^{\mathsf{T}} \in \mathbb{R}^3$ when $X(t)$ is the mean-zero Gaussian process with covariance $c_X(t) = T_3\mathrm{e}^{-t/\ell}$. The inputs `grid1`, `grid2`, `grid3` are column vectors of coordinates x_1, x_2, x_3. The circulant embedding method with interpolation is used to approximate $X(t)$, with a grid of length `M` and padding of size `Mpad`. The input `ell` sets the length scale ℓ. The output `v` is the data array $v_M(x_1, x_2, x_3)$.

```
function v=turn_band_exp_3d(grid1, grid2, grid3, M, Mpad, ell)
[xx,yy,zz]=ndgrid(grid1,grid2,grid3); % x,y,z points
sum=zeros(size(xx(:))); % initialise
T=norm(max(abs([grid1,grid2,grid3])));
gridt=-T+(2*T/(M-1))*(0:(M+Mpad-1))';% radius T encloses all points
c=cov(gridt,ell);% evaluate covariance
for j=1:M,
    X=circulant_embed_approx(c); % sample X using Algorithm 6.10
    e=uniform_sphere; % sample e using Algorithm 4.6
    tt =[xx(:), yy(:), zz(:)]*e; % project
    Xi=interp1(gridt, X, tt); sum=sum+Xi;
end;
v=sum/sqrt(M); v=reshape(v,length(grid1), length(grid2), length(grid3));

function f=cov(t, ell) % covariance given by turning bands operator
f= (1-t/ell).*exp(-t/ell);
```

Uniformly spaced bands in $\mathbb{R}^2$

Choosing $\boldsymbol{e}_j$ randomly is not practical, however, as the rate of convergence — typically $\mathcal{O}(M^{-1/2})$ — is too slow. We seek a faster rate of convergence by replacing $\boldsymbol{e}_j$ by uniformly spaced points $\boldsymbol{z}_j$ on S^{d-1}. We focus on the case $d = 2$.

In place of the random variables $\boldsymbol{e}_j$ used in (7.36), we use uniformly spaced vectors $\boldsymbol{z}_j = [\cos\theta_j, \sin\theta_j]^{\mathsf{T}} \in S^1$ for $\theta_j = \pi j/M$. See Figure 7.12. For a mean-zero Gaussian process $X(t)$ with stationary covariance $T_2 c^0(t)$, define

$$v_M(\boldsymbol{x}) := \frac{1}{\sqrt{M}} \sum_{j=1}^{M} X_j(\boldsymbol{x}^{\mathsf{T}}\boldsymbol{z}_j), \qquad \boldsymbol{x} \in \mathbb{R}^2, \tag{7.37}$$

where $X_j(t)$ are again *iid* copies of $X(t)$. Algorithm 7.9 evaluates (7.37) in the case where $X(t)$ is given by (7.34). Figure 7.13 shows realisations of $v_M(\boldsymbol{x})$ with $M = 1, 2, 10$ bands. In contrast to (7.36), $v_M(\boldsymbol{x})$ is a simple linear combination of Gaussian random variables

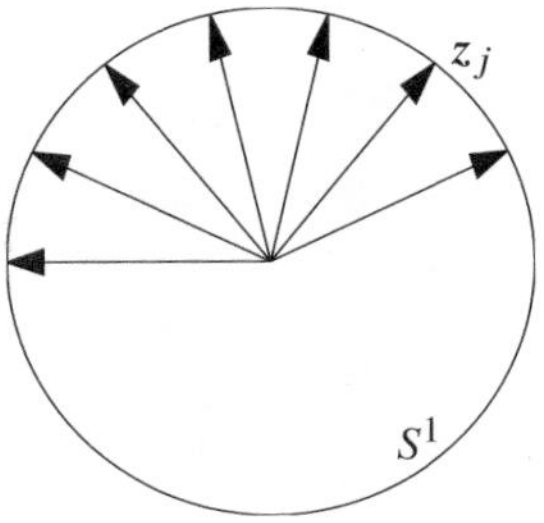

Figure 7.12 Uniformly spaced $z_j = [\cos(\theta_j), \sin(\theta_j)]^\mathsf{T}$ for $\theta_j = j\pi/M$ for $j = 1, \ldots, M$.

$X_j(\boldsymbol{x}^\mathsf{T} z_j)$ and $v_M(\boldsymbol{x})$ is now a Gaussian random field for any M. This time, however, the covariance function $c_M(\boldsymbol{x})$ of $v_M(\boldsymbol{x})$ does not equal the target covariance $c^0(\|\boldsymbol{x}\|_2)$ and is not even isotropic. Nevertheless, $v_M(\boldsymbol{x})$ is stationary and has covariance

$$c_M(\boldsymbol{x}) = \frac{1}{M}\sum_{j=1}^{M} c_X(\boldsymbol{x}^\mathsf{T} z_j), \qquad \boldsymbol{x} \in \mathbb{R}^2. \tag{7.38}$$

We show that $c_M(\boldsymbol{x})$ converges to $c^0(\|\boldsymbol{x}\|_2)$ in the limit $M \to \infty$.

Algorithm 7.9 Code to sample $v_M(\boldsymbol{x})$ in (7.37) for $X(t)$ given by (7.34). The inputs and outputs are as in Algorithm 7.7, with one additional input `M`, to specify the number of bands.

```
function v=turn_band_simple2(grid1, grid2, M)
[xx, yy] =ndgrid(grid1, grid2); sum=zeros(size(xx(:)));
for j=1:M,
    xi=randn(2,1); theta=pi*j/M; %
    e=[cos(theta); sin(theta)]; tt=[xx(:), yy(:)]*e; % project
    v=sqrt(1/2)*[cos(tt),sin(tt)]*xi; sum=sum+v; % cumulative sum
end;
v=sum/sqrt(M); % compute sample mean for v
v=reshape(v,length(grid1), length(grid2));
```

Theorem 7.49 (convergence) *Let $c^0(r)$ be an isotropic covariance in $\mathbb{R}^2$ and $\{X(t) : t \in \mathbb{R}\}$ be a mean-zero Gaussian process with stationary covariance $c_X(t) = T_2 c^0(t)$. If $c_X \in \mathrm{C}^{2p}(\mathbb{R})$ some $p \in \mathbb{N}$, there exists $K > 0$ such that*

$$\left|c^0(\|\boldsymbol{x}\|_2) - c_M(\boldsymbol{x})\right| \le K \|g\|_{\mathrm{C}^{2p}(0,\pi)} \frac{1}{M^{2p}}, \qquad \forall \boldsymbol{x} \in \mathbb{R}^2, \tag{7.39}$$

where $g(\theta) := c_X(\|\boldsymbol{x}\|_2 \cos\theta)$.

Proof It is elementary to check that $v_M(\boldsymbol{x})$ has mean zero. For $\boldsymbol{x}, \boldsymbol{y} \in \mathbb{R}^2$,

$$\begin{aligned}
\mathbb{E}\big[v_M(\boldsymbol{x}) v_M(\boldsymbol{y})\big] &= \frac{1}{M}\sum_{j,k=1}^{M} \mathbb{E}\Big[X_j(\boldsymbol{x}^\mathsf{T} z_j) X_k(\boldsymbol{y}^\mathsf{T} z_k)\Big] \\
&= \frac{1}{M}\sum_{j=1}^{M} \mathbb{E}\Big[X_j(\boldsymbol{x}^\mathsf{T} z_j) X_j(\boldsymbol{y}^\mathsf{T} z_j)\Big],
\end{aligned}$$

as $X_j(t)$, $X_k(t)$ are independent with mean zero for $j \neq k$.

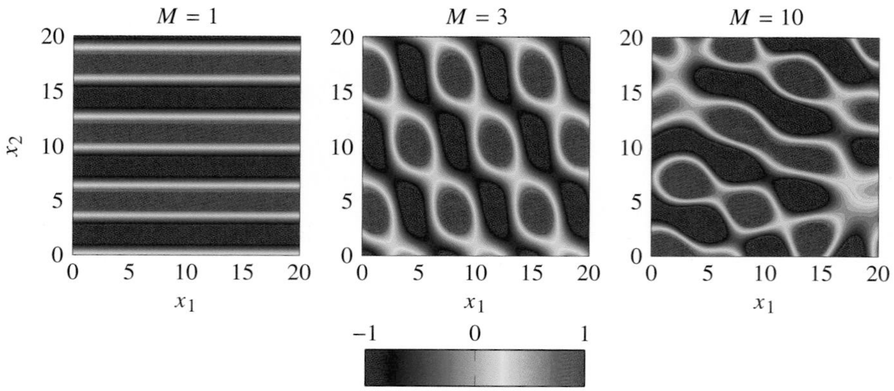

Figure 7.13 Realisations of the random field $v_M(\boldsymbol{x})$ given by (7.37) and (7.34) for $M = 1, 2, 10$. Compare to Figure 7.5, which shows two well-resolved realisations of $v_M(\boldsymbol{x})$ for $M = 20$.

$X(t)$ has stationary covariance $c_X(t)$ so that $v_M(\boldsymbol{x})$ has stationary covariance

$$c_M(\boldsymbol{x}) = \frac{1}{M}\sum_{j=1}^{M} c_X(\boldsymbol{x}^{\mathsf{T}} z_j) = \frac{1}{M}\sum_{j=1}^{M} c_X(\|\boldsymbol{x}\|_2 \cos(\theta_j + \phi))$$

for some $\phi \in [0, 2\pi]$. As c_X is even, $c_X(\|\boldsymbol{x}\|_2 \cos(\theta + \phi))$ is a π-periodic function of θ and the right-hand side is the trapezium rule for the integral

$$\frac{1}{\pi}\int_0^{\pi} g(\theta + \phi)\, d\theta = \frac{1}{\pi}\int_0^{\pi} g(\theta)\, d\theta$$

where $g(\theta) := c_X(\|\boldsymbol{x}\|_2 \cos\theta)$. The integral equals $\mathbb{E}\big[c_X(\boldsymbol{x}^{\mathsf{T}}\boldsymbol{e})\big]$, where $\boldsymbol{e} \sim \mathrm{U}(S^1)$, which by Proposition 7.47 equals $c^0(\|\boldsymbol{x}\|_2)$, and we expect $c_M(\boldsymbol{x}) \to c^0(\|\boldsymbol{x}\|_2)$.

We can get detailed estimates on the convergence of $c_M(\boldsymbol{x})$ to $c^0(\|\boldsymbol{x}\|_2)$ by applying Theorem A.6. The integrand g is periodic and $2p$ times differentiable in θ. Hence, the trapezium rule converges in the limit $M \to \infty$ with $\mathcal{O}(1/M^{2p})$ (see Theorem A.6). In particular, for some $K > 0$,

$$\big|c_M(\boldsymbol{x}) - c^0\big(\|\boldsymbol{x}\|_2\big)\big| \le K\|g\|_{\mathrm{C}^{2p}(0,\pi)}\frac{1}{M^{2p}}. \qquad \square$$

In general, we must use the methods of Chapter 6 to approximate $X(t)$. For example, we use the circulant embedding method in Algorithm 7.8 and, because $t = \mathbf{e}^{\mathsf{T}}\boldsymbol{x}_j$ are not uniformly spaced even if the sample points $\boldsymbol{x}_j$ are, we use linear interpolation and thereby approximate $X(t)$. We look at the resulting approximation error for the random field. Let $\{Y(t)\colon t \in \mathbb{R}\}$ be a mean-zero Gaussian process with stationary covariance $c_Y(t)$, where we think of $Y(t)$ as the stochastic process defined as the linear interpolant of the samples generated by the circulant embedding or quadrature methods for $X(t)$. Define

$$\tilde{v}_M(\boldsymbol{x}) := \frac{1}{\sqrt{M}}\sum_{j=1}^{M} Y_j(\boldsymbol{x}^{\mathsf{T}} z_j), \qquad \boldsymbol{x} \in \mathbb{R}^2,$$

where $Y_j(t)$ are *iid* copies of $Y(t)$, and note that $\tilde{v}_M(\boldsymbol{x})$ is a stationary Gaussian random field.

Corollary 7.50 *Let the assumptions of Theorem 7.49 hold for an isotropic covariance $c^0(r)$. There exists $K > 0$ such that*

$$\left|c^0(\|x\|_2) - \tilde{c}_M(x)\right| \le K\|g\|_{C^{2p}(0,\pi)}\frac{1}{M^{2p}} + \sup_{|t|\le\|x\|_2}\left|c_Y(t) - T_2c^0(t)\right|, \qquad \forall x \in \mathbb{R}^2,$$

where $\tilde{c}_M(x)$ is the stationary covariance of the random field $\tilde{v}_M(x)$.

Proof In the notation of Lemma 7.49, write $|c^0 - \tilde{c}_M| = |c^0 - c_M| + |c_M - \tilde{c}_M|$. The first term is bounded by $K\|g\|_{C^{2p}(0,\pi)}M^{-2p}$ using (7.39). For the second term, note

$$c_M(x) = \frac{1}{M}\sum_{j=1}^{M} c_X(x^{\mathsf{T}}z_j), \qquad \tilde{c}_M(x) = \frac{1}{M}\sum_{j=1}^{M} c_Y(x^{\mathsf{T}}z_j),$$

and $|c_M(x) - \tilde{c}_M(x)| \le \sup_{|t|\le\|x\|_2}|c_X(t) - c_Y(t)|$. The covariance of X is given by the turning bands random field and $c_X = T_2c^0$. This gives the required inequality. □

Algorithm 7.10 Code to approximate the mean-zero Gaussian random field $u(x)$ with isotropic covariance $c_q^0(r)$ using the turning bands method with M equally spaced bands and the quadrature approximation to $X(t)$. The inputs and outputs are like Algorithm 7.9 with additional inputs `q` and `ell` to specify the Whittle–Matérn parameter q and correlation length ℓ. Algorithm 6.4 (the command `squad`) is used to approximate $X(t)$.

```
function u=turn_band_wm(grid1, grid2, M, q, ell)
[xx, yy] =ndgrid(grid1, grid2);
sum=zeros(size(xx(:)));
% choose radius T to contain all grid points
T=norm([norm(grid1,inf),norm(grid2,inf)]);
for j=1:M,
    theta=j*pi/M; e=[cos(theta); sin(theta)]; % uniformly spaced
    tt=[xx(:), yy(:)]*e; % project
    [gridt, Z]=squad(2*T, 64, 64, @(s) f(s, q, ell));
    Xi=interp1(gridt-T, real(Z), tt); % interpolate
    sum=sum+Xi; % cumulative sum
end;
u=sum/sqrt(M);
u=reshape(u, length(grid1), length(grid2));

function f=f(s, q, ell) % spectral density
f=gamma(q+1)/gamma(q)*(ell^2*s)/(1+(ell*s)^2)^(q+1);
```

Example 7.51 (Whittle–Matérn covariance) The Whittle–Matérn covariance $c_q^0(r)$ was introduced in Example 7.17. Its spectral density is given by (7.11) and, by Definition 7.43, $T_dc_q^0(t)$ has spectral density

$$f^0(s)\frac{1}{2}\omega_d s^{d-1} = \frac{\Gamma(q + d/2)\omega_d}{2\Gamma(q)\pi^{d/2}}\frac{\ell^d s^{d-1}}{(1 + (\ell s)^2)^{d/2+q}}.$$

The corresponding process $X(t)$ is easily approximated using the quadrature method of §6.4. The turning bands method with uniformly spaced bands is implemented in Algorithm 7.10 for dimension $d = 2$.

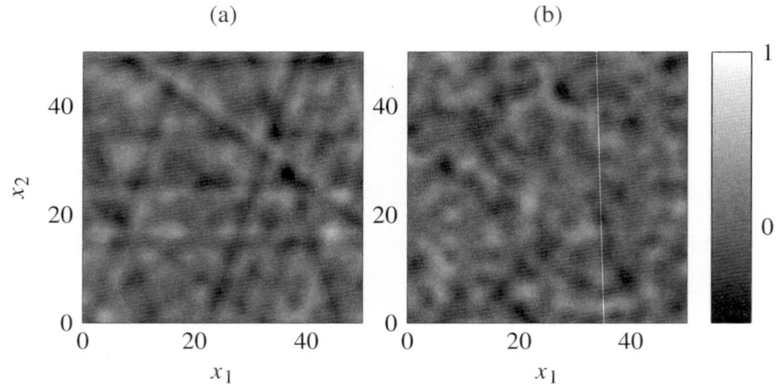

Figure 7.14 Approximate realisations of the mean-zero Gaussian random field with covariance $c_2^0(r/\ell)$ for $\ell = 1/10$ generated by Algorithm 7.10 with (a) $M = 5$ and (b) $M = 50$. Notice the streaking effect in (a) caused by choosing $M = 5$ too small.

Notice that $c_q^0(r) = \frac{1}{2^{q-1}\Gamma(q)}(r/\ell)^q K_q(r/\ell)$ and $c_q^0 \in \mathrm{C}^{2q}(\mathbb{R})$ (see Lemma 6.11). Then,

$$g(\theta) = \frac{1}{2^{q-1}\Gamma(q)}\left(\frac{r}{\ell}\cos\theta\right)^q K_q\left(\frac{r}{\ell}\cos\theta\right)$$

and Lemma 7.49 says we commit an error proportional to $\|g\|_{\mathrm{C}^{2q}(0,\pi)} M^{-2q}$ and so the turning bands method converges more rapidly when q is large. This is to be expected, as realisations for the exponential covariance ($q = 1/2$) have rougher realisations than cases with $q > 1/2$. Further, $\|g\|_{\mathrm{C}^{2p}(0,\pi)} = \mathcal{O}(1/\ell^{2p})$ and hence the error is larger for smaller ℓ. Indeed, we should take $M \gg 1/\ell$ to fully resolve the random field. See Figure 7.14 for a pair of realisations in the case $q = 2$ and $\ell = 1/10$, which shows a streaking effect for $M = 5$ bands.

Comparison with circulant embedding

When a non-negative definite embedding is available, circulant embedding provides an exact sampling method (as long as the sample points are uniformly spaced on a rectangular grid). Turning bands is an approximate method, and converges when the number of bands $M \to \infty$ and the process $X(t)$ is well resolved. For the isotropic exponential covariance, we observe in computations that the even circulant extension is non-negative definite for $\ell \ll 1$ and, in such cases, circulant embedding is the most effective method. Circulant embedding is preferred for stationary covariance models that are not isotropic, as the turning bands method is restricted to isotropic fields. For example, the Gaussian random field with separable exponential covariance can be simulated using circulant embedding but not turning bands. See Example 7.10 and Exercise 7.14.

In general, a non-negative definite embedding may be unavailable or may be so large as to be too expensive to compute. The comparison of the two methods then depends on how easy it is to sample the relevant process $X(t)$ and the number of required bands M. For example, as $X(t)$ is easy to compute for the Bessel covariance (see Example 7.45), the turning bands method is highly effective for this distribution, whilst for the circulant embedding method it is difficult to find non-negative definite embeddings (because the Bessel covariance decays slowly).

The turning bands method is also easier to generalise to dimension $d > 2$. All that is required is a sampling method for the distribution $\mathrm{U}(S^{d-1})$ or, for more efficiency, a quadrature method for an integral over S^{d-1}. The circulant embedding method does generalise to higher dimensions, but the structure of the covariance matrix is rather complicated.

7.4 Karhunen–Loève expansion of random fields

Second-order random fields can be represented by a Karhunen–Loève expansion, much as in §5.4, using the orthonormal basis provided by the eigenfunctions of the underlying covariance operator defined by

$$(\mathcal{C}\phi)(\boldsymbol{x}) = \int_D C(\boldsymbol{x},\boldsymbol{y})\phi(\boldsymbol{y})\,d\boldsymbol{y}, \qquad \boldsymbol{x} \in D. \tag{7.40}$$

Truncated expansions provide approximate realisations of the random field and are useful in analysis, as we show in §7.5. We restate the Karhunen–Loève theorems for random fields.

Theorem 7.52 (L^2 convergence of KL expansion) *Let $D \subset \mathbb{R}^d$. Consider a random field $\{u(\boldsymbol{x})\colon \boldsymbol{x} \in D\}$ and suppose that $u \in L^2(\Omega, L^2(D))$. Then,*

$$u(\boldsymbol{x},\omega) = \mu(\boldsymbol{x}) + \sum_{j=1}^{\infty} \sqrt{\nu_j}\,\phi_j(\boldsymbol{x})\,\xi_j(\omega) \tag{7.41}$$

where the sum converges in $L^2(\Omega, L^2(D))$,

$$\xi_j(\omega) := \frac{1}{\sqrt{\nu_j}}\langle u(\boldsymbol{x},\omega) - \mu(\boldsymbol{x}), \phi_j(\boldsymbol{x})\rangle_{L^2(D)}$$

and $\{\nu_j, \phi_j\}$ are the eigenvalues and eigenfunctions of the operator $\mathcal{C}$ in (7.40) with $\nu_1 \geq \nu_2 \geq \cdots \geq 0$. The random variables ξ_j have mean zero, unit variance and are pairwise uncorrelated. If u is Gaussian, then $\xi_j \sim \mathrm{N}(0,1)$ iid.

Next, consider the truncated expansion,

$$u_J(\boldsymbol{x},\omega) := \mu(\boldsymbol{x}) + \sum_{j=1}^{J} \sqrt{\nu_j}\,\phi_j(\boldsymbol{x})\,\xi_j(\omega). \tag{7.42}$$

Note that u_J is a random field with mean μ and covariance

$$C_J(\boldsymbol{x},\boldsymbol{y}) = \sum_{j=1}^{J} \nu_j \phi_j(\boldsymbol{x})\phi_j(\boldsymbol{y}). \tag{7.43}$$

Theorem 7.53 (uniform convergence of KL expansion) *If $D \subset \mathbb{R}^d$ is a closed and bounded set and $C \in \mathrm{C}(D \times D)$, then $\phi_j \in \mathrm{C}(D)$ and the series expansion of C converges uniformly. In particular,*

$$\sup_{\boldsymbol{x},\boldsymbol{y} \in D} \left| C(\boldsymbol{x},\boldsymbol{y}) - C_J(\boldsymbol{x},\boldsymbol{y}) \right| \le \sup_{\boldsymbol{x} \in D} \sum_{j=J+1}^{\infty} \nu_j \phi_j(\boldsymbol{x})^2 \to 0, \tag{7.44}$$

as $J \to \infty$, where C_J is (7.43) *and* $\sup_{\boldsymbol{x} \in D} \mathbb{E}\big[(u(\boldsymbol{x}) - u_J(\boldsymbol{x}))^2\big] \to 0$ *as* $J \to \infty$.

For stationary random fields, we can express $\int_D \mathrm{Var}(u(\boldsymbol{x}) - u_J(\boldsymbol{x}))\, d\boldsymbol{x}$ in terms of the J largest eigenvalues.

Corollary 7.54 *If the assumptions of Theorem 7.53 hold and $C(\boldsymbol{x},\boldsymbol{y}) = c(\boldsymbol{x} - \boldsymbol{y})$ then $\int_D \mathrm{Var}\, u(\boldsymbol{x})\, d\boldsymbol{x} = c(\boldsymbol{0})\mathrm{Leb}(D) = \sum_{j=1}^{\infty} \nu_j$ and*

$$\int_D \mathrm{Var}\big(u(\boldsymbol{x}) - u_J(\boldsymbol{x})\big)\, d\boldsymbol{x} = c(\boldsymbol{0})\,\mathrm{Leb}(D) - \sum_{j=1}^{J} \nu_j. \tag{7.45}$$

Proof We have $\mathrm{Var}(u(\boldsymbol{x})) = C(\boldsymbol{x},\boldsymbol{x}) = \sum_{j=1}^{\infty} \nu_j \phi_j(\boldsymbol{x})^2$. Integrating over D, we find

$$\int_D \mathrm{Var}(u(\boldsymbol{x}))\, d\boldsymbol{x} = c(\boldsymbol{0})\mathrm{Leb}(D) = \sum_{j=1}^{\infty} \nu_j,$$

as $\|\phi_j\|_{L^2(D)} = 1$. Now, since

$$\int_D \mathrm{Var}(u(\boldsymbol{x}) - u_J(\boldsymbol{x}))\, d\boldsymbol{x} = \mathbb{E}\left[\left\| \sum_{j=J+1}^{\infty} \sqrt{\nu_j}\, \phi_j \xi_j \right\|_{L^2(D)}^2 \right] = \sum_{j=J+1}^{\infty} \nu_j,$$

we obtain (7.45). □

In computations, we are interested in the relative error

$$E_J := \frac{\int_D \mathrm{Var}(u(\boldsymbol{x}) - u_J(\boldsymbol{x}))\, d\boldsymbol{x}}{\int_D \mathrm{Var}(u(\boldsymbol{x}))\, d\boldsymbol{x}} = \frac{c(\boldsymbol{0})\mathrm{Leb}(D) - \sum_{j=1}^{J} \nu_j}{c(\boldsymbol{0})\mathrm{Leb}(D)}. \tag{7.46}$$

It is the decay rate of the eigenvalues that determines how quickly $E_J \to 0$ and how large J needs to be to reduce the error to a desired threshold. To work with the truncated Karhunen–Loève expansion and to calculate the error in (7.46), we need to solve the eigenvalue problem

$$\mathcal{C}\phi = \nu\phi \tag{7.47}$$

for $\nu \neq 0$ and $\phi \in L^2(D)$, where $\mathcal{C}$ is given by (7.40). Analytical solutions for the eigenpairs $\{\nu_j, \phi_j\}$ are available in some special cases. We derive the eigenpairs for the separable exponential covariance in the cases $d = 1$ and $d = 2$.

Example 7.55 (exponential covariance $d = 1$) Consider the covariance

$$C(x,y) = \mathrm{e}^{-|x-y|/\ell} \qquad \text{on } D = [-a,a]. \tag{7.48}$$

We solve the eigenvalue problem for the corresponding integral operator $\mathcal{C}$. That is, we look for eigenpairs $\{\nu, \phi\}$ that satisfy

$$\int_{-a}^{a} \mathrm{e}^{-|x-y|/\ell}\, \phi(y)\, dy = \nu\, \phi(x), \qquad x \in [-a, a].$$

By differentiating twice we can show (as in Example 5.30) that

$$\frac{d^2\phi}{dx^2} + \omega^2 \phi = 0 \qquad \text{with} \qquad \omega^2 := \frac{2\ell^{-1} - \ell^{-2}\nu}{\nu}.$$

The associated boundary conditions are

$$\ell^{-1}\phi(-a) - \frac{d\phi}{dx}(-a) = 0, \qquad \ell^{-1}\phi(a) + \frac{d\phi}{dx}(a) = 0.$$

Hence,

$$\phi(x) = A\cos(\omega x) + B\sin(\omega x), \qquad \nu = \frac{2\ell^{-1}}{\omega^2 + \ell^{-2}}, \tag{7.49}$$

where $\omega > 0$ is either a root of

$$f_{\text{odd}}(\omega) := \ell^{-1} - \omega\tan(\omega a) \quad \text{or} \quad f_{\text{even}}(\omega) := \ell^{-1}\tan(\omega a) + \omega,$$

(see Exercise 7.16). If we let $\hat{\omega}_i, \tilde{\omega}_i$ be the positive roots of f_{odd} and f_{even} respectively and

$$\omega_i := \begin{cases} \hat{\omega}_{\lceil i/2 \rceil}, & i \text{ odd}, \\ \tilde{\omega}_{i/2}, & i \text{ even}, \end{cases} \tag{7.50}$$

then

$$\phi_i(x) = \begin{cases} A_i\cos(\omega_i x), & i \text{ odd}, \\ B_i\sin(\omega_i x), & i \text{ even}, \end{cases} \qquad \nu_i := \frac{2\ell^{-1}}{\omega_i^2 + \ell^{-2}}. \tag{7.51}$$

To have $\|\phi_i\|_{L^2(-a,a)} = 1$, we choose

$$A_i = \frac{1}{\sqrt{a + \sin(2\omega_i a)/2\omega_i}}, \qquad B_i = \frac{1}{\sqrt{a - \sin(2\omega_i a)/2\omega_i}}.$$

Example 7.56 (separable exponential $d = 2$) Consider the covariance

$$C(\boldsymbol{x}, \boldsymbol{y}) = \prod_{m=1}^{2} \mathrm{e}^{-|x_m - y_m|/\ell_m}, \qquad D = [-a_1, a_1] \times [-a_2, a_2]. \tag{7.52}$$

As C is separable, the eigenfunctions are $\phi_j(\boldsymbol{x}) = \phi_i^1(x_1)\phi_k^2(x_2)$ and the eigenvalues are $\nu_j = \nu_i^1 \nu_k^2$, where $\{\nu_i^1, \phi_i^1\}$ and $\{\nu_k^2, \phi_k^2\}$ are solutions to the one-dimensional problems

$$\int_{-a_m}^{a_m} \mathrm{e}^{-|x-y|/\ell_m}\, \phi^m(y)\, dy = \nu^m \phi^m(x), \qquad m = 1, 2.$$

Figure 7.15 shows the 100 largest eigenvalues of the one- and two-dimensional covariances for three values of ℓ and Figure 7.16 shows the relative error E_J in (7.46). Observe that the asymptotic rate of decay of ν_j does not depend on ℓ but, for a fixed tolerance ϵ, the number of terms needed to obtain $E_J \le \epsilon$ grows as $\ell \to 0$. See also Exercise 7.18. From Figure 7.16, we surmise that $E_J = \mathcal{O}(J^{-1})$ for both $d = 1$ and $d = 2$. Indeed, in the two-dimensional case with $\ell_1 = 1 = \ell_1$, for $J = 5$, 20, 50, 100 terms, the corresponding

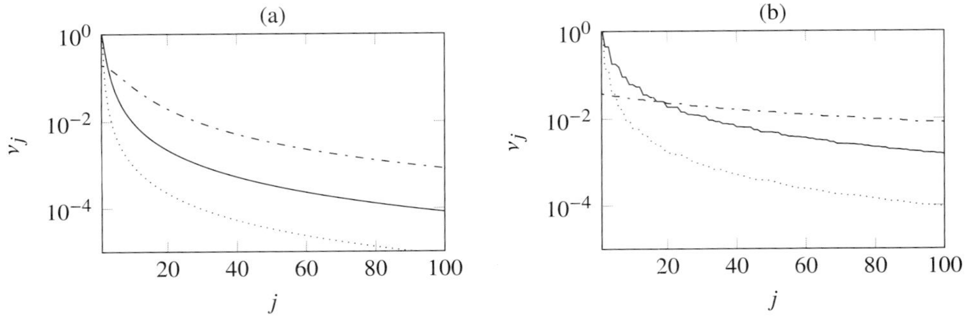

Figure 7.15 Eigenvalues ν_j for (a) the one-dimensional problem on $D = [-1, 1]$ in Example 7.55 and (b) the two-dimensional problem on $D = [-1, 1]^2$ in Example 7.56 with $\ell_1 = \ell = \ell_2$. The lines show the case $\ell = 1/10$ (dot-dash), $\ell = 1$ (solid line) and $\ell = 10$ (dotted).

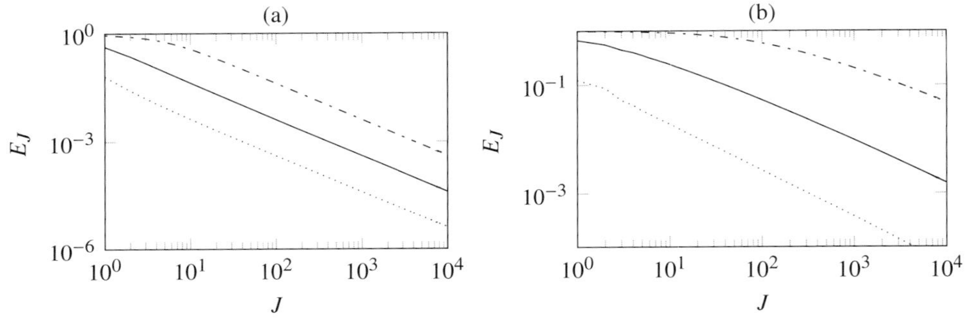

Figure 7.16 Relative error E_J in (7.46) for (a) the one-dimensional problem on $D = [-1, 1]$ in Example 7.55 and (b) the two-dimensional problem on $D = [-1, 1]^2$ in Example 7.56 with $\ell_1 = \ell = \ell_2$. The lines show the cases $\ell = 1/10$ (dot-dash), $\ell = 1$ (solid line) and $\ell = 10$ (dotted).

truncations errors are $E_J = 0.3549, 0.1565, 0.0856$ and 0.0530, respectively. We derive the theoretical rate of convergence of E_J in the next section. Finally, Figure 7.17 shows four eigenfunctions ϕ_j for the two-dimensional case. They become more oscillatory as the index increases.

Eigenvalue decay rates

How many terms J should we retain in (7.42) to accurately approximate $u(\boldsymbol{x})$? From (7.46), we see that a good approximation is obtained with fewer terms if the eigenvalues decay quickly. We now derive the asymptotic behaviour of ν_j for the separable exponential and Whittle–Matérn covariance functions and develop associated bounds on E_J in Theorems 7.60 and 7.61. We start with the following theorem of Widom. For functions $f, h \colon \mathbb{R}^+ \to \mathbb{R}$ (or sequences f_j, h_j), we write $f(s) \asymp h(s)$ as $s \to \infty$ (respectively, $f_j \asymp h_j$ as $j \to \infty$) and say f *is asymptotic to* h if $\lim f(s)/h(s) \to 1$ as $s \to \infty$ (resp., $f_j/h_j \to 1$ as $j \to \infty$).

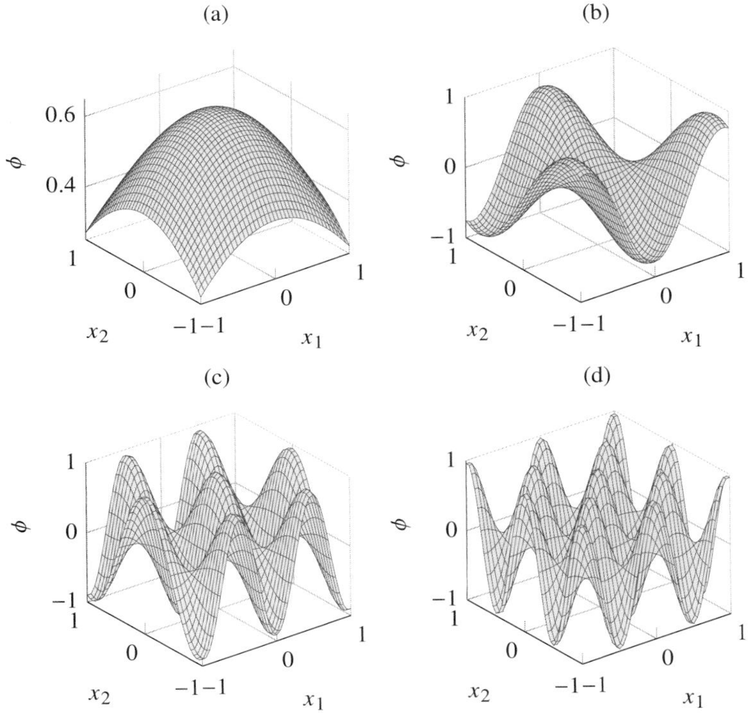

Figure 7.17 Eigenfunctions $\phi_j(\boldsymbol{x}) = \phi^1_i(x_1)\phi^2_k(x_2)$ for $(i,k) = (1,1),(2,3),(3,5),(5,5)$ (or $j = 1,10,39,79$) for $D = [-1,1]^2$ and $\ell_1 = 1 = \ell_2$ in Example 7.56.

Theorem 7.57 (Widom) *Let $c^0(r)$ be an isotropic covariance function on $\mathbb{R}^d$ with radial spectral density $f^0(s)$. Assume that*

$$f^0(s) \asymp bs^{-\rho} \qquad \text{as } s \to \infty,$$

for some $b, \rho > 0$. Let D be a bounded domain in $\mathbb{R}^d$ and consider the eigenvalues $\nu_1 \geq \nu_2 \geq \cdots \geq 0$ of the covariance operator $\mathcal{C}$ given by (7.40). Let V_d be the volume of the unit sphere in $\mathbb{R}^d$. As $j \to \infty$,

$$\nu_j \asymp K(D,d,\rho,b) j^{-\rho/d}$$

for $K(D,d,\rho,b) := (2\pi)^{d-\rho} b\,(\mathrm{Leb}(D)V_d)^{\rho/d}$.

The decay rate of the eigenvalues depends only on the dimension d and the rate of decay ρ of the spectral density f^0.

Example 7.58 (exponential covariance $d = 1$) Theorem 7.57 applies to any stationary covariance in one dimension with a spectral density $f(s) \asymp bs^{-\rho}$. For the exponential covariance $c(x) = \mathrm{e}^{-|x|/\ell}$, the spectral density is $f(s) = \ell/\pi(1 + \ell^2 s^2)$ by (6.4). Then, $f(s) \asymp 1/\pi\ell s^2$, in which case $b = 1/\pi\ell$ and $\rho = 2$. For $D = [-a,a]$, $\mathrm{Leb}(D) = 2a$ and

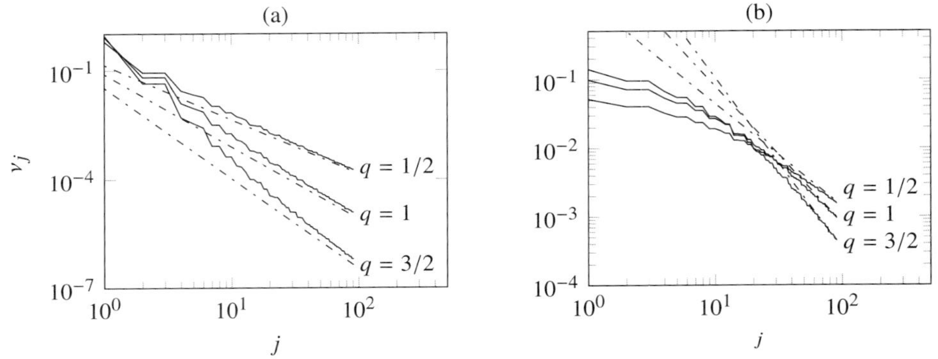

Figure 7.18 Eigenvalues (solid line) of the Whittle–Matérn covariance on $D = [0,1]^2$ with (a) $\ell = 1$ and (b) $\ell = 1/10$. Numerical approximations are given for the cases $q = 1/2$, $q = 1$, and $q = 3/2$, along with the asymptotic value of ν_j given by (7.54) (dash-dot line).

$V_1 = 2$, so that

$$\nu_j \asymp 2\pi b \left(2\pi \frac{j}{2a \times 2}\right)^{-\rho} = \frac{8a^2}{\pi^2 \ell j^2}. \tag{7.53}$$

Since the roots in (7.50) satisfy $\omega_j \asymp j\pi/2a$ and the eigenvalues in (7.49) satisfy $\nu_j \asymp 2/\ell\omega_j^2$, we find the same asymptotic formula from Example 7.55. Since ρ does not depend on ℓ, the correlation length does not affect the asymptotic convergence rate.

Example 7.59 (Whittle–Matérn covariance) The Whittle–Matérn covariance is isotropic and its radial spectral density $f^0(s)$ is given by (7.11). In this case, $f^0(s) \asymp bs^{-\rho}$ for $b = \ell^{-2q} \frac{\Gamma(q+d/2)}{\Gamma(q)\pi^{d/2}}$ and $\rho = 2q + d$. Applying Theorem 7.57, we have as $j \to \infty$

$$\nu_j \asymp (2\pi\ell)^{-2q} \frac{\Gamma(q + d/2)}{\Gamma(q)\pi^{d/2}} \left(\frac{\mathrm{Leb}(D)V_d}{j}\right)^{\rho/d}. \tag{7.54}$$

Note that ρ depends on d and on q, which controls the smoothness of the covariance function. The higher the value of q, the smoother the covariance function and the faster the eigenvalues decay.

See Figure 7.18 for a comparison of the asymptotic formula and the eigenvalues for $q = 1/2$, 1 and 3/2. In this case, we used a numerical method to approximate ν_j since analytic formulae are not available. In the particular case of the isotropic exponential covariance $c^0(r) = \mathrm{e}^{-r}$, obtained by putting $q = 1/2$ in the Whittle–Matérn covariance, we have

$$\nu_j \asymp \frac{1}{2\pi\ell} \frac{\Gamma((d+1)/2)}{\pi^{(d+1)/2}} \left(\frac{\mathrm{Leb}(D)V_d}{j}\right)^{(d+1)/d} \qquad \text{as } j \to \infty.$$

As $\Gamma(1) = 1$, this agrees with (7.53) for $d = 1$.

The asymptotics for ν_j yield the rate of convergence of $E_J \to 0$, where E_J is the Karhunen–Loève truncation error (7.46). First, consider the Whittle–Matérn covariance.

Theorem 7.60 (E_J for Whittle–Matérn) *For a closed and bounded set $D \subset \mathbb{R}^d$, let $\{u(\boldsymbol{x})\colon \boldsymbol{x} \in D\}$ be a random field with the Whittle–Matérn covariance. Then,*

$$E_J \asymp K_{q,d}\frac{1}{\ell^{2q}}\frac{1}{J^{2q/d}} \qquad \text{where } K_{q,d} := \frac{\Gamma(q+d/2)}{\Gamma(q)\pi^{d/2}(2\pi)^{2q}}\text{Leb}(D)^{2q/d}V_d^{1+2q/d}\frac{d}{2q}.$$

In particular, if $u(\boldsymbol{x})$ has the isotropic exponential covariance, then $E_J \asymp K_{1/2,d}/\ell J^{1/d}$.

Proof This follows easily from (7.46) using (7.54) because $\sum_{j=J+1}^{\infty} j^{-(1+r)} \asymp (1/r)J^{-r}$ as $J \to \infty$ for $r > 0$. □

Theorem 7.61 (E_J for separable exponential $d = 2$) *Let $\{u(\boldsymbol{x})\colon \boldsymbol{x} \in D\}$ for $D = [-a_1, a_1] \times [-a_2, a_2]$ be a random field with separable exponential covariance* (7.48) *and correlation lengths ℓ_1, ℓ_2. There exists a constant $K(a_1, a_2, \ell_1, \ell_2) > 0$ such that*

$$E_J \le K(a_1, a_2, \ell_1, \ell_2)\frac{\log^2 J}{J}.$$

Proof Let ν_i^k for $i \in \mathbb{N}$ denote the eigenvalues of the one-dimensional exponential covariance with correlation length ℓ_k on $[-a_k, a_k]$ for $k = 1, 2$. As $\text{Leb}([-a_k, a_k]) = \sum_{i=1}^{\infty} \nu_i^k$, we see that $\nu_1^k \le 2a_k$. From Example 7.56, we know $\nu_i^k = 2\ell_k^{-1}/(\omega_i^2 + \ell_k^{-2})$ and $\omega_i \ge (i-1)\pi/2a_k$. Hence, $\nu_i^k \le 2/\ell_k\omega_i^2 \le 8a_k^2/\ell_k\pi^2(i-1)^2$ for $i \ge 2$.

Taking the two together, $\nu_i^k \le K_k/i^2$ for $K_k := \max\{2a_k, 32a_k^2/\ell_k\pi^2\}$ for $i \in \mathbb{N}$. Then,

$$\nu_i^1\nu_k^2 \le \frac{K_1K_2}{i^2k^2}, \qquad i, k \in \mathbb{N}.$$

The eigenvalues of the separable exponential covariance on $[-a_1, a_1] \times [-a_2, a_2]$ with correlation lengths ℓ_1, ℓ_2 are exactly $\nu_i^1\nu_k^2$ for $i, k \in \mathbb{N}$. Order the eigenvalues in descending magnitude and denote them $\nu_j \ge \nu_{j+1}$ for $j \in \mathbb{N}$. From (7.46) with $c(0) = 1$,

$$E_J = \frac{1}{\text{Leb}(D)}\sum_{j=J+1}^{\infty} \nu_j.$$

Lemma A.15 gives a constant $K > 0$ (independent of a_1, a_2, ℓ_1, ℓ_2) such that

$$\sum_{j=J+1}^{\infty} \nu_j \le K^2K_1K_2\sum_{j=J+1}^{\infty} \mathrm{e}^{-2W(j)},$$

where $W(x)$ denotes the Lambert W function (see (A.14)). Because $W(x)$ is monotonic increasing for $x \ge 0$,

$$\sum_{j=J+1}^{\infty} \mathrm{e}^{-2W(j)} \le \int_J^{\infty} \mathrm{e}^{-2W(j)}\,dj.$$

The integral can be evaluated (see Exercise 7.17), to show

$$E_J \le \frac{2K^2K_1K_2}{\text{Leb}(D)}\frac{\log^2 J}{J}.$$

The proof is complete by choosing $K(a_1, a_2, \ell_1, \ell_2) := K^2K_1K_2/\text{Leb}(D)$. □

The constant K in Theorem 7.61 grows like $1/\ell_1\ell_2$ as $\ell_1, \ell_2 \to 0$ and J should be increased to ensure accuracy for problems with small correlation lengths.

Approximating realisations

Suppose $u(\boldsymbol{x})$ is Gaussian, with mean $\mu(\boldsymbol{x})$ and covariance $C(\boldsymbol{x},\boldsymbol{y})$. If the eigenpairs $\{\nu_j,\phi_j\}$ are known, as in Example 7.56, the truncated Karhunen–Loève expansion (7.42) can be used to generate approximate realisations of $u(\boldsymbol{x})$. Let $\boldsymbol{x}_0, \dots, \boldsymbol{x}_{N-1}$ be sample points and denote $\boldsymbol{u} = [u(\boldsymbol{x}_0),\dots,u(\boldsymbol{x}_{N-1})]^{\mathsf{T}}$. We know $\boldsymbol{u} \sim \mathrm{N}(\boldsymbol{\mu},C)$ where $\mu_{i+1} = \mu(\boldsymbol{x}_i)$ and $c_{i+1,j+1} = C(\boldsymbol{x}_i,\boldsymbol{x}_j)$, for $i,j = 0,\dots,N-1$. The vector $\boldsymbol{u}_J = [u_J(\boldsymbol{x}_0),\dots,u_J(\boldsymbol{x}_{N-1})]^{\mathsf{T}}$ has distribution $\mathrm{N}(\boldsymbol{\mu},C_J)$ where

$$[C_J]_{i+1,j+1} = C_J(\boldsymbol{x}_i,\boldsymbol{x}_j) = \sum_{m=1}^{J} \nu_m \phi_m(\boldsymbol{x}_i)\phi_m(\boldsymbol{x}_j), \qquad i,j = 0,\dots,N-1.$$

If we evaluate the eigenfunctions at the sample points, and form the vectors $\boldsymbol{\phi}_j = [\phi_j(\boldsymbol{x}_0),\dots,\phi_j(\boldsymbol{x}_{N-1})]^{\mathsf{T}}$, $j = 1,\dots,J$, we have

$$\boldsymbol{u}_J = \boldsymbol{\mu} + \Phi_J \Lambda_J^{1/2} \boldsymbol{\xi}, \qquad \boldsymbol{\xi} \sim \mathrm{N}(\boldsymbol{0},I_J), \tag{7.55}$$

where $\Phi_J := [\boldsymbol{\phi}_1,\dots,\boldsymbol{\phi}_J] \in \mathbb{R}^{N\times J}$ and $\Lambda_J \in \mathbb{R}^{J\times J}$ is diagonal with entries ν_j. We have $C_J = \Phi_J\Lambda_J\Phi_J^{\mathsf{T}}$, and if C_J is close to C then $\boldsymbol{u}_J$ is close to the target distribution. Theorem 4.63 tells us averages $\mathbb{E}[\phi(\boldsymbol{u})]$ for $\phi \in L^2(\mathbb{R}^N)$ are accurately approximated by $\mathbb{E}[\phi(\boldsymbol{u}_J)]$ if the covariance matrices C and C_J are close together.

Figure 7.19 shows a realisation of the truncated expansion (7.42) of the mean-zero Gaussian random field $u(\boldsymbol{x})$ with covariance (7.52) with $J = 5, 20, 50, 100$ terms. The realisations were generated by evaluating the exact eigenfunctions at $N = 65^2$ uniformly spaced sample points on $D = [-1,1]^2$ and forming $\boldsymbol{u}_J$ in (7.55). The relative errors in the covariance matrix

$$E_{C_J} := \frac{\|C - C_J\|_{\mathrm{F}}}{\|C\|_{\mathrm{F}}}, \tag{7.56}$$

are $E_{C_J} = 0.1715, 0.0420, 0.0159$ and 0.0078, respectively, decaying to zero as $J \to \infty$.

Now, if exact eigenpairs are not available, we could instead use approximate eigenpairs $\{\hat\nu_j,\hat\phi_j\}$ and form

$$\hat{\boldsymbol{u}}_J = \boldsymbol{\mu} + \sum_{j=1}^{J} \sqrt{\hat\nu_j}\, \hat{\boldsymbol{\phi}}_j \xi_j = \boldsymbol{\mu} + \widehat{\Phi}_J \widehat{\Lambda}_J^{1/2} \boldsymbol{\xi}, \qquad \boldsymbol{\xi} \sim \mathrm{N}(\boldsymbol{0},I_J), \tag{7.57}$$

where $\hat{\boldsymbol{\phi}}_j \in \mathbb{R}^N$ contains the values of $\hat\phi_j$, the matrix $\widehat{\Phi}_J := [\hat{\boldsymbol{\phi}}_1,\dots,\hat{\boldsymbol{\phi}}_J] \in \mathbb{R}^{N\times J}$ and $\widehat{\Lambda}_J \in \mathbb{R}^{J\times J}$ is diagonal with entries $\hat\nu_j$. The resulting vector $\hat{\boldsymbol{u}}_J$ then has distribution $\mathrm{N}(\boldsymbol{0},\widehat{C}_J)$ with $\widehat{C}_J = \widehat{\Phi}_J\widehat{\Lambda}_J\widehat{\Phi}_J^{\mathsf{T}}$. Averages $\mathbb{E}[\phi(\boldsymbol{u})]$ for $\phi \in L^2(\mathbb{R}^N)$ are accurately approximated by $\mathbb{E}[\phi(\hat{\boldsymbol{u}}_J)]$ if the covariance matrices C and $\widehat{C}_J$ are close together as described by Theorem 4.63.

Approximating eigenvalues and eigenfunctions

We briefly describe two basic numerical methods for approximating the eigenvalues and eigenfunctions. The idea is to discretise (7.47) to obtain matrix equations whose eigenvalues and eigenvectors approximate the eigenpairs of the continuous problem. Unfortunately,

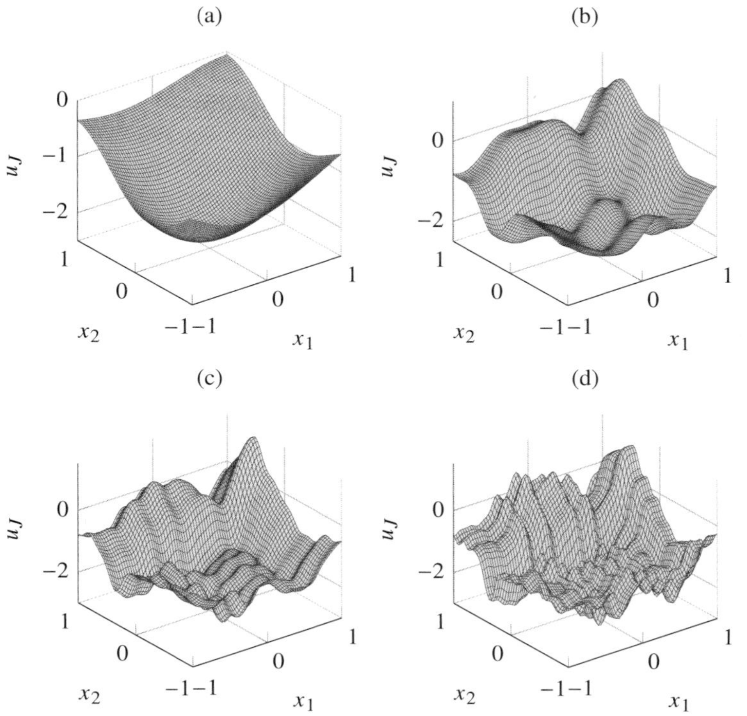

Figure 7.19 Samples of $\boldsymbol{u}_J$ in (7.55) with $N = 65^2$ uniformly spaced points on $D = [-1,1]^2$ for the Gaussian random field with separable exponential covariance, $\mu = 0$, $\ell_1 = 1 = \ell_2$ and (a) $J = 5$, (b) $J = 20$, (c) $J = 50$, (d) $J = 100$ and $\xi_j \sim \mathrm{N}(0,1)$ *iid*.

the matrices are large and dense, making computations expensive, especially in $d > 1$ dimensions.

Approximation via collocation and quadrature

Let $\boldsymbol{x}_1, \ldots, \boldsymbol{x}_P$ be points in D (not necessarily the sample points) and define

$$R_j := \int_D C(\boldsymbol{x},\boldsymbol{y})\hat{\phi}_j(\boldsymbol{y})\, d\boldsymbol{y} - \hat{\nu}_j\, \hat{\phi}_j(\boldsymbol{x}).$$

We call $\{\hat{\nu}_j, \hat{\phi}_j\}$ a collocation approximation if $R_j(\boldsymbol{x}_k) = 0$, for $k = 1, \ldots, P$. Now, suppose the $\boldsymbol{x}_k$ are quadrature points, with weights q_k, chosen so that

$$\sum_{i=1}^{P} q_i C(\boldsymbol{x}_k, \boldsymbol{x}_i)\phi(\boldsymbol{x}_i) \approx \int_D C(\boldsymbol{x}_k, \boldsymbol{y})\phi(\boldsymbol{y})\, d\boldsymbol{y}. \tag{7.58}$$

We can obtain approximations $\hat{\nu}_j$ and $\hat{\phi}_j(\boldsymbol{x}_i), i = 1, \ldots, P$, by solving

$$\sum_{i=1}^{P} q_i C(\boldsymbol{x}_k, \boldsymbol{x}_i)\hat{\phi}_j(\boldsymbol{x}_i) = \hat{\nu}_j\, \hat{\phi}_j(\boldsymbol{x}_k), \qquad k = 1, \ldots, P. \tag{7.59}$$

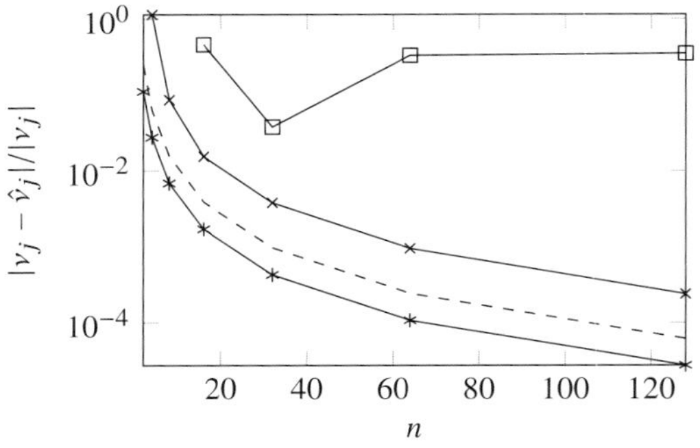

Figure 7.20 Plot of the eigenvalue error $|\nu_j - \hat{\nu}_j|/|\nu_j|$, for $j = 1(*)$, $j = 10(\times)$, and $j = 100(\square)$ against n, for Example 7.62. The dashed line is n^{-2}.

If the quadrature is exact, the collocation equations are solved exactly. In matrix notation, $CQ\hat{\boldsymbol{\phi}}_j = \hat{\nu}_j\hat{\boldsymbol{\phi}}_j$, where $C \in \mathbb{R}^{P\times P}$ is the covariance matrix for the quadrature points, $Q \in \mathbb{R}^{P\times P}$ is the diagonal matrix of weights and $\hat{\boldsymbol{\phi}}_j$ contains point evaluations of $\hat{\phi}_j$. We can convert to a symmetric problem by writing $S = Q^{1/2}CQ^{1/2}$ since $Sz_j = \hat{\nu}_j z_j$ with $z_j \coloneqq Q^{1/2}\hat{\boldsymbol{\phi}}_j$.

Example 7.62 (vertex-based quadrature) Partition $D = [-1,1]\times[-1,1]$ into $n\times n$ squares with edge length $h = 2/n$ and let $\boldsymbol{x}_k$, $k = 1,\dots,P = (n+1)^2$, be the vertices. We consider the separable exponential covariance function in (7.52) with $\ell_1 = 1 = \ell_2$ and apply the quadrature rule (7.58) with weights

$$q_i = \begin{cases} h^2/4, & \text{if } \boldsymbol{x}_i \text{ is a corner of } D,\\ h^2/2, & \text{if } \boldsymbol{x}_i \text{ lies on an edge of } D,\\ h^2, & \text{if } \boldsymbol{x}_i \text{ lies in the interior of } D. \end{cases} \tag{7.60}$$

As $n \to \infty$, the error in $\hat{\nu}_j$ for small j (corresponding to smooth eigenfunctions) converges to zero at a rate that depends on n^{-2}. However, the higher the index j, the larger n needs to be to obtain good convergence. See Figure 7.20. The eigenfunctions ϕ_j become more oscillatory as j increases and, to avoid aliasing, n must increase as the number of desired eigenpairs increases. The approximate eigenfunctions are investigated in Exercise 7.19.

If the matrix problem can be solved exactly, the accuracy depends on the choice of quadrature scheme and the regularity of $C(\boldsymbol{x},\boldsymbol{y})$. If the sample points in (7.57) are different from the quadrature points, interpolation also needs to be applied. Eigensolvers such as MATLAB's `eig` routine compute all P eigenvalues and eigenvectors of S to machine precision. However, the cost is $\mathcal{O}(P^3)$ and this quickly becomes intractable on modest PCs. Iterative solvers such as MATLAB's `eigs` routine require only matrix-vector products with S and approximate a specified number $J < P$ of eigenvalues and eigenvectors. (We usually require far fewer than P eigenpairs). However, since the cost of a matrix-vector product is $\mathcal{O}(P^2)$ for dense matrices, savings are made only if the iteration converges in much less than P iterations.

The quadrature rule in Example 7.62 can be derived by breaking the integral in (7.58) over n^2 square elements D_i and approximating each element integral by a four-point quadrature, which is exact for bilinear functions on D_i. That is,

$$\int_D f_k(\boldsymbol{y})\,d\boldsymbol{y} = \sum_{i=1}^{n^2} \int_{D_i} f_k(\boldsymbol{y})\,d\boldsymbol{y} \approx \sum_{i=1}^{n^2}\sum_{j=1}^{4} q_j^i f_k(\boldsymbol{x}_j^i), \tag{7.61}$$

where $f_k(\boldsymbol{y}) = C(\boldsymbol{x}_k, \boldsymbol{y})\phi(\boldsymbol{y})$, $q_j^i = h^2/4$ and $\boldsymbol{x}_j^i$, for $j = 1, 2, 3, 4$, are the vertices of D_i. A natural way to improve accuracy for a fixed n is to use more quadrature points in each D_i so that the rule is exact for polynomials of a higher degree.

Galerkin approximation

Another possibility is to use Galerkin approximation, as studied in Chapter 2. Given a P-dimensional subspace $V_P \subset L^2(D)$ with $V_P = \operatorname{span}\{\varphi_1(\boldsymbol{x}), \dots, \varphi_P(\boldsymbol{x})\}$, we can look for approximate eigenfunctions $\hat{\phi}_j \in V_P$ by solving

$$\int_D \left(\int_D C(\boldsymbol{x}, \boldsymbol{y}) \hat{\phi}_j(\boldsymbol{y})\,d\boldsymbol{y} \right) \varphi_i(\boldsymbol{x})\,d\boldsymbol{x} = \hat{\nu}_j \int_D \hat{\phi}_j(\boldsymbol{x}) \varphi_i(\boldsymbol{x})\,d\boldsymbol{x}, \qquad i = 1, \dots, P,$$

or $\int_D R_j(\boldsymbol{x})\varphi_i(\boldsymbol{x})\,d\boldsymbol{x} = 0$ for $i = 1, \dots P$, where R_j is the residual error

$$R_j = \int_D C(\boldsymbol{x}, \boldsymbol{y}) \hat{\phi}_j(\boldsymbol{y})\,d\boldsymbol{y} - \hat{\nu}_j \hat{\phi}_j(\boldsymbol{x}). \tag{7.62}$$

Notice that the choice $\varphi_i(\boldsymbol{x}) = \delta(\boldsymbol{x} - \boldsymbol{x}_i)$ leads us back to the collocation method. Now, since $\hat{\phi}_j \in V_P$, we have

$$\hat{\phi}_j(\boldsymbol{x}) = \sum_{k=1}^{P} a_k^j \varphi_k(\boldsymbol{x}), \tag{7.63}$$

for some coefficients a_k^j. Substituting this expansion gives

$$\sum_{k=1}^{P} a_k^j \int_D \int_D C(\boldsymbol{x}, \boldsymbol{y}) \varphi_j(\boldsymbol{y}) \varphi_i(\boldsymbol{x})\,d\boldsymbol{y}\,d\boldsymbol{x} = \hat{\nu}_j \sum_{k=1}^{P} a_k^j \int_D \varphi_j(\boldsymbol{x}) \varphi_i(\boldsymbol{x})\,d\boldsymbol{x}.$$

If we define the symmetric $P \times P$ matrices K and M via

$$[K]_{ij} = \int_D \int_D C(\boldsymbol{x}, \boldsymbol{y}) \varphi_i(\boldsymbol{y}) \varphi_j(\boldsymbol{x})\,d\boldsymbol{y}\,d\boldsymbol{x}, \qquad [M]_{ij} = \int_D \varphi_i(\boldsymbol{x}) \varphi_j(\boldsymbol{x})\,d\boldsymbol{x}, \tag{7.64}$$

and collect the coefficients in (7.63) into a vector $\boldsymbol{a}_j$, then $K\boldsymbol{a}_j = \hat{\nu}_j M \boldsymbol{a}_j$. We have a generalised matrix eigenvalue problem. K is in general dense and we still need quadrature schemes to evaluate the integrals in (7.64).

Example 7.63 (piecewise constant finite elements) Partition $D = [-1, 1]^2$ into $P = n^2$ square elements D_i as in Example 7.62 and let V_P be the space of piecewise constant functions, with $\varphi_i(\boldsymbol{x}) = 1_{D_i}(\boldsymbol{x})$. Then $\hat{\phi}_j|_{D_i} = a_k^j$ and

$$[K]_{ij} = \int_{D_i} \int_{D_j} C(\boldsymbol{x}, \boldsymbol{y})\,d\boldsymbol{y}\,d\boldsymbol{x}, \qquad [M]_{ij} = \delta_{ij} h^2, \quad h = \frac{2}{n}. \tag{7.65}$$

Note that $M = h^2 I_P$ and if we apply the midpoint rule twice then, trivially,

$$[K]_{ij} = \int_{D_i} \int_{D_j} C(\boldsymbol{x},\boldsymbol{y})\, d\boldsymbol{y}\, d\boldsymbol{x} \approx h^4 C(\boldsymbol{x}_i,\boldsymbol{x}_j), \tag{7.66}$$

where $\boldsymbol{x}_i,\boldsymbol{x}_j$ are the centres of D_i and D_j. This yields a simple matrix eigenvalue problem $Cz_j = \sigma_j z_j$, where C is the covariance matrix associated with the element centres, $\boldsymbol{a}_j = z_j/h$ and $\hat{\nu}_j = h^2\sigma_j$. See Exercise 7.20 for an implementation. The midpoint rule is exact only if $C(\boldsymbol{x},\boldsymbol{y})$ is piecewise constant. For a fixed n, accuracy can be improved by using a better quadrature method. In addition, we can choose V_P to be a space of higher-order piecewise polynomials.

7.5 Sample path continuity for Gaussian random fields

We end this chapter by studying the regularity of realisations (or 'sample paths') of Gaussian random fields and give conditions for them to be Hölder continuous. For example, we show that the Brownian sheet $\beta(\boldsymbol{x})$ has realisations that are almost surely Hölder continuous with exponent $\gamma < 1/2$ (see Definition 1.14). The method of proof is to exploit the Karhunen–Loève expansion and use $u_J(\boldsymbol{x})$, the truncated expansion in (7.42). When D is bounded and the covariance is continuous, the sequence of random variables $u_J(\boldsymbol{x})$ is Gaussian and depends continuously on $\boldsymbol{x}$. Further, $\mathbb{E}[u_J|u_{J-1}] = u_{J-1}$, which means u_J is a martingale (recall Example 4.53) with discrete parameter J. These facts have a number of important consequences that lead to the proof of the main result, Theorem 7.68.

We begin with three preliminary lemmas; the first two concern special sequences of random variables called submartingales.

Definition 7.64 (submartingale) A sequence X_J, indexed by $J \in \mathbb{N}$, of real-valued random variables on $(\Omega, \mathcal{F}, \mathbb{P})$ is a submartingale if

$$\mathbb{E}\big[|X_J|\big] < \infty \qquad \text{and} \qquad \mathbb{E}\Big[X_{J+1}\,\Big|\,X_1,\dots,X_J\Big] \ge X_J.$$

Submartingales are related to martingales (see Example 4.53). While martingales model the winnings in a fair game, submartingales model the winnings in a game where you tend to win.

Lemma 7.65 *Let X_J be a non-negative submartingale and define*

$$X_J^*(\omega) := \sup_{1\le k\le J} X_k(\omega), \qquad \omega \in \Omega.$$

If $\lambda > 0$ and $A(\lambda) := \{\omega \in \Omega\colon X_J^*(\omega) \ge \lambda\}$, then $\lambda\mathbb{P}(A(\lambda)) \le \mathbb{E}[X_J 1_{A(\lambda)}]$.

Proof We may write $A(\lambda)$ as the disjoint union of sets $A_1,\dots,A_J$ given by

$$A_k := \left\{\omega \in \Omega\colon \begin{array}{l} X_j(\omega) < \lambda \text{ for } j = 1,\dots,k-1 \\ X_k(\omega) \ge \lambda \end{array}\right\}.$$

For $k = 1,\dots,J-1$, $\mathbb{E}[X_J 1_{A_k}] = \mathbb{E}[\mathbb{E}[X_J\,|\,X_1,\dots,X_{J-1}]1_{A_k}] \ge \mathbb{E}[X_{J-1}1_{A_k}]$ by the submartingale condition. Repeating this argument, we get $\mathbb{E}[X_J 1_{A_k}] \ge \mathbb{E}[X_k 1_{A_k}]$ and this holds for $k = 1, \dots, J$. The definition of A_k gives $\mathbb{E}[X_J 1_{A_k}] \geq \lambda \mathbb{P}(A_k)$ and

$$\mathbb{E}[X_J 1_{A(\lambda)}] = \sum_{k=1}^{J} \mathbb{E}\left[X_J 1_{A_k}\right] \geq \lambda \sum_{k=1}^{J} \mathbb{P}(A_k) = \lambda\, \mathbb{P}(A(\lambda)). \qquad \square$$

Lemma 7.66 (Doob's submartingale inequality) *Let X_J be a non-negative submartingale. Then* $\mathbb{E}\left[\left(\sup_{1\le k\le J} X_k\right)^2\right] \leq 4\mathbb{E}[X_J^2]$.

Proof Using Lemma 7.65,

$$\int_0^\infty \lambda \mathbb{P}(A(\lambda))\, d\lambda \leq \int_0^\infty \mathbb{E}[X_J 1_{A(\lambda)}]\, d\lambda.$$

Using $X_J \geq 0$ and Fubini's theorem, the left-hand side becomes

$$\begin{aligned}\int_0^\infty \lambda \mathbb{P}(A(\lambda))\, d\lambda &= \int_0^\infty \int_\Omega \lambda 1_{A(\lambda)}(\omega)\, d\mathbb{P}(\omega)\, d\lambda = \int_\Omega \int_0^{X_J^*(\omega)} \lambda\, d\lambda\, d\mathbb{P}(\omega) \\ &= \frac{1}{2}\int_\Omega X_J^*(\omega)^2\, d\mathbb{P}(\omega) = \frac{1}{2}\|X_J^*\|_{L^2(\Omega)}^2. \end{aligned} \tag{7.67}$$

For the right-hand side, using the Cauchy–Schwartz inequality,

$$\begin{aligned}\int_0^\infty \mathbb{E}[X_J 1_{A(\lambda)}]\, d\lambda &= \int_0^\infty \int_{A(\lambda)} X_J(\omega)\, d\mathbb{P}(\omega)\, d\lambda \\ &= \int_\Omega \int_0^{X_J^*(\omega)} d\lambda\, X_J(\omega)\, d\mathbb{P}(\omega) = \int_\Omega X_J^*(\omega) X_J(\omega)\, d\mathbb{P}(\omega) \\ &= \mathbb{E}[X_J X_J^*] \leq \|X_J\|_{L^2(\Omega)} \|X_J^*\|_{L^2(\Omega)}. \end{aligned} \tag{7.68}$$

Together (7.67) and (7.68) give $\|X_J^*\|_{L^2(\Omega)} \leq 2\|X_J\|_{L^2(\Omega)}$. Using the definition of X_J^* gives $\mathbb{E}\left[\left(\sup_{1\le k\le J} X_k\right)^2\right] \leq 4\mathbb{E}[X_J^2]$ as required. $\square$

The third and final lemma applies to any continuous function u on a bounded domain D (see Definition 1.3). We use the notation $B(\boldsymbol{x}_0, r) := \{\boldsymbol{x} \in \mathbb{R}^d : \|\boldsymbol{x}_0 - \boldsymbol{x}\|_2 \leq r\}$, the ball of radius r in $\mathbb{R}^d$ centred at $\boldsymbol{x}_0$.

Lemma 7.67 *Let D be a bounded domain in $\mathbb{R}^d$. Fix $q > 0$. For any $\epsilon, \lambda > 0$ and any $u \in \mathrm{C}(\bar{D})$,*

$$|u(\boldsymbol{x}) - u(\boldsymbol{y})| \leq \max\left\{1, \lambda^{-1} \log M_0(u)\right\} \|\boldsymbol{x} - \boldsymbol{y}\|_2^{q-\epsilon}, \qquad \forall \boldsymbol{x}, \boldsymbol{y} \in \bar{D}, \tag{7.69}$$

where

$$M_0(u) := \sup_{\substack{B(\boldsymbol{x}_0,r)\subset D \\ B(\boldsymbol{y}_0,r)\subset D}} \int_{B(\boldsymbol{x}_0,r)} \int_{B(\boldsymbol{y}_0,r)} \exp\left(\frac{u(\boldsymbol{x}) - u(\boldsymbol{y})}{r^q}\right)^2 d\boldsymbol{x}\, d\boldsymbol{y}. \tag{7.70}$$

Proof We argue for a contradiction and suppose that (7.69) *does not* hold for some $\epsilon, \lambda > 0$ and $u \in \mathrm{C}(\bar{D})$. Let $K := \max\{1, \lambda^{-1} \log M_0(u)\}$. Then, as u is continuous and D is open, we can find balls $B(\boldsymbol{x}_0, r), B(\boldsymbol{y}_0, r) \subset D$ such that

$$|u(\boldsymbol{x}) - u(\boldsymbol{y})| > \frac{1}{2} K \|\boldsymbol{x} - \boldsymbol{y}\|_2^{q-\epsilon}, \qquad \forall \boldsymbol{x} \in B(\boldsymbol{x}_0, r),\ \boldsymbol{y} \in B(\boldsymbol{y}_0, r). \tag{7.71}$$

Now, let

$$A(\boldsymbol{x}_0, \boldsymbol{y}_0, r, K) := \int_{B(\boldsymbol{x}_0,r)} \int_{B(\boldsymbol{y}_0,r)} \exp\left(\frac{K\|\boldsymbol{x}-\boldsymbol{y}\|_2^{q-\epsilon}}{2r^q}\right)^2 d\boldsymbol{x}\, d\boldsymbol{y}. \tag{7.72}$$

Using (7.70) and (7.71), $A(\boldsymbol{x}_0, \boldsymbol{y}_0, r, K) < M_0(u)$ for all r sufficiently small. Further, $A(\boldsymbol{x}_0, \boldsymbol{y}_0, r, K) \geq \mathrm{e}^{\lambda K}$ for r sufficiently small and $K \geq 1$ (see Exercise 7.22). This is a contradiction as $\mathrm{e}^{\lambda K} \geq M_0(u)$ by definition of K. □

We are ready for the main result. If u is a mean-zero Gaussian random field on a bounded domain satisfying (7.73) then realisations of u are Hölder continuous with exponent $s/2$.

Theorem 7.68 *Let D be a bounded domain and $\{u(\boldsymbol{x})\colon \boldsymbol{x} \in \bar{D}\}$ be a mean-zero Gaussian random field such that, for some $L, s > 0$,*

$$\mathbb{E}\left[|u(\boldsymbol{x}) - u(\boldsymbol{y})|^2\right] \leq L\|\boldsymbol{x}-\boldsymbol{y}\|_2^s, \qquad \forall \boldsymbol{x}, \boldsymbol{y} \in \bar{D}. \tag{7.73}$$

For any $p \geq 1$, there exists a random variable K such that $\mathrm{e}^K \in L^p(\Omega)$ and

$$|u(\boldsymbol{x}) - u(\boldsymbol{y})| \leq K(\omega)\|\boldsymbol{x}-\boldsymbol{y}\|_2^{(s-\epsilon)/2}, \qquad \forall \boldsymbol{x}, \boldsymbol{y} \in \bar{D}, \quad a.s. \tag{7.74}$$

Proof Theorem 7.53 applies (see Exercise 7.24) and provides a truncated Karhunen–Loève expansion $u_J(\boldsymbol{x})$ defined by (7.42) such that $u_J(\cdot, \omega) \in \mathrm{C}(\bar{D})$ for all $\omega \in \Omega$. Let

$$\Delta_J(\boldsymbol{x}, \boldsymbol{y}) := \frac{u_J(\boldsymbol{x}) - u_J(\boldsymbol{y})}{\sqrt{8L\|\boldsymbol{x}-\boldsymbol{y}\|_2^s}}, \qquad \boldsymbol{x}, \boldsymbol{y} \in \bar{D}.$$

The random variables in the Karhunen–Loève expansion are *iid* with mean zero and hence $\mathbb{E}[\Delta_{J+1} \,|\, \Delta_J] = \Delta_J$. Further, the function $t \mapsto \exp(t^2)$ is convex and Jensen's inequality implies that $\mathbb{E}\left[\mathrm{e}^{\Delta_{J+1}^2} \,|\, \Delta_J\right] \geq \mathrm{e}^{\Delta_J^2}$. That is, $\mathrm{e}^{\Delta_J^2}$ is a submartingale and Lemma 7.66 gives

$$\mathbb{E}\left[\sup_{1 \leq j \leq J} \mathrm{e}^{2\Delta_j(\boldsymbol{x},\boldsymbol{y})^2}\right] \leq 4\mathbb{E}\left[\mathrm{e}^{2\Delta_J(\boldsymbol{x},\boldsymbol{y})^2}\right].$$

Note that $\mathbb{E}[\Delta_J(\boldsymbol{x}, \boldsymbol{y})^2] \leq L\|\boldsymbol{x}-\boldsymbol{y}\|_2^s / 8L\|\boldsymbol{x}-\boldsymbol{y}\|_2^s \leq 1/8$ for every $\boldsymbol{x}$, $\boldsymbol{y}$. Consequently, $\Delta_J \sim \mathrm{N}(0, \sigma^2)$ where the variance satisfies $\sigma^2 \leq 1/8$. As $1/2\sigma^2 - 2 \geq 1/4\sigma^2$,

$$\mathbb{E}\left[\mathrm{e}^{2\Delta_J(\boldsymbol{x},\boldsymbol{y})^2}\right] = \frac{1}{\sqrt{2\pi\sigma^2}} \int_{\mathbb{R}} \mathrm{e}^{2x^2} \mathrm{e}^{-x^2/2\sigma^2}\, dx \leq \frac{1}{\sqrt{2\pi\sigma^2}} \int_{\mathbb{R}} \mathrm{e}^{-x^2/4\sigma^2}\, dx = \sqrt{2}.$$

We conclude that

$$\mathbb{E}\left[\sup_{1 \leq j \leq J} \mathrm{e}^{2\Delta_j(\boldsymbol{x},\boldsymbol{y})^2}\right] \leq \sqrt{2}, \qquad \forall \boldsymbol{x}, \boldsymbol{y} \in \bar{D}.$$

Let

$$M_1(u) := \int_D \int_D \sup_{J \in \mathbb{N}} \mathrm{e}^{2\Delta_J(\boldsymbol{x},\boldsymbol{y})^2}\, d\boldsymbol{x}\, d\boldsymbol{y}.$$

As D is bounded, $M_1 \in L^1(\Omega)$. Therefore, in the notation of Lemma 7.67, we have

$$\int_{B(\boldsymbol{x}_0,r)} \int_{B(\boldsymbol{y}_0,r)} \exp\left[\frac{2|u_J(\boldsymbol{x},\omega) - u_J(\boldsymbol{y},\omega)|^2}{8Lr^s}\right] d\boldsymbol{x}\, d\boldsymbol{y} \leq M_1(\omega)$$

for any $B(\boldsymbol{x}_0,r), B(\boldsymbol{y}_0,r) \subset D$ and almost all $\omega \in \Omega$. Let $K := \max\{1,(1/p)\log M_1(u)\}$. Then $\mathrm{e}^K \in L^p(\Omega)$ and Lemma 7.67 provides

$$\frac{1}{2L^{1/2}}|u_J(\boldsymbol{x},\omega) - u_J(\boldsymbol{y},\omega)| \le K(\omega)\|\boldsymbol{x}-\boldsymbol{y}\|_2^{(s-\epsilon)/2}, \qquad \forall \boldsymbol{x},\boldsymbol{y} \in \bar{D}.$$

As $u_J(\boldsymbol{x}) \to u(\boldsymbol{x})$ as $J \to \infty$ almost surely, this gives (7.74). □

Example 7.69 (Brownian sheet) Let $\{\beta(\boldsymbol{x})\colon \boldsymbol{x} \in \mathbb{R}^+ \times \mathbb{R}^+\}$ be a Brownian sheet and consider $\boldsymbol{x} = [x_1,x_2]^\mathsf{T}, \boldsymbol{y} = [y_1,y_2]^\mathsf{T} \in \mathbb{R}^+ \times \mathbb{R}^+$. Then, $\beta(\boldsymbol{x}) - \beta(\boldsymbol{y}) = \beta(\boldsymbol{x}) - \beta(\boldsymbol{z}) + \beta(\boldsymbol{z}) - \beta(\boldsymbol{y})$ for $\boldsymbol{z} = [y_1,x_2]^\mathsf{T}$. We have $\beta(\boldsymbol{x}) - \beta(\boldsymbol{z}) \sim \mathrm{N}(0,x_2|x_1-y_1|)$ and $\beta(\boldsymbol{z}) - \beta(\boldsymbol{y}) \sim \mathrm{N}(0,y_1|x_2-y_2|)$ from Exercise 7.2. Now,

$$\mathbb{E}\left[|\beta(\boldsymbol{x}) - \beta(\boldsymbol{y})|^2\right] \le 2x_2|x_1-y_1| + 2y_1|x_2-y_2|.$$

Hence, (7.73) holds with $s = 1$ on any bounded subset of $D \subset \mathbb{R}^+ \times \mathbb{R}^+$. We conclude that realisations of the Brownian sheet (see Figure 7.1) are almost surely Hölder continuous for any exponent $\gamma < 1/2$.

In many applications (e.g., geostatistical modelling), we often encounter log-normal random fields $v(\boldsymbol{x}) = \mathrm{e}^{u(\boldsymbol{x})}$, where $u(\boldsymbol{x})$ is an underlying Gaussian random field. Our final result says, for example, that if u satisfies the conditions of Theorem 7.68, then the maximum value of $\mathrm{e}^{u(\boldsymbol{x})}$ has finite pth moments.

Corollary 7.70 *If the conditions of Theorem 7.68 hold, $\mathrm{e}^{\lambda u} \in L^p(\Omega, \mathrm{C}(\bar{D}))$ for any $p \ge 1$ and $\lambda \in \mathbb{R}$.*

Proof By Theorem 7.68, $\mathrm{e}^{\lambda u} \in \mathrm{C}(\bar{D})$ almost surely. Fix $p \ge 1$ and $\lambda \in \mathbb{R}$. Let $Z = \sup_{\boldsymbol{x}\in D} \mathrm{e}^{\lambda p(u(\boldsymbol{x})-u(\boldsymbol{0}))}$. Note that

$$\left\|\sup_{\boldsymbol{x}\in D} \mathrm{e}^{\lambda u(\boldsymbol{x})}\right\|_{L^p(\Omega)}^p = \mathbb{E}\left[Z\mathrm{e}^{\lambda p u(\boldsymbol{0})}\right] \le \|Z\|_{L^2(\Omega)}\|\mathrm{e}^{\lambda p u(\boldsymbol{0})}\|_{L^2(\Omega)}. \tag{7.75}$$

As $\mathrm{e}^{\lambda p u(\boldsymbol{0})}$ is log-normal, its mean and variance are finite and $\mathrm{e}^{\lambda p u(\boldsymbol{0})} \in L^2(\Omega)$.

Let $R \ge 1$ be such that $D \subset B(\boldsymbol{0},R)$. Choose $K(\omega)$ such that (7.74) holds and $\mathrm{e}^K \in L^q(\Omega)$ for $q = 2\lambda p|R|^{(q-\epsilon)/2} \ge 1$. Then, for $\boldsymbol{x} \in D$,

$$\mathrm{e}^{\lambda p(u(\boldsymbol{x})-u(\boldsymbol{0}))} \le \mathrm{e}^{\lambda p K(\omega)\|\boldsymbol{x}\|_2^{(q-\epsilon)/2}} \le \mathrm{e}^{\lambda p K(\omega)|R|^{(q-\epsilon)/2}} = \mathrm{e}^{qK(\omega)/2}$$

and $\mathbb{E}\left[Z^2\right] \le \mathbb{E}\left[\mathrm{e}^{qK}\right] < \infty$. As both terms in the product in (7.75) are finite, we conclude that $\sup_{\boldsymbol{x}\in D} \mathrm{e}^{\lambda u(\boldsymbol{x})}$ is in $L^p(\Omega)$. □

Derivatives of realisations of log-normal random fields are used in Chapter 9.

Theorem 7.71 *Suppose that the conditions of Theorem 7.68 hold and the covariance C of the random field $\{u(\boldsymbol{x})\colon x \in \bar{D}\}$ belongs to $\mathrm{C}^3(\bar{D} \times \bar{D})$. For almost all $\omega \in \Omega$, realisations $\mathrm{e}^{u(\cdot,\omega)}$ belong to $\mathrm{C}^1(\bar{D})$ and*

$$\left\|\frac{\partial \mathrm{e}^{u(\cdot,\omega)}}{\partial x_i}\right\|_\infty \le \left\|\frac{\partial u(\cdot,\omega)}{\partial x_i}\right\|_\infty \|\mathrm{e}^{u(\cdot,\omega)}\|_\infty, \qquad i = 1,\dots,d, \tag{7.76}$$

where $\frac{\partial u(\cdot,\omega)}{\partial x_i}$ and $\mathrm{e}^{u(\cdot,\omega)}$ belong to $L^p(\Omega, \mathrm{C}(\bar{D}))$ for any $p \ge 1$.

Proof Since the covariance $C \in \mathrm{C}^3(\bar{D} \times \bar{D})$, the Gaussian random field u is mean-square differentiable by Theorem 7.4. The mean-square derivative $\frac{\partial u}{\partial x_i}$ is then a mean-zero Gaussian random field with covariance

$$C_i(\boldsymbol{x}, \boldsymbol{y}) = \frac{\partial^2}{\partial x_i \partial y_i} C(\boldsymbol{x}, \boldsymbol{y}).$$

We have $C_i \in \mathrm{C}^1(\bar{D} \times \bar{D})$ and hence C_i is uniformly Lipschitz continuous on the bounded domain D so that Theorem 7.68 applies. Then, $|\frac{\partial z}{\partial x_i}|$ is uniformly bounded over $\boldsymbol{x} \in \bar{D}$ and (7.76) follows as $\frac{\partial \mathrm{e}^u}{\partial x_i} = \frac{\partial u}{\partial x_i}\mathrm{e}^u$. We see the two factors belong to $L^p(\Omega)$ by following the arguments in Corollary 7.70. □

7.6 Notes

Gaussian random fields and methods for their approximation are covered in (Abrahamsen, 1997; Schlather, 1999). Schlather (2001) describes a comprehensive package for simulating random fields in the programming language R. A further approach to sampling the Gaussian random fields with the Whittle–Matérn covariance using a stochastic PDE is described in Lindgren, Rue, and Lindström (2011). We have introduced a number of simple covariance models; a detailed discussion of models appropriate for earth sciences and empirical models based on experimental observations are given in Christakos (2005). Frisch (1995) develops random fields in the context of turbulence.

The two-dimensional circulant embedding method outlined in §7.2 is described in (Dietrich and Newsam, 1997; Wood and Chan, 1994) and properties of BTTB and BCCB matrices are discussed in (Davis, 1979; Vogel, 2002). The term *even circulant extension* from Definitions 7.33 and 7.35 is not standard. For brevity, we have described an embedding method that yields BCCB extension matrices for any given BTTB matrix. We have not made assumptions about symmetry or block-symmetry. If the BTTB matrix of interest is symmetric and has symmetric blocks, minimal embedding can be applied to each of the n_2 distinct blocks (as in Chapter 6) and a smaller BCCB matrix of dimension $4(n_1 - 1)(n_2 - 1)$ can be formed.

For stationary covariance functions $c(\boldsymbol{x})$ and regularly spaced grid points, the covariance matrix C is always a symmetric BTTB matrix. C only has symmetric blocks if c is even in both coordinates. The separable exponential covariance belongs to this class. For the exponential covariance in one dimension, minimal embedding always leads to a non-negative definite extension matrix for any choice of discretisation parameter and correlation length. In two dimensions, the separable exponential covariance $c(\boldsymbol{x}) = \mathrm{e}^{-|x_1|/\ell_1 - |x_2|/\ell_2}$ leads to a symmetric BTTB covariance $C = T_2 \otimes T_1$ where T_2 and T_1 are the symmetric Toeplitz covariance matrices corresponding to the one-dimensional covariances $c_2(x_2) = \mathrm{e}^{-|x_2|/\ell_2}$ and $c_1(x_1) = \mathrm{e}^{-|x_1|/\ell_1}$ respectively. If we apply minimal embedding to each Toeplitz block of C and form the corresponding minimal BCCB extension $\widetilde{C}$ of dimension $4(n_1 - 1)(n_2 - 1)$ then, since $\widetilde{C} = \widetilde{T}_2 \otimes \widetilde{T}_1$, where $\widetilde{T}_1$ and $\widetilde{T}_2$ are the minimal extensions of T_1 and T_2, it follows that $\widetilde{C}$ is always non-negative definite. Note then that the embedding we have described, with additional zero elements, always yields a BCCB matrix, but is not necessarily the most efficient embedding for a given covariance function. An alternative type of even BCCB extension which replaces the zero blocks with averages of certain other circulant blocks, is described in Dietrich and Newsam (1997).

References for the turning bands method include (Dietrich, 1995; Mantoglou and Wilson, 1982; Matheron, 1973). Gneiting (1998) provides closed-form expressions for $T_2 c^0(r)$ for some common covariance models (including the Whittle–Matérn family). The forms involve special functions (the Struve function) and are difficult to work with numerically. Often, it is more convenient to work with a random field in $d = 3$ dimensions and project down to $d = 2$ by choosing a suitable slice. The method with uniformly spaced bands that we developed for $d = 2$ can be extended to $d = 3$ with a suitable quadrature method for integrals on the sphere (see Hesse, Sloan, and Womersley (2010)). Although the version presented here gives a non-isotropic random field, offsetting the initial vector by a random angle with distribution $\mathrm{U}(0, \pi/M)$ corrects this (Schlather, 1999).

Studies of the asymptotics of the eigenvalues in (7.47) date back to Weyl (1912). Decay rates for the eigenvalues associated with a wide variety of kernels $C(\boldsymbol{x}, \boldsymbol{y})$ and domains D can be found in (Frauenfelder, Schwab, and Todor, 2005; Hille and Tamarkin, 1931; König, 1986; Reade, 1983; Weyl, 1912). In particular, a discussion of the Gaussian covariance $c^0(r) = \mathrm{e}^{-r^2}$ can be found in Frauenfelder et al. (2005).

Theorem 7.57 is a special case of one given in Widom (1963) that applies only in the isotropic case. Widom (1963) considers a family of integral operators, where the kernel $c(\boldsymbol{x} - \boldsymbol{y})$ has spectral density $f(\boldsymbol{\nu})$. Consider a bounded domain D and a function $K\colon \mathbb{R}^d \to \mathbb{R}^+$ such that $K(\boldsymbol{\nu}) \asymp (2\pi)^d f(\boldsymbol{\nu})$ as $\|\boldsymbol{\nu}\|_2 \to \infty$. If $A(\epsilon) := \mathrm{Leb}(D)\mathrm{Leb}(\{\boldsymbol{\nu} \in \mathbb{R}^d \colon K(\boldsymbol{\nu}) > \epsilon\})$ and $\phi_0(x)$ is a function such that $\mathrm{Leb}(\{x > 0\colon \phi_0(x) > \epsilon\}) = A(\epsilon)$ for $\epsilon > 0$, then the jth eigenvalue $\nu_j \asymp \phi_0((2\pi)^d j)$ as $j \to \infty$ for a large class of integral operators. We can take $\phi_0(x) = A^{-1}(x)$ if the inverse exists. Under the assumptions of Theorem 7.57, $K(\boldsymbol{\nu}) = (2\pi)^d h^0(\|\boldsymbol{\nu}\|_2)$. As V_d is the volume of the unit sphere,

$$\mathrm{Leb}\big(\{\boldsymbol{\nu} \in \mathbb{R}^d \colon h^0(\boldsymbol{\nu}) > \epsilon\}\big) = \mathrm{Leb}\big(\{\boldsymbol{\nu} \in \mathbb{R}^d \colon \|\boldsymbol{\nu}\|_2 < R(\epsilon)\}\big) = V_d R(\epsilon)^d$$

if R denotes the inverse of h^0. Then, $A(\epsilon) = \mathrm{Leb}(D) V_d R((2\pi)^{-d}\epsilon)^d$ and the inverse of A is given by

$$A^{-1}(j) = (2\pi)^d h^0\left(\left(\frac{j}{\mathrm{Leb}(D)V_d}\right)^{1/d}\right).$$

Widom's theorem states that the jth eigenvalue, which is given in Theorem 7.57, satisfies

$$\nu_j \asymp (2\pi)^d h^0\left(2\pi\left(\frac{j}{\mathrm{Leb}(D)V_d}\right)^{1/d}\right) \qquad \text{as } j \to \infty.$$

Theorem 7.61 can be generalised to the separable exponential in dimension $d > 2$.

The eigenvalue problem (7.47) is a special case of a Fredholm integral equation of the second kind, for which several classical numerical methods exist. We have briefly mentioned quadrature, collocation and Galerkin methods, and have discussed only simple implementations thereof. We applied the methods to an example with the separable exponential covariance kernel, which has multiple eigenvalues. It should be noted that superior approximations are often achieved for problems with simple, well-separated eigenvalues. More details about the numerical solution of integral eigenvalue problems, including the error analysis, can be found in (Baker, 1978; Chatelin, 1983; Delves and Walsh, 1974).

Iterative methods for solving matrix eigenvalue problems $Ax = \lambda x$, such as the Lanczos or Arnoldi iteration, require matrix-vector products with A at each step. For dense $N \times N$ matrices A, these methods are only efficient if the cost of a matrix-vector product is significantly less than $\mathcal{O}(N^2)$. If A has some special structure, this may be possible. For example, matrix-vector products with BTTB matrices can be performed via an FFT, by first applying circulant embedding. However, the matrix eigenvalue problems that result from discretisations of (7.47) have BTTB matrices A only for very simple quadrature schemes and uniformly spaced sample points. Sometimes A can be approximated well by a matrix with a special structure, for which matrix-vector products can be performed efficiently, but this introduces an additional error. Fast multipole methods are developed in Schwab and Todor (2006) which reduce the cost of matrix-vector products with certain matrices to $\mathcal{O}(N \log N)$.

Finally, for more information on the regularity of Gaussian random fields, see Adler (1981). Lemma 7.67 is a version of an inequality given in Garsia, Rodemich, and Rumsey (1971). Corollary 7.70 is implied by a much stronger theorem, known as the Fernique theorem, which says that if u is a separable Banach space-valued Gaussian random variable then $\mathbb{E}\left[e^{\alpha\|u\|^2}\right] < \infty$ for some $\alpha > 0$. See Da Prato and Zabczyk (1992).

Exercises

7.1 Let $D = \mathbb{R}^+ \times \mathbb{R}^+$ and define

$$W_N(\boldsymbol{x}) := \frac{1}{N} \sum_{j=1}^{\lfloor x_1 N \rfloor} \sum_{k=1}^{\lfloor x_2 N \rfloor} \xi_{jk}, \qquad \boldsymbol{x} = \begin{pmatrix} x_1 \\ x_2 \end{pmatrix} \in D, \tag{7.77}$$

for $\xi_{jk} \sim \mathrm{N}(0,1)$ *iid* and $N \in \mathbb{N}$.

a. Show that $\{W_N(\boldsymbol{x}) \colon \boldsymbol{x} \in D\}$ is a mean-zero Gaussian random field and that, if $N\boldsymbol{x}, N\boldsymbol{y} \in \mathbb{N}^2$, then $\mathbb{E}[W_N(\boldsymbol{x})W_N(\boldsymbol{y})] = \min\{x_1, x_2\}\min\{y_1, y_2\}$.

b. Using (7.77), show how to approximate realisations of the Brownian sheet. Implement and comment on the accuracy of the method.

7.2 Consider the Brownian sheet $\{\beta(\boldsymbol{x}) \colon \boldsymbol{x} \in \mathbb{R}^+ \times \mathbb{R}^+\}$. Show that

a. $\beta(\boldsymbol{0}) = 0$ a.s.

b. If $\boldsymbol{x}, \boldsymbol{y} \in \mathbb{R}^+ \times \mathbb{R}^+$, then

$$\beta(\boldsymbol{x}) - \beta(\boldsymbol{y}) \sim \mathrm{N}\big(0, x_1 x_2 + y_1 y_2 - 2(x_1 \wedge x_2)(y_1 \wedge y_2)\big).$$

c. For $\boldsymbol{x}, \boldsymbol{y}, \boldsymbol{z}, \boldsymbol{w} \in \mathbb{R}^+ \times \mathbb{R}^+$, the increments $\beta(\boldsymbol{y}) - \beta(\boldsymbol{x})$ and $\beta(\boldsymbol{z}) - \beta(\boldsymbol{w})$ are independent Gaussian random variables if $\{\boldsymbol{x}, \boldsymbol{y}\}$ and $\{\boldsymbol{z}, \boldsymbol{w}\}$ are the opposite corners of disjoint rectangles.

7.3 Let $\beta(\boldsymbol{x})$ denote a Brownian sheet in two dimensions and assume all realisations are continuous functions of $\boldsymbol{x}$. Fix $x > 0$ and define

$$X(t) := \frac{\beta(x,t)}{\sqrt{x}}, \qquad t \in \mathbb{R}^+. \tag{7.78}$$

Show that $X(t)$ is a Brownian motion.

7.4 Let $\{u(\boldsymbol{x}) : \boldsymbol{x} \in \mathbb{R}^d\}$ be a random field with stationary covariance $c(\boldsymbol{x})$. Show that

$$c(\boldsymbol{x}) = c\begin{pmatrix} x_1 \\ \vdots \\ x_d \end{pmatrix} = c\begin{pmatrix} -x_1 \\ \vdots \\ -x_d \end{pmatrix} = c(-\boldsymbol{x}).$$

If in addition $u(\boldsymbol{x})$ is isotropic, show that (for any combination of $\pm$)

$$c(\boldsymbol{x}) = c\left(\left[\pm x_1, \dots, \pm x_d\right]^{\mathsf{T}}\right).$$

7.5 If C_1, C_2 are covariance functions on $D_1 \subset \mathbb{R}^{d_1}$ and $D_2 \subset \mathbb{R}^{d_2}$ respectively, show that

$$C(\boldsymbol{x}, \boldsymbol{y}) = C_1(\boldsymbol{x}_1, \boldsymbol{y}_1) C_2(\boldsymbol{x}_2, \boldsymbol{y}_2), \qquad \boldsymbol{x} = \begin{pmatrix} \boldsymbol{x}_1 \\ \boldsymbol{x}_2 \end{pmatrix}, \ \boldsymbol{y} = \begin{pmatrix} \boldsymbol{y}_1 \\ \boldsymbol{y}_2 \end{pmatrix},$$

for $\boldsymbol{x}_1, \boldsymbol{y}_1 \in D_1$ and $\boldsymbol{x}_2, \boldsymbol{y}_2 \in D_2$, is a valid covariance on $D_1 \times D_2$.

7.6 Using Proposition 6.45 and Proposition 7.29 show that the eigenvalues of an $N \times N$ BCCB matrix C, where $N = n_1 n_2$, are given by

$$\boldsymbol{d}_j = \sum_{k=0}^{n_2-1} \omega_2^{-jk} \boldsymbol{\sigma}_k, \qquad j = 0, 1, \dots, n_2 - 1, \tag{7.79}$$

where $\boldsymbol{\sigma}_k$ contains the eigenvalues of the circulant block C_k.

7.7 Consider the symmetric BCCB matrix C in Example 7.24 with C_1 chosen as in (7.20) so C has symmetric blocks.

a. Compute the eigenvalues of C as in Example 7.30 using the Matlab command `ifft2`.
b. Use (7.79) to find the eigenvalues of C in terms of the eigenvalues of the individual circulant blocks C_0, C_1, and C_1^{T}. Explain why there are only four distinct eigenvalues.

7.8 Let $c \colon \mathbb{R}^2 \to \mathbb{R}$ be the stationary covariance function of a random field $u(\boldsymbol{x})$ on $D \subset \mathbb{R}^2$.

a. Show that if c is even in x_1 then it is even in x_2 and vice versa.
b. Show that if c is even in x_1 then the BTTB covariance matrix C associated with the uniformly spaced sample points in Definition 7.37 has symmetric blocks.

7.9 Investigate the Matlab command `fftshift`. Use it to generate the reduced matrix $\tilde{C}_{\text{red}}$ in (7.26) for the 24×24 BCCB extension $\tilde{C}$ of the 6×6 BTTB matrix C with reduced matrix C_{red} given in (7.25).

7.10 Using Algorithm 7.2, repeat Example 7.40 for the separable exponential covariance with $d = 2$, and $\ell_1 = 1 = \ell_2$. Show that the 5×3 reduced covariance represents a symmetric BTTB matrix with symmetric blocks.

7.11 Consider the Gaussian field $u(\boldsymbol{x})$ in two dimensions with mean zero and isotropic covariance $c^0(r) = \mathrm{e}^{-r/\ell}$.

a. Write a MATLAB routine similar to Algorithm 7.3 that uses Algorithm 7.2 to generate the associated reduced covariance matrix for a given set of uniformly spaced sample points.

b. Suppose we want to sample $u(\boldsymbol{x})$ on a uniformly spaced grid on $D = [0,1]^2$ with $n_1 = 257 = n_2$ and $\Delta x_1 = 1/256 = \Delta x_2$. Consider the cases $\ell = 1/10, 1, 10$. Is the even BCCB extension non-negative definite? If not, investigate the use of padding with Algorithm 7.6.

7.12 Let $c^0(r)$ be an isotropic covariance function in $\mathbb{R}^d$. Show that it is also a valid isotropic covariance function in $\mathbb{R}^{d-1}$. Hence, show that $J_{(n-2)/2}(r)/r^{(n-2)/2}$ is a valid isotropic covariance in $\mathbb{R}^d$ for $n \geq d$.

7.13 Let $\{u(\boldsymbol{x})\colon \boldsymbol{x} \in \mathbb{R}^d\}$ be the mean-zero Gaussian random field with stationary covariance $\mathrm{e}^{-\boldsymbol{x}^{\mathsf{T}} A \boldsymbol{x}}$ for a symmetric positive-definite $d \times d$ matrix A. By introducing a suitable change of coordinates, show how to write $u(\boldsymbol{x})$ in terms of an isotropic random field.

7.14 Implement the circulant embedding and turning bands methods for the Gaussian covariance $c(\boldsymbol{x}) = \mathrm{e}^{-\boldsymbol{x}^{\mathsf{T}} A \boldsymbol{x}}$ for a 2×2 symmetric positive-definite matrix A. Discuss the approximation errors in each case. State any extra assumptions needed for the turning band method.

7.15 Show that $c^0(r) = \sin r/r$ is a valid isotropic covariance function in $\mathbb{R}^2$. Show how to generate realisations of the mean-zero Gaussian random field in $\mathbb{R}^2$ with isotropic covariance $\sin(r)/r$ using the turning bands method.

7.16 Solve the second-order ODE in Example 7.55 to find the given expressions for the eigenvalues and normalised eigenfunctions of the exponential covariance.

7.17 Using only the property $x = W(x)\mathrm{e}^{W(x)}$ of the Lambert W function, show that

$$\int_J^\infty \mathrm{e}^{-2W(x)}\,dx = \big(2 + W(J)\big)\mathrm{e}^{-W(J)} \leq 2\frac{\log^2 J}{J}, \qquad \text{for } J \geq \mathrm{e}^2.$$

7.18 Write a MATLAB routine to compute the eigenvalues of the one-dimensional exponential covariance in Example 7.55.

a. For $\ell = 1/10, 1, 10$, and $D = [-1,1]$, how many terms needed to be retained in the Karhunen–Loève expansion to achieve $E_J \leq 10^{-2}$ in (7.46)?

b. Repeat the experiment for the two-dimensional exponential covariance with $\ell_1 = \ell_2 = \ell = 1/10, 1, 10$, and $D = [-1,1] \times [-1,1]$. How many terms need to be retained in the Karhunen–Loève expansion to achieve $E_J \leq 10^{-1}$ in (7.46)?

7.19 Write a MATLAB routine using the inbuilt function `eig` to implement the quadrature scheme in Example 7.62.

a. Apply the scheme with $n = 32$ for the separable exponential covariance with $\ell_1 = 1 = \ell_2$ on $D = [-1,1]^2$. Plot the first few eigenvectors and compare with the exact eigenfunctions, sampled at the quadrature points. What do you observe and why?

b. Modify your MATLAB code so that it approximates only the largest 20 eigenvalues and corresponding eigenvectors for the matrix problem in (a) with the iterative method `eigs`. Comment on the efficiency and accuracy of the new method.

c. Using (7.57) and your MATLAB code with $n = 64$, generate a realisation of the approximate Karhunen–Loève expansion of the Gaussian random field $u(\boldsymbol{x})$ with mean-zero and separable exponential covariance with $\ell_1 = 1 = \ell_2$ on $D = [-1,1]^2$, for $J = 5$, 20, 50, and 100. How do these realisations compare with those of the exact Karhunen–Loève expansion shown in Figure 7.19?

7.20 Write a MATLAB routine to implement the quadrature scheme in Example 7.63. Run your code with $n = 8$, 16, 32, and 64 for the separable exponential covariance as in Example 7.62 and compute the relative errors in the eigenvalues ν_1, ν_{10}, and ν_{100}.

7.21 Let $\{u(\boldsymbol{x}) \colon \boldsymbol{x} \in \mathbb{R}^d\}$ be a mean-zero Gaussian random field with isotropic covariance $c^0(r)$. Show that if $|c^0(r) - c^0(0)| \le K r^q$ for all $r \ge 0$ and some $q, K > 0$, then $u(\boldsymbol{x})$ is continuous.

7.22 With $A(\boldsymbol{x}_0, \boldsymbol{y}_0, r, K)$ defined by (7.72), prove that, for any $\lambda > 0$, there exists $r^* > 0$ such that $A(\boldsymbol{x}_0, \boldsymbol{y}_0, r, K) \ge e^{\lambda K}$ for all $K \ge 1$ and $0 < r < r^*$. Use Jensen's inequality.

7.23 Let $\{X(t) \colon t \ge 0\}$ be a fBm with Hurst exponent H. Show that $X(t)$ is almost surely Hölder continuous with exponent $0 < q < H$.

7.24 Show that if a Gaussian random field $u(\boldsymbol{x})$ satisfies (7.73), then Theorem 7.53 applies.

8
Stochastic Ordinary Differential Equations

A stochastic ordinary differential equation (SODE) is an ordinary differential equation with a random forcing, usually given by a white noise $\zeta(t)$. White noise is chosen so that the random forces $\zeta(t)$ are uncorrelated at distinct times t. For example, adding noise to the ODE $du/dt = -\lambda u$, we consider the SODE

$$\frac{du}{dt} = -\lambda u + \sigma\zeta(t), \qquad u(0) = u_0 \tag{8.1}$$

for parameters $\lambda, \sigma > 0$ and an initial condition $u_0 \in \mathbb{R}$. As we saw in §6.3, $\zeta(t) = dW(t)/dt$ for a Brownian motion $W(t)$ and we rewrite (8.1) in terms of $W(t)$ by integrating over $[0, t]$. Consider

$$u(t) = u_0 - \lambda \int_0^t u(s)\, ds + \sigma W(t), \tag{8.2}$$

which is written in short as

$$du = -\lambda u\, dt + \sigma dW(t), \qquad u(0) = u_0.$$

The solution is a stochastic process $\{u(t) \colon t \geq 0\}$ such that (8.2) holds for $t \geq 0$ and is known as the *Ornstein–Uhlenbeck process*.

More generally, we introduce a vector-valued function $\boldsymbol{f} : \mathbb{R}^d \to \mathbb{R}^d$, known as the drift, and a matrix-valued function $G \colon \mathbb{R}^d \to \mathbb{R}^{d \times m}$, known as the diffusion, and consider SODEs of the form

$$\boldsymbol{u}(t) = \boldsymbol{u}_0 + \int_0^t \boldsymbol{f}(\boldsymbol{u}(s))\, ds + \int_0^t G(\boldsymbol{u}(s))\, d\boldsymbol{W}(s), \tag{8.3}$$

also written for brevity as

$$d\boldsymbol{u} = \boldsymbol{f}(\boldsymbol{u})\, dt + G(\boldsymbol{u})\, d\boldsymbol{W}(t), \qquad \boldsymbol{u}(0) = \boldsymbol{u}_0,$$

where $\boldsymbol{u}_0 \in \mathbb{R}^d$ is the initial condition and $\boldsymbol{W}(t) = [W_1(t), \dots, W_m(t)]^\mathsf{T}$ for *iid* Brownian motions $W_i(t)$. The last term in (8.3) is a *stochastic integral* and it needs careful definition. When $G(\boldsymbol{u})$ is independent of $\boldsymbol{u}$ (such as (8.3) for $G(u) = \sigma$), we say the SODE has *additive noise* and the stochastic integral is simply $\int_0^t G\, d\boldsymbol{W}(s) = G\, \boldsymbol{W}(t)$. The case where $G(\boldsymbol{u})$ varies with $\boldsymbol{u}$, such as Example 8.4, is called *multiplicative noise* and the integrand $G(\boldsymbol{u}(s))$ is random. The stochastic integrals in Chapter 6 are for deterministic integrands and do not include this case. The proper interpretation of the stochastic integral is a delicate issue and must be understood when modelling a physical situation, as the integral $\int_0^t G(\boldsymbol{u}(s))\, d\boldsymbol{W}(s)$ and hence the solution $\boldsymbol{u}(t)$ of the SODE depends on the interpretation.

We prefer the Itô integral, which corresponds to modelling the Brownian increment $d\boldsymbol{W}(s)$ as independent of the current state $\boldsymbol{u}(s)$. The Itô integral is often used in mathematics and financial modelling as it leads to martingales (see Theorem 8.14). An alternative is to 'smooth' the differential equation, replacing $\boldsymbol{W}(t)$ by continuously differentiable approximations so that the integrals are defined by conventional calculus and a stochastic integral is defined in a limit. This approach leads to the *Stratonovich* integral, usually denoted

$$\int_0^t G(\boldsymbol{u}(s)) \circ d\boldsymbol{W}(s),$$

and to Stratonovich SODEs. Such an approach is often appropriate in physical modelling, where Brownian motion is used to model a driving noise process that is, in reality, smoother than Brownian motion.

We develop the Itô stochastic integral and calculus in §8.2. In §8.3, we introduce the Itô formula and study geometric Brownian motion, an example from financial modelling, and prove the existence and uniqueness of solutions to the initial-value problem for Itô SODEs subject to Lipschitz conditions on the drift and diffusion.

In §8.4 and §8.6, the numerical approximation by finite difference methods is addressed. The simplest such method is the *Euler–Maruyama* method, which is the stochastic version of the Euler method (3.12). We also look at a higher-order method called *Milstein's method.* We prove strong convergence for Milstein's method and examine stability of the Euler–Maruyama method. We show how to implement both methods in MATLAB and examine the $L^2(\Omega)$ approximation error numerically. For weak approximation, we describe the Monte Carlo method and a multilevel Monte Carlo method for variance reduction. A discussion of the corresponding ideas for Stratonovich SODEs is given in §8.7.

8.1 Examples of SODEs

We begin by presenting four examples of SODEs. The first three describe the dynamics of a particle in a mechanical system subject to noise. We denote the momentum by P, the position by Q, and the system energy of the particle by $H(Q,P)$. The fourth example, geometric Brownian motion, models the fluctuation of an asset price on the stock market.

Example 8.1 (Ornstein–Uhlenbeck process) Consider a particle of unit mass moving with momentum $P(t)$ at time t in a gas, subject to irregular bombardment by the ambient gas particles. The dynamics of the particle can be modelled by a dissipative force $-\lambda P(t)$, where $\lambda > 0$ is known as the dissipation constant, and a fluctuating force $\sigma\zeta(t)$, where $\zeta(t)$ is a white noise and $\sigma > 0$ is the diffusion constant. Newton's second law of motion gives the acceleration dP/dt as the sum of the two forces. We have

$$\frac{dP}{dt} = -\lambda P + \sigma\,\zeta(t), \tag{8.4}$$

which is the linear SODE (8.1) with additive noise $G = \sigma$. It is also written as

$$dP = -\lambda P\,dt + \sigma\,dW(t). \tag{8.5}$$

We usually consider the initial-value problem with $P(0) = P_0$ and the solution $P(t)$ is known as the *Ornstein–Uhlenbeck process*.

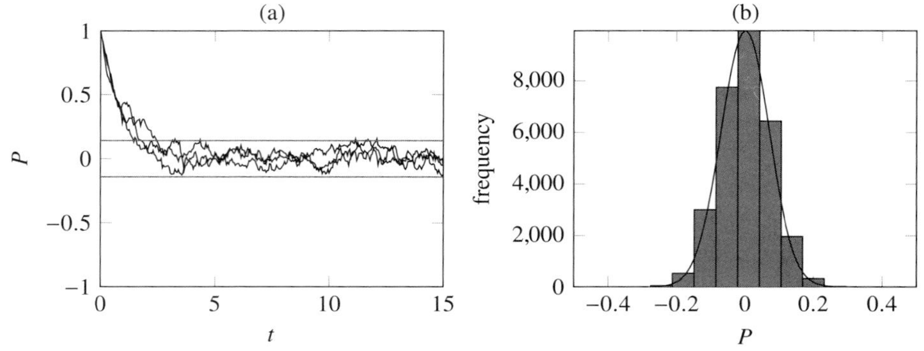

Figure 8.1 (a) Sample paths of the Euler–Maruyama approximation for (8.4) with $P_0 = 1$, $\lambda = 1$, and $\sigma = 1/2$ with $\Delta t = 0.05$. As $t \to \infty$, $P(t) \to \mathrm{N}(0, \sigma^2/2\lambda)$ in distribution and the horizontal lines mark the 90% confidence interval $\left[-2\sqrt{\sigma^2/2\lambda}, +2\sqrt{\sigma^2/2\lambda}\right]$. (b) A histogram of $P(t_n)$ for $t_n \in [10, 1000]$ and the shape is similar to the Gaussian pdf (also shown).

The variation of constants formula also applies for SODEs (see Example 8.21 and Exercise 8.7) and, in the case of (8.5), says that

$$P(t) = \mathrm{e}^{-\lambda t} P_0 + \sigma \int_0^t \mathrm{e}^{-\lambda(t-s)}\, dW(s), \tag{8.6}$$

where the second term is a stochastic integral (see Definition 6.22). We use this to derive an important relationship in statistical physics.

By definition, the expected kinetic energy per degree of freedom of a system at temperature T is given by $k_\mathrm{B} T/2$, where k_B denotes the Boltzmann constant. In particular, the temperature of a system of particles in thermal equilibrium can be determined from λ and σ. For the Ornstein–Uhlenbeck model, the expected kinetic energy $\mathbb{E}\left[P(t)^2/2\right]$ is easily calculated using Proposition 6.23:

$$\begin{aligned} \mathbb{E}\left[P(t)^2\right] &= \mathrm{e}^{-2\lambda t} P_0^2 + \sigma^2 \int_0^t \mathrm{e}^{-2\lambda(t-s)}\, ds \\ &= \mathrm{e}^{-2\lambda t} P_0^2 + \frac{\sigma^2}{2\lambda}\left(1 - \mathrm{e}^{-2\lambda t}\right) \to \frac{\sigma^2}{2\lambda} \qquad \text{as } t \to \infty. \end{aligned} \tag{8.7}$$

The expected kinetic energy $\mathbb{E}\left[\frac{1}{2}P(t)^2\right]$ converges to $\sigma^2/4\lambda$ as $t \to \infty$. Thus, $\sigma^2/4\lambda$ is the equilibrium kinetic energy and the equilibrium temperature is given by $k_\mathrm{B} T = \sigma^2/2\lambda$, which is known as the fluctuation–dissipation relation.

Taking this further, it can be shown that

$$P(t) \to \mathrm{N}(0, \sigma^2/2\lambda) \qquad \text{in distribution as } t \to \infty$$

(see Figure 8.1). In physics, the limiting distribution is known as the Gibbs canonical distribution and can be written $\mathrm{N}(0, k_\mathrm{B} T)$ with probability density function (pdf) $p(Q, P) = \mathrm{e}^{-\beta H(Q,P)}/Z$, where $\beta = 1/(k_\mathrm{B} T)$ is the inverse temperature, Z is a normalisation constant, and $H(Q, P)$ is the system energy for a particle with position Q and momentum P (in this case $H = P^2/2$).

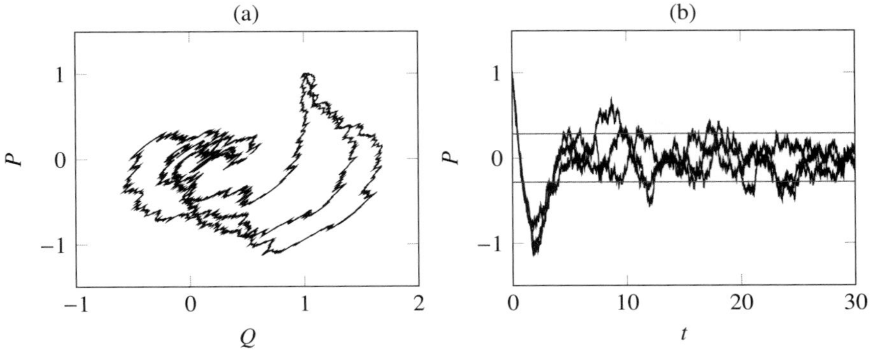

Figure 8.2 (a) Phase-plane plot of three approximate sample paths for the Langevin equation (8.8) with initial data $[Q(0), P(0)]^{\mathsf{T}} = [1, 1]^{\mathsf{T}}$ and parameters $\lambda = 1$, $\sigma = 1/5$ in the case $V(Q) = Q^2/2$. (b) The same sample paths plotted as a time series $P(t)$ against t, with horizontal lines marking the 90% confidence interval $\left[-2\sqrt{\sigma^2/2\lambda}, +2\sqrt{\sigma^2/2\lambda}\right]$ of the distribution $\mathrm{N}(0, \sigma^2/2\lambda)$. The solutions were generated by the explicit Euler–Maruyama method with time step $\Delta t = 0.0025$.

Example 8.2 (Langevin equation) If the particle in Example 8.1 has potential energy $V(Q)$ at position $Q \in \mathbb{R}$, its dynamics are described by the following SODE:

$$\begin{aligned} dQ &= P\,dt \\ dP &= \left[-\lambda P - V'(Q)\right] dt + \sigma\, dW(t) \end{aligned} \tag{8.8}$$

for parameters $\lambda, \sigma > 0$. In the notation of (8.3), $d = 2$, $m = 1$, and

$$\boldsymbol{f}(\boldsymbol{u}) = \begin{pmatrix} P \\ -\lambda P - V'(Q) \end{pmatrix}, \qquad G(\boldsymbol{u}) = \begin{pmatrix} 0 \\ \sigma \end{pmatrix},$$

for $\boldsymbol{u} = [Q, P]^{\mathsf{T}}$. The noise is additive and acts directly on the momentum P only. This SODE is known as the Langevin equation and a phase-plane plot of three approximate sample paths is shown in Figure 8.2(a) for $V(Q) = Q^2/2$.

This time, the equilibrium distribution has pdf $p(Q, P) = \mathrm{e}^{-\beta H(Q,P)}/Z$ for $H(Q, P) = P^2/2 + V(Q)$, normalisation constant Z, and inverse temperature $\beta = 1/(k_{\mathrm{B}}T) = 2\lambda/\sigma^2$. As in Example 8.1, $P(t)$ converges to a Gaussian distribution and Figure 8.2(b) shows the long-time behaviour of three sample paths of $P(t)$.

Example 8.3 (Duffing–van der Pol equation) Consider the Duffing–van der Pol SODE

$$\begin{aligned} dQ &= P\,dt \\ dP &= \left[-P(\lambda + Q^2) + (\alpha Q - Q^3)\right] dt + \sigma Q\, dW(t) \end{aligned} \tag{8.9}$$

comprising a nonlinear dissipation with parameter λ and a conservative force due to the potential $V(Q) = Q^4/4 - \alpha Q^2/2$ with parameter α. We have the following SODE for $\boldsymbol{u}(t) = [Q(t), P(t)]^{\mathsf{T}}$:

$$d\boldsymbol{u} = \begin{pmatrix} P \\ -P(\lambda + Q^2) + \alpha Q - Q^3 \end{pmatrix} dt + \begin{pmatrix} 0 \\ \sigma Q \end{pmatrix} dW(t). \tag{8.10}$$

The SODE has *multiplicative noise*, as the diffusion $G(\boldsymbol{u}) = [0, \sigma Q]^{\mathsf{T}}$ depends on $\boldsymbol{u} = [Q, P]^{\mathsf{T}}$. We explore the numerical solution of this equation in Exercise 8.9.

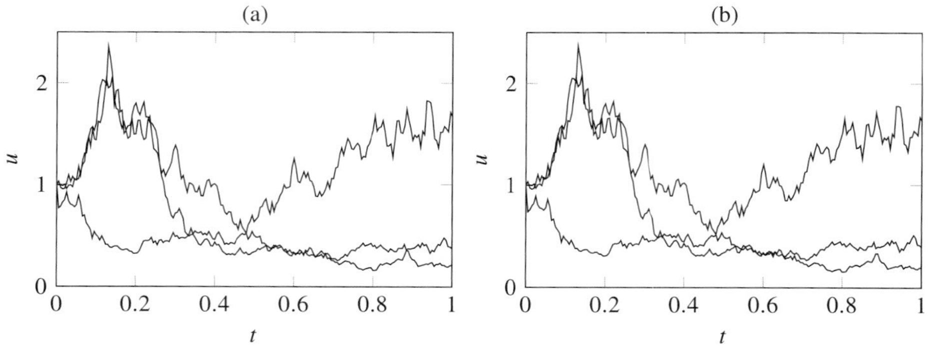

Figure 8.3 (a) Three sample paths of geometric Brownian motion, the solution of (8.11) with initial data $u_0 = 1$ and parameters $r = 1$, $\sigma = 5$ generated using the exact solution (8.34). (b) Approximations of the same sample paths given by the Euler–Maruyama method with $\Delta t = 0.005$. Geometric Brownian motion is non-negative with $u(t) \geq 0$ if $u_0 \geq 0$.

The solutions $\boldsymbol{u}(t)$ of these three examples are the same for the Itô and Stratonovich interpretations of the integral equation (8.3) (see Exercise 8.20).

Example 8.4 (geometric Brownian motion) In financial modelling, the price $u(t)$ at time t of a risk-free asset with interest rate r obeys a differential equation $du/dt = ru$. On the stock market, stock prices fluctuate rapidly and the fluctuations are modelled by replacing the risk-free interest rate r by a stochastic process $r + \sigma\zeta(t)$, for a parameter σ (known as the volatility) and white noise $\zeta(t)$. This gives the following SODE for the price $u(t)$:

$$du = r\,u\,dt + \sigma\,u\,dW(t), \qquad u(0) = u_0, \tag{8.11}$$

where $u_0 \geq 0$ is the price at time $t = 0$. In terms of (8.3), $d = m = 1$, the drift $f(u) = ru$, and the diffusion $G(u) = \sigma u$. As G depends on u, the SODE is said to have *multiplicative noise* and, for this example, the solution $u(t)$ depends on the interpretation of the integral $\int_0^t u(s)\,dW(s)$, as we show in Exercise 8.19. The Itô SODE is chosen to model stock prices, because the current price $u(t)$ of the stock is determined by past events and is independent of the current fluctuations $dW(t)$. The solution process $u(t)$ of (8.11) is called *geometric Brownian motion* and example paths are plotted in Figure 8.3(a). In Example 8.20, we find the exact solution $u(t)$ of (8.11) and hence show $u(t) \geq 0$. This is desirable in financial modelling, where stock prices are non-negative.

8.2 Itô integral

The purpose of this section is to develop the Itô integral

$$I(t) := \int_0^t X(s)\,dW(s), \tag{8.12}$$

for a class of stochastic processes $X(t)$. First, we introduce the technical conditions that ensure the integral is well defined. This subject is technical and many proofs are omitted or address only simple cases.

Stochastic processes model a time evolution and the idea of past and future is fundamental. For example, we can ask questions of a Brownian motion $W(t)$ from the point of view of a future or past time. Looking back from the future, $W(t)$ is known from our observations of the past. However, looking into the future, $W(t)$ is a random variable whose outcome is not yet known. The best we can say is the distribution of $W(t)$ at a future time t conditioned on our observations up to time $s < t$ is known to be Gaussian with mean $W(s)$ and variance $t - s$ (as the increments of Brownian motion are independent and have distribution $\mathrm{N}(0, t - s)$). Define the σ-algebra $\mathcal{F}_s := \sigma(W(s))$ (see Definition 4.18). Then,

$$\mathbb{E}\big[W(t) \,|\, \mathcal{F}_s\big] = W(s \wedge t)$$

where $s \wedge t := \min\{s,t\}$. This property of $W(t)$, known as the martingale property, extends to the stochastic integral $I(t)$, as we show in Theorem 8.14.

To handle the idea of past and future more generally, we use sub σ-algebras (see Definition 4.17) and introduce a sub σ-algebra $\mathcal{F}_t$ for each time t. Intuitively, the events in $\mathcal{F}_t$ are those observable by time t and, because we have more observations as time passes, the σ-algebras $\mathcal{F}_t$ contain more events as t increases. We make this precise with the notion of *filtration*.

Definition 8.5 (filtration) Let $(\Omega, \mathcal{F}, \mathbb{P})$ be a probability space.

(i) A *filtration* $\{\mathcal{F}_t : t \ge 0\}$ is a family of sub σ-algebras of $\mathcal{F}$ that are increasing; that is, $\mathcal{F}_s$ is a sub σ-algebra of $\mathcal{F}_t$ for $s \le t$. Each $(\Omega, \mathcal{F}_t, \mathbb{P})$ is a measure space and we assume it is complete (see Definition 1.17).

(ii) A *filtered probability space* is a quadruple $(\Omega, \mathcal{F}, \mathcal{F}_t, \mathbb{P})$, where $(\Omega, \mathcal{F}, \mathbb{P})$ is a probability space and $\{\mathcal{F}_t : t \ge 0\}$ is a filtration of $\mathcal{F}$.

Stochastic processes that conform to the notion of time described by the filtration $\mathcal{F}_t$ are known as *adapted* processes.

Definition 8.6 (adapted) Let $(\Omega, \mathcal{F}, \mathcal{F}_t, \mathbb{P})$ be a filtered probability space. A stochastic process $\{X(t) : t \in [0,T]\}$ is *$\mathcal{F}_t$-adapted* if the random variable $X(t)$ is $\mathcal{F}_t$-measurable for all $t \in [0,T]$.

For example, if $X(t)$ is $\mathcal{F}_t$-adapted, the event $\{X(s) \le a\}$ is observable at time s and should be included in $\mathcal{F}_t$ for each $t \ge s$. This is true because $\mathcal{F}_s$ is a sub σ-algebra of $\mathcal{F}_t$ and hence $X(s)$ is $\mathcal{F}_t$-measurable. Then, by definition of $\mathcal{F}_t$ measurable, the pullback set $X(s)^{-1}((-\infty, a]) = \{X(s) \le a\}$ belongs to $\mathcal{F}_t$.

To stress the relationship between Brownian motion and a filtration, we give the following definition.

Definition 8.7 ($\mathcal{F}_t$-Brownian motion) A real-valued process $\{W(t) : t \ge 0\}$ is an $\mathcal{F}_t$-Brownian motion on a filtered probability space $(\Omega, \mathcal{F}, \mathcal{F}_t, \mathbb{P})$ if

(i) $W(0) = 0$ a.s.,

(ii) $W(t)$ is continuous as a function of t,

(iii) $W(t)$ is $\mathcal{F}_t$-adapted and $W(t) - W(s)$ is independent of $\mathcal{F}_s$, $s < t$, and

(iv) $W(t) - W(s) \sim \mathrm{N}(0, t - s)$ for $0 \le s \le t$.

The $\mathcal{F}_t$-Brownian motion is a simple extension of Definition 5.11 (Brownian motion) and is easily constructed by defining an appropriate filtration. Just as each random variable X has an associated σ-algebra $\sigma(X)$ (see Definition 4.18), each process $\{X(t)\colon t \geq 0\}$ has an associated filtration, known as the *natural filtration*.

Definition 8.8 (natural filtration) If $\{X(t)\colon t \geq 0\}$ is a stochastic process, let $\mathcal{F}_t$ be the smallest σ-algebra such that $X(s)$ is measurable for all $s \leq t$. $\{\mathcal{F}_t : t \geq 0\}$ is called the *natural filtration* of $X(t)$.

Thus, $\mathcal{F}_t$-Brownian motions can be constructed from a Brownian motion $W(t)$ and its natural filtration.

Definition and properties

We turn to the definition of the Itô stochastic integral (8.12) of an $\mathcal{F}_t$-adapted process $X(t)$ with respect to an $\mathcal{F}_t$-Brownian motion $W(t)$. We start by assuming $X(t)$ has left-continuous sample paths; that is, $X(s_j,\omega) \to X(s,\omega)$ if $s_j \to s$ with $s_j \leq s$ for all $s \in [0,T]$ and $\omega \in \Omega$.

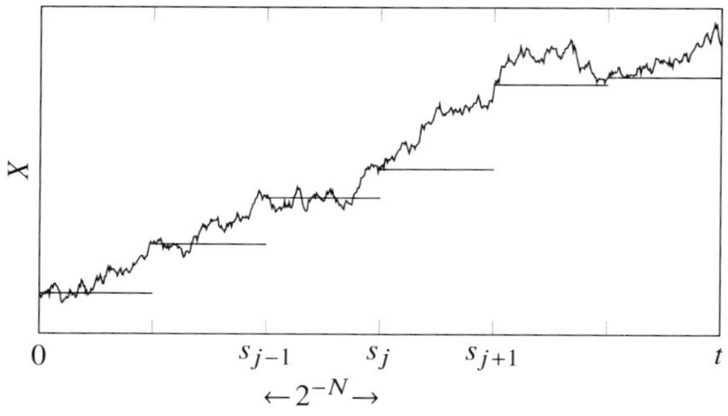

Figure 8.4 We approximate $X(t)$ on the interval $[s_j, s_{j+1})$ by $X(s_j)$, where $s_j = j2^{-N}$ for $j = 0, 1, \ldots$. The interval $[0,t]$ is partitioned into $0 = s_0 < s_1 < s_2 < \cdots < s_k < t$, where $s_k = \max\{s_j : s_j < t\}$.

The idea is to introduce a grid of points $s_j = j\Delta t$ for $j = 0, 1, 2, \ldots$, where $\Delta t = 2^{-N}$, and approximate $X(s)$ for $s \in [s_j, s_{j+1})$ by $X(s_j)$ (see Figure 8.4) and $dW(s)$ by the increment $W(s_{j+1}) - W(s_j)$. We then define the Itô stochastic integral as

$$\int_0^t X(s)\, dW(s) := \lim_{N\to\infty} S_N^X(t), \tag{8.13}$$

where we refine the grid by taking $N \to \infty$ and

$$S_N^X(t) := \sum_{s_j < t} X(s_j)\big(W(s_{j+1} \wedge t) - W(s_j)\big). \tag{8.14}$$

All but the last increment in the sum are over an interval of length $\Delta t = 2^{-N}$ (which gives a nesting property convenient for the proof of Lemma 8.9).

Surprisingly, the limit in (8.13) and hence the stochastic integral depends on the choice of $X(s_j)$ to approximate $X(t)$ on $t \in [s_j, s_{j+1})$. For example, in the case that $X(t)$ is $\mathcal{F}_t$-adapted, $X(s_j)$ is $\mathcal{F}_{s_j}$-measurable and independent of the increment $W(s_{j+1} \wedge t) - W(s_j)$

of an $\mathcal{F}_t$-Brownian motion $W(t)$. Hence, the product $X(s_j)(W(s_{j+1} \wedge t) - W(s_j))$ has mean zero. Moreover, the sum $S_N^X(t)$ and the Itô integral have mean zero. This is not true if $S_N^X(t)$ equals a sum of $X(r_j)(W(s_{j+1} \wedge t) - W(s_j))$ for some $r_j > s_j$.

The Itô integral is defined by the limit of $S_N^X(t)$ in $L^2(\Omega)$, which exists by the following lemma.

Lemma 8.9 *Let $W(t)$ be an $\mathcal{F}_t$-Brownian motion and let $\{X(t) \colon t \in [0,T]\}$ be an $\mathcal{F}_t$-adapted process with left-continuous sample paths.*

(i) $\{S_N^X(t) \colon t \in [0,T]\}$ *is $\mathcal{F}_t$-adapted and has continuous sample paths.*

(ii) $\mathbb{E}\big[S_N^X(t) \,|\, \mathcal{F}_{s_j}\big] = S_N^X(s_j)$ *for $0 \le s_j \le t$ a.s. In particular, $\mathbb{E}\big[S_N^X(t)\big] = 0$.*

(iii) *If $\{Y(t) \colon t \in [0,T]\}$ is also $\mathcal{F}_t$-adapted with left-continuous sample paths and $X, Y \in L^2(\Omega, L^2(0,T))$, then for $t_1, t_2 \in [0,T]$*

$$\mathbb{E}\Big[S_N^X(t_1)\, S_M^Y(t_2)\Big] \to \int_0^{t_1 \wedge t_2} \mathbb{E}\big[X(s)Y(s)\big]\, ds \qquad \text{as } N, M \to \infty. \tag{8.15}$$

In particular, $S_N^X(t)$ has an $L^2(\Omega)$ limit as $N \to \infty$.

Proof (i) Since $\mathcal{F}_{s_j}$ is a sub σ-algebra of $\mathcal{F}_t$ and $X(t)$ is $\mathcal{F}_t$-adapted, it holds that $X(s_j)$ is $\mathcal{F}_t$-measurable for $s_j \le t$. Similarly, $W(s)$ is $\mathcal{F}_t$-measurable for $s \le t$. Then, $S_N^X(t)$ is $\mathcal{F}_t$-measurable and the process $\{S_N^X(t) \colon t \in [0,T]\}$ is $\mathcal{F}_t$-adapted. Sample paths of $S_N^X(t)$ are continuous because $W(t)$ has continuous sample paths.

(ii) We assume for simplicity that t is integer valued, in which case

$$S_N^X(t) = \sum_{s_j < t} X(s_j)(W(s_{j+1}) - W(s_j)).$$

Now, $S_N^X(t) = S_N^X(s_k) + \big(S_N^X(t) - S_N^X(s_k)\big)$ and

$$\begin{aligned}\mathbb{E}\Big[S_N^X(t) \,\Big|\, \mathcal{F}_{s_k}\Big] &= \mathbb{E}\Big[S_N^X(s_k) \,\Big|\, \mathcal{F}_{s_k}\Big] + \sum_{s_k \le s_j < t} \mathbb{E}\Big[X(s_j)(W(s_{j+1}) - W(s_j)) \,\Big|\, \mathcal{F}_{s_k}\Big] \\ &= S_N^X(s_k) + \sum_{s_k \le s_j < t} \mathbb{E}\Big[X(s_j)\,\mathbb{E}\Big[(W(s_{j+1}) - W(s_j)) \,\Big|\, \mathcal{F}_{s_j}\Big] \,\Big|\, \mathcal{F}_{s_k}\Big],\end{aligned}$$

where we use the 'taking out what is known' and 'tower property' (see Theorem 4.52) of conditional expectation. Finally,

$$\mathbb{E}\Big[S_N^X(t) \,\Big|\, \mathcal{F}_{s_k}\Big] = S_N^X(s_k), \qquad \text{a.s.}$$

because the increment $W(s_{j+1}) - W(s_j)$ is independent of $\mathcal{F}_{s_j}$ and hence

$$\mathbb{E}\Big[(W(s_{j+1}) - W(s_j)) \,|\, \mathcal{F}_{s_j}\Big] = \mathbb{E}\big[W(s_{j+1}) - W(s_j)\big] = 0, \qquad a.s.$$

(iii) Let $s_j = j2^{-N}$ and $r_k = k2^{-M}$ for $j, k = 0, 1, \ldots$ and for simplicity assume that t_1, t_2 are integers. Write $X_j = X(s_j)$ and $Y_k = Y(r_k)$. Now,

$$\begin{aligned}\mathbb{E}[S_N^X(t_1)S_M^Y(t_2)] &= \mathbb{E}\Big[\sum_{s_j<t_1}\sum_{r_k<t_2} X_jY_k\big(W(s_{j+1})-W(s_j)\big)\big(W(r_{k+1})-W(r_k)\big)\Big]\\ &= \mathbb{E}\Big[\mathbb{E}\Big[\sum_{0\le r_k<s_j<t_1} X_jY_k(W(s_{j+1})-W(s_j))(W(r_{k+1})-W(r_k))\,\Big|\,\mathcal{F}_{s_j}\Big]\Big]\\ &\quad+\mathbb{E}\Big[\mathbb{E}\Big[\sum_{0\le s_j\le r_k<t_2} X_jY_k(W(s_{j+1})-W(s_j))(W(r_{k+1})-W(r_k))\,\Big|\,\mathcal{F}_{r_k}\Big]\Big].\end{aligned}$$

By the increment property of Definition 8.7, we have (see Exercise 8.2) almost surely

$$\mathbb{E}\Big[\big(W(r)-W(p)\big)\big(W(t)-W(s)\big)\,\Big|\,\mathcal{F}_s\Big] = \begin{cases} r-s, & p\le s\le r\le t,\\ 0, & p<r\le s<t.\end{cases} \tag{8.16}$$

We assume that $M > N$, so that $[r_k, r_{k+1})$ is either a subset of or has empty intersection with $[s_j, s_{j+1})$. For $s_j \le r_k$, both X_j and Y_k are $\mathcal{F}_{r_k}$-measurable and almost surely

$$\begin{aligned}&\mathbb{E}\Big[X_jY_k(W(s_{j+1})-W(s_j))(W(r_{k+1})-W(r_k))\,\Big|\,\mathcal{F}_{r_k}\Big]\\ &\quad= X_jY_k\mathbb{E}\Big[(W(s_{j+1})-W(s_j))(W(r_{k+1})-W(r_k))\,\Big|\,\mathcal{F}_{r_k}\Big]\\ &\quad= \begin{cases}0, & s_{j+1}\le r_k,\\ X_jY_k(r_{k+1}-r_k), & s_j\le r_k<s_{j+1}.\end{cases}\end{aligned}$$

Similarly, for $r_k < s_j$, we have $r_{k+1} \le s_j$ and

$$\begin{aligned}&\mathbb{E}\Big[X_jY_k\big(W(s_{j+1})-W(s_j)\big)\big(W(r_{k+1})-W(r_k)\big)\,\Big|\,\mathcal{F}_{s_j}\Big]\\ &\quad= X_jY_k\mathbb{E}\Big[\big(W(s_{j+1})-W(s_j)\big)\big(W(r_{k+1})-W(r_k)\big)\,\Big|\,\mathcal{F}_{s_j}\Big]=0, \qquad \text{a.s.}\end{aligned}$$

Let $\phi(r) := X_jY_k$ if $r \in [s_j, s_{j+1}) \cap [r_k, r_{k+1})$. By left continuity of the sample paths, $\phi(r) = X(s_j)Y(r_k) \to X(r)Y(r)$ as $N, M \to \infty$ with $M > N$. Then,

$$\mathbb{E}\Big[S_N^X(t_1)\,S_M^Y(t_2)\Big] = \sum_{0\le r_k<t_1\wedge t_2}\mathbb{E}[\phi(r_k)](r_{k+1}-r_k),$$

which converges to $\int_0^{t_1\wedge t_2}\mathbb{E}[X(r)Y(r)]\,dr$ as $N, M \to \infty$ with $M > N$ and hence we have (8.15). If $X \in L^2(\Omega, L^2(0,T))$, then

$$\mathbb{E}\Big[S_N^X(t)S_M^X(t)\Big] \to \|X\|^2_{L^2(\Omega, L^2(0,t))} < \infty$$

and, by Lemma 5.34, $S_N^X(t)$ has a well-defined limit in $L^2(\Omega)$. □

We have shown the Itô integral $I(t) = \int_0^t X(s)\,dW(s)$ is well defined by the $L^2(\Omega)$ limit of $S_N^X(t)$ as in (8.13), for adapted processes $X(t)$ with left-continuous sample paths. Further from (8.15), we have the so-called *Itô isometry*:

$$\Big\|\int_0^t X(s)\,dW(s)\Big\|^2_{L^2(\Omega)} = \mathbb{E}\Big[\Big|\int_0^t X(s)\,dW(s)\Big|^2\Big] = \int_0^t \mathbb{E}\big[|X(s)|^2\big]\,ds, \tag{8.17}$$

In order to conveniently apply limit theorems in Banach spaces, it is important to extend the definition to the predictable processes.

Definition 8.10 (predictable) A stochastic process $\{X(t)\colon t \in [0,T]\}$ is *predictable* if there exists $\mathcal{F}_t$-adapted and left-continuous processes $\{X_n(t)\colon t \in [0,T]\}$ such that $X_n(t) \to X(t)$ as $n \to \infty$ for $t \in [0,T]$.

Proposition 8.11 (Banach space $\mathcal{L}_2^T$) *Let $\mathcal{L}_2^T$ denote the space of predictable real-valued processes $\{X(t)\colon t \in [0,T]\}$ with*

$$\|X\|_{\mathcal{L}_2^T} := \mathbb{E}\left[\int_0^T |X(s)|^2\, ds\right]^{1/2} < \infty.$$

Then, $\mathcal{L}_2^T$ is a Banach space with norm $\|\cdot\|_{\mathcal{L}_2^T}$.

We define the Itô integral for any integrand in $\mathcal{L}_2^T$.

Definition 8.12 (Itô integral) Let $W(t)$ be an $\mathcal{F}_t$-Brownian motion and let $X \in \mathcal{L}_2^T$. By definition, X equals the limit in $\mathcal{L}_2^T$ of a sequence of left-continuous and $\mathcal{F}_t$-adapted processes $X_n(t)$, $n \in \mathbb{N}$. The stochastic integral $I_n(t) = \int_0^t X_n(s)\, dW(s)$ is well defined by Lemma 8.9 and, from (8.17) with $X = X_n - X_m$,

$$\|I_n(t) - I_m(t)\|_{L^2(\Omega)} = \mathbb{E}\left[|I_n(t) - I_m(t)|^2\right] = \int_0^t \mathbb{E}\left[|X_n(s) - X_m(s)|^2\right] ds. \tag{8.18}$$

Then, $I_n(t)$ is a Cauchy sequence in $L^2(\Omega)$ and the Itô integral

$$\int_0^t X(s)\, dW(s) := \lim_{n\to\infty} \int_0^t X_n(s)\, dW(s) \qquad \text{in } L^2(\Omega).$$

Definition 8.13 (vector-valued Itô integral) Let $\boldsymbol{W} = [W_1, \ldots, W_m]^{\mathsf{T}}$, for *iid* Brownian motions $W_j(t)$. Denote by $\mathcal{L}_2^T(\mathbb{R}^{d\times m})$ the set of $\mathbb{R}^{d\times m}$-valued processes where each component belongs to $\mathcal{L}_2^T$ and let $X \in \mathcal{L}_2^T(\mathbb{R}^{d\times m})$. The Itô integral $\int_0^t X(s)\, d\boldsymbol{W}(s)$ is the random variable in $L^2(\Omega, \mathbb{R}^d)$ with ith component

$$\sum_{j=1}^m \int_0^t X_{ij}(s)\, dW_j(s). \tag{8.19}$$

We give some useful properties of the Itô integral.

Theorem 8.14 (Itô integral process) *Let $X \in \mathcal{L}_2^T(\mathbb{R}^{d\times m})$. The following properties hold:*

(i) $\left\{\int_0^t X(s)\, d\boldsymbol{W}(s)\colon t \in [0,T]\right\}$ *is a predictable process.*
(ii) *The martingale property holds: for $0 \le r \le t \le T$,*

$$\mathbb{E}\left[\int_0^t X(s)\, d\boldsymbol{W}(s) \,\Big|\, \mathcal{F}_r\right] = \int_0^r X(s)\, d\boldsymbol{W}(s), \qquad \textit{almost surely,} \tag{8.20}$$

and in particular the integral has mean zero : $\mathbb{E}\left[\int_0^t X(s)\, d\boldsymbol{W}(s)\right] = 0.$
(iii) *For $X, Y \in \mathcal{L}_2^T(\mathbb{R}^{d\times m})$,*

$$\mathbb{E}\left[\left\langle \int_0^{t_1} X(s)\, d\boldsymbol{W}(s), \int_0^{t_2} Y(s)\, d\boldsymbol{W}(s)\right\rangle\right] = \int_0^{t_1\wedge t_2} \sum_{i=1}^m \mathbb{E}\left[\langle \boldsymbol{X}_i(s), \boldsymbol{Y}_i(s)\rangle\right] ds, \tag{8.21}$$

where X_i, Y_i denote the columns of X and Y and $\langle\cdot,\cdot\rangle$ is the $\mathbb{R}^d$ inner product. In particular, we have the Itô isometry:

$$\mathbb{E}\left[\left\|\int_0^t X(s)\,d\boldsymbol{W}(s)\right\|_2^2\right] = \int_0^t \mathbb{E}\left[\|X(s)\|_{\mathrm{F}}^2\right] ds, \qquad t \in [0,T]. \tag{8.22}$$

Proof It is clear from (8.18) that I_n is also a Cauchy sequence in $\mathcal{L}_2^T$. As $\mathcal{L}_2^T$ is Banach space, there exists a limit $I \in \mathcal{L}_2^T$ and the limit is predictable. For $d = 1 = m$, (ii)–(iii) follow from Lemma 8.9(ii)–(iii). For general d,m, the identities follow by writing out the inner products in component forms. □

Example 8.15 Consider the stochastic integral

$$I(t) = \int_0^t W(s)\,dW(s), \qquad t \geq 0.$$

The martingale property gives $\mathbb{E}[I(t)] = 0$. The Itô isometry (8.22) gives

$$\mathbb{E}\left[I(t)^2\right] = \int_0^t \mathbb{E}\left[W(s)^2\right] ds = \int_0^t s\,ds = \frac{1}{2}t^2 \tag{8.23}$$

and (8.21) gives

$$\mathbb{E}\left[I(t)I(s)\right] = \int_0^{s\wedge t} \mathbb{E}\left[W(r)^2\right] dr = \int_0^{s\wedge t} r\,dr = \frac{1}{2}(s\wedge t)^2.$$

As Itô integrals have mean zero, $\mathrm{Cov}(I(s), I(t)) = (s \wedge t)^2/2$.

8.3 Itô SODEs

Having described the Itô integral, the next step is to examine the initial-value problem for Itô SODEs of the form (8.3). First, we study the existence and uniqueness of solutions and then, using the one-dimensional Itô formula, we obtain explicit solutions for some particular SODEs in the case $d = m = 1$.

Existence and uniqueness of solutions

To show existence and uniqueness of solutions, we use the contraction mapping theorem (Theorem 1.10) on the following Banach space.

Definition 8.16 Let $(\Omega, \mathcal{F}, \mathcal{F}_t, \mathbb{P})$ be a filtered probability space. $\mathcal{H}_{2,T}$ is the set of $\mathbb{R}^d$-valued predictable processes $\{\boldsymbol{u}(t) \colon t \in [0,T]\}$ such that

$$\|\boldsymbol{u}\|_{\mathcal{H}_{2,T}} := \sup_{t\in[0,T]} \|\boldsymbol{u}(t)\|_{L^2(\Omega,\mathbb{R}^d)} = \sup_{t\in[0,T]} \mathbb{E}\left[\|\boldsymbol{u}(t)\|_2^2\right]^{1/2} < \infty.$$

This is a Banach space with the norm $\|\boldsymbol{u}\|_{\mathcal{H}_{2,T}}$ (strictly speaking it is equivalence classes of almost sure equal processes that form the Banach space).

We require the following assumption on the drift and diffusion, similar to the condition for ODEs used in Theorem 3.2.

Assumption 8.17 (linear growth/Lipschitz condition) There exists a constant $L > 0$ such that the linear growth condition holds:

$$\begin{aligned} \|f(u)\|_2^2 &\le L\big(1 + \|u\|_2^2\big), \\ \|G(u)\|_F^2 &\le L\big(1 + \|u\|_2^2\big), \qquad \forall u \in \mathbb{R}^d, \end{aligned} \tag{8.24}$$

and the global Lipschitz condition holds:

$$\begin{aligned} \|f(u_1) - f(u_2)\|_2 &\le L\|u_1 - u_2\|_2, \\ \|G(u_1) - G(u_2)\|_F &\le L\|u_1 - u_2\|_2, \qquad \forall u_1, u_2 \in \mathbb{R}^d. \end{aligned} \tag{8.25}$$

Theorem 8.18 (existence and uniqueness for SODEs) *Suppose that Assumption 8.17 holds and that $W(t)$ is an $\mathcal{F}_t$-Brownian motion on $(\Omega, \mathcal{F}, \mathcal{F}_t, \mathbb{P})$. For each $T > 0$ and $u_0 \in \mathbb{R}^d$, there exists a unique $u \in \mathcal{H}_{2,T}$ such that for $t \in [0,T]$*

$$u(t) = u_0 + \int_0^t f(u(s))\,ds + \int_0^t G(u(s))\,dW(s). \tag{8.26}$$

Proof Consider a random variable $X \in L^2(\Omega, \mathbb{R}^d)$ that is independent of $\mathcal{F}_0$ and hence of the process $W(t)$. For $u \in \mathcal{H}_{2,T}$, let

$$\mathcal{J}(u)(t) := X + \int_0^t f(u(s))\,ds + \int_0^t G(u(s))\,dW(s), \qquad t \in [0,T].$$

If $u \in \mathcal{H}_{2,T}$ is a fixed point of $\mathcal{J}$, then it satisfies the integral equation (8.26) if $X = u_0$. We show the existence of a unique fixed point by applying the contraction mapping theorem (Theorem 1.10) on the Banach space $\mathcal{H}_{2,T}$. To extend the result to any time interval, we may reapply the argument with an initial condition $X = u(kT)$ on an interval $[kT, kT + T]$ for $k = 1, 2, \ldots$.

We demonstrate the two conditions of the contraction mapping theorem and first show $\mathcal{J}$ maps into $\mathcal{H}_{2,T}$. The process $\mathcal{J}(u)(t)$ is predictable, because of Theorem 8.14(i). It remains to show the $\mathcal{H}_{2,T}$ norm of $\mathcal{J}(u)(t)$ is finite. By Jensen's inequality (A.9),

$$\mathbb{E}\left[\left\|\int_0^t f(u(s))\,ds\right\|_2^2\right] \le \mathbb{E}\left[t\int_0^t \|f(u(s))\|_2^2\,ds\right] \tag{8.27}$$

and, by the Itô isometry (8.22),

$$\mathbb{E}\left[\left\|\int_0^t G(u(s))\,dW(s)\right\|_2^2\right] = \int_0^t \mathbb{E}\left[\|G(u(s))\|_F^2\right]ds. \tag{8.28}$$

Using (A.10), (8.27), and (8.28),

$$\begin{aligned} \mathbb{E}\left[\|\mathcal{J}(u)(t)\|_2^2\right] &\le 3\mathbb{E}\left[\|X\|_2^2\right] + 3\mathbb{E}\left[\left\|\int_0^t f(u(s))\,ds\right\|_2^2\right] + 3\mathbb{E}\left[\left\|\int_0^t G(u(s))\,dW(s)\,ds\right\|_2^2\right] \\ &\le 3\mathbb{E}\left[\|X\|_2^2\right] + 3\mathbb{E}\left[t\int_0^t \|f(u(s))\|_2^2\,ds\right] + 3\int_0^t \mathbb{E}\left[\|G(u(s))\|_F^2\right]ds. \end{aligned}$$

Assumption 8.17 gives

$$\mathbb{E}\left[\|\mathcal{J}(\boldsymbol{u})(t)\|_2^2\right] \le 3\mathbb{E}\left[\|\boldsymbol{X}\|_2^2\right] + 3t \int_0^t L^2 \mathbb{E}\left[1 + \|\boldsymbol{u}(s)\|_2^2\right] ds$$
$$+ 3 \int_0^t L^2 \mathbb{E}\left[1 + \|\boldsymbol{u}(s)\|_2^2\right] ds.$$

Finally, we take the supremum over $t \in [0,T]$ in the last two terms:

$$\mathbb{E}\left[\|\mathcal{J}(\boldsymbol{u})(t)\|_2^2\right] \le 3\mathbb{E}\left[\|\boldsymbol{X}\|_2^2\right] + 3L^2(t+1)\,t\left(1 + \sup_{t\in[0,T]} \mathbb{E}\left[\|\boldsymbol{u}(t)\|_2^2\right]\right). \tag{8.29}$$

Hence, $\|\mathcal{J}(\boldsymbol{u})\|_{\mathcal{H}_{2,T}} < \infty$ and $\mathcal{J}$ maps $\mathcal{H}_{2,T}$ into $\mathcal{H}_{2,T}$.

To show $\mathcal{J}$ is a contraction, we argue similarly using the Lipschitz condition in place of the linear growth condition:

$$\begin{aligned}
&\mathbb{E}\left[\|\mathcal{J}(\boldsymbol{u}_1)(t) - \mathcal{J}(\boldsymbol{u}_2)(t)\|_2^2\right] \\
&\le 2t\,\mathbb{E}\left[\int_0^t \|\boldsymbol{f}(\boldsymbol{u}_1(s)) - \boldsymbol{f}(\boldsymbol{u}_2(s))\|_2^2\, ds\right] + 2\mathbb{E}\left[\left\|\int_0^t \left(G(\boldsymbol{u}_1(s)) - G(\boldsymbol{u}_2(s))\right) d\boldsymbol{W}(s)\right\|_2^2\right] \\
&\le 2L^2 t \int_0^t \mathbb{E}\left[\|\boldsymbol{u}_1(s) - \boldsymbol{u}_2(s)\|_2^2\right] ds + 2L^2 \int_0^t \mathbb{E}\left[\|\boldsymbol{u}_1(s) - \boldsymbol{u}_2(s)\|_2^2\right] ds \\
&\le 2L^2 t(t+1) \sup_{t\in[0,T]} \mathbb{E}\left[\|\boldsymbol{u}_1(t) - \boldsymbol{u}_2(t)\|_2^2\right].
\end{aligned}$$

Choose $T = \min\{1, 1/8L^2\}$. Then,

$$\sup_{t\in[0,T]} \mathbb{E}\left[\|\mathcal{J}(\boldsymbol{u}_1)(t) - \mathcal{J}(\boldsymbol{u}_2)(t)\|_2^2\right] \le \frac{1}{2} \sup_{t\in[0,T]} \mathbb{E}\left[\|\boldsymbol{u}_1(t) - \boldsymbol{u}_2(t)\|_2^2\right]$$

and $\|\mathcal{J}(\boldsymbol{u}_1) - \mathcal{J}(\boldsymbol{u}_2)\|_{\mathcal{H}_{2,T}} \le \frac{1}{2}\|\boldsymbol{u}_1 - \boldsymbol{u}_2\|_{\mathcal{H}_{2,T}}$. We see that $\mathcal{J}$ is a contraction on $\mathcal{H}_{2,T}$ and the contraction mapping theorem applies. Hence, there exists a unique fixed point of $\mathcal{J}$ and unique solution to (8.26). □

One-dimensional Itô formula and exact solutions

The chain rule of ordinary calculus changes in the Itô calculus to the so-called Itô formula. We discuss the Itô formula in one dimension, giving applications to geometric Brownian motion and the mean reverting Ornstein–Uhlenbeck process. The general Itô formula is given later in Lemma 8.42.

Lemma 8.19 (one-dimensional Itô formula) *Let $\Phi\colon [0,T]\times\mathbb{R} \to \mathbb{R}$ have continuous partial derivatives $\frac{\partial\Phi}{\partial t}$, $\frac{\partial\Phi}{\partial u}$ and $\frac{\partial^2\Phi}{\partial u^2}$. Let u satisfy*

$$du = f(u)\,dt + g(u)\,dW(t), \qquad u(0) = u_0$$

and suppose Assumption 8.17 holds for $d = m = 1$. Then, almost surely,

$$d\Phi = \frac{\partial\Phi}{\partial t}\,dt + \frac{\partial\Phi}{\partial u}\,du + \frac{1}{2}\frac{\partial^2\Phi}{\partial u^2} g^2\,dt \tag{8.30}$$

or written in full

$$\Phi(t,u(t)) = \Phi(0,u_0) + \int_0^t \frac{\partial \Phi}{\partial t}(s,u(s)) + \frac{\partial \Phi}{\partial u}(s,u(s))f(u(s))\,ds$$
$$+ \frac{1}{2}\int_0^t \frac{\partial^2 \Phi}{\partial u^2}(s,u(s))g(u(s))^2\,ds + \int_0^t \frac{\partial \Phi}{\partial u}(s,u(s))g(u(s))\,dW(s).$$

Proof We consider the special case $\Phi(t,u) = \phi(u)$ for a quadratic polynomial $\phi(u)$, so that (8.30) reduces to

$$d\phi(u(t)) = \phi'(u(t))\,du + \frac{1}{2}\phi''(u(t))g^2\,dt. \tag{8.31}$$

We also take $f,g \in \mathbb{R}$ to be constants, so that $u(t) = u_0 + ft + gW(t)$.

Let $s_j = j\Delta t$, where $\Delta t = 2^{-N}$ and $j = 0,1,2,\ldots$. Consider the telescoping sum

$$\phi(u(t)) - \phi(u(0)) = \sum_{s_j<t}\big(\phi(u(s_{j+1}\wedge t)) - \phi(u(s_j))\big).$$

We assume t is an integer so that $s_{j+1}\wedge t = s_{j+1}$. As ϕ is quadratic, the second-order Taylor series of $\phi(u)$ is exact and

$$\phi(u(s_{j+1})) = \phi(u(s_j)) + \phi'(u(s_j))\big(u(s_{j+1}) - u(s_j)\big) + \tfrac{1}{2}\phi''(u(s_j))\big(u(s_{j+1}) - u(s_j)\big)^2.$$

Then, $\phi(u(t)) - \phi(u(0)) = A_N + B_N$ for

$$A_N := \sum_{s_j<t}\phi'(u(s_j))\big(u(s_{j+1}) - u(s_j)\big),$$
$$B_N := \frac{1}{2}\sum_{s_j<t}\phi''(u(s_j))\big(u(s_{j+1}) - u(s_j)\big)^2. \tag{8.32}$$

Substituting $u(t) - u(s) = (f\,t - f\,s) + (g\,W(t) - g\,W(s))$,

$$A_N = \sum_{s_j<t}\phi'(u(s_j))\big(f\,s_{j+1} - f\,s_j\big) + \sum_{s_j<t}\phi'(u(s_j))\big(g\,W(s_{j+1}) - g\,W(s_j)\big)$$
$$\to \int_0^t \phi'(u(s))\,f\,ds + \int_0^t \phi'(u(s))\,g\,dW(s) \qquad \text{in } L^2(\Omega) \text{ as } N\to\infty$$

by definition of the Itô integral. As $\phi''(u)$ is constant, from (8.32),

$$B_N = \frac{1}{2}\phi''\sum_{s_j<t}\big(u(s_{j+1}) - u(s_j)\big)^2 = \frac{1}{2}\phi''\sum_{s_j<t}C_j\Delta t$$

for $C_j := (u(s_{j+1}) - u(s_j))^2/\Delta t$. Note that $C_j = (f\,s_{j+1} - f\,s_j + g\,W(s_{j+1}) - g\,W(s_j))^2/\Delta t$, which are *iid* random variables whose variance is bounded uniformly over Δt. Then,

$$\mathrm{Var}(B_N) = \frac{(\phi'')^2}{4}\sum_{s_j<t}\mathrm{Var}(C_j)\Delta t^2 \to 0 \qquad \text{as } N\to\infty.$$

Hence, $B_N \to \mathbb{E}[B_N]$ in $L^2(\Omega)$ as $N\to\infty$. Now, $\mathbb{E}[C_j] = f^2\Delta t + g^2$ and

$$\mathbb{E}[B_N] = \frac{\phi''}{2}\sum_{s_j<t}f^2\Delta t^2 + g^2\Delta t \to \frac{1}{2}\int_0^t \phi''(u(s))g\,ds, \qquad \text{as } N\to\infty.$$

Finally, $\phi(u(t)) - \phi(u(0)) = \lim_{N\to\infty}(A_N + B_N)$ and this gives (8.31). □

Consider the solution $u(t)$ of the deterministic ODE $\frac{du}{dt} = f$ and let $\phi(u) = u^2$. The standard chain rule implies that

$$\frac{d\phi(u)}{dt} = 2u\,\frac{du}{dt}. \tag{8.33}$$

However, if $u(t)$ satisfies $du = f\,dt + g\,dW(t)$, the Itô formula (8.30) says that

$$d\phi(u) = \big(2uf\,dt + 2ug\,dW(t)\big) + g^2\,dt = 2u\,du + g^2\,dt$$

and we pick up an unexpected extra term $g^2\,dt$. Since the Brownian increment $W(s_{j+1}) - W(s_j)$ is of size $|s_{j+1} - s_j|^{1/2}$, the terms arising from $(u(s_{j+1}) - u(s_j))^2$ (in B_N in the proof) that lead to $\mathbb{E}\big[(W(s_{j+1}) - W(s_j))^2\big]$ are important and the familiar rule $du^2 = 2u\,du$ does not apply in the Itô calculus.

We look at two applications of the Itô formula to solve SODEs.

Example 8.20 (geometric Brownian motion) We show the solution of the geometric Brownian motion SODE (8.11) is given by

$$u(t) = \exp\big((r - \sigma^2/2)\,t + \sigma W(t)\big)u_0. \tag{8.34}$$

For $u(0) = 0$, $u(t) = 0$ is clearly the solution to (8.11). For $u > 0$, let $\phi(u) = \log u$, so that $\phi'(u) = u^{-1}$ and $\phi''(u) = -u^{-2}$. By the Itô formula (8.31) with $\Phi(t,u) = \phi(u)$,

$$d\phi(u) = r\,dt + \sigma\,dW(t) - \frac{1}{2}\sigma^2\,dt.$$

Hence,

$$\phi(u(t)) = \phi(u_0) + \int_0^t \left(r - \frac{1}{2}\sigma^2\right) ds + \int_0^t \sigma\,dW(s)$$

and $\log u(t) = \log(u_0) + (r - \frac{1}{2}\sigma^2)t + \sigma W(t)$. Taking the exponential, we find (8.34). It is clear that, when $u_0 \ge 0$, the solution $u(t) \ge 0$ for all $t \ge 0$.

In the following example, we use the Itô formula (8.30) to examine a generalisation of the Ornstein–Uhlenbeck process (8.5) and obtain the variation of constants solution (8.6).

Example 8.21 (mean-reverting Ornstein–Uhlenbeck process) We consider the following generalisation of (8.5):

$$du = \lambda(\mu - u)dt + \sigma dW(t), \qquad u(0) = u_0,$$

for λ, μ, $\sigma \in \mathbb{R}$. We solve this SODE by using the Itô formula (8.30) with $\Phi(t,u) = \mathrm{e}^{\lambda t}\,u$. Then,

$$d\Phi(t,u) = \lambda\,\mathrm{e}^{\lambda t}u\,dt + \mathrm{e}^{\lambda t}\big(\lambda(\mu - u)dt + \sigma dW(t)\big) + 0$$

and

$$\Phi(t,u(t)) - \Phi(0,u_0) = \mathrm{e}^{\lambda t}u(t) - u_0 = \lambda\mu \int_0^t \mathrm{e}^{\lambda s}\,ds + \sigma \int_0^t \mathrm{e}^{\lambda s}\,dW(s).$$

After evaluating the deterministic integral, we find

$$u(t) = \mathrm{e}^{-\lambda t}u_0 + \mu(1 - \mathrm{e}^{-\lambda t}) + \sigma \int_0^t \mathrm{e}^{\lambda(s-t)}dW(s)$$

and this is known as the *variation of constants* solution.

Notice that $u(t)$ is a Gaussian process and can also be specified (see Corollary 5.19) by its mean $\mu(t) = \mathbb{E}[u(t)]$ and covariance $C(s,t) = \mathrm{Cov}(u(s),u(t))$. Using the mean-zero property of the Itô integral (Theorem 8.14), the mean

$$\mu(t) = \mathbb{E}\big[u(t)\big] = \mathrm{e}^{-\lambda t} u(0) + \mu\big(1 - \mathrm{e}^{-\lambda t}\big)$$

so that $\mu(t) \to \mu$ as $t \to \infty$ and the process is mean reverting. For the covariance,

$$\begin{aligned} C(s,t) = \mathrm{Cov}(u(t),u(s)) &= \mathbb{E}\Big[\big(u(s) - \mathbb{E}[u(s)]\big)\big(u(t) - \mathbb{E}[u(t)]\big)\Big] \\ &= \mathbb{E}\left[\int_0^s \sigma \mathrm{e}^{\lambda(r-s)} dW(r) \int_0^t \sigma \mathrm{e}^{\lambda(r-t)} dW(r)\right] \\ &= \sigma^2 \mathrm{e}^{-\lambda(s+t)} \mathbb{E}\left[\int_0^s \mathrm{e}^{\lambda r} dW(r) \int_0^t \mathrm{e}^{\lambda r} dW(r)\right]. \end{aligned}$$

Then, using the Itô isometry (8.21),

$$C(s,t) = \frac{\sigma^2}{2\lambda} \mathrm{e}^{-\lambda(s+t)} \big(\mathrm{e}^{2\lambda(s \wedge t)} - 1\big).$$

In particular, the variance $\mathrm{Var}(u(t)) = \sigma^2(1 - \mathrm{e}^{-2\lambda t})/2\lambda$ so that $\mathrm{Var}(u(t)) \to \sigma^2/2\lambda$ and $u(t) \to \mathrm{N}(\mu, \sigma^2/2\lambda)$ in distribution as $t \to \infty$.

8.4 Numerical methods for Itô SODEs

The geometric Brownian motion SODE (8.11) is exceptional in that the solution (8.34) is an explicit function of Brownian motion. In general, for nonlinear drift or diffusion functions, the explicit solution is not available and numerical techniques are necessary. The solution $\{\boldsymbol{u}(t) \colon t \ge 0\}$ of an SODE is a stochastic process. In this section, we examine the numerical approximation of $\boldsymbol{u}(t_n)$ by random variables $\boldsymbol{u}_n$ where $t_n = n\Delta t$, $n = 0, 1, 2, \ldots$.

One simple possibility is the Euler–Maruyama method and we sketch its derivation. From (8.3),

$$\boldsymbol{u}(t_{n+1}) = \boldsymbol{u}_0 + \int_0^{t_{n+1}} \boldsymbol{f}(\boldsymbol{u}(s))\, ds + \int_0^{t_{n+1}} G(\boldsymbol{u}(s))\, d\boldsymbol{W}(s),$$

and

$$\boldsymbol{u}(t_n) = \boldsymbol{u}_0 + \int_0^{t_n} \boldsymbol{f}(\boldsymbol{u}(s))\, ds + \int_0^{t_n} G(\boldsymbol{u}(s))\, d\boldsymbol{W}(s).$$

Subtracting, we see

$$\boldsymbol{u}(t_{n+1}) = \boldsymbol{u}(t_n) + \int_{t_n}^{t_{n+1}} \boldsymbol{f}(\boldsymbol{u}(s))\, ds + \int_{t_n}^{t_{n+1}} G(\boldsymbol{u}(s))\, d\boldsymbol{W}(s).$$

Taking both $\boldsymbol{f}$ and G constant over $[t_n, t_{n+1})$, we obtain the Euler–Maruyama method

$$\boldsymbol{u}_{n+1} = \boldsymbol{u}_n + \boldsymbol{f}(\boldsymbol{u}_n)\,\Delta t + G(\boldsymbol{u}_n)\,\Delta \boldsymbol{W}_n, \qquad \Delta \boldsymbol{W}_n := \boldsymbol{W}(t_{n+1}) - \boldsymbol{W}(t_n). \tag{8.35}$$

Since on each subinterval G is evaluated at the left-hand end point, (8.35) is consistent with our definition of the Itô integral (8.13).

We derive the Euler–Maruyama method as well as a higher-order method, known as the Milstein method, using Taylor series. In preparation for proving convergence of these methods in §8.5, we carefully identify the remainder terms. Assume that the drift $\boldsymbol{f} \in \mathrm{C}^2(\mathbb{R}^d, \mathbb{R}^d)$. Taylor's theorem (Theorem A.1) gives

$$\boldsymbol{f}(\boldsymbol{u}(r)) = \boldsymbol{f}(\boldsymbol{u}(s)) + D\boldsymbol{f}(\boldsymbol{u}(s))\big(\boldsymbol{u}(r) - \boldsymbol{u}(s)\big) + \int_0^1 (1-h) D^2 \boldsymbol{f}\big(\boldsymbol{u}(s) + h(\boldsymbol{u}(r) - \boldsymbol{u}(s))\big)\big(\boldsymbol{u}(r) - \boldsymbol{u}(s)\big)^2 \, dh,$$

which we may write as

$$\boldsymbol{f}(\boldsymbol{u}(r)) = \boldsymbol{f}(\boldsymbol{u}(s)) + \boldsymbol{R}_f(r; s, \boldsymbol{u}(s)), \qquad 0 \le s \le r,$$

for a remainder $\boldsymbol{R}_f \in \mathbb{R}^d$ given by

$$\boldsymbol{R}_f(r; s, \boldsymbol{u}(s)) := D\boldsymbol{f}(\boldsymbol{u}(s))\big(\boldsymbol{u}(r) - \boldsymbol{u}(s)\big) + \int_0^1 (1-h) D^2 \boldsymbol{f}\big(\boldsymbol{u}(s) + h(\boldsymbol{u}(r) - \boldsymbol{u}(s))\big)\big(\boldsymbol{u}(r) - \boldsymbol{u}(s)\big)^2 \, dh. \tag{8.36}$$

Similarly, if the diffusion $G \in \mathrm{C}^2(\mathbb{R}^d, \mathbb{R}^{d \times m})$,

$$G(\boldsymbol{u}(r)) = G(\boldsymbol{u}(s)) + DG(\boldsymbol{u}(s))\big(\boldsymbol{u}(r) - \boldsymbol{u}(s)\big) + R_G(r; s, \boldsymbol{u}(s)), \tag{8.37}$$

where we note that $DG(\boldsymbol{u}(s)) \in \mathcal{L}(\mathbb{R}^d, \mathbb{R}^{d \times m})$ and the remainder R_G is a $d \times m$ matrix defined by

$$R_G(r; s, \boldsymbol{u}(s)) := \int_0^1 (1-h) D^2 G\big(\boldsymbol{u}(s) + h(\boldsymbol{u}(r) - \boldsymbol{u}(s))\big)\big(\boldsymbol{u}(r) - \boldsymbol{u}(s)\big)^2 \, dh. \tag{8.38}$$

Substitute the Taylor expansions above for $\boldsymbol{f}$ and G into the integral equation

$$\boldsymbol{u}(t) = \boldsymbol{u}(s) + \int_s^t \boldsymbol{f}(\boldsymbol{u}(r)) \, dr + \int_s^t G(\boldsymbol{u}(r)) \, d\boldsymbol{W}(r), \tag{8.39}$$

to find

$$\boldsymbol{u}(t) = \boldsymbol{u}(s) + \boldsymbol{f}(\boldsymbol{u}(s))(t-s) + G(\boldsymbol{u}(s)) \int_s^t d\boldsymbol{W}(r) + \boldsymbol{R}_E(t; s, \boldsymbol{u}(s)), \tag{8.40}$$

where the remainder

$$\boldsymbol{R}_E(t; s, \boldsymbol{u}(s)) := \int_s^t \boldsymbol{R}_f(r; s, \boldsymbol{u}(s)) \, dr + \int_s^t R_G(r; s, \boldsymbol{u}(s)) \, d\boldsymbol{W}(r) + \int_s^t DG(\boldsymbol{u}(s))\big(\boldsymbol{u}(r) - \boldsymbol{u}(s)\big) \, d\boldsymbol{W}(r). \tag{8.41}$$

The Euler–Maruyama method is found by dropping the remainder from (8.40).

Definition 8.22 (Euler–Maruyama method) For a time step $\Delta t > 0$ and initial condition $\boldsymbol{u}_0 \in \mathbb{R}^d$, the Euler–Maruyama approximation $\boldsymbol{u}_n$ to the solution $\boldsymbol{u}(t_n)$ of (8.3) for $t_n = n\Delta t$ is defined by

$$\boldsymbol{u}_{n+1} = \boldsymbol{u}_n + \boldsymbol{f}(\boldsymbol{u}_n)\Delta t + G(\boldsymbol{u}_n)\,\Delta \boldsymbol{W}_n, \qquad \Delta \boldsymbol{W}_n := \int_{t_n}^{t_{n+1}} d\boldsymbol{W}(r) \tag{8.42}$$

Note also that $\Delta \boldsymbol{W}_n = \boldsymbol{W}(t_{n+1}) - \boldsymbol{W}(t_n)$.

The increments $\Delta \boldsymbol{W}_n \sim \mathrm{N}(\boldsymbol{0}, \Delta t I_m)$ *iid* and are easy to sample and the update rule (8.42) is simple to apply. The approximations $\boldsymbol{u}_n$ are random variables and, in computations, we typically generate a sample path $\boldsymbol{u}_n(\omega)$ for $n = 0, 1, \ldots, N$ that approximates a sample path $\boldsymbol{u}(\cdot, \omega)$ of the solution process $\boldsymbol{u}(t)$. Algorithm 8.1 generates sample paths $\boldsymbol{u}_n(\omega)$.

Algorithm 8.1 Code to find a sample path of the Euler–Maruyama approximation $\boldsymbol{u}_n$ for (8.3). The inputs are the initial data `u0`, final time `T`, the number of subintervals `N`, dimensions `d` and `m`, and handles `fhandle, ghandle` to the drift and diffusion functions. The outputs are a vector $[0, \Delta t, \ldots, T]^{\mathsf{T}}$ of times `t` and a matrix `u` with columns $\boldsymbol{u}_0, \ldots, \boldsymbol{u}_N$.

```
function [t, u]=EulerMaruyama(u0,T,N,d,m,fhandle,ghandle)
Dt=T/N; u=zeros(d,N+1); t=[0:Dt:T]'; sqrtDt=sqrt(Dt);
u(:,1)=u0; u_n=u0; % initial data
for n=1:N, % time loop
    dW=sqrtDt*randn(m,1); % Brownian increment
    u_new=u_n+Dt*fhandle(u_n)+ghandle(u_n)*dW;
    u(:,n+1)=u_new; u_n=u_new;
end
```

Example 8.23 (Duffing–van der Pol) The Duffing–van der Pol SODE (8.10) has drift and diffusion

$$\boldsymbol{f}(\boldsymbol{u}) = \begin{pmatrix} P \\ -P(\lambda + Q^2) + \alpha Q - Q^3 \end{pmatrix}, \qquad G(\boldsymbol{u}) = \begin{pmatrix} 0 \\ \sigma Q \end{pmatrix}$$

where $\boldsymbol{u} = [u_1, u_2]^{\mathsf{T}} = [Q, P]^{\mathsf{T}}$, for parameters λ, α, and σ. In the case $\lambda = \alpha = \sigma = 1$ and $\boldsymbol{u}_0 = [1/2, 0]^{\mathsf{T}}$, we find approximate sample paths using the Euler–Maruyama method with $\Delta t = 0.01$ on the time interval $[0, 10]$ by calling Algorithm 8.2 with the following commands:

```
>> lambda=1; alpha=1; sigma=1;
>> u0=[0.5;0]; T=10; N=1000;
>> [t, u]=vdp(u0, T, N, alpha, lambda, sigma);
```

Figure 8.5 gives plots of the resulting approximation. Exercise 8.9 investigates the Duffing–van der Pol equation as the parameters α and λ are varied.

Algorithm 8.2 Code to find a sample path of the Euler–Maruyama approximation to the Duffing–van der Pol SODE (8.10). The inputs are the initial data `u0`, final time `T`, number of time steps `N`, and parameters `alpha, lambda, sigma` for α, λ, σ. The outputs are as in Algorithm 8.1.

```
function [t,u]=vdp(u0,T,N,alpha,lambda,sigma)
[t, u]=EulerMaruyama(u0, T, N, 2, 1, @(u) vdp_f(u,lambda, alpha),...
                                     @(u) vdp_g(u,sigma));
function f=vdp_f(u, lambda, alpha) % define drift
f=[u(2); -u(2)*(lambda+u(1)^2)+alpha*u(1)-u(1)^3];
function g=vdp_g(u,sigma) % define diffusion
g=[0; sigma*u(1)];
```

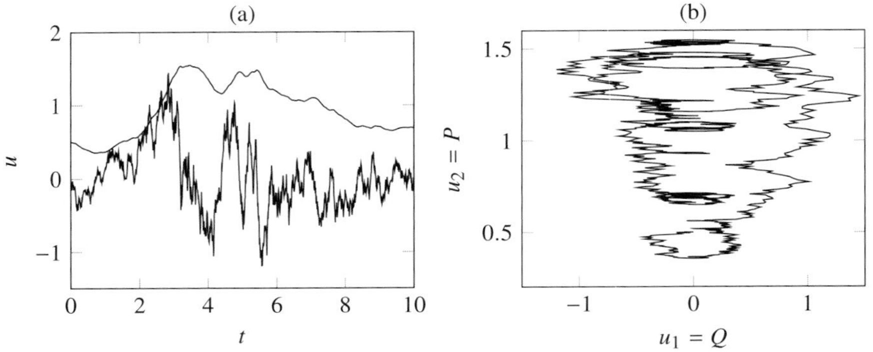

Figure 8.5 (a) A sample path of the Euler–Maruyama approximation to the Duffing–van der Pol SODE (8.10) with $\alpha = \lambda = \sigma = 1$ and initial data $\boldsymbol{u}_0 = [1/2, 0]^{\mathsf{T}}$. Note that $Q(t) = 1/2 + \int_0^t P(s)\,ds$ and is smoother that $P(t)$, which oscillates rapidly due to the Brownian motion forcing. (b) A phase-plane plot of the same sample path.

Milstein method

We develop $\boldsymbol{u}(t)$ further to derive a second numerical method, which has a smaller remainder and a better rate of convergence. From (8.40), we know

$$\boldsymbol{u}(t) = \boldsymbol{u}(s) + G(\boldsymbol{u}(s)) \int_s^t d\boldsymbol{W}(r) + \boldsymbol{R}_1(t; s, \boldsymbol{u}(s)), \tag{8.43}$$

where the remainder is

$$\boldsymbol{R}_1(t; s, \boldsymbol{u}(s)) := \boldsymbol{f}(\boldsymbol{u}(s))(t - s) + \boldsymbol{R}_E(t; s, \boldsymbol{u}(s)). \tag{8.44}$$

Using (8.43) with (8.37),

$$\begin{aligned} G(\boldsymbol{u}(r)) = G(\boldsymbol{u}(s)) + DG(\boldsymbol{u}(s))\Big(G(\boldsymbol{u}(s)) \int_s^r d\boldsymbol{W}(p)\Big) \\ + DG(\boldsymbol{u}(s))\boldsymbol{R}_1(t, s; \boldsymbol{u}(s)) + R_G(r; s, \boldsymbol{u}(s)) \end{aligned}$$

and then (8.39) gives

$$\begin{aligned} \boldsymbol{u}(t) = \boldsymbol{u}(s) + \boldsymbol{f}(\boldsymbol{u}(s))\,(t - s) + G(\boldsymbol{u}(s)) \int_s^t d\boldsymbol{W}(p) \\ + \int_s^t DG(\boldsymbol{u}(s))\Big(G(\boldsymbol{u}(s)) \int_s^r d\boldsymbol{W}(p)\Big)\, d\boldsymbol{W}(r) + \boldsymbol{R}_M(t; s, \boldsymbol{u}(s)), \end{aligned} \tag{8.45}$$

where the remainder $\boldsymbol{R}_M$ is given by

$$\begin{aligned} \boldsymbol{R}_M(t; s, \boldsymbol{u}(s)) := \int_s^t \boldsymbol{R}_f(r; s, \boldsymbol{u}(s))\, dr + \int_s^t R_G(r; s, \boldsymbol{u}(s))\, d\boldsymbol{W}(r) \\ + \int_s^t DG(\boldsymbol{u}(s))\boldsymbol{R}_1(r; s, \boldsymbol{u}(s))\, d\boldsymbol{W}(r). \end{aligned} \tag{8.46}$$

The Milstein method is found by dropping the remainder $\boldsymbol{R}_M$ from (8.45).

To simplify the presentation of Milstein's method, let $I_i(s,t) := \int_s^t dW_i(p)$ and exploit the identities

$$\int_s^t \int_s^r dW_i(p)\,dW_j(r) + \int_s^t \int_s^r dW_j(p)\,dW_i(r) = I_i(s,t)\,I_j(s,t), \qquad i \neq j,$$

and

$$\int_s^t \int_s^r dW_i(p)\,dW_i(r) = \frac{1}{2}\big(I_i(s,t)^2 - (t-s)\big).$$

See Exercises 8.4 and 8.5. Further, define

$$A_{ij}(s,t) := \int_s^t \int_s^r dW_i(p)\,dW_j(r) - \int_s^t \int_s^r dW_j(p)\,dW_i(r). \tag{8.47}$$

These are known as *Lévy areas* and the double integrals in (8.45) are written as

$$\begin{aligned} I_i I_j + A_{ij} &= 2\int_s^t \int_s^r dW_i(p)\,dW_j(r),\\ I_i I_j - A_{ij} &= 2\int_s^t \int_s^r dW_j(p)\,dW_i(r). \end{aligned} \tag{8.48}$$

Using $\boldsymbol{u}$ to denote $\boldsymbol{u}(s)$, the fourth term in (8.45) is

$$\int_s^t DG(\boldsymbol{u})\Big(G(\boldsymbol{u})\int_s^r d\boldsymbol{W}(p)\Big)\,d\boldsymbol{W}(r)$$

and has kth component (see Exercise 8.10)

$$\begin{aligned} &\frac{1}{2}\sum_{i=1}^{m}\sum_{\ell=1}^{d}\frac{\partial g_{ki}}{\partial u_\ell}(\boldsymbol{u})g_{\ell i}(\boldsymbol{u})\big(I_i(s,t)^2 - \Delta t\big)\\ &+\frac{1}{2}\sum_{i<j=1}^{m}\sum_{\ell=1}^{d}\Big(\frac{\partial g_{kj}}{\partial u_\ell}(\boldsymbol{u})g_{\ell i}(\boldsymbol{u}) + \frac{\partial g_{ki}}{\partial u_\ell}(\boldsymbol{u})g_{\ell j}(\boldsymbol{u})\Big)I_i(s,t)\,I_j(s,t)\\ &+\frac{1}{2}\sum_{i<j=1}^{m}\sum_{\ell=1}^{d}\Big(\frac{\partial g_{kj}}{\partial u_\ell}(\boldsymbol{u})g_{\ell i}(\boldsymbol{u}) - \frac{\partial g_{ki}}{\partial u_\ell}(\boldsymbol{u})g_{\ell j}(\boldsymbol{u})\Big)A_{ij}(s,t). \end{aligned}$$

Definition 8.24 (Milstein method) Consider a time step $\Delta t > 0$ and initial condition $\boldsymbol{u}_0 \in \mathbb{R}^d$, the Milstein approximation $\boldsymbol{u}_n$ to $\boldsymbol{u}(t_n)$ for $t_n = n\Delta t$ is defined by

$$\begin{aligned} u_{k,n+1} = u_{kn} &+ f_k(\boldsymbol{u}_n)\,\Delta t + \sum_{j=1}^{m} g_{kj}(\boldsymbol{u}_n)\,\Delta W_{jn} + \frac{1}{2}\sum_{i=1}^{m}\sum_{\ell=1}^{d}\frac{\partial g_{ki}}{\partial u_\ell}(\boldsymbol{u}_n)g_{\ell i}(\boldsymbol{u}_n)(\Delta W_{in}^2 - \Delta t)\\ &+\frac{1}{2}\sum_{i<j=1}^{m}\sum_{\ell=1}^{d}\Big(\frac{\partial g_{kj}}{\partial u_\ell}(\boldsymbol{u}_n)g_{\ell i}(\boldsymbol{u}_n) + \frac{\partial g_{ki}}{\partial u_\ell}(\boldsymbol{u}_n)g_{\ell j}(\boldsymbol{u}_n)\Big)\Delta W_{in}\Delta W_{jn}\\ &+\frac{1}{2}\sum_{i<j=1}^{m}\sum_{\ell=1}^{d}\Big(\frac{\partial g_{kj}}{\partial u_\ell}(\boldsymbol{u}_n)g_{\ell i}(\boldsymbol{u}_n) - \frac{\partial g_{ki}}{\partial u_\ell}(\boldsymbol{u}_n)g_{\ell j}(\boldsymbol{u}_n)\Big)A_{ij,n}, \end{aligned}$$

where $\Delta W_{in} := \int_{t_n}^{t_{n+1}} dW_i(r)$, $A_{ij,n} := A_{ij}(t_n, t_{n+1})$ and u_{kn} is the kth component of $\boldsymbol{u}_n$.

If the noise is additive then the Milstein method of Definition 8.24 and the Euler–Maruyama method of Definition 8.22 are equivalent since the derivatives of g_{kj} are zero.

A simple method for sampling the Lévy areas A_{ij} is considered in Exercise 5.3. In many cases however, the A_{ij} are not needed. When the noise is diagonal (i.e., G is a diagonal matrix), each component u_k is affected by W_k only, as $g_{kj} = 0$ when $j \neq k$, and Milstein's method reduces to

$$u_{k,n+1} = u_{kn} + f_k(\boldsymbol{u}_n)\,\Delta t + g_{kk}(\boldsymbol{u}_n)\,\Delta W_{kn} + \frac{1}{2}\frac{\partial g_{kk}}{\partial u_k}(\boldsymbol{u}_n) g_{kk}(\boldsymbol{u}_n)\big(\Delta W_{kn}^2 - \Delta t\big). \quad (8.49)$$

We implement the Milstein method for SODEs with diagonal noise in Algorithm 8.3. More generally, we say the noise is commutative when $\frac{\partial g_{kj}}{\partial u_\ell} g_{\ell i} = \frac{\partial g_{ki}}{\partial u_\ell} g_{\ell j}$ and again we do not need to sample Lévy areas.

Algorithm 8.3 Code to find a sample path of the Milstein approximation to (8.3) for a diagonal diffusion matrix G ($d = m$). Inputs and outputs are similar to those in Algorithm 8.1. However, `ghandle` specifies a vector in $\mathbb{R}^d$ for the diagonal entries of G. An additional input handle is required, `dghandle`, to specify the derivatives $\frac{\partial g_{kk}}{\partial u_k}$, $k = 1, \cdots, m$. This is also given as a vector in $\mathbb{R}^d$.

```
function [t, u]=MilsteinDiag(u0,T,N,d,m,fhandle,ghandle,dghandle)
Dt=T/N; u=zeros(d,N+1); t=[0:Dt:T]'; sqrtDt=sqrt(Dt);
u(:,1)=u0; u_n=u0; % initial data
for n=1:N, % time loop
  dW=sqrtDt*randn(m,1); gu_n=ghandle(u_n);
  u_new=u_n+Dt*fhandle(u_n)+gu_n.*dW ...
            +0.5*(dghandle(u_n).*gu_n).*(dW.^2-Dt);
  u(:,n+1)=u_new; u_n=u_new;
end
```

Example 8.25 Consider the SODE for $d = m = 2$

$$d\boldsymbol{u} = A\boldsymbol{u}\,dt + B(\boldsymbol{u})\,d\boldsymbol{W}(t), \qquad \boldsymbol{u}(0) = \boldsymbol{u}_0, \quad (8.50)$$

where A and $B(\boldsymbol{u})$ are 2×2 matrices. If $B(\boldsymbol{u})$ is diagonal, then we may apply Milstein's method using Algorithm 8.3. For the case,

$$A = \begin{pmatrix} 1 & 0 \\ 0 & -1 \end{pmatrix}, \qquad B(\boldsymbol{u}) = \begin{pmatrix} u_1 & 0 \\ 0 & 2u_2 \end{pmatrix}, \quad (8.51)$$

we have two decoupled geometric Brownian motions and, instead of forming the matrix B, we simply use the diagonal. We call Algorithm 8.3 with the following commands:

```
>> d=2; m=2; A=[1 0 ;0 -1]; sigma=[1; 2];
>> T=1; N=5000; u0=[1;1];
>> [t,u]=MilsteinDiag(u0, T, N, d, m, @(u) A*u, @(u) sigma.*u, @(u) sigma);
```

Figure 8.6 shows the resulting approximate sample path alongside an approximation by the Euler–Maruyama method.

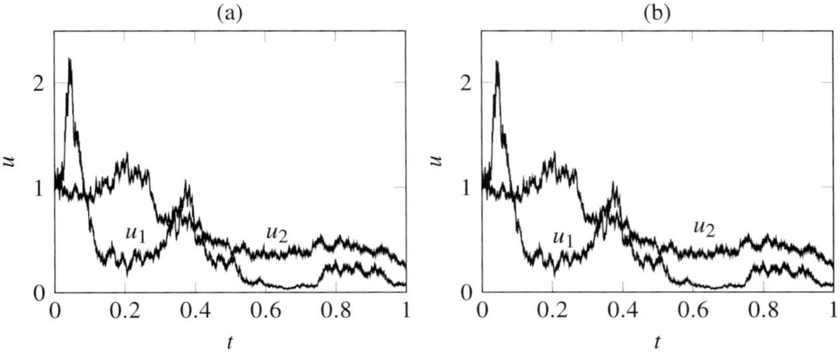

Figure 8.6 For the SODE (8.50), we plot (a) the Milstein approximation using Algorithm 8.3 and (b) the Euler–Maruyama approximation using Algorithm 8.1 to the same sample path of the solution $\boldsymbol{u}(t)$. In both cases, $\Delta t = 2 \times 10^{-4}$. Visually these two approximations appear identical.

Stability and implicit methods

To understand the stability of numerical methods for ODEs, we investigated the linear test equation in Example 3.9 and found a time step restriction is necessary for explicit methods to mimic dissipative behaviour. We now perform a stability analysis for SODEs. Instead of the linear test equation, we use the SODE (with $d = m = 1$) for geometric Brownian motion

$$du = ru\, dt + \sigma u\; dW(t), \qquad u(0) = u_0,$$

which has multiplicative noise. The solution from Example 8.20 is

$$u(t) = \exp\big((r - \sigma^2/2)\,t + \sigma\, W(t)\big)u_0$$

and, using (4.9), we see $\mathbb{E}\big[u(t)^2\big] = e^{(2r+\sigma^2)t} u_0^2$. For illustration, we consider geometric Brownian motion with parameters $r = -8$ and $\sigma = 3$. Approximate sample paths generated by the Euler–Maruyama method (using Algorithm 8.1) are shown in Figure 8.7 alongside exact sample paths. Two time steps are chosen and we see an instability for $\Delta t = 0.25$ where the numerical approximation becomes very large, while for $\Delta t = 0.0125$ the sample paths stay closer to the true solution. Stability is an issue for the numerical solution of SODEs, just as it was for the ODEs of §3.1.

Consider the case $u_0 \neq 0$. Clearly, $\mathbb{E}\big[u(t)^2\big]$ converges to zero as $t \to \infty$ if and only if $r + \sigma^2/2 < 0$. For what range of parameters does a numerical approximation u_n satisfy $\mathbb{E}\big[u_n^2\big] \to 0$ in the limit $n \to \infty$? We study this question for the Euler–Maruyama method.

Example 8.26 (Euler–Maruyama method) The Euler–Maruyama approximation u_n with time step Δt for the geometric Brownian motion SODE (8.34) is given by

$$u_{n+1} = u_n + ru_n\, \Delta t + \sigma u_n\, \Delta W_n.$$

Then, u_n is written explicitly in terms of u_0 as

$$u_n = \prod_{j=0}^{n-1} \big(1 + r\Delta t + \sigma \Delta W_j\big)\, u_0.$$

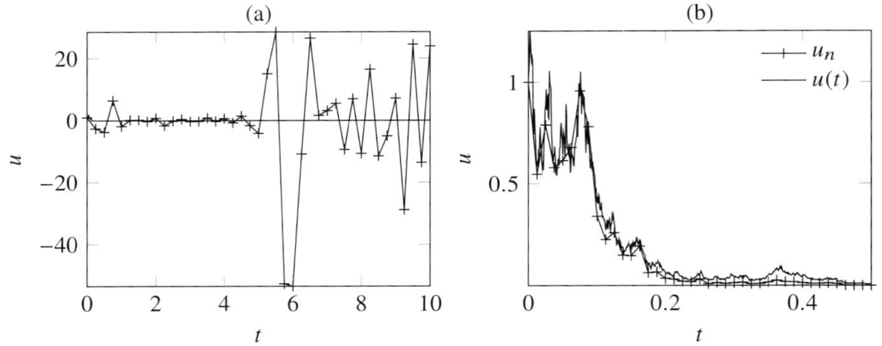

Figure 8.7 A sample path of geometric Brownian motion (8.34) with $u_0 = 1$, $r = -8$, $\sigma = 3$ plotted alongside the numerical approximation by the Euler–Maruyama method for (8.11) with (a) $\Delta t = 0.25$ and (b) $\Delta t = 0.0125$. The time step is smaller in (b) and the approximate sample path is more accurate. In (a), with the larger time step, the approximate path is unstable over a long time period and far from the exact solution.

The second moment of u_n is

$$\mathbb{E}\big[u_n^2\big] = \mathbb{E}\left[\prod_{j=0}^{n-1}\big(1 + r\Delta t + \sigma\Delta W_j\big)\right]^2 u_0^2 = \prod_{j=0}^{n-1}\mathbb{E}\Big[\big(1 + r\Delta t + \sigma\Delta W_j\big)^2\Big]u_0^2$$

$$= \prod_{j=0}^{n-1}\big((1 + r\Delta t)^2 + \sigma^2\Delta t\big)u_0^2,$$

where we exploit $\Delta W_j \sim \mathrm{N}(0, \Delta t)$ *iid*. Thus, $\mathbb{E}\big[u_n^2\big] \to 0$ as $n \to \infty$ if and only if

$$\Big|\big(1 + r\,\Delta t\big)^2 + \sigma^2\Delta t\Big| = 1 + 2\Delta t\big(r + \sigma^2/2 + \Delta t\, r^2/2\big) < 1.$$

Then, the Euler–Maruyama approximation u_n obeys $\mathbb{E}\big[u_n^2\big] \to 0$ as $n \to \infty$ if and only if

$$r + \sigma^2/2 + \Delta t\, r^2/2 < 0. \tag{8.52}$$

This is more restrictive than the stability condition $r + \sigma^2/2 < 0$ for the true solution $u(t)$. To achieve stability of the Euler–Maruyama approximation, we must choose the time step so that

$$0 < \Delta t < -2\big(r + \sigma^2/2\big)/r^2.$$

As with ODEs, time step restrictions are often too limiting in practical situations and we turn to implicit methods. We introduce an implicit version of the Euler–Maruyama method known as the θ-Euler–Maruyama method. Exercise 8.16 introduces an alternative for semilinear SODEs.

Definition 8.27 (θ-Euler–Maruyama) Consider a time step $\Delta t > 0$ and initial condition $\boldsymbol{u}_0 \in \mathbb{R}^d$, the θ-Euler–Maruyama approximation $\boldsymbol{u}_n$ to $\boldsymbol{u}(t_n)$ is defined by

$$\boldsymbol{u}_{n+1} = \boldsymbol{u}_n + \Big[(1-\theta)\boldsymbol{f}(\boldsymbol{u}_n) + \theta\boldsymbol{f}(\boldsymbol{u}_{n+1})\Big]\Delta t + G(\boldsymbol{u}_n)\Delta\boldsymbol{W}_n, \tag{8.53}$$

where θ is a parameter in $[0, 1]$ that controls the degree of implicitness in the drift term.

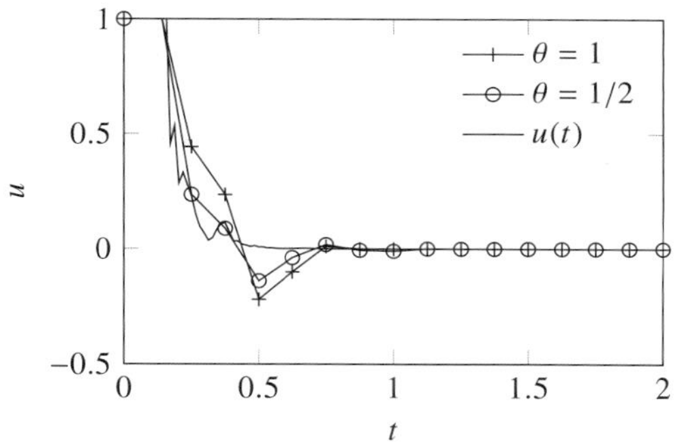

Figure 8.8 A sample path of geometric Brownian motion (8.34) with $u_0 = 1$, $r = -8$, $\sigma = 3$ and an approximation u_n of the sample path using the θ-Euler–Maruyama method on (8.11) with $\theta = 1$ and $\theta = 1/2$ with $\Delta t = 0.25$. Compare to Figure 8.7, where the Euler–Maruyama method was unstable.

The diffusion is treated explicitly, because the point of evaluation of G mimics the definition of the Itô integral.

Example 8.28 Reconsider geometric Brownian motion with parameters $r = -8$ and $\sigma = 3$. Approximate sample paths are generated by the θ-Euler–Maruyama method with $\theta = 1$ and $\theta = 1/2$ with $\Delta t = 1/4$ and are plotted in Figure 8.8(a) alongside the exact sample path. Compare to Figure 8.7(a), where the Euler–Maruyama method is unstable for the same value of Δt (i.e., $r + \sigma^2/2 + \Delta t\, r^2/2 = 4.5 > 0$). The approximation u_n is computed by the θ-Euler–Maruyama method for $\theta = 1$ and $\theta = 1/2$ by calling Algorithm 8.4 with the following commands:

```
>> u0=1; T=10; N=40; d=1; m=1; theta=1;
>> r=-8; sigma=3;
>> [t,u]=EulerMaruyamaTheta(u0,T,N,d,m,@(u)(r*u),@(u)(sigma*u),theta);
```

In both cases, the correct large n behaviour of the sample paths is observed. See Figure 8.8 for the cases $\theta = 1/2$ and $\theta = 1$. Even though the large n behaviour is correct, the numerical approximations can take negative values, which does not happen for geometric Brownian motion with a starting value $u_0 \geq 0$.

The stability calculation can be repeated to see that $\mathbb{E}\big[u_n^2\big] \to 0$ as $n \to \infty$ for the θ-Euler–Maruyama approximation u_n to geometric Brownian motion, when

$$r + \sigma^2/2 + \Delta t(1 - 2\theta)r^2/2 < 0.$$

In particular, if $\theta = 1/2$, the stability condition is $r + \sigma^2/2 < 0$, which is independent of the choice of time step and identical to the stability condition for geometric Brownian motion. See Exercise 8.13.

8.5 Strong approximation

For a given time step Δt, both the Euler–Maruyama and Milstein methods generate random variables $\boldsymbol{u}_n$ that approximate the random variables $\boldsymbol{u}(t_n)$ given by the exact solution of the SODE at $t_n = n\Delta t$. We now establish precisely how $\boldsymbol{u}_n$ converges to $\boldsymbol{u}(t_n)$ as $\Delta t \to 0$.

Algorithm 8.4 Code to find a sample path of the θ-Euler–Maruyama approximation to (8.3). The inputs and outputs are similar to those in Algorithm 8.1. An additional input `theta` is required, to specify the parameter θ.

```
function [t, u]=EulerMaruyamaTheta(u0,T,N,d,m,fhandle,ghandle,theta)
Dt=T/N; u=zeros(d,N+1); t=[0:Dt:N*Dt]'; sqrtDt=sqrt(Dt);
options=optimset('Display','Off');
u(:,1)=u0; u_n=u0; % initial data
for n=2:N+1, % time loop
    dW=sqrtDt*randn(m,1); % Brownian increment
    u_explicit=u_n+Dt*fhandle(u_n)+ghandle(u_n)*dW;
    if (theta>0) % solve nonlinear eqns for update
                % u_explicit is initial guess for nonlinear solve
       v=u_n+(1-theta)*Dt*fhandle(u_n)+ghandle(u_n)*dW;
       u_new=fzero(@(u) -u+v +theta*fhandle(u)*Dt,...
                 u_explicit, options);
    else % explicit case
       u_new=u_explicit;
    end
    u(:,n)=u_new; u_n=u_new;
end
```

As we saw in §4.3, there are many notions of convergence for random variables and the convergence of numerical methods for SODEs has to be approached with care. First, we consider *strong convergence*; that is, we care about approximating the sample path $\boldsymbol{u}(\cdot,\omega)$ for an $\omega \in \Omega$. In contrast in §8.6, we look at the *weak convergence* where only the distributions matter.

To make the notion of strong convergence definite, we use the root-mean-square or $L^2(\Omega,\mathbb{R}^d)$ error and show that

$$\sup_{0\le t_n\le T} \left\|\boldsymbol{u}(t_n) - \boldsymbol{u}_n\right\|_{L^2(\Omega,\mathbb{R}^d)} = \sup_{0\le t_n\le T} \mathbb{E}\left[\left\|\boldsymbol{u}(t_n) - \boldsymbol{u}_n\right\|_2^2\right]^{1/2} \tag{8.54}$$

is $\mathcal{O}(\Delta t)$ for the Milstein method in Theorem 8.32 and $\mathcal{O}(\Delta t^{1/2})$ for the Euler–Maruyama method in Exercise 8.11. Strong convergence is most closely linked to the numerical analysis of deterministic differential equations and makes a natural starting point. When the error (8.54) is small, sample paths $\boldsymbol{u}(\cdot,\omega)$ of the solution are well approximated (on finite time intervals) by a discrete sample path $\boldsymbol{u}_n(\omega)$ and this is useful in examining SODEs as a dynamical system. For example, the Duffing–van der Pol SODE exhibits bifurcations that can be investigated using approximate sample paths; see Exercise 8.9.

Upper bounds for the remainder terms

To prove strong convergence of the Milstein method, we find upper bounds for the remainder terms $\boldsymbol{R}_f$ in (8.36), R_G in (8.38), $\boldsymbol{R}_1$ in (8.44) and $\boldsymbol{R}_M$ in (8.46) under the following assumption.

Assumption 8.29 Let Assumption 8.17 hold, and suppose that f and G are twice continuously differentiable and the second derivatives are uniformly bounded.

First we prove a form of time regularity for the solution of the SODE (8.3).

Lemma 8.30 *Let Assumption 8.29 hold and $\boldsymbol{u}(t)$ be the solution of* (8.3). *For each $T > 0$ and $\boldsymbol{u}_0 \in \mathbb{R}^d$, there exists $K > 0$ such that for $0 \le s,t \le T$*

$$\|\boldsymbol{u}(t) - \boldsymbol{u}(s)\|_{L^2(\Omega,\mathbb{R}^d)} \le K|t-s|^{1/2}. \tag{8.55}$$

Proof From (8.3),

$$\boldsymbol{u}(t) - \boldsymbol{u}(s) = \int_s^t \boldsymbol{f}(\boldsymbol{u}(r))\,dr + \int_s^t G(\boldsymbol{u}(r))\,d\boldsymbol{W}(r).$$

Then, using (A.10),

$$\mathbb{E}\left[\|\boldsymbol{u}(t) - \boldsymbol{u}(s)\|_2^2\right] \le 2\mathbb{E}\left[\left\|\int_s^t \boldsymbol{f}(\boldsymbol{u}(r))\,dr\right\|_2^2\right] + 2\mathbb{E}\left[\left\|\int_s^t G(\boldsymbol{u}(r))\,d\boldsymbol{W}(r)\right\|_2^2\right].$$

Following the argument for deriving (8.29), we find

$$\mathbb{E}\left[\|\boldsymbol{u}(t) - \boldsymbol{u}(s)\|_2^2\right] \le 2L^2((t-s)+1)(t-s)\left(1 + \sup_{t\in[0,T]} \mathbb{E}\left[\|\boldsymbol{u}(t)\|_2^2\right]\right).$$

As $\boldsymbol{u} \in \mathcal{H}_{2,T}$, this implies (8.55). □

The following proposition gives the required estimates on the remainders.

Proposition 8.31 *Let Assumption 8.29 hold and $\boldsymbol{u}(t)$ be the solution of* (8.3). *For each $T > 0$ and $\boldsymbol{u}_0 \in \mathbb{R}^d$, there exists $K > 0$ such that the following hold for $0 \le s,t \le T$:*

(i) *For $\boldsymbol{R}_f$ and R_G given by* (8.36) *and* (8.38) *respectively,*

$$\begin{aligned}\mathbb{E}\left[\|\boldsymbol{R}_f(t;s,\boldsymbol{u}(s))\|_2^2\right] &\le K|t-s|,\\ \mathbb{E}\left[\|R_G(t;s,\boldsymbol{u}(s))\|_F^2\right] &\le K|t-s|^2.\end{aligned} \tag{8.56}$$

(ii) *For $\boldsymbol{R}_1$ given by* (8.44),

$$\mathbb{E}\left[\|\boldsymbol{R}_1(t;s,\boldsymbol{u}(s))\|_2^2\right] \le K|t-s|^2. \tag{8.57}$$

(iii) *For $\boldsymbol{R}_M$ given by* (8.46) *and $t_j = j\Delta t$ for $j = 0,1,2\ldots$,*

$$\mathbb{E}\left[\left\|\sum_{t_j<t} \boldsymbol{R}_M(t_{j+1}\wedge t; t_j, \boldsymbol{u}(t_j))\right\|_2^2\right] \le K\Delta t^2.$$

Proof (i) Under Assumption 8.29, the estimates follow from the definitions of $\boldsymbol{R}_f, R_G$ in (8.36)–(8.38) and the bound on $\|\boldsymbol{u}(t) - \boldsymbol{u}(s)\|_{L^2(\Omega,\mathbb{R}^d)}$ in (8.55).

(ii) From (8.41) and (A.10),

$$\begin{aligned}\mathbb{E}\left[\|\boldsymbol{R}_E(t;s,\boldsymbol{u}(s))\|_2^2\right] &\le 3\mathbb{E}\left[\left\|\int_s^t \boldsymbol{R}_f(r;s,\boldsymbol{u}(s))\,dr\right\|_2^2\right]\\ &+ 3\mathbb{E}\left[\left\|\int_s^t R_G(r;s,\boldsymbol{u}(s))\,d\boldsymbol{W}(r)\right\|_2^2\right] + 3\mathbb{E}\left[\left\|\int_s^t DG(\boldsymbol{u}(s))(\boldsymbol{u}(r)-\boldsymbol{u}(s))\,d\boldsymbol{W}(r)\right\|_2^2\right].\end{aligned}$$

By Jensen's inequality (A.9) and the Itô isometry (8.22),

$$\mathbb{E}\Big[\big\|\boldsymbol{R}_E(t;s,\boldsymbol{u}(s))\big\|_2^2\Big] \le 3\mathbb{E}\Big[\big\|\boldsymbol{R}_f(t;s,\boldsymbol{u}(s))\big\|_2^2\Big]|t-s|^2 + 3\int_s^t \mathbb{E}\Big[\big\|R_G(r;s,\boldsymbol{u}(r))\big\|_{\mathrm{F}}^2\Big]\,dr + 3\int_s^t \mathbb{E}\Big[\big\|DG(\boldsymbol{u}(s))(\boldsymbol{u}(r)-\boldsymbol{u}(s))\big\|_{\mathrm{F}}^2\Big]\,dr.$$

The first two terms are $\mathcal{O}(|t-s|^3)$ by (8.56). Because $DG(\boldsymbol{u})$ is a bounded linear operator uniformly over $\boldsymbol{u} \in \mathbb{R}^d$,

$$\big\|DG(\boldsymbol{u}(s))(\boldsymbol{u}(r)-\boldsymbol{u}(s))\big\|_{\mathrm{F}} \le K\big\|\boldsymbol{u}(r)-\boldsymbol{u}(s)\big\|_2$$

for some constant K and Lemma 8.30 says that the third term is $\mathcal{O}(|t-s|^2)$. Finally, taken together, we have $\|\boldsymbol{R}_E(t;s,\boldsymbol{u}(s))\|_{L^2(\Omega,\mathbb{R}^d)} = \mathcal{O}(|t-s|)$. (8.57) follows easily from the definition of $\boldsymbol{R}_1$ in (8.44).

(iii) For simplicity, assume that t is an integer multiple of Δt. Using (8.46) with (A.10) and (8.22), we have

$$\mathbb{E}\left[\left\|\sum_{t_j<t}\boldsymbol{R}_M(t_{j+1}\wedge t;t_j,\boldsymbol{u}(t_j))\right\|_2^2\right] \le 3\mathbb{E}\left[\left\|\sum_{t_j<t}\int_{t_j}^{t_{j+1}}\boldsymbol{R}_f(r;t_j,\boldsymbol{u}(t_j))\,dr\right\|_2^2\right] +3\sum_{t_j<t}\int_{t_j}^{t_{j+1}}\Big(\mathbb{E}\Big[\big\|R_G(r;t_j,\boldsymbol{u}(t_j))\big\|_{\mathrm{F}}^2\Big]+\mathbb{E}\Big[\big\|DG(\boldsymbol{u}(s))\boldsymbol{R}_1(s,\boldsymbol{u}(s))\big\|_{\mathrm{F}}^2\Big]\Big)\,dr.$$

Proposition 8.31(i) provides an $\mathcal{O}(\Delta t^2)$ estimate for the second term. For the first term, recall the definition of $\boldsymbol{R}_f$ from (8.36).

$$\begin{aligned}\boldsymbol{R}_f(r;s,\boldsymbol{u}(s)) &= D\boldsymbol{f}(\boldsymbol{u}(s))\big(\boldsymbol{u}(r)-\boldsymbol{u}(s)\big)\\ &\quad+\int_0^1(1-h)D^2\boldsymbol{f}\big(\boldsymbol{u}(s)+h(\boldsymbol{u}(r)-\boldsymbol{u}(s))\big)\big(\boldsymbol{u}(r)-\boldsymbol{u}(s)\big)^2\,dh\\ &= D\boldsymbol{f}(\boldsymbol{u}(s))\Big(G(\boldsymbol{u}(s))\int_s^r d\boldsymbol{W}(p)\Big)+D\boldsymbol{f}(\boldsymbol{u}(s))\,\boldsymbol{R}_1(s,\boldsymbol{u}(s))\\ &\quad+\int_0^1(1-h)D^2\boldsymbol{f}\big(\boldsymbol{u}(s)+h(\boldsymbol{u}(r)-\boldsymbol{u}(s))\big)\big(\boldsymbol{u}(r)-\boldsymbol{u}(s)\big)^2\,dh.\end{aligned}$$

using (8.43) gives for $r > s$. Let

$$\boldsymbol{\Theta}_j := \int_{t_j}^{t_{j+1}} D\boldsymbol{f}(\boldsymbol{u}(t_j))\Big(G(\boldsymbol{u}(t_j))\int_{t_j}^r d\boldsymbol{W}(p)\Big)\,dr$$

so that

$$\begin{aligned}\mathbb{E}\left[\left\|\sum_{j=0}^{k-1}\int_{t_j}^{t_{j+1}}\boldsymbol{R}_f(r;t_j,\boldsymbol{u}(t_j))\,dr\right\|_2^2\right] &\le 2\mathbb{E}\left[\left\|\sum_{j=0}^{k-1}\boldsymbol{\Theta}_j\right\|_2^2\right]\\ &+2\mathbb{E}\Bigg[\Bigg\|\sum_{j=0}^{k-1}\int_{t_j}^{t_{j+1}}D\boldsymbol{f}(\boldsymbol{u}(t_j))\,\boldsymbol{R}_1(t_j,\boldsymbol{u}(t_j))\\ &\quad+\int_0^1(1-h)D^2\boldsymbol{f}\big(\boldsymbol{u}(t_j)+h(\boldsymbol{u}(r)-\boldsymbol{u}(t_j))\big)\big(\boldsymbol{u}(r)-\boldsymbol{u}(t_j)\big)^2\,dh\,dr\Bigg\|_2^2\Bigg].\end{aligned}$$

The second term is $\mathcal{O}(\Delta t^2)$ by (8.57) and (8.55). For the first term, expand the sum in terms of $\langle \Theta_j, \Theta_k \rangle$ using the definition of the 2-norm. For $k > j$, we have $\mathbb{E}[\langle \Theta_j, \Theta_k \rangle] = \mathbb{E}[\langle \Theta_j, \mathbb{E}[\Theta_k \mid \mathcal{F}_{t_k}] \rangle] = 0$ a.s. Thus

$$\mathbb{E}\left[\left\| \sum_{j=0}^{k-1} \Theta_j \right\|_2^2\right] = \sum_{j=0}^{k-1} \mathbb{E}\left[\left\| \int_{t_j}^{t_{j+1}} Df(u(t_j))\Big(G(u(t_j)) \int_{t_j}^{r} dW(p)\Big)\, dr \right\|_2^2\right].$$

As $\int_{t_j}^{t_{j+1}} \int_{t_j}^{r} dW(p)\, dr \sim \mathrm{N}(\mathbf{0}, \frac{1}{3}\Delta t^3 I_m)$ (see Exercise 8.3), each term in the sum is $\mathcal{O}(\Delta t^3)$ and the sum is $\mathcal{O}(\Delta t^2)$ overall. □

Strong convergence of the Milstein method

We prove strong convergence of the Milstein method. For the proof, we define a continuous time process $u_{\Delta t}(t)$ that agrees with the approximation u_n at $t = t_n$. To do this, introduce the variable $\hat{t} := t_n$ for $t_n \le t < t_{n+1}$ and let

$$u_{\Delta t}(t) = u_{\Delta t}(\hat{t}) + f(u_{\Delta t}(\hat{t})) \int_{\hat{t}}^{t} ds + G(u_{\Delta t}(\hat{t})) \int_{\hat{t}}^{t} dW(r) + \int_{\hat{t}}^{t} DG(u_{\Delta t}(\hat{t}))\Big(G(u_{\Delta t}(\hat{t})) \int_{\hat{t}}^{r} dW(p)\Big)\, dW(r).$$

Notice that $u_{\Delta t}(\hat{t}) = u_n$ for $\hat{t} = t_n$. Furthermore, using the definition of $\hat{t}$,

$$u_{\Delta t}(t) = u_{\Delta t}(t_0) + \int_{t_0}^{t} f(u_{\Delta t}(\hat{s}))ds + \int_{t_0}^{t} G(u_{\Delta t}(\hat{s}))dW(s) + \int_{t_0}^{t} DG(u_{\Delta t}(\hat{s}))\Big(G(u_{\Delta t}(\hat{s})) \int_{\hat{s}}^{s} dW(r)\Big)\, dW(s). \tag{8.58}$$

In the following, we assume in (8.59) that $DG(u)G(u)$ is globally Lipschitz. This occurs if $G(u)$ is a linear function of u (as with geometric Brownian motion) or if $G(u)$, $DG(u)$, and $D^2G(u)$ are all uniformly bounded over $u \in \mathbb{R}^d$.

Theorem 8.32 (convergence of Milstein method) *Let Assumption 8.29 hold. Further assume, for some $L_2 > 0$, that*

$$\|DG(u_1)(G(u_1)\xi) - DG(u_2)(G(u_2)\xi)\|_{\mathrm{F}} \le L_2 \|u_1 - u_2\|_2 \, \|\xi\|_2 \tag{8.59}$$

for all $u_1, u_2 \in \mathbb{R}^d$ and $\xi \in \mathbb{R}^m$. Let u_n denote the Milstein approximation to the solution $u(t)$ of (8.3). *For $T \ge 0$, there exists $K > 0$ such that*

$$\sup_{0 \le t_n \le T} \|u(t_n) - u_n\|_{L^2(\Omega, \mathbb{R}^d)} \le K\Delta t. \tag{8.60}$$

Proof To simplify notation, let $D(s) := f(u_{\Delta t}(\hat{s})) - f(u(\hat{s}))$ and

$$M(s) := \Big[G(u_{\Delta t}(\hat{s})) - G(u(\hat{s}))\Big] + \Big[DG(u_{\Delta t}(\hat{s}))\Big(G(u_{\Delta t}(\hat{s})) \int_{\hat{s}}^{s} dW(r)\Big) - DG(u(\hat{s}))\Big(G(u(\hat{s})) \int_{\hat{s}}^{s} dW(r)\Big)\Big].$$

Then, subtracting (8.45) from (8.58),

$$\begin{aligned} \boldsymbol{u}_{\Delta t}(t) - \boldsymbol{u}(t) &= \int_0^t \boldsymbol{D}(s)\,ds + \int_0^t M(s)\,d\boldsymbol{W}(s) \\ &\quad + \sum_{t_j<t} \boldsymbol{R}_M(t_{j+1}\wedge t; t_j, \boldsymbol{u}(t_j)) + \boldsymbol{R}_M(t; t_n, \boldsymbol{u}(t_n)). \end{aligned} \tag{8.61}$$

Let $\boldsymbol{e}(\hat{t}) = \boldsymbol{u}_{\Delta t}(\hat{t}) - \boldsymbol{u}(\hat{t})$. Then,

$$\begin{aligned} \mathbb{E}\left[\|\boldsymbol{e}(\hat{t})\|_2^2\right] &\le 3\mathbb{E}\left[\left\|\int_0^{\hat{t}} \boldsymbol{D}(s)\,ds\right\|_2^2\right] + 3\mathbb{E}\left[\left\|\int_0^{\hat{t}} M(s)\,d\boldsymbol{W}(s)\right\|_2^2\right] \\ &\quad + 3\mathbb{E}\left[\left\|\sum_{t_j<\hat{t}} \boldsymbol{R}_M(t_{j+1}\wedge \hat{t}; t_j, \boldsymbol{u}(t_j))\right\|_2^2\right]. \end{aligned} \tag{8.62}$$

For the first term, the Lipschitz condition (8.25) on $\boldsymbol{f}$ and the Cauchy–Schwarz inequality give

$$\mathbb{E}\left[\left\|\int_0^{\hat{t}} \boldsymbol{D}(s)\,ds\right\|_2^2\right] \le \hat{t}\int_0^{\hat{t}} \mathbb{E}\left[\|\boldsymbol{D}(s)\|_2^2\right] ds \le \hat{t}\int_0^{\hat{t}} L^2\mathbb{E}\left[\|\boldsymbol{e}(s)\|^2\right] ds.$$

For the second term of (8.62), we have, by (8.22),

$$\mathbb{E}\left[\left\|\int_0^{\hat{t}} M(s)\,d\boldsymbol{W}(s)\right\|_2^2\right] = \int_0^{\hat{t}} \mathbb{E}\left[\|M(s)\|_{\mathrm{F}}^2\right] ds.$$

Under (8.24) and (8.59),

$$\begin{aligned} \mathbb{E}\left[\|M(s)\|_{\mathrm{F}}^2\right] &\le 2\mathbb{E}\left[\|G(\boldsymbol{u}_{\Delta t}(\hat{s})) - G(\boldsymbol{u}(\hat{s}))\|_{\mathrm{F}}^2\right] \\ + 2\mathbb{E}&\left[\left\|DG(\boldsymbol{u}_{\Delta t}(\hat{s}))\left(G(\boldsymbol{u}_{\Delta t}(\hat{s}))\int_{\hat{s}}^{s} d\boldsymbol{W}(r)\right) - DG(\boldsymbol{u}(\hat{s}))\left(G(\boldsymbol{u}(\hat{s}))\int_{\hat{s}}^{s} d\boldsymbol{W}(r)\right)\right\|_{\mathrm{F}}^2\right] \\ &\le 2(L+L_2)\,\mathbb{E}\left[\|\boldsymbol{e}(s)\|_2^2\right]. \end{aligned}$$

Consequently, for some $K_3 > 0$ independent of Δt,

$$\mathbb{E}\left[\left\|\int_0^{\hat{t}} M(s)\,d\boldsymbol{W}(s)\right\|_2^2\right] \le K_3\int_0^{\hat{t}} \mathbb{E}\left[\|\boldsymbol{e}(s)\|_2^2\right] ds.$$

For the third term of (8.62), there exists $K_2 > 0$ by Proposition 8.31(iii) again independent of Δt such that

$$\mathbb{E}\left[\left\|\sum_{j=0}^{n-1} \boldsymbol{R}_M(t_{j+1}\wedge \hat{t}; t_j, \boldsymbol{u}(t_j))\right\|_2^2\right] \le K_2\Delta t^2.$$

Putting the estimates together with (8.62), we have for $t \in [0,T]$

$$\mathbb{E}\left[\|\boldsymbol{e}(\hat{t})\|_2^2\right] \le 3(TL^2 + K_3)\int_0^{\hat{t}} \mathbb{E}\left[\|\boldsymbol{e}(s)\|_2^2\right] ds + 3K_2\Delta t^2.$$

Gronwall's inequality (Exercise 3.4) completes the proof. □

Theorem 8.32 holds for both multiplicative noise (where G depends on $\boldsymbol{u}$) and additive noise (where G is independent of $\boldsymbol{u}$). For the special case of additive noise, the Euler–Maruyama method (recall Definition 8.22) and the Milstein method are the same and Theorem 8.32 shows convergence of the Euler–Maruyama method with $\mathcal{O}(\Delta t)$. However, in general, the Euler–Maruyama method converges with $\mathcal{O}(\Delta t^{1/2})$. See Exercise 8.11.

To apply the Milstein method in dimension $d > 1$, the Lévy areas (8.47) are usually required. They are difficult to sample and, in general, must be approximated. This introduces an extra error term into the remainder defined in (8.46). To achieve the same $L^2(\Omega, \mathbb{R}^d)$ order of approximation, the error in approximating the Lévy areas must be consistent with Proposition 8.31(iii) and hence must be approximated with an error of size $\mathcal{O}(\Delta t^{3/2})$. Exercise 8.12 gives a method for approximating the Lévy areas based on applying the Euler–Maruyama method to a specific SODE. The method is slow and often the cost of approximating the Lévy areas outweighs the advantages of the higher convergence rate of the Milstein method.

Algorithm 8.5 Code to find M *iid* sample paths of the Euler–Maruyama approximation $\boldsymbol{u}_n$ to the solution $\boldsymbol{u}(t)$ of (8.3). Inputs and outputs are similar to those in Algorithm 8.1. In addition, `M` specifies the number of samples and `kappa0`$=\kappa$ determines the size of the Brownian increments used in (8.64) for Δt_{ref} `=T/(N*kappa0)`. `u` is a three-dimensional array and `u(:,j,n+1)` $= \boldsymbol{u}_n^j$ for $n = 0, \ldots, N$ and $j = 1, \ldots, M$.

```
function [t, u]=EMpath(u0,T,N,d,m,fhandle,ghandle,kappa0,M)
Dtref=T/N;  % small step
Dt=kappa0*Dtref; % large step
NN=N/kappa0;  u=zeros(d,M,NN+1); t=zeros(NN+1,1);
gdW=zeros(d,M); sqrtDtref=sqrt(Dtref); u_n=u0;
for n=1:NN+1
  t(n)=(n-1)*Dt; u(:,:,n)=u_n;
  dW=sqrtDtref*squeeze(sum(randn(m,M,kappa0),3));
  for mm=1:M
    gdW(:,mm)=ghandle(u_n(:,mm))*dW(:,mm);
  end
  u_new=u_n+Dt*fhandle(u_n)+gdW; u_n=u_new;
end
```

Calculating the L^2 error numerically

We consider strong convergence numerically by estimating the $L^2(\Omega, \mathbb{R}^d)$ norm using a sample average. That is,

$$\left\| \boldsymbol{u}(T) - \boldsymbol{u}_N \right\|_{L^2(\Omega, \mathbb{R}^d)} \approx \left(\frac{1}{M} \sum_{j=1}^{M} \left\| \boldsymbol{u}^j(T) - \boldsymbol{u}_N^j \right\|_2^2 \right)^{1/2}. \tag{8.63}$$

Here, $\boldsymbol{u}^j(T) = \boldsymbol{u}(T, \omega_j)$ are independent samples of the exact solution $\boldsymbol{u}(T)$ and $\boldsymbol{u}_N^j = \boldsymbol{u}_N(\omega_j)$ is the numerical approximation to $\boldsymbol{u}(T, \omega_j)$ for $T = N\Delta t$. There are a number of different numerical issues to consider. First, we need to simulate $\boldsymbol{u}_N^j$ using the same Brownian path as $\boldsymbol{u}^j(T)$. To do this, we choose independent sample paths $\boldsymbol{W}(\cdot, \omega_j)$ of the Brownian motion $\boldsymbol{W}(t)$ and make sure the same sample path is used for both samples.

Second, it is unusual to have explicit solutions and, in practice, $\boldsymbol{u}^j(T)$ are approximated by accurate reference solutions $\boldsymbol{u}^j_{N_{\text{ref}}}$, computed with a small time step $\Delta t_{\text{ref}} = T/N_{\text{ref}}$. To keep the same sample path, the increments $W((n+1)\Delta t_{\text{ref}}) - W(n\Delta t_{\text{ref}})$ for the reference sample paths are used to generate the required increments over a time step $\Delta t = \kappa \Delta t_{\text{ref}}$ by

$$\boldsymbol{W}((n+1)\Delta t) - \boldsymbol{W}(n\Delta t) = \sum_{j=n\kappa}^{(n+1)\kappa-1} \big(\boldsymbol{W}((j+1)\Delta t_{\text{ref}}) - \boldsymbol{W}(j\Delta t_{\text{ref}})\big). \tag{8.64}$$

Third, we need to generate a large number of samples M for (8.63) to be a good approximation. Finally, we need to do this over a range of different time steps Δt to see the error in (8.63) decay and hence observe convergence.

To start with, we compute approximations to $\boldsymbol{u}(t)$ with two different time steps using the same sample path of $\boldsymbol{W}(t)$ (taking $M = 1$). This is implemented in Algorithm 8.5 using the Euler–Maruyama method (compare to Algorithm 8.1). This algorithm uses two time steps Δt and Δt_{ref} such that $\Delta t = \kappa \Delta t_{\text{ref}}$ for some $\kappa \in \mathbb{N}$. Then, $T = N_{\text{ref}}\Delta t_{\text{ref}} = N\Delta t$ and $N_{\text{ref}} = N\kappa$. We use $\boldsymbol{u}_n$ to denote the approximation to $\boldsymbol{u}(t_n)$ with time step Δt and $\boldsymbol{u}_{n\kappa}$ to denote the approximation to $\boldsymbol{u}(t_n)$ with step Δt_{ref}. The Brownian increments for the larger step Δt are computed using steps of size Δt_{ref} using (8.64).

Example 8.33 (one sample path) Consider the two-dimensional SODE in Example 8.25 (which has independent geometric Brownian motions in each component). We examine a single sample path ($M = 1$) and generate approximations with time steps Δt and Δt_{ref} using the same Brownian path. For the time step Δt_{ref}, call Algorithm 8.5 with `N=Nref` and `kappa0=1` as follows:

```
>> d=2; m=2; T=1; Nref=5000; M=1; A=[1 0 ; 0 -1]; u0=[1;1];
>> defaultStream = RandStream.getGlobalStream;
>> savedState = defaultStream.State;
>> [tref, uref]=EMpath(u0,T,Nref,d,m,...
                      @(u) A*u,@(u)[u(1) 0 ; 0 2*u(2)],1,M);
```

To approximate the same sample path with time step $\Delta t = \kappa \Delta t_{\text{ref}}$, we reset the random number generator so the next set of Brownian increments uses the *same* random numbers as the first call. This ensures we approximate the same sample path of $\boldsymbol{u}(t)$. We then call Algorithm 8.5 with `kappa0=kappa`.

```
>> defaultStream.State = savedState; kappa=100;
>> [t, u]=EMpath(u0,T,Nref,d,m,@(u)A*u,@(u)[u(1) 0 ;0 2*u(2)],kappa,M);
>> err=squeeze(u-uref(:,:,1:kappa:end));
```

In Figure 8.9(a), we plot [`tref`, `uref`] and [`t`, `u`] for the same sample path shown in Figure 8.6. In this case, the time steps are $\Delta t_{\text{ref}} = 2 \times 10^{-4}$ and $\Delta t = \kappa \Delta t_{\text{ref}} = 0.02$ for $\kappa = 100$. The matrix `err` has columns `err(:,n)` corresponding to $\boldsymbol{u}_n - \boldsymbol{u}_{n\kappa}$ and, using this data, we plot the error $\|\boldsymbol{u}_n - \boldsymbol{u}_{n\kappa}\|_2$ in Figure 8.9(b).

We consider how to approximate $\|\boldsymbol{u}(T) - \boldsymbol{u}_N\|_{L^2(\Omega,\mathbb{R}^d)}$. We take M independent samples $\boldsymbol{u}^j(T)$ of the solution, their approximations $\boldsymbol{u}^j_N$, and estimate the error for each. Rather than use an analytical solution, we approximate $\boldsymbol{u}^j(T)$ by a reference solution $\boldsymbol{u}^j_{N_{\text{ref}}}$ obtained with $\Delta t_{\text{ref}} = T/N_{\text{ref}}$ and $N_{\text{ref}} = \kappa N$, some $\kappa > 1$. As the time step $\Delta t_{\text{ref}} = T/\kappa N$ is smaller

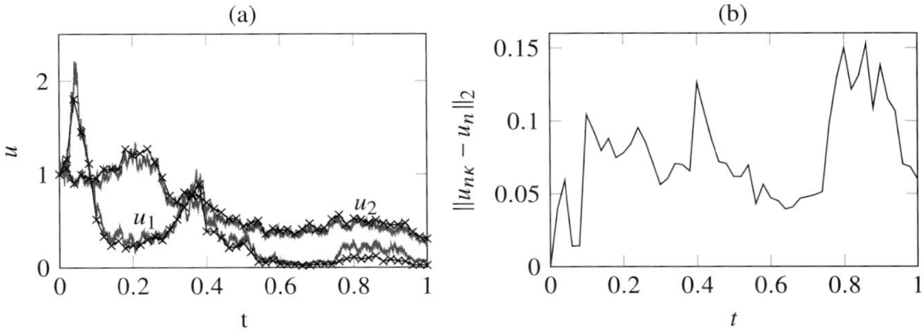

Figure 8.9 (a) Approximation to a sample path of (8.50) by the Euler–Maruyama method with $\Delta t = 0.04$ (marked by $\times$) and reference approximation computed with $\Delta t_{\text{ref}} = 2 \times 10^{-4}$ (solid line). (b) Plot of the error $\|\boldsymbol{u}_{n\kappa} - \boldsymbol{u}_n\|_2$ against t_n for $n = 0, 1, \ldots, N$.

than Δt, $\boldsymbol{u}^j_{N_{\text{ref}}}$ should be more accurate than $\boldsymbol{u}^j_N$. Then, we approximate

$$\|\boldsymbol{u}(T) - \boldsymbol{u}_N\|_{L^2(\Omega, \mathbb{R}^d)} \approx \left(\frac{1}{M} \sum_{j=1}^{M} \left\| \boldsymbol{u}^j_{N_{\text{ref}}} - \boldsymbol{u}^j_N \right\|_2^2 \right)^{1/2} \tag{8.65}$$

where we have replaced the sample $\boldsymbol{u}^j(T)$ in (8.63) by the reference solution $\boldsymbol{u}^j_{N_{\text{ref}}}$. This is different to Example 8.33 where we have one sample ($M = 1$) and we plot the error on $[0, T]$. For $M > 1$, Algorithm 8.5 computes M *iid* approximate sample paths simultaneously using the Euler–Maruyama method. Algorithm 8.6 adapts the MATLAB commands in Example 8.33 to evaluate (8.65). To deal with large M, the samples are split into blocks of size at most `Mstep`. For each block, Algorithm 8.5 is called twice and in each case Brownian paths are computed using the small time step Δt_{ref} (we reset the seed between calls to Algorithm 8.5). These blocks may be run in parallel, although here they are run in serial.

Example 8.34 (M sample paths) Continuing with Example 8.33, we estimate numerically rates of convergence of the Euler–Maruyama method using Algorithm 8.6. Although we have an exact solution of (8.50), we estimate the error using a reference solution and compute (8.65). Using Algorithm 8.6, the following commands compute (8.65) with $M = 5000$ samples and $N_{\text{ref}} = 5000$ for the six time steps $\Delta t = \kappa \Delta t_{\text{ref}}$ given by $\kappa = 5, 10, 50, 100, 500, 1000$.

```
>> d=2; m=2; T=1; N=5000; Dtref=T/N; M=5000;
>> A=[1 0 ; 0 -1]; u0=repmat([1;1],[1,M]);
>> kappa=[5,10,50,100,500,1000];
>> for k=1:length(kappa)
     [rmsErr(k)]=runEMpath(u0,T,N,d,m,@(u) A*u,...
                           @(u)[u(1) 0 ; 0 2*u(2)],kappa(k),M);
   end
```

Using this data, Figure 8.10 illustrates the rates of convergence for the Euler–Maruyama method. The figure also shows results for the Milstein method, found using a code similar to Algorithm 8.5 based on Algorithm 8.3. The rates of convergence found numerically are in agreement with the strong convergence theory of Theorem 8.32 and Exercise 8.11.

Algorithm 8.6 Code to approximate the $L^2(\Omega, \mathbb{R}^d)$ error for the Euler–Maruyama method using (8.65). Inputs are as in Algorithm 8.5, except this time `kappa` denotes κ. The output `rmsErr` gives (8.65). For the last block of samples, we also output the results `[tref, uref]` and `[t, u]` of calling Algorithm 8.5 with `kappa0=1` and `kappa0=kappa` respectively. Crucially, `u(:,j,:)` and `uref(:,j,:)` approximate the same sample path $\boldsymbol{u}^j(t)$.

```
function [rmsErr,t,u,tref,uref]=runEMpath(u0,T,Nref,d,m,...
                                  fhandle,ghandle,kappa,M)
S=0; Mstep=1000; m0=1;
for mm=1:Mstep:M
  MM=min(Mstep,M-mm+1); u00=u0(:,mm:m0+MM-1);
  defaultStream = RandStream.getGlobalStream;
  savedState = defaultStream.State;
  [tref, uref]=EMpath(u00, T, Nref, d, m, fhandle, ghandle,1,MM);
  defaultStream.State = savedState;
  [t, u]=EMpath(u00, T, Nref, d, m,fhandle, ghandle,kappa,MM);
  err=u(:,:,end)-uref(:,:,end);
  S=S+sum(sum(err.*err)); m0=m0+MM;
end
rmsErr=sqrt(S./M);
```

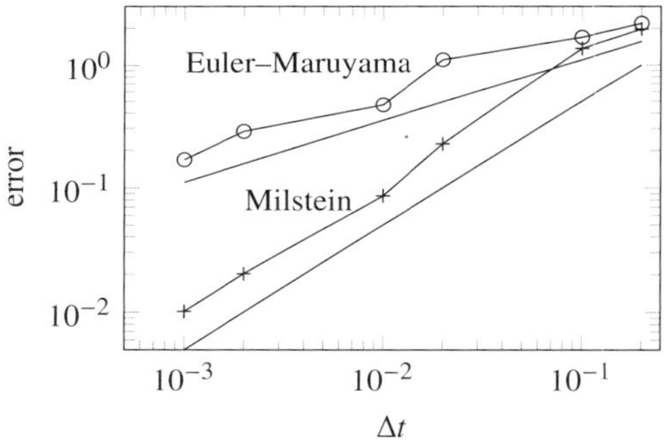

Figure 8.10 A log-log plot of the approximation (8.65) to $\|u(T) - u_N\|_{L^2(\Omega,\mathbb{R}^2)}$ at $t_N = T = 1$ against Δt, showing convergence of the Euler–Maruyama and Milstein methods for (8.50). Here $M = 5000$ samples and $\Delta t_{\text{ref}} = 2 \times 10^{-4}$. Also shown are reference lines of slope 1/2 and 1 indicating the convergence rates of the Euler–Maruyama and Milstein methods are $\mathcal{O}(\Delta t^{1/2})$ and $\mathcal{O}(\Delta t)$, respectively, in agreement with the theory.

8.6 Weak approximation

In many situations, individual sample paths are not of interest and it is average quantities, such as the temperature $\mathbb{E}\big[\frac{1}{2}P(T)^2\big]$ in Example 8.1, that are needed. The approximation of $\mathbb{E}[\phi(\boldsymbol{u}(T))]$ for a given quantity of interest or test function $\phi \colon \mathbb{R}^d \to \mathbb{R}$ is known as *weak approximation* and we now use the Monte Carlo and Euler–Maruyama methods to compute weak approximations. In other words, we generate M independent samples $\boldsymbol{u}_N^j$ for $j = 1, \ldots, M$ of the Euler–Maruyama approximation $\boldsymbol{u}_N$ to $\boldsymbol{u}(T)$ for $t_N = T$ and approximate $\mathbb{E}[\phi(\boldsymbol{u}(T))]$ by the sample average

$$\mu_M := \frac{1}{M}\sum_{j=1}^{M}\phi(\boldsymbol{u}_N^j). \tag{8.66}$$

As we saw in Example 4.72, the error can be divided as the sum of the weak discretisation error due to approximating $\boldsymbol{u}(T)$ by $\boldsymbol{u}_N$ and the Monte Carlo error due to taking M samples,

$$\mathbb{E}\big[\phi(\boldsymbol{u}(T))\big] - \mu_M = \underbrace{\big[\mathbb{E}[\phi(\boldsymbol{u}(T))] - \mathbb{E}[\phi(\boldsymbol{u}_N)]\big]}_{\text{weak discretization error}} + \underbrace{\big[\mathbb{E}[\phi(\boldsymbol{u}_N)] - \mu_M\big]}_{\text{Monte Carlo error}}. \tag{8.67}$$

From (4.34), the Monte Carlo error satisfies

$$\mathbb{P}\left(\big|\mathbb{E}[\phi(\boldsymbol{u}_N)] - \mu_M\big| < \frac{2\sigma}{\sqrt{M}}\right) > 0.95 + \mathcal{O}(M^{-1/2}), \qquad \sigma^2 := \mathrm{Var}(\phi(\boldsymbol{u}_N)).$$

By Theorem 4.58(ii) (with $r = 1/2$, $H = \mathbb{R}^d$, $X_j = \boldsymbol{u}_N$, $\Delta t = T/j$), the weak discretisation error $\mathbb{E}[\phi(\boldsymbol{u}(T))] - \mathbb{E}[\phi(\boldsymbol{u}_N)]$ for globally Lipschitz test functions $\phi\colon \mathbb{R}^d \to \mathbb{R}$ is controlled by the root-mean-square error $\|\boldsymbol{u}(T) - \boldsymbol{u}_N\|_{L^2(\Omega,\mathbb{R}^d)}$. In particular, due to Exercise 8.11, there exists $K > 0$ such that

$$\big|\mathbb{E}\big[\phi(\boldsymbol{u}_N)\big] - \mathbb{E}\big[\phi(\boldsymbol{u}(T))\big]\big| \le K\Delta t^{1/2}, \qquad t_N = T. \tag{8.68}$$

Consequently, the weak error obeys

$$\mathbb{P}\left(\big|\mathbb{E}[\phi(\boldsymbol{u}(T))] - \mu_M\big| < K\Delta t^{1/2} + \frac{2\sigma}{\sqrt{M}}\right) > 0.95 + \mathcal{O}(M^{-1/2}),$$

or informally we write $\mu_M = \mathbb{E}[\phi(\boldsymbol{u}(T))] + \mathcal{O}(\Delta t^{1/2}) + \mathcal{O}(M^{-1/2})$. In this section, we investigate weak approximation in more detail. We first look at the weak discretisation error and show, in Theorem 8.45, that the error for the Euler–Maruyama method is in fact $\mathcal{O}(\Delta t)$ when ϕ is sufficiently regular. This is smaller than the $\mathcal{O}(\Delta t^{1/2})$ error predicted by (8.68). We then turn to the Monte Carlo error and introduce the multilevel Monte Carlo method.

Markov property

We require two theoretical concepts to study weak approximation: the Markov property and the backward Kolmogorov equation. We start with the Markov property and this is easiest to describe for discrete time, such as the random variables $\boldsymbol{u}_n$ given by the Euler–Maruyama or Milstein methods. In both cases, we determine $\boldsymbol{u}_{n+1}$ from $\boldsymbol{u}_n$ and do not use $\boldsymbol{u}_{n-1}, \boldsymbol{u}_{n-2}, \dots$. In other words, the conditional distribution of the future (the random variable $\boldsymbol{u}_{n+1}$) given the present (the random variable $\boldsymbol{u}_n$) is independent of the past (the random variables $\boldsymbol{u}_{n-k}$ for $k = 1, \dots, n$). In terms of condition expectation (see Definition 4.47), we have for any bounded measurable function $\phi\colon \mathbb{R}^d \to \mathbb{R}$ that

$$\mathbb{E}\big[\phi(\boldsymbol{u}_{n+1})\,\big|\,\boldsymbol{u}_n\big] = \mathbb{E}\big[\phi(\boldsymbol{u}_{n+1})\,\big|\,\boldsymbol{u}_n, \boldsymbol{u}_{n-1}, \dots, \boldsymbol{u}_0\big].$$

This is known as the *Markov property* for a sequence $\boldsymbol{u}_n$ of random variables.

Moving to continuous time, we are primarily interested in stochastic processes with the Markov property, so-called *Markov processes*, such as the solution $\{\boldsymbol{u}(t)\colon t \ge 0\}$ of (8.3). We give the formal definition.

Definition 8.35 (transition function, Markov process) Let $(\Omega, \mathcal{F}, \mathcal{F}_t, \mathbb{P})$ be a filtered probability space. Let $B_b(\mathbb{R}^d)$ denote the set of bounded Borel measurable functions $\phi\colon \mathbb{R}^d \to \mathbb{R}$. For an $\mathbb{R}^d$-valued $\mathcal{F}_t$-adapted stochastic process $\{\boldsymbol{u}(t)\colon t \ge 0\}$, the *transition functions* are defined by

$$P_{s,t}\phi(\boldsymbol{x}) := \mathbb{E}\Big[\phi(\boldsymbol{u}(t)) \,\Big|\, \boldsymbol{u}(s) = \boldsymbol{x}\Big] \qquad \text{for } 0 \le s \le t \text{ and } \phi \in B_b(\mathbb{R}^d).$$

We say $\{\boldsymbol{u}(t)\colon t \ge 0\}$ is a *Markov process* if a.s.

$$\mathbb{E}\Big[\phi(\boldsymbol{u}(t+h)) \,\Big|\, \mathcal{F}_t\Big] = P_{t,t+h}\phi(\boldsymbol{u}(t)) \qquad \text{for } \phi \in B_b(\mathbb{R}^d) \text{ and } h,t \ge 0. \tag{8.69}$$

Further, we say the Markov process is *time homogeneous* if $P_{t,t+h}$ is independent of t and write $P_h = P_{0,h}$.

The drift $\boldsymbol{f}$ and diffusion G are functions of $\boldsymbol{u} \in \mathbb{R}^d$ and independent of time t. Moreover, the increments of Brownian motion $W(t)$ for $t > s$ are independent of $\mathcal{F}_s$. From the integral equation (8.3), we see that the conditional distribution of the solution $\boldsymbol{u}(t+h)$ of (8.3) given $\mathcal{F}_t$ depends only on $\boldsymbol{u}(t)$ and $h > 0$. As such, (8.69) holds and we see the solution $\boldsymbol{u}(t)$ of (8.3) is a time-homogeneous Markov processes.

Theorem 8.36 (Markov processes for SODEs) *Under Assumption 8.17, the solution $\boldsymbol{u}(t)$ of* (8.3) *is a time-homogeneous Markov process.*

Time-homogeneous Markov processes have an important semigroup property (see Definition 3.10) that we now introduce.

Lemma 8.37 (Chapman–Kolmogorov) *If $\boldsymbol{u}(t)$ is a $\mathbb{R}^d$-valued time-homogeneous Markov process, then*

$$P_{s+t}\,\phi = P_s P_t\,\phi, \qquad \forall s,t \ge 0, \quad \phi \in B_b(\mathbb{R}^d).$$

This is known as the Chapman–Kolmogorov *equation.*

Proof By the tower property of conditional expectation, a.s.,

$$\begin{aligned} P_{s+t}\phi(\boldsymbol{x}) &= \mathbb{E}\Big[\phi(\boldsymbol{u}(s+t)) \,\Big|\, \boldsymbol{u}(0) = \boldsymbol{x}\Big] = \mathbb{E}\Big[\mathbb{E}\big[\phi(\boldsymbol{u}(s+t)) \,\big|\, \mathcal{F}_s\big] \,\Big|\, \boldsymbol{u}(0) = \boldsymbol{x}\Big] \\ &= \mathbb{E}\Big[P_{s,s+t}\,\phi(\boldsymbol{u}(s)) \,\Big|\, \boldsymbol{u}(0) = \boldsymbol{x}\Big] = \mathbb{E}\Big[P_t\,\phi(\boldsymbol{u}(s)) \,\Big|\, \boldsymbol{u}(0) = \boldsymbol{x}\Big]. \end{aligned}$$

If $\psi(\boldsymbol{x}) := P_t\phi(\boldsymbol{x})$, then $P_{s+t}\phi(\boldsymbol{x}) = \mathbb{E}\big[\psi(\boldsymbol{u}(s)) \,\big|\, \boldsymbol{u}(0) = \boldsymbol{x}\big] = P_s\psi(\boldsymbol{x})$ and hence $P_{s+t}\phi = P_s P_t \phi$. □

Backward Kolmogorov equation

To understand the approximation of $\mathbb{E}[\phi(\boldsymbol{u}(t))]$ for the solution $\boldsymbol{u}(t)$ of (8.3) with initial condition $\boldsymbol{u}(0) = \boldsymbol{u}_0$, the function $\Phi(t, \boldsymbol{u}_0) := \mathbb{E}[\phi(\boldsymbol{u}(t))]$ plays a key role. Because $\Phi(t, \boldsymbol{u}_0) = P_t\phi(\boldsymbol{u}_0)$ (see Definition 8.35), the Chapman–Kolmogorov equation applies and we can analyse $\Phi(t, \boldsymbol{u}_0)$ via a PDE, known as the *backward Kolmogorov equation*. We start by differentiating $\boldsymbol{u}(t)$ with respect to the initial data $\boldsymbol{u}_0$. Because we are looking *back* from time $t > 0$ to see how $\boldsymbol{u}(t)$ varies with respect to the initial data $\boldsymbol{u}(0) = \boldsymbol{u}_0$, we have the name *backward* Kolmogorov equation.

Assumption 8.38 Suppose that $\boldsymbol{f} \in \mathrm{C}^\infty(\mathbb{R}^d, \mathbb{R}^d)$, $G \in \mathrm{C}^\infty(\mathbb{R}^d, \mathbb{R}^{d\times m})$, and all derivatives of $\boldsymbol{f}$ and G of order one and higher are bounded.

Assumption 8.38 implies the Lipschitz condition and hence also Assumption 8.17.

Lemma 8.39 (dependence on initial data) *Let Assumption 8.38 hold. The nth derivative $D^n\boldsymbol{u}(t)$ of $\boldsymbol{u}(t)$ with respect to the initial data $\boldsymbol{u}_0$ is a bounded linear operator for any $n \in \mathbb{N}$. Further, for each $T > 0$, there exists $K > 0$ such that for $0 \le t \le T$*

$$\mathbb{E}\Big[\big\|D^n\boldsymbol{u}(t)(\boldsymbol{\xi}_1,\dots,\boldsymbol{\xi}_n)\big\|_2^2\Big] \le K\|\boldsymbol{\xi}_1\|_2\cdots\|\boldsymbol{\xi}_n\|_2, \qquad \forall \boldsymbol{u}_0 \in \mathbb{R}^d,\ \boldsymbol{\xi}_1,\dots,\boldsymbol{\xi}_n \in \mathbb{R}^d.$$

Proof We prove this for $n = 1$; the general case is similar. Denote the matrix of first-order partial derivatives (the Jacobian) of $\boldsymbol{u}(t)$ with respect to $\boldsymbol{u}_0$ by $J(t) := d\boldsymbol{u}(t)/d\boldsymbol{u}_0$. By taking the derivative of the integral equation (8.3) in the direction $\boldsymbol{\xi} \in \mathbb{R}^d$,

$$J(t)\boldsymbol{\xi} = \boldsymbol{\xi} + \int_0^t D\boldsymbol{f}(\boldsymbol{u}(s))J(s)\boldsymbol{\xi}\,ds + \int_0^t DG(\boldsymbol{u}(s))J(s)\boldsymbol{\xi}\,d\boldsymbol{W}(s). \tag{8.70}$$

Applying the Itô isometry (8.22) to the last term,

$$\begin{aligned}\mathbb{E}\Big[\|J(t)\boldsymbol{\xi}\|_2^2\Big] \le 3\|\boldsymbol{\xi}\|_2^2 &+ 3t\int_0^t \mathbb{E}\Big[\|D\boldsymbol{f}(\boldsymbol{u}(s))J(s)\boldsymbol{\xi}\|_2^2\Big]\,ds\\ &+ 3\int_0^t \mathbb{E}\Big[\|DG(\boldsymbol{u}(s))J(s)\boldsymbol{\xi}\|_{\mathrm{F}}^2\Big]\,ds.\end{aligned}$$

Because the derivatives are bounded, for a constant $K > 0$,

$$\mathbb{E}\Big[\|J(t)\boldsymbol{\xi}\|_2^2\Big] \le 3\|\boldsymbol{\xi}\|_2^2 + 3K\,t\int_0^t \mathbb{E}\Big[\|J(s)\boldsymbol{\xi}\|_2^2\Big]\,ds + 3K\int_0^t \mathbb{E}\Big[\|J(s)\boldsymbol{\xi}\|_2^2\Big]\,ds.$$

Gronwall's inequality completes the proof. □

Definition 8.40 (polynomial growth) A function $\phi\colon \mathbb{R}^d \to \mathbb{R}$ has *polynomial growth* if $\sup_{\boldsymbol{u}_0\in\mathbb{R}^d}|\phi(\boldsymbol{u}_0)|/(1 + \|\boldsymbol{u}_0\|_2^p) < \infty$, for some $p \in \mathbb{N}$. Let $\mathrm{C}^\infty_{\mathrm{poly}}(\mathbb{R}^d)$ denote the set of $\phi \in \mathrm{C}^\infty(\mathbb{R}^d)$ such that ϕ and all its derivatives have polynomial growth. Let $C^\infty_{\mathrm{poly}}([0,T]\times\mathbb{R}^d)$ denote the set of $\Phi \in \mathrm{C}^\infty([0,T]\times\mathbb{R}^d)$ such that

$$\sup_{t\in[0,T]}\left|\frac{\partial^{|\alpha|}}{\partial^{\alpha_1}u_1\dots\partial^{\alpha_d}u_d}\Phi(t,\boldsymbol{u}_0)\right|$$

has polynomial growth for any multi-index $\boldsymbol{\alpha}$.

We know from Exercise 4.10 that every moment of a Gaussian random variable is well defined. In the next result, we show that the moments $\mathbb{E}\big[\|\boldsymbol{u}(t)\|_2^p\big]$ of solutions $\boldsymbol{u}(t)$ of SODEs are finite and this ensures $\mathbb{E}[\phi(\boldsymbol{u}(t))]$ is finite when ϕ has polynomial growth. For this reason, polynomial growing functions are a natural choice for test functions in weak approximation.

Corollary 8.41 *Let Assumption 8.38 hold. Suppose that $\phi \in C^\infty_{\mathrm{poly}}(\mathbb{R}^d)$ and let $\Phi(t,\boldsymbol{u}_0) = \mathbb{E}[\phi(\boldsymbol{u}(t))]$ where $\boldsymbol{u}(t)$ satisfies (8.3). Then $\Phi \in C^\infty_{\mathrm{poly}}([0,T]\times\mathbb{R}^d)$.*

Proof As we showed in Lemma 8.39, $J(t) = d\boldsymbol{u}(t)/d\boldsymbol{u}_0$ is uniformly bounded on $[0,T]$ by a constant K. Then, for $\boldsymbol{\xi} \in \mathbb{R}^d$ and $t \in [0,T]$,

$$\begin{aligned}\left|\nabla\Phi(t,\boldsymbol{u}_0)^{\mathsf{T}}\boldsymbol{\xi}\right| &= \left|\mathbb{E}\left[\nabla\phi(\boldsymbol{u}(t))^{\mathsf{T}} J(t)\boldsymbol{\xi}\right]\right| \\ &\le \left\|\nabla\phi(\boldsymbol{u}(t))\right\|_{L^2(\Omega,\mathbb{R}^d)}\left\|J(t)\boldsymbol{\xi}\right\|_{L^2(\Omega,\mathbb{R}^d)} \le K\left\|\nabla\phi(\boldsymbol{u}(t))\right\|_{L^2(\Omega,\mathbb{R}^d)}\left\|\boldsymbol{\xi}\right\|_2.\end{aligned}$$

As $\phi \in \mathrm{C}^\infty_{\mathrm{poly}}(\mathbb{R}^d)$, we have for some integer $p \ge 2$ that

$$\sup_{0\le t\le T}\left|\nabla\Phi(t,\boldsymbol{u}_0)^{\mathsf{T}}\boldsymbol{\xi}\right| \le \sup_{0\le t\le T} K\left(1 + \left\|\boldsymbol{u}(t)\right\|^p_{L^2(\Omega,\mathbb{R}^d)}\right)\left\|\boldsymbol{\xi}\right\|_2.$$

This is finite for $p = 2$ by Theorem 8.18 and, for $p > 2$, by Exercise 8.18. This argument shows that the first order derivatives of $\Phi(t,\boldsymbol{u}_0)$ have polynomial growth in the sense of Definition 8.40. The argument can be extended to higher-order derivatives to show $\Phi \in \mathrm{C}^\infty_{\mathrm{poly}}([0,T],\mathbb{R}^d)$. □

The final tool for deriving the backward Kolmogorov equation is the Itô formula, now for $d,m > 1$. Denote by $\mathrm{C}^{1,2}([0,T]\times\mathbb{R}^d)$ the set of functions $\Phi(t,\boldsymbol{u}_0)$ that are continuously differentiable in $t \in [0,T]$ and twice continuously differentiable in $\boldsymbol{u}_0 \in \mathbb{R}^d$.

Lemma 8.42 (Itô formula) *If $\Phi \in \mathrm{C}^{1,2}([0,T]\times\mathbb{R}^d)$ and $\boldsymbol{u}(t)$ denotes the solution of* (8.3) *under Assumption 8.17, then*

$$\Phi(t,\boldsymbol{u}(t)) = \Phi(0,\boldsymbol{u}_0) + \int_0^t \left(\frac{\partial}{\partial t} + \mathcal{L}\right)\Phi(s,\boldsymbol{u}(s))\,ds + \sum_{k=1}^m \int_0^t \mathcal{L}^k\Phi(s,\boldsymbol{u}(s))\,dW_k(s)$$

where, for $\boldsymbol{x} \in \mathbb{R}^d$ and $t > 0$,

$$\mathcal{L}\Phi(t,\boldsymbol{x}) := \boldsymbol{f}(\boldsymbol{x})^{\mathsf{T}}\nabla\Phi(t,\boldsymbol{x}) + \frac{1}{2}\sum_{k=1}^m \boldsymbol{g}_k(\boldsymbol{x})^{\mathsf{T}}\nabla^2\Phi\,(t,\boldsymbol{x})\boldsymbol{g}_k(\boldsymbol{x}), \tag{8.71}$$

$$\mathcal{L}^k\Phi(t,\boldsymbol{x}) := \nabla\Phi(t,\boldsymbol{x})^{\mathsf{T}}\boldsymbol{g}_k(\boldsymbol{x}). \tag{8.72}$$

Here $\boldsymbol{g}_k$ denotes the kth column of the diffusion matrix G. $\nabla\Phi$ is the gradient and $\nabla^2\Phi$ the Hessian matrix of second partial derivatives of $\Phi(t,\boldsymbol{x})$ with respect to $\boldsymbol{x}$.

Proposition 8.43 (backward Kolmogorov equation) *Suppose that Assumption 8.38 holds and $\phi \in \mathrm{C}^\infty_{\mathrm{poly}}(\mathbb{R}^d)$. Let $\Phi(t,\boldsymbol{u}_0) = \mathbb{E}[\phi(\boldsymbol{u}(t))]$. Then $\Phi \in \mathrm{C}^{1,2}([0,T]\times\mathbb{R}^d)$ and*

$$\frac{\partial}{\partial t}\Phi = \mathcal{L}\Phi, \qquad \Phi(0,\boldsymbol{u}_0) = \phi(\boldsymbol{u}_0), \tag{8.73}$$

where $\mathcal{L}$ is known as the generator and is defined by (8.71).

Proof $\boldsymbol{u}(t)$ is a Markov process by Theorem 8.36 and $\Phi(t,\boldsymbol{u}_0) = P_t\phi(\boldsymbol{u}_0)$ using the transition function of Definition 8.35. By the Chapman–Kolmogorov equation of Lemma 8.37,

$$\frac{\Phi(t+h,\boldsymbol{u}_0) - \Phi(t,\boldsymbol{u}_0)}{h} = \frac{P_{t+h}\phi(\boldsymbol{u}_0) - P_t\phi(\boldsymbol{u}_0)}{h} = \frac{P_h\psi(\boldsymbol{u}_0) - \psi(\boldsymbol{u}_0)}{h}$$

for $\psi(\boldsymbol{u}_0) := P_t\phi(\boldsymbol{u}_0)$. If the limit exists as $h \to 0$,

$$\frac{\partial}{\partial t}\Phi = \lim_{h\to 0}\frac{P_h\psi - \psi}{h}.$$

To show (8.73), we prove the limit is well defined and equals $\mathcal{L}\Phi$.

Note that $\psi(\boldsymbol{u}_0) = \Phi(t, \boldsymbol{u}_0)$ has two derivatives by Corollary 8.41 and hence the Itô formula applies. Then,

$$\psi(\boldsymbol{u}(h)) = \psi(\boldsymbol{u}_0) + \int_0^h \mathcal{L}\psi(\boldsymbol{u}(s))\,ds + \sum_{k=1}^m \int_0^h \mathcal{L}^k \psi(\boldsymbol{u}(s;\boldsymbol{u}_0))\,dW_k(s).$$

Rearranging,

$$\frac{\psi(\boldsymbol{u}(h)) - \psi(\boldsymbol{u}_0)}{h} = \frac{1}{h}\int_0^h \mathcal{L}\psi(\boldsymbol{u}(s))\,ds + \text{an Itô integral}.$$

Taking expectations and the limit $h \to 0$, the Itô integral vanishes and, using $\mathbb{E}[\psi(\boldsymbol{u}(h))] = P_h\psi(\boldsymbol{u}_0)$, we have

$$\lim_{h\to 0} \frac{P_h\psi(\boldsymbol{u}_0) - \psi(\boldsymbol{u}_0)}{h} = \mathbb{E}\big[\mathcal{L}\psi(\boldsymbol{u}(0;\boldsymbol{u}_0))\big] = \mathcal{L}\psi(\boldsymbol{u}_0), \tag{8.74}$$

as required. □

Example 8.44 (heat equation) Fix $T > 0$ and let

$$X^x(t) := x + W(T-t), \qquad \Psi(T,x) := \mathbb{E}\big[\phi(X^x(0))\big].$$

As we saw in Example 5.14, Ψ obeys the heat equation

$$\Psi_t = \tfrac{1}{2}\Psi_{xx} \tag{8.75}$$

with initial condition $\Psi(0,x) = \phi(x)$. In terms of Proposition 8.43, $\Psi(T,x) = \mathbb{E}[\phi(u(T))]$ for $u(T) = x + W(T)$ and $u(t)$ solves the SODE (8.3) with $f = 0$ and $G = 1$. Then, the generator is $\mathcal{L} = \frac{1}{2}\frac{d^2}{dx^2}$ and the backward Kolmogorov equation (8.73) is the same as (8.75) with $u_0 = x$.

Weak convergence of Euler–Maruyama

With the help of the backward Kolmogorov equation, we prove the weak discretisation error $\mathbb{E}[\phi(\boldsymbol{u}(T))] - \mathbb{E}[\phi(\boldsymbol{u}_N)]$ for the Euler–Maruyama approximation $\boldsymbol{u}_N$ to $\boldsymbol{u}(T)$ is $\mathcal{O}(\Delta t)$ for infinitely differentiable functions ϕ with polynomial growth.

Theorem 8.45 (Euler–Maruyama convergence) *Let Assumption 8.38 hold. For all $\phi \in C^\infty_{\text{poly}}(\mathbb{R}^d)$, $T \geq 0$, and $\boldsymbol{u}_0 \in \mathbb{R}^d$, there exists $K > 0$ independent of Δt such that*

$$\Big|\mathbb{E}\big[\phi(\boldsymbol{u}(T))\big] - \mathbb{E}\big[\phi(\boldsymbol{u}_N)\big]\Big| \leq K\Delta t, \qquad t_N = T, \tag{8.76}$$

where $\boldsymbol{u}_n$ is the Euler–Maruyama approximation to the solution $\boldsymbol{u}(t_n)$ of (8.3).

Proof Recall from §8.5 that $\hat{t} := t_n$ if $t \in [t_n, t_{n+1})$. Let $\boldsymbol{u}_{\Delta t}(t)$ denote the solution of

$$d\boldsymbol{u}_{\Delta t}(t) = \boldsymbol{f}(\boldsymbol{u}_{\Delta t}(\hat{t}))\,dt + G(\boldsymbol{u}_{\Delta t}(\hat{t}))\,d\boldsymbol{W}(t), \qquad \boldsymbol{u}_{\Delta t}(0) = \boldsymbol{u}_0. \tag{8.77}$$

Because the drift and diffusion are constant on intervals $[t_n, t_{n+1})$, $\boldsymbol{u}_{\Delta t}(t_n) = \boldsymbol{u}_n$ for $n \in \mathbb{N}$ and $\boldsymbol{u}_{\Delta t}(t)$ interpolates the numerical method.

By applying the Itô formula to (8.77) on intervals $[t_n, t_{n+1})$, we see that

$$d\Phi(t,\boldsymbol{u}_{\Delta t}(t)) = \Big(\frac{\partial}{\partial t} + \widehat{\mathcal{L}}(t)\Big)\Phi(t,\boldsymbol{u}_{\Delta t}(t))\,dt + \sum_{k=1}^{m} \widehat{\mathcal{L}}^k(t)\Phi(t,\boldsymbol{u}_{\Delta t}(t))\,dW_k(t), \tag{8.78}$$

where

$$\widehat{\mathcal{L}}(t)\Phi(t,\boldsymbol{x}) := \boldsymbol{f}(\boldsymbol{u}(\hat{t}))^{\mathsf{T}}\nabla\Phi(t,\boldsymbol{x}) + \frac{1}{2}\sum_{k=1}^{m} \boldsymbol{g}_k(\boldsymbol{u}(\hat{t}))^{\mathsf{T}}\nabla^2\Phi(t,\boldsymbol{x})\boldsymbol{g}_k(\boldsymbol{u}(\hat{t})),$$

$$\widehat{\mathcal{L}}^k(t)\Phi(t,\boldsymbol{x}) := \nabla\Phi(t,\boldsymbol{x})^{\mathsf{T}}\boldsymbol{g}_k(\boldsymbol{u}(\hat{t})), \qquad k = 1,\dots,m.$$

Let $e := \mathbb{E}[\phi(\boldsymbol{u}(T)) - \phi(\boldsymbol{u}_{\Delta t}(T))]$ and $\Phi(t,\boldsymbol{u}_0) := \mathbb{E}[\phi(\boldsymbol{u}(t))]$. Then,

$$e = \Phi(T,\boldsymbol{u}_0) - \Phi(0,\boldsymbol{u}_{\Delta t}(T))$$

and we must show $e = \mathcal{O}(\Delta t)$. By Proposition 8.43, $\Phi \in \mathrm{C}^{1,2}([0,T]\times\mathbb{R}^d)$ and the Itô formula (8.78) gives

$$\begin{aligned} e &= \Phi(T,\boldsymbol{u}_{\Delta t}(0)) - \Phi(0,\boldsymbol{u}_{\Delta t}(T)) \\ &= \mathbb{E}\bigg[\int_0^T \frac{\partial}{\partial t}\Phi(T-s,\boldsymbol{u}_{\Delta t}(s)) + \widehat{\mathcal{L}}(s)\Phi(T-s,\boldsymbol{u}_{\Delta t}(s))\,ds\bigg]. \end{aligned} \tag{8.79}$$

Changing $t \mapsto T-t$ in the backward Kolmogorov equation (8.73),

$$-\frac{\partial}{\partial t}\Phi(T-t,\boldsymbol{x}) = \mathcal{L}\Phi(T-t,\boldsymbol{x}). \tag{8.80}$$

Let $\boldsymbol{x} = \boldsymbol{u}_{\Delta t}(s)$ and substitute (8.80) into (8.79), to give

$$e = \mathbb{E}\bigg[\int_0^T \Big(\widehat{\mathcal{L}}(s)\Phi(T-s,\boldsymbol{u}_{\Delta t}(s)) - \mathcal{L}\Phi(T-s,\boldsymbol{u}_{\Delta t}(s))\Big)\,ds\bigg].$$

For simplicity, we treat only the case $\boldsymbol{f} = \boldsymbol{0}$. Then,

$$\mathcal{L}\phi = \frac{1}{2}\sum_{k=1}^{m} \boldsymbol{g}_k(\boldsymbol{x})^{\mathsf{T}}\nabla^2\phi\;\boldsymbol{g}_k(\boldsymbol{x}),$$

$$\mathcal{L}(s)\phi = \frac{1}{2}\sum_{k=1}^{m} \boldsymbol{g}_k(\boldsymbol{u}(\hat{s}))^{\mathsf{T}}\nabla^2\phi\;\boldsymbol{g}_k(\boldsymbol{u}(\hat{s})).$$

We can write

$$\begin{aligned} \widehat{\mathcal{L}}(s)\Phi(T-s,\boldsymbol{u}_{\Delta t}(s)) - \mathcal{L}\Phi(T-s,\boldsymbol{u}_{\Delta t}(s)) &= \Theta_1(s,\boldsymbol{u}_{\Delta t}(s)) - \Theta_1(\hat{s},\boldsymbol{u}_{\Delta t}(\hat{s})) \\ &\quad + \Theta_2(\hat{s},\boldsymbol{u}_{\Delta t}(\hat{s})) - \Theta_2(s,\boldsymbol{u}_{\Delta t}(s)) \end{aligned}$$

where

$$\Theta_1(s,\boldsymbol{x}) := \sum_{k=1}^{m} \boldsymbol{g}_k(\boldsymbol{u}_{\Delta t}(\hat{s}))^{\mathsf{T}}\nabla^2\Phi(T-s,\boldsymbol{x})\boldsymbol{g}_k(\boldsymbol{u}_{\Delta t}(\hat{s})), \tag{8.81}$$

$$\Theta_2(s,\boldsymbol{x}) := \sum_{k=1}^{m} \boldsymbol{g}_k(\boldsymbol{x})^{\mathsf{T}}\nabla^2\Phi(T-s,\boldsymbol{x})\boldsymbol{g}_k(\boldsymbol{x}). \tag{8.82}$$

By the Itô formula (8.78), for $i = 1,2$,

$$\Theta_i(s,\boldsymbol{u}_{\Delta t}(s)) = \Theta_i(\hat{s},\boldsymbol{u}_{\Delta t}(\hat{s})) + \int_{\hat{s}}^{s}\Big(\frac{\partial}{\partial t} + \widehat{\mathcal{L}}(r)\Big)\Theta_i(r,\boldsymbol{u}_{\Delta t}(r))\,dr + \text{Itô integral}.$$

Let

$$Q_i(r, \boldsymbol{u}_0) := \Big(\frac{\partial}{\partial t} + \widehat{\mathcal{L}}(r)\Big)\Theta_i(r, \boldsymbol{u}_0) \tag{8.83}$$

so that

$$\mathbb{E}\big[\Theta_i(s, \boldsymbol{u}_{\Delta t}(s))\big] = \mathbb{E}\big[\Theta_i(\hat{s}, \boldsymbol{u}_{\Delta t}(\hat{s}))\big] + \int_{\hat{s}}^{s} \mathbb{E}\big[Q_i(r, \boldsymbol{u}_{\Delta t}(r))\big]\,dr.$$

From (8.73),

$$\frac{\partial}{\partial s}\nabla^2\Phi(T-s,\boldsymbol{x}) = \nabla^2\frac{\partial}{\partial s}\Phi(T-s,\boldsymbol{x}) = -\nabla^2\mathcal{L}\Phi(T-s,\boldsymbol{x})$$

and hence, referring to (8.81) and (8.82), the time derivative in (8.83) can be changed to a spatial derivative. Therefore, as $\Theta_i \in \mathrm{C}^{\infty}_{\mathrm{poly}}([0,T]\times\mathbb{R}^d)$, so does Q_i and

$$\mathbb{E}\big[Q_i(r, \boldsymbol{u}_{\Delta t})\big] \le K\mathbb{E}\big[1 + \|\boldsymbol{u}_{\Delta t}(r)\|_2^p\big].$$

for some K, p independent of Δt. For $p = 2$, this is bounded uniformly for $r \in [0,T]$ because of Exercise 8.11. It also generalises to $p > 2$ as we saw in Exercise 8.18. Finally, because

$$e = \int_0^T\int_{\hat{s}}^{s} \mathbb{E}\big[Q_1(r, \boldsymbol{u}_{\Delta t}(r))\big] + \mathbb{E}\big[Q_2(r, \boldsymbol{u}_{\Delta t}(r))\big]\,dr\,ds, \tag{8.84}$$

and $|s - \hat{s}| \le \Delta t$, we gain (8.76) as required. □

Multilevel Monte Carlo

Consider the cost of computing the Monte Carlo and Euler–Maruyama approximation μ_M in (8.66). To approximate $\mathbb{E}[\phi(\boldsymbol{u}(T))]$ to an accuracy of ϵ, both the Euler–Maruyama weak discretisation error and the Monte Carlo sampling error should be $\mathcal{O}(\epsilon)$ in (8.67). Under the assumptions of Theorem 8.45, the weak discretisation error is $\mathcal{O}(\Delta t)$ so we require $\Delta t = \mathcal{O}(\epsilon)$. From (4.34), we also know the Monte Carlo error is $\mathcal{O}(1/\sqrt{M})$ so we require $M^{-1} = \mathcal{O}(\epsilon^2)$. We measure the computational cost by counting the total number of steps taken by the Euler–Maruyama method. Finding one sample of $\boldsymbol{u}_N$ requires $T/\Delta t$ steps and finding M samples requires $MT/\Delta t$ steps. Thus, by taking M proportional to ϵ^{-2} and Δt proportional to ϵ, we obtain accuracy ϵ with total cost

$$\mathrm{cost}(\mu_M) = M\frac{T}{\Delta t} = \mathcal{O}(\epsilon^{-3}). \tag{8.85}$$

A more efficient method, in many cases, is the multilevel Monte Carlo method. The idea is to use a hierarchy of time steps Δt_ℓ at levels $\ell = \ell_0, \dots, L$ rather than a fixed time step Δt. To be specific, we choose $\Delta t_\ell = \kappa^{-\ell}T$ for some $\kappa \in \{2, 3, \dots\}$. At each level, we compute the Euler–Maruyama approximation $\boldsymbol{u}_{\kappa^\ell}$ to $\boldsymbol{u}(T)$ using the time step Δt_ℓ. To simplify notation, we write $\boldsymbol{U}_\ell$ for $\boldsymbol{u}_{\kappa^\ell}$. The multilevel Monte Carlo method exploits the telescoping sum

$$\mathbb{E}\big[\phi(\boldsymbol{U}_L)\big] = \mathbb{E}\big[\phi(\boldsymbol{U}_{\ell_0})\big] + \sum_{\ell=\ell_0+1}^{L} \mathbb{E}\big[\phi(\boldsymbol{U}_\ell) - \phi(\boldsymbol{U}_{\ell-1})\big]. \tag{8.86}$$

The expectation on level L (with the smallest step Δt_L) is written as the expectation on level

ℓ_0 (with the largest step Δt_{ℓ_0}) plus a sum of corrections. Each correction is the difference of two approximations to $\boldsymbol{u}(T)$, one with time step Δt_ℓ (known as the *fine* step) and the other with time step $\Delta t_{\ell-1}$ (known as the *coarse* step). The idea is to estimate each of the expectations on the right-hand side of (8.86) separately. To estimate $\mathbb{E}[\phi(\boldsymbol{U}_{\ell_0})]$ at level ℓ_0, we compute the sample average

$$\mu_{\ell_0} := \frac{1}{M_{\ell_0}} \sum_{j=1}^{M_{\ell_0}} \phi(\boldsymbol{U}_{\ell_0}^j) \tag{8.87}$$

using M_{ℓ_0} *iid* samples $\boldsymbol{U}_{\ell_0}^j$ of $\boldsymbol{U}_{\ell_0}$. To estimate the correction at level ℓ, we compute

$$\mu_\ell := \frac{1}{M_\ell} \sum_{j=1}^{M_\ell} \Big(\phi\big(\boldsymbol{U}_\ell^{j,\text{fine}}\big) - \phi\big(\boldsymbol{U}_{\ell-1}^{j,\text{coarse}}\big)\Big), \qquad \ell = \ell_0 + 1, \dots, L, \tag{8.88}$$

where $\boldsymbol{U}_\ell^{j,\text{fine}}$ and $\boldsymbol{U}_\ell^{j,\text{coarse}}$ are *iid* samples of $\boldsymbol{U}_\ell$. The success of the method depends on coupling $\boldsymbol{U}_\ell^{j,\text{fine}}$ and $\boldsymbol{U}_{\ell-1}^{j,\text{coarse}}$ so that $\big\|\boldsymbol{U}_\ell^{j,\text{fine}} - \boldsymbol{U}_{\ell-1}^{j,\text{coarse}}\big\|_{L^2(\Omega,\mathbb{R}^d)}$ is small. This is achieved by computing $\boldsymbol{U}_\ell^{j,\text{fine}}$ and $\boldsymbol{U}_{\ell-1}^{j,\text{coarse}}$ with time steps Δt_ℓ and $\Delta t_{\ell-1}$ respectively and increments from the same Brownian sample path. The increments needed for different j and μ_ℓ are independent. Finally, $\mathbb{E}[\phi(\boldsymbol{u}(T))]$ is estimated by

$$\tilde{\mu}_M := \sum_{\ell=\ell_0}^{L} \mu_\ell. \tag{8.89}$$

For the multilevel method, the error can be divided as

$$\mathbb{E}\big[\phi(\boldsymbol{u}(T))\big] - \tilde{\mu}_M = \underbrace{\Big(\mathbb{E}\big[\phi(\boldsymbol{u}(T))\big] - \mathbb{E}\big[\phi(\boldsymbol{U}_L)\big]\Big)}_{\text{weak discretization error}} + \underbrace{\Big(\mathbb{E}\big[\phi(\boldsymbol{U}_L)\big] - \tilde{\mu}_M\Big)}_{\text{multilevel Monte Carlo error}},$$

and the multilevel Monte Carlo error can be further decomposed as

$$\Big(\mathbb{E}\big[\phi(\boldsymbol{U}_{\ell_0})\big] - \mu_{\ell_0}\Big) + \sum_{\ell=\ell_0+1}^{L} \Big(\mathbb{E}\big[(\phi(\boldsymbol{U}_\ell) - \phi(\boldsymbol{U}_{\ell-1}))\big] - \mu_\ell\Big).$$

The multilevel estimator (8.89) is more efficient than (8.66), because we achieve a Monte Carlo error of $\mathcal{O}(\epsilon)$ by computing μ_{ℓ_0} with a large number M_{ℓ_0} of samples (which are cheap to compute due to the larger time step) and by computing the correction terms with a smaller number of samples (which are more costly to compute due to the smaller time step). The error in estimating $\mathbb{E}[\phi(\boldsymbol{U}_\ell) - \phi(\boldsymbol{U}_{\ell-1})]$ by μ_ℓ is proportional to the standard deviation of $\phi(\boldsymbol{U}_\ell) - \phi(\boldsymbol{U}_{\ell-1})$. Since the Euler–Maruyama approximations $\boldsymbol{U}_\ell$ converge to $\boldsymbol{u}(T)$ in $L^2(\Omega, \mathbb{R}^d)$, the standard deviation of $\phi(\boldsymbol{U}_\ell) - \phi(\boldsymbol{U}_{\ell-1})$ decreases as ℓ increases and the time step decreases. Thus, fewer samples M_ℓ are needed to compute μ_ℓ to a given tolerance as ℓ increases. In contrast, a standard Monte Carlo simulation needs to generate a large number of samples with (fixed) smallest time step Δt_L.

In order that the smallest time step $\Delta t_L = \kappa^{-L} T \le \epsilon/2$, we put

$$L = \left\lceil \frac{\log(2T/\epsilon)}{\log(\kappa)} \right\rceil. \tag{8.90}$$

With this choice of L, under the assumptions of Theorem 8.45, the approximation $\boldsymbol{U}_L$ to $\boldsymbol{u}(T)$ has a weak discretisation error of $\mathcal{O}(\epsilon)$. We choose the number of samples M_ℓ at each level to reduce the variance of $\tilde{\mu}_M$, our estimate of $\mathbb{E}[\phi(\boldsymbol{u}(T))]$, to $\mathcal{O}(\epsilon^2)$. This guarantees that the multilevel Monte Carlo error is $\mathcal{O}(\epsilon)$ and the same as the Euler–Maruyama weak discretisation error.

Lemma 8.46 *Let Assumption 8.29 hold. For a globally Lipschitz function $\phi\colon \mathbb{R}^d \to \mathbb{R}$, consider the estimator $\tilde{\mu}_M$ of $\mathbb{E}[\phi(\boldsymbol{u}(T))]$ defined by (8.89). If $M_{\ell_0} = \lceil \epsilon^{-2} \rceil$ and $M_\ell = \lceil \epsilon^{-2}(L-\ell_0)\Delta t_\ell \rceil$ for $\ell = \ell_0+1,\dots,L$, then $\operatorname{Var}(\tilde{\mu}_M) = \mathcal{O}(\epsilon^2)$.*

Proof To simplify notation, let $Y := \phi(\boldsymbol{u}(T))$ and $Y_\ell := \phi(\boldsymbol{U}_\ell)$ for $\ell = \ell_0,\dots,L$. To compute $\boldsymbol{U}_\ell^{j,\mathrm{fine}}$ and $\boldsymbol{U}_{\ell-1}^{j,\mathrm{coarse}}$, we use increments from the same Brownain path and hence $\phi(\boldsymbol{U}_\ell^{j,\mathrm{fine}}) - \phi(\boldsymbol{U}_{\ell-1}^{j,\mathrm{coarse}})$ has the same distribution as $Y_\ell - Y_{\ell-1}$. A straightforward upper bound on the variance of a random variable $X \in L^2(\Omega)$ is given by

$$\operatorname{Var}(X) = \mathbb{E}\big[X^2\big] - \big(\mathbb{E}\big[X\big]\big)^2 = \|X\|_{L^2(\Omega)}^2 - \big(\mathbb{E}\big[X\big]\big)^2 \le \|X\|_{L^2(\Omega)}^2. \tag{8.91}$$

Using (8.91) and the triangle inequality in $L^2(\Omega)$, we find

$$\begin{aligned}\operatorname{Var}(Y_\ell - Y_{\ell-1}) &\le \|Y_\ell - Y_{\ell-1}\|_{L^2(\Omega)}^2 = \|Y_\ell - Y + Y - Y_{\ell-1}\|_{L^2(\Omega)}^2 \\ &\le \Big(\|Y_\ell - Y\|_{L^2(\Omega)} + \|Y - Y_{\ell-1}\|_{L^2(\Omega)}\Big)^2.\end{aligned}$$

Let L_ϕ denote the Lipschitz constant of ϕ. The $L^2(\Omega,\mathbb{R}^d)$ error for the Euler–Maruyama method is $\mathcal{O}(\Delta t^{1/2})$ (see Exercise 8.11) and hence

$$\|Y_\ell - Y\|_{L^2(\Omega)}^2 \le L_\phi^2 \|\boldsymbol{U}_\ell - \boldsymbol{u}(T)\|_{L^2(\Omega,\mathbb{R}^d)}^2 \le K\Delta t_\ell, \tag{8.92}$$

for some constant $K > 0$. Since $\Delta t_{\ell-1} = \kappa \Delta t_\ell$, this gives

$$\operatorname{Var}(Y_\ell - Y_{\ell-1}) \le K\big(1+\sqrt{\kappa}\big)^2 \Delta t_\ell.$$

By independence of the samples, we have

$$\begin{aligned}\operatorname{Var}(\mu_\ell) &= \frac{1}{M_\ell^2}\sum_{j=1}^{M_\ell} \operatorname{Var}(Y_\ell - Y_{\ell-1}) = \frac{1}{M_\ell}\operatorname{Var}(Y_\ell - Y_{\ell-1}) \\ &\le K\big(1+\sqrt{\kappa}\big)^2 \frac{\Delta t_\ell}{M_\ell} \le K\big(1+\sqrt{\kappa}\big)^2 \frac{\epsilon^2}{L-\ell_0}\end{aligned} \tag{8.93}$$

by the choice of M_ℓ. Notice also that $\operatorname{Var}(\mu_{\ell_0}) = \operatorname{Var}(Y_{\ell_0})/M_{\ell_0} \le \operatorname{Var}(Y_{\ell_0})\epsilon^2$ by (4.31) and the choice of M_{ℓ_0}. Since the samples are independent over different levels,

$$\begin{aligned}\operatorname{Var}(\tilde{\mu}_M) &= \operatorname{Var}(\mu_{\ell_0}) + \sum_{\ell=\ell_0+1}^{L} \operatorname{Var}(\mu_\ell) \\ &\le \operatorname{Var}(Y_{\ell_0})\epsilon^2 + K\big(1+\sqrt{\kappa}\big)^2\epsilon^2\end{aligned} \tag{8.94}$$

and we see that $\operatorname{Var}(\tilde{\mu}_M) = \mathcal{O}(\epsilon^2)$. □

The question now is: what is the computational cost of finding $\tilde{\mu}_M$ using the M_ℓ in the preceding lemma? Again, we measure the cost as the total number of steps the numerical method takes to compute $\tilde{\mu}_M$.

Lemma 8.47 *If $M_{\ell_0} = \lceil \epsilon^{-2} \rceil$, $M_\ell = \lceil \epsilon^{-2}(L-\ell_0)\Delta t_\ell \rceil$ for $\ell = \ell_0+1,\dots,L$, and L is given by* (8.90), *the computational cost of finding $\tilde{\mu}_M$ in* (8.89) *is $\mathcal{O}(\epsilon^{-2}|\log \epsilon|^2)$.*

Proof The cost of computing one sample on level ℓ_0 is $T\Delta t_{\ell_0}^{-1}$ and we compute M_{ℓ_0} samples, so $\mathrm{cost}(\mu_{\ell_0}) = M_{\ell_0} T \Delta t_{\ell_0}^{-1}$. For μ_ℓ, we need M_ℓ samples and compute both $\boldsymbol{U}_\ell^{j,\mathrm{fine}}$ (needing $T\Delta t_\ell^{-1}$ steps) and $\boldsymbol{U}_{\ell-1}^{j,\mathrm{coarse}}$ (needing $T\Delta t_{\ell-1}^{-1}$ steps). Hence,

$$\mathrm{cost}(\mu_\ell) = M_\ell T(\Delta t_\ell^{-1} + \Delta t_{\ell-1}^{-1}) = M_\ell T \Delta t_\ell^{-1}(1+\kappa^{-1}),$$

where we have used that $\Delta t_{\ell-1} = \kappa \Delta t_\ell$. Thus, the total cost of finding $\tilde{\mu}_M$ is

$$\mathrm{cost}(\tilde{\mu}_M) = M_{\ell_0} T \Delta t_{\ell_0}^{-1} + \sum_{\ell=\ell_0+1}^{L} M_\ell T \Delta t_\ell^{-1}(1+\kappa^{-1}). \tag{8.95}$$

Using the definition of M_ℓ, we see $\mathrm{cost}(\tilde{\mu}_M) = \mathcal{O}(\epsilon^{-2}) + \mathcal{O}(L^2\epsilon^{-2})$. Since $L = \mathcal{O}(|\log \epsilon|)$ by (8.90), $\mathrm{cost}(\tilde{\mu}_M) = \mathcal{O}(\epsilon^{-2}|\log \epsilon|^2)$. □

The cost of the multilevel Monte Carlo method is $\mathcal{O}(\epsilon^{-2}|\log \epsilon|^2)$ and is much lower than the cost for the naive Monte Carlo method of $\mathcal{O}(\epsilon^{-3})$ found in (8.85). For example if $\epsilon = 0.001$, we have one order of magnitude in difference in cost between the two methods.

Finally, we show how to choose M_ℓ using the variances

$$V_\ell = \begin{cases} \mathrm{Var}(\phi(\boldsymbol{U}_{\ell_0})), & \ell = \ell_0, \\ \mathrm{Var}(\phi(\boldsymbol{U}_\ell) - \phi(\boldsymbol{U}_{\ell-1})), & \ell = \ell_0+1,\dots,L, \end{cases} \tag{8.96}$$

to achieve the least Monte Carlo error (as measured by $\mathrm{Var}(\tilde{\mu}_M)$) for a given computational cost. The variances V_ℓ may be approximated by sample variances during computations and allow the algorithm to adaptively choose M_ℓ.

Lemma 8.48 *For a fixed computational cost C, there exists a $K_C > 0$ depending on C so that the variance $\mathrm{Var}(\tilde{\mu}_M)$ is minimised by taking*

$$M_{\ell_0} = \left\lceil K_C \sqrt{V_{\ell_0}\Delta t_{\ell_0}} \right\rceil, \qquad M_\ell = \left\lceil K_C \sqrt{V_\ell \Delta t_\ell/(1+\kappa^{-1})} \right\rceil, \tag{8.97}$$

where V_ℓ are defined by (8.96). *To achieve $\mathrm{Var}(\tilde{\mu}_M) = \epsilon^2/2$ for a tolerance ϵ, set*

$$K_C = 2\epsilon^{-2}\left(\sqrt{V_{\ell_0}\Delta t_{\ell_0}^{-1}} + \sum_{\ell=\ell_0+1}^{L} \sqrt{V_\ell \Delta t_\ell^{-1}(1+\kappa^{-1})} \right). \tag{8.98}$$

Proof By (8.94), $\mathrm{Var}(\tilde{\mu}_M) = \sum_{\ell=\ell_0}^{L} V_\ell/M_\ell$. From (8.95) with the notation $\delta_{\ell_0} = \Delta t_{\ell_0}$ and $\delta_\ell = \Delta t_\ell/(1+\kappa^{-1})$ for $\ell = \ell_0+1,\dots,L$, we have $\mathrm{cost}(\tilde{\mu}_M) = \sum_{\ell=\ell_0}^{L} TM_\ell/\delta_\ell$. We wish to minimise the following over $M_{\ell_0},\dots,M_L \in \mathbb{N}$:

$$\mathrm{Var}(\tilde{\mu}_M) = \sum_{\ell=\ell_0}^{L} \frac{V_\ell}{M_\ell} \qquad \text{subject to} \qquad \mathrm{cost}(\tilde{\mu}_M) = \sum_{\ell=\ell_0}^{L} T\frac{M_\ell}{\delta_\ell} = C.$$

We introduce a Lagrange multiplier λ and treat M_ℓ as real-valued variables, so that the optimality condition is

$$\frac{\partial}{\partial M_\ell}\left(\sum_{k=\ell_0}^{L} \frac{V_k}{M_k} + \lambda\left(\sum_{k=\ell_0}^{L} \frac{M_k}{\delta_k} - \frac{C}{T} \right) \right) = 0, \qquad \ell = \ell_0,\dots,L.$$

Thus, $-M_\ell^{-2} V_\ell + \lambda/\delta_\ell = 0$ and

$$M_\ell = \sqrt{V_\ell \delta_\ell / \lambda}, \qquad \ell = \ell_0, \dots, L. \tag{8.99}$$

We take $K_C = 1/\sqrt{\lambda}$ to find (8.97). To achieve $\mathrm{Var}(\tilde{\mu}_M) = \epsilon^2/2$, we need

$$\frac{1}{2}\epsilon^2 = \mathrm{Var}(\tilde{\mu}_M) = \sum_{\ell=\ell_0}^{L} \frac{V_\ell}{M_\ell} = \sqrt{\lambda} \sum_{\ell=\ell_0}^{L} \sqrt{V_\ell \delta_\ell^{-1}}$$

and hence $K_C = \frac{1}{\sqrt{\lambda}} = 2\epsilon^{-2} \sum_{\ell=\ell_0}^{L} \sqrt{V_\ell \delta_\ell^{-1}}$. □

The multilevel Monte Carlo method is implemented in Algorithms 8.7 and 8.8 for the Euler–Maruyama method. The user calls Algorithm 8.7 and specifies the underlying initial value problem, final time T, test function ϕ, multilevel parameter κ, maximum time step $\Delta t_{\max}$, and the desired accuracy ϵ. The algorithm then determines the number of levels L by (8.90) and sets the initial level ℓ_0, so that $\Delta t_{\ell_0} \le \Delta t_{\max} < \kappa \Delta t_{\ell_0}$. At each level, 10 sample paths are computed (i.e., $M_\ell = 10$) initially. Then, M_ℓ is increased if required according to (8.98) as sample variances become available to use in place of V_ℓ. The final estimate of

Algorithm 8.7 Code to implement the multilevel Monte Carlo method. Inputs are similar to those in Algorithm 8.1 with additional parameters kappa=κ, epsilon=ϵ and DTMX=$\Delta t_{\max}$. Outputs are the estimate EPu of $\mathbb{E}[\phi(\boldsymbol{u}(T))]$ and a vector M containing the number of samples taken on each level.

```
function [EPu, M]=mlmc(u0,T,d,m,fhandle,ghandle,kappa,epsilon,DTMX)
Levels=ceil(log(2*T/epsilon)/log(kappa))+1; % (=L+1)
DT=T*kappa.^(-(0:Levels-1))'; % time steps
L0=find(DT<=DTMX, 1); % coarsest level (=ell_0+1)
M=10*ones(Levels,1); % initial samples
Ckappa=(1+1./kappa); S1=zeros(Levels,1); S2=S1; ML=S1; VL=S1;
for j=L0:Levels
  N=kappa^j;
  % get samples for level j, initial pass
  [S1,S2]=getlevel(u0,T,N,d,m,fhandle,ghandle,kappa,M(j),j,L0,S1,S2);
  % estimate variance
  VL(L0:j)=S2(L0:j)./M(L0:j)-(S1(L0:j)./M(L0:j)).^2;
  KC=(2/epsilon^2)*sqrt(VL(L0)/DT(L0)); % KC coarse level only
  if j>L0, % esimate samples required (corrections)
    KC=KC+(2/epsilon^2)*sum(sqrt(VL(L0+1:j)./DT(L0+1:j)*Ckappa));
    ML(L0+1:j)=ceil(KC*sqrt(VL(L0+1:j).*DT(L0+1:j)/Ckappa));
  else % estimate sample required (coarsest level)
    ML(L0)=ceil(KC*sqrt(VL(L0)*DT(L0)));
  end
  for l=L0:j,
    dM=ML(l)-M(l);
    if dM>0 % extra samples needed
      N=kappa^l;
      M(l)=M(l)+dM; % get dM extra samples
      [S1,S2]=getlevel(u0,T,N,d,m,fhandle,ghandle,kappa,dM,l,L0,S1,S2);
    end
  end
end
EPu=sum(S1(L0:Levels)./M(L0:Levels));
```

Algorithm 8.8 Code to compute the multilevel Monte Carlo contributions at level ℓ. Inputs are as in Algorithm 8.1 with, in addition, parameters `K` $= \kappa$, the number of sample paths `MS`, and the levels `L` $= \ell$ and `L0` $= \ell_0$. `S1` contains $M_\ell \mu_\ell$ and `S2` contains $\sum_{j=1}^{M_\ell} \left(\phi(\boldsymbol{U}_\ell^{j,\text{fine}}) - \phi(\boldsymbol{U}_{\ell-1}^{j,\text{coarse}})\right)^2$. The outputs are the new values of `S1` and `S2`.

```
function [S1,S2]=getlevel(u00,T,N,d,m,fhandle,ghandle,...
                          kappa,MS,L,L0,S1,S2)
S(1)=0; S(2)=0;
Mstep=10000; % compute samples in blocks
for M=1:Mstep:MS
  MM=min(Mstep,MS-M+1);
  u0=bsxfun(@times,u00,ones(d,MM));
  if L==L0
    % compute Euler-Maruyama samples on the coarsest level
    [t,u]=EMpath(u0, T, N, d, m,fhandle,ghandle,1,MM);
    u=squeeze(u(:,:,end));
    S(1)=S(1)+sum(phi(u));
    S(2)=S(2)+sum(phi(u).^2);
  else % fine levels
    defaultStream = RandStream.getGlobalStream;
    % save state of random number generator
    savedState = defaultStream.State;
    % compute Euler-Maruyama samples
    [t,uu]=EMpath(u0, T, N, d, m,fhandle,ghandle,1,MM);
    uref=squeeze(uu(:,:,end));
    % reset random number generator
    defaultStream.State = savedState;
    % recompute the same samples with large time step
    [t,uu]=EMpath(u0, T, N, d, m,fhandle,ghandle,kappa,MM);
    u=squeeze(uu(:,:,end));
    X=(phi(uref)-phi(u));
    S(1)= S(1)+sum(X);
    S(2)= S(2)+sum(X.^2);
  end
end
S1(L)=S1(L)+S(1);
S2(L)=S2(L)+S(2);

% define the quantity of interest phi
function phiv=phi(v)
phiv=v(end,:);
```

$\tilde{\mu}_M$, given by the output `EPu` of Algorithm 8.7, is found from (8.89). Algorithm 8.7 calls Algorithm 8.8 in order to evaluate $M_\ell \mu_\ell$ and $\sum_{j=1}^{M_\ell} \left(\phi(\boldsymbol{U}_\ell^{j,\text{fine}}) - \phi(\boldsymbol{U}_{\ell-1}^{j,\text{coarse}})\right)^2$. For $\ell > \ell_0$, Algorithm 8.6 is used to compute the Euler–Maruyama approximations $\boldsymbol{U}_\ell^{j,\text{fine}}$ and $\boldsymbol{U}_{\ell-1}^{j,\text{coarse}}$ of the same sample of $\boldsymbol{u}(T)$ with time steps Δt_ℓ and $\Delta t_{\ell-1}$. For level ℓ_0, Algorithm 8.5 is used to compute the necessary Euler–Maruyama approximations. Notice that the computations in Algorithm 8.8 are split into blocks of size `Mstep`, taken here to be 10,000.

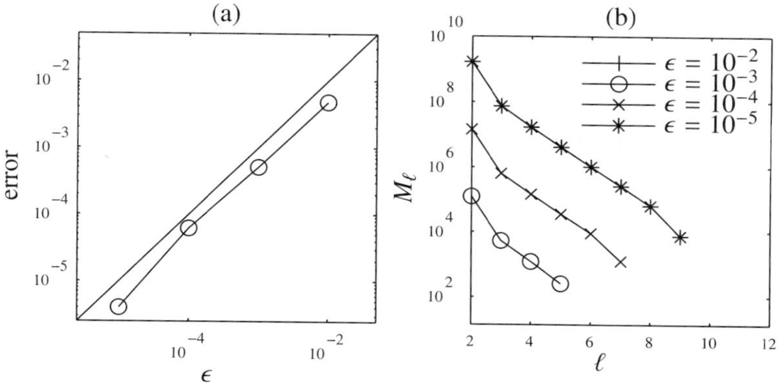

Figure 8.11 The multilevel Monte Carlo method is applied to (8.11) to estimate $\mathbb{E}[u(1)]$ for an accuracy $\epsilon = 10^{-2}, 10^{-3}, 10^{-4}$ and 10^{-5}. In (a), we plot the root-mean-square error in $\mathbb{E}[u(1)]$, averaged over 10 calls of Algorithm 8.7. In (b), the number of samples M_ℓ is plotted against ℓ (for a single call of Algorithm 8.7).

Example 8.49 Consider the geometric Brownian motion SODE (8.11) with $r = -1$ and $\sigma = 0.5$ and suppose we want to estimate $\mathbb{E}[u(T)]$ at $T = 1$. We use the multilevel Monte Carlo and Euler–Maruyama methods with $\kappa = 4$ and $\Delta t_{\max} = 0.2$. The following commands call Algorithm 8.7 and estimate $\mathbb{E}[u(1)]$ with an accuracy of $\epsilon = 10^{-3}$:

```
>> d=1; m=1; r=-1; sigma=0.5; u0=1;
>> kappa=4; T=1; DTMX=0.2; epsilon=1e-3;
>> [EPu, M]=mlmc(u0,T,d,m,@(u) r*u,@(u) sigma*u,kappa,epsilon,DTMX);
```

On a single execution of these commands, we recorded an estimate of $\mathbb{E}[u(1)] \approx 0.3680$ (returned as EPu). From (8.34) and (4.9), the true value of $\mathbb{E}[u(1)] = e^{-1} \approx 0.3679$ and so the error in the estimate is approximately 1.5×10^{-4}. Note, however, that each run of the multilevel Monte Carlo method produces a different estimate. With the stated choices for κ, ϵ, and $\Delta t_{\max}$, we have $L = 5$ and $\ell_0 = 2$. On level 2, the time step is $\Delta t_2 = 0.0625$ and 113,089 samples of the Euler–Maruyama approximation were computed. In contrast, on level 5, where the time step is much smaller, only 96 samples were taken.

In Figure 8.11(a), we examine the root-mean-square error in the multilevel Monte Carlo approximation for accuracies $\epsilon = 10^{-2}, 10^{-3}, 10^{-4}, 10^{-5}$, by averaging over 10 sample runs of Algorithm 8.7. We observe that the estimate is more accurate than the desired accuracy ϵ in each case, as the computed errors all lie below the solid diagonal line. In Figure 8.11(b), we plot the required numbers of samples M_ℓ against the level ℓ, for $\epsilon = 10^{-2}, 10^{-3}, 10^{-4}, 10^{-5}$; in each case, fewer samples are required as ℓ is increased. Now, for the standard Monte Carlo method, we would expect to have to take $M = 10^4, 10^6, 10^8, 10^{10}$ sample paths, respectively. Thus, we observe roughly one order of magnitude fewer sample paths are required using the multilevel method.

8.7 Stratonovich integrals and SODEs

In many physical applications, Brownian motion is too irregular a process and the modelling situation is better served by approximating $W(t)$ by a process $W_J(t)$ with a well-defined time derivative $W_J'(t) := dW_J(t)/dt$. We develop this idea for the case $d = 1 = m$ corresponding to the Itô SODE

$$du = f(u)\,dt + g(u)dW(t), \qquad u(0) = u_0, \tag{8.100}$$

for $f, g\colon \mathbb{R} \to \mathbb{R}$ and a one-dimensional Brownian motion $W(t)$. We replace the differential $dW(t)$ by $W_J'(t)\,dt$ and consider the ODE

$$\frac{dv_J}{dt} = f(v_J) + g(v_J)W_J'(t), \qquad v_J(0) = v_0. \tag{8.101}$$

Notice that $u(t)$ denotes the solution of the Itô SODE (8.100) and $v_J(t)$ denotes the solution of the ODE (8.101). To make our discussion definite, we work on the time interval $[0, 1]$ and choose the truncated Karhunen–Loève expansion

$$W_J(t) := \sum_{j=0}^{J} \frac{\sqrt{2}}{(j+\frac{1}{2})\pi}\xi_j \sin((j+\tfrac{1}{2})\pi t), \qquad \xi_j \sim \mathrm{N}(0,1)\ iid, \tag{8.102}$$

for the smooth approximation $W_J(t)$ to $W(t)$ (see Exercise 5.12). For fixed J, the time derivative of any sample path is given by

$$W_J'(t) = \sum_{j=0}^{J} \sqrt{2}\,\xi_j \cos((j+\tfrac{1}{2})\pi t). \tag{8.103}$$

Example 8.50 We must be careful with quantities that depend on $W_J'(t)$ as their limits as $J \to \infty$ may not exist or not converge to the obvious thing. As an example, consider the limit of

$$\int_0^1 W(t)W_J'(t)\,dt \qquad \text{as } J \to \infty.$$

Substituting $W_J'(t)\,dt = \frac{dW_J(t)}{dt}\,dt = dW_J(t)$, it is tempting to believe that $\int_0^1 W(t)\,dW_J(t)$ converges to the Itô integral $\int_0^1 W(t)\,dW(t) = \frac{1}{2}\big(W(1)^2 - 1\big)$ (see Exercise 8.4). However, we show that

$$\int_0^1 W(t)W_J'(t)\,dt \to \frac{1}{2}W(1)^2 \qquad \text{as } J \to \infty.$$

First note from (8.102) and (8.103) that

$$\lim_{J\to\infty}\int_0^1 W(s)W_J'(s)\,ds = 2\sum_{j,k=0}^{\infty}\int_0^1 \frac{\sin((k+\frac{1}{2})\pi t)\xi_k}{(k+\frac{1}{2})\pi}\xi_j\cos((j+\tfrac{1}{2})\pi t)\,dt$$

$$= \sum_{j,k=0}^{\infty}\int_0^1 \frac{\sin((k+\frac{1}{2})\pi t)\xi_k}{(k+\frac{1}{2})\pi}\xi_j\cos((j+\tfrac{1}{2})\pi t) + \frac{\sin((j+\frac{1}{2})\pi t)\xi_j}{(j+\frac{1}{2})\pi}\xi_k\cos((k+\tfrac{1}{2})\pi t)\,dt.$$

Integrating by parts,

$$\int_0^1 \frac{\sin((k+\frac{1}{2})\pi t)}{(k+\frac{1}{2})\pi}\cos((j+\tfrac{1}{2})\pi t) + \frac{\sin((j+\frac{1}{2})\pi t)}{(j+\frac{1}{2})\pi}\cos((k+\tfrac{1}{2})\pi t)\,dt = \frac{\sin((k+\frac{1}{2})\pi)\sin((j+\frac{1}{2})\pi)}{(k+\frac{1}{2})(j+\frac{1}{2})\pi^2}.$$

Since

$$\sum_{j,k=0}^{\infty} \frac{\sqrt{2}\sin((j+\frac{1}{2})\pi t)\xi_j}{(j+\frac{1}{2})\pi}\,\frac{\sqrt{2}\sin((k+\frac{1}{2})\pi t)\xi_k}{(k+\frac{1}{2})\pi} = W(1)^2,$$

we have

$$\lim_{J\to\infty}\int_0^1 W(s)W_J'(s)\,ds = \frac{1}{2}W(1)^2.$$

The message here is that limits involving $W_J'(t)$ must be carefully examined and we do not expect the solution $v_J(t)$ of (8.101) to converge to the solution of the Itô SODE (8.100). We now identify the limit of $v_J(t)$ as $J \to \infty$ as the solution of an Itô SODE with a modified drift $\tilde{f}$.

Theorem 8.51 (Wong–Zakai) *Suppose that $f, g\colon \mathbb{R} \to \mathbb{R}$ and Assumption 8.29 holds. Further suppose the function $u \mapsto f(u)/g(u)$ is globally Lipschitz continuous and $g(u) \geq g_0 > 0$ for all $u \in \mathbb{R}$. Let $v_J(t)$ be the solution of* (8.101) *and $v(t)$ be the solution of*

$$dv = \tilde{f}(v)\,dt + g(v)\,dW(t), \qquad v(0) = v_0 \tag{8.104}$$

for the modified drift $\tilde{f}(v) := f(v) + \frac{1}{2}g(v)g'(v)$. Then,

$$\sup_{0\le t\le 1}\left|v(t) - v_J(t)\right| \to 0 \qquad \text{as } J \to \infty.$$

Proof Let

$$\phi(x) := \int_0^x \frac{1}{g(y)}\,dy$$

and note that $\phi'(x) = 1/g(x)$ and $\phi''(x) = -g'(x)/g(x)^2$. By Itô's formula (Lemma 8.19),

$$d\phi(v) = \left[\phi'(v)\tilde{f}(v) + \frac{1}{2}\phi''(v)g(v)^2\right]dt + \phi'(v)g(v)\,dW(t) = \frac{f(v)}{g(v)}\,dt + dW(t),$$

as $\tilde{f}/g + (-g'/g^2)(g^2/2) = f/g$. By the chain rule,

$$\frac{d\phi(v_J)}{dt} = \phi'(v_J)\frac{dv_J}{dt} = \frac{f(v_J)}{g(v_J)} + W_J'(t).$$

In integral form, we have

$$\phi(v(t)) - \phi(v_0) = \int_0^t \frac{f(v(s))}{g(v(s))}\,ds + W(t),$$

$$\phi(v_J(t)) - \phi(v_0) = \int_0^t \frac{f(v_J(s))}{g(v_J(s))}\,ds + W_J(t).$$

By assumption, $f(u)/g(u)$ is Lipschitz continuous and

$$|\phi(x)-\phi(y)| = \left|\int_x^y \frac{1}{g(s)}\,ds\right| \ge \frac{1}{g_0}|x-y|.$$

For some constant L,

$$\left|v(t)-v_J(t)\right| \le g_0\left|\phi(v(t))-\phi(v_J(t))\right| \le \int_0^t L\left|v(s)-v_J(s)\right|ds + \left|W(t)-W_J(t)\right|.$$

Gronwall's inequality (see Exercise 3.4) implies that

$$\left|v(t)-v_J(t)\right| \le \sup_{0\le s\le 1}\left|W(s)-W_J(s)\right| \mathrm{e}^{Lt}. \tag{8.105}$$

Finally, the Karhunen–Loève expansion of Brownian motion converges uniformly, because

$$\mathbb{E}\left[\left|\sum_{j=0}^{\infty}\frac{\xi_j}{(j+1/2)\pi}\right|^2\right] = \sum_{j=0}^{\infty}\frac{1}{(j+1/2)^2\pi^2} < \infty$$

and hence $\sum_{j=0}^{\infty}\frac{\xi_j}{(j+1/2)\pi} < \infty$ a.s. This implies

$$\begin{aligned}\sup_{0\le t\le 1}\left|W_J(t)-W(t)\right| &= \sup_{0\le t\le 1}\left|\sum_{j=J+1}^{\infty}\frac{\xi_j}{(j+1/2)\pi}\sin((j+1/2)\pi t)\right|\\ &\le \left|\sum_{j=J+1}^{\infty}\frac{\xi_j}{(j+1/2)\pi}\right| \to 0, \qquad J\to\infty.\end{aligned}$$

With (8.105), this completes the proof. □

Stratonovich SODEs for $d = 1 = m$

Rather than using $W_J(t)$ and $v_J(t)$ to define a limiting process $v(t)$, it is usual to define the *Stratonovich integral* and consider $v(t)$ as the solution of a so-called *Stratonovich SODE*.

Definition 8.52 (Stratonovich integral) Let $W(t)$ be an $\mathcal{F}_t$-Brownian motion (see Definition 8.7) and let $X \in \mathcal{L}_2^T$ (see Definition 8.10). For $t \in [0,T]$, the *Stratonovich integral* $\int_0^t X(s)\circ dW(s)$ is defined by

$$\int_0^t X(s)\circ dW(s) := \int_0^t X(s)\,dW(s) + \frac{1}{2}[X,W]_t$$

where on the right we have an Itô integral and the quadratic variation $[X,W]_t$, defined by the $L^2(\Omega)$ limit

$$[X,W]_t := \lim_{\Delta t\to 0}\sum_{0<t_{n+1}\le t}\big(X(t_{n+1})-X(t_n)\big)\big(W(t_{n+1})-W(t_n)\big), \qquad t_n := n\Delta t.$$

To see the connection with Theorem 8.51, we compute the quadratic variation when $X(t) = \phi(u(t))$.

Proposition 8.53 *If $\phi \in \mathrm{C}^2(\mathbb{R})$ and $u(t)$ is the solution of* (8.100),

$$[\phi(u), W]_t = \int_0^t \phi'(u(s))g(u(s))\,ds$$

and

$$\int_0^t \phi(u(s)) \circ dW(s) = \int_0^t \phi(u(s))\,dW(s) + \frac{1}{2}\int_0^t \phi'(u(s))g(u(s))\,ds. \tag{8.106}$$

Proof This is a calculation using Taylor expansions of $\phi(u(t_{n+1}) - \phi(u(t_n))$, similar to the calculations in Lemma 8.30 and Proposition 8.31. See Exercise 8.22. □

We know $v(t)$ satisfies the Itô SODE (8.104), so that

$$v(t) = v_0 + \int_0^t \tilde{f}(v(s))\,ds + \int_0^t g(v(s))\,dW(s).$$

Replacing the Itô integral with a Stratonovich integral using (8.106), we have

$$v(t) = v_0 + \int_0^t \left[\tilde{f}(v(s)) - \tfrac{1}{2}g'(v(s))g(v(s))\right] ds + \int_0^t g(v(s)) \circ dW(s).$$

Since $\tilde{f} = f + g'g/2$, we find

$$v(t) = v_0 + \int_0^t f(v(s))\,ds + \int_0^t g(v(s)) \circ W(s). \tag{8.107}$$

That is, $v(t)$ is the solution of an SODE with the original drift f and diffusion g *if* we treat the stochastic integral in the Stratonovich sense. Equation (8.107) is known as a Stratonovich SODE and also written as

$$dv = f(v)\,dt + g(v) \circ dW(t), \qquad v(0) = v_0. \tag{8.108}$$

To numerically approximate the Stratonovich SODE (8.108), we can transform the equation to an Itô SODE (8.104) with a modified drift $\tilde{f}$ and apply the Euler–Maruyama and Milstein methods derived in §8.4. It is also possible to approximate solutions of Stratonovich SODEs directly and we discuss two basic methods, known as the Stratonovich–Milstein method and Heun's method.

The Stratonovich–Milstein method is derived in the same way as the Milstein method of Definition 8.24, except we interpret the stochastic integrals in the Stratonovich sense. Alternatively, we may write down the Milstein method for the Itô SODE (8.104) with modified drift $\tilde{f}$. For example, the $d = 1$ Milstein method for the Itô SODE (8.104) is

$$v_{n+1} = v_n + \tilde{f}(v_n)\,\Delta t + g(v_n)\,\Delta W_n + \frac{1}{2}g'(v_n)g(v_n)(\Delta W_n^2 - \Delta t),$$

where $\Delta W_n \sim \mathrm{N}(0, \Delta t)$ *iid*. Substituting $\tilde{f} = f + \frac{1}{2}g'g$ gives the Stratonovich–Milstein method in terms of the coefficients of the Stratonovich SODE (8.108). That is,

$$v_{n+1} = v_n + f(v_n)\,\Delta t + g(v_n)\,\Delta W_n + \frac{1}{2}g'(v_n)g(v_n)\,\Delta W_n^2. \tag{8.109}$$

The approximation v_n converges to $v(t_n)$ for $t_n = T$ fixed and the $L^2(\Omega)$ error is $\mathcal{O}(\Delta t)$.

Heun's method is the generalisation of the Euler–Maruyama method to Stratonovich SODEs. It is found by examining the last two terms of the Milstein method. First notice that, for $\delta \in \mathbb{R}$,

$$g(v) + g'(v)g(v)\delta = g\big(v + g(v)\delta\big) - R$$

where the remainder $R = \frac{1}{2}g''(v)\xi^2 g(v)^2\delta^2$, for some $\xi \in [0,1]$. Let $\tilde{v}_{n+1} = v_n + g(v_n)\,\Delta W_n$; then (8.109) becomes

$$v_{n+1} = v_n + f(v_n)\,\Delta t + g(v_n)\,\Delta W_n + \frac{1}{2}\big(g(\tilde{v}_{n+1}) - g(v_n)\big)\,\Delta W_n + R.$$

If we drop the remainder, we have Heun's method

$$v_{n+1} = v_n + f(v_n)\,\Delta t + \frac{1}{2}\big(g(v_n) + g(\tilde{v}_{n+1})\big)\,\Delta W_n,$$

where again $\tilde{v}_{n+1} = v_n + g(v_n)\,\Delta W_n$. It does not depend on the derivative of g and is easier to apply than the Stratonovich–Milstein method. The $L^2(\Omega)$ error is only $\mathcal{O}(\Delta t^{1/2})$ in general.

Stratonovich SODEs for d, m > 1

To end this chapter, we briefly review Stratonovich SODEs in dimension $d > 1$ with forcing by m Brownian motions. Consider the initial-value problem

$$d\boldsymbol{v} = \boldsymbol{f}(\boldsymbol{v})\,dt + G(\boldsymbol{v}) \circ d\boldsymbol{W}(t), \qquad \boldsymbol{v}(0) = \boldsymbol{v}_0, \tag{8.110}$$

for $f\colon \mathbb{R}^d \to \mathbb{R}^d$, $G\colon \mathbb{R}^d \to \mathbb{R}^{d\times m}$, and $\boldsymbol{W} = [W_1,\dots,W_m]^{\mathsf{T}}$ for independent Brownian motions $W_i(t)$. Then, $\boldsymbol{v}(t) = [v_1(t),\dots,v_d(t)]^{\mathsf{T}}$ is a solution to (8.110) if the components $v_i(t)$ for $i = 1,\dots,d$ obey

$$v_i(t) = v_{0,i} + \int_0^t f_i(\boldsymbol{v}(s))\,ds + \sum_{j=1}^m \int_0^t g_{ij}(\boldsymbol{v}(s)) \circ dW_j(s), \qquad t > 0,$$

and the last term is the Stratonovich integral of Definition 8.52. The Milstein method for the Stratonovich SODE (8.110) can be derived by applying the Milstein method of Definition 8.24 to the Itô SODE with an appropriate drift term. In this general case, define $\tilde{\boldsymbol{f}}(\boldsymbol{v}) := \big[\tilde{f}_1(\boldsymbol{v}),\dots,\tilde{f}_d(\boldsymbol{v})\big]^{\mathsf{T}}$, where

$$\tilde{f}_i(\boldsymbol{v}) = f_i(\boldsymbol{v}) + \frac{1}{2}\sum_{k=1}^d \sum_{j=1}^m g_{kj}(\boldsymbol{v})\frac{\partial g_{ij}(\boldsymbol{v})}{\partial u_k}, \qquad i = 1,\dots,d, \quad \boldsymbol{v} \in \mathbb{R}^d. \tag{8.111}$$

Then the solution $\boldsymbol{v}(t)$ of (8.110) satisfies the Itô SODE

$$d\boldsymbol{v} = \tilde{\boldsymbol{f}}(\boldsymbol{v})\,dt + G(\boldsymbol{v})\,d\boldsymbol{W}(t), \qquad \boldsymbol{v}(0) = \boldsymbol{v}_0.$$

This leads to the following version of Milstein's method for Stratonovich SODEs. We also define Heun's method.

Definition 8.54 (numerical methods for Stratonovich SODEs) For a time step $\Delta t > 0$, the *Stratonovich–Milstein approximation* $\boldsymbol{v}_n = [v_{1n}, \ldots, v_{dn}]^{\mathsf{T}}$ to $\boldsymbol{v}(t_n)$ for $t_n = n\Delta t$ is defined by

$$v_{k,n+1} = v_{kn} + f_k(\boldsymbol{v}_n)\,\Delta t + \sum_{j=1}^{m} g_{kj}(\boldsymbol{v}_n)\,\Delta W_{jn} + \frac{1}{2}\sum_{i=1}^{m}\sum_{\ell=1}^{d} \frac{\partial g_{ki}}{\partial v_\ell}(\boldsymbol{v}_n) g_{\ell i}(\boldsymbol{v}_n)\,\Delta W_{in}^2$$
$$+ \frac{1}{2}\sum_{i<j=1}^{m}\sum_{\ell=1}^{d}\left(\frac{\partial g_{kj}}{\partial u_\ell}(\boldsymbol{v}_n) g_{\ell i}(\boldsymbol{v}_n) + \frac{\partial g_{ki}}{\partial u_\ell}(\boldsymbol{v}_n) g_{\ell j}(\boldsymbol{v}_n)\right)\Delta W_{in}\,\Delta W_{jn}$$
$$+ \frac{1}{2}\sum_{i<j=1}^{m}\sum_{\ell=1}^{d}\left(\frac{\partial g_{kj}}{\partial v_\ell}(\boldsymbol{v}_n) g_{\ell i}(\boldsymbol{v}_n) - \frac{\partial g_{ki}}{\partial v_\ell}(\boldsymbol{v}_n) g_{\ell j}(\boldsymbol{v}_n)\right) A_{ij,n},$$

where $\Delta W_{in} := \int_{t_n}^{t_{n+1}} dW_i(r)$ and $A_{ij,n} := A_{ij}(t_n, t_{n+1})$ are defined by (8.47) (and are independent of the interpretation of the integral as Stratonovich or Itô). The *Heun approximation* $\boldsymbol{v}_n$ to $\boldsymbol{v}(t_n)$ for $t_n = n\Delta t$ is defined by

$$\boldsymbol{v}_{n+1} = \boldsymbol{v}_n + \boldsymbol{f}(\boldsymbol{v}_n)\,\Delta t + \tfrac{1}{2}\big(G(\boldsymbol{v}_n) + G(\tilde{\boldsymbol{v}}_{n+1})\big)\,\Delta \boldsymbol{W}_n$$

and $\tilde{\boldsymbol{v}}_{n+1} = \boldsymbol{v}_n + G(\boldsymbol{v}_n)\,\Delta \boldsymbol{W}_n$.

Finally, the following lemma corresponds to the Itô formula (Lemma 8.42) and shows that the chain rule for Stratonovich calculus resembles the classical chain rule.

Lemma 8.55 (Stratonovich rule) *Suppose that* $\Phi \in \mathrm{C}^{1,2}([0,T] \times \mathbb{R}^d)$ *and* $\boldsymbol{v}(t)$ *is the solution of* (8.110). *Then*

$$\Phi(t, \boldsymbol{v}(t)) = \Phi(0, \boldsymbol{v}_0) + \int_0^t \left(\frac{\partial}{\partial t} + \mathcal{L}_{\text{strat}}\right)\Phi(s, \boldsymbol{v}(s))\,ds + \sum_{k=1}^{m}\int_0^t \mathcal{L}^k \Phi(s, \boldsymbol{v}(s)) \circ dW_k(s),$$

where $\mathcal{L}_{\text{strat}}\Phi := \boldsymbol{f}^{\mathsf{T}}\nabla\Phi$ *and* $\mathcal{L}^k\Phi := \nabla\Phi^{\mathsf{T}}\boldsymbol{g}_k$ *(cf.* (8.71)*).*

8.8 Notes

SODEs and their mathematical theory are described in many texts, including (Gīhman and Skorohod, 1972; Karatzas and Shreve, 1991; Mao, 2008; Øksendal, 2003; Protter, 2005) and from a physical perspective in (Gardiner, 2009; Nelson, 1967; Öttinger, 1996; van Kampen, 1997). Numerical methods and their analysis are covered in (Graham and Talay, 2013; Kloeden and Platen, 1992; Milstein and Tretyakov, 2004).

Filtered probability spaces are often considered under the *usual conditions*, which is important for working with jump processes. Formally, a filtration $\{\mathcal{F}_t : t \geq 0\}$ is said to be *right continuous* if $\mathcal{F}_t = \mathcal{F}_{t+}$, for $\mathcal{F}_{t+} := \{F \in \mathcal{F}_0 : F \in \mathcal{F}_s \text{ for all } s > t\}$. Then, a filtered probability space $(\Omega, \mathcal{F}, \mathcal{F}_t, \mathbb{P})$ satisfies the *usual conditions* if $(\Omega, \mathcal{F}, \mathbb{P})$ is a *complete* measure space (see Definition 1.17), the filtration $\{\mathcal{F}_t : t \geq 0\}$ is right continuous, and $\mathcal{F}_0$ contains all sets $F \in \mathcal{F}$ with $\mathbb{P}(F) = 0$. A filtration is easily enlarged to satisfy the usual conditions.

In Definition 8.12, the stochastic integral is defined for square-integrable predictable processes, where predictable processes are the limits of left-continuous $\mathcal{F}_t$-adapted processes.

Equivalently, we may define predictable processes as functions $[0,T] \times \Omega \to \mathbb{R}$ that are measurable with respect to the so-called predictable σ-algebra $\mathcal{P}_T$, which is the smallest σ-algebra containing the sets

$$\{0\} \times F, \text{ for } F \in \mathcal{F}_0, \qquad (s,t] \times F \text{ for } 0 \le s < t < T \text{ and } F \in \mathcal{F}_s.$$

Note also $\mathcal{L}_2^T = L^2([0,T] \times \Omega, \mathcal{P}_T, \text{Leb} \times \mathbb{P})$. The definition of a $\mathbb{R}^{d\times m}$-valued predictable process is given in terms of the components, which is equivalent to the process being $\mathcal{P}_T$-measurable as a function $[0,T] \times \Omega \to \mathbb{R}^{d\times m}$. $\mathcal{P}_T$ is a sub σ-algebra of $\mathcal{B}([0,T]) \times \mathcal{F}$ and it is clear that predictable processes are jointly measurable as functions of (t,ω). In fact, the stochastic integral may also be developed under the assumption that the process is $\mathcal{F}_t$-adapted and jointly measurable (e.g., Kloeden and Platen, 1992; Øksendal, 2003). It is enough to consider the smaller class of predictable processes in this book.

We have shown convergence of the Euler–Maruyama and Milstein methods in $L^2(\Omega, \mathbb{R}^d)$ subject to smoothness conditions on the drift $\boldsymbol{f}$ and diffusion G (in particular, the global Lipschitz condition). There are a number of extensions to this result. First, the error in the $L^p(\Omega, \mathbb{R}^d)$ norm for any $p \ge 2$ converges to zero with the same rate. The key additional tool required is the Burkholder–Davis–Gundy inequality (Mao, 2008, Theorem 7.3), which says for $p \ge 2$ and a constant $C_p > 0$ that, for any $X \in \mathcal{L}_2^T(\mathbb{R}^{d\times m})$,

$$\left\| \sup_{0\le t\le T} \int_0^t X(s)\, d\boldsymbol{W}(s) \right\|_{L^p(\Omega,\mathbb{R}^d)}^2 \le C_p \left\| \int_0^T \|X(s)\|_{\mathrm{F}}^2\, ds \right\|_{L^{p/2}(\Omega)} \tag{8.112}$$

Theorem 4.58(iii) then implies convergence in probability and, for $\epsilon > 0$, there exists a random variable $K(\omega) > 0$ independent of Δt such that

$$\left|\boldsymbol{u}(T,\omega) - \boldsymbol{u}_N(\omega)\right| \le K(\omega)\Delta t^{r-\epsilon}, \qquad T = t_N,$$

where $\boldsymbol{u}_n$ is the Euler–Maruyama (respectively, Milstein) approximation and $r = 1/2$ (resp. $r = 1$). These results may be proved with fewer assumptions on the drift $\boldsymbol{f}$ and diffusion function G and, in particular, the global Lipschitz conditions can be relaxed. See (Gyöngy, 1998a; Higham, Mao, and Stuart, 2002; Jentzen, Kloeden, and Neuenkirch, 2009a,b; Kloeden and Neuenkirch, 2007). For further work on the stability of numerical methods, see (Buckwar and Sickenberger, 2011; Burrage, Burrage, and Mitsui, 2000; Higham, 2000; Higham, Mao, and Stuart, 2003).

The Euler–Maruyama and Milstein methods are basic time-stepping methods and limited in many situations. In particular, their accuracy is poor due to the order one convergence and the Euler–Maruyama method may diverge for SODEs with non-globally Lipschitz coefficients (Hutzenthaler, Jentzen, and Kloeden, 2010). More sophisticated solution strategies include, for example, Runge–Kutta methods, linear multi-step methods, variable step size control methods, and asymptotically efficient methods; see (Buckwar and Winkler, 2006; Müller-Gronbach and Ritter, 2008; Rößler, 2010). The development of higher-order methods is limited by sampling methods for the iterated integrals. Even for the Milstein methods, the Lévy area is difficult to sample and the existing methods are not quick enough to make the Milstein method efficient in all cases. This is especially true in higher dimensions. For forcing by $m = 2$ Brownian motions, (Gaines and Lyons, 1994; Ryden and Wiktorsson, 2001) provide very efficient methods, but correlations between pairs of Brownian motions are significant and Lévy areas cannot be generated pairwise for $m > 2$. General methods using the Karhunen–Loève expansion are developed by Kloeden, Platen, and Wright (1992) and an interesting method for improving the accuracy is given in Wiktorsson (2001).

Weak convergence is reviewed in Talay (1996) and the extrapolation method developed in Talay and Tubaro (1990). Long-time numerical approximation is not usually accurate in the sense of strong approximation. However, weak approximation errors over long times are small in many problems where the underlying SODE is ergodic. See for example (Mattingly, Stuart, and Tretyakov, 2010; Shardlow and Stuart, 2000; Talay, 2002; Talay and Tubaro, 1990).

The multilevel Monte Carlo method for SODEs was first popularised in Giles (2008b). In our presentation, we used time steps $\Delta t_\ell = \kappa^{-\ell} T$ for $\ell = \ell_0, \dots, L$. The smallest level ℓ_o can be chosen to avoid stability issues arising from large step sizes and the maximum number of levels L is fixed. In Giles (2008b), a value of $\kappa = 4$ is suggested as a compromise between an estimate of the optimal $\kappa \approx 7$ and the decrease in the number of levels resulting increasing κ in (8.90). In other presentations, often $\ell_0 = 0$ (so $\Delta t_0 = T$) and L is not fixed. Numerical estimates of $\mathbb{E}[\phi(\boldsymbol{U}_\ell) - \phi(\boldsymbol{U}_{\ell-1})]$ can be combined with Richardson extrapolation to reduce the error (as in Giles (2008b)). The multilevel Monte Carlo method has received a great deal of interest recently. It can readily be extended to Milstein or other methods and also to other noise processes (Dereich, 2011; Giles, 2008a). It has been examined for non-globally Lipschitz functions ϕ (Giles, Higham, and Mao, 2009) and combined with other variance reduction techniques (Giles and Waterhouse, 2009).

The Wong–Zakai theorem is described in (Twardowska, 1996; Wong and Zakai, 1965). Definition 8.52 of the Stratonovich (also called the Fisk–Stratonovich) integral is given in (Karatzas and Shreve, 1991; Protter, 2005). In analogue to (8.13), the Stratonovich integral may also be defined by

$$\int_0^t X(s) \circ dW(s) := \lim S_N^X(t) \qquad \text{as } N \to \infty \tag{8.113}$$

for $S_N^X(t) := \sum_{s_{j+1/2} \le t} X(s_{j+1/2})(W(s_{j+1} \wedge t) - W(s_j))$, $s_j = j2^{-N}$, and $s_{j+1/2} := (s_j + s_{j+1})/2$. That is, we choose the midpoint of the interval $[s_j, s_{j+1}]$ for the Stratonovich integral and the left-hand point for the Itô integral.

Many SODE models are posed on a bounded domain $D \subset \mathbb{R}^d$ and boundary conditions specified on ∂D. For example, the process may be killed or reflected at the boundary and such problems are intimately connected to elliptic boundary-value problems. For a discussion of numerical solution of SODEs with boundary conditions, see (Gobet, 2000, 2001; Milstein and Tretyakov, 1999, 2004).

In place of a Brownian motion $\boldsymbol{W}(t)$, other types of stochastic process can be used to force the SODE and this is currently an active research topic. For example, we may consider forcing by a fractional Brownian motion $B_H(t)$ in place of $W(t)$. See, for example, Mishura (2008) and numerical methods for such equations in Neuenkirch (2008). More generally, Lyon's theory of rough paths (Davie, 2007; Friz and Victoir, 2010) allows very general driving terms to be considered.

Exercises

8.1 Show that an $\mathcal{F}_t$-Brownian motion $\{W(t)\colon t \geq 0\}$ has mean function $\mu(t) = 0$ and covariance function $C(s,t) = \min\{s,t\}$ and hence satisfies the conditions of Definition 5.11.

8.2 Let $W(t)$ be an $\mathcal{F}_t$-Brownian motion. Prove that (8.16) holds.

8.3 Show that the distribution of both $\int_0^t W(s)\,ds$ and $\int_0^t s\,dW(s)$ is $\mathrm{N}(0,t^3/3)$.

8.4 Using the Itô formula for $W(t)^2$, show that

$$\int_0^t W(s)\,dW(s) = \frac{1}{2}W(t)^2 - \frac{1}{2}t. \tag{8.114}$$

8.5 For *iid* Brownian motions $W(t), W_1(t), W_2(t)$, prove that

$$\int_0^t s\,dW(s) + \int_0^t W(s)\,ds = t\,W(t)$$

$$\int_0^t W_1(s)\,dW_2(s) + \int_0^t W_2(s)\,dW_1(s) = W_1(t)W_2(t).$$

8.6 For $T > 0$, let $s_j := j2^{-N}$ for $j = 0,1,\dots$ and

$$\widetilde{S}_N(t) := \sum_{s_j<t} X\Big(\frac{s_j + s_{j+1}}{2}\Big)\big(W(s_{j+1} \wedge t) - W(s_j)\big), \qquad t \in [0,T],$$

for $X \in \mathcal{L}^2_T$. In the case $X = W$, prove that $\mathbb{E}\big[\widetilde{S}_N(t)\big] \to t/2$ as $N \to \infty$.

8.7 Consider the semilinear Itô SODE

$$du = \big[-\lambda u + f(u)\big]\,dt + G(u)\,dW(t), \qquad u(0) = u_0 \in \mathbb{R} \tag{8.115}$$

for $f \in \mathrm{C}^2(\mathbb{R},\mathbb{R})$, $G \in \mathrm{C}^2(\mathbb{R},\mathbb{R})$, and $\lambda > 0$. Using the Itô formula (8.30), show that the variation of constants formula holds: that is,

$$u(t) = \mathrm{e}^{-\lambda t}u_0 + \int_0^t \mathrm{e}^{-\lambda(t-s)}f(u(s))\,ds + \int_0^t \mathrm{e}^{-\lambda(t-s)}G(u(s))\,dW(s). \tag{8.116}$$

8.8 a. Let u_n denote the Euler–Maruyama approximation (8.42) of (8.2) for $\lambda,\sigma > 0$. In the case that $u_0 = 0$ and $T > 0$, prove that there exists $K > 0$ such that

$$\big|\mathbb{E}\big[\phi(u_n)\big] - \mathbb{E}\big[\phi(u(T))\big]\big| \leq K\|\phi\|_{L^2(\mathbb{R})}\Delta t,$$

for $t_n = T$ and $\phi \in L^2(\mathbb{R})$. Further, show that

$$\mathbb{E}\big[u_n^2\big] \to \frac{\sigma^2}{2\lambda + \lambda^2\Delta t} \qquad \text{as } n \to \infty.$$

b. Let u_n satisfy the $\theta = 1/2$ Euler–Maruyma rule

$$u_{n+1} = u_n - \frac{1}{2}\lambda(u_n + u_{n+1})\Delta t + \sigma\Delta W_n.$$

Prove that u_n^2 has the correct long term average as defined by (8.7):

$$\mathbb{E}\big[u_n^2\big] \to \frac{\sigma^2}{2\lambda} \qquad \text{as } n \to \infty.$$

8.9 Consider the Duffing–van der Pol SODE (8.10) and, for the following range of parameters, approximate and plot a sample path of the solution $\boldsymbol{u}(t)$ on the (u_1,u_2) phase-plane and describe how the dynamics change.

a. For $\lambda = 1$ and $\sigma = 0.5$, vary α from $\alpha = -0.1$ up to $\alpha = 0.2$.

b. For $\alpha = -1$ and $\sigma = 0.3$, vary λ from $\lambda = -0.1$ up to $\lambda = 0.1$.

8.10 a. From (8.45), show that, with $d = m = 1$, the Milstein method for (8.2) can be written as

$$u_{n+1} = u_n + f(u_n)\,\Delta t + g(u_n)\,\Delta W_n + \frac{1}{2}\frac{\partial g}{\partial u}(u_n)g(u_u)(\Delta W_n^2 - \Delta t).$$

b. Show that with $m = 1$ and $d \in \mathbb{N}$, the Milstein method can be written as

$$u_{k,n+1} = u_{k,n} + f_k(\boldsymbol{u}_n)\,\Delta t + g_k(\boldsymbol{u}_n)\,\Delta W_{k,n} + \frac{1}{2}\sum_{\ell=1}^{d}\frac{\partial g_k}{\partial u_\ell}(\boldsymbol{u}_n)g_\ell(\boldsymbol{u}_n)(\Delta W_n^2 - \Delta t).$$

c. Derive the form for the Milstein method given in Definition 8.24 with $m = d = 2$.

8.11 Let $\boldsymbol{u}_n$ denote the Euler–Maruyama approximation with time step Δt to (8.3) under Assumption 8.29. Prove the following: for $T > 0$, there exists $K_1, K_2 > 0$ independent of $\Delta t > 0$ such that for $0 \le t_n = n\Delta t \le T$

$$\left\|\boldsymbol{u}_n - \boldsymbol{u}(t_n)\right\|_{L^2(\Omega,\mathbb{R}^d)} \le K_1(\mathrm{e}^{K_2 t_n} - 1)^{1/2}\Delta t^{1/2}. \tag{8.117}$$

8.12 Consider the following Itô SODE with $d = 3$ and $m = 2$

$$d\boldsymbol{u} = \begin{pmatrix} 1 & 0 \\ 0 & 1 \\ -u_2 & u_1 \end{pmatrix} d\boldsymbol{W}(t), \qquad \boldsymbol{u}(0) = \boldsymbol{0} \in \mathbb{R}^3.$$

Show that $u_3(t)$ equals the Lévy area $A_{12}(0,t)$ defined in (8.47). Show how to approximate $u_3(t)$ when $t \ll 1$ by an Euler–Maruyama approximation $u_{3,n}$ and determine a condition on Δt to achieve $\|u_3(t) - u_{3,n}\|_{L^2(\Omega)} = \mathcal{O}(t^{3/2})$ for $t = n\Delta t$.

8.13 Write down the θ-Milstein method for (8.11) and show that the mean-square stability condition is

$$\left(r + \tfrac{1}{2}\sigma^2\right) + \left(\left(\tfrac{1}{2} - \theta\right)r^2 + \frac{1}{4}\sigma^4\right)\Delta t < 0.$$

8.14 Show that the Itô formula (Lemma 8.42) can be rewritten as follows: In component form, where $\boldsymbol{f} = [f_1,\dots,f_d]^{\mathsf{T}}$ and G has entries g_{ij},

$$d\Phi = \left(\frac{\partial \Phi}{\partial t} + \sum_{i=1}^{d}\frac{\partial \Phi}{\partial u_i}f_i + \frac{1}{2}\sum_{i,j=1}^{d}\sum_{k=1}^{m}\frac{\partial^2 \Phi}{\partial u_i \partial u_j}g_{ik}\,g_{jk}\right)dt + \sum_{i=1}^{d}\sum_{j=1}^{m}\frac{\partial \Phi}{\partial u_i}g_{ij}\,dW_j(t).$$

and, in the notation of vector calculus,

$$d\Phi = \left(\frac{\partial \Phi}{\partial t} + \nabla\Phi \cdot \boldsymbol{f} + \tfrac{1}{2}\,\mathrm{Tr}\,\nabla^2\Phi\, G\, G^{\mathsf{T}}\right)dt + \nabla\Phi \cdot G\, d\boldsymbol{W}(t),$$

where $\mathrm{Tr}\,\nabla^2\Phi\, G\, G^{\mathsf{T}}$ may be equivalently written as $\mathrm{Tr}\, G\, G^{\mathsf{T}}\, \nabla^2\Phi$ or $\mathrm{Tr}\, G^{\mathsf{T}}\, \nabla^2\Phi\, G$.

8.15 Consider the Itô SODE

$$du = Au\,dt + \sum_{i=1}^{m} B_i u\,dW_i(t), \qquad u(0) = u_0 \in \mathbb{R}^d, \tag{8.118}$$

where $A, B_i \in \mathbb{R}^{d\times d}$ and $W_i(t)$ are *iid* Brownian motions for $i = 1,\dots,m$. If the matrices A, B_i all commute (so that $AB_i = B_iA$ and $B_iB_j = B_jB_i$), show that

$$u(t) = \exp\left(\left(A - \frac{1}{2}\sum_{i=1}^{m} B_i^2\right)t + \sum_{i=1}^{m} B_i W_i(t)\right)u_0, \tag{8.119}$$

where $\exp(A)$ is the matrix exponential defined by $\exp(A) := \sum_{k=0}^{\infty} A^k/k!$.

8.16 Consider the semilinear SODE

$$du = \left[-Au + f(u)\right]dt + G(u)\,dW(t), \qquad u(0) = u_0, \tag{8.120}$$

where $A \in \mathbb{R}^{d\times d}$, $f\colon \mathbb{R}^d \to \mathbb{R}^d$, and $G\colon \mathbb{R}^d \to \mathbb{R}^{d\times m}$, and $W(t)$ is a vector of m *iid* Brownian motions. The semi-implicit Euler–Maruyama method to approximate the solution $u(t_n)$ of (8.120) at $t_n = n\Delta t$ is given by

$$u_{n+1} = u_n + \Delta t\left(-Au_{n+1} + f(u_n)\right)dt + G(u_n)\,\Delta W_n \tag{8.121}$$

where the increments are $\Delta W_n = W(t_{n+1}) - W(t_n)$.

a. For $d = m = 1$. Let $f(u) = \frac{\sigma}{2}u - u^3$ and $G(u) = \sqrt{\sigma}u$. Implement the semi-implicit Euler–Maruyama method with $\sigma = 2$, $M = -1$, $u_0 = 1$, $t \in [0,1]$ and plot a sample path with $\Delta t = 0.001$. The exact solution (Kloeden and Platen, 1992, p.125) is given by

$$u(t) = \frac{u_0 e^{-tA+\sqrt{\sigma}W(t)}}{\sqrt{1 + 2u_0^2\int_0^t e^{-2sA+2\sqrt{\sigma}W(s)}ds}}. \tag{8.122}$$

Compare the numerical solution with the exact solution using the trapezium rule to approximate the integral in (8.122).

b. For $d = m = 4$, let $f(u) = u - u^3$ (so $f_j = u_j - u_j^3$), $G(u) = \mathrm{diag}(u)$ and

$$A = r\begin{pmatrix} 2 & -1 & 0 & 0\\ -1 & 2 & -1 & 0\\ 0 & -1 & 2 & -1\\ 0 & 0 & -1 & 2 \end{pmatrix}.$$

Implement the semi-implicit Euler–Maruyama method for this system. Choose $r = 4$ and investigate strong convergence numerically. Compare the stability properties of the semi-implicit to the explicit Euler–Maruyama method when r is large. Choose $r = 40$ and consider time steps $\Delta t = 0.002$ and $\Delta t = 0.02$.

8.17 Consider the semilinear SODE (8.115). Using the variation of constants formula (8.116), derive the numerical method

$$u_{n+1} = e^{-\Delta t\lambda}u_n + \phi_1(-\Delta t\lambda)f(u_n)\,\Delta t + e^{-\Delta t\lambda}G(u_n)\,\Delta W_n, \tag{8.123}$$

where $\phi_1(z) := z^{-1}(e^z - 1)$. This is an example of an *exponential integrator*.

Implement (8.123) for (8.115) with $\lambda = 1$, $f(u) = u - u^3$ and $G(u) = \sqrt{2}u$ with

$u(0) = 1$. Test your sample path against the solution found using the trapezium rule to approximate the integral in (8.122) and take $T = 1$, $\Delta t = 0.001$. Compare the approximate sample to path to that found by the Euler–Maruyama method.

8.18 Show that the solution $\boldsymbol{u}(t)$ of (8.3) under Assumption 8.17 satisfies the following: for any $\boldsymbol{u}_0 \in \mathbb{R}^d$, $p \geq 2, T > 0$, there exists $K > 0$ such that

$$\sup_{0 \leq t \leq T} \mathbb{E}\left[\|\boldsymbol{u}(t)\|_2^p\right] \leq K.$$

Use Gronwall's inequality and (8.112).

8.19 For $r, \sigma > 0$, let $u(t), v(t)$ be the Itô and Stratonovich geometric Brownian motions, defined by the SODEs

$$du = ru\,dt + \sigma u\,dW(t), \qquad u(0) = u_0,$$
$$dv = rv\,dt + \sigma v \circ dW(t), \qquad v(0) = u_0.$$

By giving the exact form for $u(t)$ and $v(t)$ in terms of $W(t)$, show that $u(t,\omega) \leq v(t,\omega)$ for any $t \geq 0$ for almost all $\omega \in \Omega$. Give an example of parameters r, σ where $u(t,\omega) \to 0$ and $v(t,\omega) \to \infty$ as $t \to \infty$ a.s.

8.20 Consider $\lambda, \sigma > 0$ and $f : \mathbb{R} \to \mathbb{R}$. Using the modified drift (8.111), find the Itô form of the following Stratonovich SODEs:

$$d\begin{pmatrix} q \\ p \end{pmatrix} = \begin{pmatrix} p \\ -p\lambda - f(q) \end{pmatrix} dt + \begin{pmatrix} 0 \\ \sigma q \end{pmatrix} \circ dW(t)$$

$$d\begin{pmatrix} q \\ p \end{pmatrix} = \begin{pmatrix} p \\ -p\lambda - f(q) \end{pmatrix} dt + \begin{pmatrix} 0 \\ \sigma p \end{pmatrix} \circ dW(t).$$

8.21 Show that if $W_1(t)$, $W_2(t)$ are *iid* Brownian motions, then the quadratic variation $[W_1, W_2]_t = 0$. Hence, show the Lévy areas are equal for the Itô and Stratonovich interpretation of the integral.

8.22 Let $u(t)$ be the solution of the initial value problem (8.3) in the case $d = m = 1$. For $\phi : \mathbb{R} \to \mathbb{R}$, show that the quadratic variation

$$[\phi(u), W]_t = \int_0^t \phi'(u(s))G(u(s))\,ds.$$

State any smoothness conditions you need on f, G, and ϕ. Explain why the quadratic variation is unchanged if the Stratonovich interpretation of the integral is used to define the process $u(t)$.

8.23 Convert the following Stratonovich SODE to the equivalent Itô SODE.

$$du = \left(u - u^3\right) dt + \sigma u \circ dW, \qquad u(0) = 1.$$

Implement Heun's method and compute an approximate sample path for the SODE with $\sigma = 2$, $T = 2$, and $\Delta t = 0.002$. Compare this approximate sample path to that found by the Euler–Maruyama method for the equivalent Itô SODE.

9

Elliptic PDEs with Random Data

We now return to the elliptic boundary-value problem (BVP) on a domain $D \subset \mathbb{R}^2$

$$-\nabla \cdot \big(a(\boldsymbol{x})\nabla u(\boldsymbol{x})\big) = f(\boldsymbol{x}), \qquad \boldsymbol{x} \in D, \tag{9.1}$$

$$u(\boldsymbol{x}) = g(\boldsymbol{x}), \qquad \boldsymbol{x} \in \partial D, \tag{9.2}$$

and assume that $\{a(\boldsymbol{x})\colon \boldsymbol{x} \in D\}$ and $\{f(\boldsymbol{x})\colon \boldsymbol{x} \in D\}$ are second-order random fields (recall Definition 7.3). The PDE (9.1) is a *stochastic differential equation* in the sense that the solution $u(\boldsymbol{x})$ is a random field. However, to make a distinction between (9.1) and the stochastic differential equations (SDEs) studied in Chapters 8 and 10 — which require special calculus — we use the terminology *PDE with (correlated) random data*.

The simplest way to interpret (9.1)–(9.2) is by saying that realisations of the data give rise to examples of the deterministic BVP studied in §2.2. Given realisations $a(\cdot,\omega)$ and $f(\cdot,\omega)$, existence and uniqueness of weak solutions $u(\cdot,\omega) \in H^1_g(D)$ is assured under the conditions of Theorem 2.42. In particular, each $a(\cdot,\omega)$ must be positive on D. If so, the Galerkin finite element method from §2.3 can be used to approximate individual realisations of the solution.

Example 9.1 Consider the one-dimensional problem,

$$-\frac{d}{dx}\left(a(x)\frac{d}{dx}\right)u(x) = 1, \quad 0 < x < 1, \qquad u(0) = 0, \quad u(1) = 0, \tag{9.3}$$

where the diffusion coefficient is given by the Karhunen–Loève expansion

$$a(x,\omega) = \mu + \sum_{k=1}^{P} \frac{\sigma}{k^2\pi^2}\cos(\pi k x)\xi_k(\omega), \qquad \xi_k \sim \mathrm{U}(-1,1)\ iid \tag{9.4}$$

(see §5.4 and §7.4). The ODE (9.3) is a model for steady-state heat diffusion in a bar of unit length with uncertain spatially varying thermal conductivity coefficient. At each $x \in (0,1)$, $a(x)$ is a uniform random variable with mean μ. Finite element approximations of $u(x)$ corresponding to 20 realisations of $a(x)$ are shown in Figure 9.1(b). See also Exercise 9.1. Observe in Figure 9.1(a) that each realisation of $a(x)$ is positive on $D = (0,1)$.

Example 9.2 Let $D = (0,1)\times(0,1)$ and $a(\boldsymbol{x}) = \mathrm{e}^{z(\boldsymbol{x})}$, where $z(\boldsymbol{x})$ is a mean-zero Gaussian random field with Gaussian isotropic covariance $c^0(r) = \mathrm{e}^{-r^2/\ell^2}$ (see Example 7.13 with $A = \ell^{-2}I_2$). Two realisations of $z(\boldsymbol{x})$ for the case $\ell = 0.25$ and corresponding finite element approximations of the solution to (9.1)–(9.2) with $g = 0$ and $f = 1$ are shown in Figure 9.2. See also Exercise 9.2. The realisations of $z(\boldsymbol{x})$ were generated using the circulant embedding method (see Chapter 7). Since $z(\boldsymbol{x})$ is Gaussian, its realisations may take negative values on D. However, realisations of $a(\boldsymbol{x}) = \mathrm{e}^{z(\boldsymbol{x})}$ are always positive.

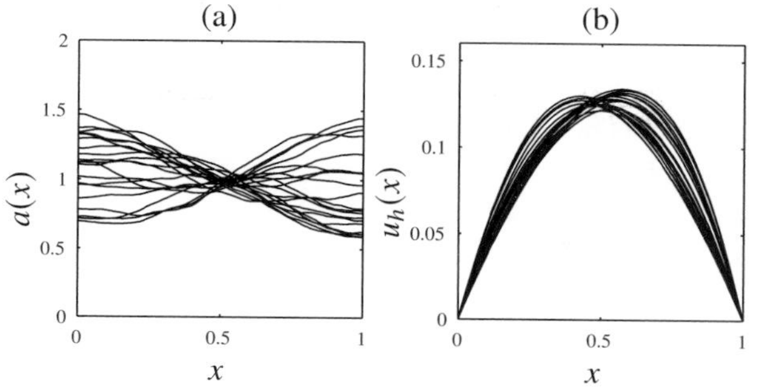

Figure 9.1 (a) Twenty realisations of $a(x)$ in (9.4) with $\mu = 1, \sigma = 4$ and $P = 50$. (b) Piecewise linear finite element approximations u_h to the corresponding solutions to (9.3).

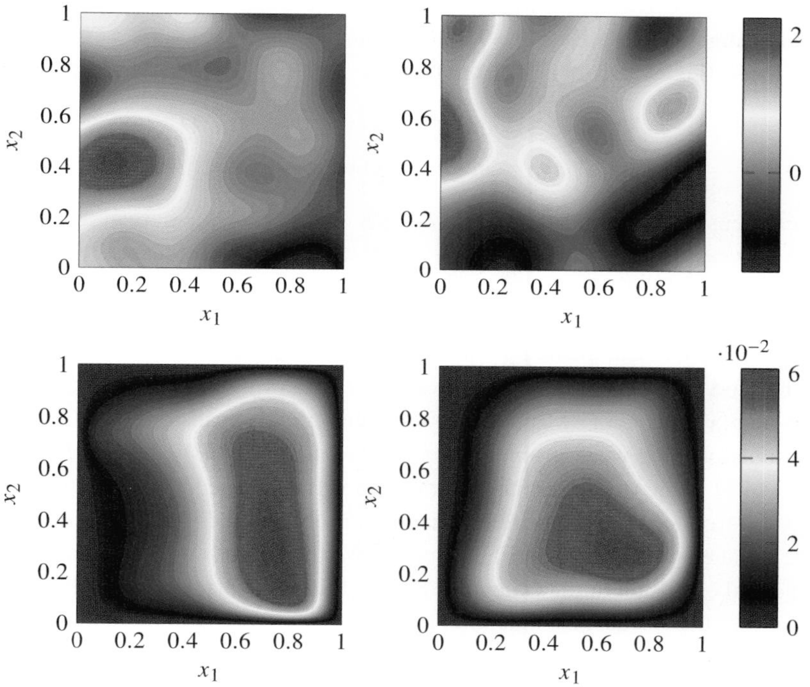

Figure 9.2 `contourf` plots of (top) realisations of $z(\boldsymbol{x})$ in Example 9.2 and (bottom) piecewise linear FEM approximations u_h to the corresponding solutions of the BVP.

Studying *realisations* of the solution to (9.1)–(9.2) is not the only possibility. We can also study weak solutions $u \in L^2(\Omega, H_g^1(D))$ (recall Definition 1.26) by setting up a single variational problem on $D \times \Omega$. In this chapter, we discuss sampling and non-sampling techniques for approximating $\mathbb{E}[u(\boldsymbol{x})]$ and $\mathrm{Var}(u(\boldsymbol{x}))$. We use finite elements for the discretisation on D and it would be useful to review the theory in §2.2 and §2.3 before reading further.

Model problem

Let $(\Omega, \mathcal{F}, \mathbb{P})$ be a probability space and let $D \subset \mathbb{R}^2$ be a bounded domain (see Definition 1.3). The σ-algebra $\mathcal{F}$ is assumed to be rich enough to support any random variable we encounter. Throughout this chapter, we work with real-valued second-order random fields $\{v(\boldsymbol{x})\colon \boldsymbol{x} \in D\}$ that are $L^2(D)$-valued random variables (recall Definition 7.3 and Example 7.8). That is, functions $v : D \times \Omega \to \mathbb{R}$ belonging to $L^2(\Omega, \mathcal{F}, L^2(D))$ that satisfy

$$\|v\|^2_{L^2(\Omega, L^2(D))} := \int_\Omega \int_D v^2(\boldsymbol{x}, \omega)\, d\boldsymbol{x}\, d\mathbb{P}(\omega) = \mathbb{E}\left[\|v\|^2_{L^2(D)}\right] < \infty.$$

We often simply write $v(\boldsymbol{x})$ for $v(\boldsymbol{x}, \omega)$ and it should be understood that $v(\boldsymbol{x})\colon \Omega \to \mathbb{R}$ for each $\boldsymbol{x} \in D$. For brevity, we also write $L^2(\Omega, X(D))$ instead of $L^2(\Omega, \mathcal{F}, X(D))$ (see Definition 4.44), where $X(D)$ is a Hilbert space of functions on D.

We assume $a, f \in L^2(\Omega, L^2(D))$ are specified random fields such as (9.4) or else have specified distributions (as in Example 9.2). Given such data, we want to find $u\colon \bar{D} \times \Omega \to \mathbb{R}$ such that $\mathbb{P}$-a.s. (i.e., with probability one),

$$-\nabla \cdot \big(a(\boldsymbol{x}, \omega) \nabla u(\boldsymbol{x}, \omega)\big) = f(\boldsymbol{x}, \omega), \qquad \boldsymbol{x} \in D, \tag{9.5}$$

$$u(\boldsymbol{x}, \omega) = g(\boldsymbol{x}), \qquad \boldsymbol{x} \in \partial D, \tag{9.6}$$

in the weak sense. The boundary data $g\colon D \to \mathbb{R}$ is deterministic here and, to ensure the existence of weak solutions, we assume $g \in H^{1/2}(\partial D)$ (see Definition 2.38). In §9.1, we study realisations of the solution to (9.5)–(9.6) corresponding to realisations of the data, review the variational formulation of the resulting deterministic problems on D and apply the Galerkin finite element method. Following this, in §9.2, we apply the Monte Carlo finite element method (MCFEM). In §9.3 we derive a weak form of (9.5)–(9.6) on $D \times \Omega$ and approximate $a(\boldsymbol{x})$ and $f(\boldsymbol{x})$ by truncated Karhunen–Loève expansions (see Chapter 7). In §9.4, we transform the weak problem on $D \times \Omega$ to a *deterministic* one on $D \times \Gamma$, where $\Gamma \subset \mathbb{R}^M$ and M is the (finite) number of random variables used to approximate the data. In §9.5, we introduce a semi-discrete variational problem on $D \times \Gamma$ that arises when we introduce finite element methods (FEMs) on D. We then focus on the stochastic Galerkin FEM, a natural extension of the Galerkin FEM discussed in Chapter 2. Finally, in §9.6, we briefly consider the stochastic collocation FEM.

9.1 Variational formulation on D

In this section and in §9.2, we assume for simplicity that $f \in L^2(D)$ and $g = 0$ (as in Examples 9.1 and 9.2). Now, given realisations $a(\cdot, \omega)$ of the diffusion coefficient, consider the random field $u(\boldsymbol{x})$ with realisations $u(\cdot, \omega) \in V := H^1_0(D)$ satisfying the weak form

$$\int_D a(\boldsymbol{x}, \omega) \nabla u(\boldsymbol{x}, \omega) \cdot \nabla v(\boldsymbol{x})\, d\boldsymbol{x} = \int_D f(\boldsymbol{x}) v(\boldsymbol{x})\, d\boldsymbol{x}, \qquad \forall v \in V. \tag{9.7}$$

Existence and uniqueness of $u(\cdot, \omega)$ is ensured by Theorem 2.42 if $a(\cdot, \omega)$ satisfies Assumption 2.34. To establish well-posedness for almost all $\omega \in \Omega$ (or, $\mathbb{P}$-a.s.), we need to make an extra assumption on $a(\boldsymbol{x})$. Consider, first, Assumption 9.3.

Assumption 9.3 (diffusion coefficients) The diffusion coefficient $a(\boldsymbol{x})$ satisfies

$$0 < a_{\min} \le a(\boldsymbol{x},\omega) \le a_{\max} < \infty, \qquad \text{a.e. in } D \times \Omega,$$

for some real constants $a_{\min}$ and $a_{\max}$. In particular, $a \in L^\infty(\Omega, L^\infty(D))$.

Example 9.4 Consider the random field (stochastic process) $a(x)$ in Example 9.1. Assuming $\mu > 0$, for any $x \in (0,1)$ and $\omega \in \Omega$, we have

$$\mu - \frac{\sigma}{\pi^2}\sum_{k=1}^{P}\frac{1}{k^2} \le a(x,\omega) \le \mu + \frac{\sigma}{\pi^2}\sum_{k=1}^{P}\frac{1}{k^2}.$$

If $\sigma < 6\mu$ then, since $\sum_{k=1}^{\infty} k^{-2} = \pi^2/6$, Assumption 9.3 holds for any P with

$$a_{\min} = \mu - \sigma/6, \qquad a_{\max} = \mu + \sigma/6.$$

The advantage in working with Assumption 9.3 is that $a_{\min}$ and $a_{\max}$ are constants and the results from §2.2 follow over to our stochastic BVP straightforwardly. The disadvantage is that log-normal random fields (such as $a(\boldsymbol{x})$ in Example 9.2) are not bounded on $D \times \Omega$ (since Gaussian random variables are unbounded). To make progress, we consider the following weaker assumption.

Assumption 9.5 (diffusion coefficients) For almost all $\omega \in \Omega$, realisations $a(\cdot,\omega)$ of the diffusion coefficient belong to $L^\infty(D)$ and satisfy

$$0 < a_{\min}(\omega) \le a(\boldsymbol{x},\omega) \le a_{\max}(\omega) < \infty, \qquad \text{a.e. in } D, \tag{9.8}$$

where

$$a_{\min}(\omega) := \operatorname*{ess\,inf}_{\boldsymbol{x}\in D} a(\boldsymbol{x},\omega), \qquad a_{\max}(\omega) := \operatorname*{ess\,sup}_{\boldsymbol{x}\in D} a(\boldsymbol{x},\omega). \tag{9.9}$$

In contrast to random fields satisfying Assumption 9.3, $a(\boldsymbol{x},\omega)$ may not be uniformly bounded over $\omega \in \Omega$. We now show that Assumption 9.5 is satisfied by many log-normal random fields, such as $a(\boldsymbol{x})$ in Example 9.2.

Assumption 9.6 Let $D \subset \mathbb{R}^2$ be a bounded domain and let $a(\boldsymbol{x}) = e^{z(\boldsymbol{x})}$, where $z(\boldsymbol{x})$ is a mean-zero Gaussian random field such that, for some $L, s > 0$,

$$\mathbb{E}\left[\left|z(\boldsymbol{x}) - z(\boldsymbol{y})\right|^2\right] \le L\|\boldsymbol{x}-\boldsymbol{y}\|_2^s, \qquad \forall \boldsymbol{x},\boldsymbol{y} \in \bar{D}. \tag{9.10}$$

Theorem 9.7 *If Assumption 9.6 holds, $a(\boldsymbol{x})$ satisfies Assumption 9.5.*

Proof Given (9.10), Theorem 7.68 implies that $z(\boldsymbol{x})$ has Hölder continuous realisations for an exponent $\gamma < s/2$ and, further, is uniformly bounded over $\boldsymbol{x} \in \bar{D}$ almost surely. As $a(\boldsymbol{x},\omega) = e^{z(\boldsymbol{x},\omega)}$ is always positive, we conclude that $a(\boldsymbol{x})$ satisfies Assumption 9.5. □

Corollary 9.8 *Let $z(\boldsymbol{x})$ be a Gaussian random field with isotropic covariance $c^0(r) = e^{-r^2/\ell^2}$ and mean zero. Then, $a(\boldsymbol{x}) = e^{z(\boldsymbol{x})}$ satisfies Assumption 9.5.*

Proof Since $\mathbb{E}[|z(\boldsymbol{x}) - z(\boldsymbol{y})|^2] = 2\left(c^0(0) - c^0(r)\right)$ for $r := \|\boldsymbol{x}-\boldsymbol{y}\|_2$,

$$\mathbb{E}\left[\left|z(\boldsymbol{x}) - z(\boldsymbol{y})\right|^2\right] = 2\left(e^0 - e^{-r^2/\ell^2}\right) \le 2\left|0 - \frac{r^2}{\ell^2}\right| = \frac{2}{\ell^2}\|\boldsymbol{x}-\boldsymbol{y}\|_2^2.$$

Hence, (9.10) holds with $L = 2/\ell^2$ and $s = 2$ and Theorem 9.7 gives the result. □

The Whittle–Matérn isotropic covariance c_q^0 with parameter q (see Example 7.17) is also investigated in Exercise 9.3.

Working with Assumption 9.5, we now prove existence and uniqueness of weak solutions satisfying (9.7), almost surely.

Theorem 9.9 *Let Assumption 9.5 hold, $f \in L^2(D)$ and $g = 0$. Then, $\mathbb{P}$-a.s., (9.7) has a unique solution $u(\cdot,\omega) \in V = H_0^1(D)$.*

Proof Fix $\omega \in \Omega$ and follow the proof of Theorem 2.42 (with $g = 0$), replacing the constants $a_{\min}$ and $a_{\max}$ with $a_{\min}(\omega)$ and $a_{\max}(\omega)$ from (9.9). □

Theorem 2.43 now provides an a priori estimate for $\|u(\cdot,\omega)\|_{H_0^1(D)}$, for almost all $\omega \in \Omega$. To study $u(\boldsymbol{x})$, we take pth moments and examine

$$\|u\|_{L^p(\Omega,H_0^1(D))} := \mathbb{E}\left[\|u\|_{H_0^1(D)}^p\right]^{1/p}.$$

Note that in (9.9) (in Assumption 9.5), $a_{\min}$ and $a_{\max}$ are random variables. In the analysis below, we will need powers of $a_{\min}$ and $a_{\max}$ to belong to certain $L^p(\Omega)$ spaces. We present two useful auxiliary results.

Lemma 9.10 *Let Assumption 9.6 hold. Then, $a_{\min}^{\lambda_1}, a_{\max}^{\lambda_2} \in L^p(\Omega)$ for any $p \ge 1$ and $\lambda_1, \lambda_2 \in \mathbb{R}$.*

Proof Corollary 7.70 tells us that $a^{\lambda} \in L^p(\Omega, C(\bar{D}))$ for any $p \ge 1$ and $\lambda \in \mathbb{R}$. The result follows by definition of $a_{\min}$ and $a_{\max}$ in (9.9). □

Lemma 9.11 *Let Assumption 9.6 hold. Then, the product $a_{\min}^{\lambda_1} a_{\max}^{\lambda_2} \in L^p(\Omega)$ for any $p \ge 1$ and $\lambda_1, \lambda_2 \in \mathbb{R}$.*

Proof See Exercise 9.4. □

We can now establish an upper bound for $\|u\|_{L^p(\Omega,H_0^1(D))}$.

Theorem 9.12 *Let the conditions of Theorem 9.9 hold so that $u(\cdot,\omega) \in V$ satisfies (9.7), $\mathbb{P}$-a.s. If Assumption 9.6 holds, then $u \in L^p(\Omega, H_0^1(D))$ for any $p \ge 1$ and*

$$\|u\|_{L^p(\Omega,H_0^1(D))} \le K_{\mathrm{p}} \|a_{\min}^{-1}\|_{L^p(\Omega)} \|f\|_{L^2(D)}. \tag{9.11}$$

Proof By Theorem 2.43 (with $g = 0$), we have

$$\|u(\cdot,\omega)\|_{H_0^1(D)} \le K_{\mathrm{p}} a_{\min}^{-1}(\omega) \|f\|_{L^2(D)},$$

for almost all $\omega \in \Omega$ where we recall K_{p} is the Poincaré constant. Hence,

$$\mathbb{E}\left[\|u\|_{H_0^1(D)}^p\right] \le K_{\mathrm{p}}^p \, \mathbb{E}\left[a_{\min}^{-p}\right] \|f\|_{L^2(D)}^p.$$

Taking pth roots gives (9.11). Since $a_{\min}^{-1} \in L^p(\Omega)$ for any $p \ge 1$ by Lemma 9.10, we have $\mathbb{E}\left[\|u\|_{H_0^1(D)}^p\right] < \infty$ and so $u \in L^p(\Omega, H_0^1(D))$. □

Galerkin finite element approximation

Next, given finite element spaces $V^h \subset H_0^1(D)$ associated with an admissible, shape-regular sequence of meshes $\mathcal{T}_h$ (see Definitions 2.53 and 2.54), consider the random field $u_h(\boldsymbol{x})$ with realisations $u_h(\cdot,\omega) \in V^h$ satisfying the weak form

$$\int_D a(\boldsymbol{x},\omega)\nabla u_h(\boldsymbol{x},\omega)\cdot\nabla v(\boldsymbol{x})\,d\boldsymbol{x} = \int_D f(\boldsymbol{x})v(\boldsymbol{x})\,d\boldsymbol{x}, \qquad \forall v \in V^h. \tag{9.12}$$

A unique solution $u_h(\cdot,\omega)$ exists, almost surely, under the same conditions as Theorem 9.9. The next result says that $u_h(\boldsymbol{x})$ also has finite pth moments.

Lemma 9.13 *Let the conditions of Theorem 9.9 hold so that $u_h(\cdot,\omega) \in V^h$ satisfies (9.12), $\mathbb{P}$-a.s. If Assumption 9.6 holds then $u_h \in L^p(\Omega, H_0^1(D))$ for any $p \geq 1$ and*

$$\|u_h\|_{L^p(\Omega,H_0^1(D))} \leq K_{\mathrm{p}}\|a_{\min}^{-1}\|_{L^p(\Omega)}\|f\|_{L^2(D)}. \tag{9.13}$$

Proof See Exercise 9.5. □

To establish a bound for the error $\|u - u_h\|_{L^2(\Omega,H_0^1(D))}$ (note, we now choose $p = 2$), we assume that $u \in L^4(\Omega, H^2(D))$ (recall, also, Assumption 2.64).

Assumption 9.14 ($L^4(\Omega, H^2(D))$-regularity) There exists a constant $K_2 > 0$, independent of $\omega \in \Omega$, such that, for every $f \in L^2(D)$, we have $u \in L^4(\Omega, H^2(D))$ and

$$|u|_{L^4(\Omega,H^2(D))} := \mathbb{E}\Big[|u|_{H^2(D)}^4\Big]^{1/4} \leq K_2\|f\|_{L^2(D)}.$$

If Assumption 9.14 holds then we can show that the error associated with piecewise linear finite element approximation is $\mathcal{O}(h)$.

Theorem 9.15 *Let the conditions of Theorem 9.9 hold and suppose Assumptions 9.6 and 9.14 hold. Let $V^h \subset H_0^1(D)$ denote a piecewise linear finite element space and let $\mathcal{T}_h$ be shape-regular. Then,*

$$\|u - u_h\|_{L^2(\Omega,H_0^1(D))} \leq K\,h\,\Big\|a_{\min}^{-1/2}a_{\max}^{1/2}\Big\|_{L^4(\Omega)}\|f\|_{L^2(D)}, \tag{9.14}$$

where $K > 0$ is a constant independent of h.

Proof Following the proof of Corollary 2.68, for each $u(\cdot,\omega) \in V$ and corresponding finite element approximation $u_h(\cdot,\omega) \in V^h$, we have

$$\|u(\cdot,\omega) - u_h(\cdot,\omega)\|_{H_0^1(D)} \leq K\,h\,a_{\min}^{-1/2}(\omega)\,a_{\max}^{1/2}(\omega)\,|u(\cdot,\omega)|_{H^2(D)}, \tag{9.15}$$

where K is a constant that does not depend on h (see Exercise 2.23). Hence,

$$\begin{aligned}\mathbb{E}\Big[\|u - u_h\|_{H_0^1(D)}^2\Big] &\leq K^2h^2\,\mathbb{E}\Big[a_{\min}^{-1}a_{\max}\,|u|_{H^2(D)}^2\Big]\\ &\leq K^2h^2\,\mathbb{E}\Big[\big(a_{\min}^{-1/2}a_{\max}^{1/2}\big)^4\Big]^{1/2}\,\mathbb{E}\Big[|u|_{H^2(D)}^4\Big]^{1/2}.\end{aligned}$$

Taking square roots and using Assumption 9.14 gives the result, with a constant K depending on K_2. Note that $a_{\min}^{-1/2}a_{\max}^{1/2} \in L^4(\Omega)$ by Lemma 9.11. □

Galerkin approximation with approximate data

In practice, we only have access to approximate samples of the diffusion coefficient or else to exact samples at a discrete set of points. To understand the implications for the approximation of $u(\boldsymbol{x})$, suppose that $\tilde{a}(\cdot,\omega)$ is a realisation of an approximation $\tilde{a}(\boldsymbol{x})$ to $a(\boldsymbol{x})$ and consider the random field $\tilde{u}_h(\boldsymbol{x})$ with realisations $\tilde{u}_h(\cdot,\omega) \in V^h$ satisfying the perturbed weak form

$$\int_D \tilde{a}(\boldsymbol{x},\omega)\,\nabla\tilde{u}_h(\boldsymbol{x},\omega)\cdot\nabla v(\boldsymbol{x})\,d\boldsymbol{x} = \int_D f(\boldsymbol{x})v(\boldsymbol{x})\,d\boldsymbol{x}, \qquad \forall v \in V^h. \tag{9.16}$$

A unique $\tilde{u}_h(\cdot,\omega)$ exists $\mathbb{P}$-a.s. if $f \in L^2(D)$ and $\tilde{a}(\boldsymbol{x})$ satisfies Assumption 9.5.

Theorem 9.16 *Let $\tilde{a}(\boldsymbol{x})$ satisfy Assumption 9.5 with random variables $\tilde{a}_{\min}$ and $\tilde{a}_{\max}$. Let the conditions of Theorem 9.9 hold so that $u_h(\cdot,\omega) \in V^h$ satisfies* (9.12) *and $\tilde{u}_h(\cdot,\omega) \in V^h$ satisfies* (9.16), *$\mathbb{P}$-a.s. Then,*

$$\|u_h - \tilde{u}_h\|_{L^2(\Omega,H_0^1(D))} \le \|\tilde{a}_{\min}^{-1}\|_{L^8(\Omega)}\|a-\tilde{a}\|_{L^8(\Omega,L^\infty(D))}\|u_h\|_{L^4(\Omega,H_0^1(D))}.$$

Proof From (2.73) (with no approximation for f), we have

$$\|u_h(\cdot,\omega) - \tilde{u}_h(\cdot,\omega)\|_{H_0^1(D)} \le \tilde{a}_{\min}^{-1}(\omega)\|a(\cdot,\omega)-\tilde{a}(\cdot,\omega)\|_{L^\infty(D)}\|u_h(\cdot,\omega)\|_{H_0^1(D)},$$

for almost all $\omega \in \Omega$. Hence,

$$\begin{aligned}\mathbb{E}\Big[\|u_h - \tilde{u}_h\|_{H_0^1(D)}^2\Big] &\le \mathbb{E}\Big[\tilde{a}_{\min}^{-4}\|a-\tilde{a}\|_{L^\infty(D)}^4\Big]^{1/2}\,\mathbb{E}\Big[\|u_h\|_{H_0^1(D)}^4\Big]^{1/2}\\ &\le \mathbb{E}\big[\tilde{a}_{\min}^{-8}\big]^{1/4}\mathbb{E}\Big[\|a-\tilde{a}\|_{L^\infty(D)}^8\Big]^{1/4}\,\mathbb{E}\Big[\|u_h\|_{H_0^1(D)}^4\Big]^{1/2}.\end{aligned}$$

Taking square roots gives the result. □

Consider the bound in Theorem 9.16. If $a(\boldsymbol{x})$ is log-normal and satisfies Assumption 9.6 then $u_h \in L^4(\Omega, H_0^1(D))$ by Lemma 9.13. For the chosen $\tilde{a}(\boldsymbol{x})$, we also need $\tilde{a}_{\min}^{-1} \in L^8(\Omega)$ and $\|a-\tilde{a}\|_{L^\infty(D)} \in L^8(\Omega)$. In (9.17) below, we construct a very simple approximation but there are many possibilities.

For the chosen mesh $\mathcal{T}_h$ with parameter h, let $\boldsymbol{v}_j^k$, $j = 1,2,3$ denote the vertices of element $\triangle_k$. Now, consider the random field $\tilde{a}(\boldsymbol{x})$ defined by

$$\tilde{a}(\boldsymbol{x},\omega)\big|_{\bar{\triangle}_k} := \frac{1}{3}\sum_{j=1}^{3} a(\boldsymbol{v}_j^k,\omega), \qquad \forall\,\triangle_k \in \mathcal{T}_h. \tag{9.17}$$

Realisations of $\tilde{a}(\boldsymbol{x})$ are piecewise constant and are generated by sampling $a(\boldsymbol{x})$ at the mesh vertices and averaging over triangles. Working with (9.17), we now establish an upper bound for the random variable $\|a-\tilde{a}\|_{L^\infty(D)}$ by exploiting a result from Chapter 7 concerning the derivatives of realisations of $a(\boldsymbol{x})$.

Lemma 9.17 *Let Assumption 9.6 hold and suppose the Gaussian random field $z(\boldsymbol{x})$ has covariance $C_z \in \mathrm{C}^3(\bar{D}\times\bar{D})$. Define $\tilde{a}(\boldsymbol{x})$ as in* (9.17)*. Then,*

$$\|a-\tilde{a}\|_{L^\infty(D)} \le a_{\max}\,K\,h, \tag{9.18}$$

for some random variable $K \in L^p(\Omega)$ independent of h. In addition, $\|a-\tilde{a}\|_{L^\infty(D)} \in L^p(\Omega)$ and $\tilde{a}_{\min}^{-1} \in L^p(\Omega)$, for any $p \ge 1$.

Proof Fix an element $\triangle_k$. For any $\boldsymbol{x} \in \bar{\triangle}_k$, note that $\|\boldsymbol{x} - \boldsymbol{v}_j^k\|_2 \le h$ for each $j = 1,2,3$. Using the triangle inequality, we have

$$|a(\boldsymbol{x},\omega) - \tilde{a}(\boldsymbol{x},\omega)| \le \frac{1}{3}\sum_{j=1}^{3}|a(\boldsymbol{x},\omega) - a(\boldsymbol{v}_j^k,\omega)|.$$

By Theorem 7.71, $a(\cdot,\omega)$ is continuously differentiable for almost all ω (i.e., realisations belong to $C^1(\bar{D})$) and the first partial derivatives are bounded by $a_{\max}(\omega)K(\omega)$, for $K := \|\nabla z\|_{C(\bar{D},\mathbb{R}^2)}$. Then,

$$|a(\boldsymbol{x},\omega) - \tilde{a}(\boldsymbol{x},\omega)| \le \frac{1}{3}\sum_{j=1}^{3} a_{\max}(\omega)K(\omega)\|\boldsymbol{x} - \boldsymbol{v}_j^k\|_2 \le a_{\max}(\omega)\, K(\omega)\, h.$$

Applying this result over all triangles gives (9.18). Taking pth moments,

$$\|a - \tilde{a}\|_{L^p(\Omega,L^\infty(D))} \le \|K\|_{L^{2p}(\Omega)}\|a_{\max}\|_{L^{2p}(\Omega)}h < \infty, \tag{9.19}$$

since K belongs to $L^p(\Omega)$ for any $p \ge 1$. In addition, $\tilde{a}_{\min}(\omega) \ge a_{\min}(\omega)$ by definition and so $\tilde{a}_{\min}^{-1}(\omega) \le a_{\min}^{-1}(\omega)$. Since $a_{\min}^{-1} \in L^p(\Omega)$ for any $p \ge 1$ by Lemma 9.10, we have $\|\tilde{a}_{\min}^{-1}\|_{L^p(\Omega)} < \infty$. □

Using the approximation (9.17), we now have the following result.

Corollary 9.18 *Let the assumptions of Theorem 9.9 and Lemma 9.17 hold and define* $\tilde{a}(\boldsymbol{x})$ *as in* (9.17). *Then, for some constant* $C > 0$,

$$\|u_h - \tilde{u}_h\|_{L^2(\Omega,H_0^1(D))} \le C\, h\, \|u_h\|_{L^4(\Omega,H_0^1(D))}. \tag{9.20}$$

Proof Applying (9.19) in the proof of Lemma 9.17 with $p = 8$ gives

$$\|a - \tilde{a}\|_{L^8(\Omega,L^\infty(D))} \le \|a_{\max}\|_{L^{16}(\Omega)}\|K\|_{L^{16}(\Omega)}h.$$

All the assumptions of Theorem 9.16 are satisfied and so (9.20) follows with

$$C := \|\tilde{a}_{\min}^{-1}\|_{L^8(\Omega)}\|a_{\max}\|_{L^{16}(\Omega)}\|K\|_{L^{16}(\Omega)}.$$ □

The approximation $\tilde{a}(\boldsymbol{x})$ in (9.17) requires *exact* samples of $a(\boldsymbol{x})$ at the vertices of the finite element mesh. If the vertices are uniformly spaced, the circulant embedding method (see Chapter 7) is a possibility. Using (9.17) together with piecewise linear finite elements now gives

$$\begin{aligned}\|u - \tilde{u}_h\|_{L^2(\Omega,H_0^1(D))} &\le \|u - u_h\|_{L^2(\Omega,H_0^1(D))} + \|u_h - \tilde{u}_h\|_{L^2(\Omega,H_0^1(D))}\\ &= \mathcal{O}(h) + \mathcal{O}(h),\end{aligned}$$

(provided the assumptions of Theorem 9.15 and Corollary 9.18 hold). This means the discretisation and the data approximation errors are balanced. Note however that the constants appearing in the above bounds depend on the correlation length ℓ (e.g., see the constant C in (9.20) and Exercise 9.6). Given ℓ, the mesh parameter should be chosen so that $\ell h^{-1} \ge K$ for some fixed positive constant K (say, $K = 5$ or 10). This ensures that the diffusion coefficient does not vary on a length scale that cannot be captured by the spatial discretisation. If not, then for a fixed h, the data error grows as $\ell \to 0$.

We now combine the finite element method with the Monte Carlo method to estimate $\mathbb{E}[u(\boldsymbol{x})]$ and $\mathrm{Var}(u(\boldsymbol{x}))$. It is interesting to note that when $a(\boldsymbol{x})$ is deterministic and $f(\boldsymbol{x})$ is a random field, we can set up a weak problem on D and approximate $\mathbb{E}[u(\boldsymbol{x})]$ without sampling; see Exercise 9.7.

9.2 Monte Carlo FEM

Given *iid* samples $\tilde{a}_r := \tilde{a}(\cdot,\omega_r)$ for $r = 1,\ldots,Q$ of the approximate diffusion coefficient $\tilde{a}(\boldsymbol{x})$ (defined, e.g., as in (9.17)), we can generate *iid* samples $\tilde{u}_h^r(\boldsymbol{x}) := \tilde{u}_h(\boldsymbol{x},\omega_r)$ of the finite element solution $\tilde{u}_h(\boldsymbol{x})$ by solving Q variational problems of the form (9.16). From Chapter 2, we know that the coefficients of each $\tilde{u}_h^r \in V^h$ are found by solving a $J \times J$ linear system of the form (2.85) (with $\boldsymbol{w}_{\mathrm{B}} = \boldsymbol{0}$ for zero boundary conditions). Here, the Galerkin matrix $A \in \mathbb{R}^{(J+J_b)\times(J+J_b)}$ and the vector $\boldsymbol{b} \in \mathbb{R}^{J+J_b}$ associated with the rth sample are defined by

$$a_{ij} := \int_D \tilde{a}_r(\boldsymbol{x})\,\nabla\phi_i(\boldsymbol{x})\cdot\nabla\phi_j(\boldsymbol{x})\,d\boldsymbol{x}, \qquad i,j = 1,\ldots,J+J_b,$$

$$b_i := \int_D f(\boldsymbol{x})\phi_i(\boldsymbol{x})\,d\boldsymbol{x}, \qquad i = 1,\ldots,J+J_b,$$

where ϕ_i is a (piecewise linear) finite element basis function. In the Monte Carlo finite element method (MCFEM), we estimate $\mathbb{E}[\tilde{u}(\boldsymbol{x})]$ by

$$\mu_{Q,h}(\boldsymbol{x}) := \frac{1}{Q}\sum_{r=1}^{Q}\tilde{u}_h^r(\boldsymbol{x}),$$

and similarly (recall (4.29)), we estimate $\mathrm{Var}(\tilde{u}(\boldsymbol{x}))$ by

$$\sigma_{Q,h}^2(\boldsymbol{x}) := \frac{1}{Q-1}\sum_{r=1}^{Q}\left(\tilde{u}_h^r(\boldsymbol{x}) - \mu_{Q,h}(\boldsymbol{x})\right)^2 = \frac{1}{Q-1}\left(\sum_{r=1}^{Q}\tilde{u}_h^r(\boldsymbol{x})^2 - Q\,\mu_{Q,h}(\boldsymbol{x})^2\right).$$

Example 9.19 Consider the BVP in Example 9.1 with $f = 1$ and

$$a(x,\omega) = \mu + \sum_{k=1}^{P}\frac{\sigma}{k^2\pi^2}\cos(\pi k x)\xi_k(\omega), \qquad \xi_k \sim \mathrm{U}(-1,1)\ iid. \tag{9.21}$$

Algorithm 9.1 generates piecewise constant approximations of realisations of $a(x)$ (using *iid* samples of $\boldsymbol{\xi} := [\xi_1,\ldots,\xi_P]^{\mathsf{T}}$) and calls Algorithm 2.1 to implement the finite element method. Suppose that $\mu = 1, \sigma = 4$ and $P = 10$ and we use a mesh of 512 elements. We can generate realisations of $\mu_{Q,h}(x)$ and $\sigma_{Q,h}^2(x)$ with $Q = 100$ using Algorithm 9.1 with the following MATLAB command.

```
>> [mean_u, var_u]=oned_MC_FEM(512,4,1,10,100);
```

Realisations of $\mu_{Q,h}(x)$ and $\sigma_{Q,h}^2(x)$ for $Q = 10$,10^2, 10^3 and 10^4 are shown in Figure 9.3. The sample mean and variance both converge as $Q \to \infty$. To interpret Figure 9.3(b), note that $\mathrm{Var}(u(x))$ is zero at $x = 0$ and $x = 1$ (on the Dirichlet boundary) because $u(x)$ is not a random variable at those points. By construction, $\tilde{u}_h(x)$ satisfies the same boundary conditions and so $\mathrm{Var}(\tilde{u}_h(x)) = 0$ at $x = 0$ and $x = 1$. In addition, $\mathrm{Var}(a(x))$ has a minimum at $x = 1/2$ (see Exercise 9.8) and this feature is inherited by the estimate of $\mathrm{Var}(\tilde{u}_h(x))$.

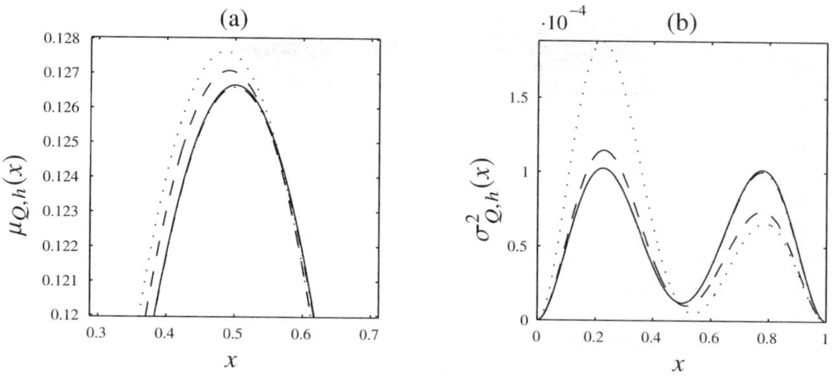

Figure 9.3 (a) MCFEM sample mean $\mu_{Q,h}(x)$ (zoomed) and (b) sample variance $\sigma^2_{Q,h}(x)$ for Example 9.19 with $\mu = 1, \sigma = 4$, $P = 10$, mesh width $h = 1/512$ and $Q = 10$ (dotted), $Q = 10^2$ (dashed), $Q = 10^3$ (dash-dot) and $Q = 10^4$ (solid).

Algorithm 9.1 Code to implement the MCFEM for Example 9.19. The inputs are `ne`, the number of finite elements; `sigma`, `mu`, `P`, the parameters σ, μ and P for the diffusion coefficient; and `Q`, the number of samples. The outputs `mean_u`, `var_u` are vectors containing the values of $\mu_{Q,h}(x)$ and $\sigma^2_{Q,h}(x)$ at the vertices.

```
function [mean_u,var_u]=oned_MC_FEM(ne,sigma,mu,P,Q)
h=1/ne; x=[(h/2):h:(1-h/2)]';
sum_us=zeros(ne+1,1); sum_sq=zeros(ne+1,1);
for j=1:Q
    xi=-1+2.*rand(P,1); a=mu.*ones(ne,1);
    for i=1:P
        a=a+sigma.*((i.*pi).^(-2)).*cos(pi.*i.*x).*xi(i);
    end
    [u,A,b]=oned_linear_FEM(ne,a,zeros(ne,1),ones(ne,1)); hold on;
    sum_us=sum_us+u; sum_sq=sum_sq+(u.^2);
end
mean_u=sum_us./Q;
var_u=(1/(Q-1)).*(sum_sq-(sum_us.^2./Q));
```

Example 9.20 Consider the BVP in Example 9.2 with $a(\boldsymbol{x}) = \mathrm{e}^{z(\boldsymbol{x})}$ where $z(\boldsymbol{x})$ is a mean-zero Gaussian random field with isotropic covariance $c^0(r) = \mathrm{e}^{-r^2/\ell^2}$ and correlation length $\ell = 0.25$. The conditions of Lemma 9.17 are satisfied in this case. Algorithm 9.2 first calls Algorithm 2.4 to create a uniform finite element mesh. It then calls Algorithm 7.6 to apply circulant embedding to generate samples of $z(\boldsymbol{x})$ at the vertices (which are uniformly spaced) and creates piecewise constant approximations of realisations of $a(\boldsymbol{x})$ using (9.17). Finally, it calls Algorithm 2.7 to implement the finite element method. Note that (9.17) and the subsequent analysis assume that *exact* samples of $a(\boldsymbol{x})$ are available. Circulant embedding may only return *approximate* samples (see Example 7.42). To control the sampling error, we must use padding.

Suppose we choose $h = 1/50$. This is sensible since $\ell h^{-1} = 12.5$. If we call Algorithm 7.6

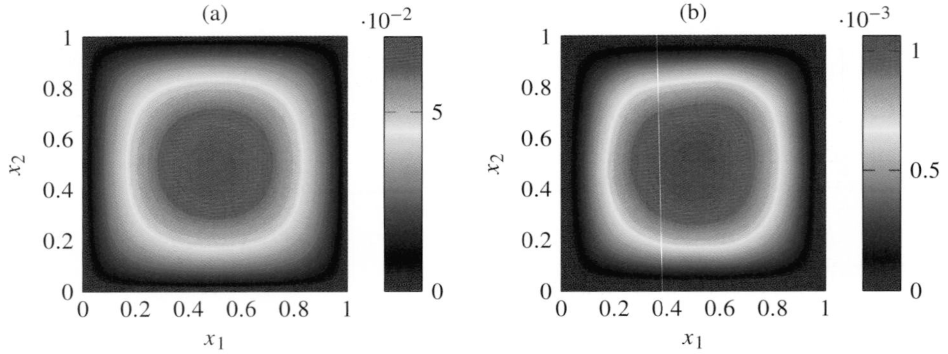

Figure 9.4 `contourf` plots of (a) MCFEM sample mean $\mu_{Q,h}(\boldsymbol{x})$ and (b) sample variance $\sigma^2_{Q,h}(\boldsymbol{x})$ for Example 9.20 with $h = 1/50$, $\ell = 0.25$ and $Q = 10^4$ samples.

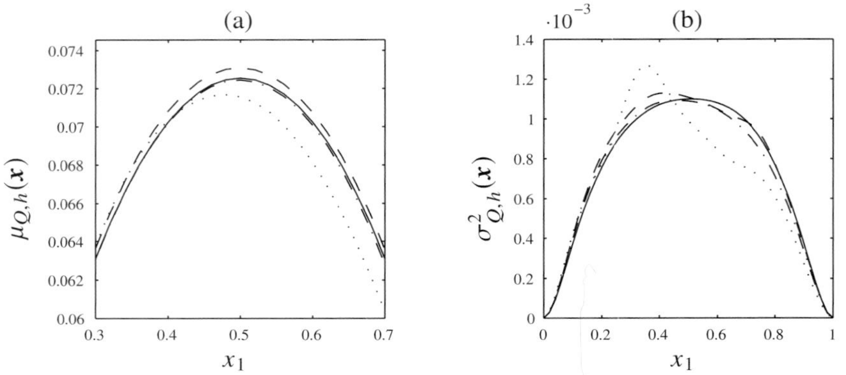

Figure 9.5 Cross-section (through $x_2 = 0.5$) of (a) MCFEM sample mean $\mu_{Q,h}(\boldsymbol{x})$ (zoomed) and (b) sample variance $\sigma^2_{Q,h}(\boldsymbol{x})$ for Example 9.20 with $h = 1/50$ and $Q = 10^2$ (dotted), $Q = 10^3$ (dashed), $Q = 10^4$ (dash-dot) and $Q = 10^5$ (solid).

without padding (by calling Algorithm 9.2 with `alpha` $= 0$) then the following warning message indicates that the samples are not exact.

```
>> [mean_u,var_u]=twod_MC_FEM(50,1000,0.25,0);
Invalid covariance;rho(D_minus)=2.9693e-06
```

The returned samples are from a Gaussian distribution with a modified covariance and the norm of the covariance error is $\mathcal{O}(10^{-6})$ (see Chapter 6 and Exercise 6.19). However, when $\ell = 0.25$, setting `alpha` $= 1$ and applying padding yields a covariance error that is $\mathcal{O}(10^{-14})$ (see also Exercise 9.2). Realisations of the Monte Carlo estimates $\mu_{Q,h}(\boldsymbol{x})$ and $\sigma^2_{Q,h}(\boldsymbol{x})$ can be generated using Algorithm 9.2 with $Q = 10^3$ approximate samples of $z(\boldsymbol{x})$ on a mesh with $h = 1/50$, using the following MATLAB command.

```
>> [mean_u,var_u]=twod_MC_FEM(50,1000,0.25,1);
Invalid covariance;rho(D_minus)=5.2286e-14
```

Algorithm 9.2 Code to implement the MCFEM for Examples 9.20–9.21 with circulant embedding. The inputs are `ns`, the number of squares to define the finite element mesh; `Q`, the number of Monte Carlo samples; `ell`, the correlation length; and α, the padding parameter. The outputs `mean_u`, `var_u` are vectors containing the values of $\mu_{Q,h}(\boldsymbol{x})$ and $\sigma^2_{Q,h}(\boldsymbol{x})$ at the mesh vertices.

```
function [mean_u, var_u]=twod_MC_FEM(ns,Q,ell,alpha)
[xv,yv,elt2vert,nvtx,ne,h]=uniform_mesh_info(ns);
xv=reshape(xv,ns+1,ns+1)'; yv=reshape(xv,ns+1,ns+1)';
b_nodes=find((xv==0)|(xv==1)|(yv==0)|(yv==1));
int_nodes=1:nvtx; int_nodes(b_nodes)=[];
% specify covariance
fhandle1=@(x1,x2)gaussA_exp(x1,x2,ell^(-2),ell^(-2),0);
n1=ns+1; n2=ns+1; m1=alpha*n1; m2=alpha*n2;
C_red=reduced_cov(n1+m1,n2+m2,1/ns,1/ns,fhandle1);
% initialise
sum_us=zeros(nvtx,1); sum_sq=zeros(nvtx,1); Q2=floor(Q/2);
for i=1:Q2
    % two realisations of a - with padding
    [z1,z2]=circ_embed_sample_2dB(C_red,n1,n2,m1,m2);
    v1=exp(z1);v2=exp(z2);
    % piecewise constant approximation
    a1=(1/3).*(v1(elt2vert(:,1))+v1(elt2vert(:,2))+v1(elt2vert(:,3)));
    a2=(1/3).*(v2(elt2vert(:,1))+v2(elt2vert(:,2))+v2(elt2vert(:,3)));
    % two realisations of FEM solution & zero bcs
    [u1_int,A1,rhs1]=twod_linear_FEM(ns,xv,yv,elt2vert,...
                                     nvtx,ne,h,a1,ones(ne,1));
    [u2_int,A2,rhs2]=twod_linear_FEM(ns,xv,yv,elt2vert,...
                                     nvtx,ne,h,a2,ones(ne,1));
    u1=zeros(nvtx,1); u1(int_nodes)=u1_int;
    u2=zeros(nvtx,1); u2(int_nodes)=u2_int;
    sum_us=sum_us+u1+u2; sum_sq=sum_sq+(u1.^2+u2.^2);
end
Q=2*Q2;
mean_u=sum_us./Q;
var_u=(1/(Q-1)).*(sum_sq-((sum_us.^2)./Q));
```

Realisations obtained with $Q = 10^4$ samples are shown in Figure 9.4 and cross-sections, for varying numbers of samples Q, are shown in Figure 9.5.

Example 9.21 Consider Example 9.20 and now fix $\ell = 0.1$. Choosing a smaller correlation length has two consequences. First, we need to choose a smaller value of h to ensure the data error does not dominate the finite element error. For example, $h = 1/128$ gives $\ell h^{-1} = 12.8$. Second, padding is not necessary as the covariance error is $\mathcal{O}(10^{-14})$ when we set `alpha` $= 0$.

Realisations of $\mu_{Q,h}(\boldsymbol{x})$ and $\sigma^2_{Q,h}(\boldsymbol{x})$ can be generated using Algorithm 9.2 (e.g., with $Q = 10^3$ samples) by using the following MATLAB command.

```
>> [mean_u,var_u]=twod_MC_FEM(128,1000,0.1,0);
Invalid covariance;rho(D_minus)=1.0417e-14
```

Two samples of $z(\boldsymbol{x})$ and the corresponding finite element approximations are shown in Figure 9.6. Notice that the realisations of $z(\boldsymbol{x})$ appear rougher than those in Figure 9.2 (when $\ell = 0.25$). Cross-sections of $\mu_{Q,h}(\boldsymbol{x})$ and $\sigma^2_{Q,h}(\boldsymbol{x})$, for varying numbers of samples Q, are shown in Figure 9.7. We observe that $\sigma^2_{Q,h}(\boldsymbol{x})$ has a local minimum at $x = 1/2$.

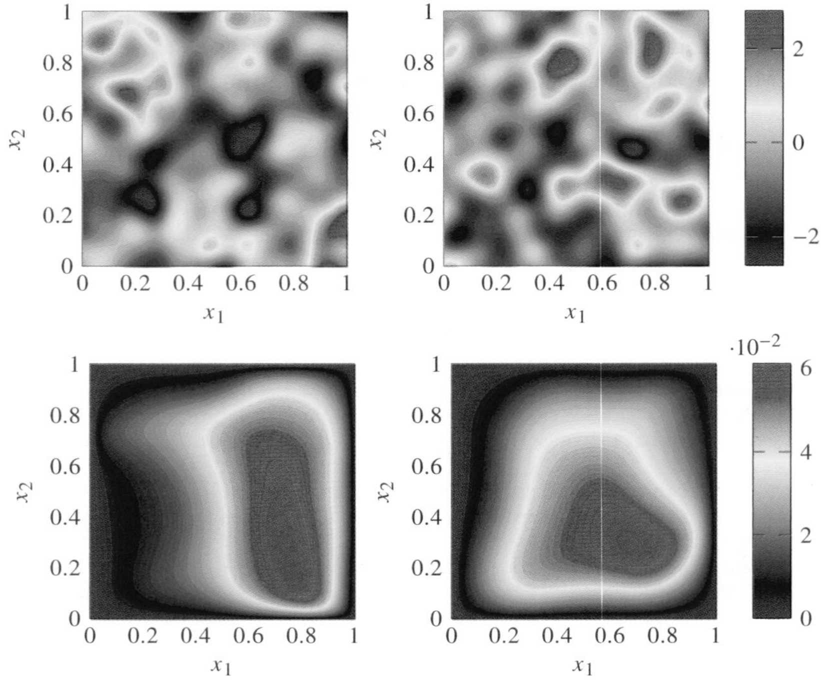

Figure 9.6 contourf plots of (top) two realisations of $z(\boldsymbol{x})$ in Example 9.21 with $\ell = 0.1$ and (bottom) corresponding piecewise linear finite element approximations u_h.

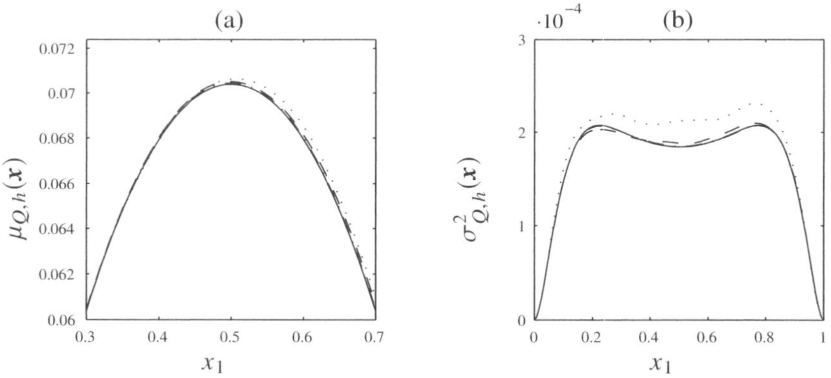

Figure 9.7 Cross-section (through $x_2 = 0.5$) of (a) sample mean $\mu_{Q,h}(\boldsymbol{x})$ (zoomed) and (b) sample variance $\sigma^2_{Q,h}(\boldsymbol{x})$ for Example 9.21 with $h = 1/128$ and $Q = 10^3$ (dotted), $Q = 10^4$ (dashed), $Q = 10^5$ (dash-dot) and $Q = 10^6$ (solid).

See the Notes at the end of this chapter for further comments about this. The relationship between the integral of $\text{Var}(u(\boldsymbol{x}))$ and ℓ is investigated in Exercise 9.9 and an example with non-zero boundary conditions is considered in Exercise 9.10.

Error analysis

We now study the MCFEM error in approximating $\mathbb{E}[u(\boldsymbol{x})]$. Working in the $H_0^1(D)$ norm (the $|\cdot|_{H^1(D)}$ semi-norm) and using the triangle inequality gives

$$\|\mathbb{E}[u]-\mu_{Q,h}\|_{H_0^1(D)} \le \underbrace{\|\mathbb{E}[u]-\mathbb{E}[\tilde{u}_h]\|_{H_0^1(D)}}_{=:\,E_{h,a}} + \underbrace{\|\mathbb{E}[\tilde{u}_h]-\mu_{Q,h}\|_{H_0^1(D)}}_{=:\,E_{\mathrm{MC}}}. \tag{9.22}$$

The first term $E_{h,a}$ on the right-hand side is the error associated with the finite element discretisation using approximate data and satisfies

$$\begin{aligned} E_{h,a} &\le \|\mathbb{E}[u]-\mathbb{E}[u_h]\|_{H_0^1(D)} + \|\mathbb{E}[u_h]-\mathbb{E}[\tilde{u}_h]\|_{H_0^1(D)} \\ &\le \|u-u_h\|_{L^2(\Omega,H_0^1(D))} + \|u_h-\tilde{u}_h\|_{L^2(\Omega,H_0^1(D))} \end{aligned}$$

(by Jensen's inequality). We obtained bounds for these errors in §9.1 (see Theorem 9.15 and Corollary 9.18). The second term E_{MC} in the bound in (9.22) is the statistical error associated with taking Q samples. We now study $\mathbb{E}[E_{\mathrm{MC}}^2]$.

Theorem 9.22 *Let the conditions of Lemma 9.13 hold and define the random variable E_{MC} as in (9.22). Then, for some constant $K>0$ independent of h,*

$$\mathbb{E}[E_{\mathrm{MC}}^2] \le KQ^{-1}. \tag{9.23}$$

Proof First note that

$$\mathbb{E}[E_{\mathrm{MC}}^2] = \mathbb{E}\Big[\|\mathbb{E}[\tilde{u}_h]-\mu_{Q,h}\|_{H_0^1(D)}^2\Big] = \frac{1}{Q^2}\mathbb{E}\left[\left\|\sum_{r=1}^{Q}\left(\mathbb{E}[\tilde{u}_h]-\tilde{u}_h^r\right)\right\|_{H_0^1(D)}^2\right].$$

Writing $X_r = \mathbb{E}[\tilde{u}_h]-\tilde{u}_h^r$ and noting that the X_r are *iid* $H_0^1(D)$-valued random variables and are hence uncorrelated (see Definition 4.35) gives

$$\mathbb{E}[E_{\mathrm{MC}}^2] = \frac{1}{Q^2}\sum_{r=1}^{Q}\mathbb{E}\Big[\|X_r\|_{H_0^1(D)}^2\Big].$$

Now,

$$\mathbb{E}\Big[\|X_r\|_{H_0^1(D)}^2\Big] \le \mathbb{E}\Big[\|\tilde{u}_h^r\|_{H_0^1(D)}^2\Big] = \mathbb{E}\Big[\|\tilde{u}_h\|_{H_0^1(D)}^2\Big],$$

for each $r=1,\ldots,Q$, since $\tilde{u}_h^r$ are *iid* samples of $\tilde{u}_h$. Hence, we have

$$\mathbb{E}[E_{\mathrm{MC}}^2] \le \frac{1}{Q}\mathbb{E}\Big[\|\tilde{u}_h\|_{H_0^1(D)}^2\Big] = \frac{1}{Q}\|\tilde{u}_h\|_{L^2(\Omega,H_0^1(D))}^2.$$

Applying Lemma 9.13 with $p=2$ now gives the result with

$$K := K_{\mathrm{p}}\|a_{\min}^{-1}\|_{L^2(\Omega)}\|f\|_{L^2(D)}. \tag{9.24}$$

□

Corollary 9.23 *Let the conditions of Lemma 9.13 hold and define the random variable E_{MC} as in (9.22). Then, for any $\epsilon>0$,*

$$\mathbb{P}\left(E_{\mathrm{MC}} \ge Q^{-1/2+\epsilon}\right) \le LQ^{-2\epsilon}, \tag{9.25}$$

for a constant $L>0$ independent of h.

Proof Note that (9.23) holds and apply Theorem 4.58(i) with $p = 2$ and $r = 1/2$. The result follows with $L = K^2$, where K is the constant in (9.24). □

The above result tells us that the statistical error is $\mathcal{O}(Q^{-1/2})$ and so

$$\|\mathbb{E}[u] - \mu_{Q,h}\|_{H_0^1(D)} = \mathcal{O}(h) + \mathcal{O}(Q^{-1/2}). \tag{9.26}$$

Since obtaining each sample of $\tilde{u}_h(\boldsymbol{x})$ requires the numerical solution of a PDE, the convergence rate with respect to Q is undesirable. Moreover, the constant L in the error bound depends on $\|a_{\min}^{-1}\|_{L^2(\Omega)}$ (see (9.24)) and convergence may be extremely slow. The cost of implementing the MCFEM amounts to the cost of generating Q samples of the Gaussian random field $z(\boldsymbol{x})$ plus the cost of solving Q linear systems of dimension $J \times J$ where $J = \mathcal{O}(h^{-2})$ (in two space dimensions). For small correlation lengths ℓ, we need fine meshes and this increases the dimension of, and the cost of solving, each system. The optimal cost of solving one system is $\mathcal{O}(J)$ (see Exercise 2.19).

Suppose $J = (\texttt{ns}-1)^2$ (as in Algorithm 9.2). Circulant embedding requires one fast Fourier transform (FFT) to factorise a $4\hat{J} \times 4\hat{J}$ circulant matrix, where $\hat{J} := (1+\alpha)^2(\texttt{ns}+1)^2$ and α is the padding parameter, followed by $Q/2$ FFTs to generate $Q/2$ pairs of samples of $z(\boldsymbol{x})$. Hence, if optimal solvers are available, the total computational cost is $\mathcal{O}(QJ) + \mathcal{O}(Q\hat{J}\log\hat{J})$, or, equivalently,

$$\mathcal{O}(Qh^{-2}) + \mathcal{O}(Qh^{-2}\log h). \tag{9.27}$$

To approximate $\mathbb{E}[u(\boldsymbol{x})]$ in the $H_0^1(D)$ norm to an accuracy of ϵ, it follows from (9.26) that we need to choose $Q = \mathcal{O}(\epsilon^{-2})$ and $h = \mathcal{O}(\epsilon)$. The cost of implementing the MCFEM is then disappointingly $\mathcal{O}(\epsilon^{-4})$. Compare this with the estimate (8.85), obtained for an SDE in one dimension.

9.3 Variational formulation on $D \times \Omega$

Following the ideas in §2.2, we now derive an alternative variational formulation of (9.5)–(9.6) by working on $D \times \Omega$ and seeking weak solutions $u\colon D \times \Omega \to \mathbb{R}$. It is natural to search for u in $L^2(\Omega, H_g^1(D))$ so that realisations belong to the standard solution space for (2.1)–(2.2). Working in $L^2(\Omega)$ guarantees that the weak solution has finite pth moments for $p = 1, 2$. Hence, the mean and variance are well defined. Many of the proofs from §2.2 follow over straightforwardly if we use Assumption 9.3 and treat $a_{\min}$ and $a_{\max}$ as constants. For simplicity, we do not consider log-normal coefficients now. However, we do allow f to be a random field and $g \neq 0$.

Homogeneous boundary conditions

As in Chapter 2, we start by considering $g = 0$ and define $V := L^2(\Omega, H_0^1(D))$. Multiplying both sides of (9.5) by a test function $v \in V$, integrating over D and then taking expectations leads to the variational problem (9.28).

Definition 9.24 (weak solution on $D \times \Omega$) A *weak solution* to the BVP (9.5)–(9.6) with $g = 0$ is a function $u \in V := L^2(\Omega, H_0^1(D))$ that satisfies

$$a(u, v) = \ell(v), \qquad \forall v \in V, \tag{9.28}$$

where $a(\cdot,\cdot): V \times V \to \mathbb{R}$ and $\ell: V \to \mathbb{R}$ are defined by

$$a(u,v) := \mathbb{E}\left[\int_D a(\boldsymbol{x},\cdot)\nabla u(\boldsymbol{x},\cdot) \cdot \nabla v(\boldsymbol{x},\cdot)\, d\boldsymbol{x}\right], \tag{9.29}$$

$$\ell(v) := \mathbb{E}\left[\int_D f(\boldsymbol{x},\cdot) v(\boldsymbol{x},\cdot)\, d\boldsymbol{x}\right]. \tag{9.30}$$

The next result establishes well-posedness of (9.28).

Theorem 9.25 *If $f \in L^2(\Omega, L^2(D))$, $g = 0$ and Assumption 9.3 holds then* (9.28) *has a unique solution $u \in V$. Further, if $a(\boldsymbol{x},\cdot)$ and $f(\boldsymbol{x},\cdot)$ are $\mathcal{G}$-measurable, for some sub σ-algebra $\mathcal{G} \subset \mathcal{F}$ then $u(\boldsymbol{x},\cdot)$ is $\mathcal{G}$-measurable.*

Proof Let $v \in V = L^2(\Omega, H_0^1(D))$ and define the norm

$$\|v\|_V := \|v\|_{L^2(\Omega, H_0^1(D))} = \mathbb{E}\left[|v|_{H^1(D)}^2\right]^{1/2}. \tag{9.31}$$

Assumption 9.3 guarantees that $a(\cdot,\cdot)$ is bounded on $V \times V$ since

$$|a(u,v)| \le a_{\max}\mathbb{E}\left[|u|_{H^1(D)}^2\right]^{1/2}\mathbb{E}\left[|v|_{H^1(D)}^2\right]^{1/2} = a_{\max}\|u\|_V\|v\|_V,$$

for all $u, v \in V$. Coercivity also holds since

$$a(v,v) = \mathbb{E}\left[\int_D a(\boldsymbol{x},\cdot)\,\nabla v(\boldsymbol{x},\cdot) \cdot \nabla v(\boldsymbol{x},\cdot)\, d\boldsymbol{x}\right] \ge a_{\min}\|v\|_V^2.$$

The Cauchy–Schwarz inequality gives

$$|\ell(v)| \le \|f\|_{L^2(\Omega, L^2(D))}\|v\|_{L^2(\Omega, L^2(D))}. \tag{9.32}$$

Poincaré's inequality (see Theorem 1.48) also gives

$$\|v(\cdot,\omega)\|_{L^2(D)} \le K_{\mathrm{p}}|v(\cdot,\omega)|_{H^1(D)}, \qquad \forall \omega \in \Omega$$

and, taking second moments, we obtain

$$\|v\|_{L^2(\Omega, L^2(D))} \le K_{\mathrm{p}}\|v\|_V.$$

Substituting into (9.32) then yields $|\ell(v)| \le K_{\mathrm{p}}\|f\|_{L^2(\Omega, L^2(D))}\|v\|_V$. Hence, existence and uniqueness of $u \in V$ is proved by the Lax–Milgram lemma (see Lemma 1.59). Since $u \in V := L^2(\Omega, \mathcal{F}, H_0^1(D))$, $u(\boldsymbol{x},\cdot)$ is $\mathcal{F}$-measurable. To prove that $u(\boldsymbol{x},\cdot)$ is also $\mathcal{G}$-measurable, for any sub σ-algebra $\mathcal{G} \subset \mathcal{F}$, when $a(\boldsymbol{x},\cdot)$ and $f(\boldsymbol{x},\cdot)$ are $\mathcal{G}$-measurable, see Exercise 9.11. □

Non-homogeneous boundary conditions

If $g \neq 0$, we choose $V = L^2(\Omega, H_0^1(D))$ as the test space and derive (9.28) as before. However, we now want to find $u \in W = L^2(\Omega, H_g^1(D)) \neq V$.

Definition 9.26 (weak solution on $D \times \Omega$) A *weak solution* to the BVP (9.5)–(9.6) is a function $u \in W := L^2(\Omega, H_g^1(D))$ that satisfies

$$a(u,v) = \ell(v), \qquad \forall v \in V, \tag{9.33}$$

where $a(\cdot,\cdot): W \times V \to \mathbb{R}$ and $\ell: V \to \mathbb{R}$ are defined in (9.29)–(9.30).

Since $\|\cdot\|_V$ in (9.31) is not a norm on $L^2(\Omega, H^1(D))$, we now write

$$|v|_W := \mathbb{E}\left[|v|^2_{H^1(D)}\right]^{1/2}.$$

Writing (9.33) as an equivalent problem with homogeneous boundary conditions, as in §2.2, leads to the following analogues of Theorems 2.42 and 2.43.

Theorem 9.27 *If Assumption 9.3 holds, $f \in L^2(\Omega, L^2(D))$ and $g \in H^{1/2}(\partial D)$ then (9.33) has a unique solution $u \in W$. Further, if $a(\boldsymbol{x}, \cdot)$ and $f(\boldsymbol{x}, \cdot)$ are $\mathcal{G}$-measurable, for some sub σ-algebra $\mathcal{G} \subset \mathcal{F}$ then $u(\boldsymbol{x}, \cdot)$ is $\mathcal{G}$-measurable.*

Theorem 9.28 *Let $u \in W$ satisfy (9.33). If Assumption 9.3 holds, $g \in H^{1/2}(\partial D)$ and $f \in L^2(\Omega, L^2(D))$ then*

$$|u|_W \le K\left(\|f\|_{L^2(\Omega, L^2(D))} + \|g\|_{H^{1/2}(\partial D)}\right)$$

where $K := \max\left(\frac{K_p}{a_{\min}}, K_\gamma\left(1 + \frac{a_{\max}}{a_{\min}}\right)\right)$ and K_γ is the constant from (2.54).

Proof See Exercise 9.12. □

We now assess the effect of approximating $a(\boldsymbol{x})$ and $f(\boldsymbol{x})$ (the data) on the accuracy of the weak solution $u \in W$ in (9.33).

Perturbed weak form on $D \times \Omega$

Replacing $a(\boldsymbol{x})$ and $f(\boldsymbol{x})$ in (9.5) by approximate random fields $\tilde{a}, \tilde{f} : D \times \Omega \to \mathbb{R}$ leads to the boundary-value problem: find $\tilde{u} : \bar{D} \times \Omega \to \mathbb{R}$ such that $\mathbb{P}$-a.s.

$$-\nabla \cdot (\tilde{a}(\boldsymbol{x}, \omega) \nabla \tilde{u}(\boldsymbol{x}, \omega)) = \tilde{f}(\boldsymbol{x}, \omega), \qquad \boldsymbol{x} \in D, \tag{9.34}$$

$$\tilde{u}(\boldsymbol{x}, \omega) = g(\boldsymbol{x}), \qquad \boldsymbol{x} \in \partial D. \tag{9.35}$$

The corresponding weak problem is: find $\tilde{u} \in W$ such that

$$\tilde{a}(\tilde{u}, v) = \tilde{\ell}(v), \qquad \forall v \in V \tag{9.36}$$

where $\tilde{a} : W \times V \to \mathbb{R}$ and $\tilde{\ell} : V \to \mathbb{R}$ are defined by

$$\tilde{a}(u, v) := \mathbb{E}\left[\int_D \tilde{a}(\boldsymbol{x}, \cdot) \nabla u(\boldsymbol{x}, \cdot) \cdot \nabla v(\boldsymbol{x}, \cdot)\, d\boldsymbol{x}\right],$$

$$\tilde{\ell}(v) := \mathbb{E}\left[\int_D \tilde{f}(\boldsymbol{x}, \cdot) v(\boldsymbol{x}, \cdot)\, d\boldsymbol{x}\right].$$

If the following assumption holds, establishing well-posedness of (9.36) is straightforward.

Assumption 9.29 (diffusion coefficients) The approximate diffusion coefficient $\tilde{a}(\boldsymbol{x})$ belongs to $L^\infty(\Omega, L^\infty(D))$ and satisfies

$$0 < \tilde{a}_{\min} \le \tilde{a}(\boldsymbol{x}, \omega) \le \tilde{a}_{\max} < \infty, \qquad \text{a.e. in } D \times \Omega,$$

for some real constants $\tilde{a}_{\min}$ and $\tilde{a}_{\max}$.

Theorem 9.30 *If Assumption 9.29 holds, $\tilde{f} \in L^2(\Omega, L^2(D))$ and $g \in H^{1/2}(\partial D)$ then there exists a unique $\tilde{u} \in W = L^2(\Omega, H_g^1(D))$ satisfying (9.36). Further, if $\tilde{a}(\boldsymbol{x},\cdot)$ and $\tilde{f}(\boldsymbol{x},\cdot)$ are $\mathcal{G}$-measurable, for some sub σ-algebra $\mathcal{G} \subset \mathcal{F}$, then $\tilde{u}(\boldsymbol{x},\cdot)$ is $\mathcal{G}$-measurable.*

Proof Follow the proof of Theorem 9.27 and the solution of Exercise 9.11, replacing the random fields $a(\boldsymbol{x})$ and $f(\boldsymbol{x})$ with $\tilde{a}(\boldsymbol{x})$ and $\tilde{f}(\boldsymbol{x})$. □

The next result gives an upper bound for the weak solution error

$$|u - \tilde{u}|_W = \mathbb{E}\left[|u - \tilde{u}|^2_{H^1(D)}\right]^{1/2},$$

in terms of the data errors $\|a - \tilde{a}\|_{L^\infty(\Omega, L^\infty(D))}$ and $\|f - \tilde{f}\|_{L^2(\Omega, L^2(D))}$.

Theorem 9.31 *Let the conditions of Theorems 9.27 and 9.30 hold so that $u \in W$ and $\tilde{u} \in W$ are the unique solutions to (9.33) and (9.36), respectively. Then,*

$$|u - \tilde{u}|_W \le K_{\mathrm{p}} \tilde{a}_{\min}^{-1} \|f - \tilde{f}\|_{L^2(\Omega, L^2(D))} + \tilde{a}_{\min}^{-1} \|a - \tilde{a}\|_{L^\infty(\Omega, L^\infty(D))} |u|_W. \tag{9.37}$$

Proof See Exercise 9.13. □

Truncated Karhunen–Loève expansions

We now consider choosing the data approximations to be truncated Karhunen–Loève expansions and examine (9.37). Denote the means of $a(\boldsymbol{x})$ and $f(\boldsymbol{x})$ by $\mu_a(\boldsymbol{x}) := \mathbb{E}[a(\boldsymbol{x})]$ and $\mu_f(\boldsymbol{x}) := \mathbb{E}[f(\boldsymbol{x})]$, and the covariance functions by

$$C_a(\boldsymbol{x}_1, \boldsymbol{x}_2) := \mathbb{E}\left[(a(\boldsymbol{x}_1) - \mu_a(\boldsymbol{x}_1))(a(\boldsymbol{x}_2) - \mu_a(\boldsymbol{x}_2))\right], \tag{9.38}$$

$$C_f(\boldsymbol{x}_1, \boldsymbol{x}_2) := \mathbb{E}\left[(f(\boldsymbol{x}_1) - \mu_f(\boldsymbol{x}_1))(f(\boldsymbol{x}_2) - \mu_f(\boldsymbol{x}_2))\right]. \tag{9.39}$$

Then, we may expand the original data in (9.5) as follows:

$$a(\boldsymbol{x}, \omega) = \mu_a(\boldsymbol{x}) + \sum_{k=1}^{\infty} \sqrt{\nu_k^a}\, \phi_k^a(\boldsymbol{x})\, \xi_k(\omega), \tag{9.40}$$

$$f(\boldsymbol{x}, \omega) = \mu_f(\boldsymbol{x}) + \sum_{k=1}^{\infty} \sqrt{\nu_k^f}\, \phi_k^f(\boldsymbol{x})\, \eta_k(\omega), \tag{9.41}$$

where (ν_k^a, ϕ_k^a) are the eigenpairs of C_a, (ν_k^f, ϕ_k^f) are the eigenpairs of C_f and ξ_k, η_k are random variables. For simplicity, we assume that ξ_k and η_k have the same distribution and write $\eta_k = \xi_k$. Now, given $P, N \in \mathbb{N}$, consider

$$\tilde{a}(\boldsymbol{x}, \omega) := \mu_a(\boldsymbol{x}) + \sum_{k=1}^{P} \sqrt{\nu_k^a}\, \phi_k^a(\boldsymbol{x})\, \xi_k(\omega), \tag{9.42}$$

$$\tilde{f}(\boldsymbol{x}, \omega) := \mu_f(\boldsymbol{x}) + \sum_{k=1}^{N} \sqrt{\nu_k^f}\, \phi_k^f(\boldsymbol{x})\, \xi_k(\omega). \tag{9.43}$$

Random variables in Karhunen–Loève expansions are always uncorrelated, though not necessarily independent. We will assume they are independent. To ensure that there exists a unique $\tilde{u} \in W$ satisfying (9.36), $\tilde{a}(\boldsymbol{x})$ must satisfy Assumption 9.29. If we choose (9.42), it

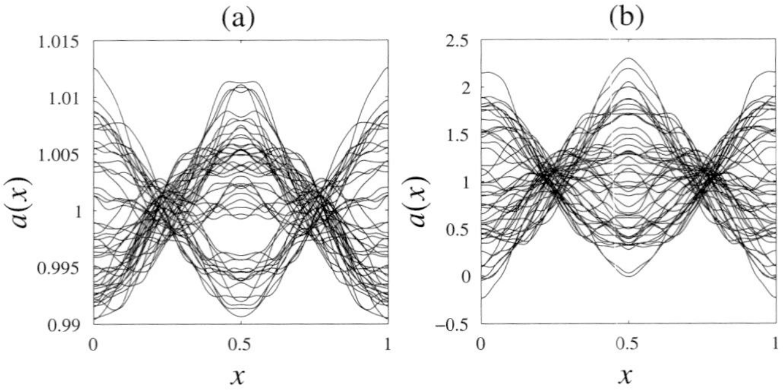

Figure 9.8 Fifty realisations of the stochastic process $a(x)$ in (9.4) with $\mu = 1$, $P = 20$ and (a) $\sigma = 0.1$ and (b) $\sigma = 10$. In (b) some realisations of $a(x)$ are negative on $(0, 1)$.

is clear that we'll need to limit our study to uniformly bounded random variables. Bounded random variables are not actually required for $\tilde{f}(\boldsymbol{x})$ since we only need $\tilde{f} \in L^2(\Omega, L^2(D))$. However, we will make the following assumption.

Assumption 9.32 (random variables) The random variables ξ_k in the Karhunen–Loève expansions (9.40) and (9.41) are independent and bounded with $\xi_k \in [-\gamma, \gamma]$, for some constant γ such that $0 < \gamma < \infty$.

Assumption 9.32 alone does not guarantee that $\tilde{a}(\boldsymbol{x})$ in (9.42) satisfies Assumption 9.29. The following example illustrates that care must be taken when choosing parameters.

Example 9.33 (well-posedness) The stochastic process $a(x)$ in (9.4) is an expansion of the form (9.42) with bounded random variables $\xi_k \in [-1, 1]$ and

$$\nu_k^a = \frac{\sigma^2}{k^4 \pi^4}, \qquad \phi_k^a(x) = \cos(\pi k x).$$

Figure 9.8 shows realisations of $a(x)$ for $\mu = 1, P = 20$ and for two choices of the parameter σ. When $\sigma = 10$, some realisations are negative on D. In that case, $a(x)$ does not satisfy Assumption 9.29, and hence (9.36) is not well posed. To guarantee that $a(x)$ is a valid diffusion coefficient, for *any* truncation parameter P, we need $\sigma < 6\mu$ (see Example 9.4).

We now consider the error bound (9.37). Theorem 7.52 gives

$$\|f - \tilde{f}\|^2_{L^2(\Omega, L^2(D))} = \mathbb{E}\left[\|f - \tilde{f}\|^2_{L^2(D)}\right] \to 0, \qquad \text{as } N \to \infty,$$

and so the first term can be made arbitrarily small by including sufficiently many terms in (9.43). However, we'll also need $\|a - \tilde{a}\|_{L^\infty(\Omega, L^\infty(D))} \to 0$ as $P \to \infty$. If $C_a \in \mathrm{C}(\bar{D} \times \bar{D})$, Theorem 7.53 only tells us that

$$\sup_{\boldsymbol{x} \in D} \mathbb{E}\left[|a(\boldsymbol{x}) - \tilde{a}(\boldsymbol{x})|^2\right] \to 0, \qquad \text{as } P \to \infty, \tag{9.44}$$

so further investigation is needed. Observe that

$$\|a - \tilde{a}\|_{L^\infty(\Omega, L^\infty(D))} = \operatorname*{ess\,sup}_{(\boldsymbol{x}, \omega) \in D \times \Omega} \left| \sum_{k=P+1}^{\infty} \sqrt{\nu_k^a}\, \phi_k^a(\boldsymbol{x})\, \xi_k(\omega) \right|$$

and if Assumption 9.32 holds then

$$\|a - \tilde{a}\|_{L^\infty(\Omega, L^\infty(D))} \le \gamma \sum_{k=P+1}^{\infty} \sqrt{\nu_k^a} \|\phi_k^a\|_\infty. \tag{9.45}$$

Note that since D is bounded, the eigenfunctions are continuous by Theorem 7.53 and it is appropriate to use the norm $\|\cdot\|_\infty$.

In Example 9.33, $\gamma = 1$ and $\sqrt{\nu_k^a}\|\phi_k^a\|_\infty \le \sigma\pi^{-2}k^{-2}$ so $\|a - \tilde{a}\|_{L^\infty(\Omega, L^\infty(D))} \to 0$ as $P \to \infty$. More generally, $\nu_k^a \to 0$ at a rate that depends on the regularity of the covariance C_a (see Example 7.59). In Karhunen–Loève expansions, $\|\phi_k^a\|_{L^2(D)} = 1$ (the eigenfunctions are normalised) and so we expect $\|\phi_k^a\|_\infty$ to grow as $k \to \infty$. Uniform bounds on $\|a - \tilde{a}\|_{L^\infty(\Omega, L^\infty(D))}$ can be established if this growth is balanced against the decay of the eigenvalues. The next result provides a bound for $\|\phi_k^a\|_\infty$ in terms of the eigenvalues of C_a.

Proposition 9.34 *Suppose that $C \in C^{2q}(D \times D)$ for a bounded domain $D \subset \mathbb{R}^d$ and denote the eigenfunctions and eigenvalues by ϕ_k and ν_k, respectively. For $2q > r > d/2$, there exists a constant $K > 0$ such that*

$$\|\phi_k\|_\infty \le K \nu_k^{-r/(2q-r)}, \qquad \text{for } k = 1, 2, \dots.$$

Proof First consider the one-dimensional domain $D = (-\pi, \pi)$. The Sobolev embedding theorem (Theorem 1.51) says that, for any $r > 1/2$, there exists $K > 0$ such that

$$\|w\|_\infty \le K\|w\|_{H^r(D)}, \qquad \forall w \in H^r(D).$$

It is convenient here to work with fractional power norms. Recall from Proposition 1.93 that the $H^r(D)$ norm is equivalent to the $\mathcal{D}(A^{r/2})$ norm for the operator A satisfying $A\phi = \phi - \phi_{xx}$ with periodic boundary conditions. Thus, for a possibly larger constant K, $\|w\|_\infty \le K\|w\|_{r/2}$. Writing $w(x) = \frac{1}{\sqrt{2\pi}} \sum_{k \in \mathbb{Z}} w_k e^{\mathrm{i}kx}$, we have

$$\|w\|_{r/2}^2 = \sum_k (1 + k^2)^r |w_k|^2. \tag{9.46}$$

Now, for any $\lambda, \nu > 0$ and $n > r$, it is easy to show that

$$\lambda^r \le \nu^2 \lambda^n + \nu^{2r/(r-n)}.$$

Hence, taking $\lambda = 1 + k^2$ in (9.46) gives

$$\|w\|_{r/2}^2 \le \nu^2 \sum_k (1 + k^2)^n w_k^2 + \nu^{2r/(r-n)} \sum_k w_k^2$$

and so

$$\|w\|_{r/2} \le \nu \|w\|_{n/2} + \nu^{r/(r-n)} \|w\|_0. \tag{9.47}$$

Thus, we have shown that for any $w \in H^n(-\pi, \pi)$ and any $\nu > 0$,

$$\|w\|_\infty \le K\|w\|_{r/2} \le K\left(\nu\|w\|_{H^n(D)} + \nu^{-r/(n-r)}\|w\|_{L^2(D)}\right).$$

This argument can be modified for any bounded domain $D \subset \mathbb{R}^d$ if $r > d/2$ (a condition which comes from the Sobolev embedding theorem), to give

$$\|w\|_\infty \le K\left(\nu\|w\|_{H^n(D)} + \nu^{-r/(n-r)}\|w\|_{L^2(D)}\right), \qquad \forall w \in H^n(D). \tag{9.48}$$

Now, the eigenfunction ϕ_k of C corresponding to the eigenvalue ν_k satisfies

$$\nu_k \phi_k(\boldsymbol{x}) = \int_D C(\boldsymbol{x},\boldsymbol{y})\phi_k(\boldsymbol{y})\,d\boldsymbol{y}$$

and so

$$\nu_k \mathcal{D}^\alpha \phi_k(\boldsymbol{x}) = \int_D \mathcal{D}^\alpha C(\boldsymbol{x},\boldsymbol{y})\phi_k(\boldsymbol{y})\,d\boldsymbol{y}, \qquad |\alpha| \le 2q.$$

The derivatives of C are uniformly bounded up to order $2q$ and hence

$$\|\phi_k\|_{H^{2q}(D)} \le \frac{1}{\nu_k} K_{2q} \|\phi_k\|_{L^2(D)},$$

for a constant K_{2q}. Choosing $w = \phi_k$, $\nu = \nu_k$, and $n = 2q$ in (9.48) gives

$$\|\phi_k\|_\infty \le K\big(K_{2q} + \nu_k^{-r/(2q-r)}\big). \qquad \square$$

Combining Proposition 9.34 with decay rates for the eigenvalues, we can find conditions for the expansion of $a(\boldsymbol{x})$ to converge in $L^\infty(\Omega, L^\infty(D))$. We now consider the isotropic Whittle–Matérn covariance of Example 7.17.

Corollary 9.35 (Karhunen–Loève $L^\infty(\Omega, L^\infty(D))$ convergence) *Let $a(\boldsymbol{x})$ and $\tilde{a}(\boldsymbol{x})$ be the Karhunen–Loève expansions* (9.40) *and* (9.42) *respectively and let $C_a(\boldsymbol{x},\boldsymbol{y}) = c_q^0(\|\boldsymbol{x}-\boldsymbol{y}\|_2)$ be the Whittle–Matérn covariance with parameter q. If Assumption 9.32 holds, then $\|a - \tilde{a}\|_{L^\infty(\Omega, L^\infty(D))} \to 0$ as $P \to \infty$, if $q \ge 2d$.*

Proof The Whittle–Matérn covariance is $2q$ times differentiable (see Lemma 6.11) and Proposition 9.34 gives $\|\phi_k^a\|_\infty = \mathcal{O}\big((\nu_k^a)^{-r/(2q-r)}\big)$ for $r > d/2$. In addition, we know that $\nu_k^a = \mathcal{O}\big(k^{-(2q+d)/d}\big)$ by Example 7.59. Now, using $q \ge 2d$ gives $\nu_k^a = \mathcal{O}(k^{-5})$ and $r/(2q-r) \le 1/7$. Thus,

$$\sqrt{\nu_k^a}\|\phi_k^a\|_\infty = \mathcal{O}\big(k^{-5/2}k^{5(1/7)}\big)$$

in (9.45) and convergence is ensured because $(-5/2) + (5)(1/7) < -1$. $\square$

If $d = 2$, we require $q > 4$ in Corollary 9.35. Note, however, that the condition $q \ge 2d$ is sufficient and not necessary.

In general, the eigenvalues and eigenfunctions of C_a and C_f are not known explicitly and we must apply numerical methods (e.g., see §7.4) to approximate them. Analysing the error bound (9.37) in that situation is more difficult. To avoid this, we introduce two Karhunen–Loève expansions with explicit eigenpairs that lead to random fields with covariance functions close to a target covariance (whose eigenpairs are not known explicitly).

Example 9.36 (explicit eigenpairs) Let $D = (0,1)$ and consider

$$a(x,\omega) = \mu_a(x) + \sum_{k=0}^{\infty} \sqrt{\nu_k^a}\,\phi_k^a(x)\,\xi_k(\omega), \qquad \xi_k \sim \mathrm{U}\big(-\sqrt{3}, \sqrt{3}\big) \textit{ iid},$$

where $\nu_0^a := 1/2$, $\phi_0^a(x) := 1$ and

$$\nu_k^a := \frac{1}{2}\exp\big(-\pi k^2 \ell^2\big), \qquad \phi_k^a(x) := \sqrt{2}\cos(k\pi x), \qquad k \ge 1.$$

Since $\mathbb{E}[\xi_k] = 0$ and $\mathrm{Var}(\xi_k) = 1$, it follows from Example 6.10 that $a(x)$ has mean μ_a and, when ℓ is small, its covariance is close to the stationary covariance

$$c(x) = \frac{1}{2\ell} \exp\left(\frac{-\pi x^2}{4\ell^2}\right),$$

Since Assumption 9.32 holds, it is easy to see that

$$\tilde{a}(x,\omega) := \mu_a(x) + \sum_{k=0}^{P} \sqrt{\nu_k^a}\, \phi_k^a(x)\, \xi_k(\omega) \tag{9.49}$$

converges to a in $L^\infty(\Omega, L^\infty(D))$ as $P \to \infty$. Realisations of $\tilde{a}(x)$ are also positive, provided μ_a is large enough compared to ℓ (see Exercise 9.14).

Example 9.37 (explicit eigenpairs) Let $D = (0,1) \times (0,1)$ and now consider

$$a(\boldsymbol{x},\omega) = \mu_a(\boldsymbol{x}) + \sum_{i=0}^{\infty}\sum_{j=0}^{\infty} \sqrt{\nu_{ij}^a}\, \phi_{ij}^a(\boldsymbol{x})\, \xi_{ij}(\omega), \qquad \xi_{ij} \sim \mathrm{U}\big(-\sqrt{3}, \sqrt{3}\big)\ iid,$$

with $\phi_{ij}^a(\boldsymbol{x}) = \phi_i(x_1)\phi_j(x_2)$ and $\nu_{ij}^a = \nu_i \nu_j$. Suppose the one-dimensional functions ϕ_k and eigenvalues ν_k are defined as in Example 9.36, so that

$$\phi_{ij}^a(\boldsymbol{x}) := 2\cos(i\pi x_1)\cos(j\pi x_2), \quad \nu_{ij}^a := \frac{1}{4}\exp\left(-\pi\left(i^2 + j^2\right)\ell^2\right), \qquad i,j \geq 1.$$

Since the ξ_{ij} are *iid* with mean zero and unit variance, $a(\boldsymbol{x})$ is a random field with covariance close to the isotropic covariance

$$c^0(r) = \frac{1}{4\ell^2} \exp\left(\frac{-\pi r^2}{4\ell^2}\right)$$

(see Example 6.10). Reordering in terms of a single index k (so that the eigenvalues ν_k^a appear in descending order) gives

$$\tilde{a}(\boldsymbol{x},\omega) = \mu_a(\boldsymbol{x}) + \sum_{k=1}^{P} \sqrt{\nu_k^a}\phi_k^a(\boldsymbol{x})\, \xi_k(\omega), \qquad \xi_k \sim \mathrm{U}\big(-\sqrt{3}, \sqrt{3}\big)\ iid, \tag{9.50}$$

which converges to $a(\boldsymbol{x})$ in $L^\infty(\Omega, L^\infty(D))$ as $P \to \infty$. See also Exercise 9.15.

We use the random field $\tilde{a}(\boldsymbol{x})$ in (9.50) in numerical experiments in §9.5.

9.4 Variational formulation on $D \times \Gamma$

The perturbed weak form (9.36) is not a convenient starting point for Galerkin approximation as the integrals involve an abstract set Ω and probability measure $\mathbb{P}$. That is,

$$\tilde{a}(u,v) = \int_\Omega \int_D \tilde{a}(\boldsymbol{x},\omega)\, \nabla u(\boldsymbol{x},\omega) \cdot \nabla v(\boldsymbol{x},\omega)\, d\boldsymbol{x}\, d\mathbb{P}(\omega),$$

and similarly for $\tilde{\ell}(v)$. We now show that if $\tilde{a}(\boldsymbol{x})$ and $\tilde{f}(\boldsymbol{x})$ depend on a finite number M of random variables $\xi_k : \Omega \to \Gamma_k \subset \mathbb{R}$ (e.g., are Karhunen–Loève expansions), then we can make a change of variable and replace (9.36) with an equivalent weak form on $D \times \Gamma$ where $\Gamma := \Gamma_1 \times \cdots \times \Gamma_M \subset \mathbb{R}^M$. Random fields that are functions of a finite number of random variables are known as *finite-dimensional noise*. We use the following definition.

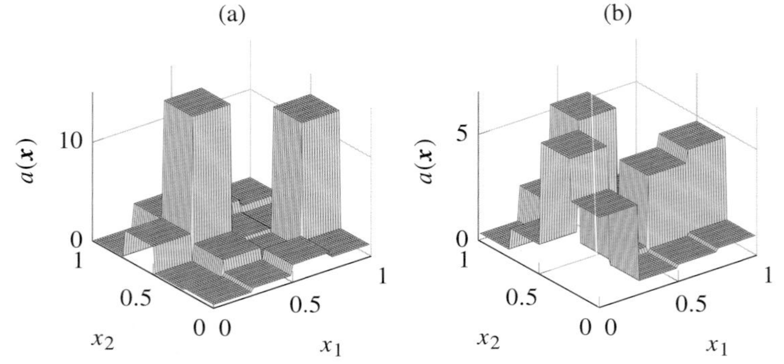

Figure 9.9 Two realisations of $a(\boldsymbol{x})$ in Example 9.39 on $D = (0,1) \times (0,1)$ with $a_0 = 0.1$ and $M = 16$ random variables $\xi_k \sim \mathrm{N}(0,1)$. D is partitioned into 16 squares and on each one $a(\boldsymbol{x})$ is log-normal with $\mathbb{E}[a(\boldsymbol{x})] \approx 1.7487$ and $\mathrm{Var}(a(\boldsymbol{x})) \approx 8.0257$.

Definition 9.38 (finite-dimensional noise) Let $\xi_k : \Omega \to \Gamma_k$ for $k = 1, \dots, M$ be real-valued random variables with $M < \infty$. A function $v \in L^2(\Omega, L^2(D))$ of the form $v(\boldsymbol{x}, \boldsymbol{\xi}(\omega))$ for $\boldsymbol{x} \in D$ and $\omega \in \Omega$, where $\boldsymbol{\xi} = [\xi_1, \dots, \xi_M]^{\mathsf{T}} : \Omega \to \Gamma \subset \mathbb{R}^M$ and $\Gamma := \Gamma_1 \times \cdots \times \Gamma_M$, is called *finite-dimensional* or *M-dimensional noise*.

A truncated Karhunen–Loève expansion of a second-order random field $v \in L^2(\Omega, L^2(D))$ clearly satisfies Definition 9.38. However, this is not the only possibility. Consider the following example.

Example 9.39 (piecewise random fields) Suppose the domain D is partitioned into M non-overlapping subdomains D_k so that $\bar{D} = \bigcup_{k=1}^M \bar{D}_k$ and let

$$\log\big(a(\boldsymbol{x}, \omega) - a_0\big) = \sum_{k=1}^{M} \xi_k(\omega)\, 1_{D_k}(\boldsymbol{x}),$$

where $a_0 > 0$, $\xi_k \sim \mathrm{N}(\mu_k, \sigma_k^2)$ are pairwise independent, and 1_{D_k} is the indicator function on D_k. This model often arises in studies of fluid flows in porous media. On each subdomain D_k, the diffusion coefficient $a(\boldsymbol{x})|_{D_k} = a_0 + \mathrm{e}^{\xi_k}$ is a log-normal random variable with

$$\mathbb{E}\Big[a(\boldsymbol{x})|_{D_k}\Big] = a_0 + \mathrm{e}^{\mu_k + \sigma_k^2/2}, \qquad \mathrm{Var}\Big(a(\boldsymbol{x})|_{D_k}\Big) = \Big(\mathrm{e}^{\sigma_k^2} - 1\Big)\mathrm{e}^{2\mu_k + \sigma_k^2},$$

(see (4.9)). The shift $a_0 > 0$ ensures that no realisation of $a(\boldsymbol{x})$ has a value less than a_0 at any $\boldsymbol{x} \in D$. It is clear that $a \in L^2(\Omega, L^2(D))$ and satisfies Definition 9.38. Realisations for the case $a_0 = 0.1$, $M = 16$, $\mu_k = 0$ and $\sigma_k = 1$ are shown in Figure 9.9.

If $v = v(\boldsymbol{x}, \boldsymbol{\xi})$ is finite-dimensional noise then we can perform a change of variable when computing expectations (recall (4.5)). For example, if p is the joint density of $\boldsymbol{\xi}$,

$$\|v\|_{L^2(\Omega, L^2(D))}^2 = \mathbb{E}\Big[\|v\|_{L^2(D)}^2\Big] = \int_{\Gamma} p(\boldsymbol{y}) \|v(\cdot, \boldsymbol{y})\|_{L^2(D)}^2 d\boldsymbol{y}.$$

Hence, by defining $y_k := \xi_k(\omega)$ and $\boldsymbol{y} = [y_1, \dots, y_M]^{\mathsf{T}}$, we can associate any random field

$v \in L^2(\Omega, L^2(D))$ satisfying Definition 9.38 with a function $v = v(\boldsymbol{x}, \boldsymbol{y})$ belonging to the weighted L^2 space,

$$L^2_p(\Gamma, L^2(D)) := \left\{ v : D \times \Gamma \to \mathbb{R} : \int_\Gamma p(\boldsymbol{y}) \|v(\cdot, \boldsymbol{y})\|^2_{L^2(D)} d\boldsymbol{y} < \infty \right\}.$$

Returning to the weak form (9.36), we have the following result.

Lemma 9.40 *Let $\tilde{a}(\boldsymbol{x})$ and $\tilde{f}(\boldsymbol{x})$ be finite-dimensional noise. Then, $\tilde{u} \in W$ satisfying* (9.36) *is also finite-dimensional noise.*

Proof First, note that $\tilde{u} \in W \subset L^2(\Omega, L^2(D))$. Since $\tilde{a}(\boldsymbol{x}, \cdot)$ and $\tilde{f}(\boldsymbol{x}, \cdot)$ are functions of $\boldsymbol{\xi} : \Omega \to \Gamma$ they are $\sigma(\boldsymbol{\xi})$-measurable (see Definition 4.18). Theorem 9.30 says that $\tilde{u}(\boldsymbol{x}, \cdot)$ is then $\sigma(\boldsymbol{\xi})$-measurable (for every $\boldsymbol{x} \in D$) and the Doob–Dynkin lemma (Lemma 4.46) tells us that $\tilde{u}(\boldsymbol{x}, \cdot)$ is a function of $\boldsymbol{\xi}$. □

We assume that $\tilde{a}(\boldsymbol{x})$ and $\tilde{f}(\boldsymbol{x})$ are given by (9.42) and (9.43). In that case, Lemma 9.40 says that $\tilde{u} \in W$ satisfying (9.36) is a function of $\boldsymbol{\xi}$ where $\boldsymbol{\xi} = [\xi_1, \dots, \xi_M]^\mathsf{T} : \Omega \to \Gamma$ contains the random variables in (9.42)–(9.43) and $M := \max\{P, N\}$. By the above argument, we can search for an equivalent weak solution $\tilde{u} : D \times \Gamma \to \mathbb{R}$ satisfying the following definition.

Definition 9.41 (weak solution on $D \times \Gamma$) Let $\tilde{a}(\boldsymbol{x})$ and $\tilde{f}(\boldsymbol{x})$ be the truncated Karhunen–Loève expansions (9.42) and (9.43) and define

$$L^2_p(\Gamma, H^1_g(D)) := \left\{ v : D \times \Gamma \to \mathbb{R} : \int_\Gamma p(\boldsymbol{y}) \|v(\cdot, \boldsymbol{y})\|^2_{H^1(D)} d\boldsymbol{y} < \infty \text{ and } \gamma v = g \right\},$$

and similarly for $L^2_p(\Gamma, H^1_0(D))$. An equivalent *weak solution* to the BVP (9.34)–(9.35) on $D \times \Gamma$ is a function $\tilde{u} \in W := L^2_p(\Gamma, H^1_g(D))$ satisfying

$$\tilde{a}(\tilde{u}, v) = \tilde{\ell}(v), \qquad \forall v \in V := L^2_p(\Gamma, H^1_0(D)), \tag{9.51}$$

where $\tilde{a} : W \times V \to \mathbb{R}$ and $\tilde{\ell} : V \to \mathbb{R}$ are defined by

$$\tilde{a}(u, v) := \int_\Gamma p(\boldsymbol{y}) \int_D \tilde{a}(\boldsymbol{x}, \boldsymbol{y}) \nabla u(\boldsymbol{x}, \boldsymbol{y}) \cdot \nabla v(\boldsymbol{x}, \boldsymbol{y}) \, d\boldsymbol{x} \, d\boldsymbol{y}, \tag{9.52}$$

$$\tilde{\ell}(v) := \int_\Gamma p(\boldsymbol{y}) \int_D \tilde{f}(\boldsymbol{x}, \boldsymbol{y}) v(\boldsymbol{x}, \boldsymbol{y}) \, d\boldsymbol{x} \, d\boldsymbol{y}. \tag{9.53}$$

The weight $p : \Gamma \to \mathbb{R}^+$ is the joint density of $\boldsymbol{\xi} = [\xi_1, \dots, \xi_M]^\mathsf{T}$ where $M := \max\{P, N\}$ and $\boldsymbol{\xi} : \Omega \to \Gamma \subset \mathbb{R}^M$.

Observe that (9.51) is an $(M+2)$-dimensional *deterministic* problem on $D \times \Gamma$ and the equivalent strong formulation is find $\tilde{u} : \bar{D} \times \Gamma \to \mathbb{R}$ such that

$$-\nabla \cdot \big(\tilde{a}(\boldsymbol{x}, \boldsymbol{y}) \nabla \tilde{u}(\boldsymbol{x}, \boldsymbol{y})\big) = \tilde{f}(\boldsymbol{x}, \boldsymbol{y}), \qquad (\boldsymbol{x}, \boldsymbol{y}) \in D \times \Gamma, \tag{9.54}$$

$$\tilde{u}(\boldsymbol{x}, \boldsymbol{y}) = g(\boldsymbol{x}), \qquad (\boldsymbol{x}, \boldsymbol{y}) \in \partial D \times \Gamma, \tag{9.55}$$

(cf. (9.34)–(9.35)). Working on Γ instead of Ω, we now have

$$\tilde{a}(\boldsymbol{x}, \boldsymbol{y}) = \mu_a(\boldsymbol{x}) + \sum_{k=1}^{P} \sqrt{\nu_k^a} \phi_k^a(\boldsymbol{x}) y_k, \qquad \boldsymbol{x} \in D, \ \boldsymbol{y} \in \Gamma, \tag{9.56}$$

and similarly for $\tilde{f}$. Well-posedness of the weak form (9.51) is established under the following assumption (cf. Assumption 9.29).

Assumption 9.42 (diffusion coefficients) The approximate diffusion coefficient $\tilde{a}$ satisfies

$$0 < \tilde{a}_{\min} \le \tilde{a}(\boldsymbol{x}, \boldsymbol{y}) \le \tilde{a}_{\max} < \infty, \qquad \text{a.e. in } D \times \Gamma,$$

for some real constants $\tilde{a}_{\min}$ and $\tilde{a}_{\max}$. In particular, $\tilde{a} \in L^\infty(\Gamma, L^\infty(D))$.

Theorem 9.43 *If Assumption 9.42 holds, $\tilde{f} \in L^2_p(\Gamma, L^2(D))$ and $g \in H^{1/2}(\partial D)$ then there exists a unique $\tilde{u} \in W = L^2_p(\Gamma, H^1_g(D))$ satisfying* (9.51).

Proof See Theorem 9.30 and replace Ω with Γ and $L^2(\Omega)$ with $L^2_p(\Gamma)$. □

9.5 Stochastic Galerkin FEM on $D \times \Gamma$

Starting from (9.51), we can now apply Galerkin approximation on $D \times \Gamma$. First, we replace $H^1_g(D)$ and $H^1_0(D)$ with the finite element spaces W^h and V^h from §2.3 and study weak solutions $\tilde{u}_h \in L^2_p(\Gamma, W^h)$.

Semi-discrete problem

Definition 9.44 (semi-discrete weak solution) A *semi-discrete weak solution* on $D \times \Gamma$ to the BVP (9.54)–(9.55) is a function $\tilde{u}_h \in L^2_p(\Gamma, W^h)$ that satisfies

$$\tilde{a}\left(\tilde{u}_h, v\right) = \tilde{\ell}\left(v\right), \qquad \forall v \in L^2_p(\Gamma, V^h), \tag{9.57}$$

where $\tilde{a}(\cdot,\cdot)$ and $\tilde{\ell}(\cdot)$ are as in (9.52)–(9.53), $V^h \subset H^1_0(D)$ and $W^h \subset H^1_g(D)$.

Theorem 9.45 *If the conditions of Theorem 9.43 hold, $V^h \subset H^1_0(D)$ and $W^h \subset H^1_g(D)$, then* (9.57) *has a unique solution $\tilde{u}_h \in L^2_p(\Gamma, W^h)$.*

The semi-discrete solution $\tilde{u}_h$ will be used for error analysis later. We are not obliged to choose finite element spaces for W^h and V^h but we will do so. We focus on triangular elements and piecewise linear polynomials. Since $L^2_p(\Gamma, V^h) \subset L^2_p(\Gamma, H^1_0(D)) =: V$, combining (9.51) and (9.57) gives

$$\tilde{a}\left(\tilde{u} - \tilde{u}_h, v\right) = 0, \qquad \forall v \in L^2_p(\Gamma, V^h) \tag{9.58}$$

and, if the finite element spaces are compatible, in the sense that

$$v - w \in L^2_p(\Gamma, V^h), \qquad \forall v, w \in L^2_p(\Gamma, W^h), \tag{9.59}$$

then $\tilde{u}_h$ is the best approximation.

Theorem 9.46 (best approximation) *Let $\tilde{u} \in L^2_p(\Gamma, H^1_g(D))$ be the unique solution to* (9.51) *and let $\tilde{u}_h \in L^2_p(\Gamma, W^h)$ be the unique solution to* (9.57). *If* (9.59) *holds then*

$$\left|\tilde{u} - \tilde{u}_h\right|_E = \inf_{v \in L^2_p(\Gamma, W^h)} \left|\tilde{u} - v\right|_E.$$

Proof See the proof of Theorem 2.49. □

Following our change of variable from $\omega \in \Omega$ to $\boldsymbol{y} \in \Gamma$, we now have

$$|v|_{\mathrm{E}}^2 := \int_\Gamma p(\boldsymbol{y}) \int_D \tilde{a}(\boldsymbol{x},\boldsymbol{y}) \nabla v(\boldsymbol{x},\boldsymbol{y}) \cdot \nabla v(\boldsymbol{x},\boldsymbol{y})\, d\boldsymbol{x}\, d\boldsymbol{y}, \tag{9.60}$$

$$|v|_W^2 := \int_\Gamma p(\boldsymbol{y})\, |v(\cdot,\boldsymbol{y})|_{H^1(D)}^2\, d\boldsymbol{y}. \tag{9.61}$$

As $\tilde{u}_h \in L_p^2(\Gamma, W^h)$ is the best approximation in the energy norm, $|\tilde{u} - \tilde{u}_h|_{\mathrm{E}} = |\tilde{u} - P_h \tilde{u}|_{\mathrm{E}}$ where $P_h \colon W \to L_p^2(\Gamma, W^h)$ is the Galerkin projection (recall Definition 2.19) satisfying

$$\tilde{a}\left(P_h \tilde{u}, v\right) = \tilde{a}\left(\tilde{u}, v\right), \qquad \forall v \in L_p^2(\Gamma, V^h). \tag{9.62}$$

Note that P_h provides a mapping from $H_g^1(D) \to W^h$ and effectively acts only on the spatial components of $\tilde{u} \in W$. To establish a bound for the error $|\tilde{u} - \tilde{u}_h|_W$ in terms of the mesh parameter h, we now make the following assumption.

Assumption 9.47 ($L_p^2(\Gamma, H^2(D))$-regularity) There exists a constant $K_2 > 0$, independent of $\boldsymbol{y} \in \Gamma$, such that for every $\tilde{f} \in L_p^2(\Gamma, L^2(D))$, the weak solution $\tilde{u} \in L_p^2(\Gamma, H^2(D))$ and satisfies

$$|\tilde{u}|_{L_p^2(\Gamma, H^2(D))} \le K_2 \|\tilde{f}\|_{L_p^2(\Gamma, L^2(D))}.$$

Theorem 9.48 *Let the conditions of Theorem 9.46 hold and suppose Assumptions 9.42 and 9.47 hold. Let $V^h \subset H_0^1(D)$ and $W^h \subset H_g^1(D)$ denote piecewise linear finite element spaces and let $\mathcal{T}_h$ be shape-regular. Then*

$$|\tilde{u} - \tilde{u}_h|_W \le K \sqrt{\frac{\tilde{a}_{\max}}{\tilde{a}_{\min}}}\, h\, \|\tilde{f}\|_{L_p^2(\Gamma, L^2(D))}, \tag{9.63}$$

where $K > 0$ is a constant independent of h.

Proof Following the proof of Corollary 2.68, for each sample $\tilde{u}(\cdot, \boldsymbol{y})$ and corresponding finite element approximation $\tilde{u}_h(\cdot, \boldsymbol{y})$, we have

$$|\tilde{u}(\cdot,\boldsymbol{y}) - \tilde{u}_h(\cdot,\boldsymbol{y})|_{H^1(D)} \le K\, h\, \tilde{a}_{\min}^{-1/2} \tilde{a}_{\max}^{1/2} |u(\cdot,\boldsymbol{y})|_{H^2(D)},$$

where K is a constant that does not depend on h (see Exercise 2.23). Hence,

$$\begin{aligned} \int_\Gamma p(\boldsymbol{y}) |\tilde{u}(\cdot,\boldsymbol{y}) - \tilde{u}_h(\cdot,\boldsymbol{y})|_{H^1(D)}^2\, d\boldsymbol{y} &\le K^2\, h^2\, \tilde{a}_{\min}^{-1}\, \tilde{a}_{\max} \int_\Gamma p(\boldsymbol{y}) |u(\cdot,\boldsymbol{y})|_{H^2(D)}^2\, d\boldsymbol{y} \\ &= K^2\, h^2\, \tilde{a}_{\min}^{-1}\, \tilde{a}_{\max} |u|_{L_p^2(\Gamma, H^2(D))}^2. \end{aligned}$$

Taking square roots and using Assumption 9.47 gives the result. □

Notice that Assumption 9.47 is slightly weaker than Assumption 9.14, which was required in Theorem 9.15 to derive a similar $\mathcal{O}(h)$ error estimate for the MCFEM when the diffusion coefficient only satisfies Assumption 9.5.

Galerkin approximation

We now derive a fully discrete, finite-dimensional problem on $D \times \Gamma$.

Definition 9.49 (stochastic Galerkin solution) A *stochastic Galerkin solution* to (9.54)–(9.55) is a function $\tilde{u}_{hk} \in W^{hk} \subset W := L^2_p(\Gamma, H^1_g(D))$ satisfying

$$\tilde{a}\left(\tilde{u}_{hk}, v\right) = \tilde{\ell}\left(v\right), \qquad \forall v \in V^{hk} \subset V := L^2_p(\Gamma, H^1_0(D)), \tag{9.64}$$

where $\tilde{a}\colon W \times V \to \mathbb{R}$ and $\tilde{\ell}\colon V \to \mathbb{R}$ are defined as in (9.52).

Applying the analysis from Chapter 2, we obtain the following results.

Theorem 9.50 *If the conditions of Theorem 9.43 hold, $V^{hk} \subset V$ and $W^{hk} \subset W$, then* (9.64) *has a unique solution $\tilde{u}_{hk} \in W^{hk}$.*

Theorem 9.51 (best approximation) *Let $\tilde{u} \in W$ and $\tilde{u}_{hk} \in W^{hk}$ be the unique solutions to* (9.51) *and* (9.64)*, respectively. If $v - w \in V^{hk}$ for all $v, w \in W^{hk}$, then*

$$\left|\tilde{u} - \tilde{u}_{hk}\right|_{\mathrm{E}} = \inf_{v \in W^{hk}} \left|\tilde{u} - v\right|_{\mathrm{E}}. \tag{9.65}$$

Furthermore, if Assumption 9.42 holds then

$$\left|\tilde{u} - \tilde{u}_{hk}\right|_{W} \leq \sqrt{\frac{\tilde{a}_{\max}}{\tilde{a}_{\min}}} \left|\tilde{u} - v\right|_{W}, \qquad \forall v \in W^{hk}. \tag{9.66}$$

To compute $\tilde{u}_{hk}$ we need to select finite-dimensional subspaces of W and V. Let $S^k \subset L^2_p(\Gamma)$ denote *any* Q-dimensional subspace with

$$S^k = \operatorname{span}\left\{\psi_1, \psi_2, \dots, \psi_Q\right\}. \tag{9.67}$$

One way to construct a finite-dimensional subspace of $V = L^2_p(\Gamma, H^1_0(D))$ is by tensorising the basis functions ψ_j in (9.67) with the previously selected (piecewise linear) finite element basis functions for V^h. Recall,

$$V^h = \operatorname{span}\left\{\phi_1, \phi_2, \dots, \phi_J\right\} \subset H^1_0(D). \tag{9.68}$$

Given a basis for the chosen S^k, the tensor product space is defined as

$$V^h \otimes S^k := \operatorname{span}\left\{\phi_i \psi_j : i = 1, \dots, J,\ j = 1, \dots, Q\right\} \tag{9.69}$$

and has dimension JQ. This space will serve as our test space V^{hk} in (9.64). To impose the boundary condition (9.55), we must take care in constructing W^{hk}. Augmenting the basis for $V^{hk} := V^h \otimes S^k$ with the J_b finite element functions ϕ_i associated with the Dirichlet boundary vertices gives

$$W^{hk} := V^{hk} \oplus \operatorname{span}\left\{\phi_{J+1}, \dots, \phi_{J+J_b}\right\}. \tag{9.70}$$

Any $w \in W^{hk}$ then has the form

$$w(\boldsymbol{x}, \boldsymbol{y}) = \sum_{i=1}^{J} \sum_{j=1}^{Q} w_{ij} \phi_i(\boldsymbol{x}) \psi_j(\boldsymbol{y}) + \sum_{i=J+1}^{J+J_b} w_i \phi_i(\boldsymbol{x}) =: w_0(\boldsymbol{x}, \boldsymbol{y}) + w_g(\boldsymbol{x}) \tag{9.71}$$

where $w_0 \in V^{hk}$ is zero on $\partial D \times \Gamma$. We fix $w_{J+1}, \dots, w_{J+J_b}$ in (9.71) so that w_g interpolates

g at the boundary vertices (for each $\boldsymbol{y} \in \Gamma$) and assume that g is simple enough that, with the above construction, $W^{hk} \subset L^2_p(\Gamma, H^1_g(D))$.

It remains to choose the space S^k. One possibility is to partition Γ into elements and use *piecewise* polynomials of a fixed degree (as we have done on D). Alternatively, we can choose *global* polynomials. In this approach, accuracy is controlled by increasing the polynomial degree, k. The approximation error will only decay as k is increased, however, if $\tilde{u}$ is a smooth function of $\boldsymbol{y} \in \Gamma$. The higher the polynomial degree, the more regularity we need. In fact, for the elliptic BVP (9.54)–(9.55), with data $\tilde{a}$ and $\tilde{f}$ given by truncated Karhunen–Loève expansions, it can be shown that $\tilde{u}\colon D \to C^\infty(\Gamma)$. Hence, we focus on global polynomial approximation.

Stochastic basis functions

We now discuss specific spaces S^k of global polynomials on $\Gamma \subset \mathbb{R}^M$ and outline a general method for constructing the basis functions ψ_j in (9.67). Let S_i^k denote the set of *univariate* polynomials of degree k or less in y_i on the interval $\Gamma_i \subset \mathbb{R}$ and suppose that

$$S_i^k = \operatorname{span}\left\{P^i_{\alpha_i}(y_i)\colon \alpha_i = 0, 1, \ldots, k\right\}, \qquad i = 1, \ldots, M, \tag{9.72}$$

where $P^i_{\alpha_i}(y_i)$ is some polynomial of degree α_i in y_i. Clearly, one possibility is $P^i_{\alpha_i}(y_i) = y_i^{\alpha_i}$ but we delay a specific choice until later. We can use the one-dimensional spaces S_i^k to construct multivariate polynomials on Γ. When $M > 1$, we may choose between so-called tensor product and complete polynomials.

First, suppose that we choose S^k to be the set of polynomials of degree k in *each* of $y_1, \ldots, y_M$ (*tensor product* polynomials). Then, $S^k = S_1^k \otimes S_2^k \otimes \cdots \otimes S_M^k$ and using (9.72), we have

$$S^k = \operatorname{span}\left\{\prod_{i=1}^M P^i_{\alpha_i}(y_i)\colon \alpha_i = 0, 1, \ldots, k, \, i = 1, \ldots, M\right\}. \tag{9.73}$$

By considering all the possibilities for the multi-indices $\boldsymbol{\alpha} := (\alpha_1, \ldots, \alpha_M)$ in (9.73), we find

$$Q := \dim(S^k) = (k+1)^M. \tag{9.74}$$

Similarly, given $\boldsymbol{k} = (k_1, k_2, \ldots, k_M)$, the set of tensor product polynomials of distinct degrees k_i in each y_i can be written as

$$S^{\boldsymbol{k}} = \operatorname{span}\left\{\prod_{i=1}^M P^i_{\alpha_i}(y_i)\colon \alpha_i = 0, 1, \ldots, k_i, \, i = 1, \ldots, M\right\}. \tag{9.75}$$

Example 9.52 (tensor product polynomials) Let $M = 2$ and $k = 1$ and consider the set of bivariate polynomials of degree 1 or less in each of y_1 and y_2. The associated set of multi-indices is $\{(0,0), (1,0), (0,1), (1,1)\}$ and so $Q = 4$. Given bases $\{P_0^1, P_1^1\}$, $\{P_0^2, P_1^2\}$ for the univariate spaces S_1^1, S_2^1, respectively, we have $S^k = \operatorname{span}\{\psi_1, \psi_2, \psi_3, \psi_4\}$ where

$$\psi_1(\boldsymbol{y}) = P_0^1(y_1)P_0^2(y_2), \qquad \psi_2(\boldsymbol{y}) = P_1^1(y_1)P_0^2(y_2),$$
$$\psi_3(\boldsymbol{y}) = P_0^1(y_1)P_1^2(y_2), \qquad \psi_4(\boldsymbol{y}) = P_1^1(y_1)P_1^2(y_2).$$

Finally, let S^k be the set of polynomials of *total* degree k or less (*complete* polynomials). Then,

$$S^k = \text{span}\left\{\prod_{i=1}^{M} P^i_{\alpha_i}(y_i)\colon \alpha_i = 0,1,\ldots,k,\ i = 1,\ldots,M, |\boldsymbol{\alpha}| \le k\right\}, \tag{9.76}$$

where $|\boldsymbol{\alpha}| = \sum_{i=1}^{M} \alpha_i$. In this case,

$$Q := \dim(S^k) = \frac{(M+k)!}{M!k!}. \tag{9.77}$$

For a fixed k and M, note that S^k in (9.76) is always a subset of S^k in (9.73) and the dimension Q may be substantially smaller. See Exercise 9.16.

Example 9.53 (complete polynomials) Let $M = 2$ and $k = 1$ and consider the set of bivariate polynomials of total degree 1 or less. The set of permitted multi-indices is now $\{(0,0),(1,0),(0,1)\}$ and $Q = 3$. The multi-index $(1,1)$ is not included as it corresponds to a polynomial of total degree 2. This time, we obtain $S^k = \text{span}\{\psi_1,\psi_2,\psi_3\}$ where

$$\psi_1(\boldsymbol{y}) = P^1_0(y_1)P^2_0(y_2), \qquad \psi_2(\boldsymbol{y}) = P^1_1(y_1)P^2_0(y_2), \qquad \psi_3(\boldsymbol{y}) = P^1_0(y_1)P^2_1(y_2).$$

In summary, tensor product and complete spaces of global polynomials on Γ can be constructed from bases for the M univariate spaces S^k_i in (9.72). Indeed, each of the spaces (9.73), (9.75) and (9.76) can be written as (9.67) with

$$\psi_j(\boldsymbol{y}) = \psi_{j(\boldsymbol{\alpha})}(\boldsymbol{y}) := \prod_{i=1}^{M} P^i_{\alpha_i}(y_i), \qquad j = 1,\ldots,Q. \tag{9.78}$$

Here, each scalar index j is assigned to some multi-index $\boldsymbol{\alpha} = (\alpha_1,\ldots,\alpha_M)$ whose components provide the degrees of the univariate basis polynomials. Only the set of multi-indices and hence the dimension Q change in each case. We will now write S^k to denote any of the spaces (9.73), (9.75), and (9.76). Since the construction of V^{hk} and W^{hk} in (9.64) is not affected, an explicit choice for S^k will be stated only where necessary. It will be convenient to choose the $P^i_{\alpha_i}$ in (9.78) to be orthonormal and so we now recall some classical results about orthogonal polynomials.

Univariate orthogonal polynomials

Given an interval $\Gamma \subset \mathbb{R}$ and a weight function $p\colon \Gamma \to \mathbb{R}^+$, define the inner product

$$\langle v,w\rangle_p := \int_\Gamma p(y)v(y)w(y)\,dy \tag{9.79}$$

for functions $v,w\colon \Gamma \to \mathbb{R}$. Let $P_{-1} = 0$ and $P_0 = 1$. It is well known that we can construct a sequence of polynomials P_j (of degree j) on Γ that are orthogonal with respect to the weighted inner product $\langle\cdot,\cdot\rangle_p$ and satisfy a three-term recurrence

$$P_j(y) = \left(a_j y + b_j\right)P_{j-1}(y) - c_j P_{j-2}(y), \qquad j = 1,2,\ldots. \tag{9.80}$$

The coefficients a_j, b_j, c_j depend on p. If we normalise the polynomials so that $\langle P_i, P_j \rangle_p = \delta_{ij}$, where δ_{ij} is the Kronecker delta function, then $\frac{c_j}{a_j} = \frac{1}{a_{j-1}}$ (see Exercise 9.17) and

$$b_j = -a_j \langle y P_{j-1}, P_{j-1} \rangle_p. \tag{9.81}$$

Now, if $\Gamma = [-\gamma, \gamma]$ for some $\gamma > 0$ and p is even, then $b_j = 0$ (see Exercise 9.18) and the three-term recurrence becomes

$$\frac{1}{a_j} P_j(y) = y P_{j-1}(y) - \frac{1}{a_{j-1}} P_{j-2}(y), \qquad j = 1, 2, \dots. \tag{9.82}$$

The three-term recurrence (9.82) (or, more generally, (9.80)) provides an important link between roots of orthogonal polynomials and eigenvalues of tridiagonal matrices. Indeed, applying (9.82) for $j = 1, \dots, N-1$ gives

$$y \begin{pmatrix} P_0(y) \\ P_1(y) \\ \vdots \\ P_{N-1}(y) \end{pmatrix} = T_N \begin{pmatrix} P_0(y) \\ P_1(y) \\ \vdots \\ P_{N-1}(y) \end{pmatrix} + \begin{pmatrix} 0 \\ 0 \\ \vdots \\ \frac{1}{a_N} P_N(y) \end{pmatrix} \tag{9.83}$$

where T_N is the symmetric tridiagonal matrix

$$T_N := \begin{pmatrix} 0 & \frac{1}{a_1} & & & \\ \frac{1}{a_1} & 0 & \frac{1}{a_2} & & \\ & \ddots & \ddots & \ddots & \\ & & \frac{1}{a_{N-2}} & 0 & \frac{1}{a_{N-1}} \\ & & & \frac{1}{a_{N-1}} & 0 \end{pmatrix}. \tag{9.84}$$

Writing (9.83) compactly as $y\boldsymbol{p}(y) = T_N \boldsymbol{p}(y) + \frac{1}{a_N} P_N(y) \boldsymbol{e}_N$, where $\boldsymbol{e}_N$ is the last column of the $N \times N$ identity matrix, we see that any value of y for which $P_N(y) = 0$ is an eigenvalue of the $N \times N$ matrix T_N in (9.84).

Many standard families of polynomials are orthogonal with respect to even weight functions on symmetric intervals and satisfy (9.82).

Example 9.54 (Legendre polynomials) Let $\Gamma = \left[-\sqrt{3}, \sqrt{3}\right]$ and $p = 1/(2\sqrt{3})$. Note that Γ is the image of a uniform random variable with mean zero and variance one and p is the associated density. Define

$$L_j(y) := \hat{L}_j\left(\frac{y}{\sqrt{3}}\right), \qquad j = 0, 1, \dots, \tag{9.85}$$

where

$$\hat{L}_j(y) := \frac{\sqrt{2j+1}}{2^j\, j!} \frac{d^j}{dy^j} \left(y^2 - 1\right)^j.$$

The resulting Legendre polynomials $L_0 = 1$, $L_1 = y$, $L_2 = \frac{\sqrt{5}}{2}\left(y^2 - 1\right), \dots$ are orthonormal and satisfy

$$\langle L_i, L_j \rangle_p = \frac{1}{2\sqrt{3}} \int_{-\sqrt{3}}^{\sqrt{3}} L_i(y) L_j(y)\, dy = \delta_{ij}.$$

It can also be shown that they satisfy the three-term recurrence (9.82) with

$$\frac{1}{a_j} = \frac{j\sqrt{3}}{\sqrt{2j-1}\sqrt{2j+1}}. \tag{9.86}$$

The associated tridiagonal matrix is investigated in Exercise 9.19. Hermite polynomials on $\Gamma = \mathbb{R}$ are also studied in Exercise 9.20.

Orthonormal stochastic basis functions

Returning to (9.64), we construct W^{hk} as in (9.70) and $V^{hk} = V^h \otimes S^k$ where S^k is some Q-dimensional space of global polynomials on Γ with basis functions ψ_j of the form (9.78). Recall now that

$$L^2_p(\Gamma) := \left\{v : \Gamma \to \mathbb{R} : \|v\|_{L^2_p(\Gamma)} < \infty\right\}, \qquad \|v\|^2_{L^2_p(\Gamma)} := \langle v, v\rangle_p,$$

and

$$\langle v, w\rangle_p := \int_\Gamma p(\boldsymbol{y})v(\boldsymbol{y})w(\boldsymbol{y})\,d\boldsymbol{y}. \tag{9.87}$$

We need $S^k \subset L^2_p(\Gamma)$ where $\Gamma \subset \mathbb{R}^M$ is the image of the random variable $\boldsymbol{\xi} = [\xi_1, \ldots, \xi_M]^\mathsf{T}$ associated with (9.42)–(9.43) and the weight p is the joint density. Let p_i denote the density of ξ_i. If the ξ_i are independent, then

$$p(\boldsymbol{y}) = p_1(y_1)p_2(y_2)\cdots p_M(y_M). \tag{9.88}$$

The next result tells us how to choose $P^i_{\alpha_i}$ in (9.78) so that the polynomials ψ_j are orthonormal with respect to the inner product $\langle\cdot,\cdot\rangle_p$ in (9.87).

Theorem 9.55 *Let p satisfy* (9.88). *Suppose that the univariate polynomials $\{P^i_{\alpha_i}(y_i)\}$ are orthonormal with respect to $\langle\cdot,\cdot\rangle_{p_i}$ on Γ_i for $i = 1, \ldots, M$. Then, the polynomials ψ_j in* (9.78) *are orthonormal with respect to $\langle\cdot,\cdot\rangle_p$ on Γ.*

Proof Consider two polynomials ψ_r and ψ_s with $r = r(\boldsymbol{\alpha})$ and $s = s(\boldsymbol{\beta})$. Then,

$$\psi_r(\boldsymbol{y}) := \prod_{i=1}^M P^i_{\alpha_i}(y_i), \qquad \psi_s(\boldsymbol{y}) := \prod_{i=1}^M P^i_{\beta_i}(y_i).$$

Since the joint density function p and the polynomials ψ_r, ψ_s are separable,

$$\begin{aligned}\langle\psi_r, \psi_s\rangle_p &:= \int_\Gamma p(\boldsymbol{y})\psi_r(\boldsymbol{y})\psi_s(\boldsymbol{y})\,d\boldsymbol{y} = \prod_{i=1}^M \int_{\Gamma_i} p_i(y_i)P^i_{\alpha_i}(y_i)P^i_{\beta_i}(y_i)\,dy_i \\ &=: \prod_{i=1}^M \left\langle P^i_{\alpha_i}, P^i_{\beta_i}\right\rangle_{p_i} = \prod_{i=1}^M \delta_{\alpha_i\beta_i} = \delta_{rs}.\end{aligned}$$

□

Constructing an orthonormal basis is straightforward. Once the distributions of the random variables ξ_i in (9.42) and (9.43) are known, the densities p_i are known and the appropriate orthogonal polynomials $P^i_{\alpha_i}$ can be selected. If all the ξ_i have the *same* distribution (as we have assumed), the M sets of univariate polynomials are identical. Hence, we can drop the superscript and simply write P_{α_i}. For elliptic BVPs, since we need Assumption 9.42

to hold, we have to work with bounded random variables (recall Assumption 9.32). We focus then on the uniform distribution. If each ξ_i has mean zero and variance one, we have $\xi_i \sim \mathrm{U}(-\sqrt{3}, \sqrt{3})$, $\Gamma_i = [-\sqrt{3}, \sqrt{3}]$ and $p_i(y_i) = 1/(2\sqrt{3})$. Theorem 9.55 now tells us that we should choose $P_{\alpha_i} = L_{\alpha_i}$ in (9.78) where L_{α_i} is the Legendre polynomial of degree α_i from Example 9.54.

Example 9.56 (uniform random variables) Let $M = 2$ and $\xi_1, \xi_2 \sim \mathrm{U}(-\sqrt{3}, \sqrt{3})$ *iid*. This situation could arise, for example, if $P = N = 2$ in (9.42)–(9.43) or if $P = 2$ and f is deterministic (so $\tilde{f} = \mu_f = f$ and $N = 0$). If we choose complete polynomials of degree $k \le 2$, $S^k = \mathrm{span}\{\psi_j\}_{j=1}^{6}$ where

$$
\begin{aligned}
\psi_1(\boldsymbol{y}) &= L_0(y_1)L_0(y_2), & \psi_2(\boldsymbol{y}) &= L_1(y_1)L_0(y_2), & \psi_3(\boldsymbol{y}) &= L_0(y_1)L_1(y_2),\\
\psi_4(\boldsymbol{y}) &= L_2(y_1)L_0(y_2), & \psi_5(\boldsymbol{y}) &= L_1(y_1)L_1(y_2), & \psi_6(\boldsymbol{y}) &= L_0(y_1)L_2(y_2),
\end{aligned}
$$

and $L_0(y) = 1$, $L_1(y) = y$ and $L_2(y) = (\sqrt{5}/2)(y^2 - 1)$ are the univariate Legendre polynomials of degrees $0, 1, 2$ on $[-\sqrt{3}, \sqrt{3}]$.

Stochastic Galerkin mean and variance

Assuming an *orthonormal* basis for $S^k \subset L^2_p(\Gamma)$ is selected, we now explain how to compute the mean and variance of the stochastic Galerkin solution. The multi-indices $\boldsymbol{\alpha}$ associated with the chosen S^k will need to be ordered in a convenient way for efficient computations. We assume only that the first one is $\boldsymbol{\alpha} = (0, 0, \ldots, 0)$. That way, since $\langle P_0, P_0\rangle_{p_i} = 1$, $P_0 = 1$ and $\psi_1(\boldsymbol{y}) = 1$.

Since $\tilde{u}_{hk} \in W^{hk}$, using (9.71) gives

$$
\tilde{u}_{hk}(\boldsymbol{x}, \boldsymbol{y}) = \sum_{i=1}^{J}\sum_{j=1}^{Q} u_{ij}\phi_i(\boldsymbol{x})\psi_j(\boldsymbol{y}) + w_g(\boldsymbol{x}),
$$

where the coefficients u_{ij} are found by solving a linear system of dimension JQ. Rearranging (and using $\psi_1 = 1$) gives

$$
\begin{aligned}
\tilde{u}_{hk}(\boldsymbol{x}, \boldsymbol{y}) &= \sum_{j=1}^{Q}\left(\sum_{i=1}^{J} u_{ij}\phi_i(\boldsymbol{x})\right)\psi_j(\boldsymbol{y}) + w_g(\boldsymbol{x}) =: \sum_{j=1}^{Q} u_j(\boldsymbol{x})\psi_j(\boldsymbol{y}) + w_g(\boldsymbol{x})\\
&= \big(u_1(\boldsymbol{x}) + w_g(\boldsymbol{x})\big)\psi_1(\boldsymbol{y}) + \sum_{j=2}^{Q} u_j(\boldsymbol{x})\psi_j(\boldsymbol{y}).
\end{aligned}
\tag{9.89}
$$

Expansions in terms of orthogonal polynomials are often referred to as *polynomial chaos* (PC) expansions. Notice that in (9.89) the PC coefficients u_j are finite element functions. The first one belongs to W^h and satisfies the Dirichlet boundary condition. For $j \ge 2$, however, $u_j \in V^h$ and is zero on the boundary.

Now, since the polynomials ψ_j are orthonormal, we have

$$\mathbb{E}[\tilde{u}_{hk}] = \int_\Gamma p(\boldsymbol{y})\,\tilde{u}_{hk}(\cdot,\boldsymbol{y})\cdot 1\,d\boldsymbol{y} = \big\langle (u_1 + w_g)\psi_1, \psi_1 \big\rangle_p + \sum_{j=2}^{Q} \big\langle u_j\psi_j, \psi_1\big\rangle_p$$
$$= (u_1 + w_g)\langle \psi_1, \psi_1\rangle_p + \sum_{j=2}^{Q} u_j \langle \psi_j, \psi_1\rangle_p = u_1 + w_g. \tag{9.90}$$

Hence, given the vectors $\boldsymbol{u}_1 := [u_{11}, \dots, u_{J1}]^\mathsf{T}$ and $\boldsymbol{w}_\mathrm{B} := [w_{J+1}, \dots, w_{J+J_b}]^\mathsf{T}$ associated with the first PC coefficient $u_1 + w_g$ in (9.89), we can compute the mean of the stochastic Galerkin solution. Similarly, the variance is

$$\mathrm{Var}(\tilde{u}_{hk}) = \mathbb{E}[\tilde{u}_{hk}^2] - \mathbb{E}[\tilde{u}_{hk}]^2 = \mathbb{E}[\tilde{u}_{hk}^2] - (u_1 + w_g)^2.$$

Expanding $\tilde{u}_{hk}$ and using the orthonormality of the ψ_j once more gives

$$\mathbb{E}[\tilde{u}_{hk}^2] = \int_\Gamma p(\boldsymbol{y})\,\tilde{u}_{hk}(\cdot,\boldsymbol{y})^2\,d\boldsymbol{y} = \langle \tilde{u}_{hk}, \tilde{u}_{hk}\rangle_p$$
$$= (u_1 + w_g)^2 \langle \psi_1, \psi_1\rangle_p + \sum_{j=2}^{Q} u_j^2 \langle \psi_j, \psi_j\rangle_p = (u_1 + w_g)^2 + \sum_{j=2}^{Q} u_j^2.$$

Hence, we have

$$\mathrm{Var}(\tilde{u}_{hk}) = \sum_{j=2}^{Q} u_j^2. \tag{9.91}$$

To compute the variance, we need the vectors $\boldsymbol{u}_j := [u_{1j}, u_{2j}, \dots, u_{Jj}]^\mathsf{T}$ associated with the PC coefficients u_j, for $j = 2, \dots, Q$.

Stochastic Galerkin linear system

We now discuss the linear system of equations that needs to be solved to compute (9.90) and (9.91). Owing to our choice of V^{hk} in (9.69), there are JQ equations, where Q is the dimension of the polynomial space S^k and J is the dimension of the finite element space V^h. If $\tilde{a}(\boldsymbol{x})$ is a truncated Karhunen–Loève expansion, we can show that choosing an orthonormal basis for S^k leads to a block-sparse coefficient matrix.

Starting from (9.64), expanding $\tilde{u}_{hk}$ as in (9.89) and setting $v = \phi_r\psi_s$, for $r = 1, \dots, J$ and $s = 1, \dots, Q$ yields a matrix equation $A\boldsymbol{u} = \boldsymbol{b}$ with block structure,

$$A = \begin{pmatrix} A_{11} & A_{12} & \dots & A_{1Q} \\ A_{21} & A_{22} & \dots & A_{2Q} \\ \vdots & \vdots & \ddots & \vdots \\ A_{Q1} & A_{Q2} & \dots & A_{QQ} \end{pmatrix}, \qquad \boldsymbol{u} = \begin{pmatrix} \boldsymbol{u}_1 \\ \boldsymbol{u}_2 \\ \vdots \\ \boldsymbol{u}_Q \end{pmatrix}, \qquad \boldsymbol{b} = \begin{pmatrix} \boldsymbol{b}_1 \\ \boldsymbol{b}_2 \\ \vdots \\ \boldsymbol{b}_Q \end{pmatrix}.$$

The jth block of the solution vector,

$$\boldsymbol{u}_j := [u_{1j}, u_{2j}, \dots, u_{Jj}]^\mathsf{T}, \qquad j = 1, \dots, Q,$$

is associated with the coefficient u_j in (9.89) and each matrix A_{sj} has the form

$$A_{sj} = \langle \psi_j, \psi_s \rangle_p K_0 + \sum_{\ell=1}^{P} \langle y_\ell \psi_j, \psi_s \rangle_p K_\ell, \qquad s, j = 1, \dots, Q, \tag{9.92}$$

where K_0 and K_ℓ, $\ell = 1, \dots, P$ are finite element matrices defined by

$$[K_0]_{ir} := \int_D \mu_a(\boldsymbol{x}) \nabla \phi_i(\boldsymbol{x}) \cdot \nabla \phi_r(\boldsymbol{x}) \, d\boldsymbol{x}, \qquad i, r = 1, \dots, J, \tag{9.93}$$

$$[K_\ell]_{ir} := \int_D \left(\sqrt{\nu_\ell^a} \phi_\ell^a(\boldsymbol{x}) \right) \nabla \phi_i(\boldsymbol{x}) \cdot \nabla \phi_r(\boldsymbol{x}) \, d\boldsymbol{x}, \qquad i, r = 1, \dots, J. \tag{9.94}$$

Note that the finite element basis functions for the boundary vertices are not included here. K_0 is the *mean diffusion matrix* and K_ℓ is a diffusion matrix associated with the ℓth eigenpair of the covariance function C_a. On the right-hand side, the vectors $\boldsymbol{b}_s$ for $s = 1, \dots, Q$ are defined by

$$\boldsymbol{b}_s := \langle \psi_1, \psi_s \rangle_p \left(\boldsymbol{f}_0 - K_{0,\mathrm{B}}^{\mathsf{T}} \boldsymbol{w}_{\mathrm{B}} \right) + \sum_{\ell=1}^{N} \langle y_\ell, \psi_s \rangle_p \boldsymbol{f}_\ell - \sum_{\ell=1}^{P} \langle y_\ell, \psi_s \rangle_p K_{\ell,\mathrm{B}}^{\mathsf{T}} \boldsymbol{w}_{\mathrm{B}},$$

where, recall, $\boldsymbol{w}_{\mathrm{B}} \in \mathbb{R}^{J_b}$ contains the boundary data. Using (9.43), we have

$$[\boldsymbol{f}_0]_r := \int_D \mu_f(\boldsymbol{x}) \phi_r(\boldsymbol{x}) \, d\boldsymbol{x}, \qquad r = 1, \dots, J, \tag{9.95}$$

$$[\boldsymbol{f}_\ell]_r := \int_D \left(\sqrt{\nu_\ell^f} \phi_\ell^f(\boldsymbol{x}) \right) \phi_r(\boldsymbol{x}) \, d\boldsymbol{x}, \qquad r = 1, \dots, J. \tag{9.96}$$

$K_{0,\mathrm{B}} \in \mathbb{R}^{J_b \times J}$ and $K_{\ell,\mathrm{B}} \in \mathbb{R}^{J_b \times J}$ are finite element diffusion matrices that account for the coupling between interior and boundary degrees of freedom. That is, $[K_{0,\mathrm{B}}]_{ir}$ and $[K_{\ell,\mathrm{B}}]_{ir}$ are defined in the same way as $[K_0]_{ir}$ and $[K_\ell]_{ir}$ in (9.93) and (9.94), but with $i = J+1, \dots, J+J_b$.

The Galerkin system can be neatly expressed using Kronecker products (recall Definition 7.25). Define $G_0 \in \mathbb{R}^{Q \times Q}$ and $G_\ell \in \mathbb{R}^{Q \times Q}$ for $\ell = 1, \dots, M$ by

$$[G_0]_{js} = \langle \psi_j, \psi_s \rangle_p, \qquad j, s = 1, \dots, Q, \tag{9.97}$$

$$[G_\ell]_{js} = \langle y_\ell \psi_j, \psi_s \rangle_p, \qquad j, s = 1, \dots, Q, \tag{9.98}$$

and the vectors $\boldsymbol{g}_0$ and $\boldsymbol{g}_\ell$, $\ell = 1, \dots, M$ by

$$[\boldsymbol{g}_0]_s = \langle \psi_s, \psi_1 \rangle_p, \qquad s = 1, \dots, Q, \tag{9.99}$$

$$[\boldsymbol{g}_\ell]_s = \langle y_\ell, \psi_s \rangle_p, \qquad s = 1, \dots, Q. \tag{9.100}$$

Recall that P and N are the numbers of random variables representing $\tilde{a}(\boldsymbol{x})$ and $\tilde{f}(\boldsymbol{x})$. $M := \max\{P, N\}$ is the total number of distinct variables. Now, using (9.97)–(9.100), we can write the Galerkin system as $A\boldsymbol{u} = \boldsymbol{b}$ where

$$A := G_0 \otimes K_0 + \sum_{\ell=1}^{P} G_\ell \otimes K_\ell, \tag{9.101}$$

$$\boldsymbol{b} := \boldsymbol{g}_0 \otimes \left(\boldsymbol{f}_0 - K_{0,\mathrm{B}}^{\mathsf{T}} \boldsymbol{w}_{\mathrm{B}} \right) + \sum_{\ell=1}^{N} \boldsymbol{g}_\ell \otimes \boldsymbol{f}_\ell - \sum_{\ell=1}^{P} \boldsymbol{g}_\ell \otimes K_{\ell,\mathrm{B}}^{\mathsf{T}} \boldsymbol{w}_{\mathrm{B}}. \tag{9.102}$$

When f is deterministic (so $\tilde{f} = \mu_f = f$) and $g = 0$, $\boldsymbol{b}$ simplifies considerably; see Exercise 9.21.

Solving the Galerkin system of JQ equations seems daunting. However, we need to solve only *one* linear system (unlike the MCFEM). Moreover, due to (9.101) and (9.102), we do not need to assemble A. Iterative solvers can be implemented using only the component matrices and vectors. See Exercises 9.22 and 9.23.

Algorithm 9.3 Code to compute the finite element components of the stochastic Galerkin system. The first seven inputs are defined in Algorithm 2.4, `mu_a`, `nu_a` and `phi_a` are the mean, a vector of eigenvalues and a matrix of sampled eigenfunctions for $\tilde{a}(\boldsymbol{x})$ and `mu_f`, `nu_f` and `phi_f` are similar for $\tilde{f}(\boldsymbol{x})$. `P` and `N` are the numbers of random variables representing the data.

```
function [K_mats,KB_mats,f_vecs,wB]=fem_blocks(ns,xv,yv,elt2vert,...
            nvtx,ne,h,mu_a,nu_a,phi_a,mu_f,nu_f,phi_f,P,N)

[Jks,invJks,detJks]=get_jac_info(xv,yv,ne,elt2vert);
b_nodes=find((xv==0)|(xv==1)|(yv==0)|(yv==1));
int_nodes=1:nvtx; int_nodes(b_nodes)=[];
wB=feval('g_eval',xv(b_nodes),yv(b_nodes));
M=max(P,N); % total no. variables
for ell=0:M
    if ell==0
        a=mu_a.*ones(ne,1); f=mu_f.*ones(ne,1);
    else
        if ell<=P
            a=sqrt(nu_a(ell))*phi_a(:,ell);
        else
            a=zeros(ne,1);
        end
        if ell<=N
            f=sqrt(nu_f(ell))*phi_f(:,ell);
        else
            f=zeros(ne,1);
        end
    end
    [Aks,bks]=get_elt_arrays2D(xv,yv,invJks,detJks,ne,elt2vert,a,f);
    A_ell = sparse(nvtx,nvtx); b_ell = zeros(nvtx,1);
    for row_no=1:3
        nrow=elt2vert(:,row_no);
        for col_no=1:3
           ncol=elt2vert(:,col_no);
           A_ell=A_ell+sparse(nrow,ncol,Aks(:,row_no,col_no),nvtx,nvtx);
        end
        b_ell = b_ell + sparse(nrow,1,bks(:,row_no),nvtx,1);
    end
    f_vecs{ell+1}=b_ell(int_nodes);
    KB_mats{ell+1}=A_ell(int_nodes,b_nodes);
    K_mats{ell+1}=A_ell(int_nodes,int_nodes);
end
end
function g=g_eval(x,y)
g=zeros(size(x)); % boundary condition
end
```

Finite element components

An advantage of choosing V^{hk} as in (9.69) is that each block of the Galerkin matrix is a linear combination of K_0, K_ℓ, $\ell = 1, \dots, P$. We can use existing finite element code to generate these, as well as $\boldsymbol{f}_0, \boldsymbol{f}_\ell$, for $\ell = 1, \dots, N$. Indeed, we can call Algorithm 2.7 with the mean data μ_a, μ_f to generate the components $K_0, \boldsymbol{f}_0$. Similarly, we can generate $K_\ell, \boldsymbol{f}_\ell$ using the ℓth eigenpairs of C_a and C_f as data. See Algorithm 9.3. Note that if $P = N$, the number of Karhunen–Loève terms for $\tilde{a}(\boldsymbol{x})$ and $\tilde{f}(\boldsymbol{x})$ is the same and the code simplifies.

Example 9.57 Let $D = (0,1) \times (0,1)$, $\tilde{f} = 1$ (so $N = 0$), $g = 0$ and suppose $\tilde{a}(\boldsymbol{x})$ is (9.50) with $\mu_a = 5$, $\ell = 1$ and $P = 6$. Note that $\tilde{a}(\boldsymbol{x})$ is a valid diffusion coefficient. We call Algorithm 2.4 to generate a uniform mesh with $h = 1/64$ and use the MATLAB code `twoDeigs.m` developed in Exercise 9.15(a) to compute the eigenvalues and evaluate the eigenfunctions for $\tilde{a}(\boldsymbol{x})$, as follows.

```
>> [xv,yv,elt2vert,nvtx,ne,h]=uniform_mesh_info(64);
>> xc=(xv(elt2vert(:,1))+xv(elt2vert(:,2))+xv(elt2vert(:,3)))/3;
>> yc=(yv(elt2vert(:,1))+yv(elt2vert(:,2))+yv(elt2vert(:,3)))/3;
>> [nu_a,phi_a]=twoDeigs(6,1,xc,yc);
```

Notice that we evaluate the eigenfunctions at the element centres (found by averaging the coordinates of the vertices). All the finite element components in (9.101) and (9.102) can now be generated by calling Algorithm 9.3 with the following MATLAB command.

```
>> [K_mats, KB_mats, f_vecs,w_B]=fem_components(64,xv,yv,...
                         elt2vert,nvtx,ne,h,5,nu_a,phi_a,1,[],[],6,0);
```

Since $\tilde{f} = 1$ here, we set `mu_f = 1` and do not supply eigenvalues and eigenfunctions for f. The outputs `K_mats,KB_mats` and `f_vecs` are cell arrays containing the matrices K_ℓ, $K_{\ell,\mathrm{B}}$ and vectors $\boldsymbol{f}_\ell$, for $\ell = 0, 1, \dots, M$. Note that if $P < M$ then the variables y_ℓ for $\ell = P+1, \dots, M$ do not appear in $\tilde{a}(\boldsymbol{x}, \boldsymbol{y})$ and the corresponding matrix components $K_\ell, K_{\ell,\mathrm{B}}$ are zero. Similarly, if $N < M$, $\boldsymbol{f}_\ell = \boldsymbol{0}$ for $\ell = N+1, \dots, M$. The output `wB` contains the boundary data $\boldsymbol{w}_\mathrm{B}$.

It remains to study the so-called stochastic components of the Galerkin system, that is, the left Kronecker factors $G_0, G_\ell, \boldsymbol{g}_0$ and $\boldsymbol{g}_\ell$ in (9.101)–(9.102).

Stochastic components

Setting $j = 1$ in (9.97)–(9.98) and assuming $\psi_1 = 1$, we see that $\boldsymbol{g}_0$ is the first column of G_0. Similarly, $\boldsymbol{g}_\ell$ is the first column of G_ℓ. The matrices G_0 and G_ℓ are clearly symmetric but only G_0 is guaranteed to be positive definite. The next result is an immediate consequence of (9.97) and Theorem 9.55.

Theorem 9.58 *Let the conditions of Theorem 9.55 hold and assume $\psi_1 = 1$. Then, $G_0 = I$ is the $Q \times Q$ identity matrix.*

The G_ℓ matrices, although not diagonal (like G_0), are still highly sparse when we use orthonormal bases for S^k.

Theorem 9.59 *Let the conditions of Theorem 9.55 hold. Assume $\Gamma_i = [-\gamma,\gamma]$ for some $\gamma > 0$ and p_i is even, for each $i = 1,\dots,M$. Then, each matrix G_ℓ, $\ell = 1,\dots,M$ in (9.98) has at most two non-zero entries per row. In particular,*

$$[G_\ell]_{js} = \begin{cases} \dfrac{1}{a_{\alpha_\ell+1}}, & \text{if } \alpha_\ell = \beta_\ell - 1 \text{ and } \alpha_i = \beta_i \text{ for each } i \in \{1,\dots,M\}\setminus\{\ell\},\\ \dfrac{1}{a_{\alpha_\ell}}, & \text{if } \alpha_\ell = \beta_\ell + 1 \text{ and } \alpha_i = \beta_i \text{ for each } i \in \{1,\dots M\}\setminus\{\ell\},\\ 0, & \text{otherwise,}\end{cases} \tag{9.103}$$

where $j = j(\boldsymbol{\alpha})$, $s = s(\boldsymbol{\beta})$ and a_{α_ℓ} is the coefficient from (9.82).

Proof Using (9.98) with $j = j(\boldsymbol{\alpha})$ and $s = s(\boldsymbol{\beta})$ gives

$$\begin{aligned}[G_\ell]_{js} &:= \langle y_\ell \psi_{j(\boldsymbol{\alpha})}, \psi_{s(\boldsymbol{\beta})}\rangle_p = \langle y_\ell P_{\alpha_\ell}, P_{\beta_\ell}\rangle_{p_\ell} \prod_{i=1, i\neq\ell}^{M} \langle P_{\alpha_i}, P_{\beta_i}\rangle_{p_i}\\ &= \langle y_\ell P_{\alpha_\ell}, P_{\beta_\ell}\rangle_{p_\ell} \prod_{i=1, i\neq\ell}^{M} \delta_{\alpha_i\beta_i}\end{aligned}$$

which is zero unless $\boldsymbol{\alpha}$ and $\boldsymbol{\beta}$ differ only in their ℓth components. Under the stated assumptions, the recurrence (9.82) applies and so

$$\begin{aligned}\langle y_\ell P_{\alpha_\ell}, P_{\beta_\ell}\rangle_{p_\ell} &= \left\langle \left(a_{\alpha_\ell+1}^{-1} P_{\alpha_\ell+1} + a_{\alpha_\ell}^{-1} P_{\alpha_\ell-1}\right), P_{\beta_\ell}\right\rangle_{p_\ell}\\ &= a_{\alpha_\ell+1}^{-1}\,\delta_{(\alpha_\ell+1)\beta_\ell} + a_{\alpha_\ell}^{-1}\,\delta_{(\alpha_\ell-1)\beta_\ell},\end{aligned}$$

which gives (9.103). Hence (with $s = s(\boldsymbol{\beta})$), we have $[G_\ell]_{js} \neq 0$ only when $j = j(\boldsymbol{\alpha})$ with $\boldsymbol{\alpha} = (\beta_1,\dots,\beta_\ell - 1,\dots,\beta_M)$ or $\boldsymbol{\alpha} = (\beta_1,\dots,\beta_\ell + 1,\dots,\beta_M)$. □

Example 9.60 (uniform random variables) Suppose $P = M = 2$ and consider $\xi_1,\xi_2 \sim$ U$(-\sqrt{3},\sqrt{3})$ *iid*. If we choose S^k as in Example 9.56 then $Q = \dim(S^k) = 6$ and we need to compute $G_1, G_2 \in \mathbb{R}^{6\times 6}$. Using Theorem 9.59 with $s = s(\boldsymbol{\alpha})$ and $j = j(\boldsymbol{\beta})$ where $\boldsymbol{\alpha} = (\alpha_1,\alpha_2)$ and $\boldsymbol{\beta} = (\beta_1,\beta_2)$ gives

$$[G_1]_{js} = \begin{cases} \dfrac{1}{a_{\alpha_1+1}}, & \text{if } \alpha_1 = \beta_1 - 1 \text{ and } \alpha_2 = \beta_2,\\ \dfrac{1}{a_{\alpha_1}}, & \text{if } \alpha_1 = \beta_1 + 1 \text{ and } \alpha_2 = \beta_2,\\ 0, & \text{otherwise,}\end{cases}$$

where, for Legendre polynomials, $1/a_{\alpha_1}$ is given by (9.86) (with $i = \alpha_1$). If we order the multi-index set as in Example 9.56 so that $\psi_1,\dots,\psi_6$ correspond to $(0,0)$, $(0,1)$, $(1,0)$, $(0,2)$, $(1,1)$, $(2,0)$, respectively, we obtain

$$G_1 = \begin{pmatrix} 0 & 0 & \frac{1}{a_1} & 0 & 0 & 0\\ 0 & 0 & 0 & 0 & \frac{1}{a_1} & 0\\ \frac{1}{a_1} & 0 & 0 & 0 & 0 & \frac{1}{a_2}\\ 0 & 0 & 0 & 0 & 0 & 0\\ 0 & \frac{1}{a_1} & 0 & 0 & 0 & 0\\ 0 & 0 & \frac{1}{a_2} & 0 & 0 & 0\end{pmatrix} = \begin{pmatrix} 0 & 0 & 1 & 0 & 0 & 0\\ 0 & 0 & 0 & 0 & 1 & 0\\ 1 & 0 & 0 & 0 & 0 & 2/\sqrt{5}\\ 0 & 0 & 0 & 0 & 0 & 0\\ 0 & 1 & 0 & 0 & 0 & 0\\ 0 & 0 & 2/\sqrt{5} & 0 & 0 & 0\end{pmatrix}.$$

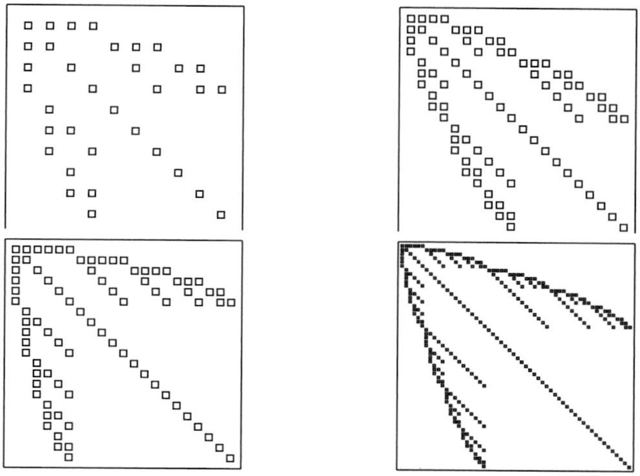

Figure 9.10 Block sparsity of the Galerkin matrix A when an orthonormal basis is used for S^k. Here, S^k contains complete polynomials with (top line) $M = 3$ and (bottom line) $M = 5$ for (left column) $k = 2$ and (right column) $k = 3$.

The calculation for G_2 is similar.

Theorem 9.58 tells us that

$$A := I \otimes K_0 + \sum_{\ell=1}^{P} G_\ell \otimes K_\ell \tag{9.104}$$

and we see the first term is block-diagonal. In addition, since (9.92) gives

$$A_{sj} = [G_0]_{js} K_0 + \sum_{\ell=1}^{P} [G_\ell]_{js} K_\ell, \qquad s, j = 1, \ldots, Q,$$

$G_0 = I$ and Theorem 9.59 gives $[G_\ell]_{ss} = 0$ for $\ell = 1, \ldots, P$, we have

$$A_{ss} = K_0, \qquad A_{sj} = \sum_{\ell=1}^{P} [G_\ell]_{js} K_\ell, \quad s \neq j.$$

Theorem 9.59 also tells us that A has at most $2P + 1$ non-zero blocks per row. Since the total number Q of blocks per row is given by (9.74) or (9.77), it is clear that A is block sparse. To investigate the block sparsity pattern of the Galerkin matrix, we consider

$$S := G_0 + \sum_{\ell=1}^{P} G_\ell. \tag{9.105}$$

The block A_{sj} of the matrix A is non-zero precisely when $S_{sj} \neq 0$.

Example 9.61 (complete polynomials) Suppose $P = M$ and consider complete polynomials of degree k. MATLAB `spy` plots of S in (9.105) for two choices of M and k are shown in Figure 9.10. Each square represents a matrix of dimension $J \times J$. Observe that most of the blocks are zero. When $M = 5$ and $k = 3$, only 266 out of $Q^2 = 56^2 = 3{,}136$ (see (9.77)) blocks are non-zero.

Finite element diffusion matrices are ill-conditioned with respect to h (see Exercise 2.5). Since the stochastic Galerkin matrix is composed of $P + 1$ such matrices, it becomes more ill-conditioned — and the system becomes harder to solve — as we refine the mesh. To develop solution schemes, we also need to determine whether the Galerkin matrix is ill-conditioned with respect to k and M. For this, we must study the eigenvalues of the G_ℓ matrices. We focus on complete polynomials and show that for a fixed M and k, G_ℓ is a permutation of a block tridiagonal matrix, whose eigenvalues are known explicitly.

Theorem 9.62 *Let the conditions of Theorem 9.59 hold and assume S^k is the set of complete polynomials of degree $\le k$. Then, each matrix G_ℓ defined in* (9.98) *has $k + 1$ eigenvalues of multiplicity one given by the roots of P_{k+1}, and s eigenvalues of multiplicity*

$$\sum_{r=1}^{k-s+1} \binom{M-1}{r}, \tag{9.106}$$

given by the roots of P_s, for $s = 1, \dots, k$.

Proof For complete polynomials, the set of multi-indices associated with S^k is

$$\Lambda := \left\{ \boldsymbol{\alpha} = (\alpha_1, \dots, \alpha_M) \colon \alpha_i = 0, 1, \dots, k,\ i = 1, \dots, M, \text{ with } |\boldsymbol{\alpha}| \le k \right\}.$$

Suppose the first $k + 1$ multi-indices, corresponding to $\psi_1, \dots, \psi_{k+1}$, are

$$(0, 0 \dots, 0), (1, 0 \dots, 0), (2, 0 \dots, 0), \dots, (k, 0, \dots, 0).$$

Since they differ only in the *first* component, Theorem 9.59 tells us that the leading $(k + 1) \times (k + 1)$ block of G_1 is the matrix T_{k+1} defined in (9.84) (with $N = k + 1$). There are no other non-zero entries in the first $k + 1$ rows of G_1 because there is no other $\boldsymbol{\alpha} \in \Lambda$ for which $\alpha_2 = \cdots = \alpha_M = 0$ and α_1 differs by one from the first component of one of the multi-indices already chosen.

Next, we list the multi-indices $\boldsymbol{\alpha}$ for which α_1 ranges from 0 to $k - 1$ and $\sum_{s=2}^{M} \alpha_s = 1$. These can be arranged into $(M - 1)$ groups, in each of which every component of $\boldsymbol{\alpha}$ except α_1 is identical. That is,

$$\begin{array}{cccc}
(0,0,\dots 0,1) & (0,0,\dots 1,0) & \cdots & (0,1,\dots 0,0) \\
(1,0,\dots 0,1) & (1,0,\dots 1,0) & \cdots & (1,1,\dots 0,0) \\
\vdots & \vdots & & \vdots \\
(k-1,0\dots 0,1) & (k-1,0,\dots 1,0) & \cdots & (k-1,1,\dots 0,0).
\end{array}$$

With this ordering, the leading $(Mk + 1) \times (Mk + 1)$ block of G_1 is

$$\begin{pmatrix} T_{k+1} & & & \\ & T_k & & \\ & & \ddots & \\ & & & T_k \end{pmatrix}.$$

Next, we list all $\boldsymbol{\alpha} \in \Lambda$ for which $\alpha_1 = 0, \dots, k - 2$ and $\sum_{s=2}^{M} \alpha_s = 2$, grouped so that every component of $\boldsymbol{\alpha}$ except α_1 is identical. This contributes

$$\binom{M-1}{1} + \binom{M-1}{2} \quad \text{tridiagonal blocks } T_{k-1} \text{ to } G_1.$$

We continue in this fashion until, finally, we list the

$$\sum_{r=1}^{k} \binom{M-1}{r}$$

groups of multi-indices with $\alpha_1 = 0$ and $\sum_{s=2}^{M} \alpha_s = k$. G_1 is block tridiagonal by Theorem 9.59. There is one T_{k+1} block and the number of T_s blocks, for $s = 1, \ldots, k$, is given by (9.106). Now, regardless of the ordering of the set Λ, there exists a permutation matrix P_1 (corresponding to a reordering of the ψ_j) such that $P_1 G_1 P_1^{\mathrm{T}}$ is the block tridiagonal matrix described. Since the eigenvalues of G_1 are unaffected by permutations, and the eigenvalues of T_s are the roots of P_s, the result holds for G_1. For $\ell > 1$, the proof is similar. □

Example 9.63 (uniform random variables) The matrices G_1 and G_2 in Example 9.60 are permutations of the block tridiagonal matrix

$$T = \left(\begin{array}{ccc|cc|c} 0 & 1 & 0 & 0 & 0 & 0 \\ 1 & 0 & 2/\sqrt{5} & 0 & 0 & 0 \\ 0 & 2/\sqrt{5} & 0 & 0 & 0 & 0 \\ \hline 0 & 0 & 0 & 0 & 1 & 0 \\ 0 & 0 & 0 & 1 & 0 & 0 \\ \hline 0 & 0 & 0 & 0 & 0 & 0 \end{array}\right) = \begin{pmatrix} T_3 & 0 & 0 \\ 0 & T_2 & 0 \\ 0 & 0 & T_1 \end{pmatrix}$$

whose eigenvalues are given by the roots of the Legendre polynomials L_3, L_2 and L_1, which are $\{-3/\sqrt{5}, 0, 3/\sqrt{5}\}$, $\{-1, 1\}$ and $\{0\}$, respectively.

For tensor product polynomials (9.73), the result is similar. The multiplicities of the eigenvalues change, however, and there are $(k+1)^M$ in total. Observe now that if we fix the polynomial degree k then G_ℓ does not become ill-conditioned as the number of variables $M \to \infty$.

Corollary 9.64 *Let the conditions of Theorem 9.62 hold. Then, for a fixed polynomial degree k, the matrix G_ℓ in (9.98) has the same eigenvalues, for any $M > 1$, for $\ell = 1, 2, \ldots, M$.*

Proof Apply Theorem 9.62 and note that only the multiplicities of the eigenvalues change as M is increased. □

For Legendre and Hermite polynomials, we have the following result.

Corollary 9.65 *Let the conditions of Theorem 9.62 hold. If the polynomials P_{α_i} are Legendre polynomials on $[-\sqrt{3}, \sqrt{3}]$ then the eigenvalues of G_ℓ belong to $[-\sqrt{3}, \sqrt{3}]$, for each $\ell = 1, \ldots, M$. If the P_{α_i} are Hermite polynomials on $\mathbb{R}$ then the eigenvalues belong to $[-2\sqrt{k}, 2\sqrt{k}]$, for each $\ell = 1, \ldots, M$.*

Proof Apply Theorem 9.62 and see Exercises 9.19 and 9.20. □

The eigenvalues of the G_ℓ matrices associated with Legendre polynomials — needed to construct an orthonormal basis when working with uniform random variables — are bounded independently of k and M. Hence, if we solve the Galerkin system using an iterative method (e.g., the conjugate gradient method), the number of iterations required to satisfy a fixed

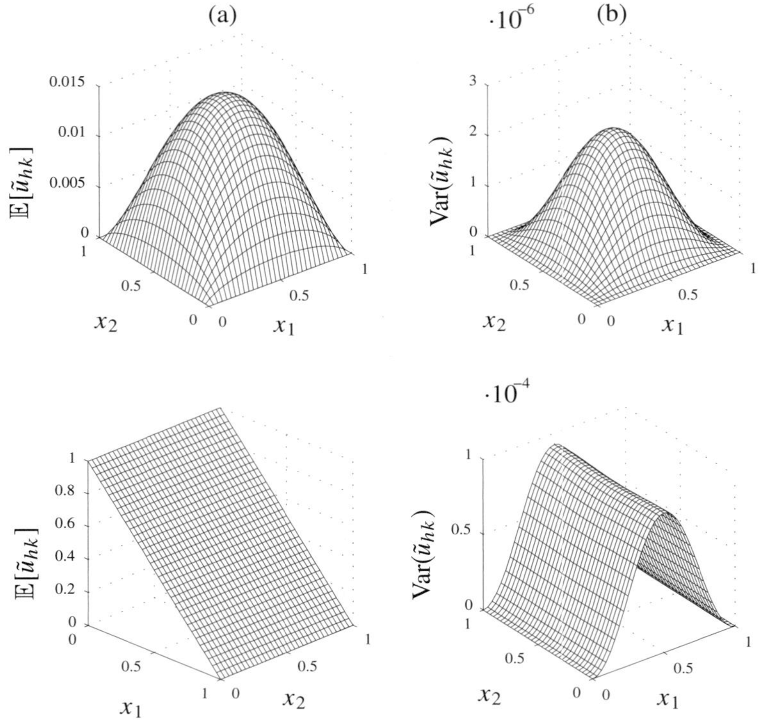

Figure 9.11 Plot of (a) $\mathbb{E}[\tilde{u}_{hk}]$ and (b) $\mathrm{Var}(\tilde{u}_{hk})$ for Example 9.66 with $P = 6$, $h = 1/32$ and $k = 3$ for (top) $f = 1$ and $g = 0$ and (bottom) $f = 0$ and mixed boundary conditions.

error tolerance, should remain bounded as $k, M \to \infty$. See also Exercise 9.23. On the other hand, the eigenvalues of the G_ℓ matrices associated with Hermite polynomials — needed to construct an orthonormal basis when working with Gaussian random variables — are *not* bounded independently of k. Of course, for our model elliptic BVP, we are not permitted to use Gaussian random variables in (9.42). However, such variables do appear in many other interesting problems.

Numerical results

We now implement the stochastic Galerkin finite element method (SGFEM) for test problems on $D = (0,1) \times (0,1)$ with a deterministic f. In this case, $\tilde{f} = f$, $N = 0$ and we have $M := \max\{P, N\} = P$ random variables. The diffusion coefficient $a(\boldsymbol{x})$ is chosen as in Example 9.37 and is approximated by $\tilde{a}(\boldsymbol{x})$ in (9.50). Note that the eigenvalues ν_k^a depend on the correlation length. When $\ell = 1$, if $P = 6$, then (9.45) gives

$$\|a - \tilde{a}\|_{L^\infty(\Omega, L^\infty(D))} \leq 2\sqrt{3} \sum_{k=7}^{\infty} \sqrt{\nu_k^a} \approx 1.4 \times 10^{-3}. \qquad (9.107)$$

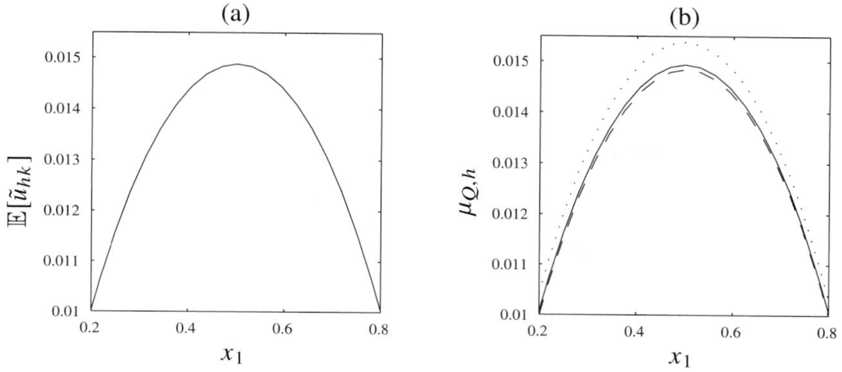

Figure 9.12 Cross-section (zoomed) of (a) SGFEM estimate $\mathbb{E}[\tilde{u}_{hk}]$ and (b) MCFEM estimate $\mu_{Q,h}$ for Example 9.67. In (a), $k = 1$ (dotted), $k = 2$ (dashed) and $k = 3$ (solid). In (b), $Q = 7$ (dotted), $Q = 28$ (dashed) and $Q = 84$ (solid).

On the other hand,

$$\mathrm{Var}\left[\|a - \tilde{a}\|_{L^2(D)}\right] = \sum_{k=7}^{\infty} \nu_k^a \approx 7.5 \times 10^{-8}. \tag{9.108}$$

We need to retain more terms in the Karhunen–Loève expansion of $a(\boldsymbol{x})$ to bound the $L^\infty(\Omega, L^\infty(D))$ norm of the data error by a fixed tolerance than to bound the variance of the $L^2(D)$ norm. Recall that the former error appears in (9.37). If we reduce the correlation length and don't increase P then the data error grows. For $\ell = 0.5$ then, even with $P = 15$,

$$\|a - \tilde{a}\|_{L^\infty(\Omega, L^\infty(D))} \leq 1.42 \times 10^{-2}, \qquad \mathrm{Var}\left[\|a - \tilde{a}\|_{L^2(D)}\right] \approx 2.8 \times 10^{-6}.$$

The number of terms P required to ensure that both data errors are small increases as $\ell \to 0$. For this reason, using Karhunen–Loève expansions to approximate random fields with small correlation lengths is usually infeasible.

Below, we apply piecewise linear finite elements on uniform meshes of D and use complete polynomials of degree k on $\Gamma = \left[-\sqrt{3}, \sqrt{3}\right]^P$. We compute the FEM components of the Galerkin system using Algorithm 9.3, compute the stochastic components similarly, and then apply the conjugate gradient method (without assembling A; see Exercise 9.22). Finally, $\mathbb{E}[\tilde{u}_{hk}]$ and $\mathrm{Var}(\tilde{u}_{hk})$ are computed via (9.90) and (9.91).

Example 9.66 As in Example 9.57, we start by choosing $\mu_a = 5$, $\ell = 1$ and $P = 6$. Plots of $\mathbb{E}[\tilde{u}_{hk}]$ and $\mathrm{Var}(\tilde{u}_{hk})$ obtained with $h = 1/32$ and $k = 3$ are shown in Figure 9.11. We consider unit forcing ($f = 1$) and homogeneous Dirichlet boundary conditions ($g = 0$) and then $f = 0$ with the mixed boundary conditions from Exercise 2.20. In the second case, note that the Dirichlet boundary is $\partial D_{\mathrm{D}} = \{0, 1\} \times [0, 1]$ and $\mathrm{Var}(\tilde{u}_{hk})$ is zero only on ∂D_{D}.

In the next section, we investigate the theoretical convergence of $\mathbb{E}[\tilde{u}_{hk}]$ to $\mathbb{E}[\tilde{u}]$. We can use Theorem 9.48 to show that the error associated with the FEM discretisation is $\mathcal{O}(h)$ but we will also need to study the error associated with the polynomial approximation on Γ. In the next two examples, we fix h and study the convergence of $\mathbb{E}[\tilde{u}_{hk}]$ and $\mathrm{Var}(\tilde{u}_{hk})$ numerically when we vary k. We focus on the case $f = 1$ and $g = 0$ (see Figure 9.11).

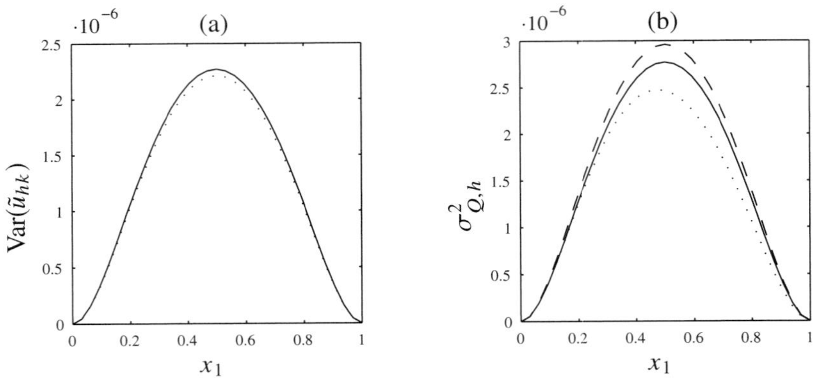

Figure 9.13 Cross-section of (a) SGFEM estimate $\mathrm{Var}(\tilde{u}_{hk})$ and (b) MCFEM estimate $\sigma^2_{Q,h}$ for Example 9.67. In (a), $k = 1$ (dotted), $k = 2$ (dashed) and $k = 3$ (solid). In (b), $Q = 7$ (dotted), $Q = 28$ (dashed) and $Q = 84$ (solid).

Example 9.67 (SGFEM vs MCFEM) In Figure 9.12, we compare SGFEM and MCFEM approximations of $\mathbb{E}[\tilde{u}(\boldsymbol{x})]$ obtained with $h = 1/32$. Again, we choose $\mu_a = 5$, $\ell = 1$ and $P = 6$. For the stochastic Galerkin method, we vary the polynomial degree k (and hence the dimension Q of S^k). For Monte Carlo, we choose the number of samples to match the dimension Q of S^k. Hence, in both cases, we have to solve JQ equations where $J = 961$. The curves in Figure 9.12(a) are indistinguishable and the SGFEM approximation converges rapidly with respect to k (and hence Q). However, we see that the MCFEM estimate converges more slowly. In Figure 9.13, we plot the approximations to $\mathrm{Var}(\tilde{u}(\boldsymbol{x}))$ and here the SGFEM approximation also converges more rapidly than the MCFEM estimate.

Example 9.68 (SGFEM k-convergence) To gain more insight into the *rate* of convergence of $\mathbb{E}[\tilde{u}_{hk}]$, we now consider the error

$$E_{2,h} := \left\| \mathbb{E}[\tilde{u}_{hk}] - \mathbb{E}[\tilde{u}_{h,\mathrm{ref}}] \right\|_{2,h}, \tag{9.109}$$

where, for $v \in V^h$, we define

$$\|v\|^2_{2,h} := \sum_{j=1}^{J} v_j^2 h^2, \qquad \text{with } v(x) = \sum_{j=1}^{J} v_j \phi_j(x).$$

Note that $\|\cdot\|_{2,h}$ is a discrete approximation of the $L^2(D)$ norm on V^h (see also (3.53) for one space dimension). We compute $E_{2,h}$ for a fixed h and vary k. Results are shown in Figure 9.14. Here, the reference solution $\tilde{u}_{h,\mathrm{ref}}$ is computed with $k = 8$ (giving $Q = 3{,}003$ when $P = 6$ and a total of $2{,}885{,}883$ equations). Note that the vertical axis is on a logarithmic scale and so $E_{2,h}$ converges to zero *exponentially* as $k \to \infty$.

When $\ell = 1$, choosing $P = 6$ yields a data error satisfying (9.107) and (9.108). If we reduce ℓ without and don't increase P then we have seen that the data error grows.

Our final example illustrates an additional side effect of retaining too few terms in the Karhunen–Loève expansion of $a(\boldsymbol{x})$.

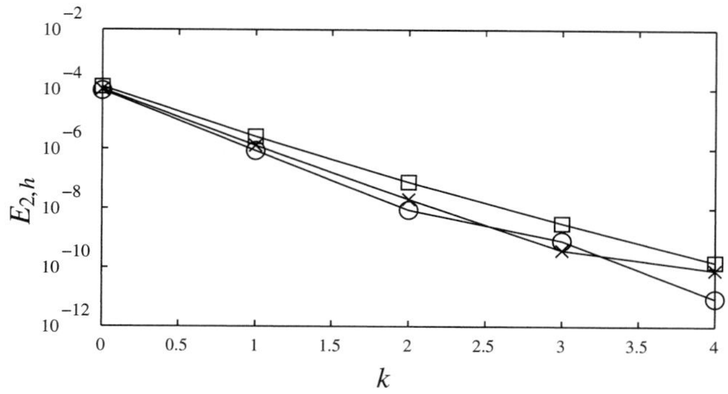

Figure 9.14 Plot of the error $E_{2,h}$ in Example 9.68 for varying k with $\mu_a = 5$, $P = 6$, $h = 1/32$ and correlation lengths $\ell = 1$ (circles), $\ell = 0.8$ (crosses), $\ell = 0.6$ (squares).

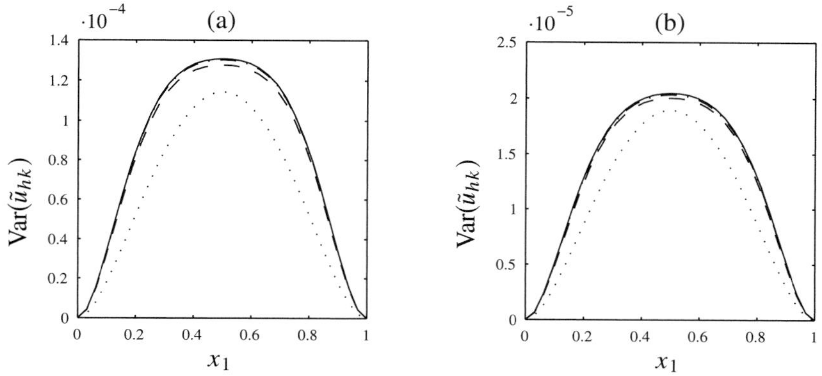

Figure 9.15 Cross-section of $\mathrm{Var}(\tilde{u}_{hk})$ in Example 9.69 with (a) $\ell = 0.75$, $\mu_a = 2$ and (b) $\ell = 0.5$, $\mu_a = 3$. $P = 1$ (dotted line), $P = 3$ (dashed), $P = 5$ (dot-dash) and $P = 7$ (solid).

Example 9.69 (varying ℓ and P) We fix $k = 3$, $h = 1/32$ and now reduce the correlation length. Since decreasing ℓ increases $\mathrm{Var}(\tilde{a}(\boldsymbol{x}))$, we need to choose μ_a large enough that $\tilde{a}(\boldsymbol{x})$ has positive realisations. In Figure 9.15, we plot $\mathrm{Var}(\tilde{u}_{hk})$ for $\ell = 0.75$ with $\mu_a = 2$ and for $\ell = 0.5$ with $\mu_a = 3$, for varying P. We see that choosing P to be too small (for a fixed ℓ) causes us to under estimate the solution variance. This fits with our earlier observation that retaining too few terms in the expansion of $a(\boldsymbol{x})$ causes us to under estimate $\mathrm{Var}(a(\boldsymbol{x}))$.

Error analysis

We now investigate the error in approximating $\mathbb{E}[u(\boldsymbol{x})]$ by $\mathbb{E}[\tilde{u}_{hk}]$, where $u(\boldsymbol{x})$ is the weak solution (a random field) associated with the original BVP (9.5)–(9.6) and $\tilde{u}_{hk} \colon D \times \Gamma \to \mathbb{R}$ is the stochastic Galerkin solution satisfying (9.64). The first contribution to the error arises from the data approximations (9.42) and (9.43). Recall then that $\tilde{u}(\boldsymbol{x})$ is the weak solution associated with the BVP (9.34)–(9.35). Working on $D \times \Omega$, we see that the error in estimating $\mathbb{E}[u(\boldsymbol{x})]$ caused by approximating the data satisfies

$$\left|\mathbb{E}[u] - \mathbb{E}[\tilde{u}]\right|_{H^1(D)} \leq \mathbb{E}\left[|u - \tilde{u}|^2_{H^1(D)}\right]^{1/2} := |u - \tilde{u}|_W.$$

We obtained an upper bound for this in Theorem 9.31. For the specific approximations (9.42) and (9.43), the bound depends on the truncation parameters P and N. To see this, recall that

$$\|f - \tilde{f}\|^2_{L^2(\Omega, L^2(D))} = \sum_{i=N+1}^{\infty} \nu_f^i$$

and by Assumption 9.32, (9.45) holds. Since their upper bounds depend on the eigenvalues, both data errors in (9.37) converge to zero as $P, N \to \infty$ with rates that depend on the regularity of C_a and C_f (see Corollary 9.35 and the Notes in Chapter 7). The smoother the covariances, the fewer Karhunen–Loève terms we require to ensure $|u - \tilde{u}|_W = \mathcal{O}(\epsilon)$ (for a given tolerance ϵ).

We now focus on the error $|\mathbb{E}[\tilde{u}] - \mathbb{E}[\tilde{u}_{hk}]|_{H^1(D)}$. For truncated Karhunen–Loève approximations $\tilde{a}(\boldsymbol{x})$ and $\tilde{f}(\boldsymbol{x})$, which are finite-dimensional noise and functions of the random vector $\boldsymbol{\xi}\colon \Omega \to \Gamma$, we can work on $D \times \Gamma$ and treat $\tilde{u}$ as the weak solution associated with the equivalent BVP (9.54)–(9.55). Hence,

$$\left|\mathbb{E}[\tilde{u}] - \mathbb{E}[\tilde{u}_{hk}]\right|_{H^1(D)} \le \left|\tilde{u} - \tilde{u}_{hk}\right|_W,$$

where we now define $|\cdot|_W$ as in (9.61) and $\tilde{u} \in L^2_p(\Gamma, H^1_g(D))$. To exploit our earlier FEM error analysis, we introduce the semi-discrete approximation $\tilde{u}_h \in L^2_p(\Gamma, W^h)$ satisfying (9.57). Using the triangle inequality then gives

$$\left|\tilde{u} - \tilde{u}_{hk}\right|_W \le \left|\tilde{u} - \tilde{u}_h\right|_W + \left|\tilde{u}_h - \tilde{u}_{hk}\right|_W. \tag{9.110}$$

If Assumption 9.47 holds, Theorem 9.48 tells us that the first term (the spatial error) is $\mathcal{O}(h)$. When computing $\tilde{u}_h$ in practice, we often approximate the data further to allow fast computation of the FEM matrices via quadrature. We do not quantify that error here but note that integration should be performed in a way that is consistent with the accuracy of the spatial discretisation.

If we work with the splitting (9.110) then, to fully analyse the SGFEM error, we need to study $|\tilde{u}_h - \tilde{u}_{hk}|_W$. This error depends on the polynomial space S^k. Note that we have not yet chosen a specific one. First, define $\mathbb{N}_0 := \mathbb{N} \cup \{0\}$ and the set of multi-indices

$$\mathbb{N}_0^M := \left\{\boldsymbol{\alpha} = (\alpha_1, \dots, \alpha_M) : \alpha_s \in \mathbb{N}_0,\ s = 1, \dots, M\right\}$$

and suppose that S^k is any space of global polynomials ψ_j on Γ of the form

$$S^k = \operatorname{span}\left\{\psi_{j(\boldsymbol{\alpha})}(\boldsymbol{y}) : \boldsymbol{\alpha} \in \Lambda \text{ for some } \Lambda \subset \mathbb{N}_0^M \text{ with } |\Lambda| = Q < \infty\right\}. \tag{9.111}$$

We have seen that each of the spaces (9.73), (9.75) and (9.76) can be written this way. Only the set Λ and the dimension Q change, where Q depends on the number of variables M and the polynomial degree k (or vector of polynomial degrees $\boldsymbol{k} = (k_1, \dots, k_M)$ in the case of (9.75)). For example, for (9.76) (complete polynomials of degree k), $\Lambda = \{\boldsymbol{\alpha} \in \mathbb{N}_0^M : |\boldsymbol{\alpha}| \le k\}$ and Q is given by (9.77). We now introduce a projection associated with S^k, denoted P_Λ.

Definition 9.70 ($L^2_p(\Gamma)$-orthogonal projection) $P_\Lambda\colon L^2_p(\Gamma) \to S^k$, where S^k is a polynomial space of the form (9.111), is the $L^2_p(\Gamma)$-orthogonal projection defined by

$$\left\langle P_\Lambda u, v\right\rangle_p = \left\langle u, v\right\rangle_p, \qquad \forall u \in L^2_p(\Gamma),\ \forall v \in S^k, \tag{9.112}$$

where the inner product $\langle \cdot, \cdot \rangle_p$ is defined in (9.87).

Using P_Λ, we can obtain an alternative splitting of the error $\|\tilde{u} - \tilde{u}_{hk}\|_W$.

Lemma 9.71 *Let $P_\Lambda \colon L^2_p(\Gamma) \to S^k$ be the $L^2_p(\Gamma)$ projection defined in (9.112). Let $\tilde{u} \in L^2_p(\Gamma, H^1_g(D))$, $\tilde{u}_h \in L^2_p(\Gamma, W^h)$ and $\tilde{u}_{hk} \in W^h \otimes S^k$ satisfy (9.51), (9.57) and (9.64) respectively and suppose Assumption 9.42 holds. Then,*

$$\|\tilde{u} - \tilde{u}_{hk}\|_W \le K_{\tilde{a}}\big(\|\tilde{u} - \tilde{u}_h\|_W + \|\tilde{u} - P_\Lambda \tilde{u}\|_W\big), \qquad K_{\tilde{a}} := \sqrt{\frac{\tilde{a}_{\max}}{\tilde{a}_{\min}}}. \tag{9.113}$$

Proof Define the Galerkin projection P_h as in (9.62). Since P_Λ acts only on the $\boldsymbol{y}$-components of $P_h\tilde{u} \in L^2_p(\Gamma, W^h)$, it follows that $P_\Lambda(P_h\tilde{u}) \in W^h \otimes S^k = W^{hk}$. Now, using the optimality property (9.65) gives

$$\begin{aligned}\|\tilde{u} - \tilde{u}_{hk}\|_E &\le \|\tilde{u} - P_\Lambda P_h \tilde{u}\|_E = \|\tilde{u} - P_h\tilde{u} + P_h\tilde{u} - P_\Lambda P_h\tilde{u}\|_E \\ &\le \|\tilde{u} - P_h\tilde{u}\|_E + \|P_h(\tilde{u} - P_\Lambda\tilde{u})\|_E \\ &\le \|\tilde{u} - \tilde{u}_h\|_E + \|\tilde{u} - P_\Lambda\tilde{u}\|_E.\end{aligned}$$

Using Assumption 9.42 and (9.60)–(9.61) gives the result. □

Comparing the bounds (9.113) and (9.110) we see that the spatial error is now multiplied by the constant $K_{\tilde{a}}$. The advantage in working with (9.113) is that the second contribution to the error (a projection error) is uncoupled from the finite element approximation.

We now examine the error $\|\tilde{u} - P_\Lambda\tilde{u}\|_W$ in (9.113) when the ξ_i in (9.42)–(9.43) are uniform *iid* random variables with $\mathbb{E}[\xi_i] = 0$ and $\mathrm{Var}(\xi_i) = 1$. In that case, $\tilde{u} \in L^2_p(\Gamma, H^1_g(D))$ with $p = 1/(2\sqrt{3})^M$ and $\Gamma = [-\sqrt{3}, \sqrt{3}]^M$ and Legendre polynomials provide a countable orthonormal basis for $L^2_p(\Gamma)$. That is,

$$L^2_p(\Gamma) = \mathrm{span}\big\{L_{j(\boldsymbol{\alpha})}(\boldsymbol{y}) \colon \boldsymbol{\alpha} \in \mathbb{N}_0^M\big\}, \qquad L_{j(\boldsymbol{\alpha})}(\boldsymbol{y}) := \prod_{s=1}^{M} L_{\alpha_s}(y_s), \tag{9.114}$$

where L_{α_s} is the polynomial of degree α_s on $[-\sqrt{3}, \sqrt{3}]$ defined in Example 9.54. Using these specific basis functions, (9.111) then becomes

$$S^k = \mathrm{span}\big\{L_{j(\boldsymbol{\alpha})}(\boldsymbol{y}) \colon \boldsymbol{\alpha} \in \Lambda \text{ for some set } \Lambda \subset \mathbb{N}_0^M \text{ with } |\Lambda| = Q < \infty\big\}. \tag{9.115}$$

The next result shows that when we expand $\tilde{u}$ in the orthonormal basis for $L^2_p(\Gamma)$, the projection error in (9.113) may be bounded using a truncated sum of the associated (Fourier) coefficients.

Lemma 9.72 *Let $p = 1/(2\sqrt{3})^M$, $\Gamma = [-\sqrt{3}, \sqrt{3}]^M$ and define S^k as in (9.115) and $P_\Lambda \colon L^2_p(\Gamma) \to S^k$ as in (9.112). Let $\tilde{u} \in L^2_p(\Gamma, H^1_g(D))$ satisfy (9.51). Then,*

$$\|\tilde{u} - P_\Lambda\tilde{u}\|_W^2 \le \sum_{j=Q+1}^{\infty} |\tilde{u}_j|^2_{H^1(D)}, \qquad \tilde{u}_j := \langle \tilde{u} - w_g, L_j \rangle_p,$$

where $w_g \in H^1_g(D)$ is defined as in (9.71) and L_j is defined in (9.114).

Proof Since $\tilde{u} \in L^2_p(\Gamma, H^1_g(D))$, we have $\tilde{u}(\boldsymbol{x},\boldsymbol{y}) = \tilde{u}_0(\boldsymbol{x},\boldsymbol{y}) + w_g(\boldsymbol{x})$ for some $\tilde{u}_0 \in L^2_p(\Gamma, H^1_0(D))$ and $w_g \in H^1_g(D)$. Using the orthonormal basis of Legendre polynomials from (9.114), we have

$$\tilde{u}_0(\boldsymbol{x},\boldsymbol{y}) = \sum_{\boldsymbol{\alpha} \in \mathbb{N}_0^M} \tilde{u}_{j(\boldsymbol{\alpha})}(\boldsymbol{x}) L_{j(\boldsymbol{\alpha})}(\boldsymbol{y}) = \sum_{j=1}^{\infty} \tilde{u}_j(\boldsymbol{x}) L_j(\boldsymbol{y}), \qquad \tilde{u}_j := \langle \tilde{u}_0, L_j \rangle_p.$$

Note that the coefficients $\tilde{u}_j \in H^1_0(D)$. Now, using (9.115), it can be shown that

$$P_\Lambda \tilde{u}_0 := \sum_{\boldsymbol{\alpha} \in \Lambda} \tilde{u}_{j(\boldsymbol{\alpha})} L_{j(\boldsymbol{\alpha})} = \sum_{j=1}^{Q} \tilde{u}_j L_j, \tag{9.116}$$

(see Lemma 1.41). Since $P_\Lambda w_g = w_g$ and $\langle L_i, L_j \rangle_p = \delta_{ij}$, we have

$$|\tilde{u} - P_\Lambda \tilde{u}|_W^2 = \left| \sum_{j=Q+1}^{\infty} \tilde{u}_j L_j \right|_W^2 \le \sum_{j=Q+1}^{\infty} |\tilde{u}_j|^2_{H^1(D)}. \qquad \square$$

For other types of random variables ξ_i, a similar result holds. However, if the density p and the interval Γ change, so do the orthonormal basis functions for $L^2_p(\Gamma)$ and S^k. Lemma 9.72 holds for any polynomial space S^k of the form (9.115). To make progress, we now consider a specific case. We focus on the tensor product space

$$S^k = S^{\boldsymbol{k}} := S_1^{k_1} \otimes S_2^{k_2} \otimes \cdots \otimes S_M^{k_M} \tag{9.117}$$

from (9.75), where $\boldsymbol{k} := (k_1, \dots, k_M)$ and $S_i^{k_i}$ is the set of polynomials of degree k_i or less in y_i on $[-\sqrt{3}, \sqrt{3}]$. The set of multi-indices in (9.115) is then

$$\Lambda := \left\{ \boldsymbol{\alpha} = (\alpha_1, \dots, \alpha_M) \colon \alpha_i = 0, 1, \dots, k_i, \text{ for } i = 1, \dots, M \right\},$$

and $Q = \dim(S^{\boldsymbol{k}}) = |\Lambda| = \prod_{i=1}^M (k_i + 1)$. Note that this includes (9.73) as a special case (when $k_1 = \cdots = k_M = k$). We now define projection operators associated with each of the spaces $S_i^{k_i}$.

Definition 9.73 ($L^2_{p_i}(\Gamma_i)$-orthogonal projection) $P_{k_i} \colon L^2_{p_i}(\Gamma_i) \to S_i^{k_i}$ is the $L^2_{p_i}(\Gamma_i)$-orthogonal projection and is defined by

$$\left\langle P_{k_i} u, v \right\rangle_{p_i} = \langle u, v \rangle_{p_i}, \qquad \forall u \in L^2_{p_i}(\Gamma_i), \forall v \in S_i^{k_i}. \tag{9.118}$$

Note that P_{k_i} denotes an *operator* now and not a polynomial (as in the previous section). For $\xi_i \sim \mathrm{U}(-\sqrt{3}, \sqrt{3})$, we have $\Gamma_i = [-\sqrt{3}, \sqrt{3}]$ and $p_i = (1/2\sqrt{3})$ in (9.118). For the tensor product polynomial spaces (9.73) and (9.75), we can use the operators from Definition 9.73 to bound the error $|\tilde{u} - P_\Lambda \tilde{u}|_W$ in Lemma 9.71 by a sum of one-dimensional projection errors.

Corollary 9.74 *Let the conditions of Lemma 9.72 hold and let S^k be the tensor product polynomial space* (9.117). *Then, with P_{k_i} is defined as in* (9.118),

$$|\tilde{u} - P_\Lambda \tilde{u}|_W \le \sum_{i=1}^{M} \left| \tilde{u} - P_{k_i} \tilde{u} \right|_W, \tag{9.119}$$

Proof First note that $P_\Lambda \tilde{u} = \left(P_{k_1} P_{k_2} \cdots P_{k_M}\right) \tilde{u}$. Since $\tilde{u} \in L_p(\Gamma, H_g^1(D))$ and $L_p^2(\Gamma) := L_{p_1}^2(\Gamma_1) \otimes \cdots \otimes L_{p_M}^2(\Gamma_M)$, it should be understood here that P_{k_i} acts only on the y_i-component of $\tilde{u}$. If we define $P_{k_0} := I$ then,

$$(I - P_\Lambda)\tilde{u} = \sum_{i=1}^{M} P_{k_0} P_{k_1} \cdots P_{k_{i-1}} (I - P_{k_i})\tilde{u},$$

and hence,

$$\left|\tilde{u} - P_\Lambda \tilde{u}\right|_W \le \sum_{i=1}^{M} \left|P_{k_0} P_{k_1} \cdots P_{k_{i-1}} (I - P_{k_i})\tilde{u}\right|_W \le \sum_{i=1}^{M} \left|(I - P_{k_i})\tilde{u}\right|_W. \qquad \square$$

It remains to bound the projection errors $|\tilde{u} - P_{k_i}\tilde{u}|_W$ in (9.119). For any $v = v(y_i) \in L_{p_i}^2(\Gamma_i)$, since P_{k_i} is an orthogonal projection, we have

$$\left\|v - P_{k_i} v\right\|_{L_{p_i}^2(\Gamma_i)} = \min_{w \in S_i^{k_i}} \left\|v - w\right\|_{L_{p_i}^2(\Gamma_i)}.$$

If $v \in \mathrm{C}^{k_i+1}(\Gamma_i)$, we can use standard approximation theory in one dimension (see Exercise 1.13) to obtain, for any $k_i \ge 0$,

$$\left\|v - P_{k_i} v\right\|_{L_{p_i}^2(\Gamma_i)} \le \left\|\frac{d v^{k_i+1}}{d y^{k_i+1}}\right\|_{L_{p_i}^2(\Gamma_i)} \frac{\sqrt{3}^{(k_i+1)}}{(k_i+1)!}. \tag{9.120}$$

Now, for multivariate functions $v \colon \Gamma \to \mathbb{R}$ we have

$$\left\|v - P_{k_i} v\right\|_{L_p^2(\Gamma)}^2 = \int_{\Gamma^*} p(\boldsymbol{y}_*) \left\|v(\boldsymbol{y}_*, \cdot) - P_{k_i} v(\boldsymbol{y}_*, \cdot)\right\|_{L_{p_i}^2(\Gamma_i)}^2 d\boldsymbol{y}_*,$$

where $\boldsymbol{y}_* \in \Gamma_* := \prod_{s=1, s \ne i}^{M} \Gamma_i \subset \mathbb{R}^{M-1}$ and

$$p(\boldsymbol{y}_*) := p_1(y_1) \cdots p_{i-1}(y_{i-1}) p_{i+1}(y_{i+1}) \cdots p_M(y_M).$$

Hence, if v is sufficiently differentiable, it follows from (9.120) that

$$\left\|v - P_{k_i} v\right\|_{L_p^2(\Gamma)} \le \left\|\frac{\partial v^{k_i+1}}{\partial y^{k_i+1}}\right\|_{L_p^2(\Gamma)} \frac{\sqrt{3}^{(k_i+1)}}{(k_i+1)!}. \tag{9.121}$$

Lemma 9.75 *Let the conditions of Corollary 9.74 hold. For $k_i \ge 0$, if $\tilde{u} \in \mathrm{C}^\infty(\Gamma, H_g^1(D))$ then,*

$$\left|\tilde{u} - P_\Lambda \tilde{u}\right|_W \le \sum_{i=1}^{M} \left|\frac{\partial \tilde{u}^{k_i+1}}{\partial y_i^{k_i+1}}\right|_W \frac{\sqrt{3}^{(k_i+1)}}{(k_i+1)!}.$$

Proof For functions $v \in L_p^2(\Gamma, H_g^1(D))$ that are sufficiently differentiable with respect to the y_i variables we have, for each $\boldsymbol{x} \in D$,

$$\left\|v(\boldsymbol{x}, \cdot) - P_{k_i} v(\boldsymbol{x}, \cdot)\right\|_{L_p^2(\Gamma)} \le \left\|\frac{\partial v^{k_i+1}(\boldsymbol{x}, \cdot)}{\partial y_i^{k_i+1}}\right\|_{L_p^2(\Gamma)} \frac{\sqrt{3}^{(k_i+1)}}{(k_i+1)!}. \tag{9.122}$$

Here, we used (9.121) noting that $v(\boldsymbol{x}, \cdot)$ is a function of $\boldsymbol{y} \in \Gamma$. We now need to relate the

$L^2_p(\Gamma)$ norm to $|\cdot|_W$. Let $v_s := \frac{\partial \tilde{u}}{\partial x_s}$, $s = 1,2$ and assume that $v_1(\boldsymbol{x},\cdot), v_2(\boldsymbol{x},\cdot) \in \mathrm{C}^\infty(\Gamma)$ a.e. in D. By definition,

$$\left|\tilde{u} - P_{k_i}\tilde{u}\right|_W^2 = \int_D \left\|v_1(\boldsymbol{x},\cdot) - P_{k_i} v_1(\boldsymbol{x},\cdot)\right\|^2_{L^2_p(\Gamma)} + \left\|v_2(\boldsymbol{x},\cdot) - P_{k_i} v_2(\boldsymbol{x},\cdot)\right\|^2_{L^2_p(\Gamma)} d\boldsymbol{x}.$$

Combining this with (9.122) and swapping the order of integration on the right-hand side gives

$$\left|\tilde{u} - P_{k_i}\tilde{u}\right|_W \le \left|\frac{\partial^{k_i+1}\tilde{u}}{\partial y_i^{k_i+1}}\right|_W \frac{\sqrt{3}^{(k_i+1)}}{(k_i+1)!}.$$

The result is now proved by applying Corollary 9.74. □

To draw out more information about the dependence of the error on the polynomial degrees k_i, we need to obtain bounds for the norms of the derivatives of $\tilde{u}$ with respect to the y_i variables. See Exercise 9.24 for technical details about this. Combining Exercise 9.24 with the above result, we now give a final bound for the projection error $|\tilde{u} - P_\Lambda \tilde{u}|_W$ associated with the tensor product space S^k in (9.117).

Theorem 9.76 (Babuška, Tempone, and Zouraris, 2004) *Let Assumption 9.42 and the conditions of Lemma 9.75 hold. Suppose (for simplicity) that $g = 0$ and let $\tilde{a}(\boldsymbol{x})$ and $\tilde{f}(\boldsymbol{x})$ be given by* (9.42)–(9.43). *If $\tilde{u}$ satisfies the strong form* (9.54) *then,*

$$\left|\tilde{u} - P_\Lambda \tilde{u}\right|_W \le \frac{\sqrt{3}K_{\mathrm{p}}}{\tilde{a}_{\min}} \sum_{i=1}^M \mathrm{e}^{-r_i k_i} \left(\sqrt{\nu_f^i} + \alpha_i \|\tilde{f}\|_{L^2_p(\Gamma, L^2(D))}\right),$$

where r_i and α_i are constants depending on the ith eigenvalue ν_i^a and eigenfunction ϕ_i^a of the covariance C_a of $a(\boldsymbol{x})$.

Proof Since $\tilde{u} \in \mathrm{C}^\infty(\Gamma, H^1_g(D))$ satisfies (9.54), combining the results of Exercise 9.24 gives, for all $\boldsymbol{y} \in \Gamma$,

$$\left|\frac{\partial^{k_i+1}\tilde{u}(\cdot,\boldsymbol{y})}{\partial y_i^{k_i+1}}\right|_{H^1(D)} \le (k_i+1)!\, \alpha_i^{k_i} \frac{K_{\mathrm{p}}}{\tilde{a}_{\min}} \left(\sqrt{\nu_f^i} + \alpha_i \|f(\cdot,\boldsymbol{y})\|_{L^2(D)}\right),$$

for each $i = 1,\dots,M$ and any $k_i \ge 0$ where

$$\alpha_i := \frac{\sqrt{\nu_a^i}\|\phi_a^i\|_{L^\infty(D)}}{\tilde{a}_{\min}} > 0.$$

Taking the $L^2_p(\Gamma)$ norm of both sides then gives

$$\frac{1}{(k_i+1)!}\left|\frac{\partial^{k_i+1}\tilde{u}}{\partial y_i^{k_i+1}}\right|_W \le \alpha_i^{k_i} \frac{K_{\mathrm{p}}}{\tilde{a}_{\min}} \left(\sqrt{\nu_f^i} + \alpha_i \|f\|_{L^2_p(\Gamma, L^2(D))}\right).$$

Hence, Lemma 9.75 gives

$$\left|\tilde{u} - P_\Lambda \tilde{u}\right|_W \le \frac{\sqrt{3}K_{\mathrm{p}}}{\tilde{a}_{\min}} \sum_{i=1}^M \left(\sqrt{3}\alpha_i\right)^{k_i} \left(\sqrt{\nu_f^i} + \alpha_i \|f\|_{L^2_p(\Gamma, L^2(D))}\right),$$

and the result follows since $\left(\sqrt{3}\alpha_i\right)^{k_i} = \mathrm{e}^{-k_i r_i}$ with $r_i := \log\left(\sqrt{3}\alpha_i\right)^{-1}$. □

Combining Theorem 9.76 and Theorem 9.48 with Lemma 9.71 now gives

$$\left|\mathbb{E}[\tilde{u}] - \mathbb{E}[\tilde{u}_{hk}]\right|_{H^1(D)} = \mathcal{O}(h) + \sum_{i=1}^{M} \mathcal{O}(\mathrm{e}^{-r_i k_i}).$$

Theorem 9.76 tells us that each of the one-dimensional polynomial errors decays exponentially fast to zero as the polynomial degree $k_i \to \infty$ (provided $r_i > 0$). In fact, the bound stated in Theorem 9.76 is not optimal. It can be improved to increase the rate k_i to $k_i + 1$ using more advanced techniques from approximation theory that are outside the scope of this book. See the Notes at the end of this chapter. Notice however that the constant r_i could be very small and depends on both $\tilde{a}_{\min}$ and C_a (and hence, on the correlation length ℓ). The bound also depends on the total number of random variables, M.

On the other hand, the MCFEM error in approximating $\mathbb{E}[\tilde{u}(\boldsymbol{x})]$ by the sample average $\mu_{Q,h} := \frac{1}{Q}\sum_{r=1}^{Q} \tilde{u}_h(\cdot, \boldsymbol{y}_r)$ satisfies

$$\left|\mathbb{E}[\tilde{u}] - \mu_{Q,h}\right|_{H^1(D)} \le \left|\tilde{u} - \mu_{Q,h}\right|_W \le \left|\tilde{u} - \tilde{u}_h\right|_W + \left|\tilde{u}_h - \mu_{Q,h}\right|_W,$$

and we see that the first contribution to the error is the same as for the SGFEM. Combining this with the analysis from §9.2 then gives

$$\left|\mathbb{E}[\tilde{u}] - \mu_{Q,h}\right|_{H^1(D)} = \mathcal{O}(h) + \mathcal{O}(Q^{-1/2}). \tag{9.123}$$

Although we did not use truncated Karhunen–Loève expansions to approximate the data in §9.2, this does not affect the analysis of the sampling error. In (9.123), Q is the number of samples $\tilde{a}(\cdot, \boldsymbol{y}_r)$, generated by sampling $\boldsymbol{y} \in \Gamma \subset \mathbb{R}^M$. Note that the bound does not depend on M.

We do not attempt a full cost comparison of the SGFEM and MCFEM here, but it is clear that for smooth covariances — for which the eigenvalues decay quickly and few Karhunen–Loève terms are required to control the data error — the SGFEM approximation to $\mathbb{E}[\tilde{u}(\boldsymbol{x})]$ is likely to converge quicker than the MCFEM approximation (recall Example 9.67).

9.6 Stochastic collocation FEM on $D \times \Gamma$

To end this chapter, we give a brief outline of a third popular family of stochastic FEMs for solving the BVP (9.54)–(9.55). We assume once again that $\tilde{a}(\boldsymbol{x})$ and $\tilde{f}(\boldsymbol{x})$ are truncated Karhunen–Loève expansions. The idea is to construct approximations by combining *collocation* on Γ with Galerkin finite element approximation on D. So-called stochastic collocation FEMs are like Monte Carlo FEMs in that we solve decoupled problems by *sampling* the random data. However, the sample points are not randomly chosen and the final approximation is constructed via interpolation.

Consider once again the strong form (9.54)–(9.55) and assume for simplicity that $g = 0$. Now select a set $\Theta := \{\boldsymbol{y}_1, \ldots, \boldsymbol{y}_Q\}$ of Q points in Γ and force the residual error to be zero at each one. That is, for $r = 1, 2, \ldots, Q$.

$$-\nabla \cdot \left(\tilde{a}\left(\boldsymbol{x}, \boldsymbol{y}_r\right) \nabla \tilde{u}(\boldsymbol{x}, \boldsymbol{y}_r)\right) - \tilde{f}(\boldsymbol{x}, \boldsymbol{y}_r) = 0, \qquad \boldsymbol{x} \in D, \tag{9.124}$$

$$\tilde{u}(\boldsymbol{x}, \boldsymbol{y}_r) = 0, \qquad \boldsymbol{x} \in \partial D. \tag{9.125}$$

If a solution to each of these decoupled problems exists, a continuous approximation on Γ can be constructed via

$$\tilde{u}_{\Theta}(\boldsymbol{x},\boldsymbol{y}) := \sum_{r=1}^{Q} \tilde{u}(\boldsymbol{x},\boldsymbol{y}_r) L_r(\boldsymbol{y}), \tag{9.126}$$

where $L_r : \Gamma \to \mathbb{R}$ now denotes an M-variate Lagrange polynomial satisfying

$$L_r(\boldsymbol{y}_s) = \delta_{rs}.$$

By construction, $\tilde{u}_{\Theta}(\boldsymbol{x},\boldsymbol{y}_r) = \tilde{u}(\boldsymbol{x},\boldsymbol{y}_r)$ for $r = 1,2,\ldots,Q$ and we may also write $\tilde{u}_{\Theta} := I_{\Theta}\tilde{u}$ where I_{Θ} is an *interpolation* operator. Each of the conditions (9.124)–(9.125) is a deterministic elliptic BVP on D. If we replace each one with a weak form and apply the Galerkin finite element method from Chapter 2, we arrive at the stochastic collocation finite element method (SCFEM).

Definition 9.77 (stochastic collocation finite element solution) Given a set of Q collocation points $\Theta := \{\boldsymbol{y}_1,\ldots,\boldsymbol{y}_Q\} \subset \Gamma \subset \mathbb{R}^M$ and a finite element space $V^h \subset H_0^1(D)$, the *stochastic collocation finite element solution* to the BVP (9.54)–(9.55) (with $g = 0$) is defined by

$$\tilde{u}_{h,\Theta}(\boldsymbol{x},\boldsymbol{y}) := \sum_{r=1}^{Q} \tilde{u}_h^r(\boldsymbol{x}) L_r(\boldsymbol{y}) \tag{9.127}$$

where each $\tilde{u}_h^r := \tilde{u}_h(\cdot,\boldsymbol{y}_r) \in W^h$ satisfies the variational problem

$$\int_D \tilde{a}_r(\boldsymbol{x})\nabla\tilde{u}_h^r(\boldsymbol{x}) \cdot \nabla v(\boldsymbol{x})\,d\boldsymbol{x} = \int_D \tilde{f}_r(\boldsymbol{x}) v(\boldsymbol{x})\,d\boldsymbol{x}, \quad \forall v \in V^h. \tag{9.128}$$

By construction (e.g., see (2.78) in §2.3), we have

$$\tilde{u}_{h,\Theta}(\boldsymbol{x},\boldsymbol{y}) = \sum_{r=1}^{Q} \tilde{u}_h^r(\boldsymbol{x}) L_r(\boldsymbol{y}) = \sum_{r=1}^{Q}\left(\sum_{i=1}^{J} u_{i,r}\phi_i(\boldsymbol{x})\right) L_r(\boldsymbol{y}) \in V^h \otimes P^{\Theta},$$

where we recall $V^h = \mathrm{span}\{\phi_1,\ldots,\phi_J\}$ and we now define

$$P^{\Theta} := \mathrm{span}\left\{L_r(\boldsymbol{y}) \colon \boldsymbol{y}_r \in \Theta\right\}.$$

P^{Θ} is a set of multivariate polynomials in $\boldsymbol{y}$ on Γ but (depending on the choice of Θ) may not be a standard set like complete or tensor product polynomials.

The mean of the SCFEM solution is given by

$$\mathbb{E}\left[\tilde{u}_{h,\Theta}(\boldsymbol{x})\right] := \int_{\Gamma} p(\boldsymbol{y})\tilde{u}_{h,\Theta}(\boldsymbol{x},\boldsymbol{y})\,d\boldsymbol{y} = \sum_{r=1}^{Q} \tilde{u}_h^r(\boldsymbol{x})\langle L_r, 1\rangle_{L_p^2(\Gamma)},$$

and similarly for the variance. To compute statistical information about the SCFEM solution, we therefore need to calculate integrals of Lagrange polynomials. Unfortunately, these polynomials are not orthogonal. Given an appropriate set of R quadrature weights w_s and quadrature points $\boldsymbol{y}_s$, we can always approximate the integrals. For example,

$$\langle L_r, 1\rangle_{L_p^2(\Gamma)} \approx \sum_{s=1}^{R} w_s L_r(\boldsymbol{y}_s), \qquad r = 1,\ldots,Q.$$

This is not as cumbersome as it sounds since it often happens — depending on the choice of Θ and the degree of the polynomial being integrated — that the previously selected interpolation points $\boldsymbol{y}_r, r = 1, \ldots, Q$ (together with appropriate weights) provide a quadrature scheme that is exact.

To implement the stochastic collocation FEM and obtain the coefficients of the finite element functions $\tilde{u}_h^r$ in (9.127), we have to solve Q linear systems of dimension $J \times J$ of the form (2.85). By exploiting (9.56), the Galerkin matrix $A \in \mathbb{R}^{(J+J_b)\times(J+J_b)}$ associated with $\boldsymbol{y}_r$ can be written as

$$A = K_0 + \sum_{\ell=1}^{P} y_{r,\ell} K_\ell,$$

where K_0 and K_ℓ are defined as in (9.93) and (9.94) and $y_{r,\ell}$ denotes the ℓth component of $\boldsymbol{y}_r$. Hence, each Galerkin matrix is a linear combination of the $P + 1$ finite element diffusion matrices that we encountered in the stochastic Galerkin FEM. We can use Algorithm 9.3 to generate these, as well as the components of the right-hand side vector $\boldsymbol{b}$.

Observe that (9.127) means that $\tilde{u}_{h,\Theta}$ is the interpolant of a semi-discrete weak solution $\tilde{u}_h : \Gamma \to V^h$ for which $\tilde{u}_h(\cdot, \boldsymbol{y})$ satisfies

$$\int_D \tilde{a}(\boldsymbol{x}, \boldsymbol{y}) \nabla \tilde{u}_h(\boldsymbol{x}, \boldsymbol{y}) \cdot \nabla v(\boldsymbol{x})\, d\boldsymbol{x} = \int_D \tilde{f}(\boldsymbol{x}, \boldsymbol{y}) v(\boldsymbol{x})\, d\boldsymbol{x}, \quad \forall v \in V^h, \forall \boldsymbol{y} \in \Gamma.$$

That is, $\tilde{u}_{h,\Theta}(\boldsymbol{x}, \boldsymbol{y}) = I_\Theta \tilde{u}_h(\boldsymbol{x}, \boldsymbol{y})$ where $I_\Theta : L^2(\Gamma) \to P^\Theta(\Gamma)$ satisfies

$$I_\Theta v(\boldsymbol{y}_r) = v(\boldsymbol{y}_r), \qquad \forall\, \boldsymbol{y}_r \in \Theta.$$

To perform error analysis, we can use the triangle inequality to obtain

$$\left|\tilde{u} - \tilde{u}_{h,\Theta}\right|_W \le \left|\tilde{u} - \tilde{u}_h\right|_W + \left|\tilde{u}_h - \tilde{u}_{h,\Theta}\right|_W =: \left|\tilde{u} - \tilde{u}_h\right|_W + \left|\tilde{u}_h - I_\Theta \tilde{u}_h\right|_W.$$

The first contribution to the error is the spatial error associated with the FEM discretisation. This is the same as for the MCFEM and SGFEM and is $\mathcal{O}(h)$ if, in particular, Assumption 9.47 holds. The second error is an interpolation error and must be analysed for a specific choice of the set Θ. In general, it depends on the number of random variables M (in a similar way as the SGFEM projection error in Theorem 9.76). The challenge lies in finding the *smallest* set Θ such that $|\tilde{u}_h - I_\Theta \tilde{u}_h|_W = \mathcal{O}(\epsilon)$ for a given tolerance, ϵ.

9.7 Notes

In many physical applications, input data is not truly random in nature. A commonly stated application of the elliptic BVP (9.1)–(9.2) is groundwater flow in a porous medium. In that scenario, the diffusion coefficient and source terms associated with a specific flow site and porous medium (rock, soil, etc.), are not random. The required input data is simply inaccessible and the decision to model it using correlated random fields reflects the inherent *uncertainty*. At the time of writing, random fields are increasingly being used to represent uncertain data with an underlying covariance (material coefficients, boundary conditions, source terms, and domain geometries) in a wide variety of PDE models.

Stochastic FEMs have been used in the engineering community (e.g., Ghanem and Spanos, 1991) for many years, to solve a variety of practical problems with uncertain parameters. Such methods are still a relatively new phenomenon in numerical analysis circles, however, and terminology is still evolving. Some of the terms we have used (e.g., Definition 9.38) are not standard. Readers should also be aware that the name stochastic FEM is given to *any* numerical method that relies on finite elements for the spatial discretisation.

In §9.1, we discussed well-posedness of deterministic weak formulations on D (corresponding to realisations of the data). We mentioned in the Notes for Chapter 2 that, for the deterministic BVP, the weak solution $u \in H^2(D)$ if $a \in \mathrm{C}^{0,1}(\bar{D})$ and D is a convex polygon. In this chapter, we introduced Assumption 9.47 as a generalisation of Assumption 2.13. If each $u(\cdot,\omega) \in H^2(D)$ and $|u(\cdot,\omega)|_{H^2(D)} \in L^4(\Omega)$ then Assumption 9.47 holds. Notice also that in Theorem 9.17, we work with diffusion coefficients with realisations in $\mathrm{C}^1(\bar{D})$ that do satisfy the aforementioned $\mathrm{C}^{0,1}(\bar{D})$ assumption.

In Chapter 4, we studied real-valued random variables and used the Berry–Esséen inequality to establish a confidence interval for the Monte Carlo estimate of the mean. Using the Monte Carlo FEM in §9.2, we investigated the sample mean of finite element solutions and encountered $H_0^1(D)$-valued random variables. For *iid* sequences of Hilbert space-valued random variables with finite third moments, the following analogue of (4.26) holds.

Theorem 9.78 (Yurinskiĭ, 1982) *Let H be a separable Hilbert space with norm $\|\cdot\|$ and X_i be iid H-valued random variables with mean zero and $\mathbb{E}\big[\|X_i\|^3\big] < \infty$. Denote the covariance operator of X_i by $\mathcal{C}$ (see Definition 4.35). There exists $c > 0$ such that for all $z \geq 0$ and $M = 1,2,\dots$*

$$\big|\mathbb{P}(\|X_M^*\| < z) - \mathbb{P}(\|Z\| < z)\big| \leq \frac{c}{\sqrt{M}},$$

where $X_M^* := (X_1 + \cdots + X_M)/\sqrt{M}$ and $Z \sim \mathrm{N}(0,\mathcal{C})$.

This result can be applied alongside Corollary 9.23 to find an $r > 0$ such that

$$\left|\mathbb{P}\left(E_{\mathrm{MC}} \leq \frac{r}{\sqrt{Q}}\right)\right| \geq 0.95 - \frac{c}{\sqrt{Q}}, \qquad Q = 1,2,\dots$$

where E_{MC} is the Monte Carlo error in (9.22). We can then deduce a confidence interval similar to (4.34), this time with respect to the Hilbert space norm.

In §9.2, we discussed only the *basic* Monte Carlo FEM. The cost, for a fixed Q and h, is given by (9.27). When Q is very large and h is very small (e.g., due to a small correlation length), the method requires significant computing resources. The systems can be solved in parallel (which can easily be done in Algorithm 9.2). However, new variants of the Monte Carlo method, including the multilevel Monte Carlo method (see Chapter 8) can significantly lower the computational cost. Recent studies of multilevel Monte Carlo FEMs for elliptic PDEs with random data can be found for example in (Barth, Schwab, and Zollinger, 2011; Cliffe et al., 2011) and quasi-Monte Carlo FEMs are investigated in Graham et al. (2011).

When $a(\boldsymbol{x})$ has a stationary covariance function of the form $c(\boldsymbol{x}/\ell)$, solutions of (9.1)–(9.2) are of special interest for small ℓ. The limiting solution can be identified with the solution of the *homogenised PDE*

$$\begin{aligned} -a_*\nabla\cdot\nabla\bar{u}(\boldsymbol{x}) &= f(\boldsymbol{x}), && \boldsymbol{x} \in D,\\ \bar{u}(\boldsymbol{x}) &= g(\boldsymbol{x}), && \boldsymbol{x} \in \partial D, \end{aligned}$$

where $a_* = 1/\mathbb{E}\big[a^{-1}\big]$ (Kozlov, 1979). The correction $u(\boldsymbol{x}) - \bar{u}(\boldsymbol{x})$ can be analysed in some cases. When $d = 1$ (Bal et al., 2008), the solution of

$$-\big(a(x/\ell)u_x\big)_x = 1, \qquad u(0) = u(1) = 0,$$

converges to a deterministic limit $\bar{u}(x)$ as $\ell \to 0$, for a large class of stationary processes $a(x)$ with rapidly decaying covariance functions. Furthermore, in distribution

$$\frac{u(x) - \bar{u}(x)}{\sqrt{\ell}} \to d(x) := \left(\int_{-\infty}^{\infty} R(y)\,dy\right)^{1/2} \int_0^1 K(x,y)\,dW(y),$$

where $R(x)$ is the stationary covariance of $a^{-1}(x)$, $W(y)$ is a Brownian motion, and

$$K(x,y) := \begin{cases} (x-1)(y-1/2), & 0 \le x \le y, \\ x(y-1/2), & y < x \le 1. \end{cases}$$

Notice that the variance is given by

$$\mathrm{Var}(d(x)) = cx(1-x)(8x^2 - 8x + 3), \qquad c := \frac{1}{12}\int_{-\infty}^{\infty} R(t)\,dt,$$

and has a minimum at $x = 1/2$. We observed similar behaviour for a two-dimensional example in Figure 9.7.

The eigenfunctions $\phi_k(x)$ of the Karhunen–Loève expansion in Example 9.1 are trigonometric functions and are uniformly bounded (in k and x). In general, we cannot expect eigenfunctions normalised in the $L^2(D)$ sense to have such behaviour. For example, consider the Haar basis $\{\phi_{jk} : k = 0,1,\dots,2^j - 1, j = 0,1,\dots\}$ for $L^2(0,1)$ given by $\phi_{jk}(x) = 2^{j/2}h(2^j x - k)$ for

$$h(x) := \begin{cases} 1, & 0 < x \le 1/2, \\ -1, & 1/2 < x \le 1, \\ 0, & \text{otherwise.} \end{cases}$$

We have $\|\phi_{jk}\|_{L^2(0,1)} = 1$ but $\|\phi_{jk}\|_\infty = 2^{j/2}$ and is unbounded. In special cases, we can control the growth in the supremum norm, as in Proposition 9.34 (where we assume smoothness of the covariance). See also Schwab and Todor (2006). Note that (9.47) is a special case of the Gagliardo–Nirenberg inequality.

In §9.5, we followed Golub and Welsch (1969) for the three-term recurrence (9.80). Standard results about orthogonal polynomials can be found in Gautschi (2004). We chose S^k to be a set of global polynomials on Γ. Piecewise polynomials are considered in (Deb, Babuška, and Oden, 2001; Elman et al., 2005a; Le Maître and Knio, 2010). We also focused on orthonormal bases but there is another possibility. Suppose we can find bases $\{P_0^i, \dots, P_k^i\}$ for each of the univariate spaces S_i^k in (9.72) such that

$$\big\langle P_r^i, P_s^i \big\rangle_{p_i} = \delta_{r,s} \quad \text{and} \quad \big\langle y_i P_r^i, P_s^i \big\rangle_{p_i} = c_{r,s}^i \delta_{r,s}, \qquad i = 1,\dots,M, \tag{9.129}$$

for $r, s = 0,1,\dots,k$. Now, given two polynomials of the form

$$\psi_i(\boldsymbol{y}) = \psi_{i(\boldsymbol{\alpha})}(\boldsymbol{y}) = \prod_{s=1}^M P_{\alpha_s}^s(y_s), \qquad \psi_j(\boldsymbol{y}) = \psi_{j(\boldsymbol{\beta})}(\boldsymbol{y}) = \prod_{s=1}^M P_{\beta_s}^s(y_s),$$

it is easy to show (follow Theorem 9.55) that $\langle \psi_i, \psi_j \rangle_p = \delta_{ij}$ and $\langle y_\ell \psi_i, \psi_j \rangle_p = c_{\alpha_\ell, \beta_\ell}^\ell \delta_{ij}$, for $\ell = 1, \ldots, M$. We say that the functions ψ_j are *doubly orthogonal* (Babuška et al., 2004). If we can find a doubly orthogonal basis for S^k, then $G_0, G_1, \ldots, G_M$ are *all* diagonal. The stochastic Galerkin matrix A is block-diagonal and solving the linear system reduces to solving Q decoupled systems of dimension J, with coefficient matrices

$$K_0 + \sum_{\ell=1}^{P} c^{\ell}_{\alpha_\ell,\alpha_\ell} K_\ell, \qquad j = 1, \ldots, Q \text{ with } j = j(\boldsymbol{\alpha}).$$

This is the same matrix that would be obtained if we applied the MCFEM and chose the specific sample $\boldsymbol{\xi} = [\xi_1, \ldots, \xi_P]^{\mathsf{T}} = [c^1_{\alpha_1,\alpha_1}, \ldots, c^P_{\alpha_P,\alpha_P}]^{\mathsf{T}}$ in (9.42). If S^k is the tensor product polynomial space (9.73), we can construct a doubly orthogonal basis of polynomials ψ_j. Details about how to compute the coefficients $c^{\ell}_{\alpha_\ell,\beta_\ell}$ can be found in Ernst and Ullmann (2010). Unfortunately, there is no such basis when S^k is given by (9.76) (again, see Ernst and Ullmann, 2010). For a fixed k, using (9.73) instead of (9.76) does not usually result in improved accuracy. Since the dimension Q may be substantially smaller for complete polynomials, working with standard orthonormal bases is often more efficient.

We have not provided code for a full implementation of the SGFEM. However, we have provided Algorithm 9.3 to compute the FEM components of the Galerkin system and an efficient iterative solver is developed in Exercises 9.22 and 9.23. It remains to write code to generate the matrices $G_1, \ldots, G_M$. If an orthonormal basis is selected for S^k, then, depending on the reader's specific choice of S^k and random variables in (9.42)–(9.43), these may be computed using Theorem 9.59. The necessary ingredients are a set of multi-indices Λ and the tridiagonal matrix (matrices) in (9.84) associated with the chosen family (families) of univariate orthogonal polynomials.

In §9.5, our discussion of the SGFEM system matrix centred on the assumption that $\tilde{a}(\boldsymbol{x})$ is given by (9.42). That is, $\tilde{a}(\boldsymbol{x})$ is a *linear* function of the random variables ξ_i. In this case, the Galerkin matrix A is block sparse. For further discussion on the properties of A and solvers, see (Pellissetti and Ghanem, 2000; Powell and Elman, 2009). Unfortunately, whenever the weak form leads to $[G_\ell]_{ij} = \langle \psi_\ell \psi_j, \psi_k \rangle_p$, the Galerkin system matrix is *block dense*. This can arise if we approximate $a(\boldsymbol{x})$ using an expansion of the form $\tilde{a}(\boldsymbol{x}, \boldsymbol{y}) = \sum_{j=1}^{N} a_j(\boldsymbol{x}) \psi_j(\boldsymbol{y})$, where the ψ_j belong to the same family of polynomials used to construct the basis for S^k, or if the underlying PDE is nonlinear.

In (9.90) and (9.91), we explained how to compute the mean and variance of the stochastic Galerkin solution. Since we have a functional representation of $\tilde{u}_{hk}$, a wealth of other statistics may be approximated. (Although we do not quantify the errors here). Consider $\mathbb{P}\left(\{\omega \in \Omega \colon |\tilde{u}(\boldsymbol{x}_*, \omega)| > \epsilon\}\right)$, where $\boldsymbol{x}_* \in D$ and ϵ is a tolerance. The probability that $|\tilde{u}(\boldsymbol{x}_*, \cdot)|$ exceeds the tolerance at $\boldsymbol{x} = \boldsymbol{x}_*$ could correspond to an extreme event or a design criterion. Writing the random field $\tilde{u}(\boldsymbol{x})$ a function of $\boldsymbol{y} \in \Gamma$ and approximating it by $\tilde{u}_{hk}$ gives

$$|\tilde{u}_{hk}(\boldsymbol{x}_*, \boldsymbol{y})| = \left| \sum_{i=1}^{J} \sum_{j=1}^{Q} u_{ij} \phi_i(\boldsymbol{x}_*) \psi_j(\boldsymbol{y}) + w_g(\boldsymbol{x}_*) \right|. \tag{9.130}$$

If $\boldsymbol{x}_*$ is one of the interior mesh vertices, say $\boldsymbol{x}_* = \boldsymbol{x}_k$. Then, since $\phi_j(\boldsymbol{x}_k) = \delta_{j,k}$,

$$|\tilde{u}_{hk}(\boldsymbol{x}_*, \boldsymbol{y})| = \left| \sum_{j=1}^{Q} u_{ik} \psi_j(\boldsymbol{y}) \right|.$$

By sampling $\boldsymbol{y} \in \Gamma$, we can now estimate the distribution of $|\tilde{u}(\boldsymbol{x}_*, \cdot)|$.

We referred to (9.89) as a *polynomial chaos* expansion. Originally (see Wiener, 1938) this term was used to describe a series expansion in terms of Hermite polynomials H_j in Gaussian random variables. More generally, any second-order random variable $\xi \in L^2(\Omega)$ (and hence second-order random field on $D \times \Omega$) may be expanded using polynomials P_j in ξ that are orthogonal with respect to the underlying density function. For convergence analysis, see Cameron and Martin (1947). Some authors still reserve *polynomial chaos* for the special case of Hermite polynomials and Gaussian random variables and refer to expansions associated with other distributions as *generalised* polynomial chaos. Note that even after transferring the weak problem from $D\times\Omega$ to $D\times\Gamma$, so that the chosen polynomials are functions of real-valued parameters y_i, we still often call the expansion a chaos. For a discussion of the Askey scheme of hypergeometric polynomials — which contains many families of polynomials that are orthogonal with respect to weight functions coinciding with probability densities of standard distributions — see Xiu and Karniadakis (2002).

For the analysis of the SGFEM error (with tensor product polynomials), we followed Babuška et al. (2004) to obtain Theorem 9.76. The analysis used to establish that result requires only standard properties of orthogonal projections and one-dimensional polynomial approximation theory. Even though the result can be slightly improved (in terms of the rate of convergence with respect to the polynomial degrees k_i) we included only Theorem 9.76, as the proof is more accessible for the intended audience of this book. The solution to the model elliptic problem is *extremely smooth* in $\mathbf{y} \in \Gamma$. In fact, it is *analytic* and this is key for the error analysis. To improve the bound, we must extend the solution $\tilde{u}$ (as a function of each y_i) to an analytic function in the complex plane and then use techniques from complex analysis (see Babuška et al., 2004 for full details). In contrast, the white-noise driven SDEs studied in Chapter 8 feature solutions with low regularity.

Lemma 9.72 provides a bound for the projection error $|\tilde{u} - P_\Lambda \tilde{u}|_W$ that can be analysed for *any* choice of polynomial space S^k. However, the characterisation of the error in terms of Fourier coefficients tells us that the *best* polynomial space of a fixed dimension Q is the one that yields the smallest $|\tilde{u}_{j(\alpha)}|_{H^1(D)}$, $j = 1, \ldots, Q$. Identifying a polynomial space of the lowest possible dimension that yields an acceptable error then amounts to finding a particular index set Λ. Recently, this observation has been exploited to derive adaptive algorithms for constructing lower-dimensional polynomial spaces than the standard ones we have considered. See (Bieri, Andreev, and Schwab, 2010; Gittelson, 2013; Schwab and Gittelson, 2011). Error analysis for stochastic Galerkin *mixed* finite element approximations of elliptic PDEs with random data (using standard tensor product polynomials) is provided in Bespalov, Powell, and Silvester (2012).

In §9.6, we briefly outlined stochastic collocation FEMs but did not comment on the choice of Θ. So-called *full-tensor* SCFEMs (Babuška, Nobile, and Tempone, 2007; Xiu and Hesthaven, 2005) use Cartesian products of M sets of interpolation points on the one-dimensional intervals Γ_i. Possibilities include Clenshaw–Curtis (CC) points (Clenshaw and Curtis, 1960) and Gauss points. If $d_i + 1$ points are selected on Γ_i, the total number of interpolation points is $Q = \prod_{i=1}^{M}(d_i + 1)$ and this quickly becomes intractable as $M \to \infty$. An interesting observation is that applying the stochastic Galerkin FEM with the tensor product polynomial space S^k in (9.73) and choosing doubly orthogonal basis functions ψ_j is equivalent to applying a full-tensor SCFEM. More sophisticated *sparse grid* SCFEMs (see Nobile, Tempone, and Webster, 2008a,b; Xiu and Hesthaven, 2005) are based on interpolation and cubature rules for high-dimensional problems (see Barthelmann, Novak, and Ritter, 2000; Bungartz and Griebel, 2004; Novak and Ritter, 1996) and are derived from the work of Smolyak (1963). The challenge lies in choosing as few sample points as possible in the M-dimensional hypercube Γ to minimise the interpolation error. Finally, it should be noted that SCFEMs (unlike SGFEMs) always lead to sparse Galerkin matrices, irrespective of the type of expansion used to approximate $a(\boldsymbol{x})$, and, from that point of view, are often more flexible. Solvers for SCFEM linear systems are discussed in Gordon and Powell (2012) and references therein.

Exercises

9.1 Write a MATLAB routine to generate a realisation of the stochastic process in (9.4) and use Algorithm 2.1 to generate a piecewise linear finite element approximation to the corresponding realisation of the solution to the BVP in (9.3).

9.2 Write a MATLAB routine to generate a pair of realisations of the isotropic Gaussian random field in Example 9.2 (hint: use Algorithm 7.6) and then use Algorithm 2.7 to generate finite element approximations to the corresponding realisations of the solution to the BVP.

9.3 Let $z(\boldsymbol{x})$ be a mean-zero Gaussian random field with the Whittle–Matérn covariance $c_q^0(r)$. Show that (9.10) holds if $s \le 2q$ and $s \in (0,2)$ and hence that realisations of $z(\boldsymbol{x})$ are almost surely Hölder continuous with any exponent $\gamma < \min\{q,1\}$.

9.4 Prove Lemma 9.11.

9.5 Prove Lemma 9.13.

9.6 Let $c \in \mathrm{C}^2(\mathbb{R}^2)$ be a stationary covariance function and consider a random field $z(\boldsymbol{x})$ on $\bar{D} = [0,1] \times [0,1]$ with mean zero and covariance $c(\boldsymbol{x}/\ell)$, for a correlation length ℓ. Show that

$$\|z\|_{L^2(\Omega, H_0^1(D))} = \left(-\frac{1}{\ell^2}\sum_{i=1}^{2}\frac{\partial^2 c(\boldsymbol{0})}{\partial x_i^2}\right)^{1/2}.$$

Discuss the error bounds in Corollary 9.17 as $\ell \to 0$.

9.7 Consider the BVP (9.5)–(9.6) where a is a deterministic function of $\boldsymbol{x} \in D$ and $f(\boldsymbol{x})$ is a random field. By choosing test functions $v \in H_0^1(D)$, derive a deterministic weak problem on D involving $\mathbb{E}[f(\boldsymbol{x})]$ that can be solved for a weak approximation to $\mathbb{E}[u(\boldsymbol{x})]$.

9.8 Let $v(x) := \lim_{P\to\infty} \mathrm{Var}(a(x))$ where $a(x)$ is defined by (9.21). Assuming that $v(x)$ is a polynomial of degree 4, determine $v(x)$ and sketch the graph to show that it has a minimum at $x = 1/2$.

9.9 Let $u(\boldsymbol{x})$ be the solution of (9.1)–(9.2) with $g = 0$, $f = 1$ and $D = (0,1) \times (0,1)$. Consider $a(\boldsymbol{x}) = \mathrm{e}^{z(\boldsymbol{x})}$ where $z(\boldsymbol{x})$ is a mean-zero Gaussian random field with isotropic covariance $c^0(r) = \mathrm{e}^{-r^2/\ell^2}$. Define

$$\theta(\ell) := \int_D \mathrm{Var}(u(\boldsymbol{x}))\, d\boldsymbol{x}.$$

Using Algorithm 9.2, approximate $\theta(\ell)$ and give a log-log plot of the approximation against ℓ. Do you observe a power law relationship between the estimate of $\theta(\ell)$ and ℓ, as $\ell \to 0$?

9.10 Let $z(\boldsymbol{x})$ be a Gaussian random field with mean-zero and isotropic covariance $c^0(r) = \mathrm{e}^{-r^2/\ell^2}$. Consider the BVP in Exercise 2.20 with mixed, non-homogeneous boundary conditions. Modify Algorithm 9.2 and implement the MCFEM to estimate the mean and variance of the solution when $f = 0$, $a(\boldsymbol{x}) = \mathrm{e}^{z(\boldsymbol{x})}$ and $\ell = 0.1$.

9.11 Let $a(\boldsymbol{x},\cdot)$ and $f(\boldsymbol{x},\cdot)$ be $\mathcal{F}$-measurable and consider the variational problem: find $u \in V := L^2(\Omega, \mathcal{F}, H_0^1(D))$ satisfying

$$\mathbb{E}\left[\int_D a(\boldsymbol{x})\nabla u(\boldsymbol{x}) \cdot \nabla v(\boldsymbol{x})\, d\boldsymbol{x}\right] = \mathbb{E}\left[\int_D f(\boldsymbol{x}) v(\boldsymbol{x})\, d\boldsymbol{x}\right], \qquad \forall v \in V.$$

Let $\mathcal{G} \subset \mathcal{F}$ be a sub σ-algebra and assume $a(\boldsymbol{x},\cdot)$ and $f(\boldsymbol{x},\cdot)$ are also $\mathcal{G}$-measurable. Consider the alternative variational problem: find $u^* \in V^* := L^2(\Omega, \mathcal{G}, H_0^1(D))$ satisfying

$$\mathbb{E}\left[\int_D a(\boldsymbol{x})\nabla u^*(\boldsymbol{x}) \cdot \nabla v^*(\boldsymbol{x})\, d\boldsymbol{x}\right] = \mathbb{E}\left[\int_D f(\boldsymbol{x}) v^*(\boldsymbol{x})\, d\boldsymbol{x}\right], \qquad \forall v^* \in V^*.$$

Show that $u = u^*$ and hence that $u(\boldsymbol{x},\cdot)$ is $\mathcal{G}$-measurable.

9.12 Prove Theorem 9.28. Hint: follow the proof of Theorem 2.43.

9.13 Prove Theorem 9.31. Hint: follow the proof of Theorem 2.46.

9.14 Derive a condition on the mean μ_a and the correlation length ℓ that guarantees that the random field $\tilde{a}(x)$ in Example 9.36 has positive realisations, for any choice of truncation parameter P.

9.15 Consider the random field $a(\boldsymbol{x})$ in Example 9.37.

a. Write a MATLAB routine `twoDeigs.m` which can be called via:

```
>>[nu2D, phi2D]=twoDeigs(P,ell,x,y)
```

to compute the P largest eigenvalues ν_k and evaluate the corresponding eigenfunctions ϕ_k in (9.50). The code should accept as inputs, the number of terms `P`, the correlation length `ell` and vectors `x` and `y` of coordinates at which to evaluate the eigenfunctions. The output `nu2D` should be a vector containing the eigenvalues and `phi2D` should be a matrix whose columns contain the sampled eigenfunctions.

b. Generate a uniform finite element mesh of triangular elements of width $h = 1/128$ on $\bar{D} = [0,1] \times [0,1]$ using Algorithm 2.4 and write a MATLAB routine to generate realisations of the truncated random field $\tilde{a}(\boldsymbol{x})$ using the mesh vertices as sample points.

9.16 Investigate the dimension Q of the tensor product polynomial space (9.73) and the complete polynomial space (9.76) for $M = 1,5$ and 10 variables. Increase the polynomial degree k and compare Q.

9.17 Prove that if we normalise the orthogonal polynomials P_j in (9.80) then $\frac{c_j}{a_j} = \frac{1}{a_{j-1}}$ and (9.81) holds.

9.18 Working with $\langle \cdot,\cdot \rangle_p$ in (9.79) and normalised polynomials P_j, show that if $\Gamma \subset \mathbb{R}$ is symmetric about zero and p is even, then $b_j = 0$.

9.19 Consider the rescaled Legendre polynomials L_j in Example 9.54. Show that the eigenvalues of the tridiagonal matrix T_{j+1} associated with the polynomial L_{j+1} lie in the interval $\left[-\sqrt{3}, \sqrt{3}\right]$.

9.20 Consider $\xi \sim \mathrm{N}(0,1)$. Identify the family of polynomials $P_j, i = 0,1,\ldots,$ that are orthonormal with respect to the density p of ξ. As in Exercise 9.19, derive the tridiagonal matrix T_{j+1} associated with P_{j+1} and find an upper bound for the largest eigenvalue.

9.21 Suppose f is deterministic and the boundary data $g = 0$. Simplify the right-hand side vector $\boldsymbol{b}$ in (9.102).

9.22 Let $\boldsymbol{v} \in \mathbb{R}^{JQ}$ and use an orthonormal basis for S^k.

a. Using (9.92), investigate how to compute $A\boldsymbol{v}$ for the Galerkin matrix A in (9.104) given only $K_0, K_\ell, \ell = 1,\ldots,P$ and the non-zero entries of the G_ℓ matrices. Can the operation be parallelised?

b. Denote consecutive blocks of $\boldsymbol{v}$ of length J by $\boldsymbol{v}_1,\ldots,\boldsymbol{v}_Q$ and define

$$V := \begin{bmatrix} \boldsymbol{v}_1 & \boldsymbol{v}_2 & \cdots & \boldsymbol{v}_Q \end{bmatrix} \in \mathbb{R}^{J\times Q}.$$

Suppose $B_1 \in \mathbb{R}^{Q\times Q}$ and $B_2 \in \mathbb{R}^{J\times J}$. Given that $(B_1 \otimes B_2)\mathrm{vec}(V) = \mathrm{vec}(B_2 V B_1^{\mathsf{T}})$, where $\mathrm{vec}(\cdot)$ is the matrix operator that stacks consecutive columns, write a MATLAB routine that accepts $\boldsymbol{v}, K_0, K_\ell$ and G_ℓ, for $\ell = 1,\ldots,P$ as inputs and returns $A\boldsymbol{v}$.

c. Investigate implementing the conjugate gradient method using the MATLAB routine `pcg.m` to solve the Galerkin system in Example 9.67 without assembling A.

9.23 Consider the bilinear form $\tilde{a}(\cdot,\cdot)$ in (9.52) and assume that μ_a is constant. Define a new bilinear form $\tilde{a}_0(\cdot,\cdot)\colon V\times V \to \mathbb{R}$, as follows:

$$\tilde{a}_0(u,v) := \int_\Gamma p(\boldsymbol{y}) \int_D \mu_a \,\nabla u(\boldsymbol{x},\boldsymbol{y}) \cdot \nabla v(\boldsymbol{x},\boldsymbol{y})\, d\boldsymbol{x}\, d\boldsymbol{y}.$$

Find constants α, β such that

$$\alpha\, \tilde{a}_0(v,v) \le \tilde{a}(v,v) \le \beta\, \tilde{a}_0(v,v), \qquad \forall v \in V.$$

Determine whether the matrix $I \otimes K_0$ is an effective preconditioner for A in (9.104) when $\xi_k \sim \mathrm{U}(-\sqrt{3},\sqrt{3})$ *iid*, $i = 1,\ldots,M$ and an orthonormal basis of Legendre polynomials is used for S^k.

9.24 Suppose $\tilde{u}$ satisfies (9.54), $g = 0$ and Assumption 9.42 holds. In addition, let $\tilde{a}(\boldsymbol{x})$ and $\tilde{f}(\boldsymbol{x})$ be given by (9.42)–(9.43).

a. For all $\boldsymbol{y} \in \Gamma$ and for $i = 1,\ldots,M$, show that

$$\left\| \frac{\partial \tilde{u}(\cdot,\boldsymbol{y})}{\partial y_i} \right\|_{H^1(D)} \le \frac{K_{\mathrm{p}}}{\tilde{a}_{\min}} \Big(\sqrt{\nu_f^i} + \alpha_i \,\|f(\cdot,\boldsymbol{y})\|_{L^2(D)} \Big),$$

where $\alpha_i := \frac{1}{\tilde{a}_{\min}} \sqrt{\nu_a^i}\,\|\phi_a^i\|_{L^\infty(D)}$.

b. For all $\boldsymbol{y} \in \Gamma$ and for $i = 1,\ldots,M$, show that for any $k_i \ge 1$,

$$\left\| \frac{\partial^{k_i+1} \tilde{u}(\cdot,\boldsymbol{y})}{\partial y_i^{k_i+1}} \right\|_{H^1(D)} \le (k_i+1)!\, \alpha_i^{k_i} \left\| \frac{\partial \tilde{u}(\cdot,\boldsymbol{y})}{\partial y_i} \right\|_{H^1(D)}.$$

10
Semilinear Stochastic PDEs

Stochastic partial differential equations (SPDEs) are PDEs that include random fluctuations, which occur in nature and are missing from deterministic PDE descriptions. As a simple example, add a random term $\zeta(t,\boldsymbol{x})$ to the deterministic heat equation to get

$$u_t = \Delta u + \zeta(t,\boldsymbol{x}), \qquad t > 0, \quad \boldsymbol{x} \in D,$$

as we did for stochastic ODEs in (8.1) except this time ζ depends on both space and time. We choose $\zeta = dW/dt$, where $W(t,\boldsymbol{x})$ is the Wiener process introduced in §10.2, and study SPDEs as Itô integral equations. For example, we consider the stochastic heat equation

$$du = \Delta u\, dt + dW(t,\boldsymbol{x}).$$

More generally, this chapter addresses semilinear stochastic PDEs such as the stochastic reaction–diffusion equation

$$du = \left[\Delta u + f(u)\right] dt + G(u)\, dW(t,\boldsymbol{x}).$$

Like the SODEs in Chapter 8, we distinguish between noise that is *additive* (G is independent of u) and *multiplicative* ($G(u)$ depends on u). The Itô integral is described in §10.3 and the existence and regularity of solutions are considered in §10.4.

In Chapter 3, we wrote semilinear PDEs as ODEs on a Hilbert space. We do the same here and consider SPDEs as semilinear SODEs on a Hilbert space H. For example, we study the stochastic semilinear evolution equation

$$du = \left[-Au + f(u)\right] dt + G(u)\, dW(t) \tag{10.1}$$

where $-A$ is a linear operator that generates a semigroup $S(t) = \mathrm{e}^{-tA}$ (see Definition 3.10). To understand (10.1), we use semigroups extensively and it is useful to review §3.2 before reading further. We again suppress the dependence on space and write $u(t)$ for $u(t,\boldsymbol{x})$ and $W(t)$ for $W(t,\boldsymbol{x})$. The solution $u(t,\boldsymbol{x})$ is a family of real-valued random variables and can be viewed as a random field. We prefer to view $u(t)$ as an H-valued stochastic process (recall Definition 5.2), where H contains functions of $\boldsymbol{x}$ in D. We talk about *sample paths* of u to mean $u(\cdot,\cdot,\omega)$ for a fixed sample ω, which varies in both space and time. For a fixed time t, however, $u(t,\cdot,\cdot)$ is a random field and we talk about *realisations* $u(t,\cdot,\omega)$ (recall Definition 7.2).

§10.5–§10.8 discuss the discretisation of (10.1) by finite difference and Galerkin methods, alongside MATLAB implementations. Theorem 10.34 is the main convergence result and applies to a class of Galerkin methods with semi-implicit Euler time stepping.

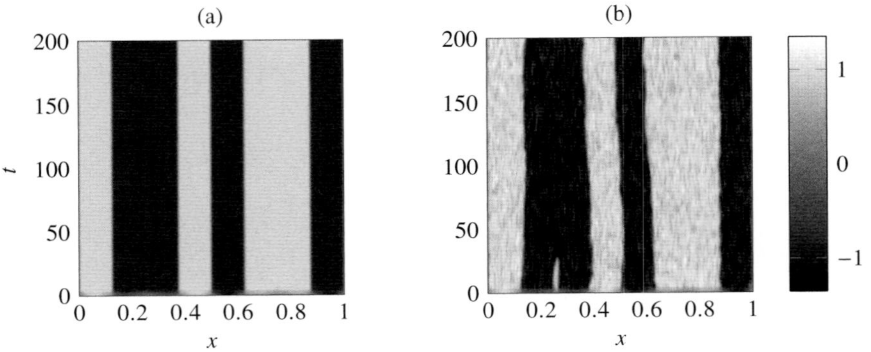

Figure 10.1 (a) A solution of the deterministic Allen–Cahn equation (3.3) with $\varepsilon = 10^{-5}$. (b) One sample path of the solution to the stochastic Allen–Cahn equation (10.2) with $\sigma = 0.02$, and the same ε and initial data as in (a). The kinks now move.

10.1 Examples of semilinear SPDEs

We present some SPDEs that arise in applications. Examples 10.1–10.3 look at models from material science, mathematical biology, and fluid dynamics, respectively. The noise term is included to capture phenomena not present in the corresponding deterministic model. In the final example, Example 10.4, the SPDE arises from a filtering problem where the aim is to estimate the state of a system from noisy observations.

Example 10.1 (phase-field models) A phase-field model describes the interface between two phases of a material, say a solid and liquid phase. A variable u distinguishes between the two phases and, for example, we might have $u \geq 1$ indicating a liquid phase, $u \leq -1$ a solid phase, and $u \in (-1, 1)$ a mix of the two phases. Equations for the space and time evolution of $u(t, \boldsymbol{x})$ may be derived from the minimization of a free energy functional. Phase-field models were originally used to investigate quenching of a binary alloy and the microstructures that arise. Noise plays a significant role and experimental results show the formation of small scale structures that result from thermal noise and are not present in the deterministic model. The simplest such phase-field model is the Allen–Cahn equation, which we first saw in its deterministic form in (3.3). We now consider the equation with external fluctuations:

$$du = \left[\varepsilon \, \Delta u + u - u^3\right] dt + \sigma \, dW(t), \tag{10.2}$$

where $\varepsilon > 0$ controls the rate of diffusion and σ controls the strength of the noise. The SPDE (10.2) has additive noise. For the deterministic case ($\sigma = 0$), there are two spatially homogeneous solutions $u = \pm 1$ that are fixed points. Transitions between the states $u = \pm 1$ are called kinks. Figure 10.1(a) shows an example solution and we see regions of space where $u \approx 1$ and $u \approx -1$ and kinks at transitions between these states.

In the stochastic case ($\sigma \neq 0$), the additive noise changes the properties of the solutions. For example, there are no longer any classical fixed points. In Figure 10.1(b), we plot one sample path of the solution to (10.2) with $\sigma = 0.02$ (with space–time white noise; see Definition 10.14). We keep the same initial data for both deterministic and stochastic simulations and show the solutions side by side in Figure 10.1. The kinks are no longer static and can interact and even annihilate each other. Similarly, new kinks may arise, induced by

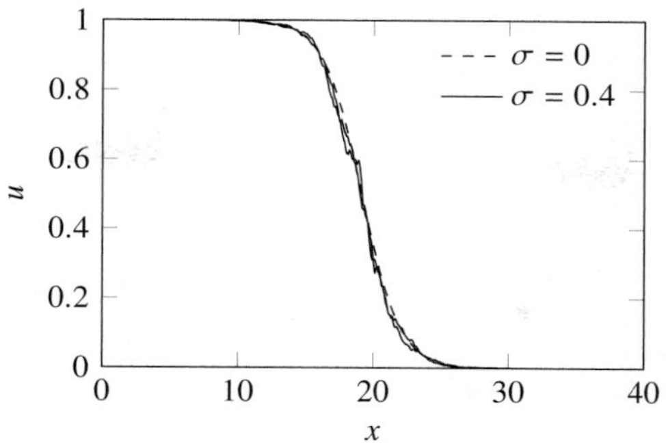

Figure 10.2 For $\alpha = -1/2$, $\varepsilon = 1$, and homogeneous Neumann boundary conditions on $[0, 40]$, one solution to the deterministic ($\sigma = 0$) and two realisations of the solution to the stochastic Nagumo equation (10.3) (with $W(t)$ as in Example 10.30 with $\sigma = 0.4$ and $\ell = 0.05$) at time $t = 10$ computed with the same initial data. The travelling waves move from left to right and both stochastic waves are behind the deterministic one.

the noise. In two spatial dimensions, stochastic forcing also changes the spatio-temporal structure of the solution. Compare, for example, the deterministic case in Figure 3.5(b) with Figure 10.11.

Example 10.2 (mathematical biology) The Nagumo equation is given by

$$u_t = \varepsilon u_{xx} + u(1 - u)(u - \tilde{\alpha}),$$

and we consider initial data $u(0, x) = u_0(x)$ on $D = (0, a)$ with homogeneous Neumann boundary conditions. This is a reduced model for wave propagation of the voltage u in the axon of a neuron. There are two model parameters $\tilde{\alpha} \in \mathbb{R}$ and $\varepsilon > 0$. The parameter $\tilde{\alpha}$ determines the speed of a wave travelling down the length of the axon and ε controls the rate of diffusion. If the parameter $\tilde{\alpha}$ varies both in space and time in a random way, we obtain an SPDE with multiplicative noise (also called internal noise). In particular, for $\tilde{\alpha} = \alpha + \sigma dW/dt$ and $\sigma > 0$, we find

$$du = \left[\varepsilon u_{xx} + u(1 - u)(u - \alpha)\right] dt + \sigma u(u - 1)\, dW(t). \tag{10.3}$$

Unlike the additive noise in the stochastic Allen–Cahn equation (10.2), this perturbation preserves the homogeneous fixed points $u = 0$ and $u = 1$ of the Nagumo equation (as the noise has no effect at $u = 0, 1$). As a consequence, if the initial data $u_0(x) \in [0, 1]$ for all $x \in (0, a)$, the solution $u(t, x) \in [0, 1]$ for all $x \in (0, a)$ and $t > 0$.

Figure 10.2 shows both a solution of the PDE and two realisations of the solution of the SPDE at $t = 10$ (using the Itô interpretation and the $W(t)$ given in Example 10.30). In each case, there is a travelling wave between $u = 0$ and $u = 1$ that travels from the left to the right. In the realisations shown, the stochastic waves are behind the deterministic one, though this varies from realisation to realisation. The behaviour depends on the interpretation of the stochastic integral: in this case, the stochastic waves travel near to the deterministic wave speed; if a Stratonovich interpretation is used, a system shift in wave speed results. We also observe that $u(t, x)$ is no longer monotonically decreasing.

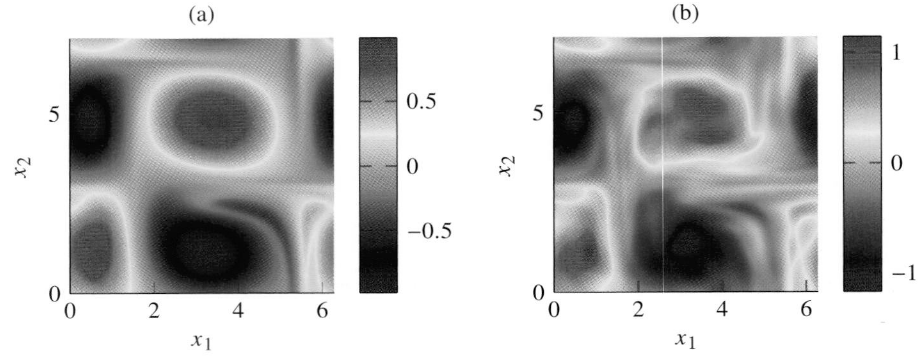

Figure 10.3 (a) Solution $u(t,\boldsymbol{x})$ at $t = 10$ of the deterministic ($\sigma = 0$) vorticity equation (10.5) and in (b) a realisation of a solution to the stochastic equation (10.6) with $\sigma = \sqrt{\varepsilon}$. The same initial data is used in both (a) and (b) and $\varepsilon = 0.001$

Example 10.3 (fluid flow) Semilinear SPDEs often arise in models of turbulent fluid flow and there is a great deal of interest in the traditional equations of fluid dynamics, such as the Navier–Stokes, Burgers, and Euler equations with stochastic forcing. The Navier–Stokes equations for the velocity $\boldsymbol{v}$ of an incompressible two-dimensional flow on a domain $D \subset \mathbb{R}^2$ are given by

$$\boldsymbol{v}_t = \varepsilon\,\Delta\boldsymbol{v} - \nabla p - (\boldsymbol{v}\cdot\nabla)\boldsymbol{v}, \qquad \nabla\cdot\boldsymbol{v} = 0, \tag{10.4}$$

where p is the pressure and $\varepsilon = 1/\mathrm{Re}$ for the Reynolds number Re. In two dimensions, the vorticity $u := \nabla\times\boldsymbol{v}$ satisfies the PDE

$$u_t = \varepsilon\,\Delta u - (\boldsymbol{v}\cdot\nabla)u, \tag{10.5}$$

where $\boldsymbol{v} = (\psi_y, -\psi_x)$ and the scalar stream function $\psi(t,\boldsymbol{x})$ satisfies $\Delta\psi = -u$. We consider the PDE for the vorticity with additive noise:

$$du = \Big[\varepsilon\,\Delta u - (\boldsymbol{v}\cdot\nabla)u\Big]\,dt + \sigma\,dW(t). \tag{10.6}$$

This is a model for large-scale flows, for example, in climate modelling or in the evolution of the red spot on Jupiter. The additive noise is a random forcing term that captures small scale perturbations. Consider $\varepsilon = 10^{-3}$, initial data $u_0(\boldsymbol{x}) = \sin(2\pi x_1/a_1)\cos(2\pi x_2/a_2)$, and periodic boundary conditions on $D = (0,a_1)\times(0,a_2)$ with $a_1 = 2\pi$ and $a_2 = 9\pi/4$. In Figure 10.3(a), we show the solution $u(t,\boldsymbol{x})$ of (10.5) at $t = 10$, where there is no noise in the system. In Figure 10.3(b), we plot one realisation of the solution of (10.6) with $\sigma = \sqrt{\varepsilon}$ also at $t = 10$ (with $W(t)$ defined in Example 10.12 with $\alpha = 0.05$).

Example 10.4 (filtering and sampling) Classic applications of filtering are estimating the state of the stock market or tracking missiles from noisy observations of part of the system. These types of problem also give rise to semilinear SPDEs. We consider a simple example. Suppose we have a signal $Y(x)$, $x \geq 0$, that satisfies the SODE

$$dY = f(Y(x))\,dx + \sqrt{\sigma}\,d\beta_1(x), \qquad Y(0) = 0, \tag{10.7}$$

where $f : \mathbb{R} \to \mathbb{R}$ is a given forcing term, $\beta_1(x)$ is a Brownian motion, and σ describes the strength of the noise. Rather than observing $Y(x)$ directly, we have noisy observations $Z(x)$ of the signal $Y(x)$. We assume these observations satisfy the SODE

$$dZ = Y(x)\,dx + \sqrt{\gamma}\,d\beta_2(x), \qquad Z(0) = 0, \tag{10.8}$$

where $\beta_2(x)$ is also a Brownian motion (independent of β_1) and γ determines the strength of the noise in the observation. If $\gamma = 0$, we observe the signal exactly. The goal is to estimate the signal $Y(x)$ given observations $Z(x)$ for $x \in [0, b]$. This can be achieved by finding the conditional distribution $\tilde{u}$ of the signal Y given the observations Z (see Example 4.51 for conditional distribution). Under some assumptions, it can be shown that $\tilde{u}$ is the stationary distribution (see §6.1) of the following SPDE:

$$du = \left[\frac{1}{\sigma}\left(u_{xx} - f(u)f'(u) - \frac{\sigma}{2}f''(u)\right)\right] dt + \frac{1}{\gamma}\left[\frac{dY}{dx} - u\right] dt + \sqrt{2}\,dW(t) \tag{10.9}$$

where $t > 0$, $x \in [0, b]$, and $W(t)$ is space–time white noise (see Definition 10.14). Since $Y(x)$ is only Hölder continuous with exponent less than $1/2$, the derivative $\frac{dY}{dx}$ and the SPDE (10.9) require careful interpretation. The SPDE is supplemented with the following boundary conditions at $x = 0$ and $x = b$:

$$\frac{\partial u}{\partial x}(t, 0) = 0, \qquad \frac{\partial u}{\partial x}(t, b) = f(u(t, b)),$$

and initial data $u(0, x) = u_0(x)$. We are interested in long-time solutions (i.e., large t) in order to approximate the stationary distribution and hence the conditional distribution $\tilde{u}$.

As a concrete example, consider the case $f(Y) = Y$ and $\sigma = 1$ in the signal equation (10.7) and $\gamma = 1/16$ in the observation equation (10.8). Then (10.9) gives the following SPDE:

$$du = \left[u_{xx} - u + 16\frac{dY}{dx} - 16u\right] dt + \sqrt{2}\,dW(t). \tag{10.10}$$

In Figure 10.4, we plot in (a) a sample signal and in (b) the corresponding observation. We use the SPDE (10.10) to reconstruct Y from the observations. In Figure 10.4(c), we show one realisation of the solution $u(t, x)$ of (10.10) at $t = 100$ and the similarity between (c) and (a) is apparent.

10.2 Q-Wiener process

In Definition 5.11, we introduced Brownian motion and, in Chapter 8, we studied ODEs forced by the derivative of Brownian motion. Our aim is to construct a generalisation of Brownian motion suitable for stochastically forcing the heat equation. That is, we want to introduce a spatial variable to the Brownian motion $W(t)$. One possibility is the Brownian sheet $\{\beta(t, x) : (t, x) \in \mathbb{R}^+ \times D\}$ (see Example 7.6) and we could study the stochastic heat equation

$$du = u_{xx}\,dt + d\beta(t, x), \qquad x \in D \subset \mathbb{R}.$$

We prefer, however, following §3.3, to treat the stochastic heat equation as a semilinear equation in $L^2(D)$ and work with Gaussian processes taking values in $L^2(D)$ (rather than the random field $\beta(t, x)$). To this end, we introduce the Q-Wiener process.

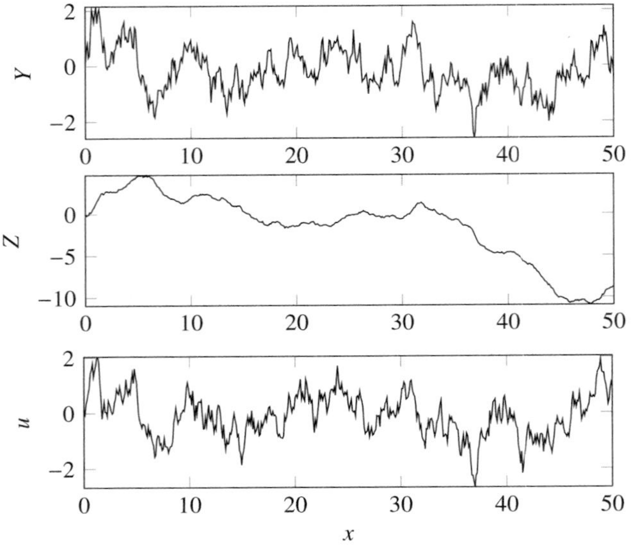

Figure 10.4 The upper plot shows the signal $Y(x)$ found from (10.7), the middle plot shows the observation $Z(x)$ from (10.8), and the lower plot is a reconstruction of the signal. The reconstructed signal is an approximate realisation of the solution $u(t, x)$ of the SPDE (10.10) at $t = 100$, which approximately samples the conditional distribution $\tilde{u}$.

In place of $L^2(D)$, we develop the theory on a separable Hilbert space U with norm $\|\cdot\|_U$ and inner product $\langle\cdot,\cdot\rangle_U$ and define the Q-Wiener process $\{W(t): t \geq 0\}$ as a U-valued process. Each $W(t)$ is a U-valued Gaussian random variable and each has a well-defined covariance operator (see Definition 4.35). The covariance operator at $t = 1$ is denoted Q and, by Proposition 4.37, it must satisfy the following assumption.

Assumption 10.5 $Q \in \mathcal{L}(U)$ is non-negative definite and symmetric. Further, Q has an orthonormal basis $\{\chi_j : j \in \mathbb{N}\}$ of eigenfunctions with corresponding eigenvalues $q_j \geq 0$ such that $\sum_{j\in\mathbb{N}} q_j < \infty$ (i.e., Q is of trace class).

Let $(\Omega, \mathcal{F}, \mathcal{F}_t, \mathbb{P})$ be a filtered probability space (see Definition 8.5). Recall the definition of $\mathcal{F}_t$-adapted (Definition 8.6) and of the Gaussian distribution $\mathrm{N}(0, Q)$ on a Hilbert space (Corollary 4.41). We define the Q-Wiener process. The name *Wiener process* distinguishes it from the real-valued Brownian motion.

Definition 10.6 (Q-Wiener process) A U-valued stochastic process $\{W(t): t \geq 0\}$ is a *Q-Wiener process* if

(i) $W(0) = 0$ a.s.,
(ii) $W(t)$ is a continuous function $\mathbb{R}^+ \to U$, for each $\omega \in \Omega$,
(iii) $W(t)$ is $\mathcal{F}_t$-adapted and $W(t) - W(s)$ is independent of $\mathcal{F}_s$ for $s < t$, and
(iv) $W(t) - W(s) \sim \mathrm{N}(0, (t-s)Q)$ for all $0 \leq s \leq t$.

In analogy to the Karhunen–Loève expansion, the Q-Wiener process can be represented as a linear combination of the eigenfunctions χ_j of Q. Instead of the uncorrelated random variables in the Karhunen–Loève expansion, we employ *iid* Brownian motions $\beta_j(t)$.

Theorem 10.7 *Let Q satisfy Assumption 10.5. Then, $W(t)$ is a Q-Wiener process if and only if*

$$W(t) = \sum_{j=1}^{\infty} \sqrt{q_j}\, \chi_j\, \beta_j(t), \qquad a.s., \tag{10.11}$$

where $\beta_j(t)$ are iid $\mathcal{F}_t$-Brownian motions and the series converges in $L^2(\Omega, U)$. Moreover, (10.11) converges in $L^2(\Omega, \mathrm{C}([0,T], U))$ for any $T > 0$.

Proof Let $W(t)$ be a Q-Wiener process and suppose without loss of generality that $q_j > 0$ for all j. Since $\{\chi_j : j \in \mathbb{N}\}$ is an orthonormal basis for U,

$$W(t) = \sum_{j=1}^{\infty} \langle W(t), \chi_j \rangle_U\, \chi_j.$$

Let $\beta_j(t) := \frac{1}{\sqrt{q_j}} \langle W(t), \chi_j \rangle_U$, so that (10.11) holds. Clearly then, $\beta_j(0) = 0$ a.s. and $\beta_j(t)$ is $\mathcal{F}_t$-adapted and has continuous sample paths. The increment

$$\beta_j(t) - \beta_j(s) = \frac{1}{\sqrt{q_j}} \langle W(t) - W(s), \chi_j \rangle_U, \qquad 0 \le s \le t,$$

is independent of $\mathcal{F}_s$. As $W(t) - W(s) \sim \mathrm{N}(0, (t-s)Q)$, Definition 4.35 gives

$$\begin{aligned}\operatorname{Cov}\big(\beta_j(t) - \beta_j(s), \beta_k(t) - \beta_k(s)\big) &= \frac{1}{\sqrt{q_j q_k}} \mathbb{E}\Big[\langle W(t) - W(s), \chi_j \rangle_U \langle W(t) - W(s), \chi_k \rangle_U\Big] \\ &= \frac{1}{\sqrt{q_j q_k}} \langle (t-s) Q \chi_j, \chi_k \rangle_U = (t-s)\, \delta_{jk}.\end{aligned}$$

Then, $\beta_j(t) - \beta_j(s) \sim \mathrm{N}(0, t-s)$ and $\beta_j(t)$ is a $\mathcal{F}_t$-Brownian motion. Any pair of increments forms a multivariate Gaussian and hence β_j and β_k are independent for $j \ne k$.

To show $W(t)$ as defined by (10.11) is a Q-Wiener process, we first show the series converges in $L^2(\Omega, U)$ for any fixed $t \ge 0$. Consider the finite sum approximation

$$W^J(t) := \sum_{j=1}^{J} \sqrt{q_j} \chi_j \beta_j(t) \tag{10.12}$$

and the difference $W^J(t) - W^M(t)$, for $M < J$. By the orthonormality of the eigenfunctions χ_j, we have, using Parseval's identity (1.43), that

$$\|W^J(t) - W^M(t)\|_U^2 = \sum_{j=M+1}^{J} q_j \beta_j(t)^2. \tag{10.13}$$

Each $\beta_j(t)$ is a Brownian motion and taking the expectation gives

$$\mathbb{E}\Big[\|W^J(t) - W^M(t)\|_U^2\Big] = \sum_{j=M+1}^{J} q_j\, \mathbb{E}\big[\beta_j(t)^2\big] = t \sum_{j=M+1}^{J} q_j.$$

As Q is trace class, $\sum_{j=1}^{\infty} q_j < \infty$ and the right-hand side converges to zero as $M, J \to \infty$. Hence the series (10.11) is well defined in $L^2(\Omega, U)$.

From (10.12), it is easily seen that $W^J(t)$ is a mean-zero $\mathcal{F}_t$-adapted Gaussian process with independent increments and these properties also hold in the limit $J \to \infty$. Similarly, from (10.11),

$$\operatorname{Cov}\Big(\big\langle W(t), \chi_j \big\rangle_U, \big\langle W(t), \chi_k \big\rangle_U\Big) = t\, q_j\, \delta_{jk}.$$

Hence $W(t) \sim \mathrm{N}(0, t\,Q)$. This observation extends in a natural way to the increments and we see $W(t)$ satisfies Definition 10.6.

To show the series converges in $L^2(\Omega, \mathrm{C}([0,T], U))$, note that (10.13) gives

$$\mathbb{E}\bigg[\sup_{0 \le t \le T} \|W^J(t) - W^M(t)\|_U^2\bigg] = \mathbb{E}\bigg[\sup_{0 \le t \le T} \sum_{j=M+1}^{J} q_j\, \beta_j(t)^2\bigg] \le \sum_{j=M+1}^{J} q_j\, \mathbb{E}\bigg[\sup_{0 \le t \le T} \beta_j(t)^2\bigg].$$

Let $C(T) := \mathbb{E}\big[\sup_{0 \le t \le T} \beta_j(t)^2\big]$, which is independent of j and finite by the Doob maximal inequality (Corollary A.12). As Q is trace class,

$$\mathbb{E}\bigg[\sup_{0 \le t \le T} \|W^J(t) - W^M(t)\|_U^2\bigg] \le C(T) \sum_{j=M+1}^{J} q_j \to 0 \qquad \text{as } M, J \to \infty.$$

Thus, $W \in L^2(\Omega, \mathrm{C}([0,T], U))$ and its sample paths belong to $\mathrm{C}([0,T], U)$ almost surely. We may therefore choose a version (see Definition 5.39) of $W(t)$ that is continuous and hence satisfies Definition 10.6. □

Consider $U = L^2(D)$ for a domain D. We seek a Q-Wiener process $W(t)$ taking values in $H^r(D)$ for a given $r \ge 0$. This can be achieved by choosing the eigenfunctions χ_j and eigenvalues q_j of the covariance operator with an appropriate rate of decay; see also Example 6.10.

Example 10.8 ($H^r_{\mathrm{per}}(0,a)$-valued process) Recall that $H^r_{\mathrm{per}}(0,a)$ is equivalent to the fractional power space $\mathcal{D}(A^{r/2})$ where $Au = u - u_{xx}$ and $\mathcal{D}(A) = H^2_{\mathrm{per}}(0,a)$ (Proposition 1.93). Denote the eigenvalues of A by λ_j and the normalised eigenfunctions by ϕ_j for $j \in \mathbb{N}$ as given by (1.28) (in real-valued notation). Recall the norm $\|u\|_r = \big(\sum_{j=1}^{\infty} \lambda_j^{2r} u_j^2\big)^{1/2}$ for coefficients $u_j = \langle u, \phi_j \rangle_{L^2(0,a)}$.

For $U = L^2(0,a)$ and a given $r \ge 0$, we define a trace class operator $Q \in \mathcal{L}(U)$ and hence a Q-Wiener process $W(t)$, so that $W(t)$ takes values in $H^r_{\mathrm{per}}(0,a) \subset U$. To do this, choose $\chi_j = \phi_j$ and an appropriate rate of decay for q_j. Exercise 1.25 suggests $q_j = \mathcal{O}\big(j^{-(2r+1+\epsilon)}\big)$ for an $\epsilon > 0$. In order that the eigenfunctions $\sin(\ell x/a)$ and $\cos(\ell x/a)$ receive the same eigenvalue (and hence that $\mathrm{e}^{2\pi \mathrm{i} \ell x/a}$ is a complex-valued eigenfunction of Q with eigenvalue $q_{2|\ell|+1}$), we choose

$$q_j = \begin{cases} \ell^{-(2r+1+\epsilon)}, & j = 2\ell + 1 \text{ or } j = 2\ell \text{ for } \ell \in \mathbb{N}, \\ 0, & j = 1. \end{cases}$$

Then, for this choice of q_j,

$$\mathbb{E}\left[\|W(t)\|_{r/2}^2\right] = \sum_{j=1}^{\infty} \lambda_j^r \, q_j \, \mathbb{E}\left[\beta_j(t)^2\right] < \infty,$$

as $\lambda_j = \mathcal{O}(j^2)$ and $\lambda_j^r q_j = \mathcal{O}(j^{2r} j^{-(2r+1+\epsilon)}) = \mathcal{O}(j^{-(1+\epsilon)})$. The resulting Q-Wiener process $W(t)$ therefore takes values in $H_{\text{per}}^r(0,a)$ a.s.

Example 10.9 ($H_0^r(0,a)$-valued process) Recall that the Sobolev space $H_0^r(0,a)$ equals the fractional power space $\mathcal{D}(A^{r/2})$ where $Au = -u_{xx}$ and $\mathcal{D}(A) = H^2(0,a) \cap H_0^1(0,a)$; see Example 1.90. Similarly to Example 10.8, the Q-Wiener process $W(t)$ defined by (10.11) with $\chi_j(x) = \sqrt{2/a}\sin(j\pi x)$ and $q_j = |j|^{-(2r+1+\epsilon)}$ leads to a $H_0^r(0,a)$-valued process.

Approximating sample paths of a Q-Wiener process

Consider an $L^2(D)$-valued Q-Wiener process $W(t)$. Numerical approximation of the sample paths of $W(t)$ is straightforward when the eigenfunctions of Q are known. First, we introduce a finite set of sample points $\boldsymbol{x}_1, \ldots, \boldsymbol{x}_K \in D$ and seek to numerically generate samples of $W(t, \boldsymbol{x}_1), \ldots, W(t, \boldsymbol{x}_K)$ for $t > 0$. This is achieved by approximating $W(t)$ by the finite sum $W^J(t)$ in (10.12) and sampling the random variables $W^J(t, \boldsymbol{x}_k)$. In Examples 10.8 and 10.9, we gave examples of $L^2(D)$-valued Q-Wiener processes where the eigenfunctions of Q are known. Conveniently, the eigenfunctions are trigonometric functions that, with careful choice of sample points $\boldsymbol{x}_k$, allow evaluation of $W^J(t, \boldsymbol{x}_k)$ for $k = 1, \ldots, K$ with a single Fourier transform. This provides an efficient method for generating approximate samples of $W(t)$.

As in (8.64), we want to sample the same sample path $W(t)$ with different time steps. To do this, we introduce a reference time step $\Delta t_{\text{ref}} = T/N_{\text{ref}}$ and construct a sample path based on the time step Δt_{ref} when computing with $\Delta t = \kappa \Delta t_{\text{ref}}$ for $\kappa \in \mathbb{N}$. To keep the same sample path, we use the increments $W^J(t_{n+1}) - W^J(t_n)$ with $t_n = n\Delta t_{\text{ref}}$, computed from (10.12) by

$$W^J(t_{n+1}) - W^J(t_n) = \sqrt{\Delta t_{\text{ref}}} \sum_{j=1}^{J} \sqrt{q_j} \chi_j \, \xi_j^n, \tag{10.14}$$

where $\xi_j^n := (\beta_j(t_{n+1}) - \beta_j(t_n))/\sqrt{\Delta t_{\text{ref}}}$. Here $\xi_j^n \sim \mathrm{N}(0,1)$ *iid* and are easily sampled. Then, increments over intervals of size $\Delta t = \kappa \Delta t_{\text{ref}}$ are computed by

$$W^J(t + \Delta t) - W^J(t) = \sum_{n=0}^{\kappa-1} \left(W^J(t + t_{n+1}) - W^J(t + t_n)\right). \tag{10.15}$$

Example 10.10 ($H_0^r(0,a)$-valued process) For the process $W(t)$ of Example 10.9, we employ the discrete sine transform to compute approximate sample paths. Choose the $K = J - 1$ sample points $x_k = ka/J$ for $k = 1, \ldots, J-1$ and take (10.14) with $\chi_j(x) = \sqrt{2/a}\sin(\pi j x/a)$ and $J - 1$ terms. Then,

$$W^{J-1}(t_{n+1}, x_k) - W^{J-1}(t_n, x_k) = \sum_{j=1}^{J-1} b_j \sin\left(\frac{\pi j k}{J}\right) \xi_j^n, \tag{10.16}$$

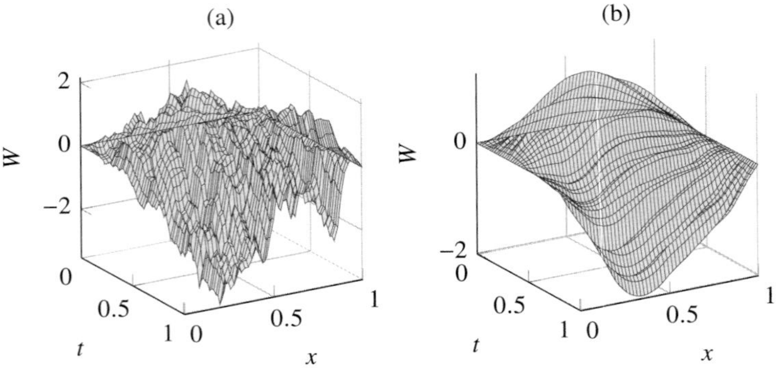

Figure 10.5 Approximate sample paths of the Q-Wiener process defined in Example 10.10 generated with $J = 128$ and $\Delta t_{\mathrm{ref}} = 0.01$ for (a) $r = 1/2$ and (b) $r = 2$. In each case $W(t,0) = W(t,1) = 0$.

Algorithm 10.1 Code to form the coefficients b_j in Example 10.10. Inputs are `dtref`$= \Delta t_{\mathrm{ref}}$, `J`$= J$, the domain size `a`, and regularity parameter `r`$= r$. The output is a vector `bj` of coefficients b_j, $j = 1,\dots,J-1$. Here we fix $\epsilon = 0.01$ in the definition of q_j using `myeps`.

```
function bj = get_onedD_bj(dtref,J,a,r)
jj = [1:J-1]'; myeps=0.001;
root_qj=jj.^-((2*r+1+myeps)/2); % set decay for H^r
bj=root_qj*sqrt(2*dtref/a);
```

where $b_j := \sqrt{2q_j \Delta t_{\mathrm{ref}}/a}$. Here q_j and hence b_j depend on the regularity parameter r. The vector of increments is the discrete sine transform (DST-1; see Exercise 1.27) of the vector $[b_1\xi_1^n,\dots,b_{J-1}\xi_{J-1}^n]^{\mathsf{T}}$. Algorithm 10.1 computes the coefficients b_j and Algorithm 10.2 samples ξ_j^n and applies DST-1 to sample the vector of increments. For maximum efficiency in using the FFT, J should be a power of 2. In the cases $r = 1/2$ and $r = 2$, we plot a sample path of $W^J(t)$ in Figure 10.5 for the domain width $a = 1$. These were generated with $J = 128$ and $\Delta t_{\mathrm{ref}} = 0.01$ using the following MATLAB commands:

```
>> dtref=0.01; kappa=100; r=1/2; J=128; a=1;
>> bj=get_onedD_bj(dtref,J,a,r);
>> dW=get_onedD_dW(bj,kappa,0,1);
```

Example 10.11 ($H^r_{\mathrm{per}}(0,a)$-valued Wiener process) To approximate sample paths of the Q-Wiener process defined in Example 10.8, we require the complex form of (10.14): let $\tilde{q}_\ell := q_{2|\ell|+1}$ for $\ell \in \mathbb{Z}$, so that $Q\mathrm{e}^{2\pi \mathrm{i}\ell x/a} = \tilde{q}_\ell \mathrm{e}^{2\pi \mathrm{i}\ell x/a}$. For J even, consider the truncated expansion

$$W^J(t_{n+1},x) - W^J(t_n,x) = \sqrt{\Delta t_{\mathrm{ref}}} \sum_{\ell=-J/2+1}^{J/2} \sqrt{\tilde{q}_\ell}\, \frac{\mathrm{e}^{2\pi \mathrm{i}\ell x/a}}{\sqrt{a}}\, \xi_\ell^n, \tag{10.17}$$

where $\xi_0^n, \xi_{J/2}^n \sim \mathrm{N}(0,1)$, $\xi_\ell^n \equiv \bar{\xi}_{-\ell}^n \sim \mathrm{CN}(0,1)$ for $\ell = 1,\dots,J/2-1$, and $\xi_0^n,\dots,\xi_{J/2}^n$ are pairwise independent. In this case, we choose sample points $x_k = a(k-1)/J$ for

Algorithm 10.2 Code to sample $W^{J-1}(t+\kappa\Delta t_{\mathrm{ref}}, x_k) - W^{J-1}(t, x_k)$ by (10.16). Inputs are the coefficients `bj` from Algorithm 10.1; `kappa` $= \kappa$, a flag `iFspace`; and the number `M` of independent realisations to compute. If `iFspace=1`, the output `dW` is a matrix of M columns with kth entry $W^{J-1}(t+\kappa\Delta t_{\mathrm{ref}}, x_k) - W^{J-1}(t, x_k)$ for $k = 1, \ldots, J-1$. If `iFspace=1` then the columns of `dW` are the inverse DST-1 of those for `iFspace=0`. `dst1` is developed in Exercise 1.27.

```
function dW=get_onedD_dW(bj,kappa,iFspace,M)
if(kappa==1) % generate xi_j
  nn=randn(length(bj),M);
else % sum over kappa steps
  nn=squeeze(sum(randn(length(bj),M,kappa),3));
end
X=bsxfun(@times,bj,nn);
if(iFspace==1) % return b_j xi_j
  dW=X;
else % return Wiener increments
  dW=dst1(X);
end
```

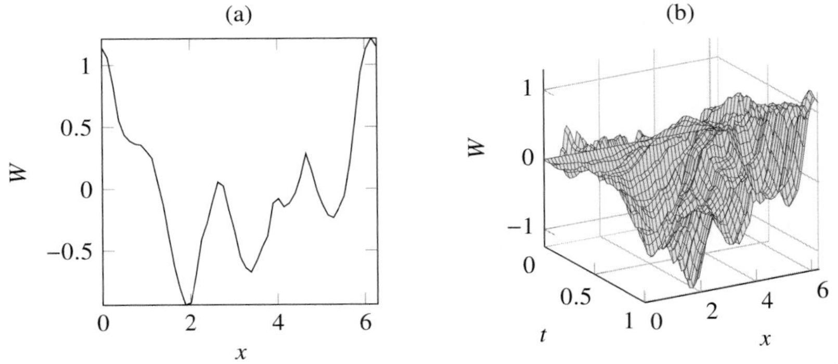

Figure 10.6 For the Q-Wiener process in Example 10.11, (a) a realisation of $W^J(1)$ computed with $\Delta t = 1$ and $\Delta t_{\mathrm{ref}} = 0.01$ and (b) a `mesh` plot of the same sample path $W^J(t_n)$ at $t_n = n\Delta t_{\mathrm{ref}}$. We see that $W^J(1)$ in (b) agrees with $W^J(1)$ in (a).

$k = 1, \ldots, J$ and (10.17) gives

$$W^J(t_{n+1}, x_k) - W^J(t_n, x_k) = \frac{\sqrt{\Delta t_{\mathrm{ref}}}}{\sqrt{a}}\left(\sum_{\ell=0}^{J/2} \sqrt{\tilde{q}_\ell}\, \mathrm{e}^{2\pi \mathrm{i}\ell x_k/a} \xi_\ell^n + \sum_{\ell=J/2+1}^{J-1} \sqrt{\tilde{q}_{\ell-J}} \mathrm{e}^{2\pi \mathrm{i}\ell x_k/a} \xi_{\ell-J}^n\right)$$

$$= \frac{1}{J}\left(\sum_{j=0}^{J/2} b_j\, \mathrm{e}^{\mathrm{i}jk/J}\, \xi_j^n + \sum_{j=J/2+1}^{J-1} b_j\, \mathrm{e}^{\mathrm{i}jk/J}\, \bar{\xi}_{J-j}^n\right), \tag{10.18}$$

where $b_j \equiv b_{J-j} := J\sqrt{\Delta t_{\mathrm{ref}} \tilde{q}_j / a}$, for $j = 0, \ldots, J/2$. This is the inverse discrete Fourier transform of $\left[b_0\xi_0^n, \ldots, b_{J/2}\xi_{J/2}^n, b_{J/2+1}\bar{\xi}_{J/2-1}^n, \ldots, b_{J-1}\bar{\xi}_1^n\right]^{\mathsf{T}}$. Algorithms 10.3 and 10.4 can be used to sample the increments and generate sample paths of $W^J(t)$. For $r = 1$ and $a = 1$, we show one sample path of $W^J(t)$ for $J = 100$ and $\Delta t_{\mathrm{ref}} = 0.01$ in Figure 10.6.

Algorithm 10.3 Code to form the coefficients b_j in (10.18). Inputs and output are as in Algorithm 10.1.

```
function bj = get_oned_bj(dtref,J,a,r)
jj=[1:J/2, -J/2+1:-1]'; myeps=0.001;
root_qj=[0; abs(jj).^-((2*r+1+myeps)/2)]; % set decay rate for H^r
bj=root_qj*sqrt(dtref/a)*J;
```

Algorithm 10.4 Code to sample $W^J(t+\kappa\Delta t_{\text{ref}}, x_k) - W^J(t, x_k)$ for $k = 1,\dots,J$ in (10.18) for $k = 1,\dots,J$. Inputs and outputs are similar to those in Algorithm 10.2.

```
function dW=get_oned_dW(bj,kappa,iFspace,M)
J=length(bj);
if(kappa==1) % generate xi_j
  nn=randn(J,M);
else % sum over kappa steps
  nn=squeeze(sum(randn(J,M,kappa),3));
end
nn2=[nn(1,:);(nn(2:J/2,:)+sqrt(-1)*nn(J/2+2:J,:))/sqrt(2);...
     nn(J/2+1,:);(nn(J/2:-1:2,:)-sqrt(-1)*nn(J:-1:J/2+2,:))/sqrt(2)];
X= bsxfun(@times,bj,nn2);
if(iFspace==1) % return b_j xi_j
  dW=X;
else % return Wiener increments
  dW=real(ifft(X));
end
```

Example 10.12 (Q-Wiener process in two dimensions) Let $D = (0,a_1)\times(0,a_2)$ and $U = L^2(D)$. Consider $Q \in \mathcal{L}(U)$ with eigenfunctions $\chi_{j_1,j_2}(\boldsymbol{x}) = \frac{1}{\sqrt{a_1 a_2}} e^{2\pi i j_1 x_1/a_1} e^{2\pi i j_2 x_2/a_2}$ and eigenvalues $q_{j_1,j_2} = e^{-\alpha\lambda_{j_1,j_2}}$, for a parameter $\alpha > 0$ and $\lambda_{j_1,j_2} = j_1^2 + j_2^2$. For even integers J_1, J_2, let

$$W^J(t,\boldsymbol{x}) := \sum_{j_1=-J_1/2+1}^{J_1/2} \sum_{j_2=-J_2/2+1}^{J_2/2} \sqrt{q_{j_1,j_2}}\, \chi_{j_1,j_2}(\boldsymbol{x})\, \beta_{j_1,j_2}(t),$$

for *iid* Brownian motions $\beta_{j_1,j_2}(t)$. Let $\boldsymbol{x}_{k_1,k_2} = [a_1(k_1-1)/J_1, a_2(k_2-1)/J_2]^{\mathsf{T}}$ for $k_i = 1,\dots,J_i$. We seek to generate two independent copies of $W^J(t,\boldsymbol{x}_{k_1,k_2})$ using a single FFT. The correct generalisation of (10.17) is

$$\Delta Z^J(t_n,\boldsymbol{x}) = \sqrt{\Delta t_{\text{ref}}} \sum_{j_1=-J_1/2+1}^{J_1/2} \sum_{j_2=-J_2/2+1}^{J_2/2} \sqrt{q_{j_1,j_2}} \frac{e^{2\pi i(j_1 x_1/a_1 + j_2 x_2/a_2)}}{\sqrt{a_1 a_2}} \xi^n_{j_1,j_2}, \tag{10.19}$$

where $\xi^n_{j_1,j_2} \sim \mathrm{CN}(0,2)$ *iid*. The increments $\Delta Z^J(t_n,\boldsymbol{x}_{k_1,k_2})$ are complex with independent real and imaginary parts with the same distribution as the increments $W^J(t_{n+1},\boldsymbol{x}_{k_1,k_2}) - W^J(t_n,\boldsymbol{x}_{k_1,k_2})$ (see Lemma 6.14). As before, (10.19) is easily evaluated at the sample points $\boldsymbol{x}_{k_1,k_2}$ by applying the two-dimensional discrete Fourier transform to the $J_1 \times J_2$ matrix with entries $b_{j_1 j_2}\xi^n_{j_1,j_2}$ for $b_{j_1,j_2} := \sqrt{\Delta t_{\text{ref}} q_{j_1,j_2}/a_1 a_2}\, J_1 J_2$. See Algorithms 10.5 and 10.6.

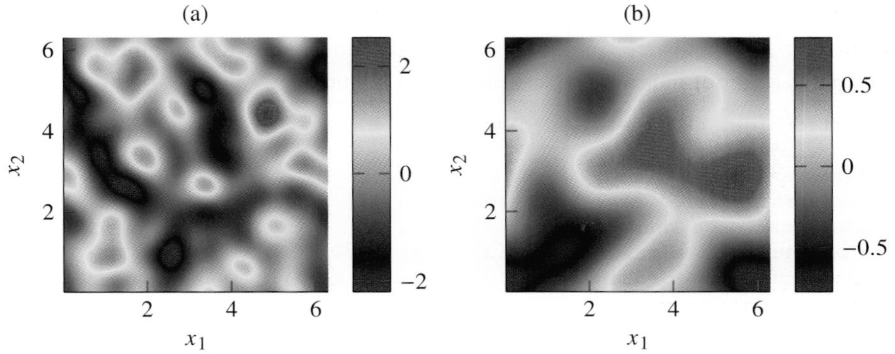

Figure 10.7 Realisations of the Q-Wiener process $W^J(t)$ defined in Example 10.12 at $t = 1$ for (a) $\alpha = 0.05$ and (b) $\alpha = 0.5$, computed with $J_1 = J_2 = 512$. Both processes take values in $H^r((0, 2\pi) \times (0, 2\pi))$ for any $r \geq 0$.

Figure 10.7(a) shows a realisation of $W^J(1)$ in the case $\alpha = 0.05$ and $a_1 = a_2 = 2\pi$, generated with $\Delta t_{\text{ref}} = 0.01$, $J_1 = J_2 = 512$ and the following MATLAB commands.

```
>> J=[512,512]; dtref=0.01; kappa=100; a=[2*pi,2*pi];
>> alpha=0.05; bj = get_twod_bj(dtref,J,a,alpha);
>> [W1,W2]=get_twod_dW(bj,kappa,1);
```

Figure 10.7(b) shows a realisation of $W^J(1)$ for $\alpha = 0.5$ and this realisation shows less variation due to the more rapid decay of q_{j_1,j_2}. Even so, $W^J(t) \in H^r_{\text{per}}(D)$ for all $r \geq 0$ in both cases due to the exponential decay of q_{j_1,j_2}.

Algorithm 10.5 Code to form the coefficients b_j in Example 10.12 for $D = (0, a_1) \times (0, a_2)$. Inputs and output are similar to those in Algorithm 10.3, with the inputs a=$[a_1, a_2]$ and J=$[J_1, J_2]$ now vectors and the additional parameter alpha=α.

```
function bj=get_twod_bj(dtref,J,a,alpha)
lambdax= 2*pi*[0:J(1)/2 -J(1)/2+1:-1]'/a(1);
lambday= 2*pi*[0:J(2)/2 -J(2)/2+1:-1]'/a(2);
[lambdaxx lambdayy]=meshgrid(lambday,lambdax);
root_qj=exp(-alpha*(lambdaxx.^2+lambdayy.^2)/2); % set decay rate noise
bj=root_qj*sqrt(dtref)*J(1)*J(2)/sqrt(a(1)*a(2));
```

Cylindrical Wiener process

We mention the important case of $Q = I$, which is not trace class on an infinite-dimensional space U (as $q_j = 1$ for all j) so that the series (10.11) does not converge in $L^2(\Omega, U)$. To extend the definition of a Q-Wiener process, we introduce the *cylindrical Wiener process*. The key point is to introduce a second space U_1 such that $U \subset U_1$ and $Q = I$ is a trace class operator when extended to U_1. Recall the Hilbert–Schmidt operators of Definition 1.60.

Algorithm 10.6 Code to sample $W^J(t + \kappa\Delta t_{\text{ref}}, \boldsymbol{x}_{k_1,k_2}) - W^J(t, \boldsymbol{x}_{k_1,k_2})$ in Example 10.12 for $k_i = 1, \ldots, J_i$. Inputs are similar to those in Algorithm 10.2, without the flag `iFspace`. The output is a pair of independent samples of the increment.

```
function [dW1,dW2]=get_twod_dW(bj,kappa,M)
J=size(bj);
if(kappa==1) % get xi_j
  nnr=randn(J(1),J(2),M); nnc=randn(J(1),J(2),M);
else % sum over kappa steps
  nnr=squeeze(sum(randn(J(1),J(2),M,kappa),4));
  nnc=squeeze(sum(randn(J(1),J(2),M,kappa),4));
end
nn2=nnr + sqrt(-1)*nnc; tmphat=bsxfun(@times,bj,nn2);
tmp=ifft2(tmphat); dW1=real(tmp); dW2=imag(tmp);
```

Example 10.13 ($L^2(0,2\pi)$ cylindrical Wiener process) Let $U = L^2(0,2\pi)$ and consider $Au = u - u_{xx}$ with domain $H^2_{\text{per}}(0,2\pi)$, normalised eigenfunctions ϕ_j, and eigenvalues λ_j. Then, $L^2(0,2\pi) \subset \mathcal{D}(A^{-r/2})$ for $r \geq 0$. The linear operator $\iota\colon L^2(0,2\pi) \to \mathcal{D}(A^{-r/2})$ defined by $\iota u = u$ (the *inclusion*) is a Hilbert–Schmidt operator if $r > 1/2$. To see this, note that

$$\|\iota\|_{\text{HS}(L^2(0,2\pi),\mathcal{D}(A^{-r/2}))} = \left(\sum_{j=1}^{\infty} \|\iota\phi_j\|^2_{-r/2}\right)^{1/2} = \left(\sum_{j=1}^{\infty} \lambda_j^{-r}\right)^{1/2} < \infty,$$

because $\lambda_j^{-r} = \mathcal{O}(j^{-2r})$ and $2r > 1$. Define Q_1 on $\mathcal{D}(A^{-r/2})$ by $Q_1\phi_j = \phi_j\|\phi_j\|^2_{-r/2}$ and note that

$$\operatorname{Tr} Q_1 = \|\iota\|^2_{\text{HS}(L^2(0,2\pi),\mathcal{D}(A^{-r/2}))}$$

and hence Q_1 is of trace class. The Q_1-Wiener process is well defined in $\mathcal{D}(A^{-r/2})$. As Q_1 has eigenvalues $\|\phi_j\|^2_{-r/2}$ and $\|\cdot\|_{-r/2}$-normalised eigenfunctions $\phi_j/\|\phi_j\|_{-r/2}$, its expansion (10.11) is effectively the case $q_j = 1$, $\chi_j = \phi_j$ and the Q_1-Wiener process makes sense of the I-Wiener process.

We present a formal definition of the cylindrical Wiener process through the sum (10.11).

Definition 10.14 (cylindrical Wiener process) Let U be a separable Hilbert space. The *cylindrical Wiener process* (also called *space–time white noise*) is the U-valued stochastic process $W(t)$ defined by

$$W(t) = \sum_{j=1}^{\infty} \chi_j \beta_j(t), \tag{10.20}$$

where $\{\chi_j\}$ is *any* orthonormal basis of U and $\beta_j(t)$ are *iid* $\mathcal{F}_t$-Brownian motions.

If $U \subset U_1$ for a second Hilbert space U_1, the series (10.20) converges in $L^2(\Omega, U_1)$ if the inclusion $\iota\colon U \to U_1$ is Hilbert–Schmidt (see Exercise 10.3). In the context of Example 10.13, we may choose $U_1 = \mathcal{D}(A^{-r/2})$ for $r > 1/2$.

The term *white noise* is used by analogy with white light since we have a homogeneous mix ($q_j = 1$) of all eigenfunctions; see (5.22). For a Q-Wiener process, we have coloured noise and the heterogeneity of the eigenvalues q_j causes correlations in space.

10.3 Itô stochastic integrals

To develop the Itô integral $\int_0^t B(s)\,dW(s)$ for a Q-Wiener process $W(s)$, our first task is to identify a suitable class of integrands $B(s)$. Recall that $W(t)$ takes values in a Hilbert space U. We treat stochastic PDEs in a Hilbert space H and wish the Itô integral to take values in H. Hence, we consider integrands B that are $\mathcal{L}(U_0, H)$-valued processes, for a subspace U_0 of U, known as the *Cameron–Martin space*. The correct set of linear operators in which to study the integrand $B(s)$ is L_0^2, which we now define. The H norm and inner product are denoted $\|\cdot\|$ and $\langle\cdot,\cdot\rangle$.

Definition 10.15 (L_0^2 space for integrands) Let $U_0 := \{Q^{1/2}u \colon u \in U\}$ for $Q^{1/2}$ defined by (1.87). L_0^2 is the set of linear operators $B\colon U_0 \to H$ such that

$$\|B\|_{L_0^2} := \left(\sum_{j=1}^{\infty} \|BQ^{1/2}\chi_j\|^2\right)^{1/2} = \|BQ^{1/2}\|_{\mathrm{HS}(U,H)} < \infty,$$

where χ_j is an orthonormal basis for U. L_0^2 is a Banach space with norm $\|\cdot\|_{L_0^2}$.

If Q is invertible, $U_0 = \mathcal{D}(Q^{-1/2})$ and, recalling that $\mathcal{D}(Q^{-1/2})$ is a Hilbert space (see Theorem 1.88), L_0^2 is then the space of Hilbert–Schmidt operators $\mathrm{HS}(U_0, H)$ from U_0 to H. See Exercise 10.5.

As in §8.2, the integrands $B(s)$ must satisfy certain measurability properties and, in this chapter, $W(t)$ is $\mathcal{F}_t$-adapted and the L_0^2-valued process $B(t)$ is predictable (see Definition 8.6). This holds if, for an orthonormal basis ϕ_k of H, the real-valued processes $\langle B(t)Q^{1/2}\chi_j, \phi_k\rangle$, for $j,k \in \mathbb{N}$, are each predictable.

We build the H-valued Itô integral on the finite-dimensional integral of Chapter 8. The truncated form $W^J(t)$ of the Q-Wiener process is finite dimensional and the integral

$$\int_0^t B(s)\,dW^J(s) = \sum_{j=1}^{J} \int_0^t B(s)\sqrt{q_j}\,\chi_j\,d\beta_j(s) \tag{10.21}$$

is well defined (see Exercise 10.6). We show the limit as $J \to \infty$ of (10.21) exists in $L^2(\Omega, H)$ and define the stochastic integral by

$$\int_0^t B(s)\,dW(s) := \sum_{j=1}^{\infty} \int_0^t B(s)\sqrt{q_j}\chi_j\,d\beta_j(s). \tag{10.22}$$

We use the notation $\{B(s)\colon s \in [0,T]\}$ to stress that $B(s)$ is a stochastic process.

Theorem 10.16 *Let Q satisfy Assumption 10.5 and suppose that $\{B(s)\colon s \in [0,T]\}$ is a L_0^2-valued predictable process such that*

$$\int_0^T \mathbb{E}\left[\|B(s)\|_{L_0^2}^2\right] ds < \infty. \tag{10.23}$$

For $t \in [0,T]$, (10.22) is well defined in $L^2(\Omega, H)$ and the following Itô isometry holds:

$$\mathbb{E}\left[\left\|\int_0^t B(s)\,dW(s)\right\|^2\right] = \int_0^t \mathbb{E}\left[\|B(s)\|_{L_0^2}^2\right] ds. \tag{10.24}$$

Further, $\left\{\int_0^t B(s)\,dW(s)\colon t \in [0,T]\right\}$ is an H-valued predictable process.

Proof To show (10.22) is well defined in $L^2(\Omega, H)$, write $I(t) := \int_0^t B(s)\,dW(s)$ and

$$\|I(t)\|^2 = \left\| \sum_{j=1}^{\infty} \int_0^t B(s)\sqrt{q_j}\,\chi_j\,d\beta_j(s) \right\|^2 = \sum_{k=1}^{\infty} \left\langle \sum_{j=1}^{\infty} \int_0^t B(s)\sqrt{q_j}\chi_j\,d\beta_j(s), \phi_k \right\rangle^2$$
$$= \sum_{k=1}^{\infty} \left(\sum_{j=1}^{\infty} \int_0^t \langle B(s)\sqrt{q_j}\,\chi_j, \phi_k \rangle\,d\beta_j(s) \right)^2.$$

For an orthonormal basis ϕ_j of H,

$$\mathbb{E}\left[\|I(t)\|^2\right] = \sum_{j,k,\ell=1}^{\infty} \mathbb{E}\left[\int_0^t \langle B(s)\sqrt{q_j}\,\chi_j, \phi_k\rangle\,d\beta_j(s) \int_0^t \langle B(s)\sqrt{q_\ell}\,\chi_\ell, \phi_k\rangle\,d\beta_\ell(s) \right].$$

The Brownian motions β_j and β_ℓ are independent and the cross terms for $j \neq \ell$ have mean zero and vanish. The Itô isometry (8.17) gives

$$\mathbb{E}\left[\|I(t)\|^2\right] = \sum_{j,k=1}^{\infty} \int_0^t \mathbb{E}\left[\langle B(s)\sqrt{q_j}\chi_j, \phi_k\rangle^2\right] ds.$$

Since $Q\chi_j = q_j\chi_j$, we have

$$\mathbb{E}\left[\|I(t)\|^2\right] = \sum_{j,k=1}^{\infty} \int_0^t \mathbb{E}\left[\langle B(s)Q^{1/2}\chi_j, \phi_k\rangle^2\right] ds = \int_0^t \mathbb{E}\left[\sum_{j,k=1}^{\infty} \langle B(s)Q^{1/2}\chi_j, \phi_k\rangle^2 \right] ds.$$

By the Parseval identity (1.43),

$$\mathbb{E}\left[\|I(t)\|^2\right] = \int_0^t \mathbb{E}\left[\sum_{j=1}^{\infty} \|B(s)Q^{1/2}\chi_j\|^2 \right] ds = \int_0^t \mathbb{E}\left[\|B(s)\|_{L_0^2}^2 \right] ds$$

and, under (10.23), $I(t)$ is well defined in $L^2(\Omega, H)$, as required.

In the basis ϕ_k, $I(t)$ has coefficients

$$\langle I(t), \phi_k\rangle = \sum_{j=1}^{\infty} \int_0^t \langle B(s)Q^{1/2}\chi_j, \phi_k\rangle\,d\beta_j(s)$$

and $\langle I(t), \phi_k\rangle$ is a mean-square limit of one-dimensional Itô integrals. Then, $\langle I(t), \phi_k\rangle \in \mathcal{L}_2^T$ and is predictable (Proposition 8.11). An H-valued process is predictable if each coefficient is predictable and hence $I(t)$ is predictable. □

The stochastic integral (10.22) for the case of cylindrical Wiener processes is defined in the same way. A cylindrical Wiener process on U is a Q-Wiener process on a second Hilbert space $U_1 \supset U$ (see Definition 10.14) for a trace class operator $Q_1 \in \mathcal{L}(U_1)$. Hence, the stochastic integral $\int_0^T B(s)\,dW(s)$ for a cylindrical Wiener process $W(s)$ is well defined for L_0^2-valued integrands $B(s)$, where $L_0^2 = \mathrm{HS}(Q_1^{1/2}U_1, H)$. It turns out that this space is independent of the choice of U_1 and in fact $L_0^2 = \mathrm{HS}(U, H)$. In other words, the Itô integral is well defined and (10.24) holds even for cylindrical Wiener processes. See Exercise 10.4.

10.4 Semilinear evolution equations in a Hilbert space

Having defined the Itô stochastic integral, we are equipped to consider Itô stochastic semilinear evolution equations of the form

$$du = \left[-Au + f(u)\right] dt + G(u)\, dW(t), \qquad u(0) = u_0, \tag{10.25}$$

given initial data $u_0 \in H$. We assume throughout this chapter that $A\colon \mathcal{D}(A) \subset H \to H$ satisfies Assumption 3.19 and hence $-A$ is the generator of a semigroup e^{-tA}; see Definition 3.11. We consider global Lipschitz nonlinearities $f\colon H \to H$ and $G\colon H \to L^2_0$ (compare with (3.41) for PDEs and Assumption 8.17 for SODEs). $W(t)$ is either a Q-Wiener process (10.11) or a cylindrical Wiener process (10.20).

Example 10.17 (stochastic heat equation with additive noise) Consider the stochastic heat equation

$$du = \Delta u\, dt + \sigma\, dW(t), \qquad u(0) = u_0 \in L^2(D), \tag{10.26}$$

where $\sigma > 0$ and homogeneous Dirichlet boundary conditions are imposed on a bounded domain D. In terms of (10.25),

$$H = U = L^2(D), \quad f = 0, \quad \text{and} \quad G(u) = \sigma I,$$

so that $G(u)v = \sigma v$ for $v \in U$ and we have additive noise. G satisfies the Lipschitz condition. In Example 3.17, we saw that $A = -\Delta$ with domain $\mathcal{D}(A) = H^2(D) \cap H^1_0(D)$ is the generator of an infinitesimal semigroup $S(t) = \mathrm{e}^{-tA}$.

In the deterministic setting of PDEs, there are a number of different concepts of solution (see §2.1). The same is true for SPDEs. We start by discussing strong and weak solutions. In all cases, we interpret (10.25) as an integral equation in time.

Definition 10.18 (strong solution) A predictable H-valued process $\{u(t)\colon t \in [0,T]\}$ is called a *strong solution* of (10.25) if

$$u(t) = u_0 + \int_0^t \left[-Au(s) + f(u(s))\right] ds + \int_0^t G(u(s))\, dW(s), \qquad \forall t \in [0,T].$$

This form of solution is too restrictive in practice as it requires $u(t) \in \mathcal{D}(A)$. Instead, we consider the weak or variational formulation of (10.25). In contrast to the deterministic case in Definition 3.36, there is no condition on du/dt, the test space is $\mathcal{D}(A)$, and we allow $u(t)$ to be H valued.

Definition 10.19 (weak solution) A predictable H-valued process $\{u(t)\colon t \in [0,T]\}$ is called a *weak solution* of (10.25) if

$$\begin{aligned} \langle u(t), v\rangle = \langle u_0, v\rangle &+ \int_0^t \left[-\langle u(s), Av\rangle + \langle f(u(s)), v\rangle\right] ds \\ &+ \int_0^t \langle G(u(s))\, dW(s), v\rangle, \qquad \forall t \in [0,T],\ v \in \mathcal{D}(A), \end{aligned} \tag{10.27}$$

where in the notation of (10.11)

$$\int_0^t \langle G(u(s))\, dW(s), v\rangle := \sum_{j=1}^{\infty} \int_0^t \langle G(u(s))\sqrt{q_j}\chi_j, v\rangle\, d\beta_j(s).$$

The term 'weak' in Definition 10.19 refers to the PDE and not to the probabilistic notion of weak solution. The weak form is convenient for developing Galerkin approximation methods as we see in §10.6.

Before looking at mild solutions, we illustrate an important link between the stochastic heat equation and the spatial dimension.

Example 10.20 (stochastic heat equation in one dimension) Consider the weak solution of (10.26) with $D = (0,\pi)$, so that $-A$ has eigenfunctions $\phi_j(x) = \sqrt{2/\pi}\sin(jx)$ and eigenvalues $\lambda_j = j^2$ for $j \in \mathbb{N}$. Suppose that $W(t)$ is a Q-Wiener process and the eigenfunctions χ_j of Q are the same as the eigenfunctions ϕ_j of A (as in Example 10.9). From (10.27) with $v \in \mathcal{D}(A)$, a weak solution satisfies

$$\begin{aligned}\langle u(t), v\rangle_{L^2(0,\pi)} &= \langle u_0, v\rangle_{L^2(0,\pi)} + \int_0^t \langle -u(s), Av\rangle_{L^2(0,\pi)}\,ds \\ &\quad + \sum_{j=1}^{\infty}\int_0^t \sigma\sqrt{q_j}\langle \phi_j, v\rangle_{L^2(0,\pi)}\,d\beta_j(s).\end{aligned} \tag{10.28}$$

Write $u(t) = \sum_{j=1}^{\infty}\hat{u}_j(t)\phi_j$ for $\hat{u}_j(t) := \langle u(t), \phi_j\rangle_{L^2(0,\pi)}$. Take $v = \phi_j$, to see

$$\hat{u}_j(t) = \hat{u}_j(0) + \int_0^t (-\lambda_j)\hat{u}_j(s)\,ds + \int_0^t \sigma\sqrt{q_j}\,d\beta_j(s).$$

Hence, $\hat{u}_j(t)$ satisfies the SODE

$$d\hat{u}_j = -\lambda_j\hat{u}_j\,dt + \sigma\sqrt{q_j}\,d\beta_j(t). \tag{10.29}$$

Each coefficient $\hat{u}_j(t)$ is an Ornstein–Uhlenbeck (OU) process (see Examples 8.1 and 8.21), which is a Gaussian process with variance

$$\operatorname{Var}\big(\hat{u}_j(t)\big) = \frac{\sigma^2 q_j}{2\lambda_j}\big(1 - \mathrm{e}^{-2\lambda_j t}\big).$$

For initial data $u_0 = 0$, we obtain, by the Parseval identity (1.43),

$$\|u(t)\|^2_{L^2(\Omega, L^2(0,\pi))} = \mathbb{E}\left[\sum_{j=1}^{\infty}|\hat{u}_j(t)|^2\right] = \sum_{j=1}^{\infty}\frac{\sigma^2 q_j}{2\lambda_j}\Big(1 - \mathrm{e}^{-2\lambda_j t}\Big). \tag{10.30}$$

The series converges if the sum $\sum_{j=1}^{\infty} q_j/\lambda_j$ is finite. For a Q-Wiener process, the sum is finite because Q is trace class and the solution $u(t)$ of (10.26) is in $L^2(0,\pi)$ a.s.

If instead $W(t)$ is a cylindrical Wiener process, $q_j = 1$ and the sum is finite only if $\lambda_j \to \infty$ sufficiently quickly. In this case, $\lambda_j = j^2$ and $\sum_{j=1}^{\infty}\lambda_j^{-1} < \infty$. Consequently, $\|u(t)\|^2_{L^2(\Omega, L^2(0,\pi))} < \infty$ and again the solution $u(t) \in L^2(0,\pi)$ a.s.

In contrast, the next example shows that $\|u(t)\|_{L^2(\Omega, L^2(D))}$ is infinite for cylindrical Wiener processes when $D \subset \mathbb{R}^2$.

Example 10.21 (stochastic heat equation in two dimensions) We repeat the calculations of Example 10.20 for (10.26) with $D = (0,\pi)\times(0,\pi)$, where the eigenvalues $\lambda_{j_1,j_2} = j_1^2 + j_2^2$ and have associated normalised eigenfunction ϕ_{j_1,j_2} for $j_1, j_2 \in \mathbb{N}$. Assume that Q also has eigenfunctions ϕ_{j_1,j_2} and eigenvalues q_{j_1,j_2}. Write $u(t) = \sum_{j_1,j_2=1}^{\infty}\hat{u}_{j_1,j_2}(t)\phi_{j_1,j_2}$. Again substituting $v = \phi_{j_1,j_2}$ into the weak form (10.27), we find each coefficient $\hat{u}_{j_1,j_2}(t)$ is an Ornstein–Uhlenbeck process:

$$d\hat{u}_{j_1,j_2} = -\lambda_{j_1,j_2}\hat{u}_{j_1,j_2}\,dt + \sigma\sqrt{q_{j_1,j_2}}\,d\beta_{j_1,j_2}(t) \tag{10.31}$$

and the variance

$$\mathrm{Var}\big(\hat{u}_{j_1,j_2}(t)\big) = \frac{\sigma^2 q_{j_1,j_2}}{2\lambda_{j_1,j_2}}\big(1 - \mathrm{e}^{-2\lambda_{j_1,j_2}t}\big).$$

If $u_0 = 0$ then $\mathbb{E}[\hat{u}_{j_1,j_2}(t)] = 0$ and

$$\|u(t)\|^2_{L^2(\Omega,L^2(D))} = \mathbb{E}\left[\sum_{j_1,j_2=1}^{\infty}\left|\hat{u}_{j_1,j_2}(t)\right|^2\right] = \sum_{j_1,j_2=1}^{\infty}\frac{\sigma^2 q_{j_1,j_2}}{2\lambda_{j_1,j_2}}\left(1 - \mathrm{e}^{-2\lambda_{j_1,j_2}t}\right). \tag{10.32}$$

When Q is trace class, the right-hand side is finite and $u(t) \in L^2(D)$ a.s. For a cylindrical Wiener process (the case $q_{j_1,j_2} = 1$), we have

$$\sum_{j_1,j_2=1}^{\infty}\frac{1}{\lambda_{j_1,j_2}} = \sum_{j_1,j_2=1}^{\infty}\frac{1}{j_1^2 + j_2^2} = \infty$$

and the solution $u(t)$ is not in $L^2(\Omega, L^2(D))$. Thus, we do not expect weak solutions of (10.26) to exist in $L^2(D)$ in two dimensions.

When A is a fourth-order differential operator and the eigenvalues grow more rapidly, (10.26) gives well-defined solutions in two dimensions for the cylindrical Wiener process; see Exercise 10.12.

Finally, we define the stochastic version of the mild solution of Definition 3.28.

Definition 10.22 (mild solution) A predictable H-valued process $\{u(t) : t \in [0,T]\}$ is called a *mild solution* of (10.25) if for $t \in [0,T]$

$$u(t) = \mathrm{e}^{-tA}u_0 + \int_0^t \mathrm{e}^{-(t-s)A} f(u(s))\,ds + \int_0^t \mathrm{e}^{-(t-s)A} G(u(s))\,dW(s), \tag{10.33}$$

where e^{-tA} is the semigroup generated by $-A$.

In general, we expect that all strong solutions are weak solutions and all weak solutions are mild solutions. Moreover, the reverse implications hold for solutions with sufficient regularity. The existence and uniqueness theory of mild solutions is easiest to develop and we do so in Theorem 10.26. In addition to the global Lipschitz condition on G, the following condition is used.

Assumption 10.23 (Lipschitz condition on G) For constants $\zeta \in (0,2]$ and $L > 0$, we have that $G : H \to L_0^2$ satisfies

$$\begin{aligned} \|A^{(\zeta-1)/2}G(u)\|_{L_0^2} &\le L(1 + \|u\|), \\ \|A^{(\zeta-1)/2}(G(u_1) - G(u_2))\|_{L_0^2} &\le L\|u_1 - u_2\|, \qquad \forall u, u_1, u_2 \in H. \end{aligned} \tag{10.34}$$

The parameter ζ controls the regularity of the stochastic forcing $G(u)\,dW(t)$. For $\zeta > 1$, the forcing must be smooth as $G(u)$ must map U_0 into $\mathcal{D}(A^{(\zeta-1)/2}) \subset H$; for $\zeta < 1$, $\mathcal{D}(A^{(\zeta-1)/2}) \supset H$ and the forcing can be rough.

Consider a bounded domain D and let $H = U = L^2(D)$ and Q satisfy Assumption 10.5. In this case, the next lemma shows that Assumption 10.23 holds for a wide class of operators G that are diagonal in the sense of Nemytskii operators. Recall from Theorem 1.65 that Q can be written as an integral operator

$$(Qu)(\boldsymbol{x}) = \int_D q(\boldsymbol{x},\boldsymbol{y})u(\boldsymbol{y})\,d\boldsymbol{y}, \qquad u \in L^2(D), \tag{10.35}$$

for a kernel $q \in L^2(D \times D)$. In terms of the Q-Wiener process, this means

$$\operatorname{Cov}(W(t,\boldsymbol{x}), W(t,\boldsymbol{y})) = t\, q(\boldsymbol{x},\boldsymbol{y})$$

and $W(1,\boldsymbol{x})$ is a mean-zero Gaussian random field (as a function of $\boldsymbol{x}$) with covariance $q(\boldsymbol{x},\boldsymbol{y})$. See Example 7.8.

The following is the analogue for G of Lemma 3.30.

Lemma 10.24 (Nemytskii G) *For a bounded domain D, let Q have kernel $q \in \mathrm{C}(\bar{D} \times \bar{D})$. For a Lipschitz continuous function $g\colon \mathbb{R} \to \mathbb{R}$, define*

$$(G(u)\chi)(\boldsymbol{x}) := g(u(\boldsymbol{x}))\,\chi(\boldsymbol{x}), \qquad \forall u,\chi \in L^2(D), \quad \boldsymbol{x} \in D.$$

Then, Assumption 10.23 holds with $\zeta = 1$ and $H = U = L^2(D)$.

Proof By Mercer's theorem, $\sum_j q_j \|\chi_j\|_\infty^2 < \infty$ (in the notation of Theorem 10.7). Then,

$$\|G(u)\|_{L_0^2} = \left(\sum_{j=1}^{\infty} \left\|G(u) q_j^{1/2} \chi_j\right\|_{L^2(D)}^2\right)^{1/2} \le \left(\sum_{j=1}^{\infty} q_j \|\chi_j\|_\infty^2\right)^{1/2} \|g(u)\|_{L^2(D)}.$$

We know $g\colon L^2(D) \to L^2(D)$ has linear growth from Lemma 3.30 and the linear growth condition in (10.34) with $\zeta = 1$ follows easily. The Lipschitz condition is similar. □

Example 10.25 (parabolic Anderson model) Corresponding to $g(u) = \sigma u$, we find the Nemytskii operator $(G(u)\chi)(\boldsymbol{x}) = \sigma\, u(\boldsymbol{x})\, \chi(\boldsymbol{x})$. When Q has a smooth kernel, this G satisfies Assumption 10.23 with $\zeta = 1$ and we have the SPDE

$$du = \Delta u\, dt + \sigma u\, dW(t),$$

which is known as the parabolic Anderson model.

As in Theorems 3.2 and 3.29, we use a fixed point argument to prove the existence and uniqueness of mild solutions for the stochastic evolution equation. In the following theorem, we allow the initial data to be random and assume that the initial data is in $L^2(\Omega, \mathcal{F}_0, H)$ (see Definition 4.44), so that the initial data is square integrable and $\mathcal{F}_0$-measurable.

Theorem 10.26 (existence and uniqueness) *Suppose that A satisfies Assumption 3.19, $f\colon H \to H$ satisfies* (3.41)*, and $G\colon H \to L_0^2$ satisfies Assumption 10.23. Suppose that the initial data $u_0 \in L^2(\Omega, \mathcal{F}_0, H)$. Then, there exists a unique mild solution $u(t)$ on $[0,T]$ to* (10.25) *for any $T > 0$. Furthermore, there exists a constant $K_T > 0$ such that*

$$\sup_{t\in[0,T]} \|u(t)\|_{L^2(\Omega,H)} \le K_T\left(1 + \|u_0\|_{L^2(\Omega,H)}\right). \tag{10.36}$$

Proof Let $\mathcal{H}_{2,T}$ denote the Banach space of H-valued predictable processes $\{u(t)\colon t \in [0,T]\}$ with norm $\|u\|_{\mathcal{H}_{2,T}} := \sup_{0\le t\le T}\|u(t)\|_{L^2(\Omega,H)}$ (compare with Definition 8.16). For $u \in \mathcal{H}_{2,T}$, define

$$(\mathcal{J}u)(t) := \mathrm{e}^{-tA}u_0 + \int_0^t \mathrm{e}^{-(t-s)A} f(u(s))\,ds + \int_0^t \mathrm{e}^{-(t-s)A} G(u(s))\,dW(s). \tag{10.37}$$

A fixed point $u(t)$ of $\mathcal{J}$ is an H-valued predictable process and obeys (10.33) and hence is a mild solution of (10.25). To show existence and uniqueness of the fixed point, we show $\mathcal{J}$ is a contraction mapping from $\mathcal{H}_{2,T}$ to $\mathcal{H}_{2,T}$ and apply Theorem 1.10.

First, we show $\mathcal{J}$ maps into $\mathcal{H}_{2,T}$. $\mathcal{J}u(t)$ is a predictable process because u_0 is $\mathcal{F}_0$-measurable and the stochastic integral is a predictable process (Theorem 10.16). To see that $\|\mathcal{J}u\|_{\mathcal{H}_{2,T}} < \infty$, we use the properties developed in Exercises 10.7 and 10.8. By Exercise 10.7, $\|\mathrm{e}^{-tA}u_0\|_{L^2(\Omega,H)} \le \|u_0\|_{L^2(\Omega,H)} < \infty$ and

$$\begin{aligned}\left\|\int_0^t \mathrm{e}^{-(t-s)A} f(u(s))\,ds\right\|_{L^2(\Omega,H)} &\le \int_0^t \|\mathrm{e}^{-(t-s)A} f(u(s))\|_{L^2(\Omega,H)}\,ds\\ &\le \int_0^t \|f(u(s))\|_{L^2(\Omega,H)}\,ds\\ &\le \int_0^t L\big(1+\|u(s)\|_{L^2(\Omega,H)}\big)\,ds.\end{aligned}$$

By the Itô isometry, Exercise 10.7, and Assumption 10.23,

$$\begin{aligned}\left\|\int_0^t \mathrm{e}^{-(t-s)A} G(u(s))\,dW(s)\right\|^2_{L^2(\Omega,H)} &= \int_0^t \mathbb{E}\left[\|A^{(1-\zeta)/2}\mathrm{e}^{-(t-s)A}A^{(\zeta-1)/2}G(u(s))\|^2_{L^2_0}\right]ds\\ \le \int_0^t \|A^{(1-\zeta)/2}\mathrm{e}^{-(t-s)A}\|^2_{\mathcal{L}(H)}ds\; L^2\Big(1+\sup_{0\le s\le t}\|u(s)\|_{L^2(\Omega,H)}\Big)^2.\end{aligned}$$

Using Exercise 10.8 with $\beta = (1-\zeta)/2$ and $0<\zeta<1$, we have for some $K>0$

$$\left\|\int_0^t \mathrm{e}^{-(t-s)A} G(u(s))\,dW(s)\right\|_{L^2(\Omega,H)} \le K\frac{T^{\zeta/2}}{\zeta^{1/2}}L\Big(1+\sup_{0\le s\le T}\|u(s)\|_{L^2(\Omega,H)}\Big).$$

Then, for $u \in \mathcal{H}_{2,T}$, all three terms in (10.37) are uniformly bounded over $t\in[0,T]$ in $L^2(\Omega,H)$ and $\|\mathcal{J}u\|_{\mathcal{H}2,T} < \infty$. Hence $\mathcal{J}$ maps into $\mathcal{H}_{2,T}$. The case $\zeta \in [1,2]$ is similar using the bound in Exercise 10.8 for $\beta \le 0$.

For the contraction property, we may show, using the Lipschitz properties of f and G, that for $\zeta \in (0,1)$

$$\|\mathcal{J}u_1(t)-\mathcal{J}u_2(t)\|^2_{L^2(\Omega,H)} \le 2t\int_0^t L^2\|u_1(s)-u_2(s)\|^2_{L^2(\Omega,H)}\,ds + 2K^2\frac{t^\zeta}{\zeta}L^2\|u_1-u_2\|^2_{\mathcal{H}_{2,T}}.$$

Then,

$$\|\mathcal{J}u_1-\mathcal{J}u_2\|^2_{\mathcal{H}_{2,T}} \le 2\Big(T^2+K^2\frac{T^\zeta}{\zeta}\Big)L^2\|u_1-u_2\|^2_{\mathcal{H}_{2,T}}$$

and $\mathcal{J}$ is a contraction on $\mathcal{H}_{2,T}$ if $(T^2+K^2T^\zeta/\zeta) < 1/2L^2$, which is satisfied for T small. The case $\zeta\in[1,2]$ is similar. Repeating the argument on $[0,T]$, $[T,2T]$,..., we find a unique mild solution for all $t>0$. The bound (10.36) is a consequence of Gronwall's inequality; see Exercise 10.9. □

The following lemma establishes regularity of the mild solution in time. The exponents θ_1, θ_2 determine rates of convergence for the numerical methods in §10.6–10.8. For simplicity, we assume u_0 takes values in $\mathcal{D}(A)$.

Lemma 10.27 (regularity in time) *Let the assumptions of Theorem 10.26 hold and let $u_0 \in L^2(\Omega, \mathcal{F}_0, \mathcal{D}(A))$. For $T > 0$, $\epsilon \in (0, \zeta)$, and $\theta_1 := \min\{\zeta/2 - \epsilon, 1/2\}$, there exists $K_{RT} > 0$ such that*

$$\|u(t_2) - u(t_1)\|_{L^2(\Omega,H)} \le K_{RT}(t_2 - t_1)^{\theta_1}, \qquad 0 \le t_1 \le t_2 \le T. \tag{10.38}$$

Further, for $\zeta \in [1,2]$ and $\theta_2 := \zeta/2 - \epsilon$, there exists $K_{RT2} > 0$ such that

$$\left\| u(t_2) - u(t_1) - \int_{t_1}^{t_2} G(u(s))\, dW(s) \right\|_{L^2(\Omega,H)} \le K_{RT2}(t_2 - t_1)^{\theta_2}. \tag{10.39}$$

Proof Write $u(t_2) - u(t_1) = \mathrm{I} + \mathrm{II} + \mathrm{III}$, where

$$\begin{aligned}
\mathrm{I} &:= \left(\mathrm{e}^{-t_2 A} - \mathrm{e}^{-t_1 A}\right) u_0,\\
\mathrm{II} &:= \int_0^{t_2} \mathrm{e}^{-(t_2-s)A} f(u(s))\, ds - \int_0^{t_1} \mathrm{e}^{-(t_1-s)A} f(u(s))\, ds,\\
\mathrm{III} &:= \int_0^{t_2} \mathrm{e}^{-(t_2-s)A} G(u(s))\, dW(s) - \int_0^{t_1} \mathrm{e}^{-(t_1-s)A} G(u(s))\, dW(s).
\end{aligned}$$

The estimation of I and II proceeds as in Proposition 3.31, except that the H norm is replaced by the $L^2(\Omega, H)$ norm. We focus on III and write $\mathrm{III} = \mathrm{III}_1 + \mathrm{III}_2$, for

$$\begin{aligned}
\mathrm{III}_1 &:= \int_0^{t_1} \left(\mathrm{e}^{-(t_2-s)A} - \mathrm{e}^{-(t_1-s)A}\right) G(u(s))\, dW(s),\\
\mathrm{III}_2 &:= \int_{t_1}^{t_2} \mathrm{e}^{-(t_2-s)A} G(u(s))\, dW(s).
\end{aligned}$$

We consider only the case $\zeta \in (0,1)$ and analyse III_1 and III_2 separately.

First, consider III_1. The Itô isometry (10.24) yields

$$\begin{aligned}
\mathbb{E}\left[\|\mathrm{III}_1\|^2\right] &= \int_0^{t_1} \mathbb{E}\left[\left\|\left(\mathrm{e}^{-(t_2-s)A} - \mathrm{e}^{-(t_1-s)A}\right) G(u(s))\right\|_{L_0^2}^2\right] ds\\
&= \int_0^{t_1} \mathbb{E}\left[\left\|A^{(1-\zeta)/2}\left(\mathrm{e}^{-(t_2-s)A} - \mathrm{e}^{-(t_1-s)A}\right) A^{(\zeta-1)/2} G(u(s))\right\|_{L_0^2}^2\right] ds.
\end{aligned}$$

Using Assumption 10.23 on G, we obtain

$$\begin{aligned}
\mathbb{E}\left[\|\mathrm{III}_1\|^2\right] \le \int_0^{t_1} &\left\|A^{(1-\zeta)/2}\left(\mathrm{e}^{-(t_2-s)A} - \mathrm{e}^{-(t_1-s)A}\right)\right\|_{\mathcal{L}(H)}^2 ds\\
&\times L^2\left(1 + \sup_{0 \le s \le t_1} \|u(s)\|_{L^2(\Omega,H)}\right)^2.
\end{aligned}$$

Note that

$$\begin{aligned}\int_0^{t_1} &\|(e^{-(t_2-s)A} - e^{-(t_1-s)A})A^{(1-\zeta)/2}\|^2_{\mathcal{L}(H)} ds \\ &= \int_0^{t_1} \|A^{1/2-\epsilon} e^{-(t_1-s)A} A^{-\zeta/2+\epsilon}(I - e^{-(t_2-t_1)A})\|^2_{\mathcal{L}(H)} ds \\ &\le \int_0^{t_1} \|A^{1/2-\epsilon} e^{-(t_1-s)A}\|^2_{\mathcal{L}(H)} \|A^{-\zeta/2+\epsilon}(I - e^{-(t_2-t_1)A})\|^2_{\mathcal{L}(H)} ds.\end{aligned}$$

By Lemma 3.22(iii) and Exercise 10.8 for $\zeta < 1$, there exists $K_1, K_2 > 0$ such that

$$\int_0^{t_1} \|(e^{-(t_2-s)A} - e^{-(t_1-s)A})A^{(1-\zeta)/2}\|^2_{\mathcal{L}(H)} ds \le \left(K_1^2 \frac{t_1^{2\epsilon}}{\epsilon}\right)(K_2^2 (t_2 - t_1)^{\zeta-2\epsilon}).$$

Then, with $C_1 := K_1 K_2 L T^\epsilon / \sqrt{\epsilon}$,

$$\|\mathtt{III}_1\|_{L^2(\Omega,H)} = \mathbb{E}\left[\|\mathtt{III}_1\|^2\right]^{1/2} \le C_1 (t_2 - t_1)^{\zeta/2-\epsilon}\left(1 + \sup_{0\le s\le T} \|u(s)\|_{L^2(\Omega,H)}\right).$$

Let us estimate $\mathbb{E}[\|\mathtt{III}_2\|^2]$. The Itô isometry (10.24) and Assumption 10.23 on G yield

$$\begin{aligned}\mathbb{E}\left[\|\mathtt{III}_2\|^2\right] &= \int_{t_1}^{t_2} \mathbb{E}\left[\|A^{(1-\zeta)/2} e^{-(t_2-s)A} A^{(\zeta-1)/2} G(u(s))\|^2_{L^2_0}\right] ds \\ &\le \int_{t_1}^{t_2} \|A^{(1-\zeta)/2} e^{-(t_2-s)A}\|^2_{\mathcal{L}(H)} ds\, L^2\left(1 + \sup_{t_1\le s\le t_2} \|u(s)\|_{L^2(\Omega,H)}\right)^2.\end{aligned}$$

By applying Exercise 10.8, we find $K_3 > 0$ such that for $C_2 := K_3 L$

$$\|\mathtt{III}_2\|_{L^2(\Omega,H)} \le C_2 (t_2 - t_1)^{\zeta/2}\left(1 + \sup_{0\le s\le T} \|u(s)\|_{L^2(\Omega,H)}\right).$$

With $C_{\mathtt{III}}^2 = (C_1 + C_2)(1 + K_T)$, we find, using (10.36), that

$$\begin{aligned}\|\mathtt{III}\|_{L^2(\Omega,H)} &\le \|\mathtt{III}_1\|_{L^2(\Omega,H)} + \|\mathtt{III}_2\|_{L^2(\Omega,H)} \\ &\le C_{\mathtt{III}} (t_2 - t_1)^{\theta_1}\left(1 + \|u_0\|_{L^2(\Omega,H)}\right).\end{aligned}$$

This holds for $\zeta \in [1,2]$ by similar methods. Now, $\|u(t_2) - u(t_1)\|_{L^2(\Omega,H)} \le \|\mathtt{I}\|_{L^2(\Omega,H)} + \|\mathtt{II}\|_{L^2(\Omega,H)} + \|\mathtt{III}\|_{L^2(\Omega,H)}$. Since $\mathcal{D}(A)$ is continuously embedded in H, the estimates on the norms of $\mathtt{I}$, $\mathtt{II}$, and $\mathtt{III}$ complete the proof of (10.38).

The proof of (10.39) is similar, except instead of $\mathtt{III}_2$, we must remove the limit $\theta_1 \le 1/2$ by considering

$$\mathtt{III}_2' := \int_{t_1}^{t_2} (e^{-A(t_2-s)} - I)\, G(u(s))\, dW(s).$$

Using Lemma 3.22(ii) with $1 \le \zeta \le 2$, there exists a constant K_3 such that

$$\begin{aligned}\|\mathtt{III}_2'\|^2_{L^2(\Omega,H)} &\le \int_0^t \|A^{(1-\zeta)/2}(e^{-A(t_2-s)} - I)\|^2_{\mathcal{L}(H)} \|A^{(\zeta-1)/2} G(u(s))\|^2_{L^2_0} ds \\ &\le \int_{t_1}^{t_2} K_3 |t_2 - s|^{\zeta-1} ds\, L^2\left(1 + \sup_{t_1\le s\le t_2} \|u(s)\|_{L^2(\Omega,H)}\right)^2.\end{aligned}$$

Using (10.36), this gives the required upper bound (10.39). □

To conclude this section, we establish spatial regularity for additive noise.

Theorem 10.28 (regularity in space for additive noise) *Let the assumptions of Theorem 10.26 hold and $u(t)$ be the mild solution to* (10.25) *for $G(u) = \sigma I$ and $\sigma \in \mathbb{R}$. If $u_0 \in L^2(\Omega, \mathcal{F}_0, \mathcal{D}(A))$, then $u(t) \in L^2(\Omega, \mathcal{D}(A^{\zeta/2}))$ for $t \in [0,T]$.*

Proof Split the mild solution (10.33) into three terms, so that $u(t) = \mathrm{I} + \mathrm{II} + \mathrm{III}$, for

$$\mathrm{I} := \mathrm{e}^{-tA} u_0, \quad \mathrm{II} := \int_0^t \mathrm{e}^{-(t-s)A} f(u(s))\, ds, \quad \mathrm{III} := \int_0^t \mathrm{e}^{-(t-s)A} \sigma\, dW(s).$$

For the first term, since $u_0 \in L^2(\Omega, \mathcal{F}_0, \mathcal{D}(A))$, $\mathbb{E}\big[\|\mathrm{e}^{-tA}u_0\|_{\zeta/2}^2\big] \le \mathbb{E}\big[\|u_0\|_1^2\big] < \infty$ and $\mathrm{I} \in L^2(\Omega, \mathcal{D}(A^{\zeta/2}))$. The second term is considered in Exercise 10.10 using (3.41) and II also belongs to $L^2(\Omega, \mathcal{D}(A^{\zeta/2}))$.

For term III, Itô's isometry (10.24) gives

$$\mathbb{E}\big[\|\mathrm{III}\|_{\zeta/2}^2\big] = \mathbb{E}\left[\left\|\int_0^t \mathrm{e}^{-(t-s)A}\sigma\, dW(s)\right\|_{\zeta/2}^2\right] = \sigma^2 \int_0^t \|A^{\zeta/2}\mathrm{e}^{-(t-s)A}\|_{L_0^2}^2\, ds.$$

Now,

$$\begin{aligned}\int_0^t \|A^{\zeta/2}\mathrm{e}^{-(t-s)A}\|_{L_0^2}^2\, ds &= \int_0^t \|A^{(\zeta-1)/2}A^{1/2}\mathrm{e}^{-(t-s)A}\|_{L_0^2}^2\, ds \\ &\le \|A^{(\zeta-1)/2}\|_{L_0^2}^2 \int_0^t \|A^{1/2}\mathrm{e}^{-(t-s)A}\|_{\mathcal{L}(H)}^2\, ds.\end{aligned}$$

The integral is finite by Exercise 10.8 and $\|A^{(\zeta-1)/2}\|_{L_0^2} < \infty$ by Assumption 10.23. Hence, III belongs to $L^2(\Omega, \mathcal{D}(A^{\zeta/2}))$. □

Example 10.29 (reaction–diffusion equation, additive noise) Consider the SPDE

$$du = \big[-Au + f(u)\big]\, dt + \sigma\, dW(t), \qquad u(0) = u_0$$

with $A = -u_{xx}$ and $\mathcal{D}(A) = H^2(0,\pi) \cap H_0^1(0,\pi)$. Choose initial data $u_0 \in \mathcal{D}(A)$. The operator A has eigenvalues $\lambda_j = j^2$. As in Proposition 1.93, the $H^r(0,\pi)$ norm is equivalent to the fractional power norm $\|u\|_{r/2}$ associated to A. For a Q-Wiener process, we take $\zeta = 1$ in Assumption 10.23. By Theorem 10.28, $u(t) \in L^2(\Omega, H^1(0,\pi))$ and this improves on the $L^2(0,\pi)$-spatial regularity given by Theorem 10.26.

For space–time white noise (see Definition 10.14), Assumption 10.23 holds for $\zeta \in (0, 1/2)$, because $\lambda_j^{(\zeta-1)} = \mathcal{O}(j^{2(\zeta-1)})$ and

$$\|A^{(\zeta-1)/2}G(u)\|_{L_0^2} = \big(\mathrm{Tr}\, A^{(\zeta-1)}\big)^{1/2} < \infty.$$

In particular, the solution $u(t)$ of the stochastic heat equation (10.26) in one dimension forced by space–time white noise takes values in $L^2(\Omega, H^{\zeta}(0,\pi))$ and has up to a half (generalised) derivatives almost surely.

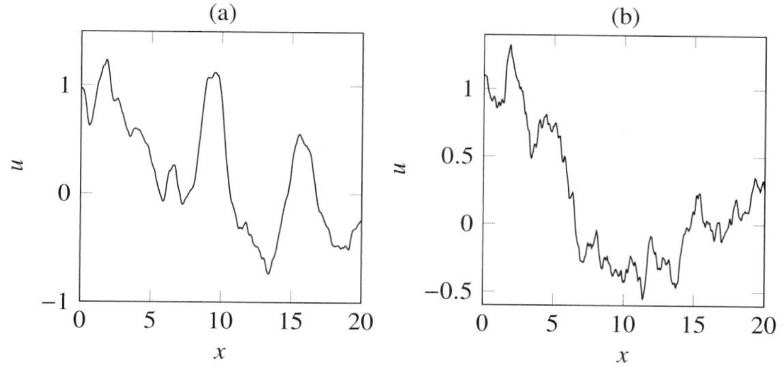

Figure 10.8 Approximate realisations of the solution to the Nagumo SPDE (10.41) at time $t = 1$ for the parameters given in Example 10.30 for (a) $\ell = 1$ and (b) $\ell = 0.1$.

10.5 Finite difference method

The finite difference method of §3.4 is easily extended to stochastic PDEs. We illustrate this for the reaction–diffusion equation with additive noise given by

$$du = \left[\varepsilon\, u_{xx} + f(u)\right] dt + \sigma\, dW(t), \qquad u(0,x) = u_0(x), \tag{10.40}$$

with homogeneous Dirichlet boundary conditions on $(0,a)$. We have parameters $\varepsilon, \sigma > 0$, a reaction term $f\colon \mathbb{R} \to \mathbb{R}$, and $W(t)$ is a Q-Wiener process on $L^2(0,a)$.

Introduce the grid points $x_j = jh$ for $h = a/J$ and $j = 0,\dots,J$. Let $\boldsymbol{u}_J(t)$ be the finite difference approximation to $[u(t,x_1),\dots,u(t,x_{J-1})]^{\mathsf{T}}$ resulting from the centred difference approximation A^{D} to the Laplacian (see (3.46)). That is, $\boldsymbol{u}_J(t)$ is the solution of

$$d\boldsymbol{u}_J = \left[-\varepsilon\, A^{\mathrm{D}} \boldsymbol{u}_J + \boldsymbol{f}(\boldsymbol{u}_J)\right] dt + \sigma\, d\boldsymbol{W}_J(t),$$

with $\boldsymbol{u}_J(0) = [u_0(x_1),\dots,u_0(x_{J-1})]^{\mathsf{T}}$ and $\boldsymbol{W}_J(t) := [W(t,x_1),\dots, W(t,x_{J-1})]^{\mathsf{T}}$. A^{D} and $\boldsymbol{f}$ are defined as in (3.48). This SODE is derived by following the same procedure as in §3.4. The finite difference method is easily adapted to other boundary conditions by changing A^{D} and $\boldsymbol{W}_J$.

To discretise in time, we may apply one of the methods discussed in Chapter 8. For example, the semi-implicit Euler–Maruyama method (8.121) with time step $\Delta t > 0$ yields an approximation $\boldsymbol{u}_{J,n}$ to $\boldsymbol{u}_J(t_n)$ at $t_n = n\Delta t$ defined by

$$\boldsymbol{u}_{J,n+1} = \left(I + \Delta t\, \varepsilon\, A^{\mathrm{D}}\right)^{-1}\left[\boldsymbol{u}_{J,n} + \boldsymbol{f}(\boldsymbol{u}_{J,n})\,\Delta t + \sigma\, \Delta \boldsymbol{W}_n\right]$$

with $\boldsymbol{u}_{J,0} = \boldsymbol{u}_J(0)$ and $\Delta \boldsymbol{W}_n := \boldsymbol{W}_J(t_{n+1}) - \boldsymbol{W}_J(t_n)$.

Suppose that Q has kernel $q(x,y)$ as in (10.35). Then, $\boldsymbol{W}_J(t) \sim \mathrm{N}(\boldsymbol{0}, t\,C)$, where C is the matrix with entries $q(x_i,x_j)$ for $i,j = 1,\dots,J-1$. The sampling algorithms of Chapters 5–7 can be used to sample $\boldsymbol{W}_J(t)$ and the increments $\Delta \boldsymbol{W}_n$. We now use the circulant embedding method to generate the increments for an example with homogeneous Neumann conditions.

Example 10.30 (Nagumo SPDE) Following Example 3.33, we consider the Nagumo equation with additive noise and consider

$$du = \left[\varepsilon\, u_{xx} + u(1-u)(u-\alpha)\right] dt + \sigma\, dW(t), \qquad u(0) = u_0, \tag{10.41}$$

on $(0,a)$ with homogeneous Neumann boundary conditions, given initial data $u_0(x) = \left(1+\exp\left(-(2-x)/\sqrt{2}\right)\right)^{-1}$. Let $W(t)$ be the Q-Wiener process on $L^2(0,a)$ with kernel $q(x,y) = \mathrm{e}^{-|x-y|/\ell}$ for a correlation length $\ell > 0$. As a function of x, $W(1,x)$ is a stochastic process with mean zero and exponential covariance as in Example 6.6.

To discretise in time, apply the semi-implicit Euler–Maruyama method to get

$$\boldsymbol{u}_{J,n+1} = \left(I + \Delta t\, \varepsilon\, A^{\mathrm{N}}\right)^{-1}\left[\boldsymbol{u}_{J,n} + \Delta t\, \boldsymbol{f}(\boldsymbol{u}_{J,n}) + \sigma\, \Delta \boldsymbol{W}_n\right],$$

where A^{N} is the $(J+1)\times(J+1)$ matrix defined in (3.51) and $\boldsymbol{u}_{J,n} \in \mathbb{R}^{J+1}$ approximates $[u(t_n,x_0),\dots,u(t_n,x_J)]^{\mathsf{T}}$. Here, $\Delta\boldsymbol{W}_n \sim \mathrm{N}(\boldsymbol{0},\Delta t\, C)$ *iid* and C has entries $q(x_i,x_j) = \mathrm{e}^{-|i-j|a/\ell J}$ for $i,j = 0,\dots,J$. The increments $\Delta\boldsymbol{W}_n$ are readily sampled with the circulant embedding method; see Algorithm 6.9. This is implemented in Algorithm 10.7. For $a = 20$, $\varepsilon = 1$, $\alpha = -1/2$, $\sigma = 1$, and $\ell = 1$, we obtain a numerical solution for $t \in [0,1]$ with $\Delta t = 10^{-3}$ and $J = 1024$ with the commands

```
>> a=20; J=1024; x=(0:a/J:a)'; u0=1./(1+exp(-(2-x)/sqrt(2)));
>> ell=1; N=1e3; T=1; epsilon=1; sigma=1;
>> [t,ut]=spde_fd_n_exp(u0,T,a,N,J,epsilon,sigma,ell,@(u)u.*(1-u).*(u+0.5));
```

Figure 10.8 shows an approximate realisation of $u(1,x)$ for noise with spatial correlation length (a) $\ell = 1$ and (b) $\ell = 0.1$. The realisation in (b), where the correlation length is smaller, shows more variation.

Algorithm 10.7 Code to approximate sample paths of the solution to (10.40) given Neumann boundary conditions and a Q-Wiener process with kernel $q(x,y) = \mathrm{e}^{-|x-y|/\ell}$. The inputs are similar to those in Algorithm 3.4, with additional arguments `sigma`, `ell` to specify the noise strength σ and spatial correlation length ℓ. The outputs are `t=` $[t_0,\dots,t_N]^{\mathsf{T}}$ and a matrix `ut`, with columns approximating one sample of $[u(t_n,x_0),\dots,u(t_n,x_J)]^{\mathsf{T}}$, $n = 0,\dots,N$.

```
function [t,ut]=spde_fd_n_exp(u0,T,a,N,J,epsilon,sigma,ell,fhandle)
Dt=T/N;  t=[0:Dt:T]'; h=a/J;
% set matrices
e = ones(J+1,1);  A = spdiags([-e 2*e -e], -1:1, J+1, J+1);
% take Neumann boundary conditions
ind=1:J+1; A(1,2)=-2; A(end,end-1)=-2;
EE=speye(length(ind))+Dt*epsilon*A/h/h;
ut=zeros(J+1,length(t)); % initialize vectors
ut(:,1)=u0; u_n=u0(ind); % set initial condition
flag=false;
for k=1:N, % time loop
  fu=fhandle(u_n); % evaluate f
  if flag==false, % generate two samples
      [x,dW,dW2]=circulant_exp(length(ind), h, ell); flag=true;
  else % use second sample from last call
      dW=dW2; flag=false;
  end;
  u_new=EE\(u_n+Dt*fu+sigma*sqrt(Dt)*dW); % update u
  ut(ind,k+1)=u_new; u_n=u_new;
end
```

We present one more example with finite differences, to illustrate again the delicacy of the case of space–time white noise.

Example 10.31 (space–time white noise) Space–time white noise is difficult to interpret and, while a reaction–diffusion equation forced by white noise in one dimension can be interpreted with mild solutions, numerical approximation remains tricky. The covariance $Q = I$ is not trace class and its kernel is the delta function. To derive an approximation to the increment $W(t_{n+1}) - W(t_n)$, we truncate the expansion (10.20) for the basis $\chi_j(t) = \sqrt{2/a}\sin(j\pi x/a)$ of $L^2(0,a)$ to J terms and write

$$W^J(t,x) = \sqrt{2/a}\sum_{j=1}^{J} \sin\Big(\frac{j\pi x}{a}\Big)\,\beta_j(t),$$

for *iid* Brownian motions $\beta_j(t)$. This is now well defined (in the previous example it was unnecessary to introduce a truncated expansion). The covariance

$$\mathrm{Cov}(W^J(t,x_i), W^J(t,x_k)) = \mathbb{E}\Big[W^J(t,x_i)W^J(t,x_k)\Big] = \frac{2t}{a}\sum_{j=1}^{J}\sin\Big(\frac{j\pi x_i}{a}\Big)\sin\Big(\frac{j\pi x_k}{a}\Big).$$

Using $x_i = ih$ and $h = a/J$ with a trigonometric identity gives

$$2\sin\Big(\frac{j\pi x_i}{a}\Big)\sin\Big(\frac{j\pi x_k}{a}\Big) = \cos\Big(\frac{j\pi(i-k)}{J}\Big) - \cos\Big(\frac{j\pi(i+k)}{J}\Big).$$

Moreover,

$$\sum_{j=1}^{J}\cos\Big(\frac{j\pi m}{J}\Big) = \begin{cases} J, & m = 0,\\ 0, & m \text{ even and } m \neq 0,\\ -1, & m \text{ odd}.\end{cases}$$

and, hence, $\mathrm{Cov}(W^J(t,x_i), W^J(t,x_k)) = (t/h)\,\delta_{ik}$ for $i,k = 1,\dots,J$.

We use $W^J(t)$ to approximate (10.40) in the case that $W(t)$ is a space–time white noise. Consider homogeneous Dirichlet boundary conditions and define an approximation $\boldsymbol{u}_J(t)$ to $[u(t,x_1),\dots,u(t,x_{J-1})]^{\mathsf{T}}$ by

$$d\boldsymbol{u}_J = \Big[-\varepsilon\, A^{\mathrm{D}}\boldsymbol{u}_J + \boldsymbol{f}(\boldsymbol{u}_J)\Big]\,dt + \sigma\, d\boldsymbol{W}^J(t)$$

for $\boldsymbol{W}^J(t) := [W^J(t,x_1),\dots,W^J(t,x_{J-1})]^{\mathsf{T}}$. We have $\boldsymbol{W}^J(t) \sim \mathrm{N}(\boldsymbol{0},(t/h)\,I)$. For a time step $\Delta t > 0$, the semi-implicit Euler–Maruyama method gives

$$\boldsymbol{u}_{J,n+1} = (I + \varepsilon\, A^{\mathrm{D}}\Delta t)^{-1}\Big[\boldsymbol{u}_{J,n} + \Delta t\,\boldsymbol{f}(\boldsymbol{u}_{J,n}) + \sigma\Delta\boldsymbol{W}_n\Big] \tag{10.42}$$

and $\Delta\boldsymbol{W}_n \sim \mathrm{N}(\boldsymbol{0},(\Delta t/h)I)$ *iid*. This method gives an approximation $\boldsymbol{u}_{J,n}$ to $[u(t_n,x_1),\dots,u(t_n,x_{J-1})]^{\mathsf{T}}$ for $t_n = n\Delta t$ and is implemented in Algorithm 10.8.

In the case $f = 0$, $\varepsilon = 1$, and $u_0 = 0$, the solution of (10.40) is a Gaussian random field and (10.30) with $q_j = 1$ gives that

$$\|u(t)\|^2_{L^2(\Omega,L^2(0,a))} = \sigma^2\sum_{j=1}^{\infty}\frac{1}{2\lambda_j}(1 - \mathrm{e}^{-2\lambda_j t}). \tag{10.43}$$

For homogeneous Dirichlet boundary conditions, the eigenvalues $\lambda_j = \varepsilon(\pi j/a)^2$ and we can evaluate $\|u\|^2_{L^2(\Omega,L^2(0,a))}$ to high precision. We use this to test the finite difference method. Specifically, we use the Monte Carlo method to approximate the expectation and the trapezium rule (A.1) to approximate the $L^2(0,a)$ norm, via

$$\|u(t_n)\|^2_{L^2(\Omega,L^2(0,a))} \approx \frac{1}{M}\sum_{m=1}^{M}\sum_{j=1}^{J}\|u^m(t_n,x_j)\|^2 h, \tag{10.44}$$

for *iid* samples $u^m(t_n,x_j)$ of the solution. The right-hand side can be approximated numerically by replacing $u^m(t_n,x_j)$ by its finite difference approximation. We do this in Algorithm 10.9, which calls Algorithm 10.8, and compare the resulting approximation to $\|u(t_n)\|_{L^2(\Omega,L^2(0,a))}$ with the exact formula (10.43). For the case $t_n = T = 0.1$, $a = 1$, $\epsilon = 0.1$, and $\sigma = 1$, (10.43) gives $\|u(t_n)\|^2_{L^2(\Omega,L^2(0,a))} \approx 0.3489$. To determine the approximation for a range of discretisation parameters, we use the following MATLAB commands:

```
>> T=0.1; epsilon=0.1; sigma=1; a=1;
>> M=1e4; J=4*2.^[2,3,4]'; N=0.25*J.^2;
   for i=1:length(N),
      u0=zeros(J(i)+1,1);
      v(i)=l2_sq_mct(T,a,N(i),J(i),M,epsilon,sigma)
   end;
```

Here, J determines the grid spacing $h = a/J$ and N determines the time step $\Delta t = T/N$. We make N/J^2 constant so that $\Delta t/h^2$ is independent of J and hence the CFL condition is satisfied. In one experiment, MATLAB computed the errors 1.562×10^{-3}, 3.906×10^{-4}, and 9.766×10^{-5} for $N = 64, 256, 1024$, which behave like $0.96\Delta t^{0.65}$ (by using `polyfit`). Theoretically, this is a type of weak convergence (see §8.6) and the error for this problem under the CFL condition is order $\Delta t^{1/2}$. This is half the rate of weak convergence for SODEs and small values of h and Δt are required for accuracy. We normally expect the rate of strong convergence to be half the rate of weak convergence (see Example 5.30 or compare Exercise 8.11 to Theorem 8.45). We therefore expect strong errors of order $\Delta t^{1/4}$ and we prove this rate occurs for the spectral Galerkin approximation with the semi-implicit Euler–Maruyama method in Theorem 10.34 (case $\zeta = 1/2$).

Algorithm 10.8 Code to generate realisations of the finite difference approximation to the solution to (10.40) with homogeneous Dirichlet boundary conditions and space–time white noise. Inputs and outputs are similar to those in Algorithm 10.7.

```
function [t,ut]=spde_fd_d_white(u0,T,a,N,J,epsilon,sigma,fhandle)
Dt=T/N;  t=[0:Dt:T]'; h=a/J;
% set matrices
e = ones(J+1,1);  A = spdiags([-e 2*e -e], -1:1, J+1, J+1);
% take zero Dirichlet boundary conditions
ind=2:J;  A=A(ind,ind);
EE=speye(length(ind))+Dt*epsilon*A/h/h;
ut=zeros(J+1,length(t)); % initialize vectors
ut(:,1)=u0; u_n=u0(ind); % set initial condition
for k=1:N, % time loop
  fu=fhandle(u_n); Wn=sqrt(Dt/h)*randn(J-1,1);
  u_new=EE\(u_n+Dt*fu+sigma*Wn);
  ut(ind,k+1)=u_new; u_n=u_new;
end
```

Algorithm 10.9 Code to approximate $\|u(T)\|_{L^2(\Omega,L^2(0,a))}$ with finite differences, assuming $f = 0$ and $u_0 = 0$. The inputs are similar to those in Algorithm 10.8 with the additional input of the number of samples M. The output gives (10.44).

```
function [out]=l2_sq_mct(T,a,N,J,M,epsilon,sigma)
v=0; u0=zeros(J+1,1);
parfor i=1:M,
    [t,ut]=spde_fd_d_white(u0,T,a,N,J,epsilon,sigma,@(u) 0);
    v=v+sum(ut(1:end-1,end).^2); % cumulative sum
end;
out= v*a/J/M; % return average
```

10.6 Galerkin and semi-implicit Euler approximation

For the Galerkin approximation of the stochastic evolution equation (10.25), we introduce a finite-dimensional subspace $\widetilde{V} \subset \mathcal{D}(A^{1/2})$ as in §3.5 and let $\widetilde{P}$ be the orthogonal projection $\widetilde{P}\colon H \to \widetilde{V}$ (see Lemma 1.41). Then, we seek the approximation $\tilde{u}(t) \in \widetilde{V}$ defined by

$$\begin{aligned}\langle \tilde{u}(t), v\rangle &= \langle \tilde{u}_0, v\rangle + \int_0^t \big[-a(\tilde{u}(s), v) + \langle f(\tilde{u}(s)), v\rangle\big]\, ds \\ &\quad + \left\langle \int_0^t G(\tilde{u}(s))\, dW(s), v\right\rangle, \qquad t \in [0,T],\ v \in \widetilde{V}\end{aligned} \tag{10.45}$$

for initial data $\tilde{u}_0 = \widetilde{P} u_0$. In other words, we take the definition of weak solution (10.27) and change $\langle u, Av\rangle$ to $a(u,v) := \langle u, v\rangle_{1/2}$ (see Lemma 1.89) and choose test functions from $\widetilde{V}$ rather than $\mathcal{D}(A)$. The Galerkin approximation $\tilde{u}(t)$ also satisfies

$$d\tilde{u} = \big[-\widetilde{A}\,\tilde{u} + \widetilde{P} f(\tilde{u})\big]\, dt + \widetilde{P}\, G(\tilde{u})\, dW(t), \qquad \tilde{u}(0) = \tilde{u}_0, \tag{10.46}$$

where $\widetilde{A}$ is defined by (3.60).

For the discretisation in time, we approximate $\tilde{u}(t_n)$ by $\tilde{u}_n$ for $t_n = n\Delta t$. Because $\widetilde{V}$ is finite dimensional, (10.46) is an SODE and we can apply the semi-implicit Euler–Maruyama method (8.121) with time step $\Delta t > 0$ to define $\tilde{u}_n$. That is, given $\tilde{u}_0$, we find $\tilde{u}_n$ by iterating

$$\tilde{u}_{n+1} = (I + \Delta t \widetilde{A})^{-1}\big(\tilde{u}_n + \widetilde{P} f(\tilde{u}_n)\,\Delta t + \widetilde{P}\, G(\tilde{u}_n)\,\Delta W_n\big) \tag{10.47}$$

for $\Delta W_n := \int_{t_n}^{t_{n+1}} dW(s)$. In practice, it is necessary to approximate G with some $\mathcal{G}\colon \mathbb{R}^+ \times H \to L_0^2$ and we study the approximation $\tilde{u}_n$ defined by

$$\tilde{u}_{n+1} = (I + \Delta t \widetilde{A})^{-1}\left(\tilde{u}_n + \widetilde{P} f(\tilde{u}_n)\,\Delta t + \widetilde{P} \int_{t_n}^{t_{n+1}} \mathcal{G}(s; \tilde{u}_n)\, dW(s)\right), \tag{10.48}$$

for initial data $\tilde{u}_0 = \widetilde{P} u_0$. For example, we may take $\mathcal{G}(s;u) = G(u)$, which gives (10.47). The difficulty here is that $\mathcal{G}(s;u)$ acts on the infinite-dimensional U-valued process $W(t)$ in (10.48) and this is difficult to implement as a numerical method. Given an orthonormal basis $\{\chi_j : j \in \mathbb{N}\}$ of U, we usually consider $\mathcal{G}(s;u) = G(u)\mathcal{P}_{J_w}$ for the orthogonal projection $\mathcal{P}_{J_w} : U \to \operatorname{span}\{\chi_1, \ldots, \chi_{J_w}\}$.

We make $\mathcal{G}$ satisfy the following assumption. Here, δ is the spatial discretisation parameter of Assumption 3.48 ($\delta = h$ for the Galerkin finite element method and $\delta = 1/\sqrt{\lambda_J}$ for the spectral Galerkin method).

Assumption 10.32 For some $\zeta \in (0,2]$, Assumption 10.23 holds. The function $\mathcal{G}\colon \mathbb{R}^+ \times H \to L_0^2$ satisfies, for some constants $K_{\mathcal{G}}, \theta, L > 0$,

$$\|\mathcal{G}(s;u_1) - \mathcal{G}(s;u_2)\|_{L_0^2} \le L\|u_1 - u_2\|, \qquad \forall s > 0,\ u_1, u_2 \in H, \tag{10.49}$$

and for $t_k \le s < t_{k+1}$

$$\|\widetilde{P}(G(u(s)) - \mathcal{G}(s;u(t_k)))\|_{L^2(\Omega,L_0^2)} \le K_{\mathcal{G}}(|s-t_k|^\theta + \delta^\zeta), \tag{10.50}$$

where $u(t)$ is the solution of (10.25).

We prefer to write $\mathcal{G}(u)$ for $\mathcal{G}(s;u)$ when it is independent of s. The assumption holds for $\mathcal{G}(s,u) \equiv \mathcal{G}(u) := G(u)\mathcal{P}_{J_w}$ for a broad class of Q-Wiener processes if the number of terms J_w is sufficiently large, as we now show.

Lemma 10.33 *Let the assumptions of Lemma 10.27 hold for some $\zeta \ge 1$. Let $W(t)$ be the Q-Wiener process on U defined by (10.11), where the eigenvalues $q_j = \mathcal{O}(j^{-(2r+1+\epsilon)})$, some $r \ge 0$ and $\epsilon > 0$. Define $\mathcal{G}(u) := G(u)\mathcal{P}_{J_w}$, where $\mathcal{P}_{J_w}$ is the orthogonal projection from U to* $\operatorname{span}\{\chi_1, \dots, \chi_{J_w}\}$*. Then, Assumption 10.32 holds for $\theta = \min\{\zeta/2 - \epsilon, 1/2\}$ if $J_w^{-r} = \mathcal{O}(\delta^\zeta)$.*

Proof First, G and hence $\mathcal{G}$ satisfy Assumption 10.32 for some $\zeta \ge 1$ so that (10.49) holds. For (10.50), first note that $\widetilde{P}(G(u(s)) - \mathcal{G}(u(t_k))) = \mathtt{I} + \mathtt{II}$, for

$$\mathtt{I} := \widetilde{P}(G(u(s)) - G(u(t_k))), \qquad \mathtt{II} := \widetilde{P}\Big(G(u(t_k)) - G(u(t_k))\mathcal{P}_{J_w}\Big). \tag{10.51}$$

The Lipschitz property gives

$$\|G(u(s)) - G(u(t_k))\|_{L^2(\Omega,L_0^2)} \le L\|u(s) - u(t_k)\|_{L^2(\Omega,H)}.$$

Using (10.38) with $\zeta \ge 1$,

$$\|\mathtt{I}\|_{L^2(\Omega,L_0^2)} \le \|G(u(s)) - G(u(t_k))\|_{L^2(\Omega,L_0^2)} \le L\,K_{RT}|s-t_k|^\theta.$$

For the second term, using (10.34) and (10.36),

$$\begin{aligned}\|\mathtt{II}\|_{L^2(\Omega,L_0^2)} &\le \|G(u(t_k))\|_{L^2(\Omega,\mathcal{L}(H))}\Big\|\Big(I - \mathcal{P}_{J_w}\Big)Q^{1/2}\Big\|_{\mathrm{HS}(U,H)}\\ &\le L(K_T+1)\left(1+\|u_0\|\right)\Big\|\Big(I - \mathcal{P}_{J_w}\Big)Q^{1/2}\Big\|_{\mathrm{HS}(U,H)}.\end{aligned}$$

Under the decay condition on q_j,

$$\|(I - \mathcal{P}_{J_w})Q^{1/2}\|_{\mathrm{HS}(U,H)} = \left(\sum_{j=J_w+1}^{\infty} q_j\right)^{1/2} = \mathcal{O}(J_w^{-r}).$$

The two bounds imply (10.50) when $J_w^{-r} = \mathcal{O}(\delta^\zeta)$. □

Strong convergence: semi-implicit Euler–Maruyama

We present a convergence theory for the semi-implicit Euler method, before demonstrating the convergence rates numerically for stochastic reaction–diffusion equations in §10.7 and §10.8. A general Galerkin subspace $\widetilde{V}$ is considered that fits into the framework of

§3.7 and we analyse convergence in the *strong* sense (see §8.5 for SODEs) for the error $\|u(t_n) - \tilde{u}_n\|_{L^2(\Omega,H)}$.

Theorem 10.34 (strong convergence) *Let the following assumptions hold:*

(i) *the assumptions of Theorem 10.26 for a unique mild solution $u(t)$ of* (10.25),
(ii) *the initial data $u_0 \in L^2(\Omega, \mathcal{F}_0, \mathcal{D}(A))$,*
(iii) *Assumption 3.48 on the Galerkin subspace $\widetilde{V}$ for discretisation parameter δ, and*
(iv) *Assumption 10.32 on $\mathcal{G}$ for some $\theta > 0$ and $\zeta \in (0,2]$.*

Let $\tilde{u}_n$ be the numerical approximation defined by (10.47). *Let $\theta_1 := \min\{\zeta/2 - \epsilon, 1/2\}$. Fix $T > 0$ and consider the limit $\Delta t, \delta \to 0$.*

Case $\zeta \in [1,2]$: *For each $\epsilon > 0$, there exists $K > 0$ independent of Δt and δ such that*

$$\max_{0 \le t_n \le T} \|u(t_n) - \tilde{u}_n\|_{L^2(\Omega,H)} \le K\big(\Delta t^{\theta_1} + \delta^{\zeta}\Delta t^{-\epsilon} + \Delta t^{\theta}\big).$$

Case $\zeta \in (0,2]$: *For each $\epsilon > 0$, there exists $K > 0$ such that*

$$\max_{0 \le t_n \le T} \|u(t_n) - \tilde{u}_n\|_{L^2(\Omega,H)} \le K\big(\Delta t^{\theta_1} + \Delta t^{\theta}\big),$$

where K can be chosen uniformly as $\Delta t, \delta \to 0$ with $\Delta t/\delta^2$ fixed (the CFL condition).

Proof Using the notation $\widetilde{S}_{\Delta t} := (I + \Delta t \widetilde{A})^{-1}$, we write the approximation obtained after n steps of (10.48) as

$$\tilde{u}_n = \widetilde{S}_{\Delta t}^{n} \widetilde{P} u_0 + \sum_{k=0}^{n-1} \widetilde{S}_{\Delta t}^{n-k} \widetilde{P} f(\tilde{u}_k)\,\Delta t + \sum_{k=0}^{n-1} \widetilde{S}_{\Delta t}^{n-k} \widetilde{P} \int_{t_k}^{t_{k+1}} \mathcal{G}(s, \tilde{u}_k)\,dW(s).$$

Subtracting from the mild solution (10.33), $u(t_n) - \tilde{u}_n = \mathtt{I} + \mathtt{II} + \mathtt{III}$ for

$$\begin{aligned}
\mathtt{I} &:= \mathrm{e}^{-t_n A} u_0 - \widetilde{S}_{\Delta t}^{n} \widetilde{P} u_0,\\
\mathtt{II} &:= \sum_{k=0}^{n-1} \left(\int_{t_k}^{t_{k+1}} \mathrm{e}^{-(t_n - s)A} f(u(s))\,ds - \widetilde{S}_{\Delta t}^{n-k} \widetilde{P} f(\tilde{u}_k)\,\Delta t \right),\\
\mathtt{III} &:= \sum_{k=0}^{n-1} \int_{t_k}^{t_{k+1}} \big(\mathrm{e}^{-(t_n - s)A} G(u(s)) - \widetilde{S}_{\Delta t}^{n-k} \widetilde{P} \mathcal{G}(s, \tilde{u}_k)\big)\,dW(s).
\end{aligned}$$

To treat $\mathtt{I}$, apply Theorem 3.54 (with $\gamma = 1$ and the $L^2(\Omega, H)$ in place of the H norm) to find $\|\mathtt{I}\|_{L^2(\Omega,H)} \le C_{\mathtt{I}}\,(\Delta t + \delta^2)$, for a constant $C_{\mathtt{I}}$ independent of δ and Δt. The second term can be treated as in Theorem 3.55, using Lemma 10.27 in place of Proposition 3.31. This leads to the following estimate, for a constant $C_{\mathtt{II}}$,

$$\|\mathtt{II}\|_{L^2(\Omega,H)} \le C_{\mathtt{II}}(\Delta t^{\theta_1} + \delta^2).$$

We break $\mathtt{III}$ into four further parts by writing

$$\mathrm{e}^{-(t_n - s)A} G(u(s)) - \widetilde{S}_{\Delta t}^{n-k} \widetilde{P} \mathcal{G}(s, \tilde{u}_k) = X_1 + X_2 + X_3 + X_4$$

for

$$X_1 := \left(\mathrm{e}^{-(t_n-s)A} - \mathrm{e}^{-(t_n-t_k)A}\right)G(u(s)), \qquad X_2 := \left(\mathrm{e}^{-(t_n-t_k)A} - \widetilde{S}_{\Delta t}^{n-k}\widetilde{P}\right)G(u(s)),$$
$$X_3 := \widetilde{S}_{\Delta t}^{n-k}\widetilde{P}\left(G(u(s)) - \mathcal{G}(s;u(t_k))\right), \qquad X_4 := \widetilde{S}_{\Delta t}^{n-k}\widetilde{P}\left(\mathcal{G}(s;u(t_k)) - \mathcal{G}(s;\tilde{u}_k)\right).$$

To estimate $\mathtt{III}$ in $L^2(\Omega,H)$, we estimate $\mathtt{III}_i = \int_0^{t_n} X_i\,dW(s)$ separately using the triangle inequality and the Itô isometry (10.24). We focus on the case $\zeta \in (0,1]$ and assume without loss of generality that $\Delta t/\delta^2 = 1$. The constants in the estimates we develop can be chosen independently of any Δt and δ satisfying $\Delta t/\delta^2 = 1$. First,

$$\begin{aligned}\|\mathtt{III}_1\|_{L^2(\Omega,H)}^2 &= \mathbb{E}\left[\left\|\sum_{k=0}^{n-1}\int_{t_k}^{t_{k+1}} X_1\,dW(s)\right\|^2\right]\\ &= \sum_{k=0}^{n-1}\int_{t_k}^{t_{k+1}} \mathbb{E}\left[\left\|\left(\mathrm{e}^{-(t_n-s)A} - \mathrm{e}^{-(t_n-t_k)A}\right)G(u(s))\right\|_{L_0^2}^2\right]ds.\end{aligned}$$

By Assumption 10.23,

$$\begin{aligned}\left\|\left(\mathrm{e}^{-(t_n-s)A} - \mathrm{e}^{-(t_n-t_k)A}\right)G(u)\right\|_{L_0^2} &= \left\|\left(\mathrm{e}^{-(t_n-s)A} - \mathrm{e}^{-(t_n-t_k)A}\right)A^{(1-\zeta)/2}A^{(\zeta-1)/2}G(u)\right\|_{L_0^2}\\ &\le \left\|\left(\mathrm{e}^{-(t_n-s)A} - \mathrm{e}^{-(t_n-t_k)A}\right)A^{(1-\zeta)/2}\right\|_{\mathcal{L}(H)}L\left(1+\|u\|\right).\end{aligned}$$

Using Lemma 3.22, there exists a constant $K_1 > 0$ such that

$$\begin{aligned}&\int_{t_k}^{t_{k+1}} \left\|\left(\mathrm{e}^{-(t_n-s)A} - \mathrm{e}^{-(t_n-t_k)A}\right)A^{(1-\zeta)/2}\right\|_{\mathcal{L}(H)}^2 ds\\ &\quad= \int_{t_k}^{t_{k+1}} \left\|A^{1/2-\epsilon}\mathrm{e}^{-(t_n-t_k)A}A^{-\zeta/2+\epsilon}\left(I - \mathrm{e}^{-(t_k-s)A}\right)\right\|_{\mathcal{L}(H)}^2 ds\\ &\quad\le K_1^2\Delta t^{\zeta-2\epsilon}\int_{t_k}^{t_{k+1}} \left\|A^{1/2-\epsilon}\mathrm{e}^{-(t_n-t_k)A}\right\|_{\mathcal{L}(H)}^2 ds, \qquad 0 \le t_k < t_{k+1} \le T.\end{aligned}$$

As in Exercise 10.8, we can find a $K_2 > 0$ so that

$$\sum_{k=0}^{n-1}\int_{t_k}^{t_{k+1}} \left\|A^{1/2-\epsilon}\mathrm{e}^{-(t_n-t_k)A}\right\|_{\mathcal{L}(H)}^2 ds \le K_2^2\frac{T^{2\epsilon}}{\epsilon}.$$

Using Theorem 10.26 to bound $\|u(s)\|_{L^2(\Omega,H)}$ for $0 \le s \le T$,

$$\|\mathtt{III}_1\|_{L^2(\Omega,H)}^2 \le C_1^2\,\Delta t^{\zeta-2\epsilon}$$

for some constant $C_1 > 0$.

For the second term, the definition of $\widetilde{T}_k$ in (3.97) gives $X_2 = \widetilde{T}_{n-k}G(u(s))$ and

$$\|\mathtt{III}_2\|_{L^2(\Omega,H)}^2 = \left\|\sum_{k=0}^{n-1}\int_{t_k}^{t_{k+1}} X_2\,dW(s)\right\|^2 = \sum_{k=0}^{n-1}\int_{t_k}^{t_{k+1}} \mathbb{E}\left[\left\|\widetilde{T}_{n-k}\,G(u(s))\right\|_{L_0^2}^2\right]ds.$$

Theorem 3.54 provides upper bounds on $\|\widetilde{T}_k v_0\|$ that depend on the regularity of v_0. For

$\zeta \in (0,1)$, we take $\alpha = (1-\zeta)/2$ and (3.98) gives a constant $K_2 > 0$ such that

$$\|\widetilde{T}_k G(u)\|_{L_0^2} \le K_2\left(\frac{\Delta t\, \delta^{\zeta-1}}{t_k} + \frac{\delta^2}{t_k^{(3-\zeta)/2}}\right)\|A^{(\zeta-1)/2}G(u)\|_{L_0^2}$$

$$= K_2\left(\frac{\delta^{\zeta-1}}{k} + \frac{\delta^2\,\Delta t^{(\zeta-3)/2}}{k^{(3-\zeta)/2}}\right)\|A^{(\zeta-1)/2}G(u)\|_{L_0^2}, \qquad k > 0.$$

Now, $\sum_{k=1}^{\infty} 1/k^r < \infty$ for $r > 1$. Using also Assumption 10.23 and $\Delta t = \delta^2$, we can find a constant $C_2 > 0$ with

$$\|\mathtt{III}_2\|_{L^2(\Omega,H)}^2 \le \frac{1}{2}C_2^2\left(\delta^{2(\zeta-1)}\Delta t + \delta^2\,\Delta t^{(\zeta-1)}\right) = C_2^2\,\Delta t^{\zeta}.$$

For $\zeta \in [1,2)$, (3.99) can be applied in place of (3.98) and we find that $\|\mathtt{III}_2\|_{L^2(\Omega,H)}^2 \le \frac{1}{2}C_2^2(\delta^{2\zeta} + \Delta t^{\zeta}) \le C_2^2\Delta t^{\zeta}$. For $\zeta = 2$, $\|\mathtt{III}_2\|_{L^2(\Omega,H)}^2 \le \frac{1}{2}C_2^2(\delta^4 + \Delta t^2)\Delta t^{-2\epsilon}$.

The necessary upper bound for $\mathtt{III}_3$ follows from (10.50) using the boundedness of $\widetilde{S}_{\Delta t}$ and $\widetilde{P}$. Then, for $C_3 := K_{\mathcal{G}}$, we have

$$\|\mathtt{III}_3\|_{L^2(\Omega,H)} \le C_3(\Delta t^{\theta} + \delta^{\zeta}).$$

For the last term, $\mathtt{III}_4$, denoting the Lipschitz constant of $\mathcal{G}$ by L,

$$\|\mathtt{III}_4\|_{L^2(\Omega,H)}^2 = \mathbb{E}\left[\left\|\sum_{k=0}^{n-1}\int_{t_k}^{t_{k+1}} X_4(s)\,dW(s)\right\|^2\right]$$

$$= \sum_{k=0}^{n-1}\int_{t_k}^{t_{k+1}} \mathbb{E}\left[\left\|\widetilde{S}_{\Delta t}^{n-k}\,\widetilde{P}(\mathcal{G}(s;u(t_k)) - \mathcal{G}(s;\tilde{u}_k))\right\|_{L_0^2}^2\right] ds$$

$$\le L^2\sum_{k=0}^{n-1}\int_{t_k}^{t_{k+1}} \mathbb{E}\left[\|u(t_k) - \tilde{u}_k\|^2\right] ds.$$

Using the triangle inequality to sum the bounds for $\mathtt{III}_1,\dots,\mathtt{III}_4$, we find that for a constant $C_{\mathtt{III}} > 0$

$$\mathbb{E}\left[\|\mathtt{III}\|^2\right] \le C_{\mathtt{III}}\left(\Delta t^{\zeta-2\epsilon} + \Delta t^{2\theta} + L^2\sum_{k=0}^{n-1}\int_{t_k}^{t_{k+1}}\mathbb{E}\left[\|u(t_k) - \tilde{u}_k\|^2\right] ds\right).$$

Combining the bounds on the norms of $\mathtt{I}$, $\mathtt{II}$, and $\mathtt{III}$, we use the discrete Gronwall lemma (Lemma A.14) as with Theorem 3.55 to complete the proof. □

Example 10.35 (reaction–diffusion) Consider the SPDE

$$du = \left[-Au + f(u)\right]dt + G(u)\,dW(t),$$

where $A = -\Delta$ with $\mathcal{D}(A) = H^2(0,1) \cap H_0^1(0,1)$ and f is globally Lipschitz. If $W(t)$ is a Q-Wiener process and Assumption 10.23 holds for some $\zeta \in (1,2)$, then $G(u)W(t)$ is smooth and in particular takes values in $\mathcal{D}(A^{(\zeta-1)/2}) = H_0^{\zeta-1}(0,1)$. Choose $\mathcal{G}(u) = G(u)\mathcal{P}_{J_w}$ so that Assumption 10.32 holds with $\theta = 1/2$ (see Lemma 10.33). Consider non-random initial data $u_0 \in H^2(0,1) \cap H_0^1(0,1)$. Theorem 10.34 gives

$$\max_{0 \le t_n \le T}\|u(t_n) - \tilde{u}_n\|_{L^2(\Omega,H)} = \mathcal{O}(\Delta t^{1/2} + \delta^{\zeta}). \tag{10.52}$$

We explore this rate of convergence numerically for the spectral and finite element Galerkin approximations in §10.7 and §10.8. For additive space–time white noise (i.e., $W(t)$ is the cylindrical Wiener process), we are restricted to $\zeta \in (0, 1/2)$ and Theorem 10.34 gives

$$\max_{0 \le t_n \le T} \left\| u(t_n) - \tilde{u}_n \right\|_{L^2(\Omega, H)} = \mathcal{O}\big(\Delta t^{1/4-\epsilon} + \Delta t^{\theta}\big), \qquad \epsilon > 0,$$

when the CFL condition holds (the dependence on Δt and δ is not given separately). The convergence rate is reduced by a factor of 2, owing to the roughness of the noise. We show how to choose $\mathcal{G}(u)$ so that the $\mathcal{O}(\Delta t^{\theta})$ term is negligible for problems with additive noise using the spectral Galerkin method in Corollary 10.38.

The Milstein method for SPDEs

We consider an alternative to the semi-implicit Euler–Maruyama time stepping that achieves a higher rate of convergence when $\zeta > 1$ in Assumption 10.32. It applies when G is Nemytskii (Lemma 10.24). In terms of SODEs, the Nemytskii assumption is like the diffusion being diagonal, in which case the Milstein method becomes simple and effective; see (8.49). We briefly discuss the Milstein method for the stochastic evolution equation (10.25). To simplify matters, we suppose that $H = U = L^2(D)$ and that Q has kernel $q(\boldsymbol{x}, \boldsymbol{y})$; see (10.35). For a smooth function $g : \mathbb{R} \to \mathbb{R}$, let G and G' be the Nemytskii operators associated with g and its derivative g'.

Discretising (10.46) with the Milstein method applied to G and keeping the drift term semi-implicit, we define an approximation $\tilde{u}_n$ to $u(t_n)$ by

$$\tilde{u}_{n+1} = \big(I + \Delta t \widetilde{A}\big)^{-1}\Big(\tilde{u}_n + \widetilde{P} f(\tilde{u}_n)\Delta t + \widetilde{P}\big[G(\tilde{u}_n)\Delta W_n + G'(\tilde{u}_n)G(\tilde{u}_n)I_{t_n,t_{n+1}}\big]\Big), \qquad (10.53)$$

where $\Delta W_n := \int_{t_n}^{t_{n+1}} dW(s) = W(t_{n+1}) - W(t_n)$ and

$$I_{s,t}(\boldsymbol{x}) := \frac{1}{2}\Big(\big(W(t, \boldsymbol{x}) - W(s, \boldsymbol{x})\big)^2 - q(\boldsymbol{x}, \boldsymbol{x})(t - s)\Big).$$

Compare with (8.49). For fixed $\boldsymbol{x} \in D$, $\beta(t) = W(t, \boldsymbol{x}) / \sqrt{q(\boldsymbol{x}, \boldsymbol{x})}$ is a standard Brownian motion. Therefore the identity (8.114) gives

$$I_{s,t}(\boldsymbol{x}) = \frac{1}{2} q(\boldsymbol{x}, \boldsymbol{x})\Big(\big(\beta(t) - \beta(s)\big)^2 - (t - s)\Big) = q(\boldsymbol{x}, \boldsymbol{x}) \int_s^t \int_s^r d\beta(p)\, d\beta(r)$$

and

$$G'(\tilde{u}_n)G(\tilde{u}_n)I_{t_n,t_{n+1}} = \int_{t_n}^{t_{n+1}} G'(\tilde{u}_n)\, G(\tilde{u}_n) \int_{t_n}^{s} dW(r)\, dW(s).$$

We understand this equation as a multiplication of the terms G', G, and $I_{t_n,t_{n+1}}$ for each $\boldsymbol{x}$.

This method fits into the framework of Theorem 10.34 with

$$\mathcal{G}(s; u) := G(u) + G'(u) \int_{t_n}^{s} dW(r), \qquad t_n \le s < t_{n+1}. \qquad (10.54)$$

We verify Assumption 10.32 for this choice of $\mathcal{G}$ and hence determine convergence of the Milstein method. In the following, θ_2 gives the rate of convergence and, for small ϵ and $\zeta > 1$, it is larger than the rate θ_1 in Theorem 10.34.

Lemma 10.36 *Let D be a bounded domain and $H = U = L^2(D)$. Suppose that*

(i) *Assumptions* (i)–(ii) *of Theorem 10.34 hold with $\zeta \geq 1$,*
(ii) *Q has kernel $q \in C(\bar{D} \times \bar{D})$,*
(iii) *G is Nemytskii (see Lemma 10.24) and associated to a function $g: \mathbb{R} \to \mathbb{R}$ with two uniformly bounded derivatives, and*
(iv) *$u(t)$ is a weak solution of* (10.25) *(see Definition 10.19).*

For $T, \epsilon > 0$, there exists $K > 0$ such that

$$\left\|G(u(s)) - \mathcal{G}(s; u(t_k))\right\|_{L^2(\Omega, L^2_0)} \leq K|t_k - s|^{\theta_2}, \qquad 0 \leq t_k \leq s \leq T.$$

where $\theta_2 := \zeta/2 - \epsilon$ and $\mathcal{G}(s,u)$ is defined by (10.54).

Proof From the weak form (10.27), for $v \in \mathcal{D}(A)$,

$$\langle u(s) - u(t_k), v\rangle_{L^2(D)} = \int_{t_k}^{s} \langle -u(r), Av\rangle_{L^2(D)} + \langle f(u(r)), v\rangle_{L^2(D)}\, dr + \left\langle \int_{t_k}^{s} G(u(r))\, dW(r), v\right\rangle_{L^2(D)}.$$

Define $R^\sharp(t_k, s) := u(s) - u(t_k) - \int_{t_k}^{s} G(u(r))\, dW(r)$. Lemma 10.27 applies and (10.39) gives $\|R^\sharp(t_k, s)\|_{L^2(\Omega, H)} \leq K_{RT2}|t_k - s|^{\theta_2}$. We have

$$\int_{t_k}^{s} \langle -u(r), Av\rangle_{L^2(D)} + \langle f(u(r)), v\rangle_{L^2(D)}\, dr = \left\langle R^\sharp(t_k, s), v\right\rangle_{L^2(D)}.$$

This holds also for $v \in L^2(D)$. Subtracting $\langle G(u(t_k))(W(s) - W(t_k)), v\rangle$, we have

$$\langle u(s) - u(t_k) - G(u(t_k))(W(s) - W(t_k)), v\rangle_{L^2(D)} = \left\langle R^\sharp(t_k, s), v\right\rangle_{L^2(D)} + \left\langle \int_{t_k}^{s} G(u(r))\, dW(r) - \int_{t_k}^{s} G(u(t_k))\, dW(r), v\right\rangle_{L^2(D)}.$$

Let $v = R^*(t_k, s) := u(s) - u(t_k) - G(u(t_k))(W(s) - W(t_k))$. The Cauchy–Schwarz inequality on $L^2(\Omega, L^2(D))$ implies that

$$\|R^*\|_{L^2(\Omega, L^2(D))} \leq \left\|R^\sharp(t_k, s)\right\|_{L^2(\Omega, L^2(D))} + \left(\int_{t_k}^{s} \mathbb{E}\left[\|G(u(r)) - G(u(t_k))\|_{L^2_0}^2\right] dr\right)^{1/2}.$$

Use the Lipschitz condition on G and (10.39):

$$\begin{aligned}\|R^*\|_{L^2(\Omega, L^2(D))} &\leq \left\|R^\sharp(t_k, s)\right\|_{L^2(\Omega, L^2(D))} + L\,|t_k - s|^{1/2} \sup_{t_k \leq r \leq s} \|u(r) - u(t_k)\|_{L^2(\Omega, L^2(D))} \\ &\leq K_{RT2}|t_k - s|^{\theta_2} + L\, K_{RT}\, |t_k - s|^{\theta_1 + 1/2}.\end{aligned}$$

Here, we find it convenient to work with the function $g: \mathbb{R} \to \mathbb{R}$ that represents the Nemytskii operator G. As $g \in C^2(\mathbb{R})$, Taylor's theorem (Theorem A.1) gives

$$g(u(s)) = g(u(t_k)) + g'(u(t_k))\big(u(s) - u(t_k)\big) + R^\dagger(t_k, s), \tag{10.55}$$

where

$$R^\dagger(t_k, s) := \int_0^1 (1 - h)\, g''\big(u(t_k) + h\,(u(s) - u(t_k))\big)\big(u(s) - u(t_k)\big)^2\, dh.$$

As g'' is uniformly bounded, for some $K_1 > 0$, we have using (10.38)

$$\|R^{\dagger}(t_k,s)\|_{L^2(\Omega,L^2(D))} \le K_1\|u(s)-u(t_k)\|^2_{L^2(\Omega,H)} \le K_1\, K_T^2\, |s-t_k|^{2\theta_1}.$$

Substitute $u(s) - u(t_k) = R^*(t_k,s) + g(u(t_k))(W(s) - W(t_k))$ into (10.55). Then, for $R(t_k,s) := R^{\dagger}(t_k,s) + g'(u(t_k))R^*(t_k,s)$.

$$g(u(s)) = g(u(t_k)) + g'(u(t_k))\, g(u(t_k))\big(W(s) - W(t_k)\big) + R(t_k,s),$$

Now, $G(u(s)) = g(u(s))$ and

$$\mathcal{G}(s,u(t_k)) = g(u(t_k)) + g'(u(t_k))g(u(t_k))\big(W(s) - W(t_k)\big).$$

Consequently, $\|(G(u(s)) - \mathcal{G}(s,u(t_k)))\chi\|_{L^2(\Omega,\mathcal{L}(H))} \le \|R\|_{L^2(\Omega,H)}\|\chi\|_\infty$ for any $\chi \in C(\bar{D})$ and

$$\begin{aligned}\|G(u(s)) - \mathcal{G}(s;u(t_k))\|^2_{L^2(\Omega,L^2_0)} &= \sum_{j=1}^{\infty}\|(G(u(s)) - \mathcal{G}(s,u(t_k)))\chi_j\|^2_{L^2(\Omega,H)}q_j\\ &\le \|R\|^2_{L^2(\Omega,H)}\sum_{j=1}^{\infty} q_j\|\chi_j\|^2_\infty.\end{aligned}$$

The sum is finite by Mercer's theorem (Theorem 1.80) and $\|R\|_{L^2(\Omega,H)} = \mathcal{O}(|t_k - s|^{\theta_2})$. □

We give the convergence behaviour of the Milstein method with $f = 0$. The convergence for a class of nonlinear functions f are considered in Exercise 10.15.

Corollary 10.37 (Milstein) *Consider* (10.25) *with* $f = 0$. *Let Assumptions* (i)–(iii) *of Theorem 10.34 and also the assumptions of Lemma 10.36 hold. Consider* $\Delta t, \delta \to 0$ *with* $\Delta t/\delta^2$ *fixed. For any* $\epsilon, T > 0$, *there exists* $K > 0$ *such that the Milstein approximation* $\tilde{u}_n$ *defined by* (10.53) *satisfies*

$$\max_{0\le t_n\le T}\|u(t_n) - \tilde{u}_n\|_{L^2(\Omega,H)} \le K\,\Delta t^{\theta_2}, \qquad \theta_2 := \zeta/2 - \epsilon.$$

Proof This is similar to the proof of Theorem 10.34. The main difference is that II does not arise and the terms Δt^{θ_1} never appear for $\theta_1 = \min\{\zeta/2 - \epsilon, 1/2\}$, only terms $\Delta t^{\zeta/2-\epsilon}$. Lemma 10.36 should be used to replace the terms in $G(u(s)) - \mathcal{G}(s;u(t_k))$. □

We complete this chapter with a number of examples concerning the spectral and finite element Galerkin methods.

10.7 Spectral Galerkin method

Suppose that A satisfies Assumption 3.19 and let ϕ_j denote the eigenfunctions of A with eigenvalues λ_j for $j \in \mathbb{N}$. For the spectral Galerkin approximation, we choose $\tilde{V} = V_J := \operatorname{span}\{\phi_1,\dots,\phi_J\}$ and we write u_J for the method of lines approximation $\tilde{u}$, $P_J : H \to V_J$ for the orthogonal projection $\tilde{P}$, and $A_J = P_J A$ for $\tilde{A}$, to emphasise the dependence on J. See §3.5. Then, (10.46) becomes

$$du_J = \big[-A_J u_J + P_J\, f(u_J)\big]\,dt + P_J\, G(u_J)\,dW(t), \qquad u_J(0) = P_J u_0, \tag{10.56}$$

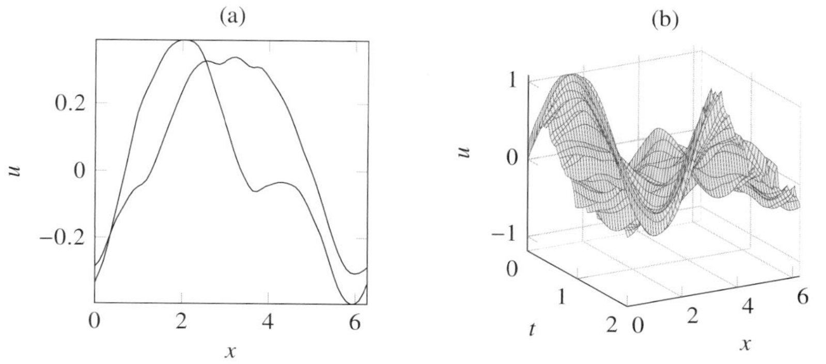

Figure 10.9 (a) Two approximate realisations at $t = 2$ and (b) an approximate sample path of the solution $u(t,x)$ to the stochastic Allen–Cahn equation (10.60) found by a spectral Galerkin and semi-implicit Euler–Maruyama approximation.

and the semi-implicit Euler–Maruyama method (10.48) becomes

$$u_{J,n+1} = \left(I + \Delta t\, A_J\right)^{-1}\left(u_{J,n} + \Delta t\, P_J\, f(u_{J,n}) + P_J\, \mathcal{G}(u_{J,n})\, \Delta W_n\right), \qquad (10.57)$$

where we choose $\mathcal{G}(u) := G(u)\mathcal{P}_{J_w}$ and initial data $u_{J,0} = P_J u_0$. This approach is especially convenient for problems with additive noise where $U = H$ and the eigenfunctions of Q and A are equal, as in Examples 10.10–10.12. In this case, $\mathcal{P}_{J_w} = P_J$ for $J_w = J$. As we now show, Assumption 10.32 holds and Theorem 10.34 gives convergence.

Corollary 10.38 (convergence, additive noise) *Suppose that $U = H$ and $\mathcal{P}_{J_w} = P_J$. Let the assumptions of Theorem 10.26 hold for $G = \sigma I$. In particular, Assumption 10.23 says that $\|A^{(1-\zeta)/2}Q^{1/2}\|_{\mathrm{HS}(U,H)} < \infty$ for some $\zeta \in (0,2]$. The spectral Galerkin and semi-implicit Euler–Maruyama approximation $u_{J,n}$ given by* (10.57) *converges to the solution $u(t_n)$ of* (10.25) *as $\Delta t \to 0$ and $J \to \infty$ with $\Delta t \lambda_J$ fixed. For $u_0 \in \mathcal{D}(A)$ and $\epsilon > 0$, there exists $K > 0$ such that*

$$\max_{0 \le t_n \le T} \left\|u(t_n) - u_{J,n}\right\|_{L^2(\Omega,H)} \le K \Delta t^{\theta_2}, \qquad \theta_2 := \zeta/2 - \epsilon. \qquad (10.58)$$

Proof Following the proof of Lemma 10.33, we show that Assumption 10.32 holds for $\mathcal{G}(u) = \sigma P_J$ with $K_{\mathcal{G}} = 0$. To do this, use (10.39) in place of (10.38) and notice that both terms in (10.51) vanish as $G = \sigma I$ and $P_J(G(u(t_k)) - \mathcal{G}(u(t_k))) = \sigma P_J(I - P_J) = 0$. Assumption 3.48 holds for $\delta = 1/\sqrt{\lambda_J}$. Here we have non-random initial data in $\mathcal{D}(A)$ and Theorem 10.34 completes the proof. □

We apply the numerical method (10.57) to the SPDE

$$du = \left[\varepsilon\, \Delta u + f(u)\right] dt + g(u)\, dW(t), \qquad u(0) = u_0 \in L^2(D), \qquad (10.59)$$

with periodic boundary conditions on the domains $D = (0,a)$ and $D = (0,a_1) \times (0,a_2)$. We consider $f, g \colon \mathbb{R} \to \mathbb{R}$ so both the reaction and noise terms are Nemytskii (see Lemmas 3.30 and 10.24). Although the Laplacian is singular in this case due to the periodic boundary conditions and Corollary 10.38 does not apply, we expect the errors to behave as in (10.58).

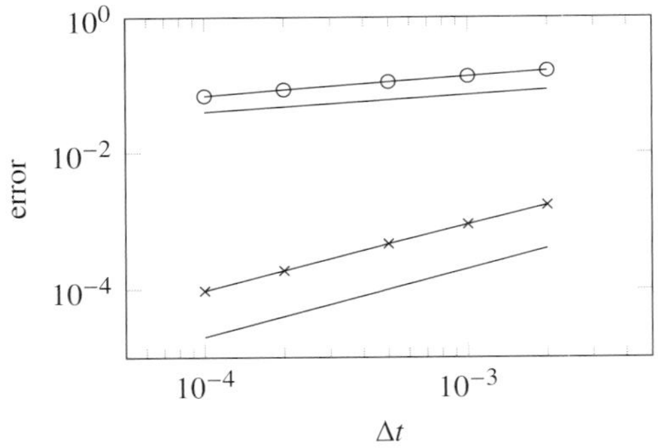

Figure 10.10 A log-log plot of the approximation to $\|u(T) - u_{J,N}\|_{L^2(\Omega, L^2(D))}$ against Δt for (10.60) with additive noise and $W(t)$ defined by Example 10.11 for $r = -1/2$ (o) and $r = 1$ (x). Reference lines of slopes 1/4 and 1 are shown and we observe errors of orders $\Delta t^{1/4}$ ($r = -1/2$) and Δt ($r = 1$).

To see this, we perform numerical experiments with the Allen–Cahn equation

$$du = \left[\varepsilon\, \Delta u + u - u^3\right] dt + \sigma\, dW(t), \qquad u(0) = u_0 \in H, \tag{10.60}$$

with periodic boundary conditions on $D = (0, 2\pi)$ and on $D = (0, 2\pi) \times (0, 16)$.

For $D = (0, a)$, Algorithm 10.10 computes samples of the vector $\boldsymbol{u}_{J,n} := [u_{J,n}(x_0), u_{J,n}(x_1), \ldots, u_{J,n}(x_J)]^{\mathsf{T}}$ for $x_j = aj/J$. We suppose $W(t)$ is the Q-Wiener process defined in Example 10.11 with parameter r. This means the eigenfunctions of Q and A are the same. The code for Algorithm 10.10 is based on Algorithm 3.7 and includes stochastic forcing by calling Algorithms 10.3 and 10.4 to compute increments and allow the simultaneous computation of M sample paths. We also pass function handles for both f and g. The code is flexible enough to examine convergence numerically.

Example 10.39 (Allen–Cahn SPDE in one dimension) Consider (10.60) on $(0, a) = (0, 2\pi)$ with $\varepsilon = 1$ and $\sigma = 1$. As initial data, we take $u_0(x) = \sin(x)$ and the Q-Wiener process defined in Example 10.11 with regularity parameter $r = 1$. We compute $M = 2$ approximate sample paths of the solution $u(t, x)$ on $[0, 2]$ using the following commands to call Algorithm 10.10. We take $J = 64$ and $\Delta t = 2/500$.

```
>> T=2; N=500; a=2*pi; J=64; r=1; epsilon=1; sigma=1;
>> M=2; kappa=1; x=[0:a/J:a]'; u0=sin(x);
>> [tref,uref,ureft]=spde_oned_Gal_MJDt(u0,T,a,N,kappa,J,J,epsilon,...
                                     @(u) u-u.^3,@(u) sigma,r,M);
```

Figure 10.9(a) illustrates two different realisations of $u(2, x)$ and Figure 10.9(b) shows one approximate sample path of $u(t, x)$. Figure 10.10 illustrates the numerical convergence of the method as $\Delta t \to 0$ for different values of r (see Example 10.43 for a description of numerical convergence for SPDEs). We plot approximate $L^2(0, a)$ errors against Δt for $r = -1/2$. Assumption 10.23 holds for $\zeta = 1 + r = 1/2$ and, by Corollary 10.38, the theoretical rate of convergence is $\theta = 1/4 - \epsilon$. We also show errors for $r = 1$, which gives $\zeta = 2$ and the convergence rate is $\theta = 1 - \epsilon$. The theoretical rate of convergence is observed in both cases. The method converges faster when the Q-Wiener process is more regular.

Algorithm 10.11 extends Algorithm 10.10 to find approximations to (10.59) on $D = (0, a_1) \times (0, a_2)$ with $W(t)$ defined by Example 10.12 (with parameter α). It returns samples

Algorithm 10.10 Code to approximate (10.59) using the semi-implicit Euler–Maruyama and spectral Galerkin method. The inputs are similar to those in Algorithm 3.5, with additional inputs to specify `ghandle` for evaluating $G(u)$, the regularity parameter `r`$=r$ (see Example 10.11), and the number of samples `M`. The outputs are `t`$= [t_0,\dots,t_N]$, a matrix `u` with columns containing M independent realisations of $[u_{J,N}(x_0),\dots,u_{J,N}(x_J)]^{\mathsf{T}}$, and one sample `ut` of the matrix with entries $u_{J,n}(x_j)$ for $j = 0,\dots,J$ and $n = 0,\dots,N$.

```
function [t,u,ut]=spde_oned_Gal_MJDt(u0,T,a,N,kappa,Jref,J,epsilon,...
                                         fhandle,ghandle,r,M)
dtref=T/N; Dt=kappa*dtref; t=[0:Dt:T]';
IJJ=J/2+1:Jref-J/2-1; % use IJJ to set unwanted modes to zero.
lambda = 2*pi*[0:Jref/2 -Jref/2+1:-1]'/a;
MM=epsilon*lambda.^2; EE=1./(1+Dt*MM); EE(IJJ)=0;% set Lin Operators
% get form of noise
iFspace=1; bj = get_oned_bj(dtref,Jref,a,r); bj(IJJ)=0;
ut(:,1)=u0; u=u0(1:Jref); uh0=fft(u);% set initial condition
uh=repmat(uh0,[1,M]); u=real(ifft(uh));
for n=1:N/kappa, % time loop
  uh(IJJ,:)=0; fhu=fft(fhandle(u)); fhu(IJJ,:)=0;
  dW=get_oned_dW(bj,kappa,iFspace,M); dW(IJJ,:)=0;
  gdWh=fft(ghandle(u).*real(ifft(dW))); gdWh(IJJ,:)=0;
  uh_new=bsxfun(@times,EE,uh+Dt*fhu+gdWh);
  uh=uh_new;  u=real(ifft(uh)); ut(1:Jref,n+1)=u(:,M);
end
ut(Jref+1,:)=ut(1,:); u=[u; u(1,:)]; % periodic
```

of the matrix $u_{J,n}$ with entries $u_{J,n}(\boldsymbol{x}_{ik})$ for $\boldsymbol{x}_{ik} = [ia_1/J_1, ka_2/J_2]$ for $i = 0,\dots,J_1$ and $k = 0,\dots,J_2$. The implementation is an extension of Algorithm 3.6 utilising samples generated from Algorithm 10.6.

Example 10.40 (stochastic Allen–Cahn equation in two dimensions) Consider (10.60) with additive noise on $D = (0,2\pi)\times(0,16)$. Let $\varepsilon = 10^{-3}$, $\sigma = 0.1$, and choose initial data $u_0(\boldsymbol{x}) = \sin(x_1)\cos(\pi x_2/8)$; see also Example 3.40. We choose the Q-Wiener process defined in Example 10.12 for a parameter α and demonstrate the effects of varying α.

For the experiments, the discretisation parameters are $J_1 = J_2 = 128$ and $\Delta t = 0.01$. We work on the time interval $[0,10]$ and call Algorithm 10.11 with the following commands to compute a single sample path for $\alpha = 0.1$:

```
>> T=10; N=1000; a=[2*pi 16]; J=[128 128];
>> alpha=0.1; epsilon=0.001; sigma=0.1; M=1; kappa=1;
>> x=[0:a(1)/J(1):a(1)]';y=[0:a(2)/J(2):a(2)]';
>> [xx yy]=meshgrid(y,x); u0=sin(xx).*cos(pi*yy/8);
>> [t,u,ut]=spde_twod_Gal(u0,T,a,N,kappa,J,epsilon,...
                         @(u) u-u.^3,@(u) sigma,alpha,M);
```

Figure 10.11(a) shows a `contourf` plot of an approximate realisation of $u(t,\boldsymbol{x})$ at $t = 10$ with $\alpha = 0.1$. In (b), we show another numerical realisation computed with $\alpha = 1$ at $t = 10$. In both (a) and (b), $W(t)$ belongs to $H^r(D)$ for any $r > 0$. When α is smaller however, the eigenvalues q_j of Q decay more slowly and $W(t)$ is more irregular, which results in the realisation (a) having more variation. Figure 3.5(b) shows the deterministic solution (case $\sigma = 0$) also at $t = 10$ with the same parameters and initial data.

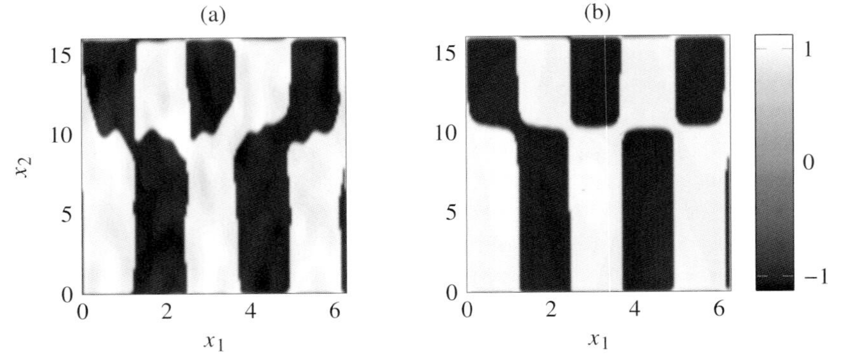

Figure 10.11 For the stochastic Allen–Cahn equation (10.60) on $(0, 2\pi) \times (0, 16)$ with the Q-Wiener process defined in Example 10.12, we show two approximate realisations of $u(t, \boldsymbol{x})$ at $t = 10$ found by the spectral Galerkin and semi-implicit Euler–Maruyama method. The parameter α controls the rate of decay of the eigenvalues of Q. The realisation is rougher for (a) $\alpha = 0.1$ than (b) $\alpha = 1$.

Exponential integrator for additive noise

The semi-implicit Euler–Maruyama method uses a basic increment ΔW_n to approximate $W(t)$. Building on the methods developed in Exercise 3.17 for ODEs and in Exercise 8.17 for SODEs, the variation of constants formula leads to an alternative method for approximating the increment for SPDEs with additive noise. Consider the method of lines SODE (10.56) and write $u_J(t) = \sum_{j=1}^{J} \hat{u}_j(t)\phi_j$. The variation of constants formula (8.116) with $t_n = n\Delta t$ gives

$$\begin{aligned} \hat{u}_j(t_{n+1}) = \mathrm{e}^{-\Delta t \lambda_j} \hat{u}_j(t_n) + \int_{t_n}^{t_{n+1}} \mathrm{e}^{-(t_{n+1}-s)\lambda_j} \hat{f}_j(u_J(s))\, ds \\ + \sigma \int_{t_n}^{t_{n+1}} \mathrm{e}^{-(t_{n+1}-s)\lambda_j} \sqrt{q_j}\, d\beta_j(s). \end{aligned} \tag{10.61}$$

To obtain a numerical method, we approximate $\hat{f}_j(u_J(s))$ by $\hat{f}_j(u_J(t_n))$ for $s \in [t_n, t_{n+1})$ and evaluate the integral, to find

$$\int_{t_n}^{t_{n+1}} \mathrm{e}^{-(t_{n+1}-s)\lambda_j} \hat{f}_j(u_J(s))\, ds \approx \frac{1 - \mathrm{e}^{-\Delta t \lambda_j}}{\lambda_j} \hat{f}_j(u_J(t_n)).$$

For the stochastic integral, we usually approximate $\mathrm{e}^{-(t_{n+1}-s)\lambda_j} \approx \mathrm{e}^{-t_{n+1}\lambda_j}$ and use a standard Brownian increment. However, from Example 8.21, we know that

$$\mathbb{E}\left[\left|\int_0^t \mathrm{e}^{-s\lambda}\, d\beta_j(s)\right|^2\right] = \frac{1 - \mathrm{e}^{-2t\lambda}}{2\lambda}.$$

The stochastic integral $\int_0^t \mathrm{e}^{-s\lambda}\, d\beta_j(s)$ has distribution $\mathrm{N}(0, (1 - \mathrm{e}^{-2t\lambda})/2\lambda)$. Hence, we can generate approximations $\hat{u}_{j,n}$ to $\hat{u}_j(t_n)$ using

$$\hat{u}_{j,n+1} = \mathrm{e}^{-\Delta t \lambda_j} \hat{u}_{n,j} + \frac{1 - \mathrm{e}^{-\Delta t \lambda_j}}{\lambda_j} \hat{f}_j(u_{J,n}) + \sigma\, b_j\, R_{j,n} \tag{10.62}$$

where $b_j := \sqrt{q_j(1-e^{-2\Delta t\lambda_j})/2\lambda_j}$ and $R_{j,n} \sim N(0,1)$ *iid*.

The method (10.62) is known as an exponential integrator and has the advantage that it samples the stochastic integral term in (10.61) exactly. See Exercise 10.17.

Algorithm 10.11 Code to approximate (10.59) using the semi-implicit Euler–Maruyama and spectral Galerkin method on $(0,a_1)\times(0,a_2)$. The inputs and outputs are similar to those in Algorithm 3.6, with additional inputs `ghandle` to define $G(u)$, regularity parameter `alpha=` α (see Example 10.12), and the number of samples `M`. The outputs are a vector `t=` $[t_0,\ldots,t_N]$, an array `u` where `u(j,k,m)` is the mth independent sample of $u_{J,N}(\boldsymbol{x}_{j-1,k-1})$, and an array `ut`, where `ut(j,k,n)` approximates one sample of $u_{J,n}(\boldsymbol{x}_{j-1,k-1})$, for $n = 0,\ldots,N$. Here $\boldsymbol{x}_{jk} = [ja_1/J_1, ka_2/J_2]^{\mathsf{T}}$.

```
function [t,u,ut]=spde_twod_Gal(u0,T,a,N,kappa,J,epsilon,...
                        fhandle,ghandle,alpha,M)
dtref=T/N; Dt=kappa*dtref; t=[0:Dt:T]'; ut=zeros(J(1)+1,J(2)+1,N);
% Set Lin Operator
lambdax= 2*pi*[0:J(1)/2 -J(1)/2+1:-1]'/a(1);
lambday= 2*pi*[0:J(2)/2 -J(2)/2+1:-1]'/a(2);
[lambdaxx lambdayy]=meshgrid(lambday,lambdax);
A=( lambdaxx.^2+lambdayy.^2);
MM=epsilon*A; EE=1./(1+Dt*MM);
bj=get_twod_bj(dtref,J,a,alpha); % get noise coeffs
u=repmat(u0(1:J(1),1:J(2)),[1,1,M]); % initial condition
uh=repmat(fft2(u0(1:J(1),1:J(2))),[1,1,M]);
% initialize
uh1=zeros(J(1),J(2),M); ut=zeros(J(1)+1,J(2)+1,N);ut(:,:,1)=u0;
for n=1:N/kappa, % time loop
  fh=fft2(fhandle(u));
  dW=get_twod_dW(bj,kappa,M);
  gudWh=fft2(ghandle(u).*dW);
  uh_new= bsxfun(@times,EE,(uh+Dt*fh+gudWh)); % update u
  u=real(ifft2(uh_new)); ut(1:J(1),1:J(2),n+1)=u(:,:,end);
  uh=uh_new;
end
u(J(1)+1,:,:)=u(1,:,:); u(:,J(2)+1,:)=u(:,1,:); % make periodic
ut(J(1)+1,:,:)=ut(1,:,:); ut(:,J(2)+1,:)=ut(:,1,:);
```

10.8 Galerkin finite element method

We consider a finite element discretisation in space (as in §3.6 for the deterministic case) for the stochastic reaction–diffusion equation (10.59). We work on the domain $(0,a)$ and fix homogeneous Dirichlet boundary conditions. Let $\widetilde{V} = V^h$ denote the space of continuous and piecewise linear functions on a uniform mesh of n_e elements with vertices $0 = x_0 < \cdots < x_{n_e} = a$ and mesh width $h = a/n_e$. We seek the finite element approximation $u_h(t) \in V^h$ to the SPDE solution $u(t)$ for $t > 0$. Substituting u_h for $\tilde{u}$ in (10.46), we obtain

$$du_h = \left[-A_h u_h + P_{h,L^2} f(u_h)\right] dt + P_{h,L^2} G(u_h)\, dW(t) \tag{10.63}$$

where A_h is defined by (3.76) and $P_{h,L^2}\colon L^2(0,a) \to V^h$ is the orthogonal projection. We choose initial data $u_h(0) = P_{h,L^2}u_0$. For a time step $\Delta t > 0$, the fully discrete method

(10.48) gives an approximation $u_{h,n}$ to $u_h(t_n)$ defined by

$$u_{h,n+1} = \left(I + \Delta t\, A_h\right)^{-1}\left(u_{h,n} + P_{h,L^2}\, f(u_{h,n})\,\Delta t + P_{h,L^2}\, \mathcal{G}(u_{h,n})\,\Delta W_n\right) \tag{10.64}$$

where $\Delta W_n = W(t_{n+1}) - W(t_n)$. We choose $\mathcal{G}(t,u) \equiv \mathcal{G}(u) := G(u)\mathcal{P}_{J_w}$ and $\mathcal{P}_{J_w}$ to be the orthogonal projection from U onto $\mathrm{span}\{\chi_1,\dots,\chi_{J_w}\}$ (for χ_j given in Assumption 10.5). The next theorem gives the conditions on J_w in terms of the mesh width h to attain convergence of this method.

Theorem 10.41 *Suppose that the following hold:*

(i) *the Assumptions of Theorem 10.26 for a unique mild solution with $\zeta \in [1,2]$,*
(ii) *the eigenvalues of Q satisfy $q_j = \mathcal{O}(j^{-(2r+1+\epsilon)})$, for some $r \ge 0, \epsilon > 0$,*
(iii) *$\mathcal{G}(u) = G(u)\mathcal{P}_{J_w}$ and $J_w \ge c\, h^{-\zeta/r}$ for a constant c, and*
(iv) *the initial data $u_0 \in \mathcal{D}(A)$.*

As $h \to 0$ and $\Delta t \to 0$, the approximation $u_{h,n}$ defined by (10.64) *to the solution $u(t)$ of* (10.59) *satisfies for $\theta_1 := \min\{\zeta/2 - \epsilon, 1/2\}$,*

$$\max_{0\le t_n \le T} \left\|u(t_n) - u_{h,n}\right\|_{L^2(\Omega, L^2(0,a))} \le K\left(\Delta t^{\theta_1} + h^{\zeta}\Delta t^{-\epsilon}\right),$$

where K is a constant independent of Δt and h.

Proof V^h satisfies Assumption 3.48 and Lemma 10.33 applies, so that Theorem 10.34 gives the result. □

For the piecewise linear finite element method (see (2.31)), we write

$$u_h(t,x) = \sum_{j=1}^{J} u_j(t)\,\phi_j(x), \tag{10.65}$$

in terms of $J = n_e - 1$ basis functions ϕ_j and coefficients $u_j(t)$ (compare with (3.78)). We choose $J_w = J$, which satisfies the condition of the theorem when $\zeta/r \le 1$. Note that $\mathcal{P}_{J_w} : U \to \mathrm{span}\{\chi_1,\dots,\chi_{J_w}\}$ and $P_J : H \to V^h$, which are distinct operators this time. Let $\boldsymbol{u}_h(t) := [u_1(t), u_2(t), \dots, u_J(t)]^{\mathsf{T}}$. Then, (10.63) can be written as

$$M\, d\boldsymbol{u}_h = \left[-K\boldsymbol{u}_h + \boldsymbol{f}(\boldsymbol{u}_h)\right] dt + \boldsymbol{G}(\boldsymbol{u}_h)\, dW(t), \tag{10.66}$$

where the vector $\boldsymbol{f}(\boldsymbol{u}_h) \in \mathbb{R}^J$ has elements $f_j = \langle f(u_h), \phi_j\rangle_{L^2(0,a)}$. M is the mass matrix with elements $m_{ij} = \langle \phi_i, \phi_j\rangle_{L^2(0,a)}$ and K is the diffusion matrix with elements $k_{ij} = a(\phi_i, \phi_j)$. Finally, $\boldsymbol{G}\colon \mathbb{R}^J \to \mathcal{L}(U, \mathbb{R}^J)$ and $\boldsymbol{G}(\boldsymbol{u}_h)\chi$ has jth coefficient $\langle G(u_h)\chi, \phi_j\rangle_{L^2(0,a)}$ for $\chi \in U$. Compare this with (3.79).

Equivalently to (10.64), we can approximate the solution $\boldsymbol{u}_h(t)$ of (10.66) at $t = t_n$ by $\boldsymbol{u}_{h,n}$, which is defined by the iteration

$$\left(M + \Delta t\, K\right)\boldsymbol{u}_{h,n+1} = M\,\boldsymbol{u}_{h,n} + \Delta t\, \boldsymbol{f}(\boldsymbol{u}_{h,n}) + G_h(\boldsymbol{u}_{h,n})\,\Delta \boldsymbol{W}_n \tag{10.67}$$

where $\boldsymbol{u}_{h,0} = \boldsymbol{u}_h(0)$ and $G_h(\boldsymbol{u}_{h,n}) \in \mathbb{R}^{J\times J_w}$ has j,k entry $\langle G(\boldsymbol{u}_{h,n})\chi_k, \phi_j\rangle_{L^2(0,a)}$ and $\Delta\boldsymbol{W}_n$ is a vector in $\mathbb{R}^{J_w}$ with entries $\langle W(t_{n+1}) - W(t_n), \chi_k\rangle_{L^2(0,a)}$ for $k = 1,\dots,J_w$. For practical computations, we write the Q-Wiener process $W(t)$ as (10.11). Then, we find $G_h(\boldsymbol{u}_{h,n})\Delta \boldsymbol{W}_n$ by multiplying the matrix $G_h(\boldsymbol{u}_{h,n})$ by the vector of coefficients $\left[\sqrt{q_1}(\beta_1(t_{n+1}) - \beta_1(t_n)), \dots, \sqrt{q_{J_w}}(\beta_{J_w}(t_{n+1}) - \beta_{J_w}(t_n))\right]^{\mathsf{T}}$.

In Algorithm 10.12, we adapt Algorithm 3.12 to the stochastic case for f and G that are Nemytskii operators (see Lemmas 3.30 and 10.24). As $J = J_w$, $G_h(\boldsymbol{u}_{h,n})$ is a diagonal matrix with entries $g(u_{h,n}^k)$, where $u_{h,n}^k$ denotes the kth entry of $\boldsymbol{u}_{h,n}$. The implementation is flexible enough to examine convergence numerically in Example 10.43. The main change is to include the stochastic forcing by calling Algorithms 10.2–10.3 to compute increments.

Algorithm 10.12 Code to approximate (10.46) using the semi-implicit Euler–Maruyama and finite element approximations. The inputs are similar to those in Algorithm 3.12 and we additionally specify `ghandle` for evaluating g, the regularity parameter r, and the number of samples M. The outputs are `t`$= [t_0, \dots, t_N]$, a matrix `u` with columns containing *iid* approximations to $[u(t_N, x_0), \dots, u(t_N, x_{n_e})]^{\mathsf{T}}$, and one approximate sample `ut` of the matrix with columns $[u(t_n, x_0), \dots, u(t_n, x_{n_e})]^{\mathsf{T}}$, $n = 0, \dots, N$.

```
function [t,u,ut]=spde_fem_MhDt(u0,T,a,Nref,kappa,neref,L,epsilon,...
                    fhandle,ghandle,r,M)
ne=neref/L; assert(mod(ne,1)==0); % require ne even
nvtx=ne+1; h=(a/ne); % set mesh width
dtref=T/Nref; Dt=kappa*dtref; t=[0:Dt:T]'; % set time steps
% set linear operator
p=epsilon*ones(ne,1); q=ones(ne,1); f=ones(ne,1);
[uh,A,b,KK,MM]=oned_linear_FEM(ne,p,q,f);
EE=MM+Dt*KK; ZM=zeros(1,M);
% get noise coeffs
bj = get_onedD_bj(dtref,neref,a,r); bj(ne:end)=0; iFspace=0;
% set initial condition
u=repmat(u0,[1,M]);ut=zeros(nvtx,Nref/kappa+1); ut(:,1)=u(:,1);
for k=1:Nref/kappa, % time loop
  dWJ=get_onedD_dW(bj,kappa,iFspace,M);
  dWL=[ZM;dWJ;ZM]; dWL=dWL(1:L:end,:);
  gdW=ghandle(u).*dWL;
  fu=fhandle(u);
  for m=1:M % set b=f(u), gdw for M samples
    b(:,m)=oned_linear_FEM_b(ne,h,fu(:,m)); % Alg 3.10
    gdw(:,m)=oned_linear_FEM_b(ne,h,gdW(:,m));
  end
  u1=EE\(MM*u(2:end-1,:)+Dt*b+gdw); % update u
  u=[ZM;u1;ZM]; ut(:,k+1)=u(:,M);
end
```

Example 10.42 (stochastic Nagumo equation) Consider the stochastic Nagumo equation (10.3) on $D = (0,1)$ with homogeneous Dirichlet boundary conditions and $\varepsilon = 10^{-3}$, $\alpha = -0.5$ and initial data $u_0(x) = \exp(-(x - 1/2)^2/\varepsilon)$. Note that $G(u) = \sigma u(1 - u)$. For $W(t)$, we take the Q-Wiener process defined in Example 10.10 with $r = 1$. We solve the SPDE using a piecewise linear finite element approximation in space with $n_e = 512$ elements. In time, we apply the semi-implicit Euler–Maruyama method with $\Delta t = 10^{-3}$.

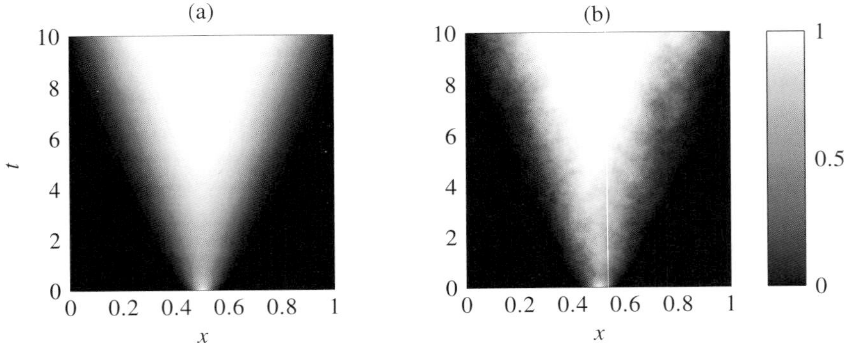

Figure 10.12 `contourf` plots of an approximate sample path of the solution to the stochastic Nagumo equation for $W(t)$ given by (a) the Q-Wiener process of Example 10.10 with $r = 1$ and (b) space–time white noise (or Example 10.10 with $r = -1/2$).

We compute an approximate sample path on $[0, 10]$ by calling Algorithm 10.12 with the following commands.

```
>> T=10; N=1e4; a=1; ne=512; h=a/ne;
>> epsilon=1e-3; r=1; M=1; sigma=0.1;
>> x=(0:h:a)'; u0=exp(-(x-0.5).^2/epsilon);
>> [t,u,ut]=spde_fem_MhDt(u0,T,a,N,1,ne,1,epsilon,...
                        @(u) u.*(1-u).*(u+0.5),@(u) sigma*u.*(1-u),r,M);
```

Figure 10.12 shows a sample path of the solution with (a) $r = 1$ and (b) $r = -1/2$. Theorem 10.41 does not apply for $r = -1/2$ as the associated covariance operator Q is not trace class. Compare Figure 10.12 to the numerical solution plotted in Figure 3.4(a) for the deterministic Nagumo equation.

Numerical illustration of convergence

With analytical solutions unavailable, we approximate $u(T)$ using two reference solutions to examine the strong rate of convergence numerically. One reference solution is computed with a small reference time step Δt_{ref} to examine convergence in the time discretisation and the other with a small reference mesh width h_{ref} to examine converge of the finite element discretisation. See, for the deterministic case, Example 3.45. Specifically, we approximate

$$\left\|u(T) - u_{h,N}\right\|_{L^2(\Omega, L^2(0,a))} \approx \left(\frac{1}{M}\sum_{m=1}^{M}\left\|u_{\text{ref}}^m - u_{h,N}^m\right\|_{L^2(0,a)}^2\right)^{1/2}, \tag{10.68}$$

where u_{ref}^m are M *iid* samples of the reference solution at time T and $u_{h,N}^m$ are their numerical approximations (see (8.65) for the SODE case). The $L^2(0, a)$ norm is approximated by the trapezium rule, as in (10.44).

The following examples consider the Nagumo equation for additive and multiplicative noise.

Example 10.43 (Q-Wiener multiplicative noise) Reconsider the Nagumo equation from Example 10.42. For the Q-Wiener process, choose the one defined in Example 10.10 with regularity parameter $r = 1$. Although f and G do not satisfy our Lipschitz assumptions, it is an illustrative example and both the Q-Wiener process and initial data are smooth.

We take $M = 100$ samples to estimate the expectation in (10.68). First we investigate convergence of the finite element discretisation with a fixed time step Δt. We compute M *iid* samples $u^m_{h_{\text{ref}},N}$, $m = 1,\dots,M$, of the reference solution with $h_{\text{ref}} = a/n_{e,\text{ref}}$. The errors are then estimated in Algorithm 10.13 by comparing $u^m_{h_{\text{ref}},N}$ with $u^m_{h,N}$ for $h = Lh_{\text{ref}}$ and a set of values of L. We make use of Algorithm 10.12, which computes $u^m_{h,N}$ to approximate $u^m_{h_{\text{ref}},N}$ by using increments of the same sample path of $W(t)$. In the following experiment, we take $T = 1$ and $N = 10^5$ so that $\Delta t = 10^{-5}$. We take a reference solution with $n_{e,\text{ref}} = 512$ elements and compare it to the solution with $n_e = 512/L$ elements with $L = 2, 4, 8, 16$ and 32. The following commands are used to call Algorithm 10.13:

```
>> T=1; N=1e5; a=1; r=1; sigma=0.5; epsilon=1e-3;
>> neref=512; href=a/neref; x=[0:href:a]';
>> u0=exp(-(x-0.5).^2/epsilon);
>> L=[2,4,8,16,32]'; M=100;
>> [h,errh]=spde_fem_convh(u0,T,a,N,neref,L,epsilon,...
                   @(u) u.*(1-u).*(u+0.5),@(u) sigma*u.*(1-u),r,M)
```

In Figure 10.13(a), we plot on a log-log scale the estimated $L^2(\Omega, L^2(0,a))$ error against the mesh width h. Comparing with the reference line of slope 2, we observe errors of order h^2 as predicted in Theorem 10.41 with $\zeta = 2$.

Now examine convergence in the time discretisation and fix a mesh width $h = a/n_e$. We compute M *iid* realisations $u^m_{h,N_{\text{ref}}}$ of the reference solution with $\Delta t_{\text{ref}} = T/N_{\text{ref}}$. The errors are then estimated in Algorithm 10.14 by comparing $u^m_{h,N_{\text{ref}}}$ to $u^m_{h,N}$ for $N = N_{\text{ref}}/\kappa$ and $\Delta t = \kappa \Delta t_{\text{ref}}$ for a set of values of κ. We expect from Theorem 10.34 errors of order $\Delta t^{1/2}$. In the following experiment, we take $n_e = 128$, $N_{\text{ref}} = 10^5$, and $T = 1$ so that $\Delta t_{\text{ref}} = 10^{-5}$. We estimate the errors with $\kappa = 5, 10, 20, 50, 100, 200, 500$ so Δt takes the values 5×10^{-5}, 10^{-4}, 2×10^{-4}, 5×10^{-4}, 10^{-3}, 2×10^{-3} and 5×10^{-3}. We use the following commands to call Algorithm 10.14:

```
>> Nref=1e5; ne=128; h=a/ne; x=[0:h:a]';
>> u0=exp(-(x-0.5).^2/epsilon);
>> kappa=[5,10,20,50,100,200,500]';
>> [dt,errT]=spde_fem_convDt(u0,T,a,Nref,kappa,ne,epsilon,...
                    @(u)u.*(1-u).*(u+0.5),@(u)sigma*u.*(1-u),r,M);
```

In Figure 10.13(b), we plot on a log-log scale the estimated $L^2(\Omega, L^2(0,a))$ error against Δt alongside a reference line of slope 1/2. This illustrates the errors are order $\Delta t^{1/2}$, which agrees with the theoretical predictions of Theorem 10.41.

For additive noise, the Euler–Maruyama method is a special case of the Milstein method and we achieve a higher rate of convergence.

Example 10.44 (Q-Wiener additive noise) Consider the Nagumo equation with additive noise on $D = (0,1)$ with $\varepsilon = 10^{-3}$ and $\alpha = -0.5$,

$$du = \left[\varepsilon u_{xx} + u(1-u)(u+1/2)\right] dt + \sigma \, dW(t). \tag{10.69}$$

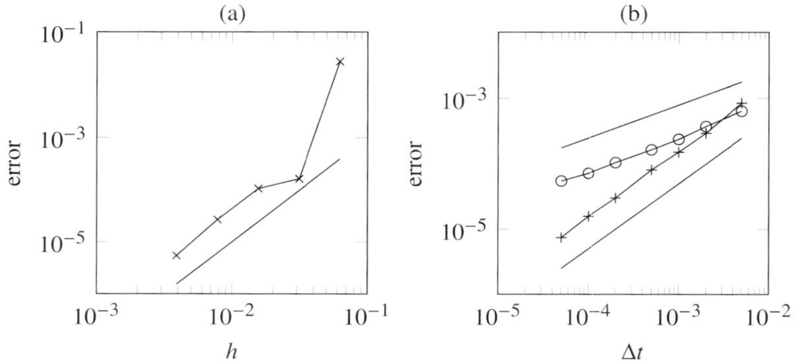

Figure 10.13 For the finite element and semi-implicit Euler–Maruyama approximation of the Nagumo equation in Examples 10.43 and 10.44, a log-log plot of the approximation to the error $\|u(T) - u_{h,N}\|_{L^2(\Omega, L^2(0,a))}$ at $T = N\Delta t = 1$ as (a) the spatial mesh size h is varied and (b) as the time step Δt is varied. In (a), we have multiplicative noise and the errors agree with the theoretical rate shown by the reference line with slope 2. In (b), we show the errors for the Nagumo equation with multiplicative noise (o) and additive noise (+), relative to reference lines of slope $1/2$ and 1. The multiplicative noise gives errors of order $\Delta t^{1/2}$, while additive noise gives errors of order Δt.

Algorithm 10.13 Code to examine convergence as $h \to 0$ of the semi-implicit Euler–Maruyama and finite element methods by calling Algorithm 10.12. The inputs `u00`, `T`, `a`, `epsilon`, `fhandle`, `ghandle`, and `M` are the same as in Algorithm 10.12. We give a vector `L` of values L_j to compute $h_j = L_j h_{\text{ref}}$ for $h_{\text{ref}} = a/n_{e,\text{ref}}$. The outputs are a vector `h` of mesh widths h_j and a vector `err` of the computed errors.

```
function [h,err]=spde_fem_convh(u00,T,a,N,neref,L,epsilon,...
                                fhandle,ghandle,r,M)
href=a/neref; xref=[0:href:a]'; kappa=1;
defaultStream=RandStream.getGlobalStream;
savedState=defaultStream.State;
[t,uref,ureft]=spde_fem_MhDt(u00,T,a,N,kappa,neref,1,...
                              epsilon,fhandle,ghandle,r,M);
for j=1:length(L)
  h(j)=href*L(j); u0=u00(1:L(j):end); % set mesh width
  defaultStream.State = savedState; % reset random numbers
  [t,u,ut]=spde_fem_MhDt(u0,T,a,N,kappa,neref,L(j),...
                          epsilon,fhandle,ghandle,r,M);
  x=[0:h(j):a]'; uinterp=interp1(x,u,xref);
  S(j)=sum(sum(uref(:,end,:)-uinterp(:,end,:)).^2)*href;
end
err=sqrt(S/M);
```

We examine convergence in the time discretisation with a finite element approximation in space. We choose the Q-Wiener process $W(t)$ defined in Example 10.10 for $r = 1$. The only difference with Example 10.43 is the additive noise.

We use Algorithm 10.14 to examine convergence as follows:

```
>> [dt,errT]=spde_fem_convDt(u0,T,a,Nref,kappa,ne,...
                   epsilon,@(u) u.*(1-u).*(u+0.5),@(u) sigma,r,M);
```

In Figure 10.13(b), we plot on a log-log scale the estimated $L^2(\Omega, L^2(0,a))$ errors against Δt. We have also plotted a reference line of slope 1. This can be compared to the multiplicative noise case from Example 10.43. We observe from the numerical experiment, as for SODEs, an improved rate of convergence. This is a consequence of the Milstein method (10.53) for additive noise being equivalent to (10.47). See Corollary 10.37.

Algorithm 10.14 Code to examine convergence as $\Delta t \to 0$ of the semi-implicit Euler–Maruyama and finite element approximations by calling Algorithm 10.12. The inputs `u00`, `T`, `a`, `epsilon`, `fhandle`, `ghandle`, and `M` are the same as in Algorithm 10.13. The remaining inputs are `Nref`$= N_{\text{ref}}$, which determines $\Delta t_{\text{ref}} = T/N_{\text{ref}}$, the number of elements `ne`, and a vector `kappa` of values κ_i to determine $\Delta t_i = \kappa_i \Delta t_{\text{ref}}$. The outputs are a vector `dt` of the time steps Δt_i and a vector `err` of the computed errors.

```
function [dt,err]=spde_fem_convDt(u00,T,a,Nref,kappa,ne,epsilon,...
                                   fhandle,ghandle,r,M)
dtref=T/Nref; h=a/ne;
defaultStream=RandStream.getGlobalStream;
savedState=defaultStream.State;
% compute the reference soln
[t,uref,ureft]=spde_fem_MhDt(u00,T,a,Nref,1,ne,1,epsilon,...
                              fhandle,ghandle,r,M);
for i=1:length(kappa)
    dt(i)=kappa(i)*dtref; % set time step
    defaultStream.State=savedState; % reset random numbers
    [t,u,ut]=spde_fem_MhDt(u00,T,a,Nref,kappa(i),ne,1,epsilon,...
                            fhandle,ghandle,r,M);
    S(i)=sum(sum((uref-u).^2)*h);
end
err=sqrt(S./M);
```

10.9 Notes

The stochastic Allen–Cahn equation (10.2) originates from Allen and Cahn (1979) and has been considered by a large number of authors. The equation is often used to test numerical methods and is an important model in physics (García-Ojalvo and Sancho, 1999). Some of the early work on SPDEs started in mathematical biology (Walsh, 1981, 1984b) and there is a growing literature on SPDEs in mathematical biology and related fields; see Laing and Lord (2010) for an overview of stochastic methods in neuroscience. We considered the stochastic Nagumo equation, though in practice the stochastic FitzHugh–Nagumo equation is preferred as a more realistic model for wave propagation in a neuron. Our example from fluid flow is the vorticity equation, which exhibits a bistability in the presence of stochastic forcing (Bouchet and Simonnet, 2009). There is, however, a great deal more interest in the stochastically forced Navier–Stokes and Euler equations as models for turbulence. Example 10.4 concerning filtering is based on Stuart, Voss, and Wiberg (2004) and is closely related to Kalman filters (Hairer et al., 2005, 2007; Voss, 2012). For reviews on filtering and related SPDEs, see (Bain and Crisan, 2009; Crisan and Rozovskiĭ, 2011; Rozovskiĭ, 1990; Stuart, 2010).

The classic reference for the theory of SPDEs and stochastic evolution equations using semigroups is Da Prato and Zabczyk (1992). The books (Chow, 2007; Jentzen and Kloeden, 2011) also take this approach and (Kruse and Larsson, 2012; van Neerven, Veraar, and Weis, 2012) give optimal regularity estimates. In contrast, the book Prévôt and Röckner (2007) takes a variational approach to establishing the existence and uniqueness of solutions to SPDEs. This approach is more widely applicable than semigroups. Prévôt and Röckner (2007, Appendix F) gives a detailed analysis of the relationship between weak and mild solutions of an SPDE. See also (Hairer, 2009; Walsh, 1984a).

We have used the fact that a separable Hilbert space-valued process $X(t)$ is predictable if its coefficients $\langle X(t), \phi_k \rangle$, for an orthonormal basis ϕ_k, are predictable. This follows from Lemma 1.20 and implies that the process $X(t)$ considered as a function $X\colon [0,T]\times\Omega \to H$ is $\mathcal{P}_T$-measurable (the predictable σ-algebra $\mathcal{P}_T$ is defined in §8.8). We have used this fact for the L^2_0-valued integrands $B(t)$ (as L^2_0 is actually a Hilbert space) and the H-valued integrals $\int_0^t B(s)\, dW(s)$. For simplicity, we assumed in Theorem 10.16 that stochastic integrals have mean-square-integrable integrands. This is not necessary and it is sufficient to know that $\mathbb{P}\left(\int_0^t \|B(s)\|^2_{L^2_0}\, ds < \infty\right) = 1$.

We have only given an introduction here to SPDEs and there are plenty of interesting directions for further study and research. Dalang, Khoshnevisan, and Rassoul-Agha (2009) covers a number of topics not included here such as the stochastic wave equation and large deviation theory. Blömker (2007) examines amplitude equations for SPDEs. Hairer and Mattingly (2011) describes the theory of invariant measures and ergodicity for semilinear SPDEs. The Itô formula (see Lemma 8.42 for SODEs) can be developed for SPDEs; see Da Prato and Zabczyk (1992) and more recently (Brzeźniak et al., 2008; Da Prato, Jentzen, and Röckner, 2010).

There is a rapidly growing literature on the numerical analysis of SPDEs. Theorem 10.34 is motivated by Yan (2005), which examined a finite element approximation in space and implicit Euler–Maruyama method in time. Our examples represent the Wiener process using eigenfunctions of the Laplacian. Piecewise constant functions (Allen, Novosel, and Zhang, 1998; Katsoulakis, Kossioris, and Lakkis, 2007; Kossioris and Zouraris, 2010) are often used to approximate the Wiener process when implementing finite element methods. The approximation of stochastic convolutions in space is considered in a more general framework in (Hausenblas, 2003a; Kovács, Lindgren, and Larsson, 2011). The spectral Galerkin approximation in space was one of the first considered for SPDEs (Grecksch and Kloeden, 1996) and has gained popularity in part because of the relative simplicity of analysis and implementation. It is the approach taken, for example, in Jentzen and Kloeden (2011). Finite differences also have a long history and strong convergence is considered, for example, in (Gyöngy, 1998b, 1999; Shardlow, 1999). Here we only examined their numerical implementation. There are also a number of general approximations results, which include wavelet based approximations — see Gyöngy and Millet (2009) (for a variational approach) or (Hausenblas, 2003a; Kovács et al., 2011) (based on semigroups).

We concentrated on the semi-implicit Euler–Maruyama time-stepping method and briefly considered the Milstein method (Jentzen and Röckner, 2010; Lang, Chow, and Potthoff, 2010, 2012). The projection $\mathcal{P}_{J_w}$ in Lemma 10.33 need not include the same number of terms J as the spatial discretisation and indeed J_w can be adapted to the smoothness of the noise (Jentzen and Kloeden, 2011; Lord and Shardlow, 2007). In (10.61), we derived an exponential time-stepping method for additive noise. See also (Jentzen and Kloeden, 2009; Kloeden et al., 2011; Lord and Rougemont, 2004). For a finite element approximation in space with an exponential integrator, see Lord and Tambue (2013). Related to exponential-based integrators are splitting methods (Cox and van Neerven, 2010; Gyöngy and Krylov, 2003). Non-uniform steps in time are considered, for example, in Müller-Gronbach and Ritter (2007) and optimal rates of convergence are obtained. The SPDEs we have considered are parabolic SPDEs. There are, however, plenty of interest and results for other types of equation. For the stochastic wave equation, see for example (Kovács, Larsson, and Saedpanah, 2010; Quer-Sardanyons and Sanz-Solé, 2006) or, for the stochastic nonlinear Schrödinger equation, see de Bouard and Debussche (2006).

For SODEs, we examined both strong and weak convergence. However, for SPDEs, we only considered strong convergence in Theorem 10.34. A number of authors (Debussche, 2011; Debussche and Printems, 2009; Geissert, Kovács, and Larsson, 2009; Hausenblas, 2003b) have examined weak convergence of numerical methods for SPDEs. Multilevel Monte Carlo methods have also been examined for SPDEs (Barth and Lang, 2012; Barth, Lang, and Schwab, 2013).

Other than Exercise 10.19, we have not considered the Stratonovich interpretation of the stochastic integrals. The book García-Ojalvo and Sancho (1999) discusses SPDEs from a physics point of view using the Stratonovich interpretation. This interpretation is often preferred as it allows for physical phenomena such as energy preservation in the nonlinear Schrödinger equation (de Bouard and Debussche, 1999) or the stochastic Landau–Lifshitz–Gilbert equation (Bañas, Brzeźniak, and Prohl, 2013). Mild solutions of the Stratonovich heat equation has been considered in Deya, Jolis, and Quer-Sardanyons (2013). The Wong–Zakai approximation (see §8.7) has also been considered in the infinite-dimensional case (Ganguly, 2013; Tessitore and Zabczyk, 2006; Twardowska, 1996; Twardowska and Nowak, 2004) and for other forms of noise (Hausenblas, 2007). As in the SODE case, it is possible to transform between the two interpretations.

Exercises

10.1 Prove that if $W(t)$ is a Q-Wiener process on $L^2(D)$ then

$$\operatorname{Tr} Q = \mathbb{E}\left[\|W(1)\|_{L^2(D)}^2\right]. \tag{10.70}$$

By performing Monte Carlo simulations with a numerical approximation to $\|\cdot\|_{L^2(D)}$, show that Algorithms 10.2, 10.4, and 10.6 produce results consistent with (10.70).

10.2 Modify Algorithm 10.5 to sample a Q-Wiener process $W(t)$ taking values in $H^r_{\text{per}}(D)$ a.s. for a given $r \geq 0$, where $D = (0, a_1) \times (0, a_2)$.

For $r = 0$, find $W(1)$ and estimate

$$\mathbb{E}\left[\|W(1)\|_{L^2(D)}^2\right], \qquad \mathbb{E}\left[\|W(1)\|_{H^1(D)}^2\right], \qquad \mathbb{E}\left[\|W(1)\|_{H^2(D)}^2\right]$$

for each $J_1 = J_2 = 16, 32, 64, 128, 256, 512$ by performing Monte Carlo simulations. Do the estimates converge as $J_1 = J_2$ is increased? Repeat for $r = 1$. Show the estimates are consistent with (10.70).

10.3 Show that (10.20) converges in $L^2(\Omega, U_1)$ if $\iota \colon U \to U_1$ is Hilbert–Schmidt.

10.4 In the context of Example 10.13, show that $\mathrm{HS}(U,H) = \mathrm{HS}(Q_1^{1/2}U_1 H)$.

10.5 If Q is invertible, show that L_0^2 is the Banach space of Hilbert–Schmidt operators from U_0 to H, where $U_0 = \mathcal{D}(Q^{-1/2})$.

10.6 Let U and H be two separable Hilbert spaces and let Q satisfy Assumption 10.5. Show that if $\|B(s)Q^{1/2}\|_{\mathrm{HS}(U,H)}$ is bounded then $\int_0^t B(s)\chi\, d\beta(s)$ is well defined, where $\chi \in U$ and $\beta(t)$ is a Brownian motion.

10.7 Show that if $u \in L^2(\Omega, H)$ then

$$\|\mathrm{e}^{-tA}u\|_{L^2(\Omega,H)} \le \|\mathrm{e}^{-tA}\|_{\mathcal{L}(H)}\|u\|_{L^2(\Omega,H)} \le \|u\|_{L^2(\Omega,H)}.$$

If $B \in L^2(\Omega, L_0^2)$ and $A \in \mathcal{L}(H)$, show that

$$\mathbb{E}\Big[\|AB\|^2_{L^2(\Omega,L_0^2)}\Big] \le \|A\|^2_{\mathcal{L}(H)}\mathbb{E}\Big[\|B\|^2_{L_0^2}\Big].$$

10.8 Let A obey Assumption 3.19. Show that if $2\beta - 1 \le 0$ then

$$\int_0^T \|A^\beta \mathrm{e}^{-tA}\|^2_{\mathcal{L}(H)}\, dt \le \frac{1}{2}\lambda_1^{2\beta-1}.$$

Show that there exists $K_\beta > 0$ such that

$$\int_0^T \|A^\beta \mathrm{e}^{-tA}\|^2_{\mathcal{L}(H)}\, dt \le \begin{cases} K_\beta^2 \dfrac{T^{1-2\beta}}{1-2\beta}, & 0 < \beta < 1/2,\\ K_\beta^2 T, & \beta \le 0. \end{cases}$$

10.9 For a Q-Wiener process $W(t)$, derive (10.36) by using Gronwall's inequality.

10.10 In the proof of Theorem 10.28, show that $\mathtt{II} \in L^2(\Omega, \mathcal{D}(A^{\zeta/2}))$.

10.11 Let $U = L^2(0, 2\pi)$ and $Q \in \mathcal{L}(U)$ be the operator with eigenfunctions $\chi_j(x) = \mathrm{e}^{\mathrm{i}jx}$ and eigenvalues $q_j = \mathrm{e}^{-\alpha|j|^2}$, $j \in \mathbb{Z}$, $\alpha > 0$. Modify Algorithm 10.3 to construct a numerical approximation to the Q-Wiener process with this covariance operator. With $J = 100$ and $\Delta t_{\mathrm{ref}} = 0.01$, plot realisations of $W(2)$ with $\alpha = 1$ and $\alpha = 0.001$.

10.12 Consider the SPDE

$$du = -\Delta^2 u\, dt + \sigma\, dW(t),$$

on $D = (0,\pi)\times(0,\pi)$ with boundary conditions $u(\boldsymbol{x}) = \Delta u(\boldsymbol{x}) = 0$ for $\in \partial D$ and initial data $u(0) = 0$. The parameter $\sigma > 0$ and $W(t)$ is the cylindrical Wiener process on $L^2(D)$. Use the weak formulation (10.27), to show that

$$\mathbb{E}\Big[\|u(t)\|^2_{L^2(D)}\Big] < \infty \qquad \text{for } t > 0.$$

10.13 Modify Algorithm 10.8 to approximate the stochastic Nagumo equation (10.3) with homogeneous Neumann boundary conditions and the cylindrical Wiener process $W(t)$ on $L^2(0,a)$. Plot an approximate sample path of the solution for $a = 20$ and $\varepsilon = 1$ with $\Delta t = 0.01$ and $h = 20/512$.

10.14 Implement the Milstein method (10.53). Use your code to approximate solutions of the stochastic Nagumo equation of Example 10.43. Take the Q-Wiener process with kernel $q(x,y) = \mathrm{e}^{-|x-y|/\ell}$ for correlation length $\ell > 0$.

10.15 Consider a Nemytskii nonlinearity $f \colon L^2(D) \to L^2(D)$ associated to a function $f \colon \mathbb{R} \to \mathbb{R}$ with two uniformly bounded derivatives. Show how to extend Corollary 10.37 to include such nonlinearities. Hint: write

$$f(u(s)) - f(u(t_k)) = \big(f(u(s)) - f(u(t_k)) - f'(u(t_k))G(u(t_k))(W(s) - W(t_k))\big) + f'(u(t_k))G(u(t_k))(W(s) - W(t_k)).$$

10.16 Consider the stochastic Allen–Cahn equation (10.2) on $D = (0, 2\pi)$ with periodic boundary conditions and $\varepsilon = 1$. Choose initial data $u_0(x) = \sin(x)$ and the Q-Wiener process $W(t)$ defined in Example 10.11. For $\Delta t = 0.004$, use Algorithm 10.10 to compute $M = 10$ approximate realisations of the solution at $t = 2$ with $J = 64$, 128, 256, 512, 1024, 2048. With $r = 0$, estimate

$$\mathbb{E}\Big[\|u(2)\|_{L^2(0,2\pi)}\Big], \qquad \mathbb{E}\Big[\|u_x(2)\|_{L^2(0,2\pi)}\Big], \qquad \mathbb{E}\Big[\|u_{xx}(2)\|_{L^2(0,2\pi)}\Big]$$

from the 10 samples for each J. Do the estimates of the norms converge as J is increased? What happens when $r = 1$? Hint: use the FFT to estimate u_x and u_{xx}.

10.17 Implement (10.62) for (10.59) by modifying Algorithm 10.3 and Algorithm 10.10. Generate sample paths for the Allen–Cahn equation (10.60) on $(0, 2\pi)$ with periodic boundary conditions.

10.18 Consider the stochastic vorticity equation (10.6) of Example 10.3 on $D = (0, a_1) \times (0, a_2)$ with $a_1 = 2\pi$ and $a_2 = 9\pi/4$. Take $\varepsilon = 10^{-3}$, $\sigma = \sqrt{\varepsilon}$, and initial data $u_0(\boldsymbol{x}) = \sin(2\pi x_1/a_1)\cos(2\pi x_2/a_2)$. Let $W(t)$ be the Q-Wiener process on $L^2(D)$ given in Example 10.12. Modify the deterministic algorithm from Exercise 3.16 to use Algorithms 10.5 and 10.6 and solve (10.6). With $J_1 = J_2 = 64$ and $\Delta t = 0.01$, plot a realisation of the solution at $t = 100$ with $\alpha = 0.1$ and $\alpha = 1$.

10.19 We may also define Stratonovich integrals with respect to the Q-Wiener process, by replacing the Itô integrals in (10.22) by Stratonovich ones. In the case $H = U = L^2(D)$ and $W(t)$ is a Q-Wiener process, write down the weak form of solution for a Stratonovich SPDE.

Use the semi-implicit Heun method (see Exercise 8.23) with the spectral Galerkin approximation, to find solutions to the following Stratonovich SPDE numerically:

$$du = \left[\varepsilon u_{xx} + u - u^3\right] dt + \sigma u \circ dW(t)$$

on $D = (0, 2\pi)$ with periodic boundary conditions and initial data $u_0(x) = \sin(x)$ for the Q-Wiener process $W(t)$ defined in Example 10.8. Take $\varepsilon = 0.001$, $\sigma = 1$, $T = 2$, $\Delta t = 0.004$, and $J = 64$ coefficients. Compare your solution to the corresponding solution of the Itô SPDE

$$du = \left[\varepsilon u_{xx} + u - u^3\right] dt + \sigma u dW(t).$$

Appendix A

A.1 Taylor's theorem

Taylor's theorem provides polynomial approximations to smooth functions and an expression for the error. We consider a function $\boldsymbol{u}\colon \mathbb{R}^d \to \mathbb{R}^m$ and use the notation $D^n\boldsymbol{u}(\boldsymbol{x})$ to denote the nth derivative of $\boldsymbol{u}$ at $\boldsymbol{x} \in \mathbb{R}^d$ for $\boldsymbol{u} \in \mathrm{C}^n(\mathbb{R}^d, \mathbb{R}^m)$. Note that $D^n\boldsymbol{u}(\boldsymbol{x})$ is a linear operator in $\mathcal{L}(\mathbb{R}^d \times \cdots \times \mathbb{R}^d, \mathbb{R}^m)$. The notation $D^n\boldsymbol{u}(\boldsymbol{x})[\boldsymbol{h}_1, \dots, \boldsymbol{h}_n]$ is used to denote the action of the linear operator on $[\boldsymbol{h}_1, \dots, \boldsymbol{h}_n] \in \mathbb{R}^d \times \cdots \times \mathbb{R}^d$. Further, the abbreviation $[\boldsymbol{h}]^n$ is used for $[\boldsymbol{h}, \dots, \boldsymbol{h}]$, so that, if $\boldsymbol{\phi}(s) = \boldsymbol{u}(\boldsymbol{x} + s\boldsymbol{h})$ for $s \in \mathbb{R}$ and $\boldsymbol{h} \in \mathbb{R}^d$, then $D^n\boldsymbol{u}(\boldsymbol{x})[\boldsymbol{h}]^n = \boldsymbol{\phi}^{(n)}(0)$, where $\phi^{(n)}$ denotes the nth derivative of $\boldsymbol{\phi}$.

Theorem A.1 (Taylor) *If $\boldsymbol{u} \in \mathrm{C}^{n+1}(\mathbb{R}^d, \mathbb{R}^m)$, then*

$$\boldsymbol{u}(\boldsymbol{x} + \boldsymbol{h}) = \boldsymbol{u}(\boldsymbol{x}) + D\boldsymbol{u}(\boldsymbol{x})[\boldsymbol{h}] + \cdots + \frac{1}{n!}D^n\boldsymbol{u}(\boldsymbol{x})[\boldsymbol{h}]^n + \boldsymbol{R}_n$$

where $\boldsymbol{x}, \boldsymbol{h} \in \mathbb{R}^d$ and the remainder

$$\boldsymbol{R}_n = \frac{1}{n!}\int_0^1 (1-s)^n D^{n+1}\boldsymbol{u}(\boldsymbol{x} + s\boldsymbol{h})[\boldsymbol{h}]^{n+1}\, ds.$$

Definition A.2 (analytic) Let D be a domain in $\mathbb{R}^d$. A function $f \in \mathrm{C}^\infty(D, \mathbb{R}^m)$ is *analytic* if, for all $\boldsymbol{x} \in D$, there is an $\epsilon > 0$ such that the Taylor series converges (i.e., $\boldsymbol{R}_n \to 0$ as $n \to \infty$) for $\|\boldsymbol{h}\|_2 < \epsilon$.

We frequently use the big-$\mathcal{O}$ notation and '$\boldsymbol{R} = \mathcal{O}(h)$ as $h \to 0$' means that there exists a constant $K > 0$ such that $\|\boldsymbol{R}\| \le Kh$ for all h sufficiently small.

Example A.3 For $u(x) = \mathrm{e}^x$, Theorem A.1 gives the classical Taylor series

$$\mathrm{e}^{x+h} = \mathrm{e}^x + \mathrm{e}^x h + \frac{1}{2!}\mathrm{e}^x h^2 + \cdots + \frac{1}{n!}\mathrm{e}^x h^n + R_n, \qquad x, h \in \mathbb{R},$$

with remainder

$$R_n = \frac{1}{n!}\int_0^1 (1-s)^n \mathrm{e}^{x+sh} h^{n+1}\, ds = \mathcal{O}(h^{n+1}) \qquad \text{as } h \to 0.$$

In this example, $R_n \to 0$ as $n \to \infty$ and e^x is analytic. This example should be contrasted with the infinitely differentiable function

$$u(x) = \begin{cases} 0, & x = 0, \\ \mathrm{e}^{-1/x^2}, & x \neq 0, \end{cases}$$

which is not analytic due to its behaviour at $x = 0$.

Example A.4 For $\boldsymbol{u} \in \mathrm{C}^1(\mathbb{R}^d, \mathbb{R}^m)$, Theorem A.1 leads to $\boldsymbol{u}(\boldsymbol{x}+\boldsymbol{h}) = \boldsymbol{u}(\boldsymbol{x}) + \boldsymbol{R}_0$ where

$$\boldsymbol{R}_0 = \int_0^1 D^1\boldsymbol{u}(\boldsymbol{x}+s\boldsymbol{h})[\boldsymbol{h}]\,ds.$$

This implies $\boldsymbol{u}$ is Lipschitz continuous if $D^1\boldsymbol{u}$ is bounded.

Example A.5 (centred differences) Let $u \in \mathrm{C}^{n+1}(\mathbb{R})$. Now, $D^n u(x)[h]^n$ is simply $u^{(n)}(x)h^n$ and, in the case $n = 2$,

$$u(x+h) = u(x) + u'(x)h + \frac{1}{2}u''(x)h^2 + R_2, \qquad R_2 = \frac{1}{2}\int_0^1 (1-s)^2 u'''(x+sh)h^3\,ds.$$

Taking $u(x+h)$ and $u(x-h)$ together, we obtain the centred difference approximation

$$u'(x) \approx \frac{u(x+h)-u(x-h)}{2h} \qquad \text{with } \mathcal{O}(h^2) \text{ error.}$$

A.2 Trapezium rule

One important application of polynomial approximation is quadrature. If we approximate a function $u\colon [a,b] \to \mathbb{R}$ by a polynomial $p(x)$, then

$$\int_a^b u(x)\,dx \approx \int_a^b p(x)\,dx,$$

where the right-hand integral is easy to evaluate. For example, the linear interpolant $p(x) = (u(a)(b-x) + u(b)(x-a))/(b-a)$ of $u(x)$ gives rise to the trapezium rule

$$\int_a^b u(x)\,dx \approx \frac{1}{2(b-a)}(u(a)+u(b))$$

and Taylor's theorem can be used to show that the error is $\mathcal{O}(|b-a|^3)$. To approximate the integral accurately, it is usual to divide $[a,b]$ into subintervals $[x_j, x_{j+1}]$ at $x_j = a + jh$ for $h = (b-a)/J$ and $j = 0,\dots,J$ and sum the trapezium rule approximations to $\int_{x_j}^{x_{j+1}} u(x)\,dx$. This gives

$$\int_a^b u(x)\,dx \approx T_J(u) := h\left[\frac{u(a)}{2} + \sum_{j=1}^{J-1} u(x_j) + \frac{u(b)}{2}\right]. \tag{A.1}$$

This is known as the *composite trapezium rule* and the approximation error is $\mathcal{O}(h^2)$. More detailed information is available in special cases.

Theorem A.6 (trapezium rule) *If $u \in \mathrm{C}^4(a,b)$, there exists $\xi \in [a,b]$ such that*

$$\int_a^b u(x)\,dx - T_J(u) = \frac{1}{12}h^2\big(u'(a) - u'(b)\big) + \frac{(b-a)}{720}\,h^4\,u''''(\xi).$$

If $u \in \mathrm{C}^{2p}(a,b)$ for some $p \in \mathbb{N}$ and $u^{(2k-1)}(a) = u^{(2k-1)}(b)$ for $k = 1,\dots,p-1$, there exists $\xi \in [a,b]$ such that

$$\int_a^b u(x)\,dx - T_J(u) = -\frac{(b-a)B_{2p}}{(2p)!}\,h^{2p}\,u^{(2p)}(\xi),$$

where B_{2p} denotes the Bernoulli numbers *($B_2 = 1/6$, $B_4 = -1/30$, $B_6 = 1/42$, $B_8 = -1/30, \dots$).*

The first equality in Theorem A.6 is used in §6.4, where we take advantage of derivatives being small at the ends of the interval. The second equality shows that, if the integrand is periodic and smooth, the trapezium rule is convergent with any order of accuracy. This fact is exploited in §7.3.

A.3 Special functions

Gamma function $\Gamma(x)$

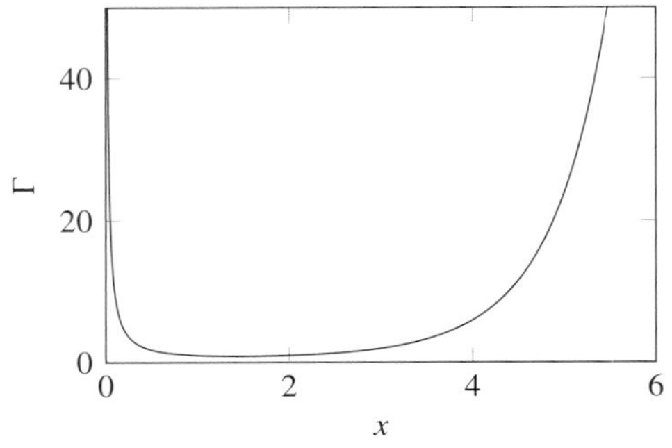

Figure A.1 The gamma function $\Gamma(x)$.

The gamma function is a generalisation of factorial defined by

$$\Gamma(x) := \int_0^\infty t^{x-1} e^{-t}\, dt, \qquad x > 0. \tag{A.2}$$

It is elementary to check that $\Gamma(x+1) = x\Gamma(x)$ using integration by parts and that $\Gamma(n+1) = n!$ for $n \in \mathbb{N}$. It is evaluated in MATLAB by `gamma(x)`. The gamma function is useful because it provides a convenient expression for the surface area and volume of the sphere in $\mathbb{R}^d$.

Lemma A.7 *Let $S^{d-1}(r) := \{\boldsymbol{x} \in \mathbb{R}^d : \|\boldsymbol{x}\|_2 = r\}$ denote the sphere of radius r in $\mathbb{R}^d$. Then, $S^{d-1}(r)$ has surface area $\omega_d r^{d-1}$ and volume $(\omega_d/d) r^d$, where $\omega_d := \dfrac{2\pi^{d/2}}{\Gamma(d/2)}$. In particular, $\omega_1 = 2$, $\omega_2 = 2\pi$, and $\omega_3 = 4\pi$.*

Bessel function $J_p(r)$

The Bessel functions frequently arise when working in polar coordinates. We are interested in the Bessel functions of the first kind, which are denoted by $J_p(r)$ and evaluated in MATLAB by `besselj(p,r)`. In some cases, the Bessel functions can be written explicitly and in particular

$$J_{-1/2}(r) := \left(\frac{2}{\pi r}\right)^{1/2} \cos(r), \qquad J_{1/2}(r) := \left(\frac{2}{\pi r}\right)^{1/2} \sin(r).$$

We also use the Bessel function $J_p(r)$ for $p > -1/2$ defined by

$$J_p(r) := \frac{2^{1-p}}{\pi^{1/2}\Gamma(p+1/2)} r^p \int_0^{\pi/2} \cos(r\cos\theta) \sin^{2p}\theta\, d\theta. \tag{A.3}$$

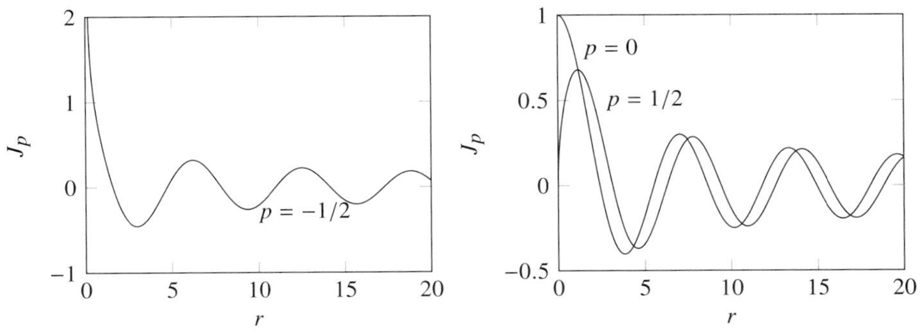

Figure A.2 Plots of the Bessel function $J_p(r)$ for $p = -1/2, 0, 1/2$

See Figure A.2. The Bessel functions provide a convenient expression for the Fourier transform of an isotropic function.

Theorem A.8 (Hankel transform) *The Fourier transform $\hat{u}$ of an isotropic function $u\colon \mathbb{R}^d \to \mathbb{C}$ is isotropic. Using $u(r)$ and $\hat{u}(\lambda)$ with $r = \|\boldsymbol{x}\|_2$ and $\lambda = \|\boldsymbol{\lambda}\|_2$, to denote $u(\boldsymbol{x})$ and $\hat{u}(\boldsymbol{\lambda})$,*

$$\hat{u}(\lambda) = \int_0^\infty \frac{J_p(\lambda r)}{(\lambda r)^p} u(r) r^{d-1}\, dr, \qquad p = \frac{d}{2} - 1. \tag{A.4}$$

(A.4) *is called the* Hankel transform.

Modified Bessel functions $K_q(r)$

The modified Bessel functions are the hyperbolic versions of the Bessel functions. They give an important class of covariance functions for random fields. The modified Bessel function of the second kind of order q is defined by $K_q(r) := \check{f}(r)/r^q$ for all $r, q > 0$, where

$$f(\lambda) := \frac{c_q}{(1+\lambda^2)^{q+1/2}}, \qquad c_q := 2^{q-1/2}\Gamma(q+1/2). \tag{A.5}$$

It is evaluated in MATLAB by `besselk(q,r)`. In terms of the Hankel transform,

$$\begin{aligned} K_q(r) &= \frac{c_q}{r^q} \int_0^\infty \frac{J_{-1/2}(\lambda r)}{(\lambda r)^{-1/2}} \frac{1}{(1+\lambda^2)^{q+1/2}}\, d\lambda \\ &= \frac{c_q}{r^q} \int_0^\infty \left(\frac{2}{\pi}\right)^{1/2} \cos(\lambda r) \frac{1}{(1+\lambda^2)^{q+1/2}}\, d\lambda. \end{aligned}$$

For $q = 1/2$, the integral can be evaluated explicitly, as follows.

Lemma A.9 *The modified Bessel function*

$$K_{1/2}(r) = \left(\frac{\pi}{2r}\right)^{1/2} e^{-r}, \qquad r \geq 0.$$

Proof From (A.5), $c_{1/2} = 1$ and $f(\lambda) = 1/(1+\lambda^2)$. By (1.44), $\hat{f}(r) = \sqrt{\pi/2}e^{-|r|}$ and hence $K_q(r) = c_{1/2}\hat{f}(r)/r^{1/2} = (\pi/2r)^{1/2}e^{-r}$ for $r \geq 0$. □

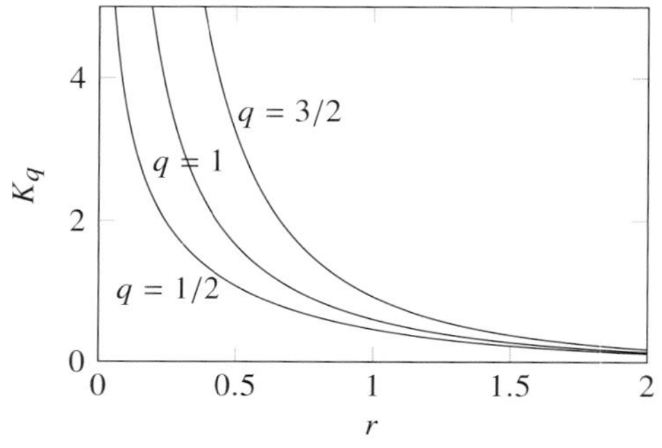

Figure A.3 Plots of the Bessel function $K_q(r)$ for $q = 1/2, 1, 3/2$

We require the following identities in §7.1: the Laplace transform of $r^p J_p(r)$ (Sneddon, 1972, p. 161) is given by

$$\int_0^\infty e^{-r\lambda} r^p J_p(r)\, dr = \frac{2^p \Gamma(p+1/2)}{\pi^{1/2}} \frac{1}{(1+\lambda^2)^{p+1/2}}. \tag{A.6}$$

The Weber–Schafheitlin integral (Watson, 1995, p. 410) for $p > -1$, $p + q > -1$, $b \geq 0$, $a > 0$ is

$$\int_0^\infty K_q(at) J_p(bt)\, t^{q+p+1}\, dt = \frac{(2a)^q (2b)^p \Gamma(p+q+1)}{(a^2+b^2)^{p+q+1}}. \tag{A.7}$$

A.4 Inequalities

The following are consequences of Jensen's inequality (Lemma 4.56).

Corollary A.10 *For a convex measurable function $\phi\colon \mathbb{R} \to \mathbb{R}$ and $f \in L^1(0,1)$,*

$$\phi\left(\int_0^1 f(s)\, ds\right) \leq \int_0^1 \phi(f(s))\, ds. \tag{A.8}$$

In particular, if $p \geq 1$,

$$\left|\int_0^t f(s)\, ds\right|^p \leq t^{p-1} \int_0^t |f(s)|^p\, ds. \tag{A.9}$$

For $a_i \in \mathbb{R}$ and $p \geq 1$,

$$\left|\sum_{i=1}^n a_i\right|^p \leq n^{p-1} \sum_{i=1}^n |a_i|^p. \tag{A.10}$$

In particular $(a_1 + a_2 + \cdots + a_n)^2 \leq n(a_1^2 + \cdots + a_n^2)$.

The following statements are well-known generalisations of Lemma 7.66 (e.g., Mao, 2008, Theorem 3.7). On a filtered probability space $(\Omega, \mathcal{F}, \mathcal{F}_t, \mathbb{P})$, a real-valued $\mathcal{F}_t$-adapted process $M(t)$ is a submartingale if $M(s) \leq \mathbb{E}[M(t) \,|\, \mathcal{F}_s]$ a.s. for $s \leq t$.

Theorem A.11 (Doob's submartingale inequality) *If $\{M(t)\colon t \ge 0\}$ is a non-negative submartingale with continuous sample paths, then*

$$\mathbb{E}\Big[\sup_{0\le t\le T} |M(t)|^2\Big] \le 4 \sup_{0\le t\le T} \mathbb{E}\big[|M(t)|^2\big]. \tag{A.11}$$

Corollary A.12 (Doob's maximal inequality) *Let $\beta(t)$ be an $\mathcal{F}_t$-Brownian motion. Then $|\beta(t)|^2$ is a submartingale and*

$$\mathbb{E}\Big[\sup_{0\le t\le T} |\beta(t)|^2\Big] \le 4 \sup_{0\le t\le T} \mathbb{E}\big[|\beta(t)|^2\big]. \tag{A.12}$$

Proof Let $M(t) = |\beta(t)|^2$. Then, for $s \le t$,

$$M(s) = |\beta(s)|^2 = \big|\mathbb{E}[\beta(t)\,|\,\mathcal{F}_s]\big|^2 \le \mathbb{E}[|\beta(t)|^2\,|\,\mathcal{F}_s] = \mathbb{E}[M(t)\,|\,\mathcal{F}_s], \qquad a.s.,$$

by Jensen's inequality (for conditional expectations) and $M(t)$ is a submartingale. As $\beta(t)$ has continuous sample paths, so does $M(t)$ and the result follows from Lemma A.11. □

Gronwall's inequality (see Exercise 3.4) also applies when b is a function (Mao, 2008, Theorem 8.1).

Lemma A.13 *Suppose that $z(t)$ satisfies*

$$0 \le z(t) \le a + \int_0^t b(s)z(s)\,ds, \qquad t \ge 0,$$

for $a \ge 0$ and a non-negative and integrable function $b(s)$. Then,

$$z(t) \le a e^{B(t)}, \qquad \text{where } B(t) := \int_0^t b(s)\,ds.$$

We also require the discrete Gronwall inequality (e.g., Stuart and Humphries, 1996, Theorem 1.1.2).

Lemma A.14 (discrete Gronwall inequality) *Consider $z_n \ge 0$ such that*

$$z_n \le a + b\sum_{k=0}^{n-1} z_k, \qquad \text{for } n = 0, 1, \dots$$

and constants $a, b \ge 0$. If $b = 1$, then $z_n \le z_0 + na$. If $b \ne 1$, then

$$z_n \le b^n z_0 + \frac{a}{1-b}(1 - b^n).$$

A.5 The Dirichlet divisor problem

For $k \in \mathbb{N}$ and $x > 0$, let

$$D_k(x) := \text{the number of } (n_1, \dots, n_k) \in \mathbb{N}^k \text{ such that } n_1 n_2 \cdots n_k \le x.$$

The study of the asymptotic behaviour of $D_k(x)$ as $x \to \infty$ is known as the *Dirichlet divisor problem* and

$$D_k(x) = x P_{k-1}(\log x) + \mathcal{O}(x^{1-1/k}), \tag{A.13}$$

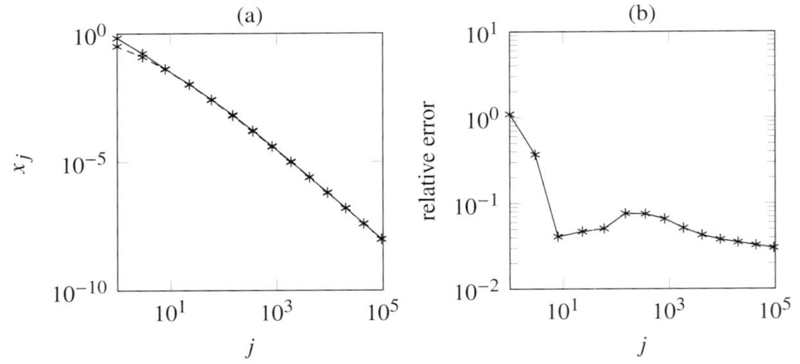

Figure A.4 (a) x_j and the asymptotic formula given by Lemma A.15 (dash-dot line) (b) the relative error in the asymptotic formula.

for a polynomial P_k of degree k. It is known that $P_1(x) = x + (2\gamma - 1)$, where the Euler constant γ is defined by $\gamma := \lim_{n\to\infty}(H_n - \log n)$ and $H_n := \sum_{k=1}^{n} \frac{1}{k}$. Its value is approximately 0.5572 and it can be evaluated in MATLAB by `-psi(1)` (where psi is a built-in function known as the polygamma function). See Ivić (2003) for more information. We use this result to find an asymptotic formula for $1/n^2m^2$ as $n, m \to \infty$, which is used in Theorem 7.61. See Figure A.4.

Define the Lambert W function for $x \geq 0$ by

$$W(x) := y, \qquad \text{where } y \in \mathbb{R} \text{ is the solution of } x = ye^y. \tag{A.14}$$

It is evaluated in MATLAB by `lambertw(x)`. Recall the $\asymp$ notation of Theorem 7.57.

Lemma A.15 *Let x_j for $j \in \mathbb{N}$ denote the numbers $1/n_1^2n_2^2$ for $n_1, n_2 \in \mathbb{N}^2$ in decreasing order, where repeat entries are included. Then, $x_j \asymp \mathrm{e}^{-2W(j)}$ as $j \to \infty$.*

Proof For $x > 0$, let $J(x)$ be the number of integer pairs $(n_1, n_2) \in \mathbb{N}^2$ such that $1/n_1^2n_2^2 \geq x$. Note that $J(x_j) = j$. Using (A.13),

$$J(x) = x^{-1/2}\log(x^{-1/2}) + x^{-1/2}(2\gamma - 1) + \mathcal{O}(x^{-1/4}), \tag{A.15}$$

where γ is Euler's constant. Let $g(y) := y \ln y$ for $y > 0$. Note that $g(y)$ has a well-defined inverse defined by $g^{-1}(x) := \mathrm{e}^{W(x)}$, because $g(\mathrm{e}^{W(x)}) = \mathrm{e}^{W(x)}W(x) = x$ by the definition of the Lambert W function in (A.14). Then,

$$J(x) = g(x^{-1/2}) + \mathcal{O}(x^{-1/2}).$$

By first showing $\log x \asymp W(x)$ as $x \to \infty$, it can be shown that $\mathrm{e}^{W(x)} \asymp \mathrm{e}^{W(y)}$ if $x \asymp y$ as $x, y \to \infty$. Hence,

$$x \asymp \left(\frac{1}{g^{-1}(J(x))}\right)^2 = \mathrm{e}^{-2W(J(x))}. \tag{A.16}$$

Replacing x by x_j and $J(x)$ by j, we have $x_j \asymp \mathrm{e}^{-2W(j)}$. □

References

P. Abrahamsen (1997), *A Review of Gaussian Random Fields and Correlation Functions*, tech. rep. 917, Norwegian Computing Centre.

R. Adams and J. Fournier (2003), *Sobolev Spaces*, 2nd ed., Pure and Applied Mathematics vol. 140, Elsevier Science.

R. J. Adler (1981), *The Geometry of Random Fields*, Chichester: John Wiley & Sons, xi+280 pp.

E. J. Allen, S. J. Novosel, and Z. Zhang (1998), "Finite element and difference approximation of some linear stochastic partial differential equations", *Stoch. Stoch. Rep.* 64: 1–2, 117–142.

S. Allen and J. W. Cahn (1979), "A microscopic theory for antiphase boundary motion and its application to antiphase domain coarsening", *Acta Metallurgica*, 27: 1085–1095.

S. Asmussen and P. W. Glynn (2007), *Stochastic Simulation: Algorithms and Analysis*, Stochastic Modelling and Applied Probability vol. 57, New York: Springer-Verlag, xiv+476 pp.

I. Babuška, R. Tempone, and G. E. Zouraris (2004), "Galerkin finite element approximations of stochastic elliptic partial differential equations", *SIAM J. Numer. Anal.* 42: 2, 800–825.

I. Babuška, F. Nobile, and R. Tempone (2007), "A stochastic collocation method for elliptic partial differential equations with random input data", *SIAM J. Numer. Anal.* 45: 3, 1005–1034.

A. Bain and D. Crisan (2009), *Fundamentals of Stochastic Filtering*, Stochastic Modelling and Applied Probability vol. 60, New York: Springer, xiv+390 pp.

C. T. H. Baker (1978), *The Numerical Treatment of Integral Equations*, Oxford: Oxford University Press.

G. Bal, J. Garnier, S. Motsch, and V. Perrier (2008), "Random integrals and correctors in homogenization", *Asymptot. Anal.* 59: 1–2, 1–26.

L. Baňas, Z. Brzeźniak, and A. Prohl (2013), "Computational studies for the stochastic Landau–Lifshitz–Gilbert Equation", *SIAM J. Sci. Comput.* 35: 1, B62–B81.

A. Barth and A. Lang (2012), "Multilevel Monte Carlo method with applications to stochastic partial differential equations", *Int. J. Comput. Math.* 89: 18, 2479–2498.

A. Barth, C. Schwab, and N. Zollinger (2011), "Multi-level Monte Carlo finite element method for elliptic PDEs with stochastic coefficients", *Numer. Math.* 119: 1, 123–161.

A. Barth, A. Lang, and C. Schwab (2013), "Multilevel Monte Carlo method for parabolic stochastic partial differential equations", *BIT*, 53: 1, 3–27.

V. Barthelmann, E. Novak, and K. Ritter (2000), "High dimensional polynomial interpolation on sparse grids", *Adv. Comput. Math.* 12: 4, 273–288.

A. Bespalov, C. E. Powell, and D. Silvester (2012), "A priori error analysis of stochastic Galerkin mixed approximations of elliptic PDEs with random data", *SIAM J. on Numer. Anal.* 50: 4, 2039–2063.

R. Bhattacharya and R. Ranga Rao (1986), *Normal Approximation and Asymptotic Expansions*, Reprint of the 1976 original, Melbourne: Robert Krieger, xiv+291 pp.

M. Bieri, R. Andreev, and C. Schwab (2010), "Sparse tensor discretization of elliptic SPDEs", *SIAM J. Sci. Comput.* 31: 6, 4281–4304.

P. Billingsley (1995), *Probability and Measure*, 3rd ed., Wiley Series in Probability and Mathematical Statistics, New York: John Wiley & Sons, xiv+593 pp.

D. Blömker (2007), *Amplitude Equations for Stochastic Partial Differential Equations*, vol. 3, Interdisciplinary Mathematical Sciences, Singapore: World Scientific.

V. I. Bogachev (2007), *Measure Theory*, vol. 1, New York: Springer-Verlag, xviii+500 pp.

F. Bouchet and E. Simonnet (2009), "Random changes of flow topology in two-dimensional and geophysical turbulence", *Phys. Rev. Lett.* 102: 9, 094504.
D. Braess (1997), *Finite Elements: Theory, Fast Solvers, and Applications in Solid Mechanics*, Cambridge: Cambridge University Press.
P. Bratley and B. L. Fox (1988), "Algorithm 659: implementing Sobol's quasirandom sequence generator.", *ACM Trans. Math. Software*, 14: 1, 88–100.
L. Breiman (1992), *Probability*, Classics in Applied Mathematics vol. 7, Corrected reprint of the 1968 original, Philadelphia: Society for Industrial and Applied Mathematics, xiv+421 pp.
S. C. Brenner and L. R. Scott (2008), *The Mathematical Theory of Finite Element Methods*, Texts in Applied Mathematics vol. 15, New York: Springer-Verlag.
F. Brezzi and M. Fortin (1991), *Mixed and Hybrid Finite Element Methods*, Springer Series in Computational Mathematics vol. 15, New York: Springer-Verlag, x+350 pp.
E. O. Brigham (1988), *Fast Fourier Transform and Its Applications*, Englewood Cliffs, NJ: Prentice Hall.
R. G. Brown and P. Y. C. Hwang (1992), *Introduction to Random Signals and Applied Kalman Filtering*, 2nd ed., New York: John Wiley & Sons.
Z. Brzeźniak, J. M. A. M. van Neerven, M. C. Veraar, and L. Weis (2008), "Itô's formula in UMD Banach spaces and regularity of solutions of the Zakai equation", *J. Differential Equations*, 245: 1, 30–58.
Z. Brzeźniak and T. Zastawniak (1999), *Basic Stochastic Processes*, Springer Undergraduate Mathematics Series, London: Springer-Verlag, x+225 pp.
E. Buckwar and T. Sickenberger (2011), "A comparative linear mean-square stability analysis of Maruyama- and Milstein-type methods", *Math. Comput. Simulat.* 81: 6, 1110–1127.
E. Buckwar and R. Winkler (2006), "Multistep methods for SDEs and their application to problems with small noise", *SIAM J. Numer. Anal.* 44: 2, 779–803.
H.-J. Bungartz and M. Griebel (2004), "Sparse grids", *Acta Numer.* 13: 147–269.
K. Burrage, P. Burrage, and T. Mitsui (2000), "Numerical solutions of stochastic differential equations", *J. Comput. Appl. Math.* 125: 1–2, 171–182.
R. E. Caflisch (1998), "Monte Carlo and quasi-Monte Carlo methods", in: Acta Numerica vol. 7, pp. 1–49, Cambridge: Cambridge University Press.
R. H. Cameron and W. T. Martin (1947), "The orthogonal development of non-linear functionals in series of Fourier–Hermite functionals", *Ann. Math.* 48: 2, 385–392.
C. Canuto, M. Y. Hussaini, A. Quarteroni, and T. A. Zang (1988), *Spectral Methods in Fluid Dynamics*, Springer Series in Computational Physics, New York: Springer-Verlag, xiv+557 pp.
F. Chatelin (1983), *Spectral Approximation of Linear Operators*, Computer Science and Applied Mathematics, New York: Academic Press, xix+458 pp.
A. J. Chorin and O. H. Hald (2006), *Stochastic Tools in Mathematics and Science*, Surveys and Tutorials in the Applied Mathematical Sciences vol. 1, New York: Springer-Verlag, viii+147 pp.
P.-L. Chow (2007), *Stochastic Partial Differential Equations*, Boca Raton, FL: Chapman & Hall CRC, x+281 pp.
G. Christakos (2005), *Random Field Models in Earth Sciences*, Mineola, NY: Dover Publications, xxv+512 pp.
P. G. Ciarlet (1978), *The Finite Element Method for Elliptic Problems*, Studies in Mathematics and its Applications vol. 4, Amsterdam: North-Holland, xix+530 pp.
C. W. Clenshaw and A. R. Curtis (1960), "A method for numerical integration on an automatic computer", *Numer. Math.* 2: 197–205.
K. A. Cliffe, M. B. Giles, R. Scheichl, and A. L. Teckentrup (2011), "Multilevel Monte Carlo methods and applications to elliptic PDEs with random coefficients", *Comput. Vis. Sci.* 14: 1, 3–15.
S. Cox and J. van Neerven (2010), "Convergence rates of the splitting scheme for parabolic linear stochastic Cauchy problems", *SIAM J. Numer. Anal.* 48: 2, 428–451.
H. Cramér and M. R. Leadbetter (2004), *Stationary and Related Stochastic Processes*, Reprint of the 1967 original, Mineola, NY: Dover Publications, xiv+348 pp.
D. Crisan and B. Rozovskiĭ (2011), *The Oxford Handbook of Nonlinear Filtering*, Oxford: Oxford: Oxford University Press.
G. Da Prato, A. Jentzen, and M. Röckner (2010), *A mild Ito formula for SPDEs*, eprint: `arXiv:1009.3526`.
G. Da Prato and J. Zabczyk (1992), *Stochastic Equations in Infinite Dimensions*, Encyclopedia of Mathematics and its Applications vol. 44, Cambridge: Cambridge University Press, xviii+454 pp.

R. C. Dalang, D. Khoshnevisan, and F. Rassoul-Agha (2009), *A Minicourse on Stochastic Partial Differential Equations*, Lecture Notes in Mathematics vol. 1962, New York: Springer-Verlag.
A. M. Davie (2007), "Differential equations driven by rough paths: an approach via discrete approximation", *Appl. Math. Res. Express.* 2: 1–40.
R. B. Davies and D. S. Harte (1987), "Tests for Hurst effect", *Biometrika*, 74: 95–102.
P. J. Davis (1979), *Circulant Matrices*, New York: John Wiley & Sons, xv+250 pp.
A. de Bouard and A. Debussche (1999), "A stochastic nonlinear Schrödinger equation with multiplicative noise", *Commun. Math. Phys.* 205: 1, 161–181.
—(2006), "Weak and strong order of convergence of a semidiscrete scheme for the stochastic nonlinear Schrödinger equation", *Appl. Math. Optim.* 54: 3, 369–399.
M. K. Deb, I. M. Babuška, and J. T. Oden (2001), "Solution of stochastic partial differential equations using Galerkin finite element techniques", *Comput. Methods Appl. Mech. Eng.* 190: 48, 6359–6372.
A. Debussche (2011), "Weak approximation of stochastic partial differential equations: the nonlinear case", *Math. Comp.* 80: 273, 89–117.
A. Debussche and J. Printems (2009), "Weak order for the discretization of the stochastic heat equation", *Math. Comp.* 78: 266, 845–863.
L. M. Delves and J. Walsh, eds. (1974), *Numerical Solution of Integral Equations*, Oxford: Oxford University Press.
S. Dereich (2011), "Multilevel Monte Carlo algorithms for Lévy-driven SDEs with Gaussian correction", *Adv. in Appl. Probab.* 21: 1, 283–311.
A. Deya, M. Jolis, and L. Quer-Sardanyons (2013), "The Stratonovich heat equation: a continuity result and weak approximations", *Electron. J. Probab.* 18: 3, 1–34.
P. Diaconis (2009), "The Markov chain Monte Carlo revolution", *Bull. Am. Math. Soc.* 46: 2, 179–205.
A. B. Dieker and M. Mandjes (2003), "On spectral simulation of fractional Brownian motion", *Probab. Engrg. Inform. Sci.* 17: 417–434.
C. R. Dietrich (1995), "A simple and efficient space domain implementation of the turning bands method", *Water Resourc. Res.* 31: 1, 147–156.
C. R. Dietrich and G. N. Newsam (1993), "A fast and exact method of multidimensional Gaussian stochastic simulations", *Water Resourc. Res.* 29: 8, 2861–2869.
—(1997), "Fast and exact simulation of stationary Gaussian processes through circulant embedding of the covariance matrix", *SIAM J. Sci. Comput.* 18: 4, 1088–1107.
R. M. Dudley (1999), *Uniform Central Limit Theorems*, Cambridge Studies in Advanced Mathematics vol. 63, Cambridge: Cambridge University Press, xiv+436 pp.
—(2002), *Real Analysis and Probability*, Cambridge Studies in Advanced Mathematics vol. 74, Cambridge: Cambridge University Press, x+555 pp.
H. C. Elman, O. G. Ernst, D. P. O'Leary, and M. Stewart (2005a), "Efficient iterative algorithms for the stochastic finite element method with application to acoustic scattering", *Comput. Methods Appl. Mech. Eng.* 194: 9-11, 1037–1055.
H. C. Elman, D. Silvester, and A. Wathen (2005b), *Finite Elements and Fast Iterative Solvers*, Oxford: Oxford University Press, xiv+400 pp.
H. C. Elman, A. Ramage, and D. J. Silvester (2007), "Algorithm 866: IFISS: A MATLAB toolbox for modelling incompressible flow", *ACM Trans. Math. Software*, 33: 14, 1–19.
O. G. Ernst and E. Ullmann (2010), "Stochastic Galerkin matrices", *SIAM J. Matrix Anal. Appl.* 31: 4, 1848–1872.
L. C. Evans (2010), *Partial Differential Equations*, 2nd ed., Graduate Studies in Mathematics vol. 19, Providence, RI: American Mathematical Society, xxii+749 pp.
R. Eymard, T. Gallouët, and R. Herbin (2000), "Finite volume methods", in: *Handbook of numerical analysis*, vol. VII, pp. 713–1020, Amsterdam: Amsterdam: North-Holland.
G. S. Fishman (1996), *Monte Carlo: Concepts, Algorithms and Applications*, Springer Series in Operations Research, New York: Springer-Verlag, xxvi+698 pp.
B. Fornberg (1996), *A Practical Guide to Pseudospectral Methods*, Cambridge Monographs on Applied and Computational Mathematics, Cambridge: Cambridge University Press, x+231 pp.
P. Frauenfelder, C. Schwab, and R. A. Todor (2005), "Finite elements for elliptic problems with stochastic coefficients", *Comput. Methods Appl. Mech. Eng.* 194: 205–228.
U. Frisch (1995), *Turbulence: The Legacy of A. N. Kolmogorov*, Cambridge: Cambridge University Press, xiv+296 pp.
B. Fristedt and L. Gray (1997), *A Modern Approach to Probability Theory*, Probability and its Applications, Basel: Birkhauser, xx+756 pp.

P. K. Friz and N. B. Victoir (2010), *Multidimensional Stochastic Processes as Rough Paths*, Cambridge Studies in Advanced Mathematics vol. 120, Cambridge: Cambridge University Press, xiv+656 pp.

H. Fujita and T. Suzuki (1991), "Evolution problems", in: *Handbook of Numerical Analysis*, vol. II, pp. 789–928, Amsterdam: North-Holland.

J. Gaines and T. Lyons (1994), "Random generation of stochastic area integrals", *SIAM J. Appl. Math.* 54: 4, 1132–1146.

A. Ganguly (2013), "Wong–Zakai type convergence in infinite dimensions", *Electron. J. Probab.* 18: 31, 1–34.

J. García-Ojalvo and J. M. Sancho (1999), *Noise in Spatially Extended Systems*, Institute for Nonlinear Science, New York: Springer-Verlag, xiv+307 pp.

C. Gardiner (2009), *Stochastic Methods: A Handbook for the Natural and Social Sciences*, 4th ed., Springer Series in Synergetics, Springer-Verlag, xviii+447 pp.

A. M. Garsia, E. Rodemich, and H. Rumsey Jr. (1971), "A real variable lemma and the continuity of paths of some Gaussian processes", *Indiana Univ. Math. J.* 20: 6, 565–578.

W. Gautschi (2004), *Orthogonal Polynomials: Computation and Approximation*, Numerical Mathematics and Scientific Computation, New York: Oxford University Press, x+301 pp.

M. Geissert, M. Kovács, and S. Larsson (2009), "Rate of weak convergence of the finite element method for the stochastic heat equation with additive noise", *BIT*, 49: 2, 343–356.

R. G. Ghanem and P. D. Spanos (1991), *Stochastic Finite Elements: A Spectral Approach*, New York: Springer-Verlag, x+214 pp.

Ĭ. Ī. Gīhman and A. V. Skorohod (1972), *Stochastic Differential Equations*, Translated from the Russian by K. Wickwire, Ergebnisse der Mathematik und ihrer Grenzgebiete, Band 72, New York: Springer-Verlag, pp. viii+354.

M. B. Giles (2008a), "Improved multilevel Monte Carlo convergence using the Milstein scheme", in: *Monte Carlo and quasi-Monte Carlo methods*, pp. 343–358, 2006, Berlin: Springer-Verlag.

—(2008b), "Multilevel Monte Carlo path simulation", *Oper. Res.* 56: 3, 607–617.

M. B. Giles and B. J. Waterhouse (2009), "Multilevel quasi-Monte Carlo path simulation", in: *Advanced Financial Modelling*, pp. 165–181, H. Abrecher, W. J. Runggaldier, and W. Schachermayer (eds.), Radon Series Computational and Applied Mathematics vol. 8, Berlin: Walter de Gruyter.

M. B. Giles, D. J. Higham, and X. Mao (2009), "Analysing multi-level Monte Carlo for options with non-globally Lipschitz payoff", *Finance Stoch.* 13: 3, 403–413.

C. J. Gittelson (2013), "An adaptive stochastic Galerkin method for random elliptic operators", *Math. Comp.* 82: 283, 1515–1541.

T. Gneiting (1998), "Closed form solutions of the two-dimensional turning bands equation", *Math. Geol.* 30: 4, 379–390.

T. Gneiting, H. Sevcíkova, D. B. Percival, M. Schlather, and Y. Jiang (2006), "Fast and exact simulation of large Gaussian lattice systems in $\mathbb{R}^2$: exploring the limits", *J. Comput. Graph. Simul.* 15: 3, 483–501.

E. Gobet (2000), "Weak approximation of killed diffusion using Euler schemes", *Stoch. Process. Appl.* 87: 2, 167–197.

—(2001), "Efficient schemes for the weak approximation of reflected diffusions", *Monte Carlo Methods Appl.* 7: 1-2, 193–202.

G. H. Golub and C. F. Van Loan (2013), *Matrix Computations*, 4th ed., Johns Hopkins Studies in the Mathematical Sciences, Baltimore, MD: Johns Hopkins University Press, xiv+756 pp.

G. H. Golub and J. H. Welsch (1969), "Calculation of Gauss quadrature rules", *Math. Comp.* 23: 106, A1–A10.

N. R. Goodman (1963), "Statistical analysis based on a certain multivariate complex Gaussian distribution (an introduction)", *Ann. Math. Statist.* 34: 1, 152–177.

A. D. Gordon and C. E. Powell (2012), "On solving stochastic collocation systems with algebraic multigrid", *IMA J. Numer. Anal.* 32: 3, 1051–1070.

D. Gottlieb and S. A. Orszag (1977), *Numerical Analysis of Spectral Methods: Theory and Applications*, CBMS-NSF Regional Conference Series in Applied Mathematics, No. 26, Philadelphia: Society for Industrial and Applied Mathematics, v+172 pp.

C. Graham and D. Talay (2013), *Stochastic Simulation and Monte Carlo Methods*, vol. 68, Stochastic Modelling and Applied Probability, Heidelberg: Springer-Verlag, pp. xvi+260.

I. G. Graham, F. Y. Kuo, D. Nuyens, R. Scheichl, and I. H. Sloan (2011), "Quasi-Monte Carlo methods for elliptic PDEs with random coefficients and applications", *J. Comput. Phys.* 230: 10, 3668–3694.

W. Grecksch and P. E. Kloeden (1996), "Time-discretised Galerkin approximations of parabolic stochastic PDEs", *Bull. Austral. Math. Soc.* 54: 1, 79–85.

G. R. Grimmett and D. R. Stirzaker (2001), *Probability and Random Processes*, 3rd ed., Oxford: Oxford University Press, xii+596 pp.

I. Gyöngy (1998a), "A note on Euler's approximations", *Potential Anal.* 8: 3, 205–216.

—(1998b), "Lattice approximations for stochastic quasi-linear parabolic partial differential equations driven by space–time white noise. I", *Potential Anal.* 9: 1, 1–25.

—(1999), "Lattice approximations for stochastic quasi-linear parabolic partial differential equations driven by space–time white noise. II", *Potential Anal.* 11: 1, 1–37.

I. Gyöngy and N. Krylov (2003), "On the splitting-up method and stochastic partial differential equations", *Ann. Probab.* 31: 2, 564–591.

I. Gyöngy and A. Millet (2009), "Rate of convergence of space time approximations for stochastic evolution equations", *Potential Anal.* 30: 1, 29–64.

W. Hackbusch (1992), *Elliptic Differential Equations: Theory and Numerical Treatment*, Springer Series in Computational Mathematics vol. 18, Translated from the author's revision of the 1986 German original by R. Fadiman and P. D. F. Ion, Berlin: Springer-Verlag, xiv+311 pp.

E. Hairer and G. Wanner (1996), *Solving Ordinary Differential Equations II*, 2nd ed., Springer Series in Computational Mathematics vol. 14, Berlin: Springer-Verlag, xvi+614 pp.

E. Hairer, S. P. Nørsett, and G. Wanner (1993), *Solving Ordinary Differential Equations I*, 2nd ed., Springer Series in Computational Mathematics vol. 8, Berlin: Springer-Verlag, xvi+528 pp.

M. Hairer, A. M. Stuart, J. Voss, and P. Wiberg (2005), "Analysis of SPDEs arising in path sampling part I: the Gaussian case", *Commun. Math. Sci.* 3: 4, 587–603.

M. Hairer, A. M. Stuart, and J. Voss (2007), "Analysis of SPDEs arising in path sampling part II: the nonlinear case", *Ann. Appl. Probab.* 17: 5-6, 1657–1706.

M. Hairer (2009), "An Introduction to Stochastic PDEs", University of Warwick Lecture Notes.

M. Hairer and J. C. Mattingly (2011), "A theory of hypoellipticity and unique ergodicity for semilinear stochastic PDEs", *Electron. J. Probab.* 16: 23.

J. M. Hammersley and D. C. Handscomb (1965), *Monte Carlo Methods*, Methuen's Monographs on Applied Probability and Statistics, London: Methuen, vii+178 pp.

E. Hausenblas (2003a), "Approximation for semilinear stochastic evolution equations", *Potential Anal.* 18: 2, 141–186.

—(2003b), "Weak approximation for semilinear stochastic evolution equations", in: *Stochastic analysis and related topics VIII*, pp. 111–128, Progress in Probability. vol. 53, Basel: Birkhäuser.

—(2007), "Wong–Zakai type approximation of SPDEs of Lévy noise", *Acta Appl. Math.* 98: 2, 99–134.

D. Henry (1981), *Geometric Theory of Semilinear Parabolic Equations*, Lecture Notes in Mathematics vol. 840, Berlin: Springer-Verlag, iv+348 pp.

K. Hesse, I. H. Sloan, and R. S. Womersley (2010), "Numerical integration on the sphere", in: *Handbook of Geomathematics*, pp. 1187–1219, W. Freeden, Z. Nashed, and T. Sonar (eds.), Berlin: Springer-Verlag.

D. J. Higham (2000), "Mean-square and asymptotic stability of the stochastic theta method", *SIAM J. Numer. Anal.* 38: 3, 753–769.

D. J. Higham, X. Mao, and A. M. Stuart (2002), "Strong convergence of Euler-type methods for nonlinear stochastic differential equations", *SIAM J. Numer. Anal.* 40: 3, 1041–1063.

—(2003), "Exponential mean-square stability of numerical solutions to stochastic differential equations", *London Math. Soc. J. Comput. Math.* 6: 297–313.

E. Hille and J. D. Tamarkin (1931), "On the characteristic values of linear integral equations", *Acta Numerica*, 57: 1–76.

W. Hundsdorfer and J. Verwer (2003), *Numerical Solution of Time-Dependent Advection–Diffusion–Reaction Equations*, Springer Series in Computational Mathematics vol. 33, Berlin: Springer-Verlag, x+471 pp.

M. Hutzenthaler, A. Jentzen, and P. E. Kloeden (2010), "Strong and weak divergence in finite time of Euler's method for stochastic differential equations with non-globally Lipschitz continuous coefficients", *Proc. R. Soc. London Ser. A*, 467: 2130, 1563–1576.

A. Iserles (1996), *A First Course in the Numerical Analysis of Differential Equations*, Cambridge Texts in Applied Mathematics, Cambridge: Cambridge University Press, xviii+378 pp.

A. Ivić (2003), *The Riemann Zeta-Function*, Reprint of the 1985 original, Mineola, NY: Dover Publications, xxii+517 pp.

A. Jentzen, P. E. Kloeden, and A. Neuenkirch (2009a), "Pathwise approximation of stochastic differential equations on domains: higher order convergence rates without global Lipschitz coefficients", *Numer. Math.* 112: 1, 41–64.

—(2009b), "Pathwise convergence of numerical schemes for random and stochastic differential equations", in: *Foundations of computational mathematics, Hong Kong 2008*, pp. 140–161, London Mathematical Society Lecture Note Series vol. 363, Cambridge: Cambridge University Press.

A. Jentzen and P. E. Kloeden (2009), "Overcoming the order barrier in the numerical approximation of stochastic partial differential equations with additive space-time noise", *Proc. R. Soc. London Ser. A*, 465: 2102, 649–667.

—(2011), *Taylor Approximations for Stochastic Partial Differential Equations*, CBMS-NSF Regional Conference Series in Applied Mathematics vol. 83, SIAM, xiv+220 pp.

A. Jentzen and M. Röckner (2010), *A Milstein scheme for SPDEs*, eprint: `arXiv:1001.2751`.

S. Joe and F. Y. Kuo (2003), "Remark on algorithm 659: implementing Sobol's quasirandom sequence generator", *ACM Trans. Math. Software*, 29: 1, 49–57.

I. Karatzas and S. E. Shreve (1991), *Brownian Motion and Stochastic Calculus*, 2nd ed., Graduate Texts in Mathematics vol. 113, New York: Springer-Verlag, xxiv+470 pp.

M. A. Katsoulakis, G. T. Kossioris, and O. Lakkis (2007), "Noise regularization and computations for the 1-dimensional stochastic Allen–Cahn problem", *Interfaces Free Bound.* 9: 1, 1–30.

P. E. Kloeden and A. Neuenkirch (2007), "The pathwise convergence of approximation schemes for stochastic differential equations", *LMS J. Comput. Math.* 10: 235–253.

P. E. Kloeden, E. Platen, and I. W. Wright (1992), "The approximation of multiple stochastic integrals", *Stochastic Anal. Appl.* 10: 4, 431–441.

P. E. Kloeden, G. J. Lord, A. Neuenkirch, and T. Shardlow (2011), "The exponential integrator scheme for stochastic partial differential equations: pathwise error bounds", *J. Comput. Appl. Math.* 235: 5, 1245–1260.

P. E. Kloeden and E. Platen (1992), *Numerical Solution of Stochastic Differential Equations*, Applications of Mathematics vol. 23, Berlin: Springer-Verlag, xxxvi+632 pp.

T. W. Koerner (1989), *Fourier Analysis*, Cambridge: Cambridge University Press.

A. N. Kolmogorov (1940), "Wienersche Spiralen und einige andere interessante Kurven Im Hilbertschen Raum", *C. R. (Doklady) Acad. URSS (N.S.)* 26: 115–118.

H. König (1986), *Eigenvalue Distribution of Compact Operators*, Operator Theory: Advances and Applications vol. 16, Basel: Birkhauser, 262 pp.

G. T. Kossioris and G. E. Zouraris (2010), "Fully-discrete finite element approximations for a fourth-order linear stochastic parabolic equation with additive space-time white noise", *M2AN Math. Model. Numer. Anal.* 44: 2, 289–322.

M. Kovács, S. Larsson, and F. Saedpanah (2010), "Finite element approximation of the linear stochastic wave equation with additive noise", *SIAM J. Numer. Anal.* 48: 2, 408–427.

M. Kovács, F. Lindgren, and S. Larsson (2011), "Spatial approximation of stochastic convolutions", *J. Comput. Appl. Math.* 235: 12, 3554–3570.

S. M. Kozlov (1979), "The averaging of random operators", *Mat. Sb. (N.S.)* 109(151): 2, 188–202.

R. Kruse and S. Larsson (2012), "Optimal regularity for semilinear stochastic partial differential equations with multiplicative noise", *Electron. J. Probab.* 17: 65, 1–19.

J. Kuelbs and T. Kurtz (1974), "Berry–Esséen estimates in Hilbert space and an application to the law of the iterated logarithm", *Ann. Probability*, 2: 387–407.

C. Laing and G. J. Lord, eds. (2010), *Stochastic Methods in Neuroscience*, Oxford: Oxford University Press, xxiv+370 pp.

J. D. Lambert (1991), *Numerical Methods for Ordinary Differential Systems*, Chichester: John Wiley & Sons, x+293 pp.

A. Lang, P.-L. Chow, and J. Potthoff (2010), "Almost sure convergence of a semidiscrete Milstein scheme for SPDEs of Zakai type", *Stochastics*, 82: 3, 315–326.

—(2012), "Erratum: Almost sure convergence of a semi-discrete Milstein scheme for SPDEs of Zakai type", *Stochastics*, 84: 4, 561–561.

P. D. Lax (2002), *Functional Analysis*, New York: John Wiley & Sons, xx+580 pp.

O. P. Le Maître and O. M. Knio (2010), *Spectral Methods for Uncertainty Quantification*, Scientific Computation, New York: Springer-Verlag, xvi+536 pp.

F. Lindgren, H. Rue, and J. Lindström (2011), "An explicit link between Gaussian fields and Gaussian Markov random fields: the stochastic partial differential equation approach", *J. R. Stat. Soc. Ser. B*, 73: 4, 423–498.

M. Loève (1977), *Probability Theory I*, 4th ed., Graduate Texts in Mathematics vol. 45, New York: Springer-Verlag, xvii+425 pp.

—(1978), *Probability Theory II*, 4th ed., Graduate Texts in Mathematics vol. 46, New York: Springer-Verlag, xvi+413 pp.

G. J. Lord and J. Rougemont (2004), "A numerical scheme for stochastic PDEs with Gevrey regularity", *IMA J. Numer. Anal.* 24: 4, 587–604.

G. J. Lord and T. Shardlow (2007), "Postprocessing for stochastic parabolic partial differential equations", *SIAM J. Numer. Anal.* 45: 2, 870–889.

G. J. Lord and A. Tambue (2013), "Stochastic exponential integrators for the finite element discretization of SPDEs for multiplicative and additive noise", *IMA J. Numer. Anal.* 33: 2, 515–543.

B. B. Mandelbrot and J. W. van Ness (1968), "Fractional Brownian motions, fractional noises, and applications", *SIAM Rev.* 10: 422–437.

A. Mantoglou and J. L. Wilson (1982), "The turning bands method for simulation of random fields using line generation by a spectral method", *Water Resourc. Res.* 18: 5, 1379–1394.

X. Mao (2008), *Stochastic Differential Equations and Applications*, 2nd ed., Chichester: Horwood, xviii+422 pp.

G. Marsaglia and W. W. Tsang (2000), "The Ziggurat method for generating random variables", *J. Statist. Software*, 8: 5, 1–7.

G. Matheron (1973), "The intrinsic random functions and their applications", *Adv. in Appl. Probab.* 5: 3, 439–468.

M. Matsumoto and T. Nishimura (1998), "Mersenne twister: a 623-dimensionally equidistributed uniform pseudorandom number generator", *ACM Trans. Model. and Comput. Simul.* 8: 1, 3–30.

J. C. Mattingly, A. M. Stuart, and M. V. Tretyakov (2010), "Convergence of numerical time-averaging and stationary measures via Poisson equations", *SIAM J. Numer. Anal.* 48: 2, 552–577.

G. N. Milstein and M. V. Tretyakov (1999), "Simulation of a space–time bounded diffusion", *Adv. in Appl. Probab.* 9: 3, 732–779.

—(2004), *Stochastic Numerics for Mathematical Physics*, Berlin: Springer-Verlag, xx+594 pp.

Y. S. Mishura (2008), *Stochastic Calculus for Fractional Brownian Motion and Related Processes*, Lecture Notes in Mathematics vol. 1929, Berlin: Springer-Verlag, xviii+393 pp.

C. Moler (1995), "Random thoughts: 10^{435} years is a very long time", *Cleve's Corner*, Fall Edition.

—(2001), "Normal behavior: Ziggurat algorithm generates normally distributed random numbers", *Cleve's Corner*, Spring Edition.

P. Mörters and Y. Peres (2010), *Brownian Motion*, Cambridge Series in Statistical and Probabilistic Mathematics, Cambridge: Cambridge University Press, xii+403 pp.

K. W. Morton and D. F. Mayers (2005), *Numerical Solution of Partial Differential Equations*, 2nd ed., Cambridge: Cambridge University Press, xiv+278 pp.

T. Müller-Gronbach and K. Ritter (2007), "An implicit Euler scheme with non-uniform time discretization for heat equations with multiplicative noise", *BIT*, 47: 2, 393–418.

—(2008), "Minimal Errors for Strong and Weak Approximation of Stochastic Differential Equations", in: *Monte Carlo and Quasi-Monte Carlo Methods 2006*, pp. 53–82, A. Keller, S. Heinrich, and H. Niederreiter (eds.), Berlin: Springer-Verlag.

E. Nelson (1967), *Dynamical Theories of Brownian Motion*, Princeton, NJ: Princeton University Press, 120 pp.

A. Neuenkirch (2008), "Optimal pointwise approximation of stochastic differential equations driven by fractional Brownian motion", *Stoch. Process. Appl.* 118: 12, 2294–2333.

G. N. Newsam and C. R. Dietrich (1994), "Bounds on the size of nonnegative definite circulant embeddings of positive definite Toeplitz matrices", *IEEE Trans. Inform. Theory*, 40: 4, 1218–1220.

H. Niederreiter (1992), *Random Number Generation and Quasi-Monte Carlo Methods*, CBMS-NSF Regional Conference Series in Applied Mathematics vol. 63, Philadelphia: Society for Industrial and Applied Mathematics, vi+241 pp.

F. Nobile, R. Tempone, and C. G. Webster (2008a), "A sparse grid stochastic collocation method for partial differential equations with random input data", *SIAM J. Numer. Anal.* 46: 5, 2309–2345.

—(2008b), "An anisotropic sparse grid stochastic collocation method for partial differential equations with random input data", *SIAM J. Numer. Anal.* 46: 5, 2411–2442.

E. Novak and K. Ritter (1996), "High-dimensional integration of smooth functions over cubes", *Numer. Math.* 75: 1, 79–97.
B. Øksendal (2003), *Stochastic Differential Equations*, 6th ed., Universitext, Berlin: Springer-Verlag, xxiv+360 pp.
H. C. Öttinger (1996), *Stochastic Processes in Polymeric Fluids*, Berlin: Springer-Verlag, xxiv+362 pp.
A. Pazy (1983), *Semigroups of Linear Operators and Applications to Partial Differential Equations*, Applied Mathematical Sciences vol. 44, New York: Springer-Verlag, viii+279 pp.
P. Z. Peebles (1993), *Probability, Random Variables, and Random Signal Principles*, New York: McGraw-Hill.
M. F. Pellissetti and R. G. Ghanem (2000), "Iterative solution of systems of linear equations arising in the context of stochastic finite elements", *Advances in Engineering Software*, 31: 607–616.
B. Picinbono (1996), "Second-order complex random vectors and normal distributions", *IEEE Trans. Signal Process*, 44: 10, 2637–2640.
C. E. Powell and H. C. Elman (2009), "Block-diagonal preconditioning for spectral stochastic finite-element systems", *IMA J. Numer. Anal.* 29: 2, 350–375.
C. E. Powell and D. J. Silvester (2007), *PIFISS: Potential (Incompressible) Flow and Iteration Software Guide*, tech. rep. 2007.14, MIMS: University of Manchester.
C. Prévôt and M. Röckner (2007), *A Concise Course on Stochastic Partial Differential Equations*, Lecture Notes in Mathematics vol. 1905, Berlin: Springer, vi+144 pp.
P. E. Protter (2005), *Stochastic Integration and Differential Equations*, 2nd ed., Stochastic Modelling and Applied Probability vol. 21, Springer-Verlag, xiv+419 pp.
A. Quarteroni and A. Valli (2008), *Numerical Approximation of Partial Differential Equations*, Springer Series in Computational Mathematics, Berlin: Springer.
L. Quer-Sardanyons and M. Sanz-Solé (2006), "Space semi-discretisations for a stochastic wave equation", *Potential Anal.* 24: 4, 303–332.
J. B. Reade (1983), "Eigenvalues of positive definite kernels", *SIAM J. Math. Anal.* 14: 1, 152–157.
M. Renardy and R. C. Rogers (2004), *An Introduction to Partial Differential Equations*, 2nd ed., Texts in Applied Mathematics vol. 13, New York: Springer-Verlag, xiv+434 pp.
F. Riesz and B. Sz.-Nagy (1990), *Functional Analysis*, Translated from the second French edition by L. F. Boron, New York: Mineola, NY: Dover Publications, xii+504 pp.
J. C. Robinson (2001), *Infinite-Dimensional Dynamical Systems*, Cambridge Texts in Applied Mathematics, Cambridge: Cambridge University Press, xviii+461 pp.
L. C. G. Rogers and D. Williams (2000), *Diffusions, Markov Processes, and Martingales*, vol. 1, Reprint of the second (1994) edition, Cambridge: Cambridge University Press, xx+386 pp.
S. M. Ross (1997), *Simulation*, 2nd ed., San Diego: Academic Press, xii+282 pp.
A. Rößler (2010), "Runge–Kutta methods for the strong approximation of solutions of stochastic differential equations", *SIAM J. Numer. Anal.* 48: 3, 922–952.
B. L. Rozovskiĭ (1990), *Stochastic Evolution Systems*, Mathematics and its Applications (Soviet Series) vol. 35, Dordrecht: Kluwer, xviii+315 pp.
W. Rudin (1987), *Real and Complex Analysis*, 3rd ed., New York: McGraw-Hill, xiv+416 pp.
T. Ryden and M. Wiktorsson (2001), "On the simulation of iterated Itô integrals", *Stoch. Process. Appl.* 91: 151–168.
M. Schlather (1999), *Introduction to Positive Definite Functions and to Unconditional Simulation of Random Fields*, tech. rep. ST-99-10, Lancaster University.
—(2001), "Simulation and analysis of random fields", *R News*, 1: 2, 18–20.
C. Schwab and R. A. Todor (2006), "Karhunen–Loève approximation of random fields by generalized fast multipole methods", *J. Comput. Phys.* 217: 100–122.
C. Schwab and C. J. Gittelson (2011), "Sparse tensor discretizations of high-dimensional parametric and stochastic PDEs", *Acta Numer.* 20: 291–467.
R. J. Serfling (1980), *Approximation Theorems of Mathematical Statistics*, New York: John Wiley & Sons, xiv+371 pp.
T. Shardlow (1999), "Numerical methods for stochastic parabolic PDEs", *Numer. Funct. Anal. Optim.* 20: 1-2, 121–145.
T. Shardlow and A. M. Stuart (2000), "A perturbation theory for ergodic properties of Markov chains", *SIAM J. Numer. Anal.* 37: 4, 1120–1137.
I. G. Shevtsova (2007), "Sharpening the upper bound for the absolute constant in the Berry–Esséen inequality", *Theory Probab. Appl.* 51: 3, 549–553.

M. Shinozuka (1971), "Simulation of multivariate and multidimensional random processes", *J. Acoust. Soc. Am.* 49: 1B, 357–368.
M. Shinozuka and C.-M. Jan (1972), "Digital simulation of random processes and its applications", *J. Sound Vibrat.* 25: 1, 111–128.
S. Smolyak (1963), "Quadrature and interpolation formulas for tensor products of certain classes of functions", *Soviet Math. Dokl.* 4: 240–243.
I. N. Sneddon (1972), *The Use of Integral Transforms*, New York: McGraw-Hill.
—(1995), *Fourier Transforms*, Reprint of the 1951 original, New York: Mineola, NY: Dover Publications, xii+542 pp.
T. Sottinen (2003), "Fractional Brownian Motion in Finance and Queueing", PhD thesis, University of Helsinki.
G. Strang and G. J. Fix (1973), *An Analysis of the Finite Element Method*, Series in Automatic Computation, Englewood Cliffs, NJ: Prentice Hall.
J. C. Strikwerda (2004), *Finite Difference Schemes and Partial Differential Equations*, 2nd ed., Philadelphia: Society for Industrial and Applied Mathematics, xii+435 pp.
A. M. Stuart (2010), "Inverse problems: a Bayesian perspective", *Acta Numer.* 19: 451–559.
A. M. Stuart and A. R. Humphries (1996), *Dynamical Systems and Numerical Analysis*, Cambridge Monographs on Applied and Computational Mathematics vol. 2, Cambridge: Cambridge University Press, xxii+685 pp.
A. M. Stuart, J. Voss, and P. Wiberg (2004), "Fast conditional path sampling of SDEs and the Langevin MCMC method", *Commun. Math. Sci.* 2: 4, 685–697.
E. Süli and D. F. Mayers (2003), *An Introduction to Numerical Analysis*, Cambridge: Cambridge University Press, x+433 pp.
D. Talay (1996), "Probabilistic numerical methods for partial differential equations: elements of analysis", in: *Probabilistic Models for Nonlinear Partial Differential Equations*, pp. 148–196, D. Talay and L. Tubaro (eds.), Lecture Notes in Math. vol. 1627, Berlin: Springer-Verlag.
—(2002), "Stochastic Hamiltonian systems: exponential convergence to the invariant measure, and discretization by the implicit Euler scheme", *Markov Process. Relat. Fields*, 8: 2, 163–198.
D. Talay and L. Tubaro (1990), "Expansion of the global error for numerical schemes solving stochastic differential equations", *Stochastic Anal. Appl.* 8: 4, 483–509.
R. Temam (1988), *Infinite-Dimensional Dynamical Systems in Mechanics and Physics*, Applied Mathematical Sciences vol. 68, New York: Springer-Verlag, xvi+500 pp.
G. Tessitore and J. Zabczyk (2006), "Wong–Zakai approximations of stochastic evolution equations", *J. Evol. Equ.* 6: 4, 621–655.
V. Thomée (2006), *Galerkin Finite Element Methods for Parabolic Problems*, 2nd ed., Springer Series in Computational Mathematics vol. 25, Berlin: Springer-Verlag, xii+370 pp.
L. N. Trefethen (2000), *Spectral Methods in MATLAB*, Software, Environments, and Tools vol. 10, Philadelphia: Society for Industrial and Applied Mathematics, xviii+165 pp.
K. Twardowska (1996), "Wong–Zakai approximations for stochastic differential equations", *Acta Appl. Math.* 43: 3, 317–359.
K. Twardowska and A. Nowak (2004), "On the relation between the Itô and Stratonovich integrals in Hilbert spaces", *Ann. Math. Sil.* 18: 49–63.
N. G. van Kampen (1997), *Stochastic Processes in Physics and Chemistry*, 2nd ed., Amsterdam: North-Holland, xiv+419 pp.
J. van Neerven, M. Veraar, and L. Weis (2012), "Stochastic maximal L^p-regularity", *Ann. Probab.* 40: 2, 788–812.
C. R. Vogel (2002), *Computational Methods for Inverse Problems*, Frontiers in Applied Mathematics vol. 23, Philadelphia: Society for Industrial and Applied Mathematics, xvi+183 pp.
J. Voss (2012), "The effect of finite element discretization on the stationary distribution of SPDEs", *Commun. Math. Sci.* 10: 4, 1143–1159.
J. B. Walsh (1981), "A stochastic model of neural response", *Adv. in Appl. Probab.* 13: 2, 231–281.
—(1984a), "An Introduction to Stochastic Partial Differential Equations", in: *École d'Été de Probabilités de Saint-Flour*, pp. 265–439, A. Dold and B. Eckmann (eds.), Springer Lecture Notes in Mathematics vol. 1180, Springer-Verlag.
—(1984b), "Regularity properties of a stochastic partial differential equation", in: *Seminar on stochastic processes*, pp. 257–290, Progress in Probability and Statistics 1983, vol. 7, Boston: Birkhäuser.

G. N. Watson (1995), *A Treatise on the Theory of Bessel Functions*, Cambridge Mathematical Library, Reprint of the second (1944) edition, Cambridge: Cambridge University Press, viii+804 pp.

H. Weyl (1912), "Das asymptotische Verteilungsgesetz der Eigenwerte linearer partieller Differentialgleichungen", *Math. Annal.* 71: 441–479.

H. Widom (1963), "Asymptotic behavior of the eigenvalues of certain integral equations", *Trans. Amer. Math. Soc.* 109: 278–295.

N. Wiener (1938), "The Homogeneous Chaos", *Am. J. Math.* 60: 4, 897–936.

M. Wiktorsson (2001), "Joint characteristic function and simultaneous simulation of iterated Itô integrals for multiple independent Brownian motions", *Adv. in Appl. Probab.* 11: 2, 470–487.

D. Williams (1991), *Probability with Martingales*, Cambridge Mathematical Textbooks, Cambridge: Cambridge University Press, xvi+251 pp.

E. Wong and M. Zakai (1965), "On the relation between ordinary and stochastic differential equations", *Internat. J. Eng. Sci.* 3: 2, 213–229.

A. T. A. Wood and G. Chan (1994), "Simulation of Stationary Gaussian Processes in $[0,1]^d$", *J. Comput. Graph. Simul.* 3: 4, 409–432.

D. Xiu and J. S. Hesthaven (2005), "High-order collocation methods for differential equations with random inputs", *SIAM J. Sci. Comput.* 27: 3, 1118–1139.

D. Xiu and G. E. Karniadakis (2002), "The Wiener–Askey polynomial chaos for stochastic differential equations", *SIAM J. Sci. Comput.* 24: 2, 619–644.

A. M. Yaglom (1962), *An Introduction to the Theory of Stationary Random Functions*, Translated and edited by R. A. Silverman, Englewood Cliffs, NJ: Prentice Hall, xiii+235 pp.

Y. Yan (2005), "Galerkin finite element methods for stochastic parabolic partial differential equations", *SIAM J. Numer. Anal.* 43: 4, 1363–1384.

K. Yosida (1995), *Functional Analysis*, Classics in Mathematics, Reprint of the sixth (1980) edition, Berlin: Springer-Verlag, xii+501 pp.

V. V. Yurinskiĭ (1982), "On the accuracy of normal approximation of the probability of hitting a ball", *Teor. Veroyatnost. i Primenen.* 27: 2, 270–278.

E. Zeidler (1995), *Applied Functional Analysis*, Applied Mathematical Sciences vol. 109, New York: Springer-Verlag, xvi+404 pp.

Index